T0295181

Wind Energy Explained

Wind Energy Explained

On Land and Offshore

Third Edition

James F. Manwell
University of Massachusetts
Amherst, United States

Emmanuel Branlard
University of Massachusetts
Amherst, United States

Jon G. McGowan
University of Massachusetts
Amherst, United States

Bonnie Ram
University of Delaware
Delaware, United States

Library of Congress Cataloging-in-Publication Data
Names: James F. Manwell, author. | Branlard, Emmanuel, author. |
    McGowan, J. G., author. | Ram, Bonnie, author.
Title: Wind energy explained : on land and offshore / James F. Manwell,
    University of Massachusetts, Amherst, United States, Emmanuel Branlard,
    University of Massachusetts, Amherst, United States, Jon G. McGowan,
    University of Massachusetts, Amherst, United States,
    Bonnie Ram, University of Delaware, Newark, United States.
Description: Third edition. | Hoboken, NJ, USA: Wiley, 2024. | Includes
    index.
Identifiers: LCCN 2023055131 (print) | LCCN 2023055132 (ebook) | ISBN
    9781119367451 (cloth) | ISBN 9781119367468 (adobe pdf) | ISBN
    9781119367475 (epub)
Subjects: LCSH: Wind power.
Classification: LCC TJ820 M374 2024 (print) | LCC TJ820 (ebook) | DDC
    621.31/2136–dc23/eng/20231221
LC record available at https://lccn.loc.gov/2023055131
LC ebook record available at https://lccn.loc.gov/2023055132

Cover Design: Wiley
Cover Images: © NREL /Alamy Stock Photo, sah/Getty Images

Set in 9.5/12.5pt STIXTwoText by Straive, Pondicherry, India

SKY10074182_050324

Contents

About the Authors

James F. Manwell is a Professor of Mechanical Engineering at the University of Massachusetts Amherst and the Founding Director of the Wind Energy Center there. He holds an MS in electrical and computer engineering and a PhD in mechanical engineering. He has been involved with a wide range of wind energy research areas since the mid-1970s. These range from wind turbine dynamics to wind hybrid power systems. His most recent research has focused on the assessment of external conditions related to the design of offshore wind turbines. He has participated in activities of the International Energy Agency (IEA), the International Electrotechnical Commission (IEC), and the International Science Panel on Renewable Energies (ISPRE). He lives in Conway, Massachusetts.

Emmanuel Branlard is an Associate Professor of Mechanical Engineering at the University of Massachusetts Amherst (UMass). He holds an MS in aerospace engineering from the French engineering school SupAero, an MS in wind energy from the Technical University of Denmark (DTU), and a PhD in wind turbine aeroelasticity from DTU. Emmanuel possesses three years of industrial experience, having worked at Orsted (Denmark) and Siemens Wind Power (USA). Additionally, he dedicated five years as a senior researcher, specializing in wind turbine multiphysics modeling at the US National Renewable Energy Laboratory, prior to his appointment at UMass.

Jon G. McGowan is a Professor of Mechanical Engineering at the University of Massachusetts Amherst. He holds an MS and a PhD in mechanical engineering. During his 40-plus years at the university, he has developed and taught a number of fundamental undergraduate/graduate engineering courses in renewable energy and energy conversion. His research and graduate student supervision at UMass have produced approximately 200 technical papers in a wide range of energy conversion applications. His recent research interests in wind engineering have been concentrated in the areas of wind system siting, hybrid systems modeling, economics, and offshore wind engineering. Professor McGowan is a Fellow of the American Society of Mechanical Engineers (ASME) and editor of the *Wind Engineering* journal. He lives in Northfield, Massachusetts.

Bonnie Ram is a Senior Researcher and Associate Director of the Center for Research in Wind at the University of Delaware and Director of Ram Consultancy. She was a strategic advisor on national environmental compliance and stakeholder engagement for the Department of Energy Wind Technology Office and the National Renewable Energy Laboratory. As a Guest Senior Researcher at the Danish Technical University, Wind Engineering Department, she created a new division on social sciences and spearheaded multidisciplinary research collaborations. She has published in peer-reviewed journals and led social and environmental science committees with the IEA, European Energy Research Alliance (EERA), Wind Energy Science Conference, North American Wind Energy Academy (NAWEA), and Research at Alpha Ventus (RAVE). She holds an MA in environmental policy and science (Clark University, Worcester, MA).

Preface

The technology of extracting energy from the wind has evolved dramatically over the past 50 years, and there have been relatively few attempts to describe that technology in a single textbook. The lack of such a text, together with a perceived need, provided the impetus for writing the first edition of this book.

The material in the original edition evolved from course notes from Wind Energy Engineering, a course that has been taught at the University of Massachusetts since the mid-1970s. These notes were later substantially revised and expanded with the support of the US Department of Energy's National Renewable Energy Laboratory (NREL). In the second edition, we added new material to reflect the rapid worldwide expansion of wind engineering in the 21st century. Now in the third edition, we have continued to update the previous texts and have added new material on offshore wind energy. We have also added material relating to the design and construction of today's larger and more sophisticated wind turbines, as well as about how the wind can provide a very significant amount of the world's energy supply.

This book provides a description of the topics that are fundamental to understanding the conversion of wind energy to electricity and its eventual use by society. These topics span a wide range, from meteorology through many fields of engineering to economics and environmental concerns. The book begins with an introduction that provides an overview of the technology and explains how it came to take the form it has today. The next chapter describes the wind resource and how it relates to energy production. Chapter 3 discusses aerodynamic principles and explains how the wind's energy will cause a wind turbine's rotor to turn. Chapter 4 delves into the dynamic and mechanical aspects of the turbine in more detail, and considers the relation of the rotor to the rest of the machine. Chapter 5 provides a summary of the electrical aspects of wind energy conversion, especially the actual generation and conversion of the electricity. Chapter 6 is about the external conditions, particularly the wind, which must be considered in the design of all turbines, and waves, which are of comparable importance for offshore wind turbines. Chapter 7 presents a summary of wind turbine materials and components. Chapter 8 discusses the design of wind turbines and how the various components are integrated into a functional machine. Chapter 9 describes wind turbine and wind farm control. Chapter 10 focuses on foundations and support structures of land-based and fixed offshore wind turbines. Chapter 11 provides an overview of floating offshore wind turbines, with a focus on their floating stability. Chapter 12 is about groups of wind turbines, known as wind farms or wind power plants, how they work together and with the electrical network to which they are connected, and how aerodynamic interactions, known as wake and blockage effects, impact their performances. Chapter 13 concerns the economics of wind energy. It describes economic analysis methods and shows how wind energy can be economically integrated with conventional forms of generation. Chapter 14 is about the wind project development process and public engagement, including permitting and environmental risk and benefits of wind energy siting. Chapter 15 discusses the installation, operation, maintenance, and decommissioning of wind turbines and wind power plants. Chapter 16 concerns high-penetration power systems, in which the wind supplies a very large fraction of the total energy supply. It includes an overview of energy storage and the emerging opportunity of "power-to-x." Finally, there are five appendices. Appendix A lists the variables used in the sixteen chapters. Appendix B gives an overview of some of the data analysis techniques that are commonly used in wind turbine design and use. Appendix C provides a summary of probability distributions of direct relevance to wind energy. Appendix D presents the key properties of some important components of support structures: cylinders, frustums, and cuboids. Finally, Appendix E informs the reader how to access problems, data files, and ancillary programs that accompany the text.

This book is intended primarily as a textbook for engineering students and for professionals in related fields who are just getting into wind energy. It is also intended to be used by anyone with a good background in mathematics and physics who wants to gain familiarity with the subject. It should be useful for those interested in wind turbine or wind power plant design *per se*. For others, it should provide enough understanding of the underlying principles of wind turbine operation and design to appreciate more fully those aspects in which they have a particular interest. These areas include turbine siting, grid integration, environmental issues, economics, social acceptance, and public policy.

The study of wind energy spans a wide range of fields. Since it is likely that many readers will not have a background in all of them, most of the chapters include some introductory material. Where appropriate, the reader is referred to other sources for more details. Some of the material is more detailed than may be of interest to many readers. The corresponding sections are marked with asterisks, and they can be bypassed without interfering with the flow of the rest of the text.

Acknowledgments

We would like to acknowledge the late Professor William Heronemus, founder of the renewable energy program at the University of Massachusetts. Without his vision and tenacity, this program would never have existed, and this book would never have been written. We are also indebted to the numerous staff and students, past and present, at the University of Massachusetts who have contributed to this program.

We would also like to acknowledge the National Wind Technology Center at the National Renewable Energy Laboratory (NREL), particularly Bob Thresher, who was instrumental in launching the first edition and contributing to subsequent editions as well.

For this third edition, we would also like to express our sincere appreciation to those who have contributed in many ways. Tony Rogers, one of the authors of the first two editions, provided a detailed review of many chapters. Others who provided very helpful reviews of full chapters include Walt Musial, Brian McNiff, and Jason Jonkman. Significant reviews were provided by Hannah Ross, Pietro Bortolotti, Ian Brownstein, Benjamin Strom, Thomas Webler, Chris Hein., and Captain Dan Kipnis.

A number of individuals, companies, and organizations have also contributed substantially, particularly with illustrations. Among these were Aidan Cronin (Siemens Gamesa), Steve Nolet (TPI, Inc.), Liz Burdock (Business Network for Offshore Wind), Andrew Henderson (Copenhagen Offshore Partners), Jan van der Tempel (Ampelmann), Zhiyu Jiang (University of Agder), Carlo Bottaso (Technical University of Munich), Fábio Espírito Santo (Principle Power), Daniela Chiaramonte (International Electrotechnical Commission), Lucy Gardiner (Osbit Ltd.), Bill Wall (LS Cable), Dieter Mechlinski (virtuellesbrueckenhofmuseum.de), Pieter Van der Avert (Scaldis Salvage & Marine Contractors) and Per Vølund (RWE Wind Services Denmark).

In addition, we appreciate the contribution of Rolf Niemeyer for arranging for one of us (JFM) to get a close-up tour of the REpower 5-MW turbine. This real turbine was the inspiration of the NREL 5-MW conceptual turbine, which is featured in examples throughout this text.

1

Introduction: Modern Wind Energy and Its Origins

The re-emergence of the wind as a significant source of the world's energy must rank as one of the significant developments of the late 20th century. The advent of the steam engine, followed by the appearance of other technologies for converting fossil fuels to useful energy, would seem to have forever relegated to insignificance the role of the wind in energy generation. In fact, by the mid-1950s that appeared to be what had already happened. By the late 1960s, however, the first signs of a reversal could be discerned, and by the early 1990s, it was becoming apparent that a fundamental reversal was underway. That decade saw a strong resurgence in the worldwide wind energy industry, with installed capacity increasing over fivefold. The 1990s were also marked by a shift to large, megawatt-sized wind turbines, a reduction and consolidation in wind turbine manufacturers, and the beginning of offshore wind power (see McGowan and Connors, 2000). Since the start of the 21st century, this trend has continued, with European countries (and manufacturers) leading the increase via government policies focused on developing domestic sustainable energy supplies and reducing pollutant emissions.

To understand what was happening, it is necessary to consider five main factors. First of all, there was a need. An emerging awareness of the finiteness of the earth's fossil fuel reserves as well as of the adverse effects of burning those fuels for energy had caused many people to look for alternatives. Second, there was the potential. Wind exists everywhere on the earth and in some places with considerable energy density. Wind had been widely used in the past, for mechanical power as well as transportation. Certainly, it was conceivable to use it again. Third, there was the technological capacity. In particular, there have been developments in other fields, which, when applied to wind turbines, could revolutionize the way they could be used. These first three factors were necessary to foster the re-emergence of wind energy, but not sufficient. There needed to be two more factors, first of all, a vision of a new way to use the wind, and second, the political will to make it happen. The vision began well before the 1960s with such individuals as Poul la Cour, Albert Betz, Palmer Putnam, and Percy Thomas. It was continued by Johannes Juul, E. W. Golding, Ulrich Hütter, and William Heronemus, but soon spread to others too numerous to mention. At the beginning of wind's re-emergence, the cost of energy from wind turbines was far higher than that from fossil fuels. Government support was required to carry out research, development, and testing; to provide regulatory reform to allow wind turbines to interconnect with electrical networks; and to offer incentives to help hasten the deployment of the new technology. The necessary political will for this support appeared at different times and to varying degrees, in a number of countries: first in the United States, Denmark, and Germany, and now in much of the rest of the world.

The purpose of this chapter is to provide an overview of wind energy technology today, so as to set a context for the rest of the book. It addresses such questions as: What does modern wind technology look like? What is it used for? How did it get this way? Where is it going?

Wind Energy Explained: On Land and Offshore, Third Edition.
James F. Manwell, Emmanuel Branlard, Jon G. McGowan, and Bonnie Ram.
© 2024 John Wiley & Sons Ltd. Published 2024 by John Wiley & Sons Ltd.

1.1 Modern Wind Turbines

A wind turbine, as described in this book, is a machine that converts the power in the wind into electricity. This is in contrast to a "windmill," which is a machine that converts the wind's power into mechanical power. As electricity generators, wind turbines are connected to some electrical network. These networks include battery-charging circuits, residential-scale power systems, isolated or island networks, and large utility grids. In terms of total numbers, the most frequently found wind turbines are actually quite small – on the order of 10 kW or less. In terms of total generating capacity, the turbines that make up the majority of the capacity are, in general, much larger – in the range of 1.5–15 MW. These larger turbines are used primarily in large utility grids, at first mostly in Europe and the United States and more recently in China, India, and elsewhere. An example of a modern wind turbine, the REpower 5M, is illustrated in Figure 1.1. This turbine was used as the conceptual basis for the NREL 5-MW reference turbine, which is used as the basis of many examples throughout this text.

To understand how wind turbines are used, it is useful to briefly consider some of the fundamental facts underlying their operation. In modern wind turbines, the actual conversion process uses the basic aerodynamic force of lift to produce a net positive torque on a rotating shaft, resulting first in the production of mechanical power and then in its transformation to electricity in a generator. Wind turbines, unlike most other generators, do not have means to control the resource available to produce energy. It is also not possible to store the wind and use it at a later time. The output of a wind turbine is thus inherently fluctuating and non-dispatchable. (The most one can do is to limit production below what the wind could produce.) Any system to which a wind turbine is connected must,

Figure 1.1 Modern utility-scale wind turbine (REpower 5M) (*Source:* J. F. Manwell (Book Author))

in some way, take this variability into account. In larger networks, the wind turbine serves to reduce the total electrical load and thus results in a decrease in either the number of conventional generators being used or in the fuel use of those that are running. In smaller networks, there may be energy storage, backup generators, and some specialized control systems. A further fact is that the wind is not transportable: it can only be converted where it is blowing. Historically, a product such as ground wheat was made at the windmill and then transported to its point of use. Today, the possibility of conveying electrical energy via power lines compensates to some extent for wind's inability to be transported. In the future, hydrogen-based energy systems may add to this possibility.

1.1.1 Modern Wind Turbine Design

Today, the most common design of wind turbines, and the type which is the primary focus of this book, is the horizontal axis wind turbine (HAWT). That is, the axis of rotation is approximatively parallel to the ground. HAWT rotors are usually classified according to the rotor orientation (upwind or downwind of the tower), hub design (rigid or teetering), rotor control (pitch vs. stall), number of blades (two or three), and how they are aligned with the wind (free yaw or active yaw). Figure 1.2 illustrates the upwind and downwind configurations.

The principal subsystems of typical land-based and offshore HAWTs are shown in Figure 1.3. In general, these include:

- The rotor, consisting of the blades and the supporting hub.
- The drivetrain, which includes the rotating parts of the wind turbine (exclusive of the rotor) and support bearings; it usually consists of shafts, gearbox, coupling, a mechanical brake, and the generator.
- The nacelle and main frame, including wind turbine housing, bedplate, and the yaw system.
- The support structure, foundation, and stationkeeping system (floating offshore).
- The machine controls.
- The balance of the electrical system, including cables, switchgear, transformers, and often electronic power converters.

The main options in wind turbine design and construction include:

- Number of blades (commonly two or three);
- Rotor orientation. Downwind or upwind of tower;
- Blade material, construction method, and profile;
- Hub design. Rigid, teetering, or hinged;
- Power control via aerodynamic control (stall control) or variable-pitch blades (pitch control);

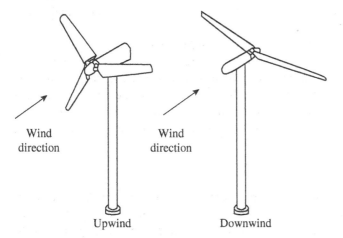

Wind
direction

Wind
direction

Upwind

Downwind

Figure 1.2 HAWT rotor configuration

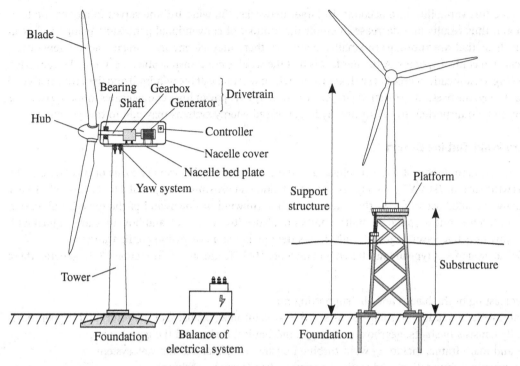

Figure 1.3 Major components of a horizontal axis wind turbine

- Fixed or variable rotor speed;
- Orientation by self-aligning action (free yaw), or direct control (active yaw);
- Synchronous or induction generator (squirrel cage or doubly fed); and
- Gearbox or direct drive generator.

A short introduction to and overview of some of the most important components follows. A more detailed discussion of the overall design aspects of these components, and other important parts of a wind power system, is contained in succeeding chapters of this book.

1.1.1.1 Rotor

The rotor consists of the hub and blades of the wind turbine. The blades produce aerodynamic forces from the wind. The hub provides for the attachment of the blades and carries the loads (forces and moments) to the main shaft.

Most turbines today have upwind rotors with three blades. There are some downwind rotors and a few designs with two blades. Single-blade turbines have been built in the past, but none are still in production. Some intermediate-sized turbines use fixed-blade pitch and stall control. Most manufacturers use pitch control, and the general trend is the increased use of pitch control, especially in larger machines. The blades on the majority of turbines are made from composites, primarily fiberglass or carbon fiber-reinforced plastics (GRP or CFRP), but sometimes wood/epoxy laminates have been used. These subjects are addressed in more detail in the aerodynamics, components, and design chapters (Chapters 3, 7, and 8).

1.1.1.2 Drivetrain

The drivetrain consists of the other rotating parts of the wind turbine downstream of the rotor. These typically include a low-speed shaft (on the rotor side), a gearbox, and a high-speed shaft (on the generator side). Other drivetrain components include the support bearings, one or more couplings, a brake, and the rotating parts of the generator (discussed separately in the next section). The purpose of the gearbox is to speed up the rate of rotation of the rotor from a low value (tens of rpm) to a rate suitable for driving a standard generator (hundreds or thousands of rpm). Two types of gearboxes are used in wind turbines: parallel shaft and planetary. For larger machines (over approximately 500 kW), the weight and size advantages of planetary gearboxes become more pronounced. A few wind turbine designs use multiple generators, and so are coupled to a gearbox with more than one output shaft. A now more common approach is to use specially designed, low-speed generators requiring no gearbox (referred to as "direct drive").

While the design of wind turbine drivetrain components usually follows conventional mechanical engineering machine design practice, the unique loading of wind turbine drivetrains requires special consideration. Fluctuating winds and the dynamics of large rotors impose significantly varying loads on drivetrain components.

1.1.1.3 Generator

Nearly all wind turbines use either induction or synchronous generators (see Chapter 5). These designs entail a constant or nearly constant rotational speed when the generator is directly connected to a utility network. If the generator is used with power electronic converters, the turbine will be able to operate at variable speed.

Many wind turbines installed in grid-connected applications use squirrel cage induction generators (SCIG). A SCIG operates within a narrow range of speeds slightly higher than its synchronous speed (a four-pole generator operating in a 60 Hz grid has a synchronous speed of 1800 rpm). The main advantages of this type of induction generator are that it is rugged, inexpensive, and easy to connect to an electrical network. A variant of the induction generator is the wound rotor induction generator (WRIG). The WRIG is often used in variable-speed applications. It is described in more detail in Chapter 5.

An increasingly popular option for utility-scale electrical power generation is the variable-speed wind turbine. There are a number of benefits that such a configuration offers, including the reduction of wear and tear on the wind turbine and the potential operation of the wind turbine at maximum efficiency over a wide range of wind speeds, yielding increased energy capture. Although there are a large number of potential hardware options for variable-speed operation of wind turbines, power electronic components are used in virtually all variable-speed machines currently being designed. When used with suitable power electronic converters, either synchronous or induction generators of either type can run at variable speeds.

1.1.1.4 Nacelle and Yaw System

This category includes the wind turbine housing, the machine bedplate or main frame, and the yaw orientation system. The main frame provides for the mounting and proper alignment of the drivetrain components. The nacelle cover protects the contents from the weather, rust, and allows for temperature regulation.

A yaw orientation system is required to keep the rotor shaft properly aligned with the wind. Its primary component is a large bearing that connects the main frame to the tower. An active yaw drive, always used with upwind wind turbines and sometimes with downwind turbines, contains one or more yaw motors, each of which drives a pinion gear against a bull gear attached to the yaw bearing. This mechanism is controlled by an automatic yaw control system with its wind direction sensor usually mounted on the nacelle of the wind turbine. Sometimes yaw brakes are used with this type of design to hold the nacelle in position, when not yawing. Free yaw systems (meaning that they can self-align with the wind) are often used on downwind wind machines.

1.1.1.5 Support Structure and Stationkeeping System

On land, a support structure includes the tower itself and the foundation. Offshore, there is an additional component known as the substructure. The tower serves to lift the rest of the turbine into the air, where the wind is. The principal types of tower design currently in use are the free-standing type using tapered steel tubes, lattice (or truss) towers, and concrete towers. For smaller turbines, guyed towers are also used. Tower height is typically 1.3–7 times the rotor radius for modern turbines (e.g., 80–150 m), but this factor is usually larger for smaller turbines. Tower

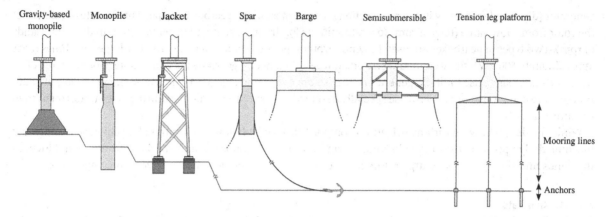

Figure 1.4 Substructures for fixed and floating offshore wind turbines

selection is greatly influenced by the characteristics of the site. The stiffness of the support structure is a major factor in wind turbine system dynamics because of the possibility of coupled vibrations between the rotor and the support structure. The tower will affect the aerodynamics of the blades as it passes in the vicinity of the tower, which will impact the turbine dynamics, resulting in cyclic loads, power fluctuations, and noise. For turbines with downwind rotors, the effect of tower shadow (the wake created by airflow behind the tower) must also be considered. For example, because of the tower shadow, downwind turbines are typically noisier than their upwind counterparts.

The substructure in an offshore turbine is the part of the support structure which is primarily in the water and mostly subject to hydrodynamic loads (waves and current). There are a variety of types of substructures (see Figure 1.4), including monopiles, multimember structures (such as the "jacket"), and gravity-based structures. Offshore wind turbines in deeper water may have floating substructures (see Figure 1.4); the principal types are the barge, spar, semisubmersible, and the tension leg platform (TLP).

The foundation provides the interface between the rest of the support structure and the soil. Floating offshore wind turbines do not have foundations in the normal sense; they do have stationkeeping systems, which consist primarily of anchors and mooring lines.

1.1.1.6 Controls

The control system for a wind turbine is important with respect to machine operation, power production, and maintaining structural integrity. A wind turbine control system includes the following components:

- Sensors. Speed, position, flow, temperature, current, voltage, etc.;
- Controllers. Mechanical mechanisms and electrical circuits;
- Power amplifiers. Switches, electrical amplifiers, hydraulic pumps, and valves;
- Actuators. Motors, pistons, magnets, and solenoids; and
- Intelligence. Computers and microprocessors.

The design of control systems for wind turbine application follows traditional control engineering practices. Many aspects, however, are quite specific to wind turbines and are discussed in Chapter 9. Wind turbine control involves the following three major aspects and the judicious balancing of their requirements:

- Setting upper bounds on and limiting the torque and power experienced by the drivetrain.
- Maximizing the fatigue life of the rotor drivetrain and other structural components in the presence of changes in the wind direction, wind speed (including gusts), and turbulence, as well as start–stop cycles of the wind turbine.
- Maximizing the energy production.

1.1.1.7 Balance of Electrical System

In addition to the generator, the wind turbine system utilizes a number of other electrical components. Some examples are cables, switchgear, transformers, power electronic converters, power factor correction capacitors, and yaw and pitch motors. Details of the electrical aspects of wind turbines themselves are given in Chapter 5. Interconnection with electrical networks is discussed in Chapter 12.

1.1.2 Power Output Prediction

The power output of a wind turbine varies with wind speed, and every wind turbine has a characteristic power performance curve. With such a curve, it is possible to predict the energy production of a wind turbine without considering the technical details of its various components. The power curve gives the electrical power output as a function of the hub height wind speed. The shape of the curve will in general differ depending on the type of power control used, the primary options being adjustable blade pitch and stall. Figure 1.5 presents examples of a power curve for a hypothetical wind turbine using the two options.

The performance of a given wind turbine can be related to three key points on the velocity scale:

- **Cut-in speed**. The minimum wind speed at which the machine will deliver useful power.
- **Rated wind speed**. the wind speed at which the rated power (generally the maximum power output of the electrical generator) is reached.
- **Cut-out speed**. the maximum wind speed at which the turbine is allowed to deliver power (usually limited by engineering design and safety constraints).

Power curves for existing machines can normally be obtained from the manufacturer. The curves are derived from field tests, using standardized testing methods. It is also possible to estimate the shape of the power curve for a given turbine using numerical tools. Such a process involves determination of the power characteristics of the wind turbine rotor and electrical generator, gearbox gear ratios, component efficiencies, and the method of power control. Chapter 8 provides details for fixed-pitch, stall-regulated turbines; Chapter 9 does the same for variable speed, pitch-controlled turbines.

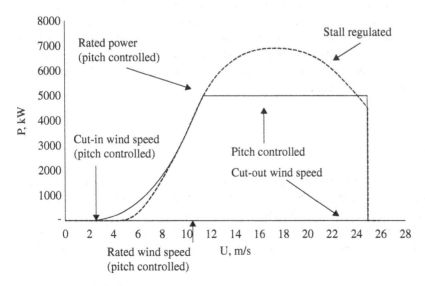

Figure 1.5 Typical wind turbine power curve

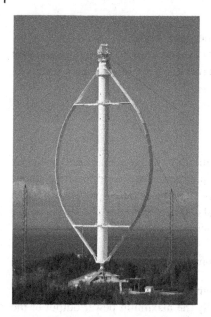

Figure 1.6 Sandia 34-m Darrieus VAWT (*Source:* Charles Jacques/Adobe Stock)

1.1.3 Other Wind Energy Concepts

The wind turbine overview provided above assumed a topology of a basic type, namely, one that employs a horizontal axis rotor, driven by lift forces. It is worth noting that a vast number of other topologies have been proposed, and in some cases built. None of these has met with the same degree of success as those with a horizontal-axis, lift-driven rotor. A few words are in order, however, to briefly summarize some of these other concepts. The closest runner-up to the HAWT is the Darrieus vertical axis wind turbine (VAWT). This concept was studied extensively in both the United States and Canada in the 1970s and 1980s. An example of a VAWT wind turbine (Sandia 34 m design) based on this concept is shown in Figure 1.6. The biggest VAWT constructed was the 110 m tall (60 m diameter), 3.8 MW, "EOLE" wind turbine, built in 1987 in Canada.

Despite some appealing features, Darrieus wind turbines had some major reliability problems, and they were never able to match corresponding HAWTs in the cost of energy. However, it is possible that the concept could emerge again for some applications. In particular, commercial small-scale solutions using the H-shaped Darrieus rotors have recently been deployed. For a summary of past work on this turbine design and other VAWT wind turbine designs the reader is referred to Paraschivoiu (2002), Price (2006), and the summary of VAWT work carried out by Sandia National Laboratories (SNL) in the United States (Sutherland *et al.*, 2012).

Another concept that appears periodically is the concentrator or diffuser augmented wind turbine (see van Bussel, 2007). In both types of design, the idea is to channel the wind to increase the productivity of the rotor. The problem is that the cost of building an effective concentrator or diffuser, which can also withstand occasional extreme winds, has always been more than the device was worth.

Finally, a number of rotors using drag instead of lift have been proposed. One concept, the Savonius rotor, has been used for some small water-pumping applications. There are two fundamental problems with such rotors: (1) they are inherently inefficient (see comments on drag machines in Chapter 3), and (2) it is difficult to protect them from extreme winds. It is doubtful whether such rotors will ever achieve widespread use in wind turbines.

The reader interested in some of the variety of wind turbine concepts may wish to consult Nelson (1996). Nelson provides a description of a number of innovative wind systems. Reviews of various types of wind machines are given in Eldridge (1980) and Le Gourieres (1982). Some of the more innovative designs are documented in work supported by the US Department of Energy (1979, 1980). A few of the many concepts are illustrated in Figures 1.7 and 1.8. It is worth noting that airborne wind energy concepts have also been investigated, using kites, gliders, and various wing concepts with energy-producing turbines.

1.2 History of Wind Energy

It is worthwhile to consider some of the history of wind energy. The history serves to illustrate the issues that wind energy systems still face today and provides insight into why turbines look the way they do. In the following summary, emphasis is given to those concepts which have particular relevance today.

The reader interested in a fuller description of the history of wind energy is referred to Park (1981), Eldridge (1980), Shepherd (1990), and Ackermann and Soder (2002). Golding (1977) presents a history of wind turbine

Horizontal axis turbines

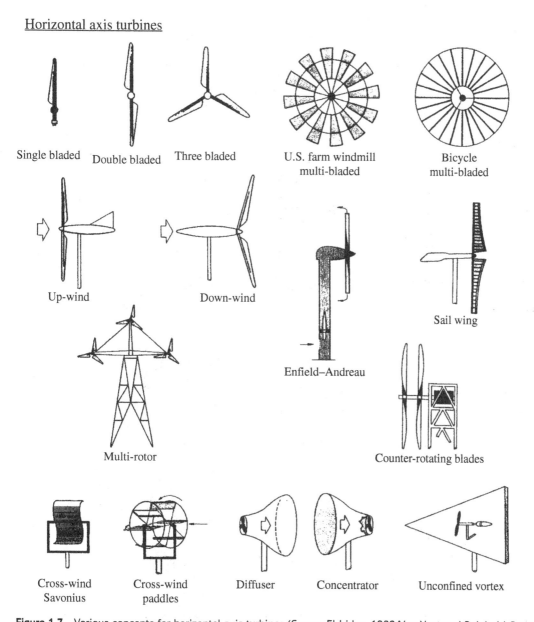

Single bladed   Double bladed   Three bladed   U.S. farm windmill   Bicycle
                                                      multi-bladed   multi-bladed

Up-wind   Down-wind   Sail wing

                                              Enfield–Andreau

Multi-rotor   Counter-rotating blades

Cross-wind   Cross-wind   Diffuser   Concentrator   Unconfined vortex
Savonius   paddles

Figure 1.7 Various concepts for horizontal axis turbines (*Source:* Eldridge, 1980/Van Nostrand Reinhold Company)

design from the ancient Persians to the mid-1950s. In addition to a summary of the historic uses of wind power, Johnson (1985) presents a history of wind electricity generation, and the US research work of the 1970–1985 period on horizontal axis, vertical axis, and innovative types of wind turbines. Historical overviews of wind energy systems and wind turbines are also contained in the books of Spera (1994), Gipe (1995), and Harrison *et al.* (2000). Eggleston and Stoddard (1987) give a historical perspective of some of the key components of modern wind turbines. Berger (1997) provides a fascinating picture of the early days of wind energy's re-emergence, particularly of the California wind farms.

Primarily drag-type

Savonius

Multi-bladed Savonius

Plates   Shield

Cupped

Primarily lift-type

φ-Darrieus

Δ-Darrieus

Giromill

Turbine

Combinations

Savonius/φ-Darrieus

Split Savonius

Magnus

Airfoil

Others

Deflector

Sunlight

Venturi

Confined vortex

Figure 1.8 Various concepts for vertical axis turbines (*Source:* Eldridge, 1980/Van Nostrand Reinhold Company)

1.2.1 A Brief History of Windmills

The first known historical reference to a windmill is from Hero (or Heron) of Alexandria, in his work *Pneumatics* (Woodcroft, 1851). Hero was believed to have lived in the 1st century A.D. His *Pneumatics* describes a device that provides air to an organ by means of a windmill. An illustration that accompanies Hero's description is shown in Figure 1.9.

There has been some debate about whether such a windmill actually existed and whether the illustration actually accompanied the original documents; see Shepherd (1990) or Drachman (1961). One of the primary scholars

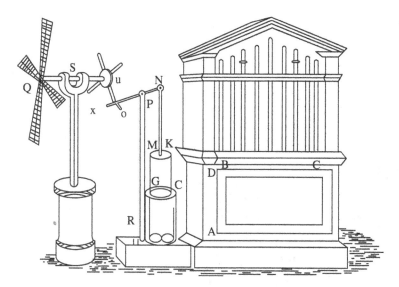

Figure 1.9 Hero's windmill (*Source:* Edward Francis Rimbault/Wikimedia Commons)

on the subject, however, H. P. Vowles, (Vowles, 1932) does consider Hero's description to be plausible. One of the arguments against the early Greeks having been familiar with windmills has to do with their presumed lack of technological sophistication. However, both mechanically driven grinding stones and gearing, which would generally be used with a wind-driven rotor, were known to exist at the time of Hero. For example, Reynolds (1983) describes water-powered grinding wheels of that time. In addition, the analysis of the Antikythera mechanism (Marchant, 2006) confirms that the early Greeks had a high degree of sophistication in the fabrication and use of gears.

Apart from Hero's windmill, the next reference on the subject dates from the 9th century A.D. (Al Masudi as reported by Vowles, 1932) Windmills were definitely in use in the Persian region of Seistan (now eastern Iran) at that time. Al Masudi also related a story indicating that windmills were in use by 644 A.D. The Seistan windmills have continued to be used up to the present time. These windmills had vertical axis rotors, as illustrated in Figure 1.10.

Windmills made their first recorded appearance in northern Europe (England) in the 12th century but probably arrived in the 10th or 11th century (Vowles, 1930). Those windmills were very different in appearance to those of Seistan, and there has been much speculation as to if and how the Seistan mills might have influenced those that appeared later in Europe. There are no definite answers here, but Vowles has suggested that the Vikings, who traveled regularly from northern Europe to the Middle East, may have brought back the concept on one of their return trips.

An interesting footnote to this early evolution concerns the change in the design of the rotor from the Seistan windmills to those of northern Europe. The Seistan rotors had vertical axes and were driven by drag forces. As such they were inherently inefficient and particularly susceptible to damage in high winds. The northern European designs had horizontal axes and were driven by lift forces. How this transition came about is not well understood, but it was to be of great significance. It can be surmised, however, that the evolution of windmill rotor design paralleled the evolution of rigging on ships during the 1st millennium A.D., which moved progressively from square sails (primarily drag devices) to other types of rigging that used lift to facilitate tacking upwind. See, for example, Casson (1991).

The early northern European windmills all had horizontal axes. They were used for nearly any mechanical task, including water pumping, grinding grain, sawing wood, and powering tools. The early mills were built on posts, so

Figure 1.10 Seistan windmill (*Source:* MorvaridiMeraj/Wikimedia Commons/CC BY-SA 4.0)

Figure 1.11 Post mill (*Source:* Paste at English Wikipedia/ Wikimedia Commons/Public domain)

that the entire mill could be turned to face the wind (or yaw) when its direction changed. These mills normally had four blades. The number and size of blades presumably were based on ease of construction as well as an empirically determined efficient solidity (ratio of blade area to swept area). An example of a post mill can be seen in Figure 1.11.

The wind continued to be a major source of energy in Europe through the period just prior to the Industrial Revolution but began to recede in importance after that time. The reason that wind energy began to disappear is primarily attributable to its non-dispatchability and its non-transportability. Coal had many advantages which the wind did not possess. Coal could be transported to wherever it was needed and used whenever it was desired. When coal was used to fuel a steam engine, the output of the engine could be adjusted to suit the load. Waterpower, which has some similarities to wind energy, was not eclipsed so dramatically. This is no doubt because waterpower is, to some extent, transportable (via canals) and dispatchable (by using ponds as storage).

Prior to its demise, the European windmill had reached a high level of design sophistication. In the later mills (or "smock mills"), such as the one shown in Figure 1.12, the majority of the structure was stationary. Only the top would be moved to face the wind. Yaw mechanisms included both manually operated arms and separate yaw rotors. Blades had acquired somewhat of an airfoil shape and included some twist. The power output of some machines could be adjusted by an automatic control system. This was the forerunner of the system used by James Watt on steam engines. In the windmill's case, a fly ball governor would sense when the rotor speed was changing. A linkage to a tentering mechanism would cause the upper millstone to move closer or farther away from the lower one, letting in more or less grain to grind. Increasing the gap would result in more grain being ground and thus a greater load on the rotor, thereby slowing it down and vice versa. See Stokhuyzen (1962) for details on this governor as well as other features of Dutch windmills.

One significant development in the 18th century was the introduction of scientific testing and evaluation of windmills. The Englishman John Smeaton, using such apparatus as illustrated in Figure 1.13, discovered three basic rules that are still applicable:

- The speed of the blade tips is ideally proportional to the speed of wind.
- The maximum torque is proportional to the speed of wind squared.
- The maximum power is proportional to the speed of wind cubed.

The 18th-century European windmills represented the culmination of one approach to using wind for mechanical power and included a number of features that were later incorporated into some early electricity-generating wind turbines.

Figure 1.12 European smock mill (Hills, 1994) (*Source:* Cambridge University Press)

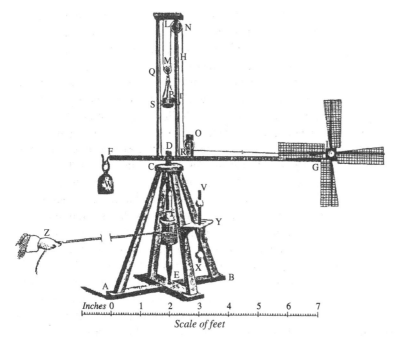

Figure 1.13 Smeaton's laboratory windmill testing apparatus (*Source:* Smeaton, 1837)

Figure 1.14 American water-pumping windmill design (*Source:* US Department of Agriculture)

As the European windmills were entering their final years, another variant of windmills came into widespread use in the United States. This type of windmill, illustrated in Figure 1.14, was most notably used for pumping water, particularly in the West. They were used on ranches for cattle and to supply water for the steam railroads. These mills were distinctive for their multiple blades and are often referred to as "fan mills." One of their most significant features was a simple but effective regulating system. This allowed the turbines to run unattended for long periods. Such regulating systems foreshadowed the automatic control systems which are now an integral part of modern wind turbines.

1.2.2 Early Wind Generation of Electricity

The initial use of wind for electricity generation, as opposed to mechanical power, included the successful commercial development of small wind generators and as well as research and experiments using large wind turbines.

When electrical generators were developed toward the end of the 19th century, it was reasonable that people would try to turn them with a windmill rotor. The first known wind generator was created in Scotland by James Blyth in 1887 (Price, 2005). The next most notable early example was built by Charles Brush in Cleveland, Ohio in 1888. These early turbines did not result in any trend, but in the following years, small electrical generators did become widespread. One of these, pioneered by Marcellus Jacobs, and illustrated in Figure 1.15, was, in some ways, the logical successor to the water-pumping fan mill. The Jacobs turbines were also significant in that their rotors had three blades with true airfoil shapes and began to resemble the turbines of today. Another feature of the Jacobs turbine was that it was typically incorporated into a complete, residential-scale power system, including battery storage. The Jacobs turbine is considered to be a direct forerunner of such modern small turbines as

Figure 1.15 Jacobs turbine (*Source:* Jacobs, Marcellus L. 1961/Reproduced from United Nations Digital Library)

those of Bergey Windpower. The expansion of the central electrical grid under the auspices of the Rural Electrification Administration during the 1930s marked the beginning of the end of the widespread use of small wind electric generators, at least for the time being.

The first half of the 20th century also saw the construction or conceptualization of a number of larger wind turbines which substantially influenced the development of today's technology. Probably the most important sequence of turbines was in Denmark. Between 1891 and 1918, Poul la Cour built more than 100 electricity-generating turbines in the 20–35 kW size range. His design was based on the latest generation of Danish smock mills. One of the more remarkable features of the turbine was that the electricity that was generated was used to produce hydrogen (by the electrolysis of water), and the hydrogen gas was then used for lighting. La Cour's turbines were followed by turbines made by Lykkegaard Ltd. and F. L. Smidth & Co prior to World War II. These ranged in size from 30 to 60 kW. Just after the war, Johannes Juul erected the 200 kW Gedser turbine, illustrated in Figure 1.16, in southeastern Denmark. This three-bladed machine was particularly innovative in that it employed aerodynamic stall for power control and used an induction generator (squirrel cage type) rather than the more conventional (at the time) synchronous generator. This type of generator is much simpler to connect to the grid than a synchronous generator. Stall is also a simple way to control power. These two concepts formed the core of the strong Danish presence in wind energy in the 1980s (Maegaard, *et al.*, 2013). One of

Figure 1.16 Danish Gedser wind turbine (*Source:* Danish Wind Turbine Manufacturers)

the pioneers in wind energy in the 1950s was Ulrich Hütter in Germany (Dörner, 2002). His work focused on applying modern aerodynamic principles to wind turbine design. Many of the concepts he worked with are still in use in some form today.

In the United States, the most significant early large turbine was the Smith–Putnam machine, built at Grandpa's Knob in Vermont in the late 1930s (Putnam, 1948). With a diameter of 53.3 m and a power rating of 1.25 MVA, this was the largest wind turbine ever built up until that time and for many years thereafter. This turbine, illustrated in Figure 1.17, was also significant in that it was the first large turbine with two blades. In this sense, it was a predecessor of the two-bladed turbines built by the US Department of Energy (DOE) in the late 1970s and early 1980s. The turbine was also notable in that the company that built it, S. Morgan Smith, had long experience in hydroelectric generation and intended to produce a commercial line of wind machines. Unfortunately, the Smith–Putnam turbine was too large, too early, given the level of understanding of wind energy engineering. It suffered a blade failure in 1945, and the project was abandoned.

Gipe and Möllerström (2022a) provide a comprehensive history of the early days of wind electric generation.

1.2.3 The Re-Emergence of Wind Energy

The re-emergence of wind energy can be considered to have begun in the late 1960s. The book *Silent Spring* (Carson, 1962) made many people aware of the environmental consequences of industrial development. *Limits to Growth* (Meadows *et al.*, 1972) followed in the same vein, arguing that unfettered growth would inevitably lead

Figure 1.17 Smith–Putnam wind turbine (*Source: Eldridge, F.R 1980/Reproduced from Scientific Research Publishing Inc.*)

to either disaster or change. Among the culprits identified were fossil fuels. The potential dangers of nuclear energy also became more public at this time. Discussion of these topics formed the backdrop for an environmental movement that began to advocate cleaner sources of energy.

In the United States, in spite of growing concern for environmental issues, not much new happened in wind energy development until the Oil Crises of the mid-1970s. Under the Carter administration, a new effort was begun to develop "alternative" sources of energy, one of which was wind energy. The US DOE sponsored a number of projects to foster the development of the technology. Most of the resources were allocated to large machines, with mixed results. These machines ranged from the 100 kW (38 m diameter) NASA MOD-0 to the 3.2 MW Boeing MOD-5B with its 98 m diameter. Much interesting data was generated but none of the large turbines led to commercial projects. DOE also supported the development of some small wind turbines and built a test facility for small machines at Rocky Flats, Colorado. A number of small manufacturers of wind turbines also began to spring up, but there was not a lot of activity until the late 1970s.

The big opportunities occurred as the result of changes in the utility regulatory structure and the provision of incentives. The US federal government, through the Public Utility Regulatory Policy Act of 1978, required utilities (1) to allow wind turbines to connect with the grid and (2) to pay the "avoided cost" for each kWh the turbines generated and fed into the grid.

The actual avoided cost was debatable, but in many states, utilities would pay enough that wind generation began to make economic sense. In addition, the federal government and some states provided investment tax credits to those who installed wind turbines. The state which provided the best incentives, and which also had regions with good winds, was California. It was now possible to install a number of small turbines together in a group ("wind farm"), connect them to the grid, and make some money.

The California wind rush was the result. Over a period of a few years, thousands of wind turbines were installed in California, particularly in the Altamont Pass, San Gorgonio Pass, and Tehachipi. A typical installation is shown in Figure 1.18. The installed capacity reached approximately 1500 MW. The early years of the California wind rush were fraught with difficulties, however. Many of the machines were essentially still prototypes, and not yet up to the task. An investment tax credit (as opposed to a production tax credit) is arguably not the best way to encourage the development and deployment of productive machines, especially when there is no means for ensuring that machines will actually perform as the manufacturer claims. When the federal tax credits were withdrawn by the Reagan administration in the early 1980s, the wind rush collapsed.

Wind turbines installed in California were not limited to those made in the United States, however. In fact, it was not long before Danish turbines began to have a major presence in the California wind farms. The Danish machines also had some teething problems in California, but in general, they were closer to production quality than their US counterparts. When all the dust had settled after the wind rush had ended, the majority of US manufacturers had gone out of business. The Danish manufacturers had been restructured or merged but had survived.

During the 1990s, a decade that saw the demise (in 1996) of the largest US manufacturer, Kenetech Windpower, the focal point of wind turbine manufacturing moved definitively to Europe, particularly Denmark

Figure 1.18   California wind farm (*Source:* Dicklyon/Wikimedia Commons/CC BY-SA 4.0)

and Germany. Concerns about climate change and continued apprehension about nuclear power resulted in a strong demand for more wind generation there and in other countries as well. The 21st century has seen some of the major European suppliers establish manufacturing plants in other countries, such as China, India, and the United States.

Since the mid-1970s, the size of the largest commercial wind turbines, as illustrated in Figure 1.19, has increased from approximately 25 kW–15 MW, with even larger machines under design. The total installed capacity in the world as of the year 2021 was about 837 GW (GWEC, 2022). Design standards and machine certification procedures have been established, so that the reliability and performance are far superior to those of the 1970s and 1980s. Offshore wind energy has become a reality; see Section 1.4. The cost of energy from wind has dropped to the point that in many sites it is on par with conventional sources, even without incentives. In those countries where incentives are in place, the rate of development is quite strong.

A comprehensive history of modern wind energy since the 1970s may be found in Gipe and Möllerström (2022b).

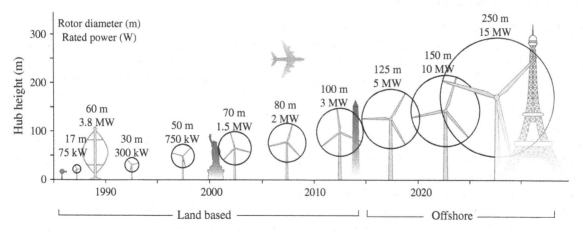

Figure 1.19   Representative size, height, and diameter of wind turbines

1.2.4 Technological Underpinnings of Modern Wind Turbines

Wind turbine technology, dormant for many years, awoke at the end of the 20th century to a world of new opportunities. Developments in many other areas of technology were adapted to wind turbines and have helped to hasten their re-emergence. A few of the many areas that have contributed to the new generation of wind turbines include materials science, computer science, aerodynamics, analytical design and analysis methods, testing and monitoring, and power electronics and controls. Materials science has brought new composites for the blades and alloys for the metal components. Developments in computer science facilitate design, analysis, monitoring, and control. Aerodynamic design methods, originally created for the aerospace industry, have now been adapted to wind turbines. Analytical design and analysis methods have now developed to the point where it is possible to have a much clearer understanding of how a new design should perform than was previously possible. Testing and monitoring using a vast array of commercially available sensors and modern data collection and analysis equipment allow designers to better understand how the new turbines actually perform. Power electronics is now widely used with wind turbines. Power electronic devices can help connect the turbine's generator smoothly to the electrical network; allow the turbine to run at variable speed, producing more energy, reducing fatigue damage, and benefit the utility in the process; facilitate operation in a small, isolated network; and transfer energy to and from storage.

1.2.5 Trends

Wind turbines have evolved a great deal over the last 50 years. They are more reliable, more cost-effective, and quieter. It cannot be concluded that the evolutionary period is over, however. It should still be possible to reduce the cost of energy at sites with lower wind speeds. Turbines for use in remote communities still remain to be made commercially viable. The world of offshore wind energy is just in its infancy. There are tremendous opportunities in offshore locations but many difficulties remain to be overcome. As wind energy comes to supply an ever-larger fraction of the world's electricity, the issues of variability, transmission, and storage must be revisited.

There will be continuing pressure on designers to improve the cost-effectiveness of turbines for all applications. Improved engineering methods for the analysis, design, and mass-produced manufacturing will be required. Opportunities also exist for the development of new materials to increase wind turbine life. Increased consideration will need to be given to the requirements of specialized applications. In all cases, the advancement of the wind industry represents an opportunity and challenge for a wide range of disciplines, especially including mechanical, electrical, materials, aeronautical, controls, ocean and civil engineering as well as computer science.

1.2.6 Wind Farms/Wind Power Plants

Wind farms or wind power plants, as they are now often called, are locally concentrated groups of wind turbines that are electrically and commercially tied together. There are many advantages to this electrical and commercial structure. Profitable wind resources are limited to distinct geographical areas. The introduction of multiple turbines into these areas increases the total wind energy produced. From an economic point of view, the concentration of repair and maintenance equipment and spare parts reduces costs. In wind farms of more than about 10 or 20 turbines, dedicated maintenance personnel can be hired, resulting in reduced labor costs per turbine and financial savings for wind turbine owners.

Wind farms were developed first in the United States in the late 1970s and then in Europe. Recently, wind farms have been developed in many other places around the world, most notably in China, but also in India, Japan, and South and Central America.

The oldest existing concentration of wind farms in the United States is in California. The California wind farms originated as a result of a number of economic factors, including tax incentives and the high cost of new conventional generation. These factors spurred a significant boom in wind turbine installation activity in California. The boom started in the late 1970s and then leveled off after 1984 when economic forces changed. The result has been the development of three main areas of California: (1) Altamont Pass, east of San Francisco, (2) the Tehachapi mountains, and (3) San Gorgonio Pass in southern California (see Figure 1.18). Many of these first wind turbines suffered from reliability problems, but in recent years older turbines have been replaced with larger, more reliable, turbines in what is known as "repowering" of the wind farms. In the 1990s, wind farms were also developed in the Midwest region of the United States. Since about 2000, development has continued at an exponential pace in the United States, with significant installations in the upper Midwest, Texas, and the state of Washington. As of the end of 2020, the United States had over 122 GW of installed wind power capacity, almost all of it in wind farms (EERE, 2021).

Wind farms in Europe started in the late 1980s in Denmark. In recent years, the number of wind turbines installed in Europe has increased tremendously as individual wind turbines, and wind farms have been developed, primarily in Denmark, Germany, Spain, the Netherlands, and Great Britain. Many of these wind turbines are in coastal areas. As available land for wind power development has become more limited in Europe, smaller installations have been built on inland mountains and in offshore wind parks, primarily in the North Sea. As of the end of 2021, Europe had over 236 GW of installed wind power capacity, most of it in wind farms (WindEurope, 2022).

The growth of wind farms in China has been especially rapid. As of 2022, China had over 217 GW of installed wind power, primarily in wind farms. See (GWEC, 2022).

1.3 Rationale for Wind Energy

In recent years, the major impetus for developing wind energy has been that it is a low-carbon alternative to fossil fuels, the burning of which is the primary source of climate change; see, for example, (IPCC, 2022). To address climate change and achieve overall social and environmental sustainability, researchers are demonstrating that rapidly decarbonizing electricity systems is both necessary and feasible; see Luderer *et al.* (2021) and Rockström *et al.* (2023).

The use of wind energy significantly reduces the amount of other pollutants that accompany the production of electricity from fossil fuels. For example, Figure 1.20, from NREL (2021), summarizes the lifecycle greenhouse gas emissions (GHGs), such as methane, nitrogen oxides, as well as carbon dioxide, from coal and combined cycle gas turbine power plants and compares them to those from wind. For fossil-fueled technologies, fuel combustion during operation of the facility emits the vast majority of GHGs. For nuclear and renewable energy technologies, most GHG emissions occur upstream of operation.

In addition, using wind energy instead of fossil fuels eliminates the cooling water that would otherwise be required by a comparable combustion-based plant or a nuclear reactor.

1.4 Offshore Wind Energy

Offshore wind has arisen much more recently than land-based wind, but it has its own history. Fundamentally, in order to be seriously considered, it was necessary that turbines themselves be large enough and reliable enough that they could generate enough electricity to pay for themselves. In addition, there needed to be the possibility to construct, install, and maintain the support structures that would be needed to hold the turbines above the sea surface. Considerable relevant experience came from the offshore oil and gas industry, which installed its first offshore platform in 1887 in a few meters of water. As of 2010, the oil and gas industry had installed a floating platform in water 2450 m deep.

Even before constructing turbines offshore was actually practical, concepts of various levels of sophistication were proposed. The first known idea for offshore wind turbines came from Hermann Honnef (1932). In the 1970s, William Heronemus proposed the use of spars for floating offshore wind turbines (Heronemus, 1972).

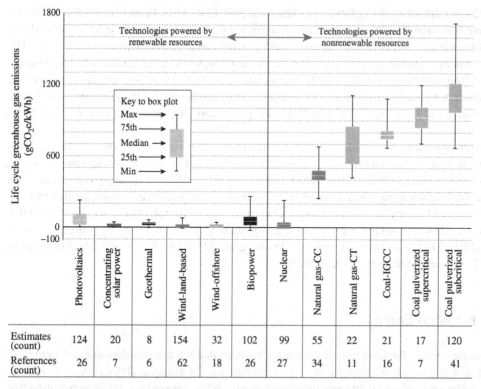

Figure 1.20 Lifecycle Greenhouse Gas Emission Estimates for Selected Electricity Generation and Storage Technologies (*Source:* Adapted from NREL, 2021)

The first group of actual offshore wind turbines was installed at Vindeby in Denmark in 1991. This wind project, which operated until 2016, consisted of eleven 450 kW machines approximately 3 km from the shore. As of 2022, 57 GW of offshore turbines had been installed worldwide, mostly in Europe and China.

As will be discussed, there are challenges to installing, operating, and maintaining wind turbines offshore, but there are good reasons for doing so. Some of these reasons are the following:

- The wind speeds offshore are generally higher offshore compared to onshore sites
- The intrinsic turbulence in the wind is lower offshore
- There is less wind shear (variation of wind speed with height) offshore
- It is possible to utilize larger turbines offshore than onshore
- There is extensive area available for siting large projects offshore
- Many promising offshore sites are relatively close to cities and load centers
- There is potential for lower environmental and societal impacts in many locations

The are many factors to consider with offshore wind energy. These include the following:

- Offshore meteorological oceanographic conditions
- Design of offshore wind turbines
- Dynamics and control of offshore wind turbines
- Support structures for fixed-bottom offshore wind turbines
- Floating offshore wind turbines

- Offshore wind electricity generation and transmission
- Offshore wind power plant design
- Installation and operation of offshore wind turbines
- Economics of offshore wind energy
- Environmental and permitting aspects of offshore wind energy
- Social acceptance and community engagement

All these topics are discussed in more detail in subsequent chapters.

1.5 Reference Wind Turbine

Throughout this text, we will have occasion to illustrate many of the concepts. This process is facilitated by the use of a "reference" wind turbine. Over the last twenty years, the National Renewable Energy Laboratory (NREL) has created four conceptual reference wind turbines that would be possible candidates. These are rated at 750 kW, 1.5, 3, and 5 MW. Since then, the Danish Technical University (DTU) developed a 10 MW conceptual turbine. More recently, an International Energy Agency working group has devised one rated at 15 MW. One of these turbines, the NREL 5-MW, has been widely used and is of such a size that it could be used on land as well as offshore. For that reason, we have elected to use the NREL 5-MW as the reference turbine throughout this book. This turbine was patterned after a commercial turbine, the REpower 5M, which had a rotor diameter of 126 m and was depicted above in Figure 1.1. The main characteristics of the NREL 5-MW are summarized in Tables 1.1 and 1.2. Note that the tower for the offshore turbine is shorter than for offshore, but the hub height is the same for both. This is because it is assumed that the offshore support structure extends 10 m above the water level. The term RNA refers to the rotor nacelle assembly, which consists of the blades, hub, drivetrain, and all other associated components on top of the tower. The turbine is described in detail in Jonkman *et al.* (2009). Other documents provide more information about that turbine on various offshore configurations. Information about all the NREL reference turbines may be found in Rinker and Dykes (2018) and about the DTU 10 MW in Bak *et al.* (2013). Additional details about the NREL 5-MW will be given elsewhere in the text as needed.

Table 1.1 Principal characteristics of NREL 5-MW reference wind turbine

Property	Value	Units
Rated power	5	MW
Hub height	90	m
Rotor diameter	$126 = 2\,(1.5 + 61.5)$	m
RNA mass	350	t
Number of blades	3	—
Gearbox gear ratio	97	—
Generator type	Variable speed	—
Power control	Collective pitch	—

Table 1.2 Characteristics of the NREL 5-MW reference wind turbine's tower, onshore and offshore

Property	Onshore	Offshore	Units
Tower height	87.6	77.6	m
Tower top diameter	3.87	3.87	m
Tower base diameter	6.0	6.5	m
Tower top wall thickness	0.0247	.019	m
Tower base wall thickness	0.0351	.027	m
Tower mass	347.4	237	t

1.6 Layout of the Book

This book consists of 16 chapters, including this introduction, and three appendices. Chapter 2 describes the wind resource, which provides the available power. Chapter 3 discusses the aerodynamic processes that convert the power in the wind to mechanical power. Chapter 4 concerns the mechanics and dynamics of the turbine, focusing on the rotor and tower. Chapter 5 introduces electrical power and generation. Chapter 6 concerns the external conditions, particularly wind and waves, that form the basis of wind turbine structural design. Chapter 7 is about the materials and important components and subsystems that make up a wind turbine. Chapter 8 discusses wind turbine design. Chapter 9 discusses wind turbine controls. Chapter 10 is about support structures for land-based and fixed offshore wind turbines. Chapter 11 concerns floating support structures and stationkeeping (mooring) systems for offshore turbines in deep water. Chapter 12 is about multiple turbine projects (wind farms) and wind power plants, wakes, and blockage effects. Chapter 13 is about wind energy economics. Chapter 14 is about wind energy project development, including permitting, environmental issues, and public engagement. Chapter 15 is about wind turbine/project installation, operation, and maintenance. Finally, Chapter 16 provides an overview of the expected wind power plants of the future which may provide a very large fraction of the electrical network in which they are installed.

It is assumed throughout the book that readers have a background in mathematics through calculus and basic physics. Familiarity with linear algebra and the fundamental engineering topics of statics, dynamics, strength of materials, and fluid mechanics would also be helpful. Within the text there are certain sections that are more advanced; they are indicated with asterisks.

The appendices provide a summary of nomenclature (A) and a more in-depth discussion of a few fundamental background topics: data processing (B), probability and statistics (C), properties of fundamental support structure components (D) and information regarding problems and associated files (E).

References

Ackermann, T. and Soder, L. (2002) An overview of wind energy-status 2002. *Renewable and Sustainable Energy Reviews*, **6**, 67–127.

Bak, C.; Zahle, F., Bitsche, R., Kim, T., Yde, A., Henriksen, L. C., Hansen, M. H., Blasques, J. P. A. A., Gaunaa, M. and Natarajan, A. (2013) *The DTU 10-MW Reference Wind Turbine, DTU Wind Energy Section*, https://backend.orbit.dtu.dk/ws/files/55645274/The_DTU_10MW_Reference_Turbine_Christian_Bak.pdf Accessed 6/10/2023.

Berger, J. J. (1997) *Charging Ahead: The Business of Renewable Energy and What it Means for America*. University of California Press, Berkeley, CA.

Carson, R. (1962) *Silent Spring*. Houghton Mifflin, New York.

Casson, L. (1991) *The Ancient Mariners: Seafarers and Sea Fighters of the Mediterranean in Ancient Times*. Princeton University Press, Princeton, NJ.

Dörner, H. (2002) *Drei Welten- ein Leben: Prof. Dr Ulrich Hütter*, Heilbronn.

Drachman, A. G. (1961) Heron's windmill. *Centaurus*, **7**(2), 145–151.

EERE (2021) *Wind Market Reports: 2021 Edition*, Office of Energy Efficiency and Renewable Energy (EERE), https://www.energy.gov/eere/wind/wind-market-reports-2021-edition Accessed 6/10/2023.

Eggleston, D. M. and Stoddard, F. S. (1987) *Wind Turbine Engineering Design*. Van Nostrand Reinhold, New York.

Eldridge, F. R. (1980) *Wind Machines*, 2nd edition. Van Nostrand Reinhold, New York.

Gipe, P. (1995) *Wind Energy Comes of Age*. John Wiley & Sons, Inc., New York.

Gipe, P. and Möllerström, E. (2022a) An overview of the history of wind turbine development: part I – the early wind turbines until the 1960s. *Wind Engineering*, **46**(6), 1973.

Gipe, P. and Möllerström, E. (2022b) An overview of the history of wind turbine development: part II – the 1907s Onward. *Wind Engineering*, **47**(1), 220.

Golding, E. W. (1977) *The Generation of Electricity by Wind Power*. E. & F. N. Spon, London.

GWEC (2022) *Global Wind Report 2022*, https://gwec.net/global-wind-report-2022/ Accessed 11/30/2023.

Harrison, R., Hau, E. and Snel, H. (2000) *Large Wind Turbines: Design and Economics*. John Wiley & Sons, Ltd, Chichester.

Heronemus, W. E. (1972) Pollution-free energy from offshore winds. *Proceedings of 8th Annual Conference and Exposition,* Marine Technology Society: Washington, DC.

Hills, R. L. (1994) *Power from Wind*. Cambridge University Press, Cambridge, UK.

Honnef, H. (1932) *Windkraftwerke*. Friedrich Vieweg & Sohn, Braunschweig, Germany.

IPCC (2022) *Climate Change 2022: Mitigation of Climate Change, Intergovernmental Panel on Climate Change*, https://www.ipcc.ch/report/ar6/wg3/downloads/report/IPCC_AR6_WGIII_FullReport.pdf Accessed 6/10/2023.

Jacobs, M. L. (1961) Experience with Jacobs wind-driven electric generating plant, 1931–1957. *Proceeding of the United Nations Conference on New Sources of Energy*, **7**, 337–339.

Johnson, G. L. (1985) *Wind Energy Systems*. Prentice Hall, Englewood Cliffs, NJ.

Jonkman, J., Butterfield, S., Musial, W. and Scott, G. (2009) *Definition of a 5-MW Reference Wind Turbine for Offshore System Development*, NREL/TP-500-38060, Golden, CO.

Le Gourieres, D. (1982) *Wind Power Plants*. Pergamon Press, Oxford.

Luderer, G., Madeddu, S., Merfort, L., Ueckerdt, F., Pehl, M., Pietzcker, R., Rottoli, M., Schreyer, F., Bauer, N., Baumstark, L., Bertram, C., Dirnaichner, A., Humpenöder, F., Levesque, A., Popp, A., Rodrigues, R., Strefler, J. and Kriegler, E. (2021) Impact of declining renewable energy costs on electrification in low-emission scenarios. *Nature Energy*, https://doi.org/10.1038/s41560-021-00937-z.

Maegaard, P., Krenz, A. and Palz, W. (2013) *Wind Power for the World: The Rise of Modern Wind Energy*. Taylor and Francis Group.

Marchant, J. (2006) In search of lost time. *Nature*, **444**, 534–538.

McGowan, J. G. and Connors, S. R. (2000) Windpower: a turn of the century review. *Annual Review of Energy and the Environment*, **25**, 147–197.

Meadows, D. H., Meadows, D. L., Randers, J. and Behrens III, W. W. (1972) *The Limits to Growth*. Universe Books, New York.

Nelson, V. (1996) *Wind Energy and Wind Turbines*. Alternative Energy Institute, Canyon, TX.

NREL (2021) *Life Cycle Greenhouse Gas Emissions from Electrcity Generation: Update*. NREL FS-6A50-80580, Golden, CO. September 2021.

Paraschivoiu, I. (2002) *Wind Turbine Design: With Emphasis on Darrieus Concept*. Polytechnic International Press, Montreal.

Park, J. (1981) *The Wind Power Book*. Cheshire Books, Palo Alto, CA.

Pice, T. J. (2006) UK Large-scale wind power programme from 1970 to 1990: the Carmarthen Bay experiments and the Musgrove vertical-axis turbines. *Wind Engineering*, **30**(3), 225–242.

Price, T.J. (2005) James Blyth – Britain's first modern wind power pioneer. *Wind Engineering*, **29**(3), 191–200.

Putnam, P. C. (1948) *Power From the Wind.* Van Nostrand Reinhold, New York.

Reynolds, T. S. (1983) *Stronger than a Hundred Men: A History of the Vertical Water Wheel.* Johns Hopkins University Press, Baltimore.

Rinker, J. and Dykes, K. (2018) *WindPACT Reference Wind Turbines*, NREL Technical Report, NREL/TP-5000-67667, available from https://www.nrel.gov/docs/fy18osti/67667.pdf. Accessed 6/10/2023.

Rockström, J., Gupta, J., Qin, D., Lade, S.J., Abrams, J. F., Andersen, L. S., Armstrong McKay, D. I., Bai, X., Bala, G., Bunn, S. E., Ciobanu, D., DeClerck, F., Ebi, K., Gifford, L., Gordon, C., Hasan, S., Kanie, N., Lenton, T. M., Loriani, S., Liverman, D. M., Mohamed, A., Nakicenovic, N., Obura, D., Ospina, D., Prodani, K., Rammelt, C., Sakschewski, B., Scholtens, J., Stewart-Koster, B., Tharammal, T., van Vuuren, D., Verburg, P. H., Winkelmann, R., Zimm, C., Bennett, E. M., Bringezu, S., Broadgate, W., Green, P. A., Huang, L., Jacobson, L., Ndehedehe, C., Pedde, S., Rocha, J., Scheffer, M., Schulte-Uebbing, L., de Vries, W., Xiao, C., Xu, C., Xu, X., Zafra-Calvo, N. and Zhang, X. (2023) Safe and just Earth system boundaries. *Nature*, https://doi.org/10.1038/s41586-023-06083-8.

Shepherd, D. G. (1990) *The Historical Development of the Windmill*, NASA Contractor Report 4337, DOE/NASA 52662.

Smeaton (1837) *Reports of the Late John Smeaton*, F.R.S., Made on Various Occasions, in the Course of His Employment as a Civil Engineer. 2nd ed. London, UK: M. Taylor.

Spera, D. A. (Ed.) (1994) *Wind Turbine Technology: Fundamental Concepts of Wind Turbine Engineering.* ASME Press, New York.

Stokhuyzen, F. (1962) *The Dutch Windmill*, Merlin Press, Dublin, Ireland.

Sutherland, H. J., Berg, D. E. and Ashwill, T. D. (2012) *A Retrospective of VAWT Technology.* Sandia Report SAND2012-0304 Sandia National Laboratories, Albuquerque, NM.

US Department of Energy (1979) *Wind Energy Innovative Systems Conference Proceedings*, Solar Energy Research Institute (SERI).

US Department of Energy (1980) *SERI Second Wind Energy Innovative Systems Conference Proceedings*, Solar Energy Research Institute (SERI).

van Bussel, G. J. W. (2007) The science of making more torque from wind: diffuser experiments and theory revisited. *Journal of Physics: Conference Series*, **75**, 1–11.

Vowles, H. P. (1930) An inquiry into the origins of the windmill. *Journal of the Newcomen Society*, **11**, 1–14.

Vowles, H. P. (1932) Early evolution of power engineering. *Isis*, **17**(2), 412–420.

WindEurope (2022) *Wind Energy in Europe: 2021 Statistics and the Outlook for 2022-2026*, https://windeurope.org/intelligence-platform/product/wind-energy-in-europe-2021-statistics-and-the-outlook-for-2022-2026/ Accessed 6/2/2023.

Woodcroft, B. (1851) *Translation from the Greek of The Pneumatics of Hero of Alexandria.* Taylor Walton and Maberly, London. Available at: https://www.loc.gov/item/07041532/ Accessed 2/7/2024.

2

Wind Characteristics and Resources

2.1   Introduction

This chapter will cover an important topic in wind energy: wind resources and characteristics – especially as they apply to wind power production. The material covered in this chapter can be of direct use to other aspects of wind energy which are discussed elsewhere in this book. For example, knowledge of the wind characteristics at a particular site is relevant to the following topics:

- **Performance evaluation**. Performance evaluation requires determining the expected energy productivity and cost-effectiveness of a particular wind energy project based on the wind resource.
- **Siting**. Siting requirements can include the assessment or prediction of the relative desirability of candidate sites for one or more wind turbines.
- **Installation/operation/maintenance**. The wind conditions have a significant effect on the installation, operation, and maintenance of wind turbines. Wind resource information is useful for load management and defining operational procedures (such as start-up and shutdown), and the prediction of maintenance or system life. Knowledge of wind conditions is important to determine "weather windows" used in the scheduling of turbine installation and maintenance, particularly offshore. Such scheduling is often affected by unusual occurrences (high winds).

The focus of this chapter is on the wind resource as the "fuel" of the wind turbine, that is, the source of the kinetic energy which the turbine converts into mechanical and then electrical form. Later on, Chapter 6 will discuss those aspects of the wind that are most relevant to turbine design. The present chapter starts with a general discussion of wind resource characteristics in Section 2.2, followed by Section 2.3 which is about the characteristics of the atmospheric boundary layer that are directly applicable to wind energy applications. Section 2.4 discusses wind data analysis and resource estimation. Section 2.5 is about wind turbine energy production estimates using statistical techniques. Section 2.6 provides an overview of available worldwide wind resource assessment data followed by a discussion of wind prediction and forecasting in Section 2.7. The next Section 2.8, reviews wind resource measurement techniques and instrumentation. The chapter concludes in Section 2.9 with a summary of some more advanced topics.

There are numerous other sources of information on wind characteristics as related to wind energy. Historically, these include the classic reference of Putnam (1948) as well as the works of Justus (1978), Hiester and Pennell (1981), and Rohatgi and Nelson (1994). More recent work on the subject includes that of Brower (2012), Landberg (2016), and Lundquist *et al.* (2019).

Wind Energy Explained: On Land and Offshore, Third Edition.
James F. Manwell, Emmanuel Branlard, Jon G. McGowan, and Bonnie Ram.
© 2024 John Wiley & Sons Ltd. Published 2024 by John Wiley & Sons Ltd.

2.2 General Characteristics of the Wind Resource

In discussing the wind resource, it is important to consider such topics as the global origins of the wind, the general characteristics of the wind, and estimates of the wind resource potential.

2.2.1 Wind Resource: Global Origins

2.2.1.1 Overall Global Patterns

The original source of the energy contained in the earth's wind resource is the sun. Global winds are caused by pressure differences across the earth's surface due to the uneven heating of the earth by solar radiation. Most obviously, the amount of solar radiation absorbed at the earth's surface is greater at the equator than at the poles. The variation in incoming energy sets up convective cells in the lower layers of the atmosphere (the troposphere). In a simple flow model, warm air rises at the equator and sinks at the poles after cooling down. The circulation of the atmosphere that results from uneven heating is greatly influenced by the effects of the rotation of the earth, resulting in a speed of about 464 m/s at the equator, decreasing to near zero at the poles. In addition, seasonal variations in the distribution of solar energy give rise to variations in the circulation.

The spatial variations in heat transfer to the earth's atmosphere create variations in the atmospheric pressure field that cause air to move from high to low pressure. There is a pressure gradient force in the vertical direction, but this is usually canceled by the downward gravitational force. Thus, the winds blow predominately in the horizontal plane, responding to horizontal pressure gradients. At the same time, there are forces that strive to mix the different temperature and pressure air masses distributed across the earth's surface. In addition to the pressure gradient and gravitational forces, inertia of the air, the earth's rotation, and friction with the earth's surface (resulting in turbulence), all affect the atmospheric winds. The influence of each of these forces on atmospheric wind systems differs depending on the scale of motion considered.

As shown in Figure 2.1, worldwide wind circulation involves large-scale wind patterns which cover the entire planet. These affect prevailing near-surface winds. It should be noted that this model is an oversimplification because it does not reflect the effect that land masses have on wind distribution.

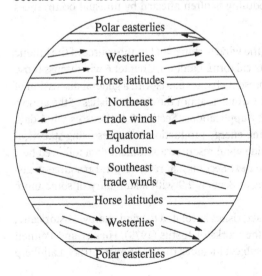

Figure 2.1 Near-surface (geostrophic) winds of worldwide circulation pattern (*Source:* Hiester and Pennell, 1981/U.S. Department of Energy/Public domain)

2.2.1.2 Mechanics of Wind Motion

In one of the simplest models for the mechanics of the atmosphere's motion, four atmospheric forces can be considered. These include pressure forces, the Coriolis force caused by the rotation of the earth, inertial forces due to large-scale circular motion, and frictional forces at the earth's surface.

The pressure force on the air (per unit mass, N/kg), F_p, is given by:

$$F_p = \frac{-1}{\rho} \frac{\partial p}{\partial n} \tag{2.1}$$

where ρ is the density of the air (kg/m^3) and n is the direction normal to lines of constant pressure. Also, $\partial p / \partial n$ is defined as the pressure gradient normal to the lines of constant pressure, or isobars. The Coriolis force (per unit mass), F_c, a fictitious force caused by measurements with respect to a rotating reference frame (the earth), is expressed as:

$$F_c = f\,U \tag{2.2}$$

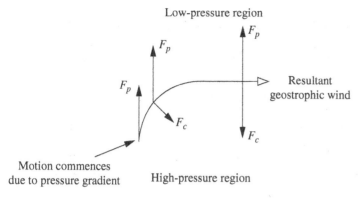

Figure 2.2 Illustration of the geostrophic wind; F_p, pressure force on the air; and F_c, Coriolis force

where U (m/s) is the wind speed and f is the Coriolis parameter [$f = 2\omega \sin(\phi)$], ϕ represents the latitude and ω the angular rotation of the earth. Thus, the magnitude of the Coriolis force depends on wind speed and latitude. The direction of the Coriolis force is perpendicular to the direction of motion of the air. The resultant of these two forces leads to an equilibrium wind, called the geostrophic wind, which tends to be parallel to isobars (see Figure 2.2).

The magnitude of the geostrophic wind, U_g, is a function of the balance of forces and is given by:

$$U_g = \frac{-1}{f\rho} \frac{\partial p}{\partial n} \tag{2.3}$$

This is an idealized case since the presence of areas of high and low pressure causes the isobars to be curved. The curvature imposes a further force on the wind, a centrifugal force. The resulting wind, called a gradient wind, U_{gr}, is shown in Figure 2.3.

The gradient wind is also parallel to the isobars and is the result of the balance of the forces:

$$\frac{U_{gr}^2}{R} = -f\,U_{gr} - \frac{1}{\rho} \frac{\partial p}{\partial n} \tag{2.4}$$

where R is the radius of curvature of the path of the air particles.

A final force on the wind is due to friction at the earth's surface. That is, the earth's surface exerts a horizontal force upon the moving air, the effect of which is to retard the flow. This force decreases as the height above the ground increases and becomes negligible above the boundary layer (defined as the near-earth region of the atmosphere where viscous forces are important). Above the boundary layer, a frictionless wind balance is established, and the wind flows with the gradient wind velocity along the isobars. The term geostrophic wind is commonly used to refer to the high-altitude wind, above the boundary layer, that drives the flow within it. In the boundary layer, friction at the surface causes the wind to be diverted more toward the low-pressure region. More details concerning the earth's boundary layer and its characteristics will also be given in later sections.

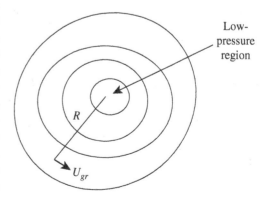

Figure 2.3 Illustration of the gradient wind U_{gr}; R, radius of curvature

2.2.1.3 Other Atmospheric Circulation Patterns

The general circulation flow pattern described previously best represents a model for a smooth spherical surface. In reality, the earth's surface varies considerably, with large ocean and land masses. These different surfaces can affect the flow of air due to variations in pressure fields, the absorption of solar radiation, and the amount of moisture available.

The oceans act as a large sink for energy. Therefore, the movement of air is often affected by the ocean circulation. All these effects lead to differential pressures which affect the global winds and many of the persistent regional winds, such as those occurring during monsoons. In addition, local heating or cooling may cause persistent local winds to occur on a seasonal or daily basis. These include sea breezes and mountain winds.

Smaller-scale atmospheric circulation can be divided into secondary and tertiary circulation (see Rohatgi and Nelson, 1994). Secondary circulation occurs if the centers of high or low pressure are caused by heating or cooling of the lower atmosphere. Secondary circulations include the following:

- tropical cyclones (e.g., hurricanes, typhoons, depending on the region);
- monsoon circulation;
- extratropical cyclones.

Tertiary circulations are small-scale, local circulations characterized by local winds. These include the following:

- land and sea breezes;
- valley and mountain winds (example: flow in California passes);
- monsoon-like flow;
- foehn winds (dry, high-temperature winds on the downwind side of mountain ranges);
- thunderstorms;
- tornadoes.

Examples of tertiary circulation, valley and mountain winds, are shown in Figure 2.4. During the day, the warmer air of the mountain slope rises and replaces the heavier cool air above it. The direction reverses at night, as cold air drains down the slopes and stagnates in the valley floor.

An understanding of these wind patterns, and other local effects, is important for the evaluation of potential wind energy sites.

2.2.2 Temporal and Spatial Characteristics of the Wind

Atmospheric motions vary in both time (seconds to months) and space (centimeters to thousands of kilometers). Figure 2.5 summarizes the time and space variations of atmospheric motion as applied to wind energy. As will be

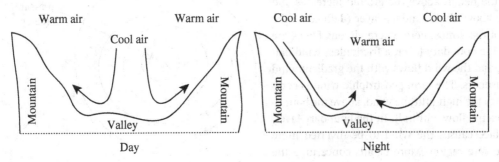

Figure 2.4 Diurnal valley and mountain wind (Rohatgi and Nelson, 1994). (*Source:* Reproduced by permission of Alternative Energy Institute)

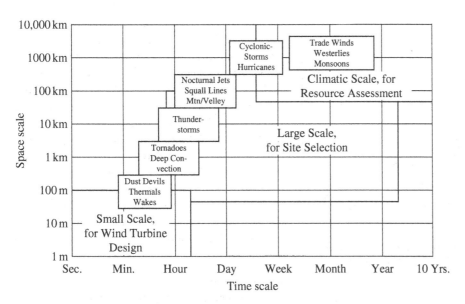

Figure 2.5 Time and space scales of atmospheric motion (Spera, 1994). (*Source:* Reproduced by permission of ASME)

discussed subsequently, spatial variations are generally dependent on height above the ground and global and local geographical conditions.

Variations in Time

Following conventional practice, variations in wind speed in time can be divided into the following categories:

- interannual;
- annual;
- diurnal;
- short-term (gusts and turbulence).

A review of each of these categories as well as comments on wind speed variation due to location and wind direction follow.

Interannual

Interannual variations in wind speed occur over time scales greater than one year. They can have a large effect on long-term average wind turbine power production. The ability to estimate the interannual variability at a given site is almost as important as estimating the long-term mean wind at a site. Meteorologists generally conclude that it takes 30 years of data to determine long-term values of weather or climate and that it takes at least five years to arrive at a reliable average annual wind speed at a given location. Nevertheless, shorter data records can be useful. Aspliden *et al.* (1986) provide a rule of thumb that one year of record data is generally sufficient to predict long-term seasonal mean wind speeds with an accuracy of ±10% with a confidence level of 90%. However, the effects of climate change can add another layer of complexity to long-term forecasting.

Annual

Significant variations in seasonal or monthly averaged wind speeds are common over most of the world. For example, for the eastern one-third of the United States, maximum wind speeds occur during the winter and early spring. Spring maxima occur over the Great Plains, the North Central States, the Texas Coast, in the basins and valleys of

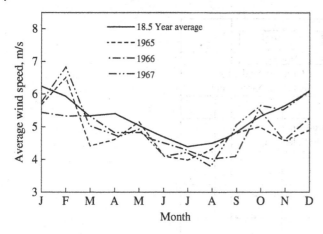

Figure 2.6 Seasonal changes in monthly average wind speeds (*Source:* Hiester and Pennell, 1981/U.S. Department of Energy/Public domain)

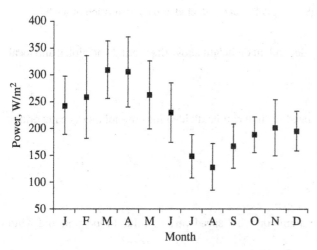

Figure 2.7 Seasonal variation in available wind power per unit area for Amarillo, Texas (Rohatgi and Nelson, 1994). (*Source:* Reproduced by permission of Alternative Energy Institute)

the West, and the coastal areas of Central and Southern California. Winter maxima occur over all US mountainous regions, except for some areas in the lower Southwest, where spring maxima occur. Spring and summer maxima occur in the wind corridors of Oregon, Washington, and California.

Figure 2.6 illustrates seasonal changes in monthly wind speed for Billings, Montana. This figure illustrates that the typical behavior of monthly variation cannot defined by a single year of data.

Similarly, Figure 2.7 provides an illustration of the importance of annual wind speed variation and its effect on available wind power (error bars show the standard deviation).

Diurnal (Time of Day)

In both tropical and temperate latitudes, large wind variations also can occur on a diurnal, or daily, time scale. This type of variation is due to differential heating of the earth's surface during the daily radiation cycle. A typical diurnal variation is an increase in wind speed during the day with the wind speeds lowest during the hours from midnight to sunrise. Daily variations in solar radiation are responsible for diurnal wind variations in temperate latitudes over relatively flat land areas. The largest diurnal changes generally occur in spring and summer, and the smallest in winter. Furthermore, the diurnal variation in wind speed may vary with location and altitude above sea level. For example, at altitudes high above surrounding terrain, e.g., mountains or ridges, the diurnal pattern may be very different. This variation can be explained by mixing, or momentum transfer, from the upper air to the lower air, and this mixing is different at different times of the day. The diurnal patterns of momentum transfer will also affect the vertical profile of the wind and the atmospheric stability (see Sections 2.3.3 and 2.3.4).

As illustrated in Figure 2.8, there may be significant year-to-year differences in diurnal behavior, even at fairly windy locations. Although gross features of the diurnal cycle can be established with a single year of data, more detailed features such as the amplitude of the diurnal oscillation and the time of day that the maximum winds occur cannot be determined precisely.

Short-term

Short-term wind speed variations of interest include turbulence and gusts. Figure 2.9, illustrating output from an anemometer (described later), shows the type of short-term wind speed variations that can be expected in the atmospheric boundary layer.

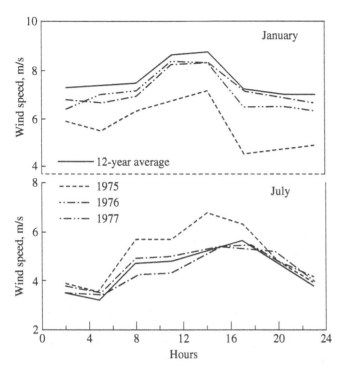

Figure 2.8 Monthly mean diurnal wind speeds for January and July for Casper, Wyoming (*Source:* Hiester and Pennell, 1981/U.S. Department of Energy/Public domain)

Turbulence can be thought of as random wind speed fluctuations imposed on the mean flow. These fluctuations occur in all three directions: in the direction of the wind, perpendicular to the average wind, and vertical. Turbulence in the atmosphere arises from the irregular and chaotic movement of air particles, due to multiple factors, such as varying temperatures, pressure fluctuations, and obstacles in the flow. The time scale of turbulent variations is on the order of seconds to minutes. Turbulence and its effects will be discussed in more detail in Chapter 6.

A gust is a discrete event within a turbulent wind field. Gusts are of particular interest in the design of wind turbines and are also discussed in more detail in Chapter 6.

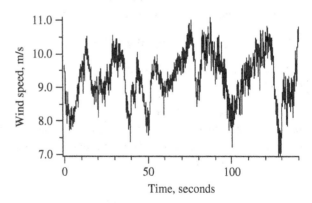

Figure 2.9 Typical plot of wind speed vs. time for a short period

Variations Due to Location

Wind speed is also very dependent on local topographical and ground cover variations. For example, as shown in Figure 2.10 differences between two sites close to each other can be significant. The graph shows monthly and five-year mean wind speeds for two sites 21 km apart. The five-year average mean wind speeds differ by about 12% (4.75 and 4.25 m/s annual averages).

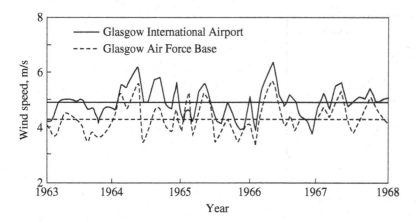

Figure 2.10 Time series of monthly wind speeds for Glasgow, Montana International Airport and Air Force Base (AFB) (*Source:* Hiester and Pennell, 1981/U.S. Department of Energy/Public domain)

Variations in Wind Direction

Wind direction also varies over the same time scales over which wind speeds vary. The occurrence of different wind directions is usually represented using a wind rose (see Section 2.8.6 and Figure 2.33). Seasonal variations may be small, on the order of 30°, or the average monthly winds may change direction by 180° over a year. Short-term direction variations are the result of the turbulent nature of the wind. These short-term variations in wind direction need to be considered in wind turbine design and siting. Other parameters, such as wind shear and turbulence intensity, are likely to be affected by the wind direction because of differences in surface roughness, terrain, etc. For offshore turbines, the relation between wind direction and fetch (the horizontal distance over which wave-generating winds blow) can also be significant. Fetch affects wave height and by extension affects surface roughness, wind shear, and turbulence; see Chapter 6.

2.2.3 Estimation of Potential Wind Resource

This section summarizes the available potential of the wind resource and how that relates to wind turbine power production.

2.2.3.1 Available Wind Power

As illustrated in Figure 2.11, one can determine the mass flow of air, $\dot{m} = \mathrm{d}m/\mathrm{d}t$, through a rotor disc of area A. From the continuity equation of fluid mechanics, the mass flow rate is a function of air density, ρ, and air velocity (assumed uniform), U, and is given by:

$$\dot{m} = \rho A U \tag{2.5}$$

From basic physics, the kinetic energy per unit time, or power (W), of the flow is given by:

$$P = \frac{1}{2}\dot{m}U^2 = \frac{1}{2}\rho A U^3 \tag{2.6}$$

The wind power per unit area, P/A or wind power density (W/m^2) is:

$$\frac{P}{A} = \frac{1}{2}\rho U^3 \tag{2.7}$$

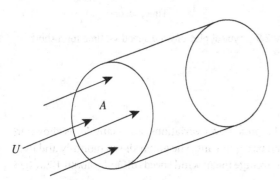

Figure 2.11 Flow of air through a rotor disc; A, area; U, wind velocity

One should note the following:

- The wind power density is proportional to the density of the air. For standard conditions (sea level, 15 °C), the density of air is 1.225 kg/m³.
- Power from the wind is proportional to the area swept by the rotor (or the rotor radius squared for a conventional horizontal axis wind machine).
- The wind power density is proportional to the cube of the wind velocity.

The actual power production potential of a wind turbine must take into account the fluid mechanics of the flow passing through a power-producing rotor, and the aerodynamics and efficiency of the rotor and drive train (including generator). In practice, a maximum of about 45% of the available wind power is harvested by the best modern horizontal-axis wind turbines (this will be discussed in Chapter 3).

One should note that due to the cubic relationship between power and wind speed, the average wind power density over the course of an extended time period, $\overline{P}/A$, is considerably higher than one would estimate based on the cube of the average wind speed.

Some sample qualitative magnitude evaluations of the wind resource, based on the average wind power density, are:

$$\overline{P}/A < 100 \text{ W/m}^2 - \text{low}$$
$$\overline{P}/A \approx 400 \text{ W/m}^2 - \text{good}$$
$$\overline{P}/A > 700 \text{ W/m}^2 - \text{great}$$

It should be noted, however, that there is now a trend toward designing turbines with low specific power rotors (i.e., large rotors in relation to generator rating) to produce energy at low wind speed speeds, so "low" wind resource areas can be utilized.

2.2.3.2 Estimates of Worldwide Resource

Based on wind resource data and an estimate of the real efficiency of actual wind turbines, numerous investigators have made estimates of the wind power potential of regions of the earth and of the entire earth itself. It will be shown in Chapter 3 that the maximum power-producing potential that can be theoretically realized from the kinetic energy contained in the wind is about 60% of the available power. Real efficiency is somewhat less than that.

Using estimates for regional wind resources, one can calculate the (electrical) power-producing potential of wind energy. It is important to distinguish between the different types of wind energy potential that can be estimated. Typical estimates (see World Energy Council, 1993; McKenna *et al.*, 2021) identify the following five categories:

1) **Meteorological potential.** This is equivalent to the available wind resource.
2) **Site or geographical potential.** This is based on the meteorological potential but is restricted to those sites that are geographically available for power production.
3) **Technical potential.** The technical potential is calculated from the site potential, accounting for the available technology.
4) **Economic potential.** The economic potential is the technical potential that can be realized economically.
5) **Implementation or feasible potential.** The implementation potential takes into account constraints and incentives to assess the wind turbine capacity that can be implemented within a certain time frame.

Historically, one of the earliest global wind energy resource assessments was carried out by Gustavson (1979). In this study, Gustavson based his resource estimate on the input of the solar energy reaching the earth and how much of this energy could be transformed into useful wind energy. On a global basis, his estimate was that the global wind resource was about 1000 PWh/yr (i.e., 10^{15} kWh/yr). In comparison, the global consumption of electricity at that time was about 5.5 PWh/yr, while the annual solar radiation incident on the earth's surface amounts to approximately 1.53 million PWh/yr.

In more recent times, there have been a number of very detailed studies that have examined the potential global wind resource on land (see Rinne *et al.*, 2018; McKenna *et al.*, 2021) and both land and offshore (see Eurek *et al.*, 2017). These studies considered the current wind turbine technology and land use constraints. It should also be noted that they calculated much higher estimates of the global wind resource than some other previous studies. For example, Eurek *et al.* (2017) estimated that the annual global onshore potential was 560 PWh/yr and offshore potential was 315 PWh/yr.

2.3 Characteristics of the Atmospheric Boundary Layer

The atmospheric boundary layer, also known as the planetary boundary layer, is the lowest part of the atmosphere and its characteristics are directly influenced by contact with the earth's surface. Here, physical quantities such as velocity, temperature, and relative humidity can change rapidly in space and time. For example, an important parameter in the characterization of the wind resource is the variation of horizontal wind speed with height above the ground. One would expect the horizontal wind speed to be zero at the earth's surface and to increase with height in the atmospheric boundary layer. This variation of wind speed with elevation is called the vertical profile of the wind speed or vertical wind shear (see Section 2.3.3.) In wind energy engineering, the determination of vertical wind shear is an important consideration since (1) it directly determines the productivity of a wind turbine with a certain hub height, and (2) it can strongly influence the lifetime of a turbine rotor blade; see Chapters 6–8.

In addition to variations due to the atmospheric stability, the variation of wind speed with height depends on surface roughness and terrain. These factors will be discussed in subsequent sections.

2.3.1 Atmospheric Density and Pressure

As is apparent in Equation 2.7, the power in the wind is proportional to air density. Air density, ρ (kg/m^3), is a function of temperature, T (°K), and pressure, p (kPa), both of which vary with height. The density of dry air can be determined by applying the ideal gas law:

$$\rho = \frac{p}{RT} \cong 3.4837 \frac{p}{T} \tag{2.8}$$

where R is the specific ideal gas constant for air (0.287 kPa m^3/kg °K). Because the power in the wind is proportional to the density, wind power production can be expected to be lower at high altitude and higher in cold climate (not accounting for icing effects). Moist air is slightly less dense than dry air, but corrections for air moisture are rarely used. When necessary, air density as a function of moisture content can be found in most texts on thermodynamics.

The international standard atmosphere assumes that the sea-level temperature and pressure are 288.15 °K and 101.325 kPa, resulting in a standard sea-level density of 1.225 kg/m^3.

Air pressure decreases with elevation above sea level. The pressure in the international standard atmosphere up to an elevation of 5000 m is very closely approximated by:

$$p = 101.29 - (0.011837)\,z + \left(4.793 \times 10^{-7}\right)z^2 \tag{2.9}$$

where z is the elevation in meters and the pressure is in kPa. Of course, the actual pressure may vary from the standard pressure as weather patterns change. In practice, at any location, the daily and seasonal temperature fluctuations have a much greater effect on air density than do daily and seasonal changes in pressure and air moisture.

2.3.2 Stability of the Atmospheric Boundary Layer

A particularly important characteristic of the atmosphere is its stability – the tendency to resist vertical motion or suppress existing turbulence. The stability of the atmospheric boundary layer is a determining factor for the wind speed gradients that occur in the first few hundred meters above the ground (e.g., wind shear and wind veer, see Section 2.3.3) and plays an important role in the diffusion of wind turbine wakes (see Chapter 12).

Atmospheric stability is usually classified as stable, neutrally stable, or unstable. The stability of the earth's atmosphere is governed by the vertical temperature distribution resulting from the radiative heating or cooling of its surface and the subsequent convective mixing of the air adjacent to the surface.

- In a stable boundary layer, the temperature is cool near the surface and warmer higher in the atmosphere (typical of nighttime, cold and calm conditions), resulting in low mixing, low turbulence, and strong vertical wind variation with height (both longitudinal, but also transversal, i.e., strong shear and veer).
- In a neutral boundary layer, the temperature does not vary much with height (typical during the day when the sun heats the ground), resulting in efficient, but moderate, vertical mixing, and moderate wind gradients with height and moderate turbulence levels.
- In an unstable boundary layer, the temperature is warmer near the surface than above (typical of sunny days), resulting in strong vertical mixing as the warm air tends to rise, low vertical wind gradients and high turbulence levels.

A summary of how the atmospheric temperature changes with elevation (assuming an adiabatic expansion) follows.

2.3.2.1 Lapse Rate

The lapse rate of the atmosphere is defined as the rate of change of temperature with height. As will be shown in the following analysis, it is easier to determine the lapse rate by calculating the change in pressure with height and using conventional thermodynamic relationships. If the atmosphere is approximated as a dry (no water vapor in the mixture) ideal gas, the relationship between a change in pressure and a change in elevation for a fluid element in a gravitational field is given by:

$$\mathrm{d}p = -\rho g\,\mathrm{d}z \tag{2.10}$$

where p = atmospheric pressure (Pa), ρ = atmospheric density (kg/m^3), z = elevation above ground (m), and g = local gravitational acceleration (m/s^2, assumed constant here).

The negative sign results from the convention that height, z, is measured positively upward, and that the pressure, p, decreases in the positive z direction.

The first law of thermodynamics for an ideal gas closed system of unit mass undergoing a quasi-static change of state is given by:

$$\mathrm{d}q = \mathrm{d}u + p\mathrm{d}v = \mathrm{d}h - v\mathrm{d}p = c_p\,\mathrm{d}T - \frac{1}{\rho}\,\mathrm{d}p \tag{2.11}$$

where T = temperature (°K), q = heat transferred (kJ/kg), u = internal energy (kJ/kg), h = enthalpy (kJ/kg), v = specific volume (m^3/kg), c_p = constant pressure specific heat (kJ/kg °K).

For an adiabatic process (no heat transfer) $\mathrm{d}q = 0$, and Equation 2.11 becomes:

$$c_p\,\mathrm{d}T = \frac{1}{\rho}\,\mathrm{d}p \tag{2.12}$$

Substitution for dp in Equation 2.10 and rearrangement gives:

$$\left(\frac{dT}{dz}\right)_{Adiabatic} = g\frac{1}{c_p} \tag{2.13}$$

The rate at which temperature decreases with an increase in height for a system with no heat transfer is about 10 °C per 1000 m and is known as the dry adiabatic lapse rate. Using conventional nomenclature, the lapse rate, Γ, is defined as the negative of the temperature gradient in the atmosphere. Therefore, the dry adiabatic lapse rate is given by:

$$\Gamma = -\left(\frac{dT}{dz}\right)_{Adiabatic} \approx \frac{10°C}{1000\ m} \tag{2.14}$$

The dry adiabatic lapse rate is extremely important in meteorological studies since a comparison of its value to the actual lapse rate in the lower atmosphere is a measure of the stability of the atmosphere.

To generalize, any atmosphere lapse rate whose dT/dz is less than (dT/dz)$_{adiabatic}$ is a stable one. One should note that the standard international lapse rate seldom occurs in nature. This explains the need for the daily balloon soundings taken worldwide to determine the actual lapse rate. Also, in order to have stability, it is not necessary for an inversion (increase of temperature with height) to exist. When one does exist, however, the atmosphere is even more stable.

2.3.2.2 Monin–Obukov Stability Length

An important parameter in quantifying atmospheric stability is the Monin–Obukov stability length, L, which is defined by Equation 2.15.

$$L = -\frac{(u^*)^3}{\kappa \beta Q_0} \tag{2.15}$$

where u^* is known as the friction velocity (see also Section 2.3.3.1)

$\kappa = 0.4$ (von Kármán's constant)

$\beta = g/T$ is a buoyancy parameter with g = gravitational constant and T = temperature, °K

$Q_0 = \dfrac{H_0}{\rho\,C_p}$ = surface kinematic heat flux, where H_0 = surface heat flux, W/m^2

ρ = air density, kg/m^3

C_p = heat capacity of air, J/kg °K

The Monin–Obukov stability length expresses the relative importance of buoyancy and turbulence in atmospheric mixing. A dimensionless parameter z/L is also frequently used, in which z indicates the height above the surface. This parameter is also related to the Richardson number, which is discussed in Section 2.3.4.

2.3.3 The Steady Wind: Wind Speed Variation with Height

As shown in Figure 2.12, real wind velocity varies in space, direction, and time. The figure shows that the mean longitudinal wind

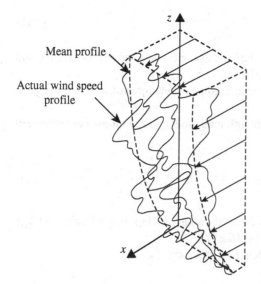

Figure 2.12 Sample wind speed profiles (*Source:* Van der Tempel, 2006/TU Delft)

speed increases with height, which defines the phenomenon called wind shear, or more strictly vertical wind shear. The transverse wind speed (not shown in the figure) can also vary with height (in particular in stable atmospheres), a phenomenon referred to as veer. The remainder of this section concerns the longitudinal wind shear.

In the field of wind energy, two mathematical models or "laws" have generally been used to model the vertical profile of wind speed over regions of homogenous, flat terrain (e.g., fields, deserts, and prairies). The first approach, the log law, has its origins in boundary layer flow in fluid mechanics and in atmospheric research. It is based on a combination of theoretical and empirical research. The second approach, most commonly used on land, is the power law. Both approaches are subject to uncertainty caused by the variable, complex nature of real, turbulent flows. A summary of each of these laws and their general application follows.

2.3.3.1 Logarithmic Profile (Log Law)

Although there are a number of ways to arrive at a prediction of a logarithmic wind profile (e.g., mixing length theory, eddy viscosity theory, and similarity theory), a mixing length type analysis given by Wortman (1982) is summarized here.

Near the surface of the earth, the momentum equation reduces to:

$$\frac{\partial p}{\partial x} = \frac{\partial}{\partial z} \tau_{xz} \tag{2.16}$$

where x and z are the horizontal and vertical coordinates, p is the pressure, and τ_{xz} is the shear stress in the direction of x whose normal coincides with z.

In this region, the pressure is independent of z and integration yields:

$$\tau_{xz} = \tau_0 + z \frac{\partial p}{\partial x} \tag{2.17}$$

where τ_0 is the surface value of the shear stress. Near the surface the pressure gradient is small, so the second term on the right-hand side may be neglected. Also, using the Prandtl mixing length theory (Schlichting, 1968), the shear stress may be expressed as:

$$\tau_{xz} = \rho \, \ell^2 \left(\frac{\partial U}{\partial z} \right)^2 \tag{2.18}$$

where ρ is the density of the air, U is the horizontal component of velocity, and ℓ is the mixing length. Note that U is used here, signifying that the effects of turbulence have been averaged out. Combining Equations 2.17 and 2.18 gives Equation 2.19:

$$\frac{\partial U}{\partial z} = \frac{1}{\ell} \sqrt{\frac{\tau_0}{\rho}} = \frac{u^*}{\ell} \tag{2.19}$$

where $u^* = \sqrt{\frac{\tau_0}{\rho}}$ is defined as the friction velocity. Typical values for u^* range from 0.01 m/s on smooth surfaces to 0.5 m/s in urban areas. But as it will be shown below, if the velocity U is known at a given height, then there is no need to determine the value of u^*.

If one assumes a mixing length, $\ell = \kappa z$, with $\kappa = 0.4$ (von Kármán's constant), then Equation 2.19 can be integrated directly from z_0 to z where z_0 is the surface roughness length, which characterizes the roughness of the ground terrain. This yields Equation 2.20, which defines the logarithmic wind profile:

$$U(z) = \frac{u^*}{\kappa} \ln\left(\frac{z}{z_0} \right) \tag{2.20}$$

Table 2.1 Values (approximate) of surface roughness length for various types of terrain

Terrain description	z_0 (mm)
Very smooth, ice or mud	0.01
Calm open sea	0.20
Blown sea	0.50
Snow surface	3.0
Lawn grass	8.0
Rough pasture	10
Fallow field	30
Crops	50
Few trees	100
Many trees, hedges, and few buildings	250
Forest and woodlands	500
Suburbs	1500
Centers of cities with tall buildings	3000

The integration of Equation 2.19 is from the lower limit of z_0 instead of 0 because natural surfaces are never uniform and smooth. Table 2.1 gives some approximate surface roughness lengths for various surface and terrain types. Note the low values for offshore situations.

Equation 2.20 can also be written as:

$$\ln(z) = \left(\frac{\kappa}{u^*}\right) U(z) + \ln(z_0) \tag{2.21}$$

This equation can be plotted as a straight line on a semilog graph. The slope of this graph is κ/u^*, and from a graph of experimental data, u^* and z_0 can be calculated.

The log law is often used to extrapolate wind speed from a reference height, z_r, to another level using Equation 2.22:

$$\frac{U(z)}{U(z_r)} = \ln\left(\frac{z}{z_0}\right) / \ln\left(\frac{z_r}{z_0}\right) \tag{2.22}$$

Sometimes, the log law is modified to consider mixing at the earth's surface, by expressing the mixing length as $\ell = \kappa(z + z_0)$. When this is used, the log profile becomes:

$$\frac{U(z)}{U(z_r)} = \ln\left(\frac{z + z_0}{z_0}\right) / \ln\left(\frac{z_r + z_0}{z_0}\right) \tag{2.23}$$

2.3.3.2 Power Law Profile

The power law is a simple model for the vertical wind speed profile. Its basic form is:

$$\frac{U(z)}{U(z_r)} = \left(\frac{z}{z_r}\right)^{\alpha} \tag{2.24}$$

Table 2.2 Effect of power-law exponent α on estimates of wind power density at higher elevations

	α = 0.1	1/7	0.3
U_{100m} (m/s)	6.29	6.95	10.0
P/A (W/m^2)	152.8	205.4	608.2
% increase over 10 m	99.5	168.3	694.3

where $U(z)$ is the wind speed at height z, $U(z_r)$ is the reference wind speed at height z_r, and α is the power law exponent.

Early work on this subject showed that under certain conditions α is equal to 1/7, indicating a correspondence between wind profiles and flow over flat plates (see Schlichting, 1968). In practice, the exponent α depends on the situation; α varies with such parameters as elevation, time of day, season, nature of the terrain, wind speed, temperature, and various thermal and mechanical mixing parameters. Various methods have been proposed for estimating the power law exponent when data is not available. One technique is based on the use of experimental data from direct measurement (see Brower, 2012).

Example: Suppose $U_0 = 5$ m/s at 10 m and $\rho = 1.225$ kg/m^3. The wind power density P/A is thus $= 75.6$ W/m^2. The wind speeds and corresponding wind power densities as predicted by values of α of 0.1, 1/7, and 0.3 at 100 m are shown in Table 2.2. It is apparent that the value of α is extremely significant in extrapolating wind speed (and wind power) from one height to another.

2.3.3.3 Vertical Wind Shear Offshore

Offshore wind shear involves a dynamic interaction between wind and waves because the surface roughness in the ocean changes with wave conditions, which are in turn affected by the ambient wind speed.

As discussed above, wind shear on land is often modeled with the power law, Equation 2.24. For offshore applications, the log law, Equation (2.22), is more common. The log law is particularly applicable for offshore application since the surface roughness changes with sea state (i.e., wave height) and so the wind shear will change accordingly. Wave height and sea state are discussed in detail in Chapter 6.

2.3.4 Wind Shear and Stability

Vertical wind shear is also a function of stability. A more general form of Equation 2.20 is:

$$U(z) = \frac{u^*}{k}\left(\ln\left(\frac{z}{z_0}\right) - \Psi\left(\frac{z}{L}\right)\right) \tag{2.25}$$

where L = Monin–Obukov stability length

An important related parameter is the Richardson number, which is a measure of the relative importance of potential energy (i.e., buoyancy) and kinetic energy. It is given by:

$$Ri = \frac{zg(T_{air} - T_{surface})}{(T_{air} + 273.15)U_{10}^2} \tag{2.26}$$

where g = gravitational constant, $z = 10$ m, T_{air} and $T_{surface}$ are air and surface temperature in °C, and U_{10} is wind speed (m/s) at 10 m above the surface.

According to Hsu (2003), the stability parameter z/L and the Richardson number (Ri) are related such that:
For unstable conditions, i.e., $T_{surface} > T_{air}$, $z/L = 7.6\ Ri$ (so $z/L < 0$)

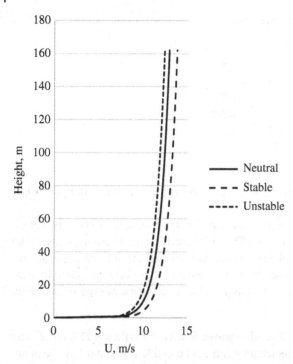

Figure 2.13 Illustration of effect of stability on wind shear

For stable conditions, i.e., $T_{air} > T_{surface}$, $z/L = 6.0\ Ri$. Using the above relations for the stability parameter, for unstable conditions we have:

$$\Psi\left(\frac{z}{L}\right) = 1.0496\left(-\frac{z}{L}\right)^{0.4591} \qquad (2.27)$$

and for stable conditions:

$$\Psi\left(\frac{z}{L}\right) = -5\frac{z}{L} \qquad (2.28)$$

Example: Consider a situation in which $z_0 = 0.00062$ m. Suppose that the wind speed at 10 m is 10 m/s. The wind speed at 80 m assuming: (1) neutral stability, (2) $T_{air} = 10\,°C$ and $T_{surface} = 2\,°C$ (stable), and (3) $T_{air} = 2\,°C$ and $T_{surface} = 10\,°C$ (unstable) can be found as follows.

For the neutral case, $U(80) = 12.2$ m/s. In the second case, $Ri = 0.02772$, $\Psi = -0.8315$, and $U(80) = 13.0$ m/s. In the third case, $Ri = -0.02852$, 0.5202, and $U(80) = 11.6$ m/s. The variation of wind speed with height using the values from this example is illustrated in Figure 2.13.

2.3.5 Effect of Terrain on Wind Characteristics

The importance of terrain features on wind characteristics is discussed in various siting handbooks for wind systems (see Hiester and Pennell, 1981; and Wegley *et al.*, 1980). Some of the effects of terrain include velocity deficits, unusual wind shear, and wind acceleration. The influence of terrain features on the energy output from a turbine may be so great that the economics of the whole project may depend on the proper selection of the site.

In the previous section, two methods were described (log profile and power law profile laws) for modeling the vertical wind speed profile. These were developed for flat and homogenous terrain. One can expect that any irregularities on the earth's surface will modify the wind flow, thus compromising the applicability of these prediction tools. This section presents a qualitative discussion of a few of the more important areas of interest regarding terrain effects.

2.3.5.1 Classification of Terrain

The most basic classification of terrain divides it into flat and non-flat terrain. Many authors define non-flat terrain as complex terrain (this is defined as an area where terrain effects are significant on the flow over the land area being considered). Flat terrain is terrain with small irregularities such as forest and shelter belts (see Wegley *et al.*, 1980). Non-flat terrain has large-scale elevations or depressions such as hills, ridges, valleys, and canyons. To qualify as flat terrain, the following conditions must hold. Note that some of these rules include wind turbine geometry:

- Elevation differences between the wind turbine site and the surrounding terrain are not greater than about 60 m anywhere in an 11.5 km diameter circle around the turbine site.
- No hill has an aspect ratio (height to width) greater than 1/50 within 4 km upstream and downstream of the site.
- The elevation difference between the lower end of the rotor disc and the lowest elevation on the terrain is greater than three times the maximum elevation difference (h) within 4 km upstream (see Figure 2.14).

Non-flat or complex terrain, according to Hiester and Pennell (1981), consists of a great variety of features, and one generally uses the following subclassifications: (1) isolated elevation or depression, and (2) mountainous terrain. Flow conditions in mountainous terrain are complex because the elevations and depressions occur in a random fashion. Flow in such terrain is divided into two classifications: small and large scales. The distinction between the two is made with a comparison to the planetary boundary layer, which is assumed to be about 1 km. That is, a hill of a height that is a small fraction of the planetary boundary layer (approximately 10%) is considered to have small-scale terrain features.

An important point to be made here is that information on wind direction should be considered when defining the terrain classification. For example, if an isolated hill (200 m high and 1000 m wide) were situated 1 km south of a proposed site, the site could be classified as non-flat. If, however, the wind blows only 5% of the time from this direction with a low average speed, say 2 m/s, then this terrain should be classified as flat.

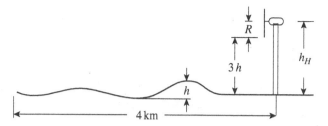

Figure 2.14 Determination of flat terrain (*Source:* Wegley *et al.*, 1980/WindBooks)

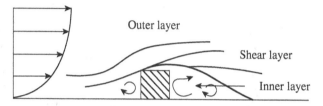

Figure 2.15 Schematic of a wake (Rohatgi and Nelson, 1994) (*Source:* Reproduced by permission of Alternative Energy Institute)

2.3.5.2 Flow Over Flat Terrain with Obstacles

Flow over flat terrain with human-made and natural obstacles has been studied extensively. Human-made obstacles include buildings, silos, etc. Natural obstacles include rows of trees, shelter belts, etc. For human-made obstacles, a common approach is to consider the obstacle to be a rectangular block and to consider the flow to be two-dimensional. This type of flow, shown in Figure 2.15, produces a wake with a velocity deficit, and, as illustrated, a free shear layer that separates from the leading edge and reattaches downwind, forming a boundary between an inner recirculating flow region (eddy) and the outer flow region.

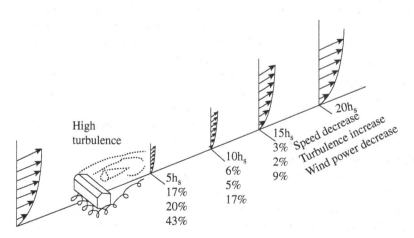

Figure 2.16 Speed, power, and turbulence effects downstream of a building (*Source:* Wegley *et al.*, 1980/WindBooks)

Figure 2.17 Effect of change in surface roughness from smooth to rough (*Source:* Wegley *et al.*, 1980/WindBooks)

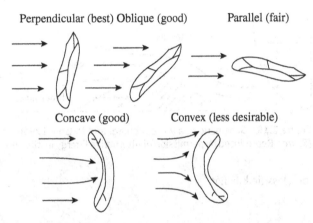

Perpendicular (best) Oblique (good)   Parallel (fair)

Concave (good)   Convex (less desirable)

Figure 2.18 Effect of ridge orientation and shape on site suitability (*Source:* Wegley *et al.*, 1980/WindBooks)

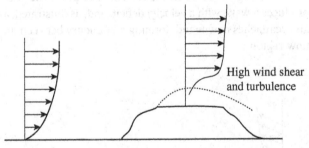

High wind shear and turbulence

Figure 2.19 Region of high wind shear over a flat-topped ridge (*Source:* Wegley *et al.*, 1980/WindBooks)

The results of an attempt to quantify data from constructed obstacles are shown in Figure 2.16, where the change in available power and turbulence is shown in the wake of a sloped-roof building. Note that the estimates in the figure apply at a level equal to one building height, h_S, above the ground, and that power losses become small downwind of the building after a distance equal to $15\,h_S$.

2.3.5.3 Flow in Flat Terrain with a Change in Surface Roughness

In most natural terrain, the surface of the earth is not uniform and changes significantly from location to location. This affects the local wind profile. For example, Figure 2.17 shows that the downwind profile changes significantly in going from a smooth to a rough surface.

2.3.5.4 Characteristics of Non-flat Terrain: Small-scale Features

Researchers (Hiester and Pennell, 1981) have divided non-flat terrain into isolated and mountainous terrain, in which the first refers to terrain of small-scale features and the latter refers to large-scale features. For small-scale flows, this classification is further divided into elevations and depressions. A summary of each follows.

Elevations

Flow over elevated terrain features resembles flow around obstacles. Characterization studies of this type of flow in water and wind tunnels, especially for ridges and small cliffs, have been carried out. Examples of the results for ridges are given below.

Ridges are elongated hills that are less than or equal to 600 m above the surrounding terrain and have little or no flat area on the summit. The ratio of length to height should be at least 10. Figure 2.18 illustrates that, for wind turbine siting, the ideal prevailing wind direction should be perpendicular to the ridge axis. When the prevailing wind is not perpendicular, the ridge will not be as attractive a site. Also, as shown in this figure, concavity in the windward direction enhances speed-up, and convexity reduces speed-up by deflecting the wind flow around the ridge.

The slope of a ridge is also an important parameter. Steeper slopes give rise to stronger wind flow, but on the lee of ridges, steeper slopes give rise to high turbulence. Furthermore, as shown in Figure 2.19, a flat-topped ridge creates a region of high wind shear due to the separation of the flow. Negative shear (flow reversal) may also occur near the surface.

Depressions

Depressions are characterized by a terrain feature lower than the surroundings. The change in speed of the wind is greatly increased if depressions can effectively channel the wind. This classification includes features such as valleys, canyons, basins, and passes. In addition to diurnal flow variations in certain depressions, there are many factors that influence the flow in depressions. These include orientation of the wind in relation to the depression, atmospheric stability, the width, length, slope, and roughness of the depression, and the regularity of the section of valley or canyon.

Shallow valleys and canyons (<50 m) are considered small-scale depressions, and other features such as basins and gaps are considered large-scale depressions. The large number of parameters affecting the wind characteristics in a valley, along with the variability of these parameters from valley to valley, make it almost impossible to draw specific conclusions valid for flow characterization.

2.3.5.5 Characteristics of Non-flat Terrain: Large-scale Features

Large-scale features are ones for which the vertical dimension is significant in relation to the planetary boundary layer. They include mountains, ridges, high passes, large escarpments, mesas, deep valleys, and gorges. The flow over these features is the most complex, and flow predictions for this category of terrain classification are the least quantified. The following types of large depressions have been studied under this terrain classification:

- valley and canyons;
- slope winds;
- prevailing winds in alignment;
- prevailing winds in nonalignment;
- gaps and gorges;
- passes and saddles;
- large basins.

An example of a large depression with the prevailing winds in alignment is shown in Figure 2.20. This occurs when moderate to strong prevailing winds are parallel to or in alignment (within about 35°) with the valley or canyon. Here, the mountains can effectively channel and accelerate the flow.

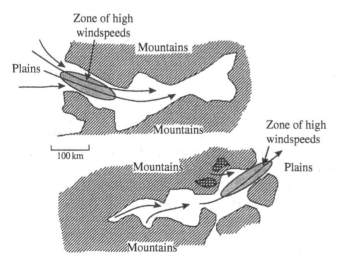

Figure 2.20 Increased wind speeds due to channeling of prevailing winds by mountains (Rohatgi and Nelson, 1994) (*Source:* Reproduced by permission of Alternative Energy Institute)

2.4 Wind Data Analysis and Resource Estimation

In this section, it is assumed that a large quantity of wind data has been collected. (Wind measurements and instrumentation are discussed later in this chapter.) This data could include direction data as well as wind speed data. There are a number of ways to summarize the data in a compact form so that one may evaluate the wind resource or wind power production potential of a particular site. These include both direct and statistical techniques. Furthermore, some of these techniques can be used with a limited amount of wind data (e.g., average wind speed only) from a given site. This section will review the following topics:

- wind turbine energy production in general;
- direct (nonstatistical) methods of data analysis and resource characterization;
- statistical analysis of wind data and resource characterization;
- statistically based wind turbine productivity estimates.

2.4.1 General Aspects of Wind Turbine Energy Production

In this section, we will determine the productivity (both maximum energy potential and machine power output) of a wind turbine at a site in which wind speed information is available in either time series format or a summary format (average wind speed, standard deviation, etc.)

The power available from wind is $P = (1/2)\rho A U^3$ as shown in Equation 2.6. In practice, the power available from a wind turbine, P_w, is given by a power curve, as was introduced in Chapter 1. A typical curve, $P_w(U)$, simplified for purposes of illustration, was shown there. As also discussed in Chapter 1, the power curve illustrates three important characteristic velocities (1) the cut-in velocity, (2) the rated velocity, and (3) the cut-out velocity. Later sections of this text, including Chapters 3, 8, and 9, will describe how such curves can be estimated. Normally, these curves are based on test or numerical data, as described in IEC (2017).

In equation form, the power curve for a pitch-controlled wind turbine (which is presently the most common type) is given as follows:

$$
\begin{aligned}
P_w(U) &= \frac{1}{2}\rho C_p \eta \pi R^2 U^3, &&\text{for } U_{in} \leq U \leq U_R \\
P_w(U) &= P_R, &&\text{for } U_R \leq U \leq U_{out} \\
P_w(U) &= 0, &&\text{for } U < U_{in} \text{ or } U > U_{out}
\end{aligned}
\tag{2.29}
$$

where U_{in}, U_R, and U_{out} are the cut-in, rated, and cut-out wind speeds, respectively.

The parameter C_p is known as the power coefficient and is a measure of the fraction of the power available in the wind that the rotor can extract. It is discussed in detail in Chapter 3. Note that C_p is usually a function of the wind speed. The parameter η is the drivetrain efficiency, it includes the combined efficiencies of the gearbox (if any) and the generator and is usually a function of the drivetrain speed.

For a stall-regulated turbine, the rated power may or may not be the maximum power and the maximum power will likely be produced at a higher wind speed than the normal rated wind speed of a comparable pitch-controlled turbine. The cut-in wind speed is also likely to be higher for a stall-regulated turbine than for a pitch-controlled turbine.

In many practical cases, the power curve is discretized for a series of wind speeds U_i and written $P_w(U_i)$. Sometimes a polynomial curve fit is used to describe the power in the wind speed ranges between cut-in and rated; see the example of the NREL 5-MW in Section 2.5.2.

In the following sections, methods for the determination of turbine production will be analyzed, as well as methods to summarize wind speed information from a given site. The following four approaches will be considered:

- direct use of data averaged over a short time interval;
- the method of bins;
- development of velocity and power curves from data;
- statistical analysis using summary measures.

The next section summarizes the use of the three nonstatistical methods.

2.4.2 Direct Methods of Data Analysis, Resource Characterization, and Turbine Productivity

2.4.2.1 Direct Use of Data

Suppose one is given a series of N wind speed observations, U_i, each averaged over the time interval Δt. These data can be used to calculate the following useful parameters:

1) The long-term average wind speed, $\overline{U}$, over the total period of data collection is:

$$\overline{U} = \frac{1}{N} \sum_{i=1}^{N} U_i \tag{2.30}$$

2) The standard deviation (sample based) of the individual wind speed averages, σ_U is:

$$\sigma_U = \sqrt{\frac{1}{N-1} \sum_{i=1}^{N} \left(U_i - \overline{U} \right)^2} = \sqrt{\frac{1}{N-1} \left\{ \sum_{i=1}^{N} U_i^2 - N\overline{U}^2 \right\}} \tag{2.31}$$

3) The average wind power density, $\overline{P}/A$, as explained previously, is the average available wind power per unit area. It is given by:

$$\overline{P}/A = \frac{1}{2} \rho \frac{1}{N} \sum_{i=1}^{N} U_i^3 \tag{2.32}$$

Similarly, the wind energy density per unit area for a given extended time period $N\Delta t$ long is given by:

$$E/A = \frac{1}{2} \rho \Delta t \sum_{i=1}^{N} U_i^3 = \left(\overline{P}/A \right) (N\Delta t) \tag{2.33}$$

4) The average wind turbine power, $\overline{P}_w$, is:

$$\overline{P}_w = \frac{1}{N} \sum_{i=1}^{N} P_w(U_i) \tag{2.34}$$

where $P_w(U_i)$ is the power as given by the discretized version of the power curve of Equation 2.29.

5) The energy from a wind turbine, E_w, is:

$$E_w = \sum_{i=1}^{N} P_w(U_i)(\Delta t) = \overline{P}_w N\Delta t \tag{2.35}$$

2.4.2.2 Method of Bins

The method of bins also provides a way to summarize wind data and determine expected turbine productivity. The data must first be separated into the wind speed intervals or bins in which they occur. A data point is considered to be in a bin if its value is greater than the lower limit of the bin and equal to or less than the upper limit. It is most convenient to use bins of the same size. Suppose that the data are separated into N_B bins of width ΔU, with midpoints m_j, and with n_j, the number of occurrences in each bin, such that:

$$N = \sum_{j=1}^{N_B} n_j \tag{2.36}$$

The values found from Equations 2.30, 2.32, 2.34, and 2.35 can be estimated from the following:

$$\overline{U} = \frac{1}{N} \sum_{j=1}^{N_B} m_j n_j \tag{2.37}$$

$$\sigma_U = \sqrt{\frac{1}{N-1}\left\{\sum_{j=1}^{N_B} m_j^2 n_j - N\left(\overline{U}\right)^2\right\}}$$

$$= \sqrt{\frac{1}{N-1}\left\{\sum_{j=1}^{N_B} m_j^2 n_j - N\left(\frac{1}{N}\sum_{j=1}^{N_B} m_j n_j\right)^2\right\}} \tag{2.38}$$

$$\overline{P}/A = \frac{\rho}{2N} \sum_{j=1}^{N_B} m_j^3 n_j \tag{2.39}$$

$$\overline{P}_w = \frac{1}{N} \sum_{j=1}^{N_B} P_w(m_j) n_j \tag{2.40}$$

$$E_w = \sum_{j=1}^{N_B} P_w(m_j) n_j \, \Delta t \tag{2.41}$$

A histogram (bar graph) showing the number of occurrences vs. bin midpoint is usually plotted when using this method. Figure 2.21 illustrates a typical histogram. This histogram was derived from one year of 10-minute data, for which the mean was 7.5 m/s and the standard deviation was 3.94 m/s. The bin width was 0.5 m/s.

2.4.2.3 Velocity and Power Duration Curves from Data

Velocity and power duration curves can be useful when comparing the energy potential of candidate wind sites. As used in this text, the velocity duration curve is a graph with wind speed on the y axis and the number of hours in the year for which the speed equals or exceeds each particular value on the x axis. An example of velocity duration

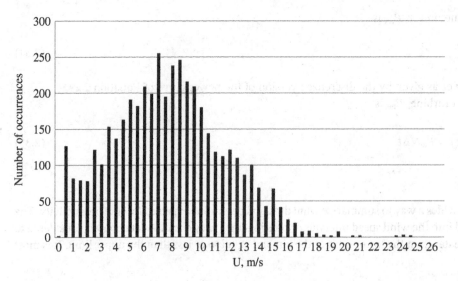

Figure 2.21 Typical histogram

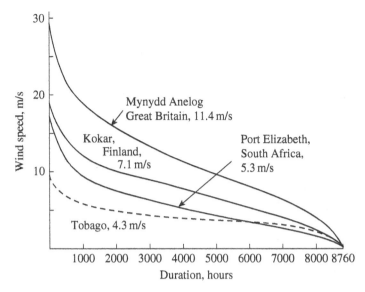

Figure 2.22 Velocity duration curve example (Rohatgi and Nelson, 1994) (*Source:* Reproduced by permission of Alternative Energy Institute)

curves (Rohatgi and Nelson, 1994) for various parts of the world (with average wind speeds varying from about 4–11 m/s) is shown in Figure 2.22. This type of figure gives an approximate idea about the nature of the wind regime at each site. The total area under the curve is a measure of the average wind speed. In addition, the flatter the curve, the more constant are the wind speeds (e.g., characteristic of the trade-wind regions of the earth). The steeper the curve, the more variable is the wind regime.

A velocity duration curve can be converted to a power duration curve by cubing the ordinates, which are then proportional to the available wind power for a given rotor-swept area. The difference between the energy potential of different sites is visually apparent because the areas under the curves are proportional to the annual energy available from the wind. The following steps must be carried out to construct velocity and power duration curves from data:

- arrange the data in bins;
- find the number of hours that a given velocity (or power per unit area) is exceeded;
- plot the resulting curves.

A turbine power duration curve for a wind turbine at a given site may be constructed using the power duration curve (adjusted by the rotor area) in conjunction with a power curve for the wind turbine. A hypothetical example of such a curve is shown in Figure 2.23. Note that the losses in energy production with the use of an actual wind turbine at this site can be identified.

2.4.3 Statistical Analysis of Wind Data

Statistical analysis can be used to determine the wind energy potential of a given site and to estimate the energy output from a wind turbine installed there. If time series measured data are available at the desired location and height, there may be little need for a data analysis in terms of probability distributions and statistical techniques. That is, the previously described techniques may be all that are needed. On the other hand, if projection of measured data from one location to another is required, or when only summary data are available, then there are distinct advantages to the use of analytical representations for the probability distribution of wind speed.

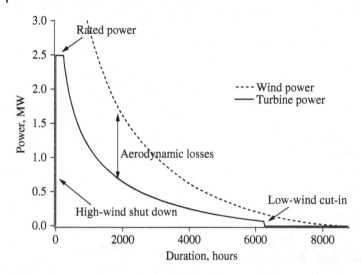

Figure 2.23 Hypothetical turbine power duration curve

For statistical analysis, a probability distribution is used. As discussed next, this is typically characterized by a probability density function or a cumulative distribution function.

2.4.3.1 Probability Density Function

The relative frequency of occurrence of wind speeds may be described by the probability density function, $f(U)$, of wind speed. The fundamentals of probability density functions (a.k.a. PDF) are summarized in Appendix C and should be reviewed. In any case, it is important to recall that the probability (Pr) of a wind speed occurring between U_a and U_b is:

$$\Pr(U_a \leq U \leq U_b) = \int_{U_a}^{U_b} f(U)\,\mathrm{d}U \tag{2.42}$$

If $f(U)$ is known, the following parameters can be calculated:
Mean wind speed, $\overline{U}$:

$$\overline{U} = \int_0^\infty U f(U)\mathrm{d}U \tag{2.43}$$

Standard deviation of wind speed, σ_U:

$$\sigma_U = \sqrt{\int_0^\infty \left(U - \overline{U}\right)^2 f(U)\,\mathrm{d}U} \tag{2.44}$$

Average wind power density, $\overline{P}/A$

$$\overline{P}/A = \frac{1}{2}\rho \int_0^\infty U^3 f(U)\mathrm{d}U = \frac{1}{2}\rho\overline{U^3} \tag{2.45}$$

where $\overline{U^3}$ is the expected value for the cube of the wind speed.

It should also be noted that the probability density function can be usefully superimposed on a wind velocity histogram if the histogram is first normalized. This is done by dividing the number of occurrences in the histogram by the total number of occurrences times the bin width, i.e., $N\Delta U$.

2.4.3.2 Cumulative Distribution Function

The cumulative distribution function $F(U)$ represents the time fraction or probability that the wind speed is <u>less</u> than or equal to a given wind speed, U. That is: $F(U) = \Pr(U' \leq U)$ where U' is a dummy variable and $\Pr()$ indicates probability. It can be shown that:

$$F(U) = \int_0^U f(U')\,\mathrm{d}U' \tag{2.46}$$

Also, the derivative of the cumulative distribution function (a.k.a. CDF) is equal to the probability density function, i.e.:

$$f(U) = \frac{\mathrm{d}F(U)}{\mathrm{d}U} \tag{2.47}$$

Note that the velocity duration curve is closely related to the CDF. In fact, the velocity duration curve $= 8760 \times (1 - F(u))$, but with the x and y axes reversed.

2.4.3.3 Commonly Used Probability Distributions

Two probability distributions are commonly used in wind data analysis (1) the Weibull and (2) the Rayleigh. The Weibull distribution is based on two parameters which may be derived from the mean and standard deviation. The Rayleigh distribution uses one parameter: the mean wind speed. Both the Rayleigh and Weibull distributions are called "skew" distributions in that they are defined only for values greater than 0. The Rayleigh is actually a Weibull distribution in which one of the parameters is fixed.

Weibull Distribution

Use of the Weibull probability density function requires knowledge of two parameters: k, a shape factor, and c, a scale factor. Both parameters are functions of $\overline{U}$ and σ_U. The Weibull probability density function and the cumulative distribution function for wind speed are given by (see also Appendix C):

$$f(U) = \left(\frac{k}{c}\right)\left(\frac{U}{c}\right)^{k-1} \exp\left[-\left(\frac{U}{c}\right)^k\right] \tag{2.48}$$

$$F(U) = 1 - \exp\left[-\left(\frac{U}{c}\right)^k\right] \tag{2.49}$$

Examples of a Weibull probability density function, for a single value of wind speed and various values of k, are illustrated in Figure 2.24. As shown, as the value of k increases, the curve has a sharper peak, indicating that there is less wind speed variation. Methods to determine k and c from $\overline{U}$ and σ_U are presented below.

It is not a completely straightforward process to get c and k in terms of $\overline{U}$ and σ_u, but there are a number of methods that can be used. For example, for $1 \leq k < 10$, a good approximation for k is (Justus, 1978):

$$k = \left(\frac{\sigma_U}{\overline{U}}\right)^{-1.086} \tag{2.50}$$

The scale parameter c is:

$$c = \frac{\overline{U}}{\Gamma(1 + 1/k)} \tag{2.51}$$

where $\Gamma(x) = $ gamma function $= \int_0^\infty e^{-t} t^{x-1}\,\mathrm{d}t$

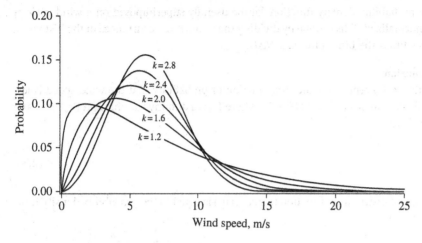

Figure 2.24 Example of Weibull probability density function for $\overline{U}$ = 6 m/s and various values of k

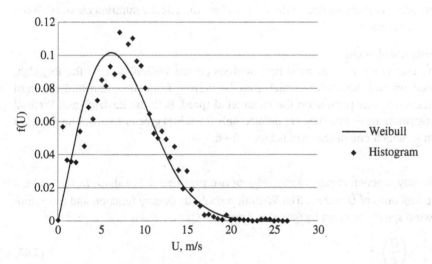

Figure 2.25 Sample normalized histogram and Weibull PDF

See Appendix C for more details.

Alternatively, c can be estimated from the following approximation (Lysen, 1983):

$$c = \overline{U}/(0.568 + 0.433/k)^{1/k} \tag{2.52}$$

An example of a Weibull PDF superimposed on a normalized version of the histogram in Figure 2.21 is shown in Figure 2.25. For this data, k and c were estimated to be 2.01 and 8.46 m/s, respectively.

Rayleigh Distribution

The Rayleigh distribution is the simplest velocity probability distribution to represent the wind resource since it requires only a knowledge of the mean wind speed, $\overline{U}$. As shown in Appendix C, the probability density function and the cumulative distribution function are given by:

$$f(U) = \frac{\pi}{2}\left(\frac{U}{\overline{U}^2}\right)\exp\left[-\frac{\pi}{4}\left(\frac{U}{\overline{U}}\right)^2\right] \tag{2.53}$$

$$F(U) = 1 - \exp\left[-\frac{\pi}{4}\left(\frac{U}{\overline{U}}\right)^2\right] \tag{2.54}$$

It should also be noted that the Rayleigh distribution is actually a Weibull distribution for which $k = 2$; see Appendix C.

2.5 Wind Turbine Energy Production Estimates Using Statistical Techniques

For a given wind regime's probability density function, $f(U)$, and a known turbine's power curve, $f_w(U)$, the average wind turbine power, $\overline{P}_w$, is given by:

$$\overline{P}_w = \int_0^\infty P_w(U)f(U)dU \tag{2.55}$$

The average power from a wind turbine whose C_p and efficiency η are known can then be found by applying Equation 2.29 in Equation 2.55 to yield Equation 2.56:

$$\overline{P}_w = \frac{1}{2}\rho\,\pi R^2\eta \int_{U_{in}}^{U_R} C_p U^3 f(U)\,dU + P_R \int_{U_R}^{U_{out}} f(U)\,dU \tag{2.56}$$

The average wind turbine power, $\overline{P}_w$, can be used to calculate a related performance parameter, the capacity factor, CF. The capacity factor of a wind turbine is defined as the ratio of the energy actually produced by the turbine to the energy that could have been produced if the machine ran at its rated power, P_R, over a given time period. Thus:

$$\text{CF} = \overline{P}_w/P_R \tag{2.57}$$

We are now in a position to use statistical methods for the estimation of the energy productivity of a specific wind turbine at a given site with a minimum of information. Two examples using the Rayleigh and Weibull distributions follow.

2.5.1 Productivity of an Idealized Turbine Using the Rayleigh Distribution

A measure of the maximum possible average power from a given rotor diameter can be calculated assuming an ideal wind turbine and using a Rayleigh probability density function. The analysis, based on the work of Carlin (1997), assumes the following:

- Idealized wind turbine, no losses, machine power coefficient, C_P, equal to the Betz limit ($C_{P,Betz} = 16/27$). As will be discussed in the next chapter, the Betz limit is the theoretical maximum possible power coefficient.
- There is no power limitation at high winds.
- Wind speed occurrences are given by a Rayleigh distribution.

The average wind machine power, $\overline{P}_w$, is given by Equation 2.55. For a Rayleigh, this is:

$$\overline{P}_w = \frac{1}{2}\rho\,\pi R^2\eta \int_0^\infty C_p(\lambda)U^3\left\{\frac{2U}{U_c^2}\exp\left[-\left(\frac{U}{U_c}\right)^2\right]\right\}dU \tag{2.58}$$

where U_c is a characteristic wind velocity given by: $U_c = 2\overline{U}/\sqrt{\pi}$.

For an ideal machine, $\eta = 1$, and the power coefficient can be replaced with the Betz value of $C_{P,Betz} = 16/27$, thus:

$$\overline{P}_w = \frac{1}{2}\rho\,\pi R^2\,U_c{}^3 C_{p,Betz}\int_0^\infty \left(\frac{U}{U_c}\right)^3\left\{\frac{2U}{U_c}\exp\left[-\left(\frac{U}{U_c}\right)^2\right]\right\}dU/U_c \tag{2.59}$$

One can now normalize the wind speed by defining a dimensionless wind speed, x, such that: $x = U/U_c$. This simplifies the previous integral as follows:

$$\overline{P}_w = \frac{1}{2}\rho\,\pi R^2\,U_c{}^3 C_{p,Betz}\int_0^\infty (x)^3\left\{2x\,\exp\left[-(x)^2\right]\right\}dx \tag{2.60}$$

Note that the wind machine constants have been removed from the integral. The integral can now be evaluated over all wind speeds. Its value is $(3/4)\sqrt{\pi}$. Thus:

$$\overline{P}_w = \frac{1}{2}\rho\,\pi R^2\,U_c{}^3(16/27)(3/4)\sqrt{\pi} \tag{2.61}$$

Using the diameter, D, divided by 2 instead of the radius and substituting for the characteristic velocity, U_c, the equation for average power is further simplified to:

$$\overline{P}_w = \rho\left(\frac{2}{3}D\right)^2\overline{U}^3 \tag{2.62}$$

This hypothetical turbine is sometimes referred to as a 1-2-3 Rayleigh–Betz turbine because of the underlying assumptions and because the average power is function of the density to the first power, the diameter to the second power, and the average wind speed to the third power.

Example: Calculate the average annual production of a 126 m diameter 1-2-3 Rayleigh–Betz turbine at sea level in a 7.5 m/s average annual wind velocity regime

$$\overline{P}_w = \left(1.225\,kg/m^3\right)\left(\frac{2}{3}\times 126\,m\right)^2(7.5\,m/s)^3 = 3.65\,MW$$

Multiplication of this by 8760 hr/yr yields an expected annual energy production of 31.9×10^3 MWh.

2.5.2 Productivity of a Realistic Turbine Using the Weibull Distribution

For a more realistic wind turbine, the average power may be calculated using Equation 2.56, but it is often more convenient to rewrite that equation using the cumulative distribution function. This may be done with the help of Equation 2.47, yielding Equation 2.63:

$$\overline{P}_w = \int_0^\infty P_w(U)\,dF(U) \tag{2.63}$$

The CDF, $F(U)$, for a Weibull distribution is given in Equation 2.49. With that equation, the integral in Equation 2.63 can be approximated with a summation over N_B bins. The resulting Equation 2.64 can then be used to find the average wind turbine power:

$$\overline{P}_w = \sum_{j=1}^{N_B} P_w\left(\frac{U_{j-1}+U_j}{2}\right)\left\{\exp\left[-\left(\frac{U_{j-1}}{c}\right)^k\right] - \exp\left[-\left(\frac{U_j}{c}\right)^k\right]\right\} \tag{2.64}$$

Note that Equation 2.64 is the statistical method's equivalent to Equation 2.40. In particular, the relative frequency, n_j/N, corresponds to the term in brackets, and the wind turbine power is calculated at the midpoint between U_{j-1} and U_j. Finally, since the rated power spans a significant wind speed range, the smaller bins can just

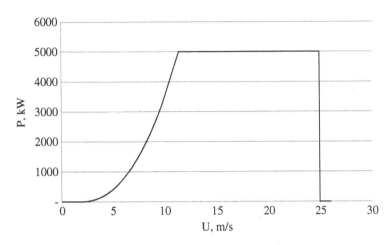

Figure 2.26 NREL 5-MW turbine power curve

be limited to the range between cut-in and rated wind speed and the last bin can span that entire range, so that Equation 2.64 can be written as Equation 2.65 (with N_B here being the number of bins below rated wind speed):

$$\overline{P}_w = \sum_{j=1}^{N_B} P_w\left(\frac{U_{j-1} + U_j}{2}\right)\left\{\exp\left[-\left(\frac{U_{j-1}}{c}\right)^k\right] - \exp\left[-\left(\frac{U_j}{c}\right)^k\right]\right\}$$
$$+ P_R\left\{\exp\left[-\left(\frac{U_R}{c}\right)^k\right] - \exp\left[-\left(\frac{U_{out}}{c}\right)^k\right]\right\}$$

(2.65)

Example: Consider the NREL 5-MW reference wind turbine introduced in Chapter 1 (Jonkman *et al.*, 2009). An approximation to its power curve (in kW) is given in Equation 2.66 and is also illustrated in Figure 2.26.

$$P(U) = \begin{cases} 5000, & \text{for } 11.4 < U \leq 25 \\ -0.0235\,U^6 + 0.8794\,U^5 - 12.872\,U^4 + 95.745\,U^3, & \text{otherwise} \\ -332.23\,U^2 + 554.63\,U - 367.99, & \text{for } 3 < U \leq 11.4 \end{cases}$$

(2.66)

Substituting Equation 2.66 into Equation 2.64 with a wind speed of 7.5 m/s and a Weibull $k = 2$ yields an average power of 1.9 MW. These effects can be seen clearly in the power duration curve of Figure 2.27, which includes the NREL 5-MW, the 1-2-3 Rayleigh–Betz turbine of the previous example, and the power in the wind of a 126 m diameter area. The average power with a realistic power curve is thus approximately one-half that of the idealized Rayleigh–Betz turbine.

2.6 Regional Wind Resource Assessment

2.6.1 Overview

As noted by Brower (2012), the acquisition of information about the wind regime in a region is a key step in a site selection process. Thus, one of the first steps required for a regional wind energy feasibility study is an estimate of the available wind resource. As summarized by Landberg *et al.* (2003), the following methods have been used for estimating the wind resource of an area:

1) Folklore.
2) Measurements only.

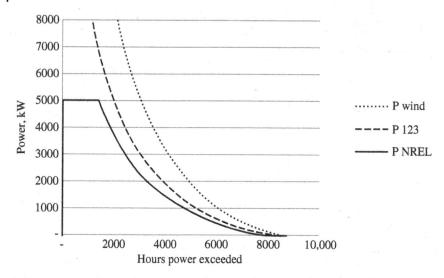

Figure 2.27 Power duration for the NREL 5-MW and idealized 126 m turbine

3) Measure–correlate–predict.
4) Global databases.
5) Wind atlas methodology.
6) Site data-based modeling.
7) Mesoscale modeling.
8) Combined meso/microscale modeling.

Nowadays, mesoscale databases are the most commonly used for wind resource assessments (see Section 2.7.2). A detailed review of all of these methods is beyond the scope of this text. In this section, a review of worldwide wind resource data that is readily available will be presented. For a general review of the subject, see Murthy and Rahi (2017).

The use of available wind resource data is an important part of any resource assessment activity. In evaluating available wind data, however, it is important to realize the data's limitations. That is, not all of this type of information has been collected for the purpose of wind energy assessment, and many data collection stations were located near or in cities, in relatively flat terrain or areas with low elevation (e.g., airports). Studies (e.g., in the United States and Europe) resulted in the completion of detailed wind atlases. These are documents that contain data on the wind speeds and direction in a region at various wind turbine heights. These original atlases provided a general description of the wind resource within a large area (mesoscale) but typically were not able to provide enough information for the detailed examination of candidate sites for wind development (microscale). In recent times, however, wind atlases with high resolution (as fine as 20–30 m) have been produced. These are designed to quantify a particular location's wind resource (also at various above-ground heights).

2.6.2 World Wind Resource Information

There are numerous technical publications that summarize wind resource information for other parts of the world. Historically, in 1981 the US Department of Energy's Pacific Northwest Laboratory (PNL) created a world resource map based on ship data, national weather data, and terrain (Cherry *et al.*, 1981). Landberg *et al.* (2003) note that the European Wind Atlas method, usually in combination with the micro-scale modeling tool WA^{S}P (Mortensen *et al.*, 2011), has been used for wind resource studies in a large portion of the world.

Worldwide wind resource information is available via the Global Wind Atlas (GWA) that was developed and maintained by DTU Wind Energy (see https://globalwindatlas.info) in partnership with the World Bank. Its version 3.0 contains data on wind resources around the world. In addition to onshore wind resource information, it provides information on offshore sites up to 200 km from the shoreline. Wind resource data was provided at heights 10, 50, 100, 150, and 200 m above ground/sea level. US-based organizations such as NREL, the Pacific Northwest National Laboratory (PNNL), Sandia National Laboratory, the US DOE, the Agency for International Development, and the American Wind Energy Association, and numerous European research and development agencies have provided technical assistance for wind resource assessment in developing countries. These have included Mexico, Indonesia, the Caribbean Islands, the former Soviet Union, and various South American countries. Resource assessments in these countries have focused on the development of rural wind power applications.

2.6.3 United States Resource Information

Historically, in the 1970s, a preliminary wind resource assessment of the United States was carried out that produced 12 regional wind energy atlases. The most recent wind resource assessment and characterization is summarized by the U.S. Department of Energy's Office of Energy Efficiency and Renewable Energy (EERE, 2020). In this work, using their WINDExchange website (2020), state-level wind resource maps based on datasets from NREL can be generated. An example of land-based state-level data (80 m hub height) is shown in Figure 2.28.

In addition to land-based wind data, this dataset includes NREL offshore wind data (Musial *et al.,* 2016). The offshore report includes the following:

- Gross resource area up to 200 nm (370 km) offshore
- Turbine hub height (100 m)
- Capacity array power density
- Energy production potential (6 MW turbine power curve)
- Technology exclusions
- Land use and environmental exclusions

An example of the available offshore resource data (Maine) is given in Figure 2.29.

One should also note that the NREL data has been used in conjunction with the Wind Integration National Dataset (WIND) Toolkit to provide a foundation for future wind integration studies (Draxl *et al.,* 2015)

2.7 Wind Forecasting and Modelling from Data

2.7.1 Wind Forecasting Overview

With the inherent variability of the wind resource, it is often valuable to be able to predict, or forecast, the wind speed for some time ahead. The importance of wind forecasting has been recognized in the past, especially as the level of penetration of wind energy into individual grids has increased (Veers *et al.*, 2019).

Wind forecasting can be divided into three-time scales:

1) **Short-term**: Minutes, hours, and days ahead
2) **Medium-term**: Energy and maintenance forecasting
3) **Long-term**: Predictions of lifetime energy production, system loads, and reliability.

Regardless of a formal definition of time scales, short-term forecasting is most important when the fraction of wind on the grid increases, affecting the grid stability depending more on a variable renewable resource. As pointed out by Veers *et al.* (2019), long-term energy forecasts depend on many of the same physical processes that are

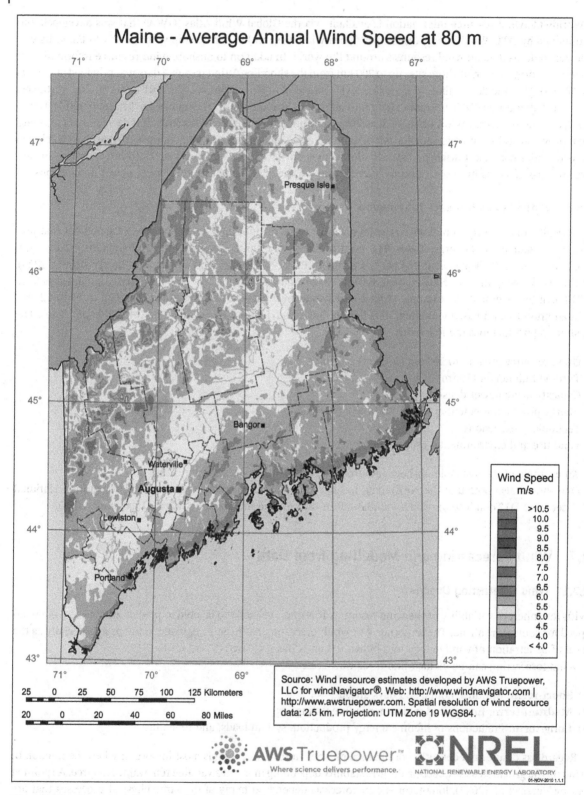

Figure 2.28 Average wind speed at 80 m for Maine (*Source:* WINDExchange, 2020/U.S. Department of Energy's Wind Energy Technologies Office/Public Domain)

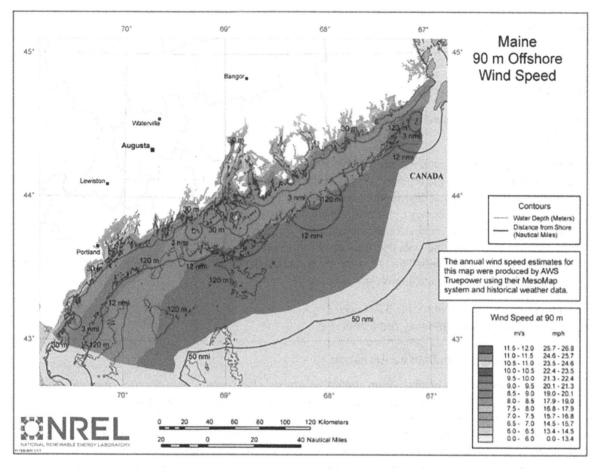

Figure 2.29 Maine average wind speed (90 m) offshore (Musial *et al.*, 2016)

important for short-term estimates, but they are also influenced by longer-term sources of variability and those influenced by climate change (for example see Pryor *et al.,* 2018).

It should be also noted that better understanding of meteorological ocean (metocean) condition over different timescales is expected to improve forecasting capabilities for offshore wind, especially through the coupling of metocean physics and wind systems. Differing from onshore wind, offshore wind forecasting requires improvements since unfavorable weather prevents installation, repair, and maintenance tasks. Offshore data-driven forecasting models for improving offshore wind turbine availability and maintenance are discussed by Pandit *et al.* (2020).

Recent work in wind energy forecasting has been concentrated on the next few hours or the one-to-two-day time frame and is generally defined as short-term wind energy forecasting. In this regard, an excellent source of information is contained in the European-based ANEMOS project work [see Kariniotakis *et al.* (2006) for a review of this project]. The overall aim of this project was to review and develop accurate and robust models that substantially outperform current state-of-the-art forecasting models for both onshore and offshore applications. Since the start of their work, the amount of literature and research on this subject has rapidly increased and reviews on the subject generally consider only the most important papers (Giebel *et al.*, 2011). A review of the early short-term forecasting work (up to 2003) was given by Landberg *et al.* (2003) who stated that most of the forecasting models consist of all or most of the following:

- A numerical weather prediction (NWP) model output;
- Input of observations;
- A numerical forecasting model and output.

Advanced wind energy prediction or forecasting methods can be divided into three main model groups (Tian, 2020):

1) **Physical models**: These consist of several submodels, which together deliver a translation from a NWP forecast at a certain grid point to a power forecast at the desired site, and at turbine hub height. Each submodel should contain an analytical description of its governing physical process. One should also note that a major factor that controls the accuracy of the NWP forecast is the resolution of the grid (Negnevitsky, 2016).
2) **Statistical models**: This approach uses meteorological predictions, historical measurements, and the generation output of statistical methods that have to be estimated from data.
3) **Combined or hybrid models**: This type of model combines the physical and statistical approaches to use the advantages of both to generate a forecast.

As summarized by Veers *et al.* (2019), it is expected that advances in wind forecasting technology will continue to improve wind power plant efficiency and output.

2.7.2 Wind Modeling from Data via Reanalysis

The use of reference data sets obtained from atmospheric models has become more common in recent times with the most typical modeled data called reanalysis data. Brower (2012) notes that all reanalysis data sets are created from historical weather observations (generally from surface, weather balloons, satellite, and aircraft-borne instruments) to drive a global or regional weather prediction model. This type of data production interpolates meteorological observations in space and time using weather prediction models. Reanalysis data is produced by numerous national weather organizations. One important example is the MERRA data set (Liléo and Petrik, 2010); see http://gmao.gsfc.nasa.gov/research/merra/. As pointed out by Rose and Apt (2015), researchers have used reanalysis data for wind resource assessment, long-term trends, long-term variability, geographic smoothing, and extreme winds. For example, most recent long-scale studies and databases for wind power potential rely on reanalysis methods (see McKenna *et al.*, 2021).

A discussion of the advantages and disadvantages of reanalysis data is beyond the scope of this text; publications by Rose and Apt (2015, 2016) discuss these subjects in detail.

2.8 Wind Measurement and Instrumentation

2.8.1 Overview

Instrumentation for wind energy applications is an important subject and has been discussed in detail by numerous authors. These include the text of Brower (2012), and a detailed DTU report (Pena *et al.*, 2015). In addition, the National Renewable Energy Laboratory (NREL) has discussed wind instrumentation in detail in their Wind Energy Instrumentation Atlas (Lundquist *et al.*, 2019).

So far, in this chapter, it has been assumed that one has sufficient and reliable meteorological wind speed data for the location of interest. In most wind energy applications, such information is not available, however, and measurements must be made specifically for determining the wind resource at the candidate location.

There are three types of instrument systems used for wind measurements:

- instruments used by national meteorological services;
- instruments designed specifically for measuring and characterizing the wind resource;
- instruments specially designed for high sampling rates for determining gust, turbulence, and inflow wind information for analyzing wind turbine response.

For each wind energy application, the type and amount of instrumentation required varies widely. In general, wind energy applications use the following types of meteorological sensors:

- anemometers to measure wind velocity;
- wind vanes to measure wind direction;
- thermometers to measure the ambient air temperature;
- barometers to measure the air pressure.

In this section, the discussion will be limited to the first two types of sensors. For more detail on the use of the third and fourth sensor types, one should refer to the wind resource assessment handbook of Bailey *et al.* (1996) and the NREL Wind Energy Instrumentation Atlas (Lundquist *et al.*, 2019). Furthermore, wind instrumentation systems consist of three major components: sensors, signal conditioners, and recorders. In the following review, these components will be discussed in more detail.

2.8.2 Wind Speed Measuring Instrumentation

The sensors of wind-measuring instrumentation can be classified according to their principle of operation via the following:

- **Momentum transfer**: Cups, propellers, and pressure plates;
- **Pressure on stationary sensors**: Pitot tubes and drag spheres;
- **Heat transfer**: Hot wires and hot films;
- **Doppler effects**: Acoustics and laser;
- **Special methods**: Ion displacement, vortex shedding, etc.

Despite the number of potential instruments available for wind speed measurements, in most wind energy applications five different systems have been used. As discussed below, they include:

- cup anemometers;
- propeller anemometers;
- sonic anemometers;
- acoustic Doppler sensors (sodar);
- laser Doppler sensors (lidar).

2.8.2.1 Cup Anemometers

The cup anemometer is probably the most common instrument for measuring the wind speed (Brower, 2012), as they are inexpensive, robust, and do not need an external power source. Cup anemometers use their rotation, which varies in proportion to the wind speed, to generate a signal. Today's most common designs feature three cups mounted on a small shaft. The rate of rotation of the cups can be measured by:

- electrical or electronic voltage changes (AC or DC);
- a photoelectric switch.

An electronic cup anemometer gives a measurement of instantaneous wind speed. The lower end of the rotating spindle is connected to a miniature AC or DC generator or a counter, and the analog output is converted to wind speed via a variety of methods.

The photoelectric switch type has a disc containing up to 120 slots and a photocell. The periodic passage of the slots produces pulses during each revolution of the cup.

The response and accuracy of a cup anemometer are determined by its weight, physical dimensions, and internal friction. By changing any of these parameters, the response of the instrument will vary. If turbulence measurements are desired, small, lightweight, low-friction sensors should be used. Typically, the most responsive cups have a distance constant (the distance that must be traveled by a cylindrical volume of air passing through the anemometer to record 63% of an instantaneous speed change) of about 1 m. Where turbulence data are not required, the cups can be larger and heavier, with distance constants from 2 to 5 m. This limits the maximum usable data sampling rate to no greater than once every few seconds. Typical accuracy values (based on wind tunnel tests) for cup anemometers are about ±2%.

Environmental factors can affect cup anemometers and reduce their reliability. These include ice or blowing dust. Dust can lodge in the bearings, causing an increase in friction and wear and reducing anemometer wind speed readings. If an anemometer ices up, its rotation will slow, or completely stop, causing erroneous wind speed signals, until the sensor thaws completely. Heated cup anemometers can be used, but they require a significant source of power. Because of these problems, the assurance of reliability for cup anemometers depends on calibration and service visits. The frequency of these visits depends on the site environment and the value of the data.

One commonly used anemometer in the US wind industry is the Maximum cup anemometer. The sensor is about 15 cm in diameter (see Figure 2.30). This anemometer has a generator that provides a sine wave voltage output. It has a Teflon® sleeve bushing bearing system that is not supposed to be affected by dust, water, or lack of lubrication. The frequency of the sine wave is proportional to the wind speed. Special anemometers based on this design (16 pole magnet) can be used for some turbulence measurements with a 1 Hz sampling rate.

A list of some of the most commonly used cup anemometers (and other wind resource equipment manufacturers) is given in Brower (2012).

Figure 2.30 Maximum cup anemometer

2.8.2.2 Propeller Anemometers

Propeller anemometers use the wind blowing into a propeller to turn a shaft that drives an AC or DC (most common) generator, or a light chopper to produce a pulse signal. The designs used for wind energy applications have a fast response and behave linearly in changing wind speeds. In a typical horizontal configuration, the propeller is kept facing the wind by a tail vane, which also can be used as a direction indicator. The accuracy of this design is about ±2%, similar to the cup anemometer. The propeller is usually made of polystyrene foam or polypropylene. The problems of reliability of propeller anemometers are similar to those discussed for cup anemometers.

When mounted on a fixed vertical arm, the propeller anemometer may be used for measuring the vertical wind component. A configuration for measuring three components of wind velocity is shown in Figure 2.31. The propeller anemometer responds primarily to wind parallel to its axis, and the wind perpendicular to the axis has no effect.

2.8.2.3 Sonic Anemometers

Sonic anemometers were initially developed in the 1970s. They use ultrasonic sound waves to measure wind speed and direction. Wind velocity is measured based on the time of flight of sonic pulses between pairs of transducers. A review of their theory of operation is given by Cuerva and Sanz-Andres (2000).

One-, two-, or three-dimensional flow can be measured via signals from multiple transducers. Typical wind engineering applications use two- or three-dimensional sonic anemometers. The spatial resolution is determined by the path length between transducers (typically 10–20 cm). Sonic anemometers can be used for turbulence measurements with fine temporal resolution (20 Hz or better). Compared to cup anemometers, sonic anemometers face challenges related to their proper installation, accurate calibration, power consumption, and cost. A review of the state of the art of sonic anemometry in wind engineering applications is given by Pedersen *et al.* (2006).

2.8.2.4 Acoustic Doppler Sensors (Sodar)

Sodar (standing for <u>so</u>und <u>d</u>etection <u>a</u>nd <u>r</u>anging) is classified as a remote sensing system, since it can make measurements without placing an active sensor at the point of measurement. Since such devices do not require tall (and expensive) towers, the potential advantages of their use are obvious. Remote sensing is used extensively for meteorological and aerospace purposes, but only in recent times has it been used for wind siting and performance measurements.

Figure 2.31 Propeller-type anemometer for measuring three wind velocity components

Sodar is based on the principle of acoustic backscattering. In order to measure the wind profile with sodar, acoustic pulses are sent vertically and at a small angle to the vertical. For measurement of three-dimensional wind velocity, at least three beams in different directions are needed. The acoustic pulse transmitted into the air experiences backscattering from particles or fluctuations in the refractive index of air. These fluctuations can be caused by wind shear as well as by temperature and humidity gradients. The acoustic energy scattered back to the ground is then collected by microphones and processed to produce useful data. Assuming that the sender and the receiver are not separated, the sodar configuration is referred to as a monostatic sodar. At the present time, all commercial sodars used for wind energy applications are monostatic (simplifying the system design and reducing its size).

Given the local speed of sound, the travel time between emission and reception determines the height the signal represents. A change in the acoustic frequency of the echo (Doppler shift) occurs if the scattering medium has a component of motion parallel to the beam motion. Thus, estimation of the speed of the wind speed parallel to the beam as a function of height can be carried out via frequency spectrum analysis of the received back-scattered signal.

Sodars have been used for both onshore and offshore wind siting studies. There has been a great deal of development on these devices over the last several years, and they are now commercially available from a number of sources (see Brower, 2012).

A detailed review and evaluation of sodar for wind energy applications was sponsored under the Wind Energy Sodar Evaluation program (WISE; see de Noord *et al.*, 2005). These authors point out that, although sodar systems can be commercially purchased, the following issues have arisen:

- For wind energy applications, stricter requirements exist for uncertainty, reliability, and validity of the data than for other sodar applications. For example, more work is needed on filtering techniques for data analysis.
- A general procedure for calibration of sodar systems has not been established.

- Low wind speeds (below 4 m/s), high wind speeds (above 18 m/s), and other atmospheric phenomena can cause difficulty with sodar measurements. This is especially important because power curve determination requires measurement of wind speeds up to the cut-out wind speed, typically around 25 m/s.
- Sodar systems have been designed for use on land. For offshore applications, problems associated with vibrations from supporting structures and an increased level of background noise have to be addressed.
- Sodar is designed to send a nearly vertical beam, and to use sodar for the detection of oncoming gusts will require a near-horizontal beam.
- Sodar systems have primarily been used at sites with easy access for maintenance and noncomplex terrain. For wind energy applications at remote sites, and in complex terrain, power supply and data communication may present challenges.

Many of these problems have been addressed, and a review of the status of remote sensing using sodar systems is contained in a comprehensive DTU Wind Energy report (Pena *et al.*, 2015). Current sodar technology is also discussed by Lundquist *et al.* (2019).

2.8.2.5 Laser Doppler Sensors (Lidar)

Lidar (<u>l</u>ight <u>d</u>etection <u>an</u>d <u>r</u>anging), similar to sodar, is also classified as a remote sensing device and can similarly be used to make measurements of a three-dimensional wind field. In this device, a beam of light is emitted, the beam interacts with the air and some of the light is scattered back to the lidar. The returned light is analyzed to determine the speed and distances to the particles from which it was scattered. In addition, the basic Lidar principle relies on the measurement of the Doppler shift of radiation scattered by natural aerosols that are carried by the wind.

Lidars have been used extensively in meteorological and aerospace applications, with the cost of meteorological lidar systems being quite high. However, developments in commercially available lidar systems have produced lower-cost systems for wind speed determination at heights of interest in wind energy applications. In addition, eye safety concerns have been overcome since most lidar lasers emit at the eye-safe wavelength of 1.5 microns. Brower (2012) notes that two different types of lidar exist for wind resource assessment: profiling and three-dimensional scanning lidars. The profiling type is most used in wind resource assessment.

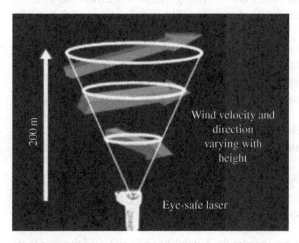

Figure 2.32 Schematic of conically scanned lidar system (*Source:* David A. Smith *et al.*, 2006/Reproduced from John Wiley & Sons, Inc.)

At the present time, there are two types of commercial profiling lidar devices available for wind engineering applications: (1) a constant wave, variable focus design, and (2) a pulsed lidar with a fixed focus. Wind speeds at heights up to 200 m have been measured by both types of lidar systems. The reader is referred to the work of Courtney *et al.* (2008) for a comparative review of these different systems.

As an example of an application of a constant wave lidar system, a portable and compact lidar system was used (Smith *et al.*, 2006) to determine horizontal and vertical wind speed and direction at heights up to 200 m. As shown in Figure 2.32, the lidar beam is offset at 30° to the vertical. The beam scans as it revolves at one revolution per second. As the beam rotates, it intercepts the wind at different angles, thereby building up a wind speed map around a disc of air. In a typical operation, three scans are performed at each height, and wind measurements are taken at five heights.

When using lidar to measure turbulence, it is important to consider the volume averaging effects. Lidar measures the average wind speed over a volume of air (over the line of sight of the beam, and by using different beam directions), not just a single point. This can affect accuracy for small-scale turbulence and introduce systematic errors; see Sathe *et al.* (2010).

A comprehensive discussion of the current state-of-the-art for wind energy lidar applications and technology is given by Lundquist *et al.* (2019). DTU Wind Energy has produced an extensive review of the use of lidar for wind energy remote sensing (Pena *et al.*, 2015) where both onshore and offshore applications of lidar have been used. In addition, the International Energy Agency (IEA) has established a Task (32) focused on identifying and mitigating barriers to the adoption of this technology for wind energy applications (Clifton *et al.*, 2018).

For offshore wind energy applications, floating lidar was introduced in 2009 as an offshore wind measurement technology focusing on the specific needs for wind resource assessment (Gottschall *et al.*, 2017). As reviewed by these authors, floating lidar systems (FLS) have been implemented to replace standard wind measurement techniques such as offshore instrumentation towers. Bishoff *et al.* (2021) note that a precise knowledge of the measurement accuracy or uncertainty of the FLS for different wind and wave conditions is very important. They also note that the lack of knowledge could be mitigated with a simulation tool for different types of FLS and variable wind and wave conditions.

Because of the need for more cost-effective wind measurement systems at the higher heights required by larger wind turbines, it is expected that the future will see extensive research and development on lidar systems for wind energy applications.

2.8.3 Wind Direction Instrumentation

Wind direction is normally measured via the use of a wind vane. A conventional wind vane consists of a broad tail that the wind keeps on the downwind side of a vertical shaft. A counterweight at the upwind end provides balance at the junction of the vane and shaft. Friction at the shaft is reduced with bearings, and so the vane requires a minimum force to initiate movement. For example, the usual threshold of this force occurs at wind speeds on the order of 1 m/s. Because of their mechanical and aerodynamic design, wind vanes will typically have a response time of a few seconds and will have slightly different readings than a sonic anemometer.

Wind vanes usually produce signals by contact closures or by potentiometers. The accuracy obtained from potentiometers is higher than that from contact closures, but the potentiometer-based wind vanes usually cost more. As with cup and propeller anemometers, environmental problems (blowing dust, salt, and ice) affect the reliability of wind vanes.

2.8.4 Instrumentation Towers

Since it is desirable to collect wind data at the hub height of turbines, one needs to use towers (often referred to as "meteorological mast" or simply "met mast") that can reach from a minimum of 20 m up to (or above) 150 m. Sometimes communications towers are available near the site under consideration. In most cases, however, towers must be installed specifically for wind measurement systems, which can be costly (from tens or hundreds of thousands of dollars onshore, to several millions offshore).

Instrumentation towers come in many styles: self-supporting, lattice or tubular towers, guyed lattice towers, and guyed tilt-up towers. Guyed tilt-up towers that can be erected from the ground are the most common type today. Many of these towers have been designed specifically for wind measurements and they are lightweight and can be moved easily. They require small foundations and can usually be installed in less than a day. It's important to exercise caution when anemometers are in the wake of a tower, as this can result in inaccuracies in the collected data.

More details on this subject are included in the wind resource assessment handbook of Bailey *et al.* (1996) and Brower (2012).

2.8.5 Data Recording Systems

In the development of a wind measurement program, one must select some type of data recording system in order to display, record, and analyze the data obtained from the sensors and transducers. The types of displays used for wind instruments are either of the analog type (meters) or of the digital type (LED, LCD) and supply one with current information. Typical displays use dials, lights, and digital counters. Recorders can provide past information and also may provide current information. Today, the recorders used in wind instrumentation systems generally are based on solid-state devices.

In general, the favored method of handling the large amount of data needed for complete analysis is the use of data loggers or data acquisition using dedicated computers or servers. A number of data logging systems are available on the market that record wind speed and direction averages and standard deviations, as well as maximum wind speed during the averaging interval. These systems often record the data on removable storage cards. Most also allow the data to be downloaded via modem and a cell phone.

The choices of methods and data recording systems are large, and each has its own advantages and disadvantages. The particular situation will define the data requirements, which, in turn, will dictate the choice of recording methods.

2.8.6 Wind Data Processing

The data produced by a wind monitoring system can be processed in a number of ways before being analyzed in detail. This involves checking the calculation and calculation of basic statistics. Results may include but are not limited to:

- average horizontal wind speeds over specified time intervals;
- variations in the horizontal wind speed over the sampling intervals (standard deviation, turbulence intensity, and maxima);
- average horizontal wind direction;
- variations in the horizontal wind direction over the sampling intervals (standard deviation);
- speed and direction distributions;
- persistence;
- determining gust parameters;
- statistical analysis, including autocorrelation, power spectral density, length and time scales, and spatial and time correlations with nearby measurements;
- steady and fluctuating u, v, w wind components;
- diurnal, seasonal, annual, interannual, and directional variations of any of the above parameters.

Some mention has been made of many of these measures of wind data, except for persistence and those of turbulence. Persistence is the duration of the wind speed within a given wind speed range. Also, histograms of the frequency of continuous periods of wind between the cut-in and cut-out wind speeds would provide information on the expected length of periods of continuous turbine operation. As noted previously, turbulence is discussed in Chapter 6.

A wind rose is a diagram showing the temporal distribution of wind direction and azimuthal distribution of wind speed at a given location. A wind rose (an example of which is shown in Figure 2.33) is a convenient tool for displaying anemometer data (wind speed and direction) for siting analysis. This figure illustrates the most common form, which consists of 16 equally spaced radial lines (each represents a compass point). The line length is proportional to the frequency of the wind from the compass point, with the circles forming a scale. The frequency of calm conditions is indicated in the center. The longest lines identify the prevailing wind directions. Wind roses generally are used to represent annual, seasonal, or monthly data.

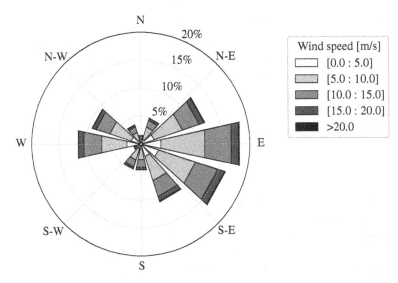

Figure 2.33 Example of a wind rose diagram

2.8.7 Overview of a Wind Monitoring Program

A wind monitoring program is an important part of wind resource assessment (see Brower, 2012). Historically, a detailed handbook on this subject was prepared by AWS Scientific (Bailey *et al.*, 1996). The handbook, which was designed for use in wind energy training seminars, contains ten chapters and an appendix. Following the approach of this handbook, a wind assessment and monitoring program includes the following components:

- review of guiding principles of a wind resource assessment program;
- determination of costs and labor requirements for a wind monitoring program;
- siting of monitoring systems;
- determination of measurement parameters;
- selection of monitoring station instrumentation;
- installation of monitoring systems;
- station operation and maintenance;
- data collection and handling;
- data validation, processing, and reporting.

2.9 Additional Topics

There are some important topics in the area of wind characterization that are beyond the scope of this chapter, but they will be summarized briefly here.

- Use of numerical or computational fluid dynamic (CFD) models for flow characterization;
- Micrositing
- Mesoscale to microscale flow modeling
- Statistically based resource assessment techniques.

A short description of each of these advanced topics follows.

2.9.1 Use of Numerical or Computational Fluid Dynamic Models for Flow Characterization

The progress in computational or numerical modeling of complex flow fields has spread to the field of wind energy. For example, since many potential locations for wind turbines involve siting in complex terrain, it is useful to have analytical tools that can characterize the wind fields in these locations.

CFD modeling is one of the most rapidly expanding areas in fluid mechanics (see Brower, 2012 for wind resource model examples). As summarized by Veers *et al.* (2019), the rapid progress in CFD models and the ability to analyze complex flows are expected to expand with the ever-increasing power of digital computers and the associated graphical routines.

2.9.2 Micrositing

Micrositing (also called siting) is defined as a resource assessment tool used to determine the exact position of one or more wind turbines on a parcel of land to optimize the power production. An objective of micrositing is to locate the wind turbines in the wind farm to maximize annual energy production or to yield the largest financial return for the owners of the wind farm.

Effective micrositing depends on a combination of detailed wind resource information for the specific site and, generally, the use of CFD models to predict the detailed flow field in the wind farm (including wake effects). The output is then combined in another model that gives a prediction of the energy output of the wind farm. Some micrositing models are even able to give the optimized location for wind turbine placement. Historically, examples of micrositing models were summarized by Rohatgi and Nelson (1994). Walls (2015) presented a simplified model for wind farm micrositing. A review of recent work on this subject is given by Gonzalez *et al.* (2014)

2.9.3 Mesoscale to Microscale Flow Modeling

As an example of the development of CFD modeling and the need for improved micrositing models, the coupling of mesoscale (atmospheric or meteorological global scale −10–1000 km in horizontal extent) and microscale flow models represents a significant challenge to wind energy research (see Heinz, 2021). Rodrigo *et al.* (2017) state that the coupling of meteorological and wind engineering flow models represents a major area of research for the next generation of wind condition assessment, wind turbine, and wind farm design tools. Heinz (2021) notes that microscale simulation methods can yield detailed information about terrain effects and interactions with structures and concludes that the atmosphere drives changes at the wind farm scale, thus it is important to model the variability at the mesoscale.

Veers *et al.* (2019) specifically observe that the mesoscale and the microscale are numerically modeled differently, thus making the study of atmospheric effects on wind farms that span these scales extremely difficult. They state that the mesoscale processes are on the order of 5–100 km in size and are generally modeled with grid spacing of 1–10 km. On the other hand, microscale processes, which drive wind turbine and wind farm processes, extend well below 1 km and have grid spacing of 5–100 m horizontally. Furthermore, atmospheric phenomena that span 0.5–1.5 km exist at the interface of mesoscale and microscale processes. Wyngaard (2004) called this zone unknown territory or "terra incognita" which spans atmospheric processes and their respective physical models with fundamentally different turbulence flow characteristics. Thus, as summarized by Heinz (2021), future work in this area will involve more comprehensive modeling of the nature of this transition region.

2.9.4 Statistically Based Resource Assessment Techniques/Measure-Correlate Predict Approach

For an estimate of the wind resource potential at a site with little or no wind resource measurement, one approach may be to link this site to a nearby site that has long-duration wind resource measurements. Landberg and Mortensen (1993) note that this link can be accomplished using either physical methods (using CFD models) or via the use of statistical methods (based on statistical correlations between the two-time series of data). A statistically based technique that has been widely used is the Measure–Correlate–Predict (MCP) approach. The basic idea behind the MCP approach is to establish relations between wind speed and direction at a potential wind turbine site and a location where wind speed and direction have been measured over a long period of time.

Many MCP methods are used in the wind industry or have been proposed for this purpose. Some determine a relationship between the concurrent data sets, which is then used with the long-term reference site data to estimate the long-term characteristics at the candidate site. Others determine the relationship between the period of concurrent reference site data and long-term reference site data and apply that to the shorter-term candidate site data in order to obtain, for example, a year of data that is representative of the long-term conditions. MCP methods may be used to estimate wind direction as well as wind speed characteristics. Finally, the data may be analyzed separately by month or by wind direction sector to improve the estimates of long-term wind characteristics.

Data sets used as long-term references include data from airports, weather balloon data, archived upper-level atmospheric data (reanalysis data), and data from meteorological towers. These towers have often been installed by government agencies to support the wind industry. When using these data sets, the quality of the data needs to be carefully evaluated.

The main difference between the various MCP methods is the form of the functional relationship that is used to relate the wind speeds and directions of the long-term data set to those of the short-term data set. This relationship may be a simple ratio, a linear relationship, or a more complicated nonlinear or probabilistic relationship. Linear wind speed relationships that have been used include linear regression and the "variance" method (Rogers *et al.*, 2005).

The variance method determines a linear relationship between hourly (or ten minutes) wind speed averages of the reference site and candidate site data sets. This relationship is used to estimate the wind speeds at the candidate site, $\hat{u}_c$ (the caret indicates an estimated value), when the wind speed at the reference site is u_r:

$$\hat{u}_c = au_r + b \tag{2.67}$$

In this relationship, the slope, a, is the ratio of the standard deviations of the hourly averages at the candidate site, σ_c, and reference site, σ_r, during the concurrent data period:

$$b = \frac{\sigma_c}{\sigma_r} \tag{2.68}$$

The offset, b, is a function of the mean wind speeds at the concurrent reference site, μ_r, and the candidate site, μ_c, as well as of the standard deviations:

$$a = \mu_c - \left(\frac{\sigma_c}{\sigma_r}\right)\mu_r \tag{2.69}$$

Equation 2.67 is then used with all the hourly averages of the reference site data set to estimate the long-term wind speeds at the candidate site. These estimated wind speeds are then used to determine the statistics of the long-term wind speeds at the candidate site.

The slope and offset in this model have been chosen to ensure that the mean and the variance of the estimated long-term wind speeds during the period of concurrent data are the same as those of the measured values. This ensures that not only the mean wind speed but also the wind speed distribution is correctly modeled. The variance method has been shown to correctly predict the statistics of long-term wind speeds at a variety of sites (see Rogers *et al.*, 2005).

Joint probability distributions can also be used to map the relationship between the reference site wind speeds and directions and the candidate site values. For example, the probability that the wind speed, u_c, and direction, θ_c, at the candidate site have specific values, $f(u_c, \theta_c)$, when the wind speed and direction at the reference site have values u_r and θ_r is $f(u_c, \theta_c | u_r, \theta_r)$. $f(u_c, \theta_c | u_r, \theta_r)$ can be determined from the concurrent data at each site. Then, the long-term probability distribution of the wind speeds and directions at the candidate site can be estimated by summing up all of the probabilities of occurrences of u_c, θ_c for all occurrences of u_r, θ_r in the long-term reference site data set. The advantage of this method is that it directly provides estimates of both long-term wind speed and long-term direction distributions at the candidate site. On the other hand, it does not provide estimates of the temporal characteristics of the winds at the site. These may be needed, for example, for studies of time-of-day pricing of the electricity generated by wind turbines at the candidate site.

The accuracy of an MCP method depends on a variety of factors (see Brower, 2012). First, it assumes that the long-term reference data and candidate data are accurate. Errors in these data sets will obviously result in erroneous predictions. Second, the longer the data set at the candidate site, the more accurately the results will represent long-term trends. In any case, the data collection at the candidate site should include representative data from all wind speeds and directions. Third, the reference site data should be representative of the climate at the candidate site. The relationship between the two data sets may be affected by diurnal effects, time delays between weather patterns arriving at the two sites, unique weather patterns, the distance between the two sites, and stability differences or topographic effects that create unique flow patterns at one site or the other. Finally, unknown stochastic or systematic changes in the relationship between the two sites at time scales greater than the period of concurrent data may introduce additional uncertainty into the final result. In any case, the results are also only good for the height of the measurements and for the specific mast location at the candidate site. More detailed modeling may still be necessary to define the specific flow patterns at candidate sites if the terrain is such that the predictions cannot be applied to nearby proposed turbine locations.

References

Aspliden, C. I., Elliot, D. L. and Wendell, L.L. (1986) Resource Assessment Methods, Siting, and Performance Evaluation, in: R. Guzzi and C.G. Justin (Eds.), *Physical Climatology for Solar and Wind Energy*, World Scientific, New Jersey.

Bailey, B. H., McDonald, S. L., Bernadett, D. W., Markus, M. J. and Elsholtz, K. V. (1996) *Wind Resource Assessment Handbook*, AWS Scientific Report. (NREL Subcontract No. TAT-5-15283-01).

Bischoff, O., Wolken-Mohlmann, G. and Cheng, P.W.(2021) Presentation and validation of a simulation environment for floating lidar systems. *Journal of Physics: Conference Series* **2018**, 012009.

Brower, M. C. (2012) *Wind Resource Assessment*. Wiley, New Jersey.

Carlin, P. W. (1997) Analytic expressions for maximum wind turbine average power in a Rayleigh wind regime. *Proc. of the 1997 ASME/AIAA Wind Symposium*, Reno, NV, pp. 255–263.

Cherry, N. J., Elliot, D. L. and Aspliden, C. I. (1981) World-wide wind resource assessment. *Proc. AWEA Wind Workshop V*, American Wind Energy Association: Washington, DC.

Clifton, A., Clive, P., Gottschall, J., Schlipf, D., Simley, E., Simmons, L., Stein, D., Trabucchi, D., Vasiljevic, N. and Wurth, I. (2018) IEA wind task 32: wind lidar identifying and mitigating barriers to the adoption of wind lidar. *Remote Sensing*, **10**, 406–428.

Courtney, M., Wagner, R. and Lindelow, P. (2008) Commercial LIDAR profilers for wind energy: a comparative guide. *Proc. of the 2008 European Wind Energy Conference. Brussels.*

Cuerva, A. and Sanz-Andres, A. (2000) On sonic anemometer measurement theory. *Journal of Wind Engineering and Industrial Aerodynamics*, **88**(1), 25–55.

de Noord, M., Curvers, A., Eecen, P., Antoniou, I., Jørgensen, H. E., Pedersen, T. F., Bradley, S., Hünerbein S. and von Kindler D. (2005) *WISE Wind Energy SODAR Evaluation Final Report*, ECN Report, ECN-C-05-044.

Draxl, C., Clifton, A., Hodge, B-M. and McCaa, J. (2015) The wind integration national dataset (WIND) toolkit. *Applied Energy*, **151**, 355–366.

EERE (2020) *Wind Resource and Characterization*, U.S. DOE Office of Energy Efficiency and Renewable Energy Internet: https://www.energy.gov/eere/wind/wind-resource-assessment-and-characterization Accessed 6/2/2020.

Eurek, K., Sullivan, P., Gleason, M, Hettinger, D., Heimiller, D. and Lopez, A. (2017) An improved global wind resource estimate for integrated assessment models. *Energy Economics*, **64**, 552–567.

Giebel, G., Draxl, C, Brownsword, R, Kariniotakis, G and Denhard, M. (2011) *The-State-of The-Art in Short-Term Prediction of Wind Power*, Risø DTU, Report no. NEI-DK-55211.

Gonzalez, J. S., Payan, M. B., Riquelme Santos, J. M. and Gonzalez-Longatt, F. (2014) A review and recent developments in the optimal wind-turbine micros-siting problem. *Renewable and Sustainable Energy Reviews*, **20**, 133–144.

Gottschall, J., Gribben, B., Stein, D. and Wurth, I. (2017) Floating lidar as an advanced offshore wind speed measurement technique: current technology status and gap analysis in regard to full maturity. *WIREs Energy and Environment*, **6**, 1–16.

Gustavson, M. R. (1979) Limits to wind power utilization. *Science*, **204**, 13–18.

Heinz, S. (2021) Theory-based mesoscale to microscale coupling for wind energy applications. *Applied Mathematical Modelling*, **98**, 563–575.

Hiester, T. R. and Pennell, W. T. (1981) *The Meteorological Aspects of Siting Large Wind Turbines*, Pacific Northwest Laboratories Report PNL-2522, NTIS, https://www.osti.gov/servlets/purl/6657537.

Hsu, H. A. (2003) "Estimating Overwater Friction Velocity and Exponent of Power-Law Wind Profile from Gust Factor during Storms," *Journal of Waterway*, Port, Coastal and Ocean Engineering. July/August.

IEC (2017) *Wind Energy Generation Systems – Part 12-1: Power Performance Measurements of Electricity Producing Wind Turbines, IEC* 61400-12-1:2017.

Jonkman, J., Butterfield, S., Musial, W., and Scott, G. (2009) *Definition of a 5-MW Reference Wind Turbine for Offshore System Development*, NREL/TP-500-38060, National Renewable Energy Laboratory, Golden, CO.

Justus, C. G. (1978) *Winds and Wind System Performance*. Franklin Institute Press, Philadelphia, PA.

Kariniotakis, G., Halliday, J., Brownsword, R., Marti, I., Palomares, A.M., Cruz, I., Madsen, H., Nielsen, T.S., Nielsen, H. A., Focken, U. and Lange, M. (2006) February. Next Generation Short-Term Forecasting of Wind Power–Overview of the ANEMOS Project. In *European Wind Energy Conference*, EWEC 2006.

Landberg, L. (2016) *Meteorology for Wind Energy: An Introduction*. Wiley, Chichester, UK.

Landberg, L. and Mortensen, N. G. (1993) A comparison of physical and statistical methods for estimating the wind resource at a site. *Proc. 15th British Wind Energy* Association Conference, York, England, pp. 119–125.

Landberg, L., Myllerup, L., Rathmann, O., Petersen, E. L., Jorgensen, B. H., Badger, J. and Mortensen, N. G. (2003) Wind resource estimation – an overview. *Wind Energy*, **6**, 261–271.

Liléo, S. and Petrik, O. (2010) Investigation on the use of NCEP/NCAR, MERRA and NCEP/CFSR reanalysis data in wind resource analysis. *Proceedings of EWEC 2010*, Warsaw, Poland.

Lundquist, J. K., Clifton, A., Dana, S., Huskey, A., Moriarty, P., van Dam, J. and Herges, T. (2019) *Wind Energy Instrumentation Atlas*, NREL Report NREL/TP-5000-68986.

Lysen, E. H. (1983) *Introduction to Wind Energy*, SWD Publication SWD 82-1, Amersfoort, NL.

McKenna, R., Pfenninger, S., Heinrichs, H., Schmidt, J., Staffell, I., Bauer, C., Gruber, K., Hahmann, A.N., Jansen, M., Klingler, M., Landwehr, N., Larsen, X. G., Lilliestam, J., Pickering, B., Robinius, M., Trondle, T., Turkovska, O., Wehrle, S. and Wohland, J. (2021) High-resolution large-scale onshore wind energy assessments: a review of potential definitions, methodologies and future research needs. *Renewable Energy*, **182**, 659–684

Mortensen, N. G., Heathfield, D. N., Rathmann, O. S. and Nielsen, M. (2011) *Wind Atlas Analysis and Application Program: Wasp 10 Help Facility*, Internet:https://backend.orbit.dtu.dk/ws/portalfiles/portal/116352660/WAsP_10_Help_Facility.pdf Accessed 6/8/2020.

Murthy, K. S. R. and Rahi, O. P. (2017) A comprehensive review of wind resource assessment. *Renewable and Sustainable Energy Reviews*, **72**, 1320–1342.

Musial, W., Heimiller, D., Beiter, P., Scott, G. and Draxl, C. (2016) 2016 *Offshore Wind Energy Resource Assessment for the United States*, NREL Report NREL/TP-5000-665599.

Negnevitsky, M. (2016) Wind Power Forecasting Techniques, Chapter 2, in: J.H. Lehr and J. Keeley (Eds.) *Alternative Energy and Shale Gas Encyclopedia* Wiley, New Jersey.

Pandit, R.K., Kolios, A. and Infield, D. (2020) Data-driven weather forecasting models performance comparison for improving offshore wind turbine availability and maintenance. *IET Renewable Power Generation*, **14**(13), 2386–2394.

Pedersen, T. F., Dahlberg, J. Å., Cuerva, A., Mouzakis, F., Busche, P., Eecen, P., Sanz-Andres, A., Franchini, S. and Petersen, S. M. (2006) *ACCUWIND-Accurate Wind Speed Measurements in Wind Energy-Summary Report.*

Pena, A., Hasager, C. B. Badger, M., Barthelmie, R. J., Bingöl, F., Cariou, J-P., Emeis, S., Frandsen, S. T., Harris, M. and Karagali, I. (2015) *Remote Sensing for Wind Energy*, DTU Wind Energy Report No. 0084(EN).

Pryor, S. C., Shepherd, T. J. and Barthelmie, R. J. (2018) Interannual variability of wind climates and wind turbine annual production. *Wind Energy Science*, **3**, 651–665.

Putnam, P. C. (1948) *Power from the Wind.* Van Nostrand Reinhold, New York.

Rinne, E., Holttinen, H., Kiviluoma, J. and Rissanen, S. (2018) Effects of turbine technology and land use on wind power resource potential. *Nature Energy*, **3**, 494–500.

Rodrigo, J. S. Arroyo, R. A. C., Moriarty, P., Churchfield, M., Kosovic, B., Rethore, P-E., Hansen, K. S., Hahmann, A., Mirocha, J. D. and Rife, D. (2017) Mesoscale to microscale wind farm flow modeling and evaluation. *WIREs Energy and Environment*, **6**, 1–30.

Rogers A. L., Rogers J. W. and Manwell, J. F. (2005) Comparison of the performance of four measure–correlate–predict algorithms. *Journal of Wind Engineering and Industrial Aerodynamics*, **97**(3), 243–264.

Rohatgi, J. S. and Nelson, V. (1994) *Wind Characteristics: An Analysis for the Generation of Wind Power.* Alternative Energy Institute, Canyon, TX.

Rose, S. and Apt, J. (2015) What can reanalysis data tell us about wind power? *Renewable Energy*, **83**, 963–969.

Rose, S. and Apt, J. (2016) Quantifying sources of uncertainty in reanalysis derived wind speed. *Renewable Energy*, **94**, 157–165.

Sathe, A., Mann, J., Gottschall, J. and Courtney, M. (2010). *Systematic Error in the Estimation of the Second Order Moments of Wind Speeds by Lidars.* In Detailed Program (online) ISARS. https://backend.orbit.dtu.dk/ws/portalfiles/portal/5125448/Systematic+error_Mann.pdf Accessed 8/19/2023.

Schlichting, H. (1968) *Boundary Layer Theory*, 6th edition. McGraw-Hill, New York.

Smith, D. A., Harris, M. and Coffey, A. (2006) Wind LIDAR evaluation at the Danish wind test site in Hovsore. *Wind Energy*, **9**, 87–93.

Spera, D. A. (Ed.) (1994) *Wind Turbine Technology: Fundamental Concepts of Wind Turbine Engineering.* ASME Press, New York.

Tian, Z. (2020) A State-of-The-Art review on wind power deterministic prediction. *Wind Engineering*, **45**(5), 1374–1392.

Van der Tempel, J. (2006) *Design of Support Structures for Offshore Wind Turbines*, PhD Thesis, TU Delft, NL.

Veers, P., Dykes, K., Lantz, E., Barth, S., Bottasso, C. L., Carlson, O., Clifton, A., Green, J., Green, P., Holttinen, H., Laird, D., Lehtomaki, V., Lundquist, J., Manwell, J., Marquis, M., Meneveau, C., Moriarty, P., Munduate, M., Muskulus, M., Naughton, J., Pao, L., Paquette, J., Peinke, J., Robertson, A., Rodrigo, J. S., Sempreviva, A. M., Smith, J. C., Tuohy, A. and Wiser, R. (2019) Grand challenges in the science of wind energy. *Science*, **366**, 6464.

Walls, E. (2015) Continuum wind flow models: introduction to model theory and case study review. *Wind Engineering*, **39**(3), 271–298.

Wegley, H. L., Ramsdell, J. V., Orgill M. M. and Drake R. L. (1980) *A Siting Handbook for Small Wind Energy Conversion Systems*, Battelle Pacific Northwest Lab., PNL-2521, Rev. 1, NTIS https://www.osti.gov/servlets/purl/5490541 Accessed 12/1/2023.

WINDExchange (2020) *Wind Energy in Maine.* https://windexchange.energy.gov/states/me Accessed 6/2/2020.

World Energy Council (1993) *Renewable Energy Resources: Opportunities and Constraints 1990–2020.* World Energy Council, London.

Wortman, A. J. (1982) *Introduction to Wind Turbine Engineering.* Butterworth, Boston, MA.

Wyngaard, J. C. (2004) Toward numerical modeling in the terra incognito. *Journal of the Atmospheric Sciences*, **61**, 1816–1826.

3

Aerodynamics of Wind Turbines

3.1 General Overview

Wind turbine power production depends on the interaction between the rotor and the wind. As discussed in Chapter 2, the wind may be considered to be a combination of the mean wind and turbulent fluctuations about that mean flow. Experience has shown that the major aspects of wind turbine performance (mean power output and mean loads) are determined by the aerodynamic forces generated by the mean wind. Periodic aerodynamic forces caused by wind shear, off-axis winds, and rotor rotation, and randomly fluctuating forces induced by turbulence and dynamic effects are the source of fatigue loads and are a factor in the peak loads experienced by a wind turbine. These are, of course, important but can only be understood once the aerodynamics of steady-state operation have been understood. Accordingly, this chapter focuses primarily on steady-state aerodynamics. An overview of the complex phenomena of unsteady aerodynamics and how they are addressed in the analysis of rotor performance is presented at the end of the chapter. Topics related to the flow about a wind turbine (wakes and blockage effects), which can affect the performance of wind turbines within a wind plant, are discussed in Chapter 12.

Practical horizontal axis wind turbine (HAWT) designs use airfoils to transform the kinetic energy in the wind into useful energy. The material in this chapter provides the background to enable the reader to understand power production with the use of airfoils, calculate an optimum blade shape for the start of a blade design, and analyze the aerodynamic performance of a rotor with a known blade shape and airfoil characteristics. A number of authors have derived methods for predicting the steady-state performance of wind turbine rotors. The classical analysis of the wind turbine was originally developed by Joukowski (1918), Betz (1926), and Glauert (1935) in the 1930s. Subsequently, the theory was expanded and adapted for solution by digital computers (see Wilson and Lissaman, 1974; Wilson et al., 1976; de Vries, 1979). In all of these methods, momentum theory and blade element theory are combined into a strip theory that enables calculation of the performance characteristics of an annular section of the rotor. The velocities and loads in each annulus are determined independently of the other annuli. The characteristics for the entire rotor are then obtained by integrating, or summing, the values obtained for each of the annular sections. This is the main approach used in this chapter.

The chapter starts with a general overview in Section 3.1 and it is followed in Section 3.2 with the analysis of an idealized wind turbine rotor using momentum theory or vorticity formulations. The discussion introduces important concepts and illustrates the general behavior of wind turbine rotors and the airflow over them. The analyses are also used to determine theoretical performance limits for wind turbines.

General aerodynamic concepts and the operation of airfoils are introduced next in Section 3.3. This information is then used to consider the advantages of using airfoils for power production over other approaches.

Wind Energy Explained: On Land and Offshore, Third Edition.
James F. Manwell, Emmanuel Branlard, Jon G. McGowan, and Bonnie Ram.
© 2024 John Wiley & Sons Ltd. Published 2024 by John Wiley & Sons Ltd.

Sections 3.4 and 3.5 detail the classical analytical approach to the steady-state analysis of HAWTs, as well as some applications and examples of its use. First, the details of momentum theory and blade element theory are developed. The results illustrate the derivation of the general blade shape used in wind turbines. The combination of the two approaches, called strip theory or blade element momentum (BEM) theory, is then used to outline a procedure for the aerodynamic design and performance analysis of a wind turbine blade design which takes the effects of wake rotation into account. Simplified analyses are presented in Section 3.6 to obtain the optimum blade shape for ideal operating conditions, and a simplified design procedure is provided. Section 3.7 presents advanced methods for rotor analyses that are used to avoid some of the limitations of the steady-state BEM analyses. These advanced methods include unsteady BEM, vortex methods, and computational fluid dynamics (CFDs). The computer codes used for analyzing rotor performance are discussed. Finally, Section 3.8 covers the basics of vertical axis aerodynamic rotor performance calculations.

Most material in this chapter should be accessible to readers without a fluid dynamics background. Nevertheless, it would be helpful to be familiar with a variety of concepts including Bernoulli's equation, streamlines, control volume analyses, and the concepts of laminar and turbulent flow. The material does require an understanding of basic physics. Starred sections are more advanced.

3.2 Idealized Wind Turbine Rotor and Actuator Disc Theory

To estimate the power produced by a wind turbine, it is necessary to estimate the aerodynamic forces that the wind exerts on the structure. The aerodynamic forces can be obtained by solving the fluid mechanics equations, which are known as the Navier–Stokes equations (Lamb, 1916; Batchelor, 1967, Bertin and Cummings, 2014; White, 2015). This complex set of equations expresses three conservation principles: the conservation of mass, Newton's 2nd law, and the first law of thermodynamics. In their general forms, these equations are complex and difficult to solve. In this section, several simplifications are used to be able to estimate the aerodynamic forces and the rotor performance. (1) The simplified form of the fluid mechanics equations, referred to as momentum theory, is used. In momentum theory, the conservation laws are expressed for a fixed geometrical volume, called a control volume. For instance, the conservation of mass is expressed by stating that the mass of fluid entering the volume must equal the mass exiting the volume. (2) The geometry of the wind turbine is simplified. Instead of modeling the individual blades and airfoils of the turbines, it is assumed that the rotor is a uniform, porous, disc, which extracts energy from the fluid that passes through it. This is called the actuator disc assumption. (3) Simplifications of the flow physics are introduced, such as steady, incompressible, inviscid flow. (4) The analyses are restricted to cases where the flow is mostly axial (called 1D momentum theory), or axisymmetric with swirl in the wake (called 2D momentum theory). The physics and modeling of "non-idealized" wakes are discussed in Chapter 12. Despite these different assumptions, the actuator disc momentum theory provides great insights on the performance of a generic rotor (sometimes called ideal rotor). Specific wind turbine designs can be studied by combining results from momentum theory (this section) with the performance results from the airfoils of each blade (Section 3.3), in what is known as the BEM theory (Section 3.5).

3.2.1 One-dimensional Momentum Theory and the Betz Limit

3.2.1.1 Assumptions

The one-dimensional (1D) momentum theory provides the basis for the simplest model of a wind turbine that can be used to predict the power produced, the thrust force from the wind, and the relationship between the aerodynamic loading and local wind speed. The theory is generally attributed to Rankine (1865) and Froude (1889), who developed the theory to predict the performance of ship propellers. Modern extensions of the theory are found in Hansen (2008), Sørensen (2016), Branlard (2017), and van Kuik (2018). In this theory, the rotor is represented by a uniform

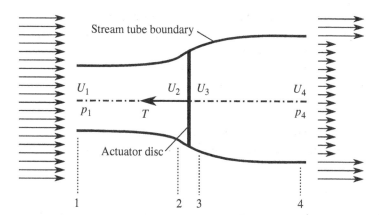

Figure 3.1 Actuator disc model of a wind turbine; U, mean air velocity; 1, 2, 3, and 4 indicate locations

"actuator disc"; therefore, the results are obtained for a generic rotor, without considering its design. Sources of losses are not considered, and for this reason, the results can be seen as the ones obtained for an ideal rotor.

The analysis uses a control volume whose boundaries are the surface of a stream tube and two cross-sections of the stream tube, labeled 1 and 4 (see Figure 3.1). A stream tube is formed by streamlines which are always tangent to the flow velocity; therefore, there is no flow across a stream tube boundary. The only flow is across the ends of the stream tube. The turbine is represented by a uniform "actuator disc" which is assumed to create a discontinuity of pressure in the stream tube of air flowing through it. On a real turbine, the change of pressure is generated by the airfoils, but in the idealized actuator disc theory, the pressure jump is simply assumed to be present without accounting for its source. The pressure jump across the disc implies that the actuator disc is extracting kinetic energy from the flow. This energy extraction is slowing down the flow which, by conservation of mass, implies that the stream tube cross sections expand from upstream to downstream of the disc. A radial velocity component is thus present at the cross Sections 3.2 and 3.3, whereas the flow is axial in Sections 3.1 and 3.4. In practice, it will be seen that the radial component drops out of the equations.

The 1D momentum theory analysis uses the following assumptions:

- homogenous, incompressible, steady-state fluid flow;
- no frictional drag;
- an infinite number of blades forming an actuator disc;
- uniform thrust over the disc;
- a nonrotating wake;
- the static pressure far upstream and far downstream of the rotor is equal to the undisturbed ambient static pressure;
- no pressure force is acting on the stream tube boundary.

3.2.1.2 Results of the 1D Momentum Theory

The main results of the 1D momentum theory are derived below. First, the conservation of linear momentum is applied to the control volume enclosing the whole system to find the net force on the contents of the control volume. That force is equal and opposite to the thrust, T, which is the force of the wind on the wind turbine. From the conservation of linear momentum, the thrust is equal and opposite to the rate of change of linear momentum of the air stream:

$$T = U_1(\rho A U)_1 - U_4(\rho A U)_4 \tag{3.1}$$

where ρ is the air density, A is the cross-sectional area, U is the air velocity, $\rho A U$ is the linear momentum, and the subscripts indicate values at the numbered cross-sections in Figure 3.1. An additional term would be present in Equation (3.1) if the assumption of no pressure force on the stream tube boundary were omitted (see Section 3.2.3). Then, the conservation of mass is applied to the whole system, which for a steady-state flow is: $(\rho A U)_1 = (\rho A U)_4 = \dot{m}$, where $\dot{m}$ is the mass flow rate. Therefore, Equation (3.1) can be re-written as:

$$T = \dot{m}(U_1 - U_4) \tag{3.2}$$

For a wind turbine (as opposed to a propeller), the thrust is typically positive so the velocity behind the rotor, U_4, is less than the free stream velocity, U_1. Another expression for the thrust can be obtained using the net sum of the pressure forces on each side of the actuator disc:

$$T = A_2(p_2 - p_3) \tag{3.3}$$

Bernoulli equation is used to find expressions for the pressure. This equation states that the sum of the static pressure ("p") and the dynamic pressure ("$\frac{1}{2}\rho U^2$") is constant along a streamline (assuming frictionless flows under conservative forces). Bernoulli equation can be used in the two control volumes on either side of the actuator disc because no work is done in each of these domains. In the stream tube upstream of the disc, the Bernoulli equation gives:

$$p_1 + \frac{1}{2}\rho U_1^2 = p_2 + \frac{1}{2}\rho(U_2^2 + U_r^2) \tag{3.4}$$

where U_r is the radial velocity at the rotor disc, assumed to be continuous across it. In the stream tube downstream of the disc:

$$p_3 + \frac{1}{2}\rho(U_3^2 + U_r^2) = p_4 + \frac{1}{2}\rho U_4^2 \tag{3.5}$$

where it is assumed that the far-upstream and far-downstream static pressures are equal ($p_1 = p_4$) and that the velocity across the disc remains the same ($U_2 = U_3$). If one solves for ($p_2 - p_3$) using Equations (3.4) and (3.5) and substitutes that into Equation (3.3), one obtains:

$$T = \frac{1}{2}\rho A_2\left(U_1^2 - U_4^2\right) \tag{3.6}$$

It is seen that the radial velocity does not affect the thrust, which justifies the appellation of "one-dimensional" momentum theory. Equating the thrust values from Equations (3.2) and (3.6) and recognizing that the mass flow rate is also $\dot{m} = \rho A_2 U_2$, one obtains:

$$U_2 = \frac{U_1 + U_4}{2} \tag{3.7}$$

Thus, the wind velocity at the rotor plane is the average of the upstream and downstream wind speeds. This is the main result of the 1D momentum theory, from which simple expressions for the thrust and power can be derived. The equations for power and thrust are typically expressed by introducing a dimensionless parameter called the axial induction factor, a (also called axial interference factor). It is defined as the fractional decrease in wind velocity between the free stream and the rotor plane velocity:

$$a = \frac{U_1 - U_2}{U_1} \tag{3.8}$$

Or equivalently

$$U_2 = U_1(1 - a) \tag{3.9}$$

With this definition, Equation (3.7) becomes:

$$U_4 = U_1(1 - 2a) \tag{3.10}$$

The quantity $u_i = -U_1 a$ is referred to as the induced velocity at the rotor. Therefore, Equation (3.9) states that the wind velocity at the rotor is a combination of the free stream velocity (U_1) and the induced wind velocity (u_i): $U_2 = U_1 + u_i$. As the axial induction factor increases from 0, the wind speed behind the rotor slows more and more. If $a = 1/2$, the wind has slowed to zero velocity behind the rotor, and the simple theory is no longer applicable.

The power, P, extracted by the 1D actuator disc is equal to the thrust times the velocity at the disc:

$$P = TU_2 = \frac{1}{2}\rho A_2\left(U_1^2 - U_4^2\right)U_2 = \frac{1}{2}\rho A_2 U_2(U_1 + U_4)(U_1 - U_4) \tag{3.11}$$

Substituting for U_2 and U_4 from Equations (3.9) and (3.10) gives:

$$P = \frac{1}{2}\rho A U^3 4a(1 - a)^2 \tag{3.12}$$

where from now on, the notations are simplified as follows: the control volume area at the rotor, A_2, is replaced by A, the rotor area, and the free stream velocity U_1 is replaced by U. With these notations, and using Equations (3.6), (3.9) and (3.10), the axial thrust on the disc is:

$$T = \frac{1}{2}\rho A U^2 [4a(1 - a)] \tag{3.13}$$

Wind turbine rotor performances are usually characterized using nondimensional coefficients, so that different designs can be compared to each other. The power coefficient, C_P, represents the fraction of the power in the wind that is extracted by the rotor. It is defined as:

$$C_P = \frac{P}{\frac{1}{2}\rho U^3 A} = \frac{\text{Rotor power}}{\text{Power in the wind}} \tag{3.14}$$

Similarly, the thrust coefficient is defined as:

$$C_T = \frac{T}{\frac{1}{2}\rho U^2 A} = \frac{\text{Thrust force}}{\text{Dynamic force}} \tag{3.15}$$

From Equation (3.12), the power coefficient obtained with the 1D momentum theory is:

$$C_P = 4a(1 - a)^2 \tag{3.16}$$

From Equation (3.13), the thrust coefficient obtained with the 1D momentum theory is:

$$C_T = 4a(1 - a) \tag{3.17}$$

A graph of the 1D momentum theory power and thrust coefficients as a function of the axial induction is given in Figure 3.2. The 1D momentum theory (sometimes called Betz theory) is not valid for axial induction factors greater than 0.5 since it would predict negative velocities in the far wake ($U_4 < 0$). In practice, as the axial induction factor approaches and exceeds 0.5, complicated flow patterns that are not represented in this simple model result in thrust coefficients that can go as high as 2.0 (Wilson et al., 1976). The details of wind turbine operation at these high axial induction factors appear in Section 3.5.1.3. It is noted that all the results presented in this section can also be obtained using a vorticity formulation (see Section 3.2.4).

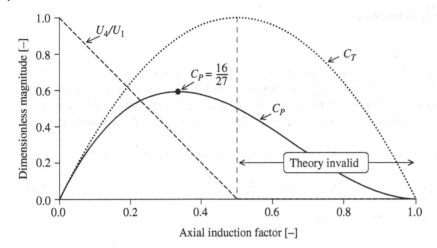

Figure 3.2 Operating parameters for the one-dimensional momentum theory of an actuator disc; U_1, velocity of undisturbed air; U_4, air velocity behind the rotor; C_P, power coefficient; C_T, thrust coefficient

3.2.1.3 Maximum Power Production – the Betz–Joukowski Limit

The maximum C_P is determined by taking the derivative of the power coefficient (Equation 3.16) with respect to a and setting it equal to zero, yielding $a = 1/3$, and thus:

$$C_{P,\mathrm{max}} = 16/27 \tag{3.18}$$

Since losses were neglected, the value of $C_{P,\mathrm{max}}$ corresponds to the maximum power extraction possible for any rotor design. This result was obtained independently by Betz and Joukowski, and the value $C_{P,\mathrm{max}} = 16/27$ is therefore referred to as the Betz–Joukowski limit. An ideal rotor would need to be operated such that the wind speed at the rotor is 2/3 of the free stream velocity to be at maximum power production. This results in an optimal axial induction factor of $a = 1/3$. Using Equation (3.17), it is found that C_T has a value of 8/9 at maximum power output ($a = 1/3$). This is lower than the maximum C_T value of 1.0 obtained when $a = 1/2$ and corresponds to a downstream velocity of zero.

For a realistic turbine, as will be discussed subsequently, three main aerodynamic effects lead to a decrease in the maximum achievable power coefficient:

- rotation of the wake behind the rotor;
- finite number of blades and associated tip losses;
- nonzero aerodynamic drag.

The electrical power generated is further reduced compared to the aerodynamic power due to additional mechanical and electrical losses, coming mostly from the gearbox and generator, respectively. The overall turbine efficiency of the wind turbine is then:

$$C_{P,\mathrm{elec}} = \frac{P_{\mathrm{elec}}}{(1/2)\rho A U^3} = \eta_{\mathrm{mech}} C_P \tag{3.19}$$

where η_{mech} is the mechanical and electrical efficiency of the turbine, and P_{elec} is the actual electrical power generated. Thus:

$$P_{\mathrm{elec}} = \eta_{\mathrm{mech}} C_P \frac{1}{2} \rho A U^3 \tag{3.20}$$

3.2.1.4 *Discussion on the Assumptions

The term "one-dimensional" momentum theory is confusing because, as it was seen, the flow has both axial and radial velocity components. The radial component arises from the conservation of mass, which implies that the flow has to expand in the radial direction if the axial velocity is reduced. It was seen that the radial component cancels out in the analysis, which explains why this component is often omitted in the presentation of this theory. Yet, the 1D momentum theory leads to several inconsistencies. The assumption that the axial velocity is constant along the span of the disc cannot hold. The axial velocity must vary with the radial coordinate, that is: $U_2 = U_2(r)$, as otherwise, it would be inconsistent with the continuity equation and the assumption of irrotationality of the flow. Nevertheless, it can be shown that if the incoming and far wake velocities U_1 and U_4 are radially constant, then the average axial velocity at the disc is the mean of these two velocities:

$$\bar{U}_2 = \frac{1}{A}\int_A U_2(r)\mathrm{d}S = \frac{1}{2}(U_1 + U_4) \tag{3.21}$$

It was further shown by van Kuik and Lignarolo (2015) that the norm of the velocity at the disc is constant, i.e., $\sqrt{U_r(r)^2 + U_2(r)^2}$ = constant, where U_r is the radial velocity at the actuator disc. The reader is directed to the following references for more details on these inconsistencies (Sørensen, 2016; Branlard, 2017; van Kuik, 2018).

3.2.2 Simplified Two-Dimensional Momentum Theory (with wake rotation)

3.2.2.1 Introduction and Main Assumptions

In Section 3.2.1, it was assumed that the flow was axial and that no rotation was imparted to the flow. In reality, the flow exerts a torque on the rotor, which in turn implies a change of angular momentum and a rotation of the flow in the opposite direction to the rotor. The two-dimensional (2D) momentum theory accounts for the axial and tangential components of the flow. In this theory, the aerodynamic forces vary along the rotor radius and therefore the momentum analyses are performed on elementary annular stream tubes. An annular stream tube illustrating the rotation of the wake is shown in Figure 3.3.

The production of rotational kinetic energy in the wake results in less energy extraction by the rotor than would be expected without wake rotation. In general, the kinetic energy in the wind turbine wake will be higher if the extracted torque is higher. Thus, as will be shown here, slow-running wind turbines (with a low rotational speed and a high torque) experience more wake rotation losses than high-speed wind machines with low torque.

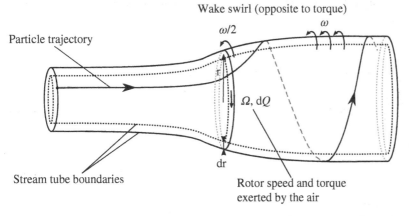

Figure 3.3 Stream tube model of flow behind rotating wind turbine rotor. The wake rotation is opposite of the torque exerted by the air on the rotor. (*Source:* Sketch inspired by a figure from Lysen (1982)/Consultancy Services Wind Energy Developing Countries)

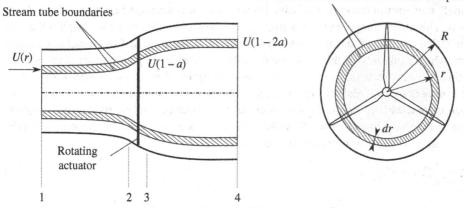

Figure 3.4 Stream tube used for two-dimensional momentum theory; U, velocity of undisturbed air; a, induction factor; r, radius

Figure 3.4 gives a schematic of the parameters involved in this analysis. Subscripts denote values at the cross-sections identified by numbers 1–4. The notations follow the ones introduced in Section 3.2.1. The angular velocity in the far wake is denoted ω and the angular velocity of the rotor, Ω. The actuator disc is assumed to induce a change in the angular velocity of the flow across the rotor, while the axial velocity remains continuous. The angular velocity is assumed to be 0 far upstream of the rotor, ω in the far wake, and $\omega/2$ at the disc. It is assumed that ω is small compared to Ω, and therefore, the pressure in the far wake is equal to the pressure in the free stream (see Wilson *et al.*, 1976). An annular stream tube, located between the radius r and $r+dr$, and resulting in a cross-sectional area equal to $2\pi r dr$ (see Figure 3.4), is considered. The method assumes that each stream tube can be analyzed independently of the others, and that no pressure force acts on the stream tube. The pressure, wake rotation, and induction factors are all assumed to be functions of radius.

3.2.2.2 Main Equations of the 2D Momentum Theory

The pressure jump across the actuator disc can be related to the change of angular velocity across it. The pressure jump can be obtained by using a control volume that moves with the angular velocity of the rotor, and by applying the energy equation in the sections before and after the rotor. The result is (see Glauert, 1935 for the derivation):

$$p_2 - p_3 = \rho\left(\Omega + \frac{1}{2}\omega\right)\omega r^2 \tag{3.22}$$

The elementary thrust, dT, resulting from the pressure difference on an annular element is:

$$dT = (p_2 - p_3)dA = \left[\rho\left(\Omega + \frac{1}{2}\omega\right)\omega r^2\right]2\pi r dr \tag{3.23}$$

The tangential induction factor, a', is defined as:

$$a' = \omega/(2\Omega) \tag{3.24}$$

As a result, the expression for the thrust becomes:

$$dT = 4a'(1 + a')\frac{1}{2}\rho\Omega^2 r^2 2\pi r dr \tag{3.25}$$

Note that when wake rotation is included in the analysis, the induced velocity at the rotor consists of not only the axial component, Ua, but also a tangential component in the rotor plane, $r\Omega a'$. Following the linear momentum analysis of Section 3.2.1, the thrust on an annular cross-section can also be determined by computing the change of linear momentum in the stream tube. The assumptions of the 2D momentum theory are such that the result from 1D momentum theory (Equation 3.7) still holds locally (for a given radius r), that is, $U_2 = U(1 - a)$, $U_4 = U(1 - 2a)$. An alternative expression for the elementary thrust is then obtained (based on Equation 3.2) as follows:

$$dT = d\dot{m}\,(U_1 - U_4) = 4a(1 - a)\frac{1}{2}\rho U^2 2\pi r dr \tag{3.26}$$

Equating the two expressions for thrust gives:

$$\frac{a(1 - a)}{a'(1 + a')} = \frac{\Omega^2 r^2}{U^2} = \lambda_r^2 \tag{3.27}$$

where λ_r is the local speed ratio, defined as the ratio of the rotor speed at some intermediate radius to the wind speed:

$$\lambda_r = \frac{\Omega r}{U} = \lambda\frac{r}{R} \tag{3.28}$$

and where λ is the tip speed ratio, defined using the blade tip speed:

$$\lambda = \frac{\Omega R}{U} \tag{3.29}$$

The tip speed ratio often occurs in the aerodynamic equations for the rotor.

Equation (3.27) indicates a relationship that must hold for the simplified 2D momentum theory to be valid. Thus, the 2D theory results in axial and tangential induction factors that are related to each other and in which the axial induction factor is not the same as that of the 1D theory. We note that this relationship might not necessarily hold for a rotor with a finite number of blades, after the tip-loss factor is introduced (see Section 3.5.1).

Next, one can derive an expression for the torque on the rotor by applying the conservation of angular momentum. For this situation, the torque exerted on the rotor, Q, must equal the change in angular momentum of the wake. On an incremental annular area element, this gives:

$$dQ = (d\dot{m})(\omega r)(r) = (\rho U_2 2\pi r dr)(\omega r)(r) \tag{3.30}$$

Since $U_2 = U(1 - a)$ and $a' = \omega/2\Omega$, this expression reduces to

$$dQ = 4a'(1 - a)\frac{1}{2}\rho U\Omega r^2 2\pi r dr \tag{3.31}$$

The power generated at each element, dP, is given by:

$$dP = \Omega dQ \tag{3.32}$$

Substituting for dQ in this expression and using the definition of the local speed ratio, λ_r, (Equation 3.28), the expression for the power generated at each element becomes:

$$dP = \frac{1}{2}\rho A U^3 \left[\frac{8}{\lambda^2}a'(1 - a)\lambda_r^3 d\lambda_r\right] \tag{3.33}$$

It can be seen that the power from any annular ring is a function of the axial and tangential induction factors and the local speed ratio. The axial and tangential induction factors determine the magnitude and direction of the airflow at the rotor plane. The local speed ratio is a function of the tip speed ratio and radius.

The incremental contribution to the power coefficient, $\mathrm{d}C_P$, from each annular ring is given by:

$$\mathrm{d}C_P = \frac{\mathrm{d}P}{\frac{1}{2}\rho A U^3} \tag{3.34}$$

Thus,

$$C_P = \frac{8}{\lambda^2} \int_0^\lambda a'(1-a)\lambda_r^3 \mathrm{d}\lambda_r \tag{3.35}$$

Similarly, the thrust coefficient is obtained as:

$$C_T = \frac{8}{\lambda^2} \int_0^\lambda a\,(1-a)\lambda_r \mathrm{d}\lambda_r \tag{3.36}$$

In order to integrate these expressions, one needs to relate the variables a, a', and λ_r (see Glauert, 1948; Sengupta and Verma, 1992). For instance, one can solve Equation (3.27) to express a' in terms of a. One gets:

$$a' = -\frac{1}{2} + \frac{1}{2}\sqrt{\left[1 + \frac{4}{\lambda_r^2}a(1-a)\right]} \tag{3.37}$$

If the axial induction along the span of the blade is known, then the tangential induction, the power coefficient, and the thrust coefficient can be calculated according to the 2D momentum theory. A special case where the axial and tangential induction factors can be determined is the case of maximum power extraction (see next section).

3.2.2.3 Maximum Power Coefficient

The aerodynamic conditions for the maximum possible power production occur when the term $a'(1-a)$ in Equation (3.35) is at its greatest value. Substituting the value for a' from Equation (3.37) into $a'(1-a)$ and setting the derivative with respect to a equal to zero yields:

$$\lambda_r^2 = \frac{(1-a)(4a-1)^2}{1-3a} \tag{3.38}$$

This equation defines the axial induction factor for maximum power as a function of the local tip speed ratio in each annular ring. An analytical solution for $a(\lambda_r)$ is given in Branlard (2017). Substituting into Equation (3.27), one finds that for maximum power in each annular ring:

$$a' = \frac{1-3a}{4a-1} \tag{3.39}$$

If Equation (3.38) is differentiated with respect to a, one obtains a relationship between $\mathrm{d}\lambda_r$ and $\mathrm{d}a$ at those conditions that result in maximum power production:

$$2\lambda_r \mathrm{d}\lambda_r = \left[6(4a-1)(1-2a)^2/(1-3a)^2\right]\mathrm{d}a \tag{3.40}$$

Now, substituting Equations (3.38–3.40) into the expression for the power coefficient (Equation 3.35) gives:

$$C_{P,\mathrm{max}} = \frac{24}{\lambda^2} \int_{a_1}^{a_2} \left[\frac{(1-a)(1-2a)(1-4a)}{(1-3a)}\right]^2 \mathrm{d}a \tag{3.41}$$

Here, the lower limit of integration, a_1, corresponds to the axial induction factor for $\lambda_r = 0$, and the upper limit, a_2, corresponds to the axial induction factor at $\lambda_r = \lambda$. Also, from Equation (3.38):

$$\lambda^2 = (1-a_2)(1-4a_2)^2/(1-3a_2) \tag{3.42}$$

Note that from Equation (3.38), $a_1 = 0.25$ gives $\lambda_r = 0$. Equation (3.42) can be solved for the values of a_2 that correspond to operation at tip speed ratios of interest. Note also from Equation (3.42), $a_2 = 1/3$ is the upper limit of the axial induction factor, a, giving an infinitely large tip speed ratio.

The definite integral can be evaluated by changing variables: substituting x for $(1-3a)$ in Equation (3.41). The optimal power coefficient is then:

$$C_{P,\max} = \frac{8}{729\lambda^2} \left\{ \frac{64}{5}x^5 + 72x^4 + 124x^3 + 38x^2 - 63x - 12[\ln(x)] - 4x^{-1} \right\}_{x = (1-3a_2)}^{x = 0.25} \tag{3.43}$$

Table 3.1 presents a summary of numerical values for $C_{P,\max}$ as a function of λ, with corresponding values for the axial induction factor at the tip, a_2 (solved using Equation 3.38).

The results of this analysis are graphically represented in Figure 3.5, which also shows the Betz–Joukowski limit of the ideal turbine based on the previous linear momentum analysis. The results show that, the higher the tip speed ratio, the closer C_P can approach the theoretical maximum.

Table 3.1 Power coefficient, $C_{P,\max}$, as a function of tip speed ratio, λ; a_2 = axial induction factor when the tip speed ratio equals the local speed ratio

λ	a_2	$C_{P,\max}$
0.5	0.2983	0.289
1.0	0.3170	0.416
1.5	0.3245	0.477
2.0	0.3279	0.511
2.5	0.3297	0.532
5.0	0.3324	0.570
7.5	0.3329	0.581
10.0	0.3331	0.585

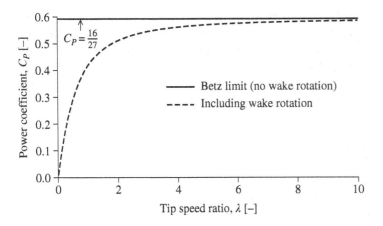

Figure 3.5 Theoretical maximum power coefficient as a function of tip speed ratio for an ideal rotor, with and without wake rotation

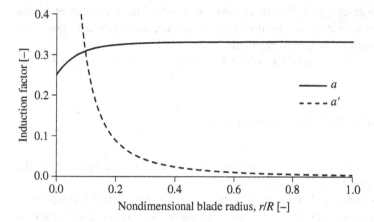

Figure 3.6 Induction factors for an ideal rotor with wake rotation; tip speed ratio, $\lambda = 7.5$; a, axial induction factor; a', tangential induction factor; r, radius; R rotor radius

Figure 3.6 shows the axial and tangential induction factors at function of spanwise location for an ideal rotor with a tip speed ratio of 7.5. It can be seen that the axial induction factors are close to the ideal of 1/3 until one gets near the hub. Tangential induction factors are close to zero in the outer parts of the rotor but increase significantly near the hub.

In the previous two sections, basic physics has been used to determine the nature of the airflow around a wind turbine rotor and theoretical limits on the maximum power that can be extracted from the wind. These developments assumed that the pressure was recovered in the far wake and that the stream tubes were independent. These assumptions are lifted in Section 3.2.3. In Section 3.2.4, vorticity formulations of the flow about an actuator disc are presented which offer a different perspective on the flow.

3.2.3 *General 2D Momentum Theory

The general 2D momentum theory removes two assumptions from the simplified 2D-momentum theory presented above: the pressure in the far wake may be different from the ambient pressure (i.e., $p_4 \neq p_1$), and the pressure force on the stream tube boundaries is accounted for. The theory can be attributed to the work of Joukowski (1918) and Glauert (1935). The inclusion of the pressure term is found in the book of Sørensen (Sørensen, 2016).

3.2.3.1 Momentum Theory Equations for Arbitrary Control Volumes

To understand the general 2D momentum theory, it is valuable to present the general momentum theory equations. For a steady and incompressible flow of a homogeneous and inviscid fluid, the conservation laws over a fixed control volume CV, with a closed surface boundary ∂CV, are:

$$\int_{\partial CV} \rho \mathbf{U} \cdot \hat{n} \mathrm{d}S = \mathbf{0}$$

$$\int_{\partial CV} \rho \mathbf{U}(\mathbf{U} \cdot \hat{n}) \mathrm{d}S = -\mathrm{d}\mathbf{T} - \int_{\partial CV} p\hat{n} \mathrm{d}S$$

$$\int_{\partial CV} (\mathbf{r} \times \rho \mathbf{U})(\mathbf{U} \cdot \hat{n}) \mathrm{d}S = -\mathrm{d}\mathbf{Q} - \int_{\partial CV} \mathbf{r} \times (p\hat{n}) \mathrm{d}S$$

$$\int_{\partial CV} \rho \left(\frac{p}{\rho} + \frac{U^2}{2} \right)(\mathbf{U} \cdot \hat{n}) \mathrm{d}S = -\mathrm{d}P$$

(3.44)

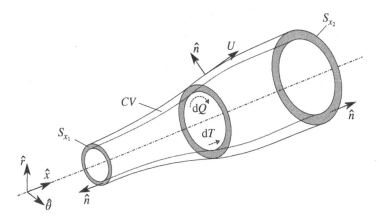

Figure 3.7 Typical control volume used in momentum analyses. The volume is delimited by two stream tubes at different radial positions and two planes normal to the x direction

where bold notations are used for vectors, $\boldsymbol{U}$ is the fluid velocity, ρ is the fluid density, p is the pressure, $\hat{\boldsymbol{n}}$ is a local unit vector pointing outward of the elementary surface area dS of the control volume surface ∂CV, and $\boldsymbol{r}$ is the position vector from the origin of the system to dS (see Figure 3.7). The total force, moment, and power within the control volume are noted d$\boldsymbol{T}$, d$\boldsymbol{Q}$, and dP, respectively. From top to bottom, the equations represent the conservation of mass, linear momentum, angular momentum, and energy, respectively.

The control volumes used in the 1D and 2D momentum theory all have similar properties: the control volumes consist of axisymmetric stream tubes delimited by two planes perpendicular to the x-axis, delimited by the coordinates x_1 and x_2, with $x_1 < x_2$. An example of such a control volume is shown in Figure 3.7.

By definition of a stream tube, the flow is along the longitudinal surface and hence orthogonal to the normal of the surface, i.e., $\boldsymbol{U} \cdot \hat{\boldsymbol{n}} = 0$, except at the ends of the stream tube. The conservations of mass and momentum in the x-direction are then:

$$\int_{S_{x_2}} \rho U_x \mathrm{d}S - \int_{S_{x_1}} \rho U_x \mathrm{d}S = 0$$

$$\int_{S_{x_2}} \rho U_x^2 \mathrm{d}S - \int_{S_{x_1}} \rho U_x^2 \mathrm{d}S = -\mathrm{d}T_x - \int_{\partial CV} p\hat{\boldsymbol{n}} \cdot \hat{\boldsymbol{x}}\mathrm{d}S \qquad (3.45)$$

$$\int_{S_{x_2}} \rho r U_\theta U_x \mathrm{d}S - \int_{S_{x_1}} \rho r U_\theta U_x \mathrm{d}S = -\mathrm{d}Q_x$$

where S_x is the surface of the stream tube at x and dT_x and dQ_x are the loads in the x direction comprised within the control volume. The pressure integral in the angular momentum equation vanishes since it only has a component along $\hat{\boldsymbol{\theta}}$ for an axisymmetric surface. On the other hand, the integral of the pressure over the control volume boundary in the linear momentum equation does not vanish. This integral was neglected in the 1D and simplified 2D momentum theory.

3.2.3.2 The General 2D Momentum Theory

The assumptions for general 2D momentum theory are: (1) steady, incompressible and axisymmetric flow; (2) homogeneous and inviscid fluid; (3) axisymmetric rotor loads concentrated on an actuator disc inducing a discontinuity of tangential velocity at the actuator disc while the axial and radial velocity remain continuous.

No tangential velocity is present upstream of the disc but a rotational motion is present in the wake. Since the pressure is not recovering in the far wake, the equations involve the rotational speed, pressure, and velocity in the far wake, respectively, noted ω_4, p_4, and U_4. Axial and tangential induction factors b and b' are defined in the far wake such that $U_4 = U(1 - b)$ and $\omega_4 = 2\Omega b'$. Detailed derivation of the equations is presented in Sørensen (2016) or Branlard (2017). Only the final results with and without the pressure forces on the stream tube are given here.

The general momentum theory expresses the axial velocity at the rotor as:

$$U_2 = \frac{U + U_4}{2} \frac{\Delta p}{\frac{1}{2}\rho(U^2 - U_4^2)}\left[1 - \frac{\mathrm{d}T_p}{\Delta p\,\mathrm{d}A}\right] \tag{3.46}$$

where $\mathrm{d}T_p$ is the contribution of the pressure force to the thrust, and $\mathrm{d}T_{p,\,Side}$ is the pressure integral over the side of the stream tube:

$$\mathrm{d}T_p = (p_0 - p_4)\,\mathrm{d}A_4 + \mathrm{d}T_{p,Side}, \qquad \mathrm{d}T_{p,Side} = \int_{Side} (p_4 - p)\boldsymbol{n}\cdot\boldsymbol{e}_z\mathrm{d}S \tag{3.47}$$

The continuity equation is used to express $= \dfrac{\mathrm{d}A_4}{\mathrm{d}A} = \dfrac{U_2}{U_4} = \dfrac{1-a}{1-b}$. The pressure differences across the rotor are given by:

$$\frac{\Delta p}{\frac{1}{2}\rho U^2} = 4\lambda_r^2 a'(1 + a') \tag{3.48}$$

The pressure difference between upstream and the far wake is:

$$\frac{p_0 - p_4}{\frac{1}{2}\rho U^2} = 4\lambda_r^2 a'(1 + b') - 2b\left(1 - \frac{b}{2}\right) \tag{3.49}$$

The main result of the general momentum equation is obtained by combining the above equations, leading to a constraint equation between the velocities and pressures:

$$b^2\frac{1-a}{1-b} = 4\lambda_r^2 a'(1 + a')\left[\frac{1 + b'}{1 + a'}\frac{1-a}{1-b} - 1\right] + \frac{\mathrm{d}T_{p,Side}}{\frac{1}{2}\rho U^2\,\mathrm{d}A} \tag{3.50}$$

In the derivations of Joukowski and Glauert (Joukowski, 1918), the term $\mathrm{d}T_{p,\,Side}$ is neglected, which effectively renders every stream tube independent. Even if this term is neglected, the general momentum theory equations do not form a closed system, and further approximations need to be introduced for their resolutions. The results presented in Section 3.2.2 assume that $p_0 = p_4$ and $b' = a'$. These assumptions together with $\mathrm{d}T_{p,\,Side} = 0$ lead to $b = 2a$, forming the "simplified" 2D momentum theory of Glauert. Wald presented momentum analyses in integral form to avoid the assumption of annuli independence (Wald, 2006). Different models are summarized in Sørensen (2016) together with the associated optimal rotor models.

3.2.4 *Actuator Disc Vorticity Formulation and Link to Momentum Theory

3.2.4.1 Velocity–Pressure and Vorticity Formulations
Many theoretical results of wind turbine aerodynamics are obtained using vorticity formulations. The analyses presented in the previous sections are based on momentum analyses, where fluid velocity and pressure are the main variables. The fluid mechanic equations can also be expressed in an alternate formulation that uses vorticity as the main variable. Mathematically, vorticity is defined as the curl of the velocity field (the curl operator is a

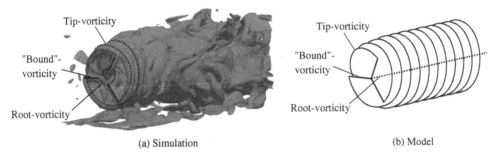

(a) Simulation   (b) Model

Figure 3.8 Vorticity about a wind turbine. (a) Vorticity contours as obtained from a wind turbine simulation with atmospheric boundary layer and turbulence. Bound, root, and tip vorticity are highlighted in the figure, (b) Example of idealization of the vorticity field onto lines and helical filaments of vorticity

cross-product of partial derivatives). Vorticity has dimensions s^{-1}. In polar coordinates (r, θ, x), the components of the vorticity vectors, noted ω, are defined as:

$$\omega_r = \frac{1}{r}\frac{\partial u_x}{\partial \theta} - \frac{\partial u_\theta}{\partial x}, \qquad \omega_\theta = \frac{\partial u_r}{\partial x} - \frac{\partial u_x}{\partial r}, \qquad \omega_x = \frac{1}{r}\left[\frac{\partial(ru_\theta)}{\partial r} - \frac{\partial u_r}{\partial \theta}\right] \tag{3.51}$$

The vorticity takes on large values in regions where the flow experiences strong gradients such as the neighborhood of a vortical structure or a shear layer. An illustration of vorticity contours about a wind turbine, as obtained from a numerical simulation, is shown in Figure 3.8 (left). Only the strong vorticity is shown in the figure, and it is seen that strong vorticity is found along the blades (bound vorticity), along the tip, and at the root. The bound vorticity originates from the lift force generated by the blade airfoils, as will be seen in Section 3.3. This complex vorticity field can be simplified by assuming a representation using helical vortex filaments (see right of Figure 3.8). Other simplified models may use vortex rings, or vortex cylinders.

The dynamics of vorticity are expressed by taking the curl of the Navier–Stokes equations, and, in many applications, the velocity–pressure and vorticity formulations can be shown to be equivalent. This is the case for the actuator disc model described in the next section. Wake dynamics and the evolution of vorticity (convection, diffusion, and meandering) for "non-idealized" wakes are discussed in Chapter 12.

3.2.4.2 Vorticity Formulation of the Ideal Actuator Disc without Wake Rotation

First, an ideal actuator disc in an inviscid fluid that generates a constant pressure drop but no wake rotation is considered. The process by which the actuator disc generates an infinitesimal vortex sheet in the wake is discussed in this paragraph. The vortex sheet is generated at the rim of the disc. It extends from the disc to infinitely far downstream and follows the streamline that passes through the rim of the disc. Figure 3.9 illustrates the vortex sheet. The vorticity surface arises from the sudden jump in streamwise velocity that the actuator disc generates between the flow that passes through the disc and the flow that goes around it (see "Velocity jump" in the figure).

In a viscous fluid, this velocity would change continuously about the rim of the disc, forming a shear layer filled with vorticity, but under the inviscid assumption this shear layer is confined to an infinitesimal vorticity sheet and a discontinuous velocity jump occurs across it. The vorticity sheet is force-free and convects at the mean of the velocity on both sides of it, noted U_+ and U_- on the figure. It follows that the vorticity surface will coincide with the streamline that passes through the rim of the disc. Following the streamline, the radius of the tubular vortex sheet increases until it reaches an equilibrium in the far wake. The intensity of the infinitesimal vortex sheet is a vector, noted γ_θ on the figure, which can be determined mathematically as:

$$\gamma_\theta = \hat{n} \times (U_+ - U_-) \tag{3.52}$$

where $\hat{n}$ is the vector normal to the sheet going in the direction from "−" to "+." Equation (3.52) indicates that the magnitude of the vortex sheet intensity is equal to the velocity jump across it, and its direction is orthogonal to the

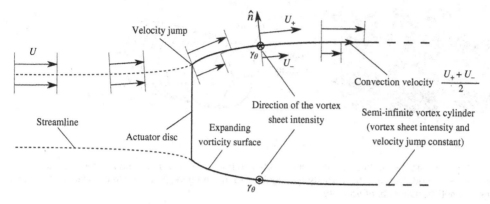

Figure 3.9 Vorticity surface generated behind an ideal actuator disc without wake rotation. The vorticity surface occurs due to the jump of velocity between the flow that passes the actuator disc and the flow outside. The surface follows the streamline and plateau

velocity difference and contained within the sheet. In this case, the vorticity is solely in the tangential direction (in and out of the plane of the figure) since no wake rotation is assumed and the flow is purely axial and radial. The magnitude of γ_θ progressively increases toward the far wake (because the velocity jump across it increases), where the wake radius reaches an equilibrium value, and from this point on, the wake corresponds to a semi-infinite vortex cylinder. The vortex sheet intensity has a dimension of ms^{-1} (vorticity times a length scale). It is a singular representation of the vorticity such that $\omega_\theta dr \sim \gamma_\theta$ as the thickness of the vorticity region dr goes to 0 to become an infinitely thin vortex sheet. The concentration of vorticity from a vortex volume to a vortex sheet is illustrated in Figure 3.42 (see Section 3.7.3). Using the axisymmetry assumption ($\partial/\partial\theta \equiv 0$) and the fact that no tangential velocity is present ($u_\theta = 0$), Equation (3.51), gives $\omega_r = 0$, $\omega_x = 0$, and solely ω_θ indeed remains.

3.2.4.3 Approximate Solution Using a Vortex Cylinder

The radius and intensity of the vorticity surface described in the previous paragraph can only be computed numerically. Methods using discrete vortex rings were, for instance, used by Øye (1990) and van Kuik and Lignarolo (2015). An approximate analytical solution may be obtained by introducing the following assumptions: the vorticity surface has a constant radius and intensity. The difference in velocity field between the constant radius cylinder ("rigid") and the "free" vortex cylinder is illustrated in Figure 3.10.

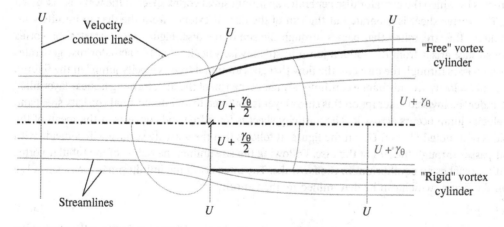

Figure 3.10 Velocity field induced by a "free" vortex cylinder (top half) and a "rigid" vortex cylinder (bottom half). γ_θ is negative as shown

The wake vorticity surface consists then of a semi-infinite vorticity surface of constant tangential intensity γ_θ. It can be shown that the velocity induced by a semi-infinite cylinder of tangential vorticity is constant and equal to $\gamma_\theta/2$ at the cylinder inlet and γ_θ at infinity downstream. The superposition of the free stream and the induced velocity field is then such that the resulting velocity is $U + \gamma_\theta/2$ at the disc and $U + \gamma_\theta$ in the far wake. Introducing the axial induction factor $a = -\gamma_\theta/(2U)$, the result from the momentum theory that the induction factor at the rotor is half the one in the far wake is directly retrieved. The fact that the approximate vortex model gives the same result as the simplified momentum theory confirms that the momentum theory result was indeed approximate. This was discussed in Section 3.2.4. The advantage of the approximate vortex cylinder model compared to the simplified momentum theory is that the velocity field may be obtained analytically in the entire domain. The velocity induced from a vorticity distribution can be obtained using the Biot–Savart law given in Equation (3.194) (see Section 3.7.3). For a vortex sheet, the Biot–Savart law is:

$$u_i(x) = \frac{-1}{4\pi} \int\int_S \frac{(x - x')}{\| x - x' \|^3} \times \gamma(x') \, dS(x')$$

(3.53)

where the integral is performed over the points x' belonging to the vorticity surface S. Applying Equation (3.53) to a semi-infinite cylinder of constant tangential intensity γ_θ, the axial induced velocity on the cylinder axis can be calculated as:

$$u_i(x) = \frac{\gamma_\theta}{2} \left(1 + \frac{x}{\sqrt{R^2 + x^2}} \right)$$

(3.54)

where R is the rotor radius, and x is the distance along the axis, positive downstream, with the actuator disc located at $x = 0$. The full velocity field is the sum of the free stream U and the induced velocity u_i. Introducing $a = -\gamma_\theta/(2U)$, the full velocity field on the axis is:

$$U(x) = U \left[1 - a \left(1 + \frac{x}{\sqrt{R^2 + x^2}} \right) \right]$$

(3.55)

From this expression, it is seen that the velocity infinitely upstream is U, the velocity at the rotor is $U(1 - a)$, and the velocity far downstream is $U(1 - 2a)$. The rigid vortex cylinder result is therefore in agreement with the simplified 1D momentum theory, but it also provides the velocity at other locations. The above expression and the velocity field at any point are given for instance in Branlard and Gaunaa (2015). It is used to compute the blockage effect in Chapter 12.

3.2.4.4 Vorticity Formulation of an Actuator Disc with Wake Rotation

When the actuator disc generates rotation of the wake, the following vorticity components are present in the wake: a longitudinal "root vortex" which induces swirl in the wake, and a vorticity surface with tangential and longitudinal vorticity. A radial "bound vorticity" component is also present at the disc to account for the change of tangential velocity across it. The bound vorticity at the disc can be thought of as coming from the bound vorticity generated by the lift of an infinite number of infinitely thin blades distributed in the rotor plane. Analytical solutions for the entire velocity field may be obtained under the same assumptions used in the previous paragraph, namely, that the wake does not expand. These velocity fields are reported in Branlard and Gaunaa (2015). The ideal actuator disc with wake rotation is yet an abstruse model since the swirl required to sustain the constant pressure drop must go to infinity at the root. To alleviate this issue, authors have used a regularized root vortex (see e.g., Sørensen and van Kuik, 2011). In practice, an actuator disc with constant pressure is never obtained.

In a realistic actuator disc configuration, the loading varies radially along the disc span. The corresponding vorticity model is then obtained using a continuous superposition of the vorticity model presented above. The varying loading along the span is associated with a varying value of the bound circulation. Vorticity is continuously trailed into the wake, and the wake consists of a continuous superposition of concentric infinitesimal vorticity surfaces. At each radial position, the infinitesimal vorticity surface has an initial intensity corresponding to the change of the bound vorticity at this location. The vorticity surface follows the streamline and consists of a component normal and tangential to the streamlines, determined by the local velocity. The wake consists of a continuous vorticity volume, with typically strong vorticity at the tip and root. The resolution of the vorticity and velocity field for the continuous actuator disc requires the development of dedicated numerical solvers.

3.2.4.5 Approximate Formulation and Relation to Strip Theory

Under some assumptions, it can be shown that the vortex formulation and the strip theory provide the same results. If wake expansion is neglected, an assumption of greater validity for lower thrust coefficients and analytical solutions of the velocity field in the entire domain can be obtained with the vortex formulation. If one further assumes that the stream tubes are independent, then the vortex model results are identical to the simplified 2D momentum theory. For a realistic flow, the different stream tubes are not independent, and the wake rotation will induce an additional pressure drop. The topic is discussed in Branlard and Gaunaa (2016).

To relate the vorticity formulations with momentum theory, the following approximate relationships can be used. The intensity of the tangential wake vorticity surface (in m/s) can be approximated as:

$$\gamma_\theta = -U\left[1 - \sqrt{1 - C_T}\right] = -2\,a\,U \tag{3.56}$$

The vorticity surface may be discretized into vortex rings, where the vorticity is condensed into an infinitely thin line. The intensity of a vortex line is referred to as circulation (in m^2/s). If the vortex rings are spaced by Δx, the circulation is obtained from Equation (3.56) as:

$$\Gamma_\theta = -U\left[1 - \sqrt{1 - C_T}\right]\Delta x \tag{3.57}$$

The circulation of the root vortex can be approximated as:

$$\Gamma_{tot} = \frac{\pi R U}{\lambda} C_T = \frac{\pi U^2}{\Omega} 4a(1 - a) \tag{3.58}$$

This value also corresponds to the total bound circulation of the rotor. The bound circulation of each blade (if assumed constant) is Γ_{tot}/B. If helical filaments are used, the helical pitch may be approximated as $h = 2\pi R (1 - a)/\lambda$ (the value with $a = 0$ is often used in propeller literature).

3.3 Airfoils and General Concepts of Aerodynamics

Airfoils are structures with specific geometric shapes that are used to generate mechanical forces due to the relative motion between the airfoil and its surrounding fluid. Wind turbine blades use airfoils to develop mechanical power and effectively introduce the "pressure drop" that was used in the actuator-disc models. The cross-sections of wind turbine blades have the shape of airfoils. The width and length of the blade are functions of the desired aerodynamic performance, the maximum desired rotor power, the assumed airfoil properties, and strength considerations. Before the details of wind turbine power production are explained, aerodynamic concepts related to airfoils need to be discussed.

3.3.1 Airfoil Terminology

A number of terms are used to characterize an airfoil, as shown in Figure 3.11. The mean camber line is the locus of points halfway between the upper and lower surfaces of the airfoil. The most forward and rearward points of the

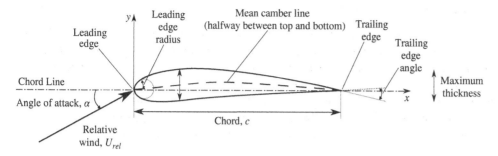

Figure 3.11 Airfoil nomenclature

mean camber line are called the leading and trailing edges, respectively. The straight line connecting the leading and trailing edges is the chord line of the airfoil, and the distance from the leading to the trailing edge measured along the chord line has designated the chord, c, of the airfoil. The camber is the distance between the mean camber line and the chord line, measured perpendicular to the chord line. The thickness is the distance between the upper and lower surfaces, also measured perpendicular to the chord line. Finally, the angle of attack, α, is defined as the angle between the relative wind, U_{rel}, and the chord line (the angle of attack is hard to define in 3D, see Branlard 2017, chapter 3). Not shown in the figure is the span of the airfoil, which is the length of the airfoil perpendicular to its cross-section. The airfoil coordinates are typically given in a x–y coordinate system such that the leading edge is at $x = 0$ and the trailing edge at $x = c$. The geometrical shape of the airfoil determines its aerodynamic performance. The classical aerodynamic theory of airfoils reveals in particular the importance of the following parameters: the leading-edge radius, mean camber line, maximum thickness and thickness distribution of the profile, and the trailing edge angle.

There are many types of airfoils (see Abbott and von Doenhoff, 1959; Althaus and Wortmann, 1981; Althaus, 1996; Miley, 1982; Tangler, 1987). Traditionally, airfoils were designed as "families" or "series" which group a set of airfoils with similar geometries. The NACA series is common amongst the aerospace community, and it has been used in the early designs of wind turbines. Manufacturers now typically design dedicated airfoils within an optimization process as part of the blade design. Typically, only a few airfoils are used along the span of the blade, each with different thicknesses. The rest of the blade geometry is then obtained by interpolation between these airfoil stations. A few examples of airfoils are shown in Figure 3.12. The NACA 0012 is a 12% thick symmetric airfoil. The other airfoils in the figure are the ones used in the NREL 5-MW turbine, displaying different thicknesses and cambers.

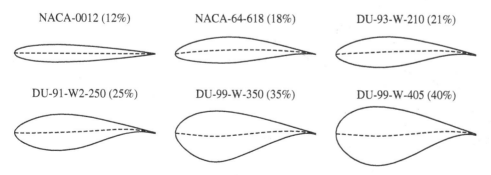

Figure 3.12 Sample airfoils. NACA 0012 airfoil and airfoils used in the NREL 5-MW turbine. The relative thickness is indicated in parenthesis

3.3.2 Lift, Drag, and Nondimensional Parameters

Airflow over an airfoil produces a distribution of forces over the airfoil surface. The flow velocity over airfoils increases over the convex surface resulting in lower average pressure on the "suction" (upper) side of the airfoil compared with the concave or "pressure" (lower) side of the airfoil. Meanwhile, viscous friction between the air and the airfoil surface slows the airflow to some extent next to the surface.

As shown in Figure 3.13, the resultant of all of these pressure and friction forces is usually resolved into two forces and a moment that acts along the chord at a distance that is approximatively $c/4$ from the leading edge (at the "quarter chord," which is location on the chord line ¼ of the distance from the leading edge):

- **Lift force**. Defined to be perpendicular to direction of the oncoming airflow. The lift force is a consequence of the unequal pressure on the upper and lower airfoil surfaces.
- **Drag force**. Defined to be parallel to the direction of the oncoming airflow. The drag force is due both to viscous friction forces at the surface of the airfoil and to unequal pressure on the airfoil surfaces facing toward and away from the oncoming flow.
- **Pitching moment**. Defined to be about an axis perpendicular to the airfoil cross-section.

The choice of the quarter chord point is a convention that has been retained because, for many airfoils, the aerodynamic moment about this point varies little with the angle of attack within the typical operating region of the airfoil (for instance, α between minus 5 and 10°). This result is also supported by the thin airfoil theory of a flat plate. The point where the aerodynamic forces are defined is referred to as the aerodynamic center. In Figure 3.13, the coordinate system of *OpenFAST* (see Chapter 8) is used, which is different from Figure 3.11.

Theory and research have shown that many flow problems can be characterized by nondimensional parameters. The most important nondimensional parameter for defining the characteristics of fluid flow conditions is the Reynolds number. The Reynolds number, Re, is defined by:

$$\text{Re} = \frac{UL}{\nu} = \frac{\rho UL}{\mu} = \frac{\text{Inertial force}}{\text{Viscous force}} \tag{3.59}$$

where ρ is the fluid density, μ is fluid viscosity, $\nu = \mu/\rho$ is the kinematic viscosity, and U and L are a velocity and length that characterize the scale of the flow. For the characterization of airfoil performances, these are taken as the incoming stream velocity, U_{rel}, and the chord length on the airfoil. For example, if U_{rel} is 65 m/s, ν is 0.000013 m²/s and the chord length is 2 m, the Reynolds number is 10 million.

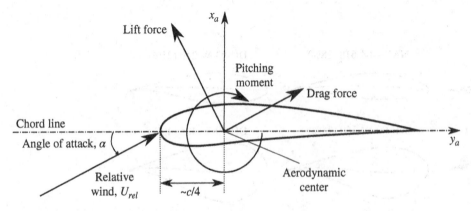

Figure 3.13 Forces and moments on an airfoil section, α, angle of attack; c, chord. The direction of positive forces and moments is indicated by the direction of the arrow

Additional nondimensionalized force and moment coefficients, which are functions of the Reynolds number, can be defined for two- or three-dimensional objects, based on wind tunnel tests or CFDs simulations. Three-dimensional airfoils have a finite span and force and moment coefficients are affected by the flow around the end of the airfoil. 2D airfoil data, on the other hand, are assumed to have an infinite span (no end effects). 2D data are measured in such a way that there is indeed no airflow around the end of the airfoil in the test section in the wind tunnel. Force and moment coefficients for flow around 2D objects are usually designated with a lowercase subscript, as in C_d for the 2D drag coefficient. In that case, the forces measured are forces per unit span. Lift and drag coefficients that are measured for flow around three-dimensional objects are usually designated with an upper case subscript, as in C_D. The three-dimensional-coefficients are rarely used in wind energy applications. Rotor design usually uses 2D coefficients, determined for a range of angles of attack and Reynolds numbers, in wind tunnel tests. The 2D lift coefficient is defined as:

$$C_l = \frac{F_L/l}{\frac{1}{2}\rho U^2 c} = \frac{\text{Lift force/unit length}}{\text{Dynamic force/unit length}} \tag{3.60}$$

The 2D drag coefficient is defined as:

$$C_d = \frac{F_D/l}{\frac{1}{2}\rho U^2 c} = \frac{\text{Drag force/unit length}}{\text{Dynamic force/unit length}} \tag{3.61}$$

and the pitching moment coefficient is:

$$C_m = \frac{M}{\frac{1}{2}\rho U^2 c^2} = \frac{\text{Pitching moment/unit length}}{\text{Dynamic moment/unit length}} \tag{3.62}$$

where ρ is the density of air, U is the velocity of undisturbed airflow, A is the projected airfoil area (chord × span), c is the airfoil chord length, and l is the airfoil span. The combination of the lift drag and moment coefficients as function of the angle of attack is commonly referred to as the polar data of the airfoils. The pressure coefficient:

$$C_p = \frac{p - p_\infty}{\frac{1}{2}\rho U^2} = \frac{\text{Static pressure}}{\text{Dynamic pressure}} \tag{3.63}$$

is used to analyze the pressure distribution around the airfoil surface. The *pressure* coefficient around an airfoil should not be confused with the *power* coefficient of a wind turbine, C_P, which is written with an upper case subscript. The two coefficients are typically discussed in very different contexts (airfoils or wind turbines). The open-source tool XFOIL (Drela, 1989) is widely used to obtain the lift, drag, and pressure coefficients for an airfoil for which the geometry is known.

3.3.3 Flow Over an Airfoil

The lift, drag, and pitching moment coefficients of an airfoil are generated by the pressure variation over the airfoil surface and the friction between the air and the airfoil.

The pressure variations are caused by changes in air velocity that can be understood using Bernoulli's equation, which was introduced in Section 3.2.1.2. As the airflow accelerates around the rounded leading edge, the pressure drops, resulting in a negative pressure gradient. When the flow approaches the trailing edge, it decelerates and the surface pressure increases, resulting in a positive pressure gradient. If, given the airfoil design and the angle of attack, the air speeds up more over the upper surface than over the lower surface of the airfoil, then there is a net lift force. Similarly, the pitching moment is a function of the integral of the moments of the pressure forces about the quarter chord over the surface of the airfoil.

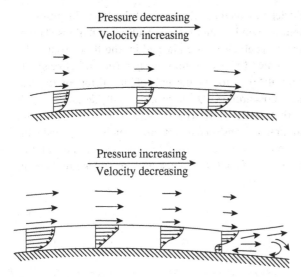

Figure 3.14 Effects of favorable (decreasing pressure) and adverse (increasing pressure) pressure gradients on the boundary layer (*Source:* Miley, 1982/National Technical Information Service/Public Domain)

Drag forces are a result of both the pressure distribution over the airfoil and the friction between the airflow and the airfoil. The component of the net pressure distribution in the direction of the airflow results in drag due to the pressure. Drag due to friction is a function of the viscosity of the fluid and dissipates energy into the flow field.

Drag also causes the development of two different regions of flow: one farther from the airfoil surface, where frictional effects are negligible and the boundary layer, immediately next to the airfoil surface, where frictional effects dominate. In the boundary layer, the velocity increases from zero at the airfoil surface to that of the friction-free flow outside of the boundary layer. The boundary layer on a wind turbine blade may vary in thickness from a millimeter to tens of centimeters. The boundary layer height decreases with increasing Reynolds number.

The flow in the boundary layer may be laminar (smooth and steady) or turbulent (irregular with three-dimensional vortices). At the leading edge of the airfoil, the flow is laminar. Usually, at some point downstream, the flow in the boundary layer becomes turbulent as the interaction between viscosity and nonlinear inertial forces causes a "transition" to chaotic, turbulent flow. Laminar boundary layers result in much lower frictional forces than turbulent boundary layers.

The pressure gradient of the flow has a significant effect on the boundary layer, as illustrated in Figure 3.14. That pressure gradient may be favorable (positive in the direction of the flow) or adverse (against the flow). Flow in the boundary layer is accelerated or decelerated by the pressure gradient. In the boundary layer, the flow is also slowed by surface friction. Thus, in an adverse pressure gradient and with the help of surface friction, the flow in the boundary layer may be stopped or it may reverse direction. This results in the flow separating from the airfoil, causing a condition called stall. Boundary layers that have already transitioned to turbulent flow are less sensitive to an adverse pressure gradient than laminar boundary layers, but once the laminar or turbulent boundary layer has separated from the airfoil, the lift drops. An airfoil can only efficiently produce lift as long as the surface pressure distributions can be supported by the boundary layer.

It is important to distinguish the effects of turbulence in the atmosphere from that in the boundary layer of the airfoil. Wind turbine airfoils operate in the turbulent planetary boundary layer but the scale of the turbulent fluctuations in the atmosphere is much larger than the scale of the turbulence in the boundary layer of a wind turbine airfoil. The flow in the boundary layer is only sensitive to fluctuations in the order of the size of the boundary layer itself. Thus, the atmospheric turbulence does not affect the airfoil boundary layer directly. It may affect it indirectly through changing angles of attack, which will change the flow patterns and pressure distributions over the blade surface.

3.3.4 Airfoil Behavior

It is useful to consider the flow around a cylinder as a starting point for looking at airfoils. The flow can be best visualized with the help of streamlines. A streamline can be thought of as the path that a particle would take if placed in a flowing fluid. A flow field can then be depicted in terms of a number of streamlines. Streamlines have a few visually interesting properties. For example, streamlines that converge indicate an increase in velocity and a

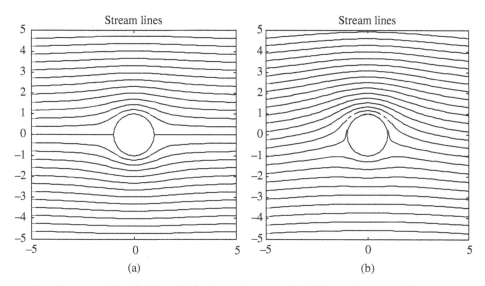

Figure 3.15 (a) Flow around stationary cylinder, (b) flow around rotating (CW) cylinder

decrease in pressure. The opposite is the case for diverging streamlines. It is also the case that Bernoulli's equation is only strictly applicable along streamlines. Figure 3.15(a) illustrates a flow around a stationary cylinder. In this case, the depiction of the flow is based on the assumption that there is no drag and no inertia. It can be seen that the streamlines move closer together as they pass the cylinder. This indicates that the velocity is increasing and the pressure is decreasing. The pattern is symmetrical on both sides of the cylinder, so there is no net lift on the cylinder. In fact, in the absence of viscous drag, there is no net force at all on the cylinder in this situation.

When there is rotational flow, however, the situation changes. The fluid rotation may be brought about either by rotation of an object in the flow, or it may result from the shape of the object (such as an airfoil), which imparts a rotational motion to the fluid.

Rotational flow is often described in terms of vorticity and circulation (see also Section 3.2.4). If an element of fluid is rotating, its angular velocity is characterized by its vorticity. In a 2D flow, only the z-component of the vorticity is present, noted, ω_z, and given by:

$$\omega_z = \frac{\partial u_x}{\partial y} - \frac{\partial u_y}{\partial x} \tag{3.64}$$

where u_x is the velocity component in the direction of the flow (x) and u_y is the component perpendicular to the flow (y). The vorticity is also equal to twice the angular velocity of the fluid element.

Circulation, Γ, is the integral of the vorticity of the elements, multiplied by their respective incremental areas, over the region of interest, as shown in Equation (3.65).

$$\Gamma = \int\int \left(\frac{\partial u_x}{\partial y} - \frac{\partial u_y}{\partial x}\right) dx dy \tag{3.65}$$

It can be shown that, in general, the lift per unit length on a body is given by $\tilde{L} = \rho \Gamma U_\infty$ (N/m), where U_∞ = free stream velocity. This relationship is referred to as the Kutta–Joukowski theorem. A static cylinder does not generate lift, but a spinning cylinder does. For such a cylinder of radius r, the value of the circulation is $\Gamma = 2\pi r u_r$ where $u_r = 2\pi r N$ is the cylinder's peripheral velocity when N is the rotational speed in rps. Figure 3.15(b) illustrates the flow of a rotating cylinder. Note that in this case, the streamlines are closer together at the top than they are at the bottom, indicating that there is a net lower pressure, and hence lift, in the vertical direction. This phenomenon is

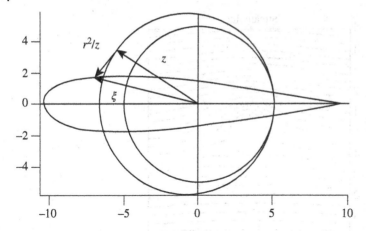

Figure 3.16 Airfoil derived from a transformed cylinder

known as the Magnus effect. The Magnus effect is the physical basis of the Flettner rotor, which has been used successfully in ship propulsion (Flettner, 1926).

The method outlined above can be used to predict pressure distributions about an airfoil. The general method is called conformal mapping. The first step in doing this is transforming the coordinate system of the flow so that the cylinder shape is transformed into a shape that resembles the airfoil of interest. Figure 3.16 illustrates a symmetrical airfoil derived from a cylinder in this way. The resulting airfoil shape is the locus of points ξ resulting from the transformation given by Equation (3.66).

$$\xi = z + r^2/z \tag{3.66}$$

where z is the vector to a circle offset from the origin. This transformation is called a conformal map, and the one presented is attributed to the work of Joukowski (1918). In the example shown, the radius of the circle is 5.5, and the offset is 0.8. A circle of radius 5 centered at the origin is included for comparison.

This method of analysis (i.e., the application of transformations of shapes, streamlines, and pressure distributions) provides the foundation of thin airfoil theory, which is used to predict the characteristics of most commonly used airfoils. Thin airfoil theory shows, for example, that the lift coefficient of a symmetrical airfoil at low angles of attack is equal to $2\pi\alpha$ when the angle α is measured in radians, as shown in Equation (3.67) The lift slope for a streamlined airfoil is often around 2π, and this value can also be exceeded. Details on the theory of lift and circulation, as well as on the use of transformations, are given in Abbott and von Doenhoff (1959) and most aerodynamics textbooks, such as Bertin and Cummings (2014).

$$C_l = 2\pi\alpha \quad \text{(with } \alpha \text{ in radians),} \qquad \text{or,} \qquad C_l = \frac{\pi^2}{90}\alpha \quad \text{(with } \alpha \text{ in}°\text{)} \tag{3.67}$$

Under ideal conditions, all symmetric airfoils of finite thickness would have similar theoretical lift coefficients. This means that lift coefficients would increase with increasing angles of attack and continue to increase until the angle of attack reached 90°. The behavior of real symmetric airfoils does indeed approximate this theoretical behavior at low angles of attack, although not for higher angles of attack. For example, typical lift and drag coefficients for a symmetric airfoil, the NACA 0012 airfoil, the profile of which was shown in Figure 3.12, are shown in Figure 3.17 as a function of angle of attack and Reynolds number. The lift coefficient for a flat plate under ideal conditions is also shown for comparison.

Note that, in spite of the very good correlation at low angles of attack, there are significant differences between actual airfoil operation and the theoretical performance at higher angles of attack. The differences are due

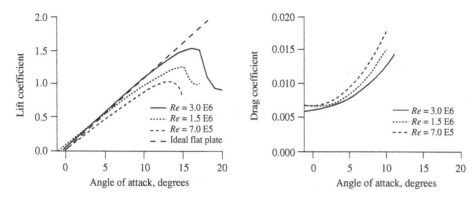

Figure 3.17 Lift and drag coefficients for the NACA 0012 symmetric airfoil (*Source:* Miley, 1982/National Technical Information Service/Public Domain.); Re, Reynolds number

primarily to the assumption, in the theoretical estimate of the lift coefficient, that air has no viscosity. As mentioned earlier, surface friction due to viscosity slows the airflow next to the airfoil surface, resulting in a separation of the flow from the surface at higher angles of attack and a rapid decrease in lift.

Airfoils for HAWTs often are designed to be used at low angles of attack, where lift coefficients are fairly high, and drag coefficients are fairly low. The lift coefficient for a symmetrical airfoil is zero at an angle of attack of zero and increases to over 1.0 before decreasing at higher angles of attack. The drag coefficient is usually much lower than the lift coefficient at low angles of attack. It increases at higher angles of attack.

Note, also, that there are significant differences in airfoil behavior at different Reynolds numbers. For example, as Reynolds numbers decrease, viscous forces increase in magnitude compared to inertial forces. This increases the effects of surface friction, affecting velocities, the pressure gradient, and the lift generated by the airfoil. Rotor designers must make sure that appropriate Reynolds number data are available for the detailed analysis of a wind turbine rotor system. Airfoil performance measurements at high Reynolds number are usually difficult to acquire or expensive since a wind tunnel of large dimension, speed, or fluid density is required for the measurement.

The lift coefficient at low angles of attack can be increased, and drag can often be decreased by using a cambered airfoil (Miley, 1982; Eggleston and Stoddard, 1987). Cambered airfoils have a nonzero lift coefficient at an angle of attack of zero, and their profile coefficients won't be odd functions of the angle of attack like the ones of symmetric airfoils. Airfoil behavior can be categorized into three flow regimes: the attached flow regime, the high lift/stall development regime, and the flat plate/fully stalled regime (Spera, 1994). For example, the DU-93-W-210 airfoil is used in some European wind turbines, and in the NREL 5-MW reference turbine (Jonkman *et al.*, 2009). Its cross-sectional profile is shown in Figure 3.12. The lift, drag, and pitching moment coefficients for this same airfoil are shown in Figure 3.18 for a Reynolds number of 3 million, together with the different flow regimes.

3.3.4.1 Attached Flow Regime

At low angles of attack (up to about 7° for the DU-93-W-210 airfoil), the flow is attached to the upper surface of the airfoil. In this attached flow regime, lift increases with the angle of attack and drag is relatively low. The lift slope in this region is close to 2π (with alpha in radians) for a typical airfoil.

3.3.4.2 High Lift/Stall Development Regime

In the high lift/stall development regime (from about 7 to 11° for the DU-93-W-210 airfoil), the lift coefficient peaks as the airfoil becomes increasingly stalled. Stall occurs when the angle of attack exceeds a certain critical value (say 10–16°, depending on the profile and the Reynolds number) and separation of the boundary layer on the upper

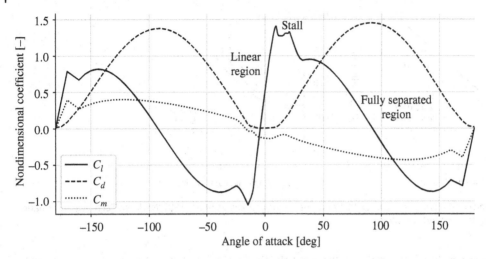

Figure 3.18 Lift, drag, and moment coefficients for the DU-93-W-210 airfoil, and the different flow regimes: linear, stall, and fully separated (flat plate theory)

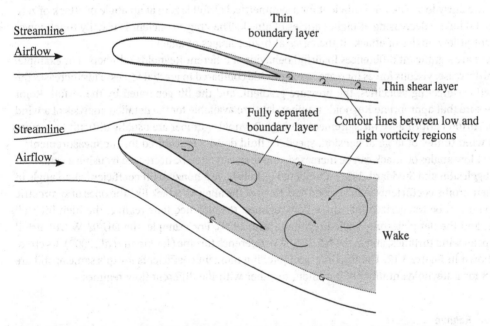

Figure 3.19 Illustration of airfoil boundary layers and wakes at low angle of attack (top) and high angle of attack (bottom)

surface takes place, as shown in Figure 3.19. This causes a wake to form above the airfoil, which reduces lift and increases drag.

Stall can occur at certain blade locations or conditions of wind turbine operation. It is sometimes used to limit wind turbine power in high winds. For example, many early wind turbine designs using fixed pitch blades relied on power regulation control via aerodynamic stall of the blades. That is, as wind speed increases, stall progresses outboard along the span of the blade (toward the tip), causing decreased lift and increased drag. In a well-designed, stall-regulated machine, this results in nearly constant power output as wind speeds increase above a certain value and a smooth reduction of power at very high wind speeds.

3.3.4.3 Flat Plate/Fully Stalled Regime

In the flat plate/fully stalled regime, at larger angles of attack up to 90°, the airfoil acts increasingly like a simple flat plate with approximately equal lift and drag coefficients at an angle of attack of 45° and zero lift at 90°.

3.3.4.4 Modeling of Post-stall Airfoil Characteristics

Measured wind turbine airfoil data are used to design wind turbine blades. Wind turbine blades may often operate in the stalled region of operation, but data at high angles of attack are sometimes unavailable. Because of the similarity of stalled behavior to flat plate behavior, models have been developed to model lift and drag coefficients for stalled operation. Information on modeling post-stall behavior of wind turbine airfoils can be found in Viterna and Corrigan (1981). Summaries of the Viterna and Corrigan model can be found in Spera (1994) and Eggleston and Stoddard (1987).

3.3.5 Airfoils for Wind Turbines

Modern wind turbines typically use different airfoils along the span of the blade (see Figure 3.12). The choice of airfoil is part of the blade design process in which the aerodynamics, structural, and aeroacoustics requirements need to be fulfilled.

The rotor torque (and hence power) is mainly generated by the outer part of the rotor due to the larger moment arm at that location. As a result of this, the tip of the blade is typically designed using high-performance airfoils: small thickness of about 15–21% and high lift-to-drag ratios of about 100–150. Aeroacoustic emission may need to be considered in the airfoil design due to the high relative wind velocities at the tip. By the same moment arm argument, the root of the blade is the part that carries the highest loads, and it is designed to provide structural support to the rest of the blade: the root region consists of a cylindrical section and higher thickness airfoils. To ensure a relative continuity of airfoil shapes and performances along the span of the blade, HAWT blades have been designed using airfoil "families," which were introduced in Section 3.3.1; see also Hansen and Butterfield (1993). The airfoil design must accommodate the higher thickness needed from a structural standpoint and cannot simply optimize the aerodynamic performances. For an overview of wind turbine blade aerodynamic design, see Section 3.4.

The airfoil designer must account for operations at off-design conditions. The blade surfaces become rougher with time due to the environment (dirt, insects, and rain erosion). Power outputs as low as 40% of their clean value were observed in the field. Airfoils whose performances are robust with changes in roughness should be preferred. Low sensitivity to surface roughness may be achieved by locating the point of transition to turbulent flow near the leading edge, as the flow approached stall. Stall behavior is also critical to airfoil design. Significant fluctuations in angle of attacks occur over a rotor revolution due to gusts or the change of wind speed. As a result of this, operation close to the maximum-lift angle of attack will intermittently fall into the stall region, resulting in damaging load oscillations for both stall- and pitch-regulated turbines. To reduce such oscillations, the designed operating angle of attack should be chosen below the highest lift point, and airfoil designs with abrupt stall behavior should be avoided. Passive aerodynamic devices such as vortex generators may be used to alter the performance of an airfoil and delay the onset of stall. Active flow controls, such as trailing edge flaps, or boundary layer control systems may be developed to dynamically change the airfoil performances.

In general, airfoil design is a compromise between the points mentioned above. For instance: midboard sections typically need higher thickness to withstand the loads, but higher thickness airfoils typically produce less lift. To deliver sufficient torque at low wind speeds, without significantly large chords, the airfoils can be optimized for high lift coefficients, but high maximum lift coefficients often accompany sensitivity to leading-edge roughness.

In the 1970s and early 1980s, wind turbine designers felt that minor differences in airfoil performance characteristics were far less important than optimizing blade twist and taper. For this reason, little attention was paid to the task of airfoil selection. Thus, airfoils that were in use by the aircraft industry were chosen because aircraft were viewed as similar applications. Aviation airfoils such as the NACA 44xx and NACA 230xx (Abbott and von Doenhoff, 1959) were popular airfoil choices because they had high maximum lift coefficients, low pitching moment, and low minimum drag. The NACA classification has 4, 5, and 6 series wing sections. For wind turbines, four-digit series were generally used, for example NACA 4415. The first integer indicates the maximum value of the mean camber line ordinate in percent of the chord. The second integer indicates the distance from the leading edge to the maximum camber in tenths of the chord. The last two integers indicate the maximum section thickness in percent of the chord.

Dedicated wind turbine airfoil designs started to emerge in the early 1980s. The new designs were selected for their reduced sensitivity to leading-edge roughness and their smooth stall behavior. Examples of airfoils series that have been used by academia and industry are the NACA 63-4xx and 63-6xx airfoils, the S8xx series, the FFA W-xxx series, the DU airfoils (Timmer and van Rooij, 2003), or the Riso-A/B airfoils (Fulgsang *et al.*, 2004).

The determination of the airfoil characteristics (lift, drag, and moment coefficients) is critical to determining the wind turbine performance. These airfoil data are required at different Reynolds numbers since this number varies along the span of the blade. Typical Reynolds numbers found in wind turbine operation are in the range of 500 000 and 10 million. A catalog of airfoil data at these low Reynolds numbers was compiled by Miley (1982). The profile data can be obtained experimentally or numerically. Experiments typically are done in wind tunnels or in the field. The high Reynolds numbers of modern large-scale wind turbines reach the limit of current wind tunnel facilities. Different airfoil design codes have been used by wind energy engineers to design airfoils specifically for HAWTs. Some of the most used codes in wind energy engineering were XFOIL (Drela, 1989), RFOIL (Timmer and van Rooij, 2003), and PROFOIL (Selig and Tangler, 1995). Modern procedures rely on CFDs software packages (see e.g. Sørensen *et al.*, 2016, and Section 3.7.4).

An important aspect of wind turbine airfoils is that three-dimensional effects will affect their performance. The onset of stall is delayed near the root of the blade, leading to higher loads. Some investigators also alter the performances of the airfoils at the tip due to the strong radial flow present there. The 2D performances obtained in a wind tunnel or with CFD are typically corrected to account for these effects. Common corrections for stall delay are the Snel (1994) or the Du and Selig (1998) corrections; see also Sant (2007).

3.3.6 Lift vs. Drag Machines

Wind energy converters that have been built over the centuries can be divided into lift machines and drag machines. Lift machines use lift forces to generate power. Drag machines use drag forces. The HAWTs that are the primary topic of this book (and almost all modern wind turbines) are lift machines, but many useful drag machines have been developed. The advantages of lift-over-drag machines are described in this section through the use of a few simple examples.

3.3.6.1 Drag Machines

As discussed in Chapter 1, drag-driven windmills, shown in Figure 3.20, were used in the Middle East over a thousand years ago. They included a vertical axis rotor consisting of flat surfaces in which half of the rotor was shielded from the wind. The simplified model on the right in Figure 3.20 is used to analyze the performance of this drag machine. For simplicity, a characteristic radius *r* is used.

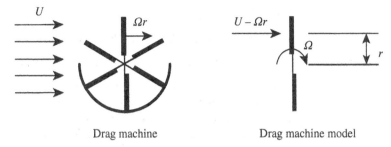

Drag machine   Drag machine model

Figure 3.20 Simple drag machine and model; U, velocity of the undisturbed airflow; Ω, angular velocity of wind turbine rotor; r, characteristic radius

The drag force, F_D, is a function of the relative wind velocity at the rotor surface, which is in turn a function of the difference between the wind speed, U, and the speed of the surface, Ωr:

$$F_D = C_d\left(\frac{1}{2}\rho A U_{rel}^2\right) = C_d\left(\frac{1}{2}\rho(U - \Omega r)^2 A\right) \tag{3.68}$$

where A is the drag surface area and where the three-dimensional drag coefficient, C_d, for a square plate is assumed to be 1.1.

The rotor power is the product of the torque due to the drag force and rotational speed of the rotor surfaces:

$$P = C_d\left(\frac{1}{2}\rho(U - \Omega r)^2 A\right)\Omega r = \left(\rho A U^3\right)\left[\frac{1}{2}C_d\lambda(1-\lambda)^2\right] \tag{3.69}$$

where $\lambda = \Omega r/U$ is the tip speed ratio for the characteristic radius r. The radius r is assumed to be representative of the turbine performance. A more rigorous approach would require the integration of the power generated at all radial positions. The power coefficient shown in Figure 3.21 was obtained for $C_d = 1.0$. It is a function of λ, the ratio of the surface velocity to the wind speed, and is based on an assumed total machine area of $2A$:

$$C_P = \frac{1}{2}C_d\lambda(1-\lambda)^2 \tag{3.70}$$

The power coefficient is zero at speed ratios of 0 (no motion) and 1.0 (the speed at which the surface moves at the wind speed and experiences no drag force). The peak power coefficient occurs at a speed ratio of 1/3 for a

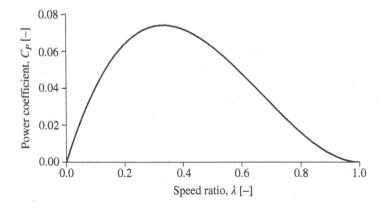

Figure 3.21 Power coefficient of flat plate drag machine for a drag coefficient of $C_d = 1.0$.

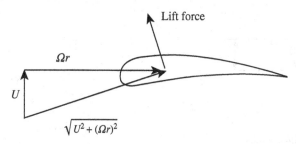

Figure 3.22 Relative velocity of a lift machine; for notation, see Figure 3.20

value $C_{P,\max} = \frac{2}{27}C_d$. This power coefficient is significantly lower than the Betz limit of $C_{P,\max} = \frac{16}{27} = 0.593$. This example also illustrates one of the primary disadvantages of a pure drag machine: the rotor surface cannot move faster than the wind speed. Thus, the wind velocity relative to the power-producing surfaces of the machine, U_{rel}, is limited to the free stream velocity:

$$U_{rel} = U(1-\lambda)\lambda < 1 \tag{3.71}$$

It should be pointed out that some drag-based machines, such as the Savonius rotor (see Chapter 1), may achieve maximum power coefficients of greater than 0.2 and may have tip speed ratios greater than 1.0. This is primarily due to the lift developed when the rotor surfaces turn out of the wind as the rotor rotates (Wilson *et al.*, 1976). Thus, the Savonius rotor and many other drag-based devices may also experience some lift forces.

3.3.6.2 Lift Machines

The forces in lift machines are also a function of the relative wind velocity and the lift coefficient:

$$F_L = C_l \left(\frac{1}{2}\rho A U_{rel}^2 \right) \tag{3.72}$$

The maximum lift and drag coefficients of airfoils are of similar magnitude. One significant difference in the performance between lift and drag machines is that much higher relative wind velocities can be achieved with lift machines. Relative velocities are always greater than the free stream wind speed, sometimes by an order of magnitude. As illustrated in Figure 3.22, the relative wind velocity at the airfoil of a lift machine is (when the induction factors are neglected):

$$U_{rel} = \sqrt{U^2 + (\Omega r)^2} = U\sqrt{1 + \lambda^2} \tag{3.73}$$

With speed ratios of up to 10 and forces that are a function of the square of the relative speed, it can be seen that the forces that can be developed by a lift machine are significantly greater than those achievable with a drag machine with the same surface area. The larger forces allow for much greater power coefficients.

3.3.7 Blade Element Theory

Blade Element Theory (BET), derived by W. Froude (Froude, 1878), can be used to compute loads on a rotor using the 2D airfoil performances at each spanwise position. The BET typically needs to be combined with another aerodynamic theory to determine the inflow velocity on the blades. Such combinations lead to so-called lifting-line methods (described in Section 3.7.1). The combination of the BET and the momentum theory presented in Section 3.2 forms the BEM method (described in Section 3.5). The combination of BET with vortex wake methods leads to lifting-line vortex wake methods (Section 3.7.3), and the combination with CFD methods leads to actuator-line methods (Section 3.7.4).

3.3.7.1 Assumptions of the Blade Element Theory

In the blade element theory, the forces on the blades of a wind turbine are expressed as a function of lift and drag coefficients and the angle of attack. As shown in Figure 3.23, for this analysis, the blade is assumed to be divided into N sections (or elements). c is the chord of the element, and dr is the radial length of the element. The figure also

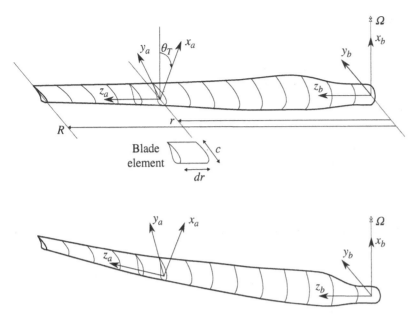

Figure 3.23 Representation of a blade with blade elements. Top: straight blade. Bottom: non-straight/deflected blade; c, airfoil chord; dr, radial length of element; r, radius; R, rotor radius; Ω, angular velocity of rotor; θ_T: geometrical twist of the section. "b" is the coordinate system attached to the blade root. "a" is the coordinate system attached to the airfoil section

shows the geometrical twist of the section, θ_T, which by convention is negative about the z axis (most twist values are then positive for a wind turbine blade).

The assumptions of the blade element theory are the following:

- There is no aerodynamic interaction between elements.
- The forces on the blades are determined solely by the (2D) lift and drag characteristics of the airfoils at each element (C_l and C_d) and the flow velocity in the airfoil coordinate system (neglecting any radial flow effect).
- The relative wind at each radial position is known (It is determined using other methods such as momentum theory, vortex methods, or CFD).

3.3.7.2 General Relations of the Blade Element Theory

The Blade Element Theory effectively assumes that each airfoil of the blade behaves as a 2D airfoil, as presented in Section 3.3.2. The coordinate system of a given airfoil, located at the radial position r, is illustrated in Figures 3.23 and 3.24.

The relative wind vector, U_{rel}, at a given radial position r, is the vector sum of the free stream velocity vector, the induced velocity vector, and the blade velocity vector (from its rotation and elastic motion). According to Blade Element Theory, the radial flow (along z_a) does not influence the blade loads. The forces acting on the blade section are the incremental lift and drag forces, noted dF_L and dF_D, respectively, expressed in Newtons. By definition, they are perpendicular and parallel to the relative wind. The incremental aerodynamic moment (in Nm) is positive about the z_a axis. In the coordinate system of this airfoil, the incremental loads are given as:

$$dF_L = C_l(\alpha)\frac{1}{2}\rho U_{rel,\perp}^2 cdr \tag{3.74}$$

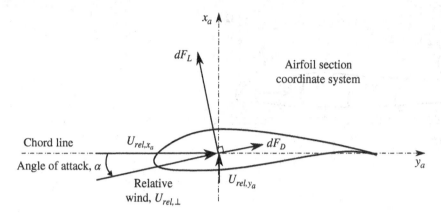

Figure 3.24 Forces and relative wind acting on a blade element. The lift and drag forces are perpendicular and parallel to the relative wind

$$dF_D = C_d(\alpha)\frac{1}{2}\rho U^2_{rel,\perp}cdr \qquad (3.75)$$

$$dM = C_m(\alpha)\frac{1}{2}\rho U^2_{rel,\perp}c^2dr \qquad (3.76)$$

Where the norm of the relative wind and the angle of attack are obtained based on the components of the relative wind in the $x_a - y_a$ plane as:

$$U_{rel,\perp} = \sqrt{U^2_{rel,x_a} + U^2_{rel,y_a}}, \quad \tan\alpha = \frac{U_{rel,x_a}}{U_{rel,y_a}} \qquad (3.77)$$

and where $C_l(\alpha)$ and $C_d(\alpha)$ are the lift and drag coefficients, obtained at the angle of attack α. These functions are usually stored as tabulated data, obtained numerically or experimentally. Equations (3.74–3.76) can be written in vector form. The vectors can then be converted into different coordinate systems (e.g., the blade system, the rotor plane system, or the global system).

3.3.7.3 Blade Element Theory for a Rotor with Straight Blades

For a rotor with straight blades (e.g., no prebend, presweep, or coning), it is common to project the relative wind into components normal and tangential to the plane of rotation, noted, respectively, U_n and U_t. The components of the relative wind and the forces acting on a blade element, for a straight blade, are illustrated in Figure 3.25. Because the blade is straight, the relative velocity norm in the $x_a - y_a$ plane is the same as the one in the "$n - t$" plane, such that $U_{rel,\perp} = \sqrt{U^2_n + U^2_t}$. In the rest of this section, $U_{rel,\perp}$ will be written U_{rel}.

The angle formed by the relative wind in the "$n - t$" plane is called the inflow angle, ϕ. The inflow angle relates the wind components as follows:

$$\tan\phi = \frac{U_n}{U_t} \qquad (3.78)$$

The total blade section pitch angle, θ_p, is the angle between the chord line and the plane of rotation. In general, it consists of the geometric twist angle, θ_T, the blade pitch angle resulting from the actuation of the blade pitch controller, θ_c, and the elastic torsion of the section, θ_e:

$$\theta_p = \theta_T + \theta_c + \theta_e \qquad (3.79)$$

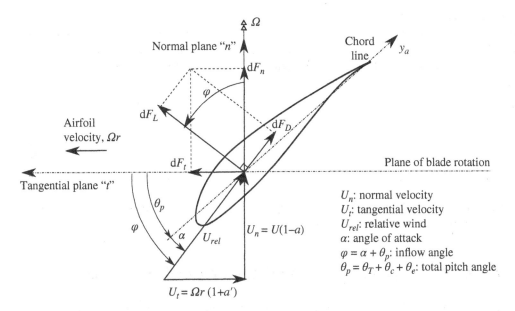

Figure 3.25 Blade section velocity triangle for application of the Blade Element Theory. Note: U_t is positive in the negative "t" direction

The angle of attack is the angle between the chord line and the relative wind. It is obtained as the difference between the inflow angle and the section pitch angle:

$$\alpha = \phi - \theta_p \tag{3.80}$$

The lift and drag (Equations (3.74) and (3.75)) can be projected onto two components: dF_n, the incremental force normal to the plane of rotation (which contributes to thrust); and dF_t, is the incremental force tangential to the circle swept by the rotor (which contributes to the torque and power). The projection gives:

$$dF_n = dF_L \cos \phi + dF_D \sin \phi \tag{3.81}$$

$$dF_t = dF_L \sin \phi - dF_D \cos \phi \tag{3.82}$$

If the rotor has B blades, the total normal force (thrust) on the sections at a distance, r, from the center is:

$$dT = B\frac{1}{2}\rho U_{rel}^2 (C_l \cos \phi + C_d \sin \phi)c dr \tag{3.83}$$

The differential torque due to the tangential force operating at a distance, r, from the center is given by:

$$dQ = B\frac{1}{2}\rho U_{rel}^2 (C_l \sin \phi - C_d \cos \phi)cr dr \tag{3.84}$$

Note that the effect of drag is to decrease torque and hence power, as well as to increase the thrust loading.

Thus, from blade element theory, one also obtains two equations (Equations 3.83 and 3.84) that define the normal force (thrust) and tangential torque on the annular rotor section as a function of the flow angles at the blades and airfoil characteristics. These equations will be used in Section 3.5 in conjunction with momentum theory to form the BEM theory, which is used to obtain the performance for any arbitrary blade shape. The results will also be used in Section 3.6, with additional assumptions, to obtain "ideal" blade shapes using simplified rotor design procedures.

3.3.7.4 Results Using the Induction Factors

The axial and tangential induction factors introduced in Section 3.2.2 are commonly understood as inductions normal and tangential to the actuator disc plane. Therefore, they can be used to define the normal and tangential velocity components as: $U_n = U(1 - a)$ and $U_t = \Omega r(1 + a')$. With these notations, the following geometrical relationships are found:

$$\tan\phi = \frac{U(1-a)}{\Omega r(1+a')} = \frac{1-a}{(1+a')\lambda_r} \tag{3.85}$$

$$U_{rel}^2 \sin^2\phi = U^2(1-a)^2 \tag{3.86}$$

$$U_{rel}^2 \sin\phi \cos\phi = U(1-a)\, \Omega r(1+a') \tag{3.87}$$

Equations (3.85–3.87) will be referred to as the "velocity triangle" equations. The blade element theory and velocity triangle equations need to be modified for non-straight blades (see Section 3.7.1).

3.4 Aerodynamic Blade Design for Modern Horizontal Axis Wind Turbines

The design of modern HAWT blades consists of a sophisticated multidisciplinary design process. This section briefly summarizes the key aspects of such process. Readers interested in the details of wind turbine rotor design can refer to Chapter 3 of Veers, 2019 and the references cited in the next sections.

3.4.1 Simple Aerodynamic Design

As a starting point, the blade designer may be given rated values of wind speed, rotational speed, and power. From this, an aerodynamic optimization can be carried out where the designer varies the main aerodynamic parameters: the number of blades, the series of airfoil shapes along the blade, and the chord and twist along the blade. The most common aerodynamic tool for blade design is the BEM method, which is described in Section 3.5. Simplified aerodynamic design procedures are presented in Section 3.6. Modern blade design requires a multidisciplinary approach and is part of the entire design process of the wind turbine, as discussed in the following paragraphs.

3.4.2 Design Requirements

Wind turbine blades are designed to convert the kinetic energy in the wind into torque while having structural properties that ensure the required ultimate and fatigue strength for the entire operational life. To be approved for commercialization, blades must respect certification standards, such as the IEC 61400-1 (see Chapter 8). The design of wind turbine blades is also subject to tight cost requirements. The figure of merit of such design process is often levelized cost of energy (LCOE; see Chapter 13. Other metrics, which account for variations of electricity prices based on availability of wind and other market conditions, are also increasingly used. Noise, manufacturing, and logistics constraints represent additional design drivers. In industrial practice, multi-objective design processes are often adopted to expose the different choices during the design phase (Bortolotti *et al.*, 2020).

3.4.3 Load Case Simulations

To ensure that a given turbine design will satisfy the design criteria, the wind turbine is simulated for a range of operating conditions using a mid-fidelity aero-servo-elastic solver (see Chapter 8). The number of design load cases (DLCs) defined by international standards is in the order of 10^3 for land-based machines and grows to $10^4/10^5$ for offshore turbines. The rotor inputs for aero-servo-elastic solver are the blade shape (airfoils, chord, and twist) and its elastic properties. The performance of the airfoils is computed from wind tunnel experiments or CFD solvers (see Section 3.3.5). The aerodynamic loads are typically computed using the BEM method.

In addition to the mid-fidelity load case simulations, additional checks are performed to: verify the elastic stability of the blade (see Chapter 4) and its integrity using a 3D solver, compute aeroacoustics noise emissions, and compute the overall LCOE for the turbine. The entire process may be automated into a multidisciplinary framework, as discussed below.

3.4.4 Multidisciplinary Design

Historically, the blade design was conducted by groups of specialists (aerodynamics and aeroacoustics, loads, structures, and controls) that iterated on a design until all requirements were met. This approach works in practice but given the conflicting constraints and objectives of blade design, it is often slow and can lead to suboptimal products. More recently, formal systems engineering approaches have become more popular to integrate the simulation models described in the previous into a single optimization framework. Thanks to multidisciplinary design analysis and optimization (MDAO) models (Bortolotti *et al.*, 2016; Zahle *et al.*, 2016), it is now possible to run design studies automatically and speed up the exploration of the design space.

3.5 Steady-State Performance Prediction: The Blade Element Momentum Method

This section presents the BEM (sometimes also referred to as strip theory) method, which is used to predict the aerodynamic performance of a given wind turbine design. The BEM method builds on the foundation presented in the previous sections, namely, it uses results from the momentum theory presented in Section 3.2.2 and the blade element theory presented in Section 3.3.7. The analysis includes wake rotation, drag, and losses due to a finite number of blades, stall, and off-design performance. In this section, it is assumed that the turbine geometry (twist, chord, and mean line) and its airfoil coefficients are known. For simplicity, the rotor blades are assumed to be straight (this assumption is relaxed in Section 3.7). For a given wind turbine design, the BEM method can be used to determine the aerodynamic force and power under different operating conditions (wind speed, rotor speed, and pitch angle). For rotor design, the BEM method may be used in an iterative approach. That is, one starts with a given blade shape and predicts its performance, tries another shape, and repeats the prediction until a suitable blade has been chosen.

The basic BEM equations are given first before presenting the solution procedures and the various corrections added to the method.

3.5.1 The Blade Element Momentum Equations

The BEM equations are obtained by equating the thrust and torque equations derived from momentum and blade element theories. First, the main results are recalled. For a realistic rotor with a finite number of blades, the results of momentum theory need to be corrected using the so-called "tip-loss" factor. Then, the thrust and torque equations are equated and the main BEM equations are obtained. These equations require iterations to be solved, which is the topic of the next sections.

3.5.1.1 Summary of Blade Element Theory Results

Blade element theory provides the elementary thrust and torque acting on the blade elements located at radius r on all blades (Equations 3.83 and 3.84), which are rewritten here as Equations (3.88) and (3.89):

$$dT = B\frac{1}{2}\rho U_{rel}^2 c_n c dr = \sigma\pi\rho\, U_{rel}^2 c_n r dr \tag{3.88}$$

$$dQ = B\frac{1}{2}\rho U_{rel}^2 c_t c r dr = \sigma\pi\rho\, U_{rel}^2\, c_t\, r^2 dr \tag{3.89}$$

where the normal and tangential coefficients were introduced, defined as:

$$c_n = C_l \cos\phi + C_d \sin\phi \tag{3.90}$$

$$c_t = C_l \sin\phi - C_d \cos\phi \tag{3.91}$$

and the local solidity, σ, defined as:

$$\sigma = \frac{Bc}{2\pi r} \tag{3.92}$$

The "velocity triangle" relations are then used to express the relative velocity (U_{rel}^2) as function of the induction factors. Equations (3.86) and (3.87) can be used indifferently. Yet, it is common to use Equation (3.86) for the thrust, and Equation (3.87) for the torque, leading to:

$$dT = \sigma\pi\rho \frac{U^2(1-a)^2}{\sin^2\phi} c_n r\, dr \tag{3.93}$$

$$dQ = \sigma\pi\rho \frac{U(1-a)\,\Omega r(1+a')}{\sin\phi \cos\phi} c_t r^2\, dr \tag{3.94}$$

3.5.1.2 Summary of Momentum Theory Results

The simplified momentum theory with wake rotation (Section 3.2.2) applies the conservation of linear momentum to the control volume of radius r and thickness dr, leading to an expression for the differential thrust (Equation 3.26), which is repeated here:

$$dT = 4a(1-a)\,\rho U^2 \pi r\, dr \tag{3.95}$$

Similarly, from the conservation of angular momentum equation, Equation (3.31), the differential torque, Q, imparted to the blades (and equally, but oppositely, to the air) is:

$$dQ = 4a'(1-a)\,\rho U \pi r^3 \Omega\, dr \tag{3.96}$$

Thus, from momentum theory, one gets two equations, Equations (3.95) and (3.96), that define the thrust and torque on an annular section of the rotor as a function of the axial and tangential induction factors (i.e., of the flow conditions).

3.5.1.3 Momentum Theory for a Finite Number of Blades – Tip Losses

The momentum results were obtained for an actuator disc, with an infinite number of blades. The flow differences between a rotor with finite and infinite number of blades are referred to as tip losses. Physically, these differences arise because the forces exerted by the blades on the flow are localized to the vicinity of the blades. Also, pressure on the suction side of a blade is lower than that on the pressure side, therefore the air tends to flow around the tip from the lower to upper surface, reducing lift and hence power production near the tip. This effect is most noticeable with fewer, wider blades, and vanishes for an infinite number of blades.

A number of methods have been suggested for including the effect of tip loss. The most straightforward approach to use is one developed by Prandtl and later extended by Glauert (see de Vries, 1979). According to this method, a correction factor, F, must be introduced into the momentum equations. This correction factor is a function of the number of blades, the angle of relative wind, and the position of the blade. Based on Prandtl's method:

$$F = \frac{2}{\pi} \cos^{-1}\left[\exp\left(-\left\{ \frac{(B/2)[1-(r/R)]}{(r/R)\sin\phi} \right\} \right) \right] \tag{3.97}$$

where B is the number of blades, and the angle resulting from the inverse cosine function is assumed to be in radians. If the inverse cosine function is in degrees, then the initial factor, $2/\pi$, is replaced by 1/90. Note, also, that

F is always between 0 and 1. This tip loss correction factor characterizes the reduction in the forces, or the increase of the flow, at a radius r along the blade that is due to the tip loss at the end of the blade.

The tip loss correction factor affects the forces derived from momentum theory. Thus, Equations (3.95) and (3.96) become:

$$dT = 4F\,a(1-a)\,\rho U^2\pi r\,dr \tag{3.98}$$
$$dQ = 4Fa'(1-a)\,\rho U\pi r^3\Omega\,dr \tag{3.99}$$

For an infinite number of blades, $F = 1$.

3.5.1.4 Blade Element Momentum Theory Equations
The BEM equations are obtained by equating the differential thrust and torque from momentum and blade element theory. By equating the normal force equations from momentum and blade element theory (Equations 3.93 and 3.98), one obtains:

$$\frac{a}{1-a} = k, \quad \text{with} \quad k = \frac{\sigma c_n}{4F\sin^2\phi} \tag{3.100}$$

Equating the torque equations (Equations 3.94 and 3.99) leads to:

$$\frac{a'}{1+a'} = k', \quad \text{with} \quad k' = \frac{\sigma c_t}{4F\sin\phi\cos\phi} \tag{3.101}$$

These equations can be solved for the induction factors as follows:

$$a = \frac{k}{1+k} = \frac{1}{1/k+1} = \left[1 + \frac{4F\sin^2\phi}{\sigma c_n}\right]^{-1} \tag{3.102}$$

$$a' = \frac{k'}{1-k'} = \frac{1}{1/k'-1} = \left[\frac{4F\sin\phi\cos\phi}{\sigma c_t} - 1\right]^{-1} \tag{3.103}$$

This set of equation needs to be solved iteratively since the coefficients c_n and c_t and depends on C_l and C_d, which depends on the angle of attack, which in turns depends on the flow angle and the induction coefficients. Solution methods for this set of equations are presented in Section 3.5.2. Once the induction factors are known, the blade loads are computed using the blade element theory results from Section 3.3.7. In that final step, the influence of drag is always included. It is sometimes omitted in the calculation of the induction factors (see the following section).

3.5.1.5 *BEM Equations with Coning and Prebend
If the blades are not straight (such as with coning angle or prebend), one approach consists in assuming that the blade element theory loads are spread over a curvilinear length ds over the blade span, while the momentum theory loads are spread over a radial distance dr in the rotor plane normal to the shaft. The local slope of the blade is noted κ, with $\cos\kappa = \frac{dr}{ds}$. For pure coning (no prebend), κ is the cone angle. A careful derivation of the BEM equations leads to similar expressions to Equations (3.100) and (3.101) but with k and k' containing a factor $1/\cos\kappa$. When computing the blade element theory loads, and the integral loads over the rotor, it is also necessary to consider the orientation of the airfoil coordinate system, because this is the one where the blade element theory loads are defined. See Section 3.7.2 for a general description of lifting-line algorithms with arbitrary blade meanline.

3.5.1.6 Blade Element Momentum Theory Equations Without Drag
In the calculation of induction factors (a and a'), some authors set C_d equal to zero (see Wilson and Lissaman, 1974). For airfoils with low drag coefficients, this simplification introduces negligible errors, but the error will

be greater toward the root where drag effects are important. Without drag, $c_n = \cos \phi \, C_l$, and $c_t = \sin \phi C_d$. After some algebraic manipulation using Equation (3.85) (which relates a, a', ϕ, and λ_r, based on geometric considerations) and Equations (3.100) and (3.101), the following useful relationships are obtained:

$$C_l = 4 F \sin \phi \, \frac{(\cos \phi - \lambda_r \sin \phi)}{\sigma(\sin \phi + \lambda_r \cos \phi)} \tag{3.104}$$

$$(1 - a)/a' = (4 F \lambda_r \, \sin \, \phi)/(\sigma C_l) \tag{3.105}$$

$$a'/(1 + a') = \sigma C_l/(4 F \, \cos \, \phi) \tag{3.106}$$

$$a/a' = \lambda_r / \tan \phi \tag{3.107}$$

$$a = 1/\left[1 + \left(4 F \sin^2 \phi/(\sigma C_l \cos \phi)\right)\right] \tag{3.108}$$

$$a' = 1/[(4 F \, \cos \, \phi/(\sigma C_l)) - 1] \tag{3.109}$$

$$U_{rel} = \frac{U(1 - a)}{\sin \phi} = \frac{U}{(\sigma C_l/4F) \cot \phi + \sin \phi} \tag{3.110}$$

3.5.2 BEM Algorithm – Solution Methods

The nonlinearity of the airfoil coefficients implies that the equations cannot be solved directly: the induction factors depend on the airfoil coefficients, which depend on the angle of attack, which in turn depend on the flow angle and therefore the axial inductions. Two solution methods will be proposed to determine the flow conditions and forces at each blade section. The first solution is an iterative numerical approach that can be used with or without drag in the induction factors, and which is most easily extended for flow conditions with large axial induction factors. It is the preferred approach for the numerical implementation of the BEM method. The second one uses the measured airfoil characteristics and the BEM equations to solve directly for C_l and a. This method only applies when drag is omitted in the calculation of the inductions. It can be solved numerically, but it also lends itself to a graphical solution that clearly shows the flow conditions at the blade and illustrates the existence of multiple solutions (see Section 3.5.1.1). A third solution method consists of using the flow angle as the main variable; it is the method used in AeroDyn (see Ning, 2014).

3.5.2.1 Method 1 – Iterative Solution for a and a'

This solution method starts with guesses for a and a', from which flow conditions and new induction factors are calculated. Specifically:

1) Guess values of a and a' (for instance $a = 0.2$, $a' = 0.01$).
2) Calculate the angle of the relative wind from Equation (3.85).
3) Calculate the angle of attack from $\phi = \alpha + \theta_p$ and then C_l and C_d.
4) Project the lift coefficients onto c_n and c_t using Equations (3.90) and (3.91).
5) Calculate a and a' from Equations (3.102) and (3.103) (or Equations (3.108) and (3.109) if drag is omitted).
6) Apply high-thrust correction (see Section 3.5.1.3).
7) (Optional: use a relaxation factor to update a and a' based on the new and old values, for instance: $a_{new} = a_{new} k_{rel} + (1 - k_{rel}) \, a_{old}$, where k_{rel} is a relaxation factor between 0 and 1).

Steps 2–5 are repeated until the newly calculated induction factors are within some acceptable tolerance of the previous ones. This method is especially useful for highly loaded rotor conditions, as described in Section 3.5.1.3. A step-by-step detail of the BEM algorithm is given in Section 3.6.4.3.

3.5.2.2 Method 2 – Solving for C_l and α

The second solution procedure only applies for the case where drag is omitted in the calculations of the induction factors. Since $\phi = \alpha + \theta_p$, for a given blade geometry and operating conditions, there are two unknowns in Equation (3.104), C_l and α at each section. In order to find these values, one can use the empirical C_l vs. α curves for the chosen airfoil (see de Vries, 1979). One then finds the C_l and α from the empirical data that satisfy Equation (3.104). This can be done either numerically or graphically (as shown in Figure 3.26). Once C_l and α have been found, a' and a can be determined from any two of Equations (3.106) through (3.109). It should be verified that the axial induction factor at the intersection point of the curves is less than 0.5 to ensure that the result is valid.

3.5.3 Computing the Rotor Performances

Once the induction factors have been obtained for each section, the aerodynamic loads can be computed using blade element theory. The flow quantities, ϕ, α, and U_{rel}^2, the aerodynamic coefficients, c_n and c_t, are known from the BEM algorithm. The total thrust and torque are obtained from the blade element theory by integrating the differential thrust and torque from Equations (3.88) and (3.89):

$$T = \int_{r_h}^{R} dT = B \int_{r_h}^{R} \frac{1}{2}\rho U_{rel}^2 c_n\, c \, dr \tag{3.111}$$

$$Q = \int_{r_h}^{R} dQ = B \int_{r_h}^{R} \frac{1}{2}\rho U_{rel}^2 c_t\, c\, r\, dr \tag{3.112}$$

where r_h is the rotor radius at the hub of the blade. The coefficients c_n and c_t must contain the influence of drag for the loads to be representative. The power is the product of the torque with the rotational speed:

$$P = \Omega Q \tag{3.113}$$

The dimensionless thrust, torque, and power coefficients are then obtained as follows:

$$C_T = \frac{T}{\frac{1}{2}\rho\pi R^2 U^2} = \frac{2}{\lambda^2}\int_{\lambda_h}^{\lambda} \frac{U_{rel}^2}{U^2}\sigma\, c_n\, \lambda_r\, d\lambda_r \tag{3.114}$$

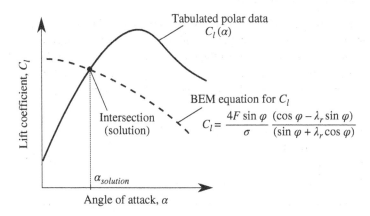

Figure 3.26 Graphical solution method to find the angle of attack where the C_l matches with the BEM equations; C_l, two-dimensional lift coefficient; α, angle of attack; λ_r, local speed ratio; ϕ, angle of relative wind; σ, local rotor solidity

$$C_Q = \frac{Q}{\frac{1}{2}\rho\pi R^2 U^2 R} = \frac{2}{\lambda^3} \int_{\lambda_h}^{\lambda} \frac{U_{rel}^2}{U^2} \sigma\, c_t\, \lambda_r^2\, \mathrm{d}\lambda_r \tag{3.115}$$

$$C_P = \frac{P}{\frac{1}{2}\rho\pi R^2 U^3} = \frac{2}{\lambda^2} \int_{\lambda_h}^{\lambda} \frac{U_{rel}^2}{U^2} \sigma\, c_t\, \lambda_r^2\, \mathrm{d}\lambda_r \tag{3.116}$$

where $\lambda_h = \Omega r_h / U$ is the local speed ratio at the hub, and λ_r is the local speed ratio (Equation 3.28). The second expression can be used if the thrust, torque, and power have not been integrated previously.

3.5.3.1 Power Coefficient without Drag in the Induction Factors

The results presented in the previous paragraph are general formulae from the blade element theory. It is possible to insert the BEM equations into it, leading to various forms. The interest of such approach is purely academic. The general formulae of the previous paragraph should preferably be used since they also work for high-thrust conditions. Brief examples are given here for the cases where drag is omitted in the computation of the induction factors because the formulae are shorter. Inserting Equation (3.105) into Equation (3.116) leads to (Wilson and Lissaman, 1974):

$$C_P = \left(8/\lambda^2\right) \int_{\lambda_h}^{\lambda} F\, \lambda_r^3 a'(1-a)[1-(C_d/C_l)\cot\phi]\, \mathrm{d}\lambda_r \tag{3.117}$$

Note that when $C_d = 0$, this equation for C_P is the same as the one derived from momentum theory, including wake rotation, Equation (3.35). Using different combination of BEM equations, the following formula can be obtained (de Vries, 1979):

$$C_P = \left(8/\lambda^2\right) \int_{\lambda_h}^{\lambda} F\, \sin^2\phi(\cos\phi - \lambda_r\sin\phi)(\sin\phi + \lambda_r\cos\phi)[1-(C_d/C_l)\cot\phi]\, \lambda_r^2 \mathrm{d}\lambda_r \tag{3.118}$$

The derivation of Equation (3.118) is algebraically complex and is left as an exercise for the interested reader. Note that even though the axial induction factors were determined assuming $C_d = 0$, the drag is included here in the power coefficient calculation.

3.5.4 Effect of Drag and Blade Number on Optimum Performance

At the beginning of the chapter, the maximum theoretically possible power coefficient for wind turbines was determined as a function of tip speed ratio. As explained earlier in this chapter, airfoil drag and tip losses reduce the power coefficients of wind turbines. The maximum achievable power coefficient for turbines with an optimum blade shape but a finite number of blades and aerodynamic drag has been calculated by Wilson et al. (1976). Their fit to the data is accurate to within 0.5% for tip speed ratios from 4 to 20, lift-to-drag ratios (C_l/C_d) from 25 to infinity, and from one to three blades (B):

$$C_{p,\max} = \left(\frac{16}{27}\right)\lambda \left[\lambda + \frac{1.32 + \left(\frac{\lambda-8}{20}\right)^2}{B^{\frac{2}{3}}}\right]^{-1} - \frac{(0.57)\lambda^2}{\frac{C_l}{C_d}\left(\lambda + \frac{1}{2B}\right)} \tag{3.119}$$

Figure 3.27 shows the maximum achievable power coefficients for a turbine with one, two, and three optimum blades and no drag, based on Equation (3.119). The performance for ideal conditions including wake rotation (an infinite number of blades) is also shown. It can be seen that at a given tip speed ratio, the power coefficient

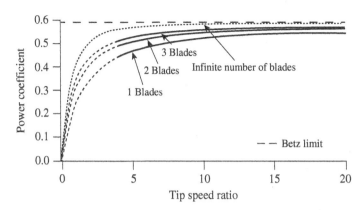

Figure 3.27 Maximum achievable power coefficients as a function of number of blades, no drag

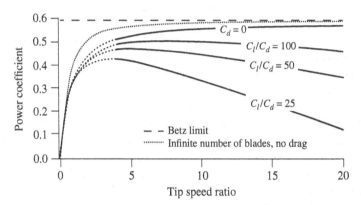

Figure 3.28 Maximum achievable power coefficients of a three-bladed optimum rotor as a function of the lift-to-drag ratio, C_l/C_d

increases C_P for an increasing number of blades. Most wind turbines use two or three blades and, in general, most two-bladed wind turbines use a higher tip speed ratio than most three-bladed wind turbines. Thus, there is little practical difference in the maximum achievable C_P between typical two- and three-bladed designs, assuming no drag. The effect of the lift-to-drag ratio on maximum achievable power coefficients for a three-bladed rotor is shown in Figure 3.28. There is clearly a significant reduction in maximum achievable power as the airfoil drag increases. For reference, the DU-93-W-210 airfoil has a maximum C_l/C_d ratio of 140 at an angle of attack of 6°, and the 19% thick LS(1) airfoil has a maximum C_l/C_d ratio of 85 at an angle of attack of 4°. It can be seen that it clearly benefits the blade designer to use airfoils with high lift-to-drag ratios. In practice, rotor power coefficients may be further reduced as a result of (1) non-optimum blade designs that are easier to manufacture, (2) the lack of airfoils at the hub, and (3) aerodynamic losses at the hub end of the blade.

3.5.5 Performance Curves for a Given Design – C_P – λ Curves

In practice, a blade design is only optimal for a single tip-speed ratio. The wind turbine will operate a large amount of time in conditions different from its design point. It is therefore necessary to evaluate the wind turbine design over a range of operating conditions, typically varying the tip speed ratio and the pitch. This can be done using the steady BEM method presented in this section. The rotor performances are computed for different operating points

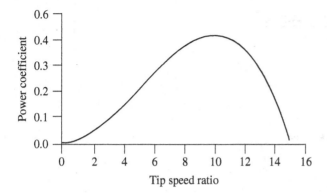

Figure 3.29 Sample $C_P - \lambda$ curve for a high tip speed ratio wind turbine

doing a parametric study. The results are usually presented as a graph of power coefficient vs. tip speed ratio, called a $C_P - \lambda$ curve, as shown in Figure 3.29.

This $C_P - \lambda$ is different from the one in Figure 3.27 since it applies to a fixed turbine design (chord, twist, and airfoil), whereas Figure 3.27 represents the power coefficient for different designs that are optimized at each tip speed ratio. $C_P - \lambda$ curves are used in wind turbine design to determine the rotor power for any combination of wind and rotor speed. It is important to note that airfoil performances vary with the Reynolds number, and therefore the $C_P - \lambda$ curves will be different for the same design but with different blade length.

Modern turbines use rotor speed and pitch regulation to optimize or limit the power output (see Chapter 9). During the rotor speed regulation, the turbine rotational speed is adapted for each wind speed so that it operates at the design tip speed ratio where the power coefficient is maximum. During the pitch regulation, the rotor speed is constant, and the pitch varies to keep the power constant. It is therefore necessary to evaluate the performance of a given design for different pitch and tip-speed ratio. $C_P - \lambda - \theta_p$ surfaces are therefore generated to design the controller and evaluate its trajectory on this surface. An example of $C_P - \lambda - \theta_p$ surface is shown in Figure 3.30 for the NREL 5-MW wind turbine.

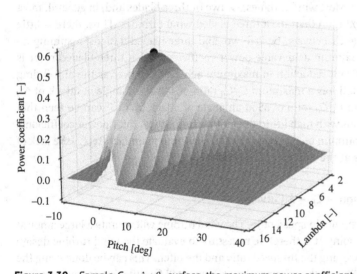

Figure 3.30 Sample $C_P - \lambda - \theta_p$ surface, the maximum power coefficient is represented with a dot

3.5.6 Advanced Flow Configurations and Aerodynamic Challenges

When a section of blade has a pitch angle or flow conditions very different from the design conditions, a number of complications can affect the analysis. These include multiple solutions in the region of transition to stall and solutions for highly loaded conditions with values of the axial induction factor approaching and exceeding 0.5.

3.5.6.1 Multiple Solutions to Blade Element Momentum Equations

In the stall region, as shown in Figure 3.31, there may be multiple solutions for C_l. Each of these solutions is possible. The correct solution should be that which maintains the continuity of the angle of attack along the blade span. The introduction of relaxation in the iterative BEM algorithm usually helps ensuring this continuity (see Hansen, 2008 and Section 3.5.2).

3.5.6.2 Wind Turbine Flow States

Measured wind turbine performance closely approximates the results of BEM theory at low values of the axial induction factor. Momentum theory is no longer valid at axial induction factors greater than 0.5 because the wind velocity in the far wake would be negative. In practice, as the axial induction factor increases above 0.5, the flow patterns through the wind turbine become much more complex than those predicted by momentum theory. A number of operating states for a rotor have been identified (see Eggleston and Stoddard, 1987). The operating states relevant to wind turbines are designated the windmill state and the turbulent wake state. The windmill state is the normal wind turbine operating state. The turbulent wake state occurs under operation in high winds. Figure 3.32 illustrates fits to measured thrust coefficients for these operating states. The windmill state is characterized by the flow conditions described by momentum theory for axial induction factors less than about 0.5. Above $a = 0.5$, in the turbulent wake state, measured data indicate that thrust coefficients increase up to about 2.0 at an axial induction factor of 1.0. This state is characterized by a large expansion of the slipstream, turbulence, and recirculation behind the rotor. While momentum theory no longer describes the turbine behavior, empirical relationships between C_T and the axial induction factor are often used to predict wind turbine behavior.

3.5.6.3 Rotor Modeling for the Turbulent Wake State

The rotor analysis discussed so far uses the equivalence of the thrust forces determined from momentum theory and from blade element theory to determine the angle of attack at the blade. In the turbulent wake state, the thrust determined by momentum theory is no longer valid.

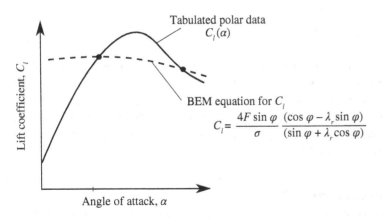

Figure 3.31 Multiple solutions: α, angle of attack; C_l, two-dimensional lift coefficient

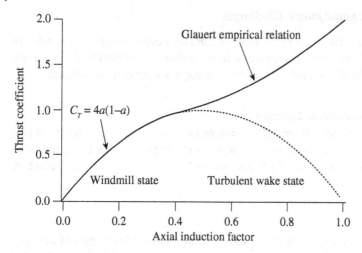

Figure 3.32 Fits to measured wind turbine thrust coefficients

In the turbulent wake state, corrections need to be implemented using an empirical relationship between the axial induction factor and the thrust coefficient, a correction referred to as high-thrust correction. The empirical relationship developed by Glauert, and shown in Figure 3.32 (see Eggleston and Stoddard, 1987), including tip losses, is:

$$a = (1/F)\left[0.143 + \sqrt{0.0203 - 0.6427(0.889 - C_T)}\right] \tag{3.120}$$

This equation is valid for $a > 0.4$ or, equivalently for $C_T > 0.96$.

The Glauert empirical relationship was determined for the overall thrust coefficient for a rotor. It is customary to assume that it applies equally to equivalent local thrust coefficients for each blade section. The local thrust coefficient, C_{T_r}, can be defined for each annular rotor section as (Wilson *et al.*, 1976):

$$C_{T_r} = \frac{dT}{(1/2)U^2\rho 2\pi r dr} \tag{3.121}$$

Using the equation for the normal force from blade element theory, Equation (3.93), the local thrust coefficient is:

$$C_{T_r} = \sigma(1-a)^2(C_l \cos \phi + C_d \sin \phi)/\sin^2 \phi \tag{3.122}$$

The solution procedure can then be modified to include heavily loaded turbines. The easiest procedure to use is the iterative procedure (Method 1) that starts with the selection of possible values for a and a'. Once the angle of attack and C_l and C_d have been determined, the local thrust coefficient can be calculated according to Equation (3.122). If $C_T < 0.96$ then the previously derived equations can be used. If $C_T > 0.96$ then the next estimate for the axial induction factor should be determined using the local thrust coefficient and Equation (3.120). The tangential induction factor, a', can be determined from Equation (3.103).

3.5.6.4 Blade Coning, Advanced Flows, and Aerodynamic Challenges
The steady BEM analysis assumes that the prevailing wind is uniform and aligned with the rotor axis and that the blades rotate in a plane perpendicular to the rotor axis. There are, however, a number of important steady-state and dynamic effects that cause increased loads or decreased power production from those expected with the steady

BEM theory presented here, especially increased transient loads. Some of these effects are listed in this section. Their modeling is discussed in Section 3.7.2.

The wind impinging on the wind turbine blades will be unsteady because of wind shear, yaw error, vertical wind components, turbulence, nacelle tilt, and elastic motion of the blade. Wind shear will result in wind speeds that vary with height across the rotor disc. Wind turbines often operate with a steady-state or transient yaw error (misalignment of the rotor axis and the wind direction about the vertical yaw axis of the turbine). Yaw error results in a flow component parallel to the rotor disc. The winds at the rotor may also have a vertical component, especially at sites in complex terrain. Turbulence results in a variety of wind conditions over the rotor. The angular position of the blade in the rotor plane is called the azimuth angle and is measured from some suitable reference; vertical is now the standard reference. Each of the effects mentioned above results in conditions at the blade varying with blade azimuth angle. Finally, blades are also often attached to the hub at a slight angle to the plane perpendicular to the rotor axis. This blade coning may be done to reduce bending moments in the blades or to keep the blades from striking the tower. In a rotor analysis, each of these situations is usually handled with appropriate geometrical transformations. Blade coning is handled by resolving the aerodynamic forces into components that are perpendicular and parallel to the rotor plane. Off-axis flow is also resolved into the flow components that are perpendicular and parallel to the rotor plane. Rotor performance is then determined for a variety of rotor azimuth angles. The axial and in-plane components of the flow that depend on the blade position result in angles of attack and aerodynamic forces that fluctuate cyclically as the blades rotate. BEM equations that include terms for blade coning were discussed in Section 3.5.1.5 (see also Wilson *et al.,* 1976). Linearized methods for dealing with small off-axis flows and blade coning are also discussed in Chapter 4. For unsteady flows, the steady BEM algorithm presented in this section is no longer suitable. It is common, however, to adapt the equations and add unsteady models (such as dynamic inflow, dynamic stall, and yaw model) to implement an unsteady BEM tool. Unsteady BEM codes are discussed in Section 3.7.2.

Steady-state effects that influence wind turbine behavior include the degradation of blade performance due to surface roughness, the effects of stall on blade performance, and blade rotation. As mentioned in Section 3.3.5, blade surfaces roughened by damage and debris can significantly increase drag and decrease the lift of an airfoil. This has been shown to decrease power production by as much as 40% on certain airfoils. The only solution is frequent blade repair and cleaning or the use of airfoils that are less sensitive to surface roughness.

Additionally, when parts of a blade operate in the stall region, fluctuating loads may result. On stall-regulated horizontal axis rotors, much of the blade may be stalled under some conditions. Stalled airfoils do not always exhibit the simple relationship between the angle of attack and aerodynamic forces that are evident in lift and drag coefficient data. The turbulent separated flow occurring during stall can induce rapidly fluctuating flow conditions and rapidly fluctuating loads on the wind turbine.

Finally, the lift and drag behavior of airfoils is measured in wind tunnels under nonrotating conditions. Investigation has shown that the same airfoils, when used on a HAWT, may exhibit delayed stall and may produce more power than expected. The resulting unexpectedly high loads at high wind speeds can reduce turbine life. This behavior has been linked to spanwise pressure gradients that result in a spanwise velocity component along the blade that helps keep the flow attached to the blade, delaying stall and increasing lift.

Because of the different challenges listed above, steady BEM tools need to be supplemented with more advanced rotor analysis tools. Unsteady BEM codes are the most common analysis tool, but they are often complemented with high-fidelity CFD tools. These advanced rotor analysis tools are discussed in Section 3.7.

3.6 Simplified Performance Analyses and Designs

The steady BEM method presented in Section 3.5 can be applied to arbitrary blade geometries, but it requires a numerical implementation. In this section, the BEM equations are exploited by introducing a variety of simplifications to be able to obtain the performance of the turbine. These analyses can be used to obtain "optimum" blade designs, with or without wake rotation. Then, the BEM method is used in a simplified rotor design procedure. The section presents four aerodynamic blade design procedures:

- The "optimum" blade design for a rotor with an infinite number of blades and no wake rotation.
- The "optimum" blade design for a rotor with an infinite number of blades, but accounting for wake rotation. This blade design can be used as the start for a general blade design analysis.
- A linearized performance calculation procedure.
- An illustration of a general blade design analysis assuming a constant design lift coefficient.

Blade design procedures used by the industry are significantly more advanced than the procedures presented in this section. In particular, this section focuses on the aerodynamics, whereas a blade design is a complex compromise between structural integrity, ease of fabrication, and aerodynamic performance over the range of wind and rotor speeds that the turbine will encounter. The procedures discussed here provide a useful starting point, however.

3.6.1 Blade Shape for Ideal Rotor without Wake Rotation

In Section 3.2.1.3, the maximum possible power coefficient from a wind turbine, assuming no wake rotation or drag, was determined to occur with an axial induction factor of 1/3. If the same simplifying assumptions are applied to the equations of momentum and blade element theory, the analysis becomes simple enough that an ideal blade shape can be determined. The blade shape approximates one that would provide maximum power at the design tip speed ratio of a real wind turbine. This will be called the "Betz optimum rotor."

In this analysis, the following assumptions will be made:

- There is no wake rotation; thus $a' = 0$.
- There is no drag; thus $C_d = 0$.
- There are no losses due to a finite number of blades (i.e., no tip loss, $F = 1$).
- The axial induction factor, a, is 1/3 in each annular stream tube.

First, a design tip-speed ratio, λ, the desired number of blades, B, the radius, R, and an airfoil with known lift and drag coefficients as a function of angle of attack need to be chosen. An angle of attack (and, thus, a lift coefficient at which the airfoil will operate) is also chosen. This angle of attack should be selected such that C_d/C_l is minimal in order to most closely approximate the assumption that $C_d = 0$. These choices allow the twist and chord distribution of a blade that would provide Betz limit power production (given the input assumptions) to be determined. With the assumption that $a = 1/3$, one gets from momentum theory (Equation 3.95):

$$dT = \rho U^2 4 \left(\frac{1}{3}\right)\left(1 - \frac{1}{3}\right)\pi r dr = \rho U^2 \left(\frac{8}{9}\right)\pi r dr \tag{3.123}$$

and, from blade element theory (Equation 3.88), with $C_d = 0$:

$$dT = B\frac{1}{2}\rho U_{rel}^2 (C_l \cos\phi) c dr \tag{3.124}$$

A third equation, Equation (3.86), can be used to express U_{rel} in terms of other known variables:

$$U_{rel} = \frac{U(1-a)}{\sin\phi} = \frac{2U}{3\sin\phi} \tag{3.125}$$

Similar to the approach used in BEM, the equations of momentum theory and blade element theory are equated. In this case, equating Equations (3.123) and (3.124) and using Equation (3.125), yields:

$$\frac{C_l Bc}{4\pi r} = \tan\phi \sin\phi \tag{3.126}$$

Equation (3.85), which relates a, a', and ϕ based on geometrical considerations, can be used to solve for the blade shape. This equation, with $a' = 0$ and $a = 1/3$, provides the flow angle:

$$\tan\phi = \frac{2}{3\lambda_r} \tag{3.127}$$

Therefore

$$\frac{C_l Bc}{4\pi r} = \left(\frac{2}{3\lambda_r}\right)\sin\phi \tag{3.128}$$

Rearranging, and noting that $\lambda_r = \lambda(r/R)$, one can determine the angle of the relative wind and the chord of the blade for each section of the ideal rotor:

$$\phi = \tan^{-1}\left(\frac{2}{3\lambda_r}\right) \tag{3.129}$$

$$c = \frac{8\pi r \sin\phi}{3 B C_l \lambda_r} \tag{3.130}$$

These relations can be used to find the chord and twist distribution of the Betz optimum blade. As an example, suppose: $\lambda = 7$, the airfoil has a lift coefficient of $C_l = 1$, C_d/C_l has a minimum at $\alpha = 7°$ and, finally, that there are three blades, so $B = 3$. Then, from Equations (3.129) and (3.130), the results shown in Table 3.2 are obtained. Here, the chord and radius have been non-dimensionalized by dividing by the rotor radius. In this process, Equations (3.79) and (3.80) are also used to relate the various blade angles to each other (see Figure 3.25), namely, $\theta_T = \phi - \alpha - \theta_c$, and $\theta_P = \theta_T + \theta_c$. If it is assumed that the twist angle is 0 at the tip, then $\theta_c = -1.6$. The chord

Table 3.2 Twist and chord distribution for the example Betz optimum blade; r/R, fraction of rotor radius; c/R, non-dimensionalized chord

r/R	c/R	Angle of rel. wind, ϕ (deg)	Twist angle, θ_T (deg)	Section pitch, θ_P (deg)
0.1	0.275	43.6	38.2	36.6
0.2	0.172	25.5	20.0	18.5
0.3	0.121	17.6	12.2	10.6
0.4	0.092	13.4	8.0	6.4
0.5	0.075	10.8	5.3	3.8
0.6	0.063	9.0	3.6	2.0
0.7	0.054	7.7	2.3	0.7
0.8	0.047	6.8	1.3	−0.2
0.9	0.042	6.0	0.6	−1.0
1	0.039	5.4	0	−1.6

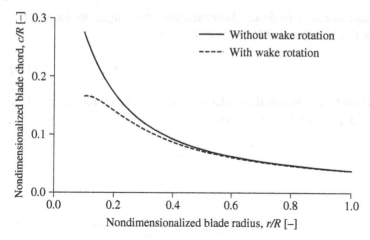

Figure 3.33 Blade chord for the example Betz optimum blade as described in Section 3.6.1 (without wake rotation) and Section 3.6.2 (with wake rotation)

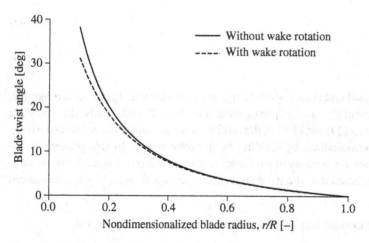

Figure 3.34 Blade twist angle for sample Betz optimum blade as described in Section 3.6.1 (without wake rotation) and Section 3.6.2 (with wake rotation)

and twist of this blade are illustrated in Figures 3.33 and 3.34. Three-dimensional representations are also shown in Figure 3.35.

It can be seen that blades designed for optimum power production have an increasingly large chord and twist angle as one gets closer to the blade root. One consideration in blade design is the cost and difficulty of fabricating the blade. An optimum blade would be very difficult to manufacture at a reasonable cost, but the design provides insight into the blade shape that might be desired for a wind turbine.

3.6.2 Blade Shape for Optimum Rotor with Wake Rotation

In the previous section, wake rotation was omitted by setting $a' = 0$ into the steady BEM equations. In this section, this assumption is lifted to obtain the blade shape for an ideal rotor that includes the effects of wake rotation. Wake rotation is included, but drag ($C_d = 0$) and tip losses ($F = 1$) are ignored as in the previous section. One can perform

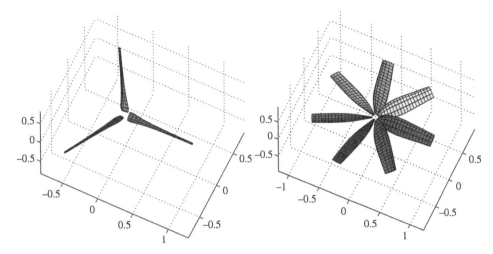

Figure 3.35 Three-dimensional representation of "ideal" rotor without wake rotation for $\lambda = 10$, $B = 3$ (left), and $\lambda = 2$, and $B = 7$ (right). The axes made dimensionless by dividing by the blade length

the maximum-C_P-optimization by taking the partial derivative of that part of the integral for C_P (Equation 3.118) which is a function of the angle of the relative wind, ϕ, and setting it equal to zero, i.e.:

$$\frac{\partial}{\partial \phi}\left[\sin^2\phi(\cos\phi - \lambda_r \sin\phi)(\sin\phi + \lambda_r \cos\phi)\right] = 0 \tag{3.131}$$

This yields:

$$\lambda_r = \sin\phi(2\cos\phi - 1)/[(1 - \cos\phi)(2\cos\phi + 1)] \tag{3.132}$$

After some algebra, the chord and flow angle for an ideal rotor with wake rotation are:

$$\phi = \frac{2}{3}\tan^{-1}\left(\frac{1}{\lambda_r}\right) \tag{3.133}$$

$$c = \frac{8\pi r}{BC_l}(1 - \cos\phi) \tag{3.134}$$

The induction factors can be calculated from Equations (3.108) and (3.39):

$$a = \left[1 + \frac{4\sin^2\phi}{\sigma C_l \cos\phi}\right]^{-1} \tag{3.135}$$

$$a' = \frac{1 - 3a}{4a - 1} \tag{3.136}$$

These results can be compared with the result for an ideal blade without wake rotation given in Equations (3.129) and (3.130). Note that the optimum values for ϕ and c, including wake rotation, are often similar to, but could be significantly different from, those obtained without assuming wake rotation. Also, as before, select α where C_d/C_l is minimal.

The total solidity of the rotor, σ_{rot}, is the ratio of the planform area of the blades to the swept area, thus:

$$\sigma_{rot} = \frac{B}{\pi R^2}\int_{r_h}^{R} c\,dr \tag{3.137}$$

Table 3.3 Three optimum rotors

r/R	λ = 1 B = 12		λ = 7 B = 3		λ = 10 B = 2	
	ϕ	c/R	ϕ	c/R	ϕ	c/R
0.1	56.2	0.093	36.7	0.166	30.0	0.168
0.2	52.5	0.164	23.7	0.141	17.7	0.119
0.3	48.9	0.215	17.0	0.110	12.3	0.086
0.4	45.5	0.250	13.1	0.087	9.4	0.067
0.5	42.3	0.273	10.6	0.072	7.5	0.054
0.6	39.4	0.285	8.9	0.061	6.3	0.046
0.7	36.7	0.290	7.7	0.053	5.4	0.039
0.8	34.2	0.290	6.7	0.046	4.8	0.035
0.9	32.0	0.287	6.0	0.041	4.2	0.031
1.0	30.0	0.281	5.4	0.037	3.8	0.028
Solidity, σ_{rot}		0.86		0.068		0.036

Note: B, number of blades; c, airfoil chord length; r, blade section radius; R, rotor radius; λ, tip speed ratio; ϕ, angle of relative wind

The optimum blade rotor solidity can be found in methods discussed above. In general, the rotor solidity decreases with increased design tip-speed ratios. When the blade is modeled as a set of N blade sections of equal span, the rotor solidity can be calculated as:

$$\sigma_{rot} \cong \frac{B}{N\pi}\left(\sum_{i=1}^{N} c_i/R\right) \cong \frac{2}{N}\left(\sum_{i=1}^{N}\sigma_i\right) \tag{3.138}$$

where the local solidity σ_i is defined in Equation (3.92).

The blade shapes for three sample optimum rotors, assuming wake rotation, are given in Table 3.3. Here, C_l is assumed to be 1.0 at the design angle of attack, and the hub radius is taken as $r_h = 0.1R$ to compute the solidity. The chord and twist for the case $B = 3$ and $\lambda = 7$ are illustrated in Figures 3.33 and 3.34 together with the case with no wake rotation. In these rotors, the blade twist is directly related to the angle of the relative wind because the angle of attack is assumed to be constant (see Equations 3.79 and 3.80). Thus, changes in blade twist would mirror the changes in the angle of the relative wind shown in Table 3.3. It can be seen that the slow 12-bladed machine would have blades that had a roughly constant chord over the outer half of the blade and smaller chords closer to the hub. The blades would also have a significant twist. The two faster machines would have blades with an increasing chord as one went from the tip to the hub. The blades would also have a significant twist but much less than the 12-bladed machine. The fastest machine would have the least twist, which is a function of local speed ratio only. It would also have the smallest chord because of the low angle of the relative wind (see Equations 3.133 and 3.134).

3.6.3 Linearized HAWT Rotor Performance Calculation Procedure

Manwell (1990) proposed a linearized method for calculating the performance of a HAWT rotor that is particularly applicable for an unstalled rotor. The method uses the previously discussed blade element theory and incorporates an analytical method for finding the blade angle of attack. Depending on whether tip losses are included, few or no iterations are required. The method assumes that two conditions apply:

- The airfoil section lift coefficient vs. angle of attack relation must be linear in the region of interest.
- The angle of attack must be small enough that the small-angle approximations may be used.

These two requirements normally apply if the section is unstalled. They may also apply under certain partially stalled conditions for moderate angles of attack if the lift curve can be linearized.

The linearized method is the same as Method 2 outlined in Section 3.5.2, with the exception of a simplification for determining the angle of attack and the lift coefficient for each blade section. The essence of the linearized method is the use of an analytical (closed form) expression for finding the angle of attack of the relative wind at each blade element. It is assumed that the lift and drag curves can be approximated by:

$$C_l = C_{l,0} + C_{l,a}\alpha \tag{3.139}$$

$$C_d = C_{d,0} + C_{d,a1}\alpha + C_{d,a2}\alpha^2 \tag{3.140}$$

where α is the angle of attack in deg and $C_{l,a}$ is in deg^{-1}.

When the lift curve is linear and when small-angle approximations can be used, it can be shown that the angle of attack (degrees) is given by:

$$\alpha = \frac{180}{\pi} \left[\frac{-q_2 \pm \sqrt{q_2^2 - 4q_1 q_3}}{2 q_3} \right] \tag{3.141}$$

where

$$q_1 = C_{l,0}\, d_2 - \frac{4F}{\sigma} d_1 \sin\theta_p \tag{3.142}$$

$$q_2 = \frac{180 C_{l,a} d_2}{\pi} + d_1 C_{l,0} - \frac{4F}{\sigma}\left(d_1 \cos\theta_p - d_2 \sin\theta_p\right) \tag{3.143}$$

$$q_3 = \frac{180 C_{l,a} d_1}{\pi} + \frac{4F}{\sigma} d_2 \cos\theta_p \tag{3.144}$$

$$d_1 = \cos\theta_p - \lambda_r \sin\theta_p \tag{3.145}$$

$$d_2 = \sin\theta_p + \lambda_r \cos\theta_p \tag{3.146}$$

Using this approach, the angle of attack can be calculated from Equation (3.141) once an initial estimate for the tip loss factor is determined. The lift and drag coefficients can then be calculated from Equations (3.139) and (3.140), using Equation (3.172). Iteration with a new estimate of the tip loss factor may be required.

The linearized method provides angles of attack very close to those of the more detailed method for many operating conditions. For example, results for the analysis of one blade of the University of Massachusetts WF-1 wind turbine are shown in Figure 3.36. This was a three-bladed turbine with a 10 m rotor, using near-optimum tapered and twisted blades. The lift curve of the NACA 4415 airfoil was approximated by $C_l = 0.368 + 0.0942\alpha$. The drag coefficient equation constants were 0.00994, 0.000259, and 0.0001055. Figure 3.36 shows the lift coefficient vs. angle of attack from Equation (3.104) and from the corresponding equation in the linearized method for one of the blade elements. The points at which these curves cross the empirical lift coefficient line determine the actual angle of attack and the lift coefficient. Also plotted in Figure 3.36 is the axial induction factor, a, for the element. Note that it is the right-hand intersection point that gives a value of $a < 1/2$, as is normally the case. In this case, the results from the Equation (3.104) are $C_l = 1.265$ at $\alpha = 9.53°$; they are very from Equation (3.141): $C_l = 1.274$ at $\alpha = 9.62°$.

3.6.4 Aerodynamic Rotor Design Procedure Using the Steady BEM

The steady BEM algorithm can be used in a generalized rotor design procedure. The procedure begins with the choice of various rotor parameters and the choice of different airfoils. For simplicity, only one airfoil is used in

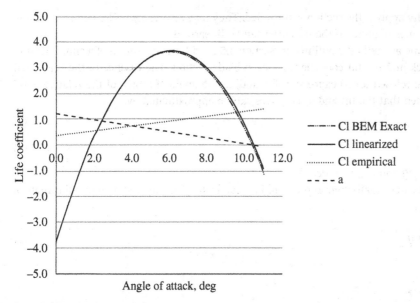

Figure 3.36 Comparison of calculation methods for one blade element; a, axial induction factor; C_l, two-dimensional lift coefficient

this section to illustrate the method. An initial blade shape is determined using the optimum blade shape assuming wake rotation presented in Section 3.6.2. The final blade shape and performance are determined iteratively considering drag, tip losses, and ease of manufacture. The steps in determining a blade design follow.

3.6.4.1 Determine Basic Rotor Parameters

1) Begin by deciding what power, P, is needed at a particular wind velocity, U. Include the effect of a probable C_P and efficiencies, η, of various other components (e.g., gearbox, generator, and pump). The radius, R, of the rotor may be estimated by using Equation (3.20) to obtain:

$$R = \sqrt{2P/C_P\eta\rho\pi U^3} \tag{3.147}$$

2) According to the type of application, choose a tip speed ratio, λ. For a water-pumping windmill, for which greater torque is needed, use $1 < \lambda < 3$. For electrical power generation, use $4 < \lambda < 10$. The higher-speed machines use less material in the blades and have smaller gearboxes but require more sophisticated airfoils.
3) Choose the number of blades, B, from Table 3.4. Note: if fewer than three blades are selected, there are a number of structural dynamic problems that must be considered in the hub design. One solution is a teetered hub (see Chapter 7).
4) Select an airfoil. If $\lambda < 3$, curved plates can be used. If $\lambda > 3$, use a more aerodynamic shape.

3.6.4.2 Define the Blade Shape

5) Obtain and examine the empirical curves for the aerodynamic properties of the airfoil at each section (the airfoil may vary from the root to the tip), i.e., C_l vs. α, C_d vs. α. Choose the design aerodynamic conditions, $C_{l,\text{design}}$ and α_{design}, such that $C_{d,\text{design}}/C_{l,\text{design}}$ is at a minimum for each blade section.

Table 3.4 Suggested blade number, B, for different tip speed ratios, λ

λ	B
1	8–24
2	6–12
3	3–6
4	3–4
>4	1–3

6) Divide the blade into N elements (usually 10–20). Use the optimum rotor theory from Section 3.6.2 to estimate the shape of the ith element with a midpoint radius of r_i (where $i = 1$ is closest to the hub):

$$\lambda_{r,i} = \frac{\lambda r_i}{R} \tag{3.148}$$

$$\phi_i = \frac{2}{3} \tan^{-1}\left(\frac{1}{\lambda_{r,i}}\right) \tag{3.149}$$

$$c_i = \frac{8\pi r_i}{BC_{l,\text{design},i}}(1 - \cos\phi_i) \tag{3.150}$$

$$\theta_{T,i} = \theta_{p,i} - \theta_c \tag{3.151}$$

$$\theta_{p,i} = \phi_i - \alpha_{\text{design},i} \tag{3.152}$$

7) Using the optimum blade shape as a guide, select a blade shape that promises to be a good approximation. For ease of fabrication, linear variations of chord, thickness, and twist might be chosen. For example, if a_1, b_1, and a_2 are coefficients for the chosen chord and twist distributions, then the chord and twist can be expressed as:

$$c_i = a_1(R - r_i) + b_1 \tag{3.153}$$

$$\theta_{T,i} = a_2(R - r_i) \tag{3.154}$$

3.6.4.3 Solve the BEM Equations

8) As outlined above, one of two methods might be chosen to solve the equations for the blade performance.

Method 1 – General Iterative Solution for a and a'

Iterating to find the axial and tangential induction factors using Method 1 requires initial guesses for their values. To find initial values, start with values from an adjacent blade section, values from the previous blade design in the iterative rotor design process, or use an estimate based on the design values from the starting optimum blade design:

$$\phi_{i,1} = \frac{2}{3} \tan^{-1}\left(\frac{1}{\lambda_{r,i}}\right) \tag{3.155}$$

$$a_{i,1} = \left[1 + \frac{4\sin^2\phi_{i,1}}{\sigma_{i,\text{design}}C_{l,\text{design}}\cos\phi_{i,1}}\right]^{-1} \tag{3.156}$$

$$a'_{i,1} = \frac{1 - 3a_{i,1}}{(4a_{i,1}) - 1} \tag{3.157}$$

Having guesses for $a_{i,1}$ and $a'_{i,1}$, start the iterative solution procedure for the jth iteration. For the first iteration $j = 1$. Calculate the angle of the relative wind and the tip loss factor:

$$\tan \phi_{i,j} = \frac{U(1 - a_{i,j})}{\Omega r(1 + a'_{i,j})} = \frac{1 - a_{i,j}}{(1 + a'_{i,j})\lambda_{r,i}} \tag{3.158}$$

$$F_{i,j} = \frac{2}{\pi} \cos^{-1}\left[\exp\left(-\left\{\frac{(B/2)[1 - (r_i/R)]}{(r_i/R)\sin \phi_{i,j}}\right\}\right)\right] \tag{3.159}$$

Determine $C_{l,i,j}$ and $C_{d,i,j}$ from the airfoil lift and drag data, using:

$$\alpha_{i,j} = \phi_{i,j} - \theta_{p,i} \tag{3.160}$$

Project the lift coefficients onto $c_{n,i,j}$ and $c_{t,i,j}$ using Equations (3.90) and (3.91):

$$c_{n,i,j} = C_{l,i,j} \cos \phi_{i,j} + C_{d,i,j} \sin \phi_{i,j} \tag{3.161}$$

$$c_{t,i,j} = C_{l,i,j} \sin \phi_{i,j} - C_{d,i,j} \cos \phi_{i,j} \tag{3.162}$$

Calculate the local thrust coefficient:

$$C_{Tr,i,j} = \frac{\sigma_i(1 - a_{i,j})^2\left(C_{l,i,j} \cos \phi_{i,j} + C_{d,i,j} \sin \phi_{i,j}\right)}{\sin^2 \phi_{i,j}} \tag{3.163}$$

Update a and a' for the next iteration. If $C_{Tr,i,j} < 0.96$:

$$a_{i,j+1} = \left[1 + \frac{4F_{i,j} \sin^2 \phi_{i,j}}{\sigma_i c_{n,i,j}}\right]^{-1} \tag{3.164}$$

If $C_{Tr,i,j} > 0.96$:

$$a_{i,j} = (1/F_{i,j})\left[0.143 + \sqrt{0.0203 - 0.6427(0.889 - C_{Tr,i,j})}\right] \tag{3.165}$$

$$a'_{i,j+1} = \left[\frac{4F_{i,j} \sin \phi_{i,j} \cos \phi_{i,j}}{\sigma_i c_{t,i,j}}\right]^{-1} \tag{3.166}$$

If the newest induction factors are within an acceptable tolerance of the previous guesses, then the other performance parameters can be calculated. If not, then the procedure starts again at Equation (3.158) with $j = j + 1$.

Method 2 – Solving for C_l and α (assuming no drag)

Find the actual angle of attack and lift coefficients for the center of each element, using the following equations and the empirical airfoil curves:

$$C_{l,i} = 4F_i \sin \phi_i \frac{(\cos \phi_i - \lambda_{r,i} \sin \phi_i)}{\sigma_i(\sin \phi_i + \lambda_{r,i} \cos \phi_i)} \tag{3.167}$$

$$\sigma_i = Bc_i/2\pi r_i \tag{3.168}$$

$$\phi_i = \theta_{p,i} + \alpha_i \tag{3.169}$$

$$F_i = \frac{2}{\pi} \cos^{-1} \left[\exp \left(- \left\{ \frac{(B/2)[1 - (r_i/R)]}{(r_i/R) \sin \phi_i} \right\} \right) \right] \tag{3.170}$$

The lift coefficient and angle of attack can be found by iteration or graphically. A graphical solution is illustrated in Figure 3.26. The iterative approach requires an initial estimate of the tip loss factor. To find a starting F_i, start with an estimate for the angle of the relative wind of:

$$\phi_{i,1} = \frac{2}{3} \tan^{-1}(1/\lambda_{r,i}) \tag{3.171}$$

For subsequent iterations, find F_i using:

$$\phi_{i,j+1} = \theta_{P,i} + \alpha_{i,j} \tag{3.172}$$

where j is the number of the iteration. Usually, only a few iterations are needed.

Finally, calculate the axial induction factor:

$$a_i = 1/\left[1 + \left(4 \sin^2 \phi_i \right) / (\sigma_i C_{l,i} \cos \phi_i) \right] \tag{3.173}$$

and the tangential induction factor:

$$a_i' = 1/[(4 \cos \phi_i)/(\sigma_i C_{l,i}) - 1] \tag{3.174}$$

If a_i is greater than 0.4, use Method 1.

3.6.4.4   Calculate Rotor Performance and Modify Blade Design

9) Having solved the equations for the performance at each blade element, the power coefficient is determined using a sum approximating the integral in Equation (3.116):

$$C_P = \sum_{i=1}^{N} \left(\frac{2}{\lambda^2} \right) \left[(1-a)^2 + \lambda_{ri}^2 (1+a')^2 \right] \sigma_i c_{t,i,j} \lambda_{ri}^2 \Delta \lambda_r \tag{3.175}$$

With $\Delta \lambda_r = \lambda_{ri} - \lambda_{r(i-1)}$ and where k is the index of the first "blade" section consisting of the actual blade airfoil. If the drag is neglected for simplicity, Equation (3.118) is used instead, leading to:

$$C_P = \sum_{i=1}^{N} \left(\frac{8 \Delta \lambda_r}{\lambda^2} \right) F_i \sin^2 \phi_i (\cos \phi_i - \lambda_{ri} \sin \phi_i)(\sin \phi_i + \lambda_{ri} \cos \phi_i) \left[1 - \left(\frac{C_d}{C_l} \right) \cot \phi_i \right] \lambda_{ri}^2 \tag{3.176}$$

10) Modify the design if necessary and repeat steps 8–10 in order to find the best design for the rotor, given the limitations of fabrication.

3.7   *Advanced Methods of Rotor Analysis

In the previous sections of this chapter, the steady BEM theory has been used to predict steady-state rotor performances. Simplified blade design procedures were presented based on this approach. As mentioned in Section 3.5.1.4, the steady BEM approach is challenged in advanced flows or when higher fidelity is needed. There are other approaches to predicting blade performance and to designing blades, in particular, approaches that can

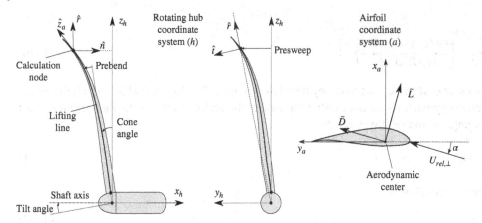

Figure 3.37 Coordinate system used for the lifting-line algorithm. Left: side view for a case with prebend and coning (no sweep). Center: front view for a case with presweep (no prebend and coning assumed). Right: airfoil coordinate system

capture the unsteadiness of the flow and the general case where blades are "out-of-plane" due to coning, prebend, presweep, and deflections (see Figure 3.37). The methods used for rotor analyses may be classified based on:

- the geometry used to represent the blades (disc, lifting-line, lifting-surface, or full-blade geometry, as described in the subsequent paragraphs)
- the main variables used by the solver: velocity–pressure for momentum-based and traditional CFD methods, vorticity for vorticity-based methods, and series of coefficients for spectral methods (see Canuto, 1988).

It is possible to develop a performance prediction tool for any combination of the geometry classes and solver classes mentioned above. Here, the focus is on the most popular combinations.

This section will begin by presenting the main principles behind lifting-line methods (Section 3.7.1), which are the most used. The most popular lifting-line methods are steady BEM, unsteady BEM, lifting-line-based vortex methods, and actuator-line CFD. Each of the subsequent subsections focuses on a solver class: unsteady BEM and engineering models (Section 3.7.2), vortex methods (Section 3.7.3), and traditional CFD (Section 3.7.4). The section ends with some discussions on the modeling challenges currently faced.

3.7.1 The General Structure of Lifting-Line-Based Algorithms

3.7.1.1 Introduction

Lifting-line methods are among the most popular performance prediction methods for wind turbines. They usually offer a trade-off between accuracy and computational expense. Lifting-line methods effectively rely on the Blade Element theory presented in Section 3.3.7. The distributed loads on the airfoil surface are lumped at each cross-section of the blade onto the mean line of the blade and do not account directly for the geometry of each cross-section. The cross-section loads are obtained using tabulated airfoil performance data. The solver does neither need to resolve the geometry of the cross-section nor the boundary layer around the structure. This last point is where the method gains in computational performance since blade-resolved tools must resolve the thickness of the boundary layer, which, for the Reynolds number at which the wind turbine airfoils are operating, requires a large number of computational cells. The lifting-line method is attributed to the independent work of Lanchester (1907) and Prandtl (1918), but it is inspired by the Blade Element theory of W. Froude (1878). The main assumption of the lifting-line approach is that the specific geometry of the blade has only a local effect in the vicinity of the blade and the flow characteristics in the wake, a few chord distances away from the blade, are primarily driven by the overall loading distribution concentrated on the mean line of the blade. The mean line of the blade is usually taken as the line formed by joining the aerodynamic center of each cross-section of the blade. The second assumption of the method is that the loading distribution along the lifting line can be determined based on the flow in the vicinity of the lifting line or on the lifting line itself. The assumptions are challenged when the flow of each cross-section is far from a 2D flow in such a way that the boundary-layer state strongly affects the performances, for instance: three-dimensional and unsteady stall, rotational effects, highly turbulent/waked flows, and fluid–structure interactions. Despite these limitations, lifting-line methods have been successfully used for the prediction of rotor aerodynamics in a large suite of tools: steady BEM, unsteady BEM, lifting-line-based vortex methods, and actuator-line CFD. The common elements of all these tools are presented in the subsequent paragraphs. Specific details for each solver follow in the next subsections.

3.7.1.2 Cross-sectional Loads

The key part of the lifting-line method is the estimation of the aerodynamic loads on a cross-section. The method assumes that the performance of each cross-section is independent of the neighboring sections, and it is only a function of the local relative wind velocity (noted U_{rel}, see Figure 3.13), consisting of the local wind velocity and the relative motion of the section. The local wind velocity is sometimes sampled at a point located at a distance of three-quarter chord from the leading edge along the mean line of the airfoil. The justifications for this choice are multiple and are open to discussion. It corresponds to the point where the induced velocity from the bound vortex balances the mean flow and satisfies the no-flow-through condition for a 2D flow (Pistolesi's theorem). Other implementations use the velocity at the aerodynamic center, which is often taken as the quarter-chord point of the airfoil. Methods that introduce a regularization parameter, such as the actuator line CFD method, may choose an optimal probing point for the velocity based on the regularization parameter. Alternative methods may use several probing points to estimate the local velocity. In general, a point away from the local pitching axis is necessary to capture aerodynamic effects due to the pitching/torsional motion of the airfoil in aero-elastic simulations. Once the local relative wind velocity is determined, the aerodynamic forces are found based on a lookup table of lift, drag, and moment coefficients (also called polar data). These coefficients are tabulated based on the angle of attack, and in some cases Reynolds number. The tabulated data are read from a file and interpolated based on the local velocity. The obtained loads are then applied at the aerodynamic center of the airfoil section. The next paragraph summarizes the lifting-line algorithm.

3.7.1.3 Lifting-line Algorithm

Most of the lifting-line tools follow the structure presented in this paragraph. The notations are illustrated in Figure 3.37.

   The blades are discretized into a set of spanwise nodes, each representing a calculation point for a given cross-section. The vector from the hub to the node is noted $\boldsymbol{r}$. For each section, the locations of the aerodynamic center or a "probing" point (e.g., the ¼ or ¾ chord point) are stored in the section coordinate system. For each spanwise positions, the following steps are taken:

1) Compute the elastic velocity of the section, $\boldsymbol{V}_{elast}$. This velocity includes the rotational velocity of the rotor ($\boldsymbol{\Omega} \times \boldsymbol{r}$) and any structural motions due to the blade flexibility (computed by an elastic solver when available).
2) Compute the local wind velocity at the "probe" location, $\boldsymbol{U}_{wind}$. This velocity includes the freestream velocity and induced velocities.
3) From the relative wind velocity as follows: $\boldsymbol{U}_{rel} = \boldsymbol{U}_{wind} - \boldsymbol{V}_{elast}$.
4) Project $\boldsymbol{U}_{rel}$ into the airfoil coordinate system (defined by the unit vectors $\hat{\boldsymbol{x}}_a$ and $\hat{\boldsymbol{y}}_a$, see Figure 3.37), compute the velocity norm in the $\hat{\boldsymbol{x}}_a - \hat{\boldsymbol{y}}_a$ plane, noted $U_{rel,\perp}$, and the angle of attack, α, angle between $\boldsymbol{U}_{rel}$ and the instantaneous chord line (includes deflections, torsion, pitch, and twist):

$$U_{rel,\perp} = \sqrt{(\boldsymbol{U_{rel}} \cdot \hat{\boldsymbol{x}}_a)^2 + (\boldsymbol{U_{rel}} \cdot \hat{\boldsymbol{y}}_a)^2}, \qquad \tan \alpha = \frac{\boldsymbol{U_{rel}} \cdot \hat{\boldsymbol{x}}_a}{\boldsymbol{U_{rel}} \cdot \hat{\boldsymbol{y}}_a} \qquad (3.177)$$

5) Retrieve the aerodynamic coefficients from tabulated data: $C_l(\alpha)$, $C_d(\alpha)$ and $C_m(\alpha)$.

6) (Optional) Apply a dynamic stall model to determine unsteady airfoil coefficients.

7) Compute the spanwise aerodynamic forces per unit length at the aerodynamic center (i.e., at the lifting line)
$\tilde{L} = \frac{1}{2}\rho U_{rel,\perp}^2 c C_l$ and $\tilde{D} = \frac{1}{2}\rho U_{rel,\perp}^2 c C_d$ (in N/m), and the aerodynamic moment $\tilde{M} = \frac{1}{2}\rho U_{rel,\perp}^2 c^2 C_m$ (in Nm/m)
(see Equations (3.74–3.76)). The forces are gathered into a vector, defined in the $\hat{x}_a - \hat{y}_a$ plane. The moment is about $\hat{z}_a$.

8) (Optional) Compute the local circulation $\Gamma = \frac{1}{2}U_{rel,\perp}^2 c C_l$

9) Project the section forces and moment, defined in the a-system into the global coordinate system.

10) (Optional) Integrate the loads along the blades to get the integrated forces and moments at the blade root and hub, and the total aerodynamic power and thrust (from all blades).

Once these steps are completed for each section, the loads are provided to the flow solver, resulting in another iteration. Once the aerodynamic loads are converged, they may be transferred to the structural solver (for further iteration), or a new time-step is carried out. In CFD actuator methods, the loads are usually smeared so that they will be continuous and cover several cells, in order to reproduce the realistic loading from a blade. In vorticity-based methods, the circulation is the main "loading-variable," and it is also distributed spatially using a regularization method similar to the one used in actuator-line methods.

3.7.2 Unsteady Blade Element Momentum Method

3.7.2.1 Introduction
Traditional CFD and vortex methods inherently account for unsteady aerodynamic effects. The Blade Element Method, on the other hand, relies on a steady-state theory. Unsteady BEM codes use the general lifting-line algorithm to compute the loads. The steady BEM equations are used to obtain quasi-steady axial and tangential induction factors (usually from loads projected onto the rotor disk plane). Generalization of the algorithm is required for non-straight blades (see Section 3.7.2.10). Ad-hoc engineering models are then used to modify these quasi-steady induction factors and extend the method for unsteady simulations. This section starts by introducing the different sources of unsteadiness of the flow, before presenting the different models used in an unsteady BEM code.

3.7.2.2 Unsteady Aerodynamic Effects
There are several unsteady aerodynamic phenomena that have a large effect on wind turbine operation. An overview of the different effects is given in this paragraph. Each of them is described in subsequent paragraphs where engineering models suitable to perform unsteady BEM simulations are given. The variation of the mean inflow with height, referred to as shear, and the variation of the mean wind direction with height, referred to as veer, will result in different wind speeds perceived by each blade node. The fact that the turbine is yawed or tilted compared to the mean inflow will also introduce differences in inflow and hence in loads. The turbulent eddies carried along with the mean wind cause rapid changes in speed and direction over the rotor disc. These changes cause fluctuating aerodynamic forces, increased peak forces, blade vibrations, and significant material fatigue. Additionally, the transient effects of tower shadow, dynamic stall, dynamic inflow, and rotational sampling (all explained below) change turbine operation in unexpected ways. Many of these effects occur at the rotational frequency of the rotor or at multiples of that frequency. Effects that occur once per revolution are often referred to as having a frequency of 1P. Similarly, effects that occur at three or n times per revolution of the rotor are referred to as occurring at a frequency of 3P or nP.

3.7.2.3 Unsteady Relative Velocity: Frozen Turbulence and Rotational Sampling
In the BEM theory, the wind speed is assumed to be steady and uniform. In a real situation, however, the relative wind speed perceived by a blade section will be highly unsteady. In unsteady BEM simulations, it is assumed that

the relative wind speed on the lifting line can be taken as the sum of the instantaneous undisturbed wind speed at this location (as if the turbine were not present), and the elastic velocity of the section originating from the blade motion. It is thus assumed that the wind turbine does not affect the turbulence or the shear. The two-way interaction between the turbine and the flow field is not accounted for in the unsteady BEM method (they are inherently present in traditional or vorticity-based CFD). A three-dimensional turbulent field is pre-generated with the characteristics of a given site, with computer codes such as *TurbSim* (see Chapter 6). These fields are then assumed to be convected with the mean wind speed, an assumption referred to as "Taylor's frozen hypothesis." As time evolves, the unsteady BEM simulation extracts points in the turbulence field at the lifting-line locations without changing the field. The velocity from the atmosphere, including turbulence effects, will be noted $\boldsymbol{U}_{atm}$. The effective turbulence seen by the rotor has a different frequency content than the turbulence field itself, due to the effect known as "rotational sampling" (see Connell, 1982): as the blades rotate, they "sample" the flow field at a different rate than what a fixed observer would measure. The rotational sampling effect is a function of the turbine dimension and the local tip speed ratio along the blade.

3.7.2.4 Yaw and Tilt: Skew Model

In common operating conditions, the mean flow across the rotor area will not be normal to the disk. Turbulence contributes to the misalignment, but the main quasi-steady contributions are coming from the yaw and tilt angle. The yaw angle (θ_{yaw}) is the difference between the horizontal wind direction and the normal to the rotor, as illustrated in Figure 3.38 (assuming $\theta_{skew} = \theta_{yaw}$).

The tilt angle is the angle formed between the normal to the rotor and the horizontal. Typically, the main shafts of wind turbines with upwind rotors are tilted upwards to increase the distance between the tower and the blades and avoid tower strikes. This distance is referred to as tower clearance.

The yawed and tilted situations are aerodynamically equivalent if the influence of the surface (ground or ocean) is neglected. Both effects are usually treated simultaneously in an unsteady BEM implementation by considering the direction of the mean flow across the rotor disk. This average flow consists of the mean undisturbed inflow and the induced velocities from the wake. The direction formed by the mean flow vector with the rotor axis is measured by the inflow skew angle θ_{skew}. In the case of pure yaw (no tilt), then $\theta_{skew} = \theta_{yaw}$. The wake will tend to follow this main direction but will have a slightly higher angle, referred to as the wake skew angle, noted χ (see Figure 3.38). The following empirical relationship can be used to relate the two:

$$\chi = (1 + 0.6\bar{a})\, \theta_{skew} \tag{3.178}$$

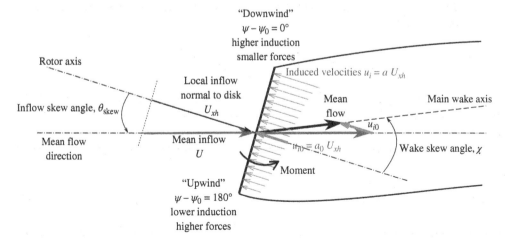

Figure 3.38 Skewed configuration of a wind turbine. Definition of the inflow skew angle and wake skew angles

where $\bar{a}$ is the average axial induction over the rotor. In the implementation presented by Hansen (2008), the inflow skew angle and the wake skew angles are determined using velocities at the radial position $r/R \approx 0.7$.

The skewed configuration of a rotor was studied by Glauert (1926) using a lifting-line analysis, and by Coleman *et al.* (1945) using a vorticity formulation. The models are sometimes attributed to the later developments of Pitt and Peters (Pitt and Peters, 1981; Goankar and Peters, 1986). There are two corrections that are needed for the BEM algorithm to account for skew.

The first correction is an empirical correction, which replaces the momentum theory expression for the thrust (Equation 3.98), with:

$$dT = 4F\,a\sqrt{(1-a)^2 + \tan^2\theta_{\text{skew}}}\,\rho U^2 \cos^2\theta_{\text{skew}}\,\pi r\,dr \tag{3.179}$$

It is readily seen that when $\theta_{\text{skew}} = 0$, Equation (3.98) is retrieved. Carrying the derivations of the BEM equations, it is seen that the factor k defined in Equation (3.100), is now multiplied by $\left[\sqrt{1 + ((\tan\theta_{\text{skew}})/(1-a))^2}\right]^{-1}$.

The second correction redistributes the axial induction factors to ensure that the induced velocities are stronger at the "downwind part" (see Figure 3.38), as follows:

$$a = a_0\left[1 + 2F_t(r)\tan\frac{\chi}{2}\cos(\psi - \psi_0)\right] \tag{3.180}$$

where a_0 is the uncorrected induction value as obtained from the BEM method (and using 3.179), ψ is the azimuthal position of the rotor, and F_t is referred to as the flow expansion function. The reference azimuthal position ψ_0 is such that $\psi - \psi_0 = 0$ at the "downwind" part of the rotor. The flow expansion function is $F_t(r) = \dfrac{r}{2R}$ in Glauert's model, which is a consistent approximation of Coleman's formulation. The model from Equation (3.180) is commonly applied in unsteady BEM codes. Extension of Coleman's model is found in Sant (2007) and Branlard *et al.* (2014).

3.7.2.5 Tower Influence (Tower Shadow and Tower Wake)

The term tower shadow refers to the change of wind speed around the tower caused by the tower obstruction. This influence is usually separated into a potential flow part, which affects the surrounding flow, and a wake part, which affects the flow downstream of the tower. Each blade will encounter the tower shadow effect once per revolution (corresponding to the frequency 1P). The rapid drops of power and vibrations in the nacelle and tower will then occur at the frequency BP, where B is the number of blades. The potential part of the tower shadow is modeled using the potential flow about a cylinder:

$$U_{dist,r} = U\left(1 - \frac{r_{twr}(z)^2}{r^2}\right)\cos\theta, \qquad U_{dist,\theta} = -U\left(1 + \frac{r_{twr}(z)^2}{r^2}\right)\sin\theta \tag{3.181}$$

where r_{twr} is the tower radius at the height z under consideration, r and θ are polar coordinates centered on the tower, and U_{dist} is the "disturbed" flow. The notations and streamlines about the tower are given on the left of Figure 3.39.

A wake model relevant for downwind wind rotors is, for instance, the model used in *AeroDyn* (Moriarty and Hansen, 2005):

$$U_{wake,x} = (1 - u_d)U, \quad \text{with} \quad u_d = \frac{C_d}{\sqrt{r}}\cos^2\left(\frac{\pi y}{2\sqrt{r}}\right) \tag{3.182}$$

where C_d is the drag coefficient of the tower (typically $C_d = 0.6$). In *AeroDyn*, both the potential flow and the wake models are combined in the variable U_{dist} (see Moriarty and Hansen, 2005).

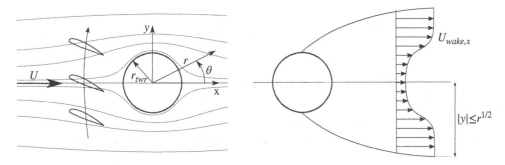

Figure 3.39 Left: Potential flow about the tower; U free stream velocity. Right: tower shadow. As the blade passes near the tower it will experience different angles of attack

3.7.2.6 Nacelle Influence
Simple models and analyses are found in the work of Anderson *et al.* (2020) and Wald (2006).

3.7.2.7 Unsteady Airfoil Aerodynamics
Unsteady airfoil aerodynamics refers to the study of airfoil loads under unsteady inflow. A particular effect is the dynamic stall, which may bring about or delay stall behavior. Rapid changes in wind speed (for example, when the blades pass through the tower shadow or turbulence) cause a sudden detachment and then reattachment of airflow along the airfoil. Such effects at the blade surface cannot be predicted with steady-state aerodynamics and will affect turbine operation. Unsteady airfoil aerodynamic effects occur on time scales of the order of the time for the relative wind at the blade to traverse the blade chord, approximately $c/\Omega r$. For large wind turbines, this might be on the order of 0.5 seconds at the blade root to 0.001 seconds at the blade tip. The most popular dynamic stall models amongst unsteady BEM codes include those of: Øye, for trailing edge stall (Øye, 1991); Leishman and Beddoes, for attached flows and trailing edge stall (Leishman and Beddoes, 1989) and its simplified state-space formulation from Hansen (2004); Theodorsen, for attached flows (Theodorsen 1935), or the Gormont model (1973). Øye's dynamic stall model is briefly presented below. The unsteady lift coefficient is computed as a linear combination of the inviscid lift coefficient, $C_{l,\text{inv}}$, and the fully separated lift coefficient $C_{l,\text{fs}}$. Both lift coefficients are determined from the steady lift coefficient, usually provided as tabulated data, noted $C_l^{st}(\alpha)$, where the superscript *st* stands for "steady." The unsteady lift coefficient is modeled as:

$$C_l(\alpha,t) = f_s(t)\,C_{l,\text{inv}}(\alpha) + (1-f_s(t))\,C_{l,\text{fs}}(\alpha) \tag{3.183}$$

where α is the instantaneous angle of attack and f_s is called the separation function which acts as a relaxation factor between the two flow situations. The inviscid lift coefficient is $C_{l,\text{inv}} = C_{l,\alpha}(\alpha - \alpha_0)$, where $C_{l,\alpha}$ is the slope of the steady lift curve about α_0, where α_0 the angle of attack for which $C_l^{st}(\alpha_0) = 0$. The fully separated lift coefficient may be modeled in different ways. In most engineering models, the slope of the fully separated lift coefficient around α_0 is $C_{l,\alpha}/2$. The approach presented below uses a steady separation function, f_s^{st} to define the fully separated lift coefficient. The steady separation function is taken as the separation point on a flat plate for a potential Kirchhoff flow (Hansen *et al.*, 2004):

$$f_s^{st}(\alpha) = \min\left\{ \left[2\sqrt{\frac{C_l^{st}(\alpha)}{C_{l,\alpha}(\alpha-\alpha_0)}} - 1 \right]^2, 1 \right\} \qquad \alpha \text{ close to } \alpha_0$$
$$f_s^{st}(\alpha) = 0 \qquad \alpha \text{ away from } \alpha_0 \tag{3.184}$$

When $\alpha = \alpha_0$, $f_s^{st}(\alpha_0) = 1$. Away from α_0, the function drops progressively to 0. As soon as the function reaches 0 on both sides of α_0, f_s^{st} is kept at the constant value of 0. The fully separated lift coefficient is then derived from the steady separation function as:

$$C_{l,fs}(\alpha) = \frac{C_l^{st}(\alpha) - C_{l,\alpha}(\alpha - \alpha_0)f_s^{st}(\alpha)}{1 - f_s^{st}(\alpha)} \text{ when } f_s^{st} \neq 1, \qquad C_{l,fs}(\alpha) = \frac{C_l^{st}(\alpha)}{2} \text{ when } f_s^{st} = 1 \tag{3.185}$$

The values of $C_{l,inv}$, $C_{l,fs}$, and f_s^{st} can be tabulated for various angles of attack and stored together with C_l^{st}. Øye's dynamic stall model uses a first-order differential system to establish the dynamics of the unsteady separation function, f_s:

$$\frac{df_s(t)}{dt} + \frac{1}{\tau}f_s(t) = \frac{1}{\tau}f_s^{st}(\alpha(t)) \tag{3.186}$$

where $\tau = \dfrac{k_\tau c}{U_{rel}}$ is the time constant of the flow separation, scaled by a value k_τ usually chosen as $k_\tau = 3$, and where U_{rel} is the relative velocity of the airfoil section. It is readily seen that f_s reaches the value f_s^{st} when the system is in a steady state (i.e., when $\dfrac{df_s(t)}{dt} = 0$). The state equation for the separation function may be integrated between two timesteps separated by a time Δt, as:

$$f_s(t + \Delta t) = f_s^{st}(\alpha(t)) + \left[f_s(t) - f_s^{st}(\alpha(t))\right]e^{-\frac{\Delta t}{\tau}} \tag{3.187}$$

The form above is used in several unsteady BEM codes. An example of application is provided in Figure 3.40, for the DU-93-W-210 airfoil (presented in Figure 3.12). Prescribed oscillations of angle of attack $\alpha(t) = 10 + 5 \sin \omega t$ (in deg) are used, where the reduced frequency is $k_\omega = \dfrac{\omega c}{2 U_0} = 0.6$, the chord is $c = 1$ m, and the inflow is $U_0 = 10$ m/s. The dynamic lift coefficient can be seen to oscillate around the steady lift coefficient, between the fully separated and inviscid lift coefficients.

3.7.2.8 Dynamic Inflow

Dynamic inflow, or dynamic wake, refers to the fact that the induced velocities at the rotor adapt progressively to a change of the rotor loading. The rotor loading varies continuously due to changes in the inflow experienced by the blade (turbulence, shear, tower shadow, etc.) or changes in control inputs (pitch angle, generator torque, and yaw

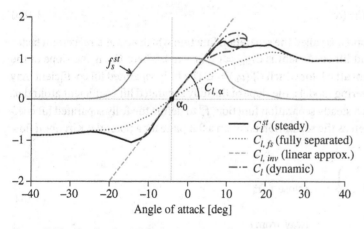

Figure 3.40 Application of Øye's dynamic stall model to the DU-93-W-210 airfoil

angle). Induced velocities at the rotor are a function of the flow field in the wake, and in particular the vorticity in the wake. Applying the steady-state BEM equations at two timesteps with different loadings would imply that the entire wake has adapted to a new equilibrium from one timestep to the next, which would be unrealistic. The wake holds a memory of the previous loading configurations, and it continuously adapts to a new loading. A change of loading implies a change of vorticity emitted into the wake. The new value of the vorticity propagates and diffuses progressively downstream. In turn, the velocity induced at the rotor by the wake vorticity also changes progressively. The time scale involved corresponds roughly to the convection velocity of the vorticity in the wake, in the order of $D/\overline{U}$, where $\overline{U}$ is the mean ambient flow (potentially accounting for axial induction). The time scale is between 5 and 20 seconds for a modern turbine. Due to the difference in axial induction, it is expected that the time constant toward the tip is shorter than toward the root (Snel and Schepers, 1995). Most engineering models are a function of the radial position and use exponential decay based on chosen time constants. Different dynamic inflow models are presented in Snel and Schepers (1991, 1993) and Pitt and Peters (1981). Recent models are found in Madsen *et al.* (2019) and Ferreira *et al.* (2022).

The model of Øye can be written into a first-order form as follows (Branlard *et al.*, 2022):

$$\frac{da_{\text{red}}}{dt} = \frac{-1}{\tau_1}a_{\text{red}} + \frac{1-k}{\tau_1}a_{\text{qs}} \tag{3.188}$$

$$\frac{da}{dt} = \frac{1}{\tau_2}a_{\text{red}} + \frac{-1}{\tau_2}a + \frac{k}{\tau_2}a_{\text{qs}} \tag{3.189}$$

where a is the unsteady axial induction factor and a_{qs} is the quasi-steady induction, which corresponds to the value obtained using a steady-state BEM algorithm. The variable a_{red} is the reduced axial induction factor, which is used as an intermediate value. The time constants are modeled as:

$$\tau_1 = \frac{1.1}{1-1.3\min(\bar{a},0.5)}\frac{R}{\overline{U}}, \qquad \tau_2 = \left[0.39 - 0.26\left(\frac{r}{R}\right)^2\right]\tau_1 \tag{3.190}$$

where $\bar{a}$ is the mean axial induction over the rotor and $\overline{U}$ is the mean free stream velocity over the rotor[1]. The time constants were tuned using a vortex ring model of the wake. It is readily seen that the steady solution of the system leads to $a = a_{\text{qs}}$. Equations (3.188) and (3.189) can be integrated using a first-order solver to obtain a based on the quasi-steady values a_{qs}. The same model is generally used for a'. The model effectively filters the high-frequency fluctuations of the quasi-steady inductions and accounts for the time history of the inductions.

An example illustrating the dynamic inflow model is shown in Figure 3.41. The NREL 5-MW is operated at a constant rotor speed of 9.161 rpm and under a uniform wind speed of 8 m/s, while undergoing a pitch step from 0 to 4 deg and back. When the dynamic inflow is active ($\tau_1 > 0$), the aerodynamic torque and induction progressively respond to the change in pitch.

3.7.2.9 Modified Steady-State BEM Equations

The steady-state BEM equations (Equations 3.100 and 3.101) need to be modified to account for general situations. As mentioned in 3.5.1.5, k and k' need to be multiplied by $1/\cos\kappa$, where κ is the local slope of the blade (containing prebend and coning). As mentioned in 3.7.2.4, the momentum balance needs to be corrected to account for skew situations. The BEM equations are modified as:

$$\frac{a}{1-a} = k, \quad \text{with} \quad k = \frac{\sigma c_n}{4F\sin^2\phi}\frac{1}{\cos\kappa}\left[\sqrt{1 + [(\tan\theta_{\text{skew}})/(1-a)]^2}\right]^{-1} \tag{3.191}$$

---

1 In some implementations, $\overline{U}$ is estimated using values at 70% radius. In others, the mean is computed from the values at each blade node.

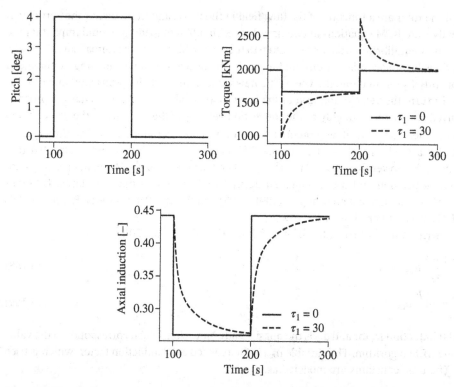

Figure 3.41 Application of the dynamic inflow model for a pitch-step simulation of the NREL 5-MW turbine.

$$\frac{a'}{1+a'} = k', \quad \text{with} \quad k' = \frac{\sigma c_t}{4F \sin \phi \cos \phi} \frac{1}{\cos \kappa} \tag{3.192}$$

where θ_{skew} is the angle between the mean flow and the normal to the rotor (either from tilt or yaw).

3.7.2.10 Overall Structure of an Unsteady BEM Algorithm

An unsteady BEM code follows the general structure of the lifting-line algorithm presented in Section 3.7.1. In the general case (prebend, coning, deflections, etc.), the blades (lifting lines) are "out-of-plane": they do not reside in the actuator disc plane that was used in the derivations of the BEM algorithm. The algorithm therefore needs to be generalized. The momentum balance in the BEM algorithm (the part that defines the axial and tangential induction factors) is performed in a "projected" actuator disc plane, normal to the rotor shaft, as if the blades were projected onto this plane (see Figure 3.37 for notations). The unit vectors of this coordinate system are written as: $\hat{r}$ (radial), $\hat{n}$ (normal to rotor disk), and $\hat{t}$ (tangential to rotor disk, opposite the direction of the rotation). For a straight blade (no prebend, no coning, and no presweep), the coordinate system is illustrated in Figure 3.24. The blade element theory part of the BEM algorithm (loads and angle of attack) is performed in the airfoil coordinate system $(\hat{x}_a - \hat{y}_a)$, where the chord line is defined according to the blade geometry (coning, prebend, and twist) and instantaneous deformations (flapwise, edgewise, and torsion). A general unsteady BEM algorithm therefore requires the definition of transformation matrices between these different coordinate systems. These considerations, and the account for unsteady effects, mainly affect step 1 of Section 3.7.1. It can be decomposed as follows:

1) Retrieve the elastic velocity of the blade nodes, V_{elast}. For a straight, stiff blade, $V_{elast} = -\Omega r \hat{t}$. In the general case, elastic motions from the entire structure need to be considered.

2) Retrieve the undisturbed wind speed at the blade nodes, U_{atm}. The undisturbed wind speed includes all atmospheric effects: turbulence (including deterministic gusts), shear, and veer. For a steady, uniform wind, without tilt or yaw: $U_{atm} = U\hat{n}$.

3) Apply flow disturbances to U_{atm} due to the influence of the tower (Equations 3.181 and 3.182) and nacelle. The disturbed flow velocity is further noted U_{dist}.

4) Compute the quasi-steady axial and tangential induction factors a_{qs}, and a'_{qs} at the blade nodes, using a steady BEM algorithm, generalized as follows:

Compute the vector of quasi-steady induced velocities, $U_{ind,qs}$, based on the axial and tangential inductions from the previous iteration, and the velocities U_{dist} and V_{elast} projected into the $\hat{n} - \hat{t}$ system:

$$U_{ind,qs} = a_{qs}\left(U_{dist,n} - V_{elast,n}\right)\hat{n} + a'_{qs}(U_{dist,t} - V_{elast,t})\hat{t} \tag{3.193}$$

a) Compute the angle of attack at each blade node using the relative wind, $U_{rel} = U_{dist} - U_{elast} + U_{ind,qs}$, projected into the airfoil coordinate system (Equation (3.177) and Figure 3.37).

b) Compute the aerodynamic coefficients in the airfoil system ($C_l(\alpha)$ and $C_d(\alpha)$) and project them into the $\hat{n} - \hat{t}$ system, to obtain c_n and c_t. Note that Equations (3.161) and (3.162) are only valid when the blades are straight. A general projection is required otherwise.

c) Apply the BEM equations to obtain the quasi-steady induction factors in the $\hat{n} - \hat{t}$ system using the modified steady BEM Equations (3.191) and (3.192). Repeat substeps a–d as part of the BEM algorithm iterations.

5) Compute the dynamic axial and induction factors, a and a', using the dynamic inflow model (Equations 3.188 and 3.189).

6) Correct the axial induction with the skew correction model (Equation 3.180).

7) Compute the vector of dynamic-induced velocities U_{ind} based on the dynamic axial and tangential induction factors (Equation 3.193, but using a and a' instead of the quasi-steady values).

8) Compute the relative velocity $U_{rel} = U_{dist} - U_{elast} + U_{ind}$, which is further used in the lifting-line algorithm given in Section 3.7.1.

A dynamic stall model, such as the ones presented in this section, is used in step 5 of the lifting-line algorithm. Several variations of the steps mentioned above may be found. The BEM implementation of the code *HAWC2* performs these steps on a planar grid instead of at the blade sections. The values on the grid are then interpolated back to the blade locations (see Madsen *et al.*, 2019). Additional details for the implementation of an unsteady BEM algorithm may be found in the book of Hansen (2008) or Branlard (2017).

3.7.3 Vortex Methods and Vortex Theory

3.7.3.1 Introduction to Vortex Methods

Vortex methods refer to a specific branch of CFDs which focuses on the tracking of vorticity (an introduction to vorticity was presented in Section 3.2.4). High-accuracy results are reachable with vortex methods, sometimes outperforming traditional CFD, and the method can be successfully applied to viscous and compressible flows. Such performances are obtained by means of the "vortex particle" methods (Winckelmans and Leonard, 1993; Papadakis, 2014). Yet, a wide range of formulations are possible, leading to compromises on accuracy. The most common vortex methods applied to wind energy are "vortex filament" methods, which are of lower fidelity than traditional CFD. Vortex theory, or vortex models, also lie on the low end of the accuracy spectrum. In such models, the vorticity distribution is usually assumed, and the corresponding velocity field is then used to derive an engineering model (see Figure 3.8). A large part of the theoretical results and engineering models used in wind energy, as developed by Betz, Prandtl, Joukowski, or Glauert, were obtained using vortex theory analyses.

In vortex methods, the continuous vorticity field is represented by a finite number of vortex elements. These elements can be vortex particles or vortex filaments, and their points convect with the local flow field. The representation (projection) of a continuous vorticity field into a finite number of elements is the main source of the inaccuracy of the method.

Viscosity diffuses the free vorticity downstream and introduces vorticity at the solid boundaries. The two effects are treated separately or may be totally or partially ignored in inviscid methods. Most viscous simulations with vortex methods rely on the assumption that the convection and diffusion steps can be done separately. This is referred to as viscous splitting. During the convection step, the vortex elements are propagated (and stretched) with the fluid velocity. The positions and intensities of the vortex elements can be projected to obtain a continuous vorticity field. The vorticity equation is solved and used to update the intensities of the vortex elements. Finally, the velocity field is obtained by inversion of the definition of vorticity $\omega = \text{curl } \boldsymbol{u}$. This is achieved either using the Biot–Savart law (in grid-free methods) or using a Poisson solver (in grid-based methods). The key aspects of vortex methods will be illustrated in the subsequent section with the presentation of the vortex lattice method. Additional details may be found in the books of Branlard (2017) and Cottet and Koumoutsakos (2000).

3.7.3.2 Velocity Fields from a Vorticity Distribution – Biot–Savart Law and Poisson Equation

The determination of velocities is needed both in the implementation of vortex methods and also for use in simple engineering models to gain quick insight into a given problem. In some situations, one can assume a vorticity distribution based on the engineering knowledge of the problem, which is usually easier than to assume a velocity and pressure distribution. Examples will be given at the end of this paragraph. Vortex methods use vorticity as a main variable, but the velocity field is needed to convect the vortex elements. The velocity reconstruction from a vorticity distribution is straightforward for incompressible flows in the absence of solid boundaries. In this case, the velocity field is determined by the vorticity distribution and a given constant velocity, $\boldsymbol{u} = \boldsymbol{u}_\omega + \boldsymbol{U}_0$ ("Helmholtz decomposition" of the velocity field). Under these assumptions, the rotational part of the velocity field may be written as the curl of a vector potential, $\boldsymbol{\psi}$, which is divergence free: $\boldsymbol{u}_\omega = \text{curl } \boldsymbol{\psi}$. Introducing the vector potential into the definition of the vorticity, $\omega = \text{curl } \boldsymbol{u}$, leads to the Poisson equation $\omega = -\Delta \boldsymbol{\psi}$, where the notation Δ refers to the Laplacian operator. The resolution of the Poisson equation is a common problem in physics. The analytical solution is referred to as the Biot–Savart law. In practice, the analytical solution requires an integration which is done numerically and may be computationally expensive. It is also possible to use numerical methods, such as finite differences, to solve the Poisson equation on a numerical grid. When the numerical approach is used, the vector potential is obtained on the solution grid, and finite differences are then employed to determine the velocity (i.e., by calculating $\boldsymbol{u}_\omega = \text{curl } \boldsymbol{\psi}$). The Biot–Savart law, on the other hand, can be applied to determine the vector potential $\boldsymbol{\psi}$ and the velocity field $\boldsymbol{u}_\omega$ directly. The remainder of this paragraph focuses on the Biot–Savart law. Many resources are available in the literature regarding the numerical resolution of the Poisson equation. It is highly instructive to implement a Poisson solver using finite differences. Optimized libraries are found for most programming languages. One thing to note is that the numerical Poisson solvers will always "smoothen" the velocity field and will require special care at the boundary of the domain.

The Biot–Savart law is written as an integral over the entire volume of vorticity, V:

$$\boldsymbol{\psi}(\boldsymbol{x}) = \frac{1}{4\pi} \iiint_V \frac{\omega(\boldsymbol{x}')}{\| \boldsymbol{x} - \boldsymbol{x}' \|} \, \mathrm{d}v(\boldsymbol{x}'), \qquad \boldsymbol{u}_\omega(\boldsymbol{x}) = \frac{-1}{4\pi} \iiint_V \frac{(\boldsymbol{x} - \boldsymbol{x}')}{\| \boldsymbol{x} - \boldsymbol{x}' \|^3} \times \omega(\boldsymbol{x}') \, \mathrm{d}v(\boldsymbol{x}') \qquad (3.194)$$

The operation is time-consuming since an integration over the entire domain needs to be performed for each point of interest $\boldsymbol{x}$. The continuous vorticity may be projected onto numerical elements, such as vortex particles and vortex segments. In such cases, the integration is replaced by a summation of the velocity fields induced by each vortex element. A vortex particle has a strength representing the amount of vorticity in a given volume, $\boldsymbol{\alpha} = \omega \mathrm{d}V$. Figure 3.42 illustrates how the vorticity within a volume may be represented by a vortex sheet, segment, or particle.

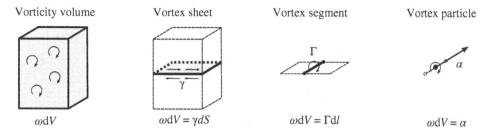

Figure 3.42 Projection of a volume of vorticity into a vortex sheet, of vortex segment and a vortex particle. Each reduction of dimension introduces higher-order singularities in the velocity field

The velocity field induced by a vortex particle located at x_0 is:

$$u(x) = \frac{-1}{4\pi} \frac{(x - x_0)}{\| x - x_0 \|^3} \times \alpha \tag{3.195}$$

A vortex segment is represented by a circulation strength such that: $\Gamma d l = \omega dV$, where $d l$ is the vector between the two segment extremities, further noted x_1 and x_2. The induced velocity by such vortex segment is:

$$u(x) = \frac{\Gamma}{4\pi} \frac{(r_1 + r_2)}{r_1 r_2 (r_1 r_2 + r_1 \cdot r_2)} r_1 \times r_2 \tag{3.196}$$

with $r_1 = x - x_1$ and $r_2 = x - x_2$ and the double bars indicate the norm.

Numerical functions that return the velocity field induced by one vortex segment or one vortex particle are easily implemented. From this, it is possible to determine the velocity field induced by a more complex distribution of vorticity by simply decomposing the distribution of vorticity into particles or segments and superposing the velocity fields from each individual segment. For instance, the velocity field induced by a vortex ring can be obtained by discretizing the ring into straight segments. The velocity field may be compared with the analytical solution of a vortex ring for validation (see e.g., Branlard, 2017). A cylindrical vorticity surface can be discretized into a set of vortex rings. For a non-expanding cylindrical surface, the results may be compared to analytical solutions as well. As seen in Section 3.2.4, such models form a good representation of the wake behind an actuator disc, and results consistent with simple momentum theory are expected. Yet, in contrast to momentum theory, it is possible to obtain the velocity field in the entire domain. More advanced representations can be used by considering a finite number of blades and representing the wake by helical tip-vortices behind each blade. The helical vortices may be discretized with vortex segments, and the velocity field is determined numerically. Valuable information is obtained using these simple analyses that assume a distribution of vorticity, and the reader is encouraged to experiment with these simple tools. The wake is said to be "rigid" since the vorticity does not dynamically convect with the local velocity field. More advanced methods lift this assumption, in which case the vorticity in the wake is referred to as "free" vorticity. Vortex lattice methods are examples of vortex methods with a free vortex wake.

3.7.3.3 Vortex Lattice Method

The vortex lattice method is probably the most common vortex method in use in the field of wind energy. This method, which relies on straight vortex filaments, can be seen as an intermediate in terms of accuracy, between BEM and traditional CFD. Figure 3.43 illustrates the vortex lattice method and its assumptions of concentration of vorticity compared to the continuous representation.

The vorticity from the boundary layer around the blade is condensed into a single bound vorticity line, usually located along the line formed by the aerodynamic center of the airfoils (see Section 3.3). The wake vorticity is assumed to be confined into a thin layer at high Reynolds number and is thus can be represented by a vortex sheet,

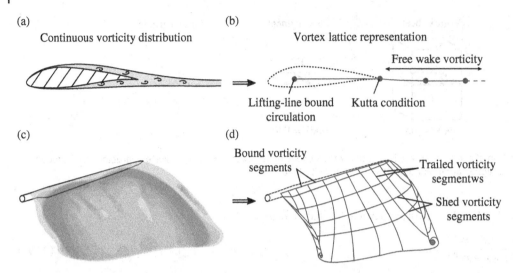

Figure 3.43 Vortex lattice representation of a wind turbine blade and its wake. Top: spanwise cross-section. Bottom: 3D perspective. (a) and (c) are the continuous vorticity representations, (b) and (d) are the vortex lattice representations

which is further discretized using vortex segments. The full vorticity field is thus approximated by a set of vortex segments. The vortex segments located along the trailing edge of the blade have zero intensity, a condition referred to as the Kutta condition. The circulation of each vortex segment along the lifting line is determined using the Kutta–Joukowski theorem, which provides circulation as a function of the velocity on the lifting line and the lift coefficient: $\Gamma(r) = \frac{1}{2}U_{rel,\perp}^2 cC_l(\alpha)$. Once the bound circulation is determined, it is emitted into the near wake and becomes "free" vorticity. The trailed vorticity has the intensity: $\Gamma_t = -\frac{d\Gamma}{dr}dr$, or, in discrete form, $\Gamma_t[i] = \Gamma[i] - \Gamma[i+1]$, where i is the index of along the blade nodes. The free vortex segments in the wake are convected based on the local velocity at their location. This velocity is computed by summing the velocity fields from each vortex segment. At any given time step, once the bound circulation has been computed, the position and circulation of all vortex segments are known. The velocity field on the lifting line may then be computed and used within the lifting-line algorithm presented in Section 3.7.1.3. The method is well documented in the report from van Garrel (van Garrel, 2003). More extensive resources are given in Katz and Plotkin (2001).

3.7.3.4 Advanced Vortex Methods
High-accuracy results may be obtained using vortex particle methods. These methods are better suited to account for viscous diffusion and viscous reconnection of vorticity lines. The implementation of the method is more involved since it often requires introducing a mesh, with projection and interpolation from the particle to the mesh and vice-versa. Cottet and Koumoutsakos (2000) describe the mathematical development of the method. The book chapter from Winckelmans (2004) offers a brief but exhaustive introduction to the method.

3.7.3.5 Advantages and Limitations of Vortex Methods
The main advantage of vortex methods and vortex theory lies in the broad-spectrum approaches, leading to a wide range of formulations and levels of complexity. Vortex methods can obtain levels of accuracy anywhere between BEM and state-of-the-art CFD: the vortex cylinder theory has an accuracy slightly higher than BEM; high-order vortex particle methods compete with actuator line CFD in terms of accuracy and computational speed. Vortex lattice methods offer a good compromise between BEM and CFD and can provide solutions where the range of

validity of BEM starts to be challenged, such as large-yaw angles, large blade deflections, and situations where the blade may interact with the wake.

Another key motivation to use vorticity-based methods is that vorticity is an intuitive variable, which offers the possibility to pinpoint driving mechanisms and separate effects within different physical phenomena. Indeed, the Biot–Savart law introduces a mathematical and causal link between vorticity and velocity. A given vorticity patch induces a velocity field in the entire domain and may be identified as the source explaining a given phenomenon. For example, one can identify the effect of the shed and trailed vorticity behind a blade separately, or the effect of tangential and longitudinal vorticity of a wind turbine wake. This approach yields a better understanding of both steady and unsteady rotor aerodynamics. With this better understanding of the physics at play, vortex-based methods can open possibilities for improving existing models and developing new models for BEM-based codes.

A drawback of vortex methods is that the mid-fidelity models, such as vortex lattice models, may be computationally expensive, unless dedicated algorithms are implemented, such as tree-codes and fast-multipole methods. Additionally, viscous diffusion is not easily accounted for in these formulations, and ad-hoc models are required. The drawback of the high-fidelity vortex particles method lies in the apparent complexity of the method. Practical textbooks and open-source libraries are still lacking to ease the implementation of such methods for newcomers. A comprehensive list of pros and cons related to vortex methods is found in Branlard (2017).

3.7.4 Computational Fluid Dynamics (CFD)

3.7.4.1 Introduction

CFDs encompasses various numerical methods to solve the fluid dynamics equation. The methods may be classified according to the flow equations, the underlying numerical methods, and the treatment of the lifting bodies. The flow solvers used for wind energy applications typically solve for the incompressible Navier–Stokes equations. The Mach number involved in wind energy is usually sufficiently low (i.e., below 0.3) to justify the incompressible assumption. Vortex methods are actually CFD methods that solve for the vorticity form of the Navier–Stokes equation. The term "vortex methods" is nevertheless preferred to avoid confusion with traditional CFD which uses a velocity–pressure formulation. The most common numerical methods are finite differences, finite elements, and finite volumes. Amongst these methods, the finite volume method is the most popular. Wind turbines may be modeled in three different ways: by an actuator-disc, by actuator-lines (e.g., using the lifting-line algorithm presented in Section 3.7.1), or by resolving the full blade geometry. Presenting the fundamentals of CFD is beyond the scope of this text; the reader is referred to dedicated references for further details, e.g., Versteeg and Malalasekera (1995). Examples of dedicated flow solvers for wind turbines are the codes Ellipsys (Sørensen, 1995) and ExaWind (Sprague *et al.*, 2020).

3.7.4.2 Challenges of CFD

One of the main challenges of CFD applied to wind turbines is the different scales involved. The airfoils of the blade operate at high Reynolds numbers, which imply that a very fine resolution, of the order of millimeters, is required near the surface to capture the boundary layer. Different models exist to moderate this requirement, but the challenge still persists. On the opposite end of the spectrum, the simulation needs to capture large atmospheric fluctuations, which have scales of the order of kilometers, hence requiring a large computational domain. Turbulence introduces an additional challenge since it involves a continuous cascade of scales, and a turbulence model has to be introduced to account for scales smaller than the grid size. Finally, numerical diffusion is inevitable due to round-off errors and the evaluation of gradients on a computational grid. The introduction of wall models, turbulence models, mesh-blending models, and other models dedicated to the reduction of numerical errors is the main source of inaccuracy of the method and a source of disparity between the results obtained with different implementations. Grid convergence studies should always accompany a CFD simulation to ensure the convergence of the method.

3.7.4.3 Actuator-Disc and Actuator-Line Methods

Actuator-disc (AD) and actuator-line (AL) methods are fairly straightforward to implement in a generic flow solver. Both methods fall under the category of "lifting-line" methods, and the algorithm presented in Section 3.7.1 applies. The interaction between the flow solver and the lifting-line algorithm is done in steps 1 and 8. During step 1, the flow solver provides the velocity at different probe locations, which typically correspond to the locations of the aerodynamic centers of each blade and each blade section. Different implementations may slightly adjust this location to account for the smearing that will be discussed shortly. Steps 1–8 proceed, leading to the determination of the lift and drag forces. These forces are introduced back into the flow solver as body forces. In the case of actuator-disc simulations, the forces are averaged over the surface of the disc. A smearing of the forces is usually applied, so that the forces spread over several grid cells. Axisymmetric Gaussian smearing is typically used, but ideally, the smearing should be inspired by the shape of the airfoil and the boundary layer height at the trailing edge. The smearing parameter, often noted ϵ, is crucial for the method. Different recommendations are found in the literature for the choice of this parameter, e.g., Martinez-Tossas *et al.* (2017) and Meyer Forsting *et al.* (2019). One advantage of the actuator-line method is that it uses the same lifting-line algorithm as unsteady BEM codes. The same underlying profile data can then be used for both tools and results are easily comparable. This makes this tool a good candidate to verify or tune unsteady BEM tools, which are still the standard tool used in the design process. The main inconvenience of the method is that it relies on blade element theory assumptions which assume that cross-sectional performances may be determined based on an angle of attack. The angle of attack is yet ill-defined in three dimensions, leading to uncertainties of the results.

3.7.4.4 Blade-Resolved CFD

Blade-resolved CFD represents the state of the art of current wind turbine simulations. Only a handful of such simulations are performed by manufacturers, though it is progressively taking greater importance in the design process, due to increases in computational capacity. The advantage of the method is that it does not rely on pre-tabulated data and assumptions about the definition of the angle of attack, as opposed to lifting-line-based methods. Blade-resolved CFD requires a complex meshing around the geometry. The overset method is commonly used to blend the complex mesh surrounding a body with the simpler background mesh. Additional models are still required to capture the different scales of turbulence, introducing potential sources of inaccuracies. As of 2022, fluid-structure simulation, using blade-resolved CFD for the aerodynamics and beam-based multibody solvers for the elasticity, required about two hours of simulation for one rotor revolution on 2000 CPUs. Direct numerical simulations (DNS) of wind turbines, which solve for the whole range of spatial and temporal scales of the flow, are still years away.

3.7.5 Validation and Modelling Challenges

Modeling challenges related to BEM, vortex methods, and traditional CFD were touched upon in the previous section. Validation against measurements is key to evaluate the different methods and assess potential model improvements.

Numerous efforts have been made to validate rotor performance models with in-field measurements. This is difficult because most test programs have instrumented wind turbines in typical operating environments with shear and turbulence. Sensors to measure the incoming wind field must be upwind of the rotor to ensure that the data is undisturbed by the wind turbine itself. By the time the turbulent flow field reaches the wind turbine, the instantaneous wind speeds across the rotor have changed from the values measured upwind. Thus, it is difficult to compare specific wind conditions with measured loads and with model outputs. In most cases, inflow conditions, measurements, and results can only be compared statistically. Nevertheless, field tests have provided important insights, and measurements have highlighted aerodynamic issues to be considered in modeling codes. The advancement of remote sensing technologies offers the opportunity to measure the inflow closer to the rotor and in several locations quasi-simultaneously. Optical technologies that perform 3D scanning of the blade geometry are also being used more and more to account for any manufacturing defects. The use of these technologies will reduce the uncertainty gap between the results from field experiments and numerical simulations. Examples of full-scale validation and modeling campaigns for wind energy are the WakeBench experiment (Doubrawa *et al.* 2020); the DanAero experiment investigated under the IEA Wind Task 29 (Boorsma *et al.*, 2023).

Wind tunnels provide a more controlled environment for testing, but large wind tunnels and large wind turbines are needed to achieve the Reynolds numbers at the operating conditions of wind turbines. One example of a test of a wind turbine with a 10 m diameter rotor which was performed at the NASA Ames wind tunnel in California (Simms *et al.*, 2001).

3.8 Aerodynamics of Vertical Axis Wind Turbines

3.8.1 General Overview

As previously noted in Chapter 1, wind turbine rotors may rotate about either a horizontal or a vertical axis. Although most wind turbines have historically been of the horizontal axis type, vertical axis machines have been used in some situations of interest. This section gives a brief overview of vertical axis turbines and summarizes key aspects of their aerodynamics.

Vertical axis wind turbines (VAWTs) may have either drag-driven or lift-driven rotors. The most common drag-driven vertical axis turbine is the Savonius rotor. It has been used for water pumping and other high-torque applications. The argument in favor of Savonius rotor turbines has been that they can be relatively inexpensive to build. In practice, by virtue of being drag-driven machines, they have intrinsically low power coefficients. In addition, they have a solidity approaching 1.0, so they are very heavy relative to the power that they produce. They are also difficult to protect from damage in high winds. See also Section 3.3.6.1.

When vertical-axis turbines have been used for electrical power generation, they have nearly always used lift-driven rotors. Typically, these rotors have had one of two types of configuration: (1) straight blades or (2) curved blades with a troposkein shape. The former type of rotor is often referred to as an H-rotor and the latter type known as the Darrieus rotor. Some rotors with straight blades have incorporated a pitching mechanism, but most lift-driven vertical-axis turbines have fixed pitch blades. Thus, power limitation at high winds is accomplished by stalling when a grid-connected generator keeps the rotational speed constant – see Chapter 9.

The principal advantage of vertical axis rotors is that they do not require any special mechanisms for yawing into wind. Another advantage is that, since the blades are generally untwisted and of constant chord, they can be made by mass production extrusion or pultrusion.

In practice, vertical-axis turbines have not been used nearly as widely as horizontal-axis turbines. The reasons have to do with the balance between some of the intrinsic benefits and limitations. By the nature of the aerodynamics of the rotor, the structural loads on each blade vary greatly during each rotation. Such loads contribute to high fatigue damage (see Chapter 7) and require that the blades and joints themselves have a very long cycle life. In addition, the vertical axis turbines do not lend themselves to being supported by a separate, tall tower. This means that a large fraction of the rotor tends to be located close to the ground in a region of relatively low wind. Productivity may then be less than that of a horizontal axis machine of equivalent rated power but on a taller tower.

3.8.2 Aerodynamics of a Straight-blade Vertical Axis Wind Turbine

The following analysis applies to a straight-blade vertical-axis wind turbine. The first section concerns a single stream tube method of analysis and follows the approach of de Vries (1979). The second section summarizes the multiple stream tube method. This is followed by a brief discussion of the double multiple stream tube method

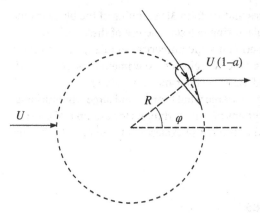

Figure 3.44 Geometry of a vertical axis turbine

(which is more widely used in practice but is more complicated). The Darrieus rotor can be modeled with a modification of the straight-blade methods. In that case, the blades are divided into sections, and the effects of the different distances of the sections from axis of rotation are accounted for; see Section 3.8.3.

3.8.2.1 Single Stream Tube Analysis

A single blade of a vertical-axis wind turbine, viewed from above, is illustrated in Figure 3.44. In the figure, the blade is shown rotating in the counter-clockwise direction, and the wind is seen impinging on the rotor from left to right. As is typical in vertical-axis wind turbines, the airfoil is symmetric. The blade is oriented so that the chord line is perpendicular to the radius of the circle of rotation. The orientation shown here implies that the blade has a zero-degree preset pitch angle. Rotating the blade about the quarter chord, however, such that the chord line is no longer perpendicular to the radius of the circle of rotation would result in a positive or negative preset pitch angle. The degree of preset pitch would have a significant effect on turbine performance, but will not be discussed here. The radius defining the angular position of the blade (normally meeting the chord line at the quarter chord) makes an angle of ϕ with the inflow (wind) direction, as shown in the figure.

Figure 3.45 illustrates the components of the inflow acting on the blade. As can be seen, a component due to rotation is tangential to the circle of rotation, and thus parallel to the chord line of the airfoil. One component of the inflow also acts tangentially. Another wind component is normal to the circle, and so perpendicular to the airfoil. An induction factor, a, accounts for the deceleration in the inflow as it passes through the rotor.

The velocity relative to the blade element is:

$$U_{rel}^2 = \{\Omega R + (1-a)U\sin(\phi)\}^2 + \{(1-a)U\cos(\phi)\}^2 \tag{3.197}$$

This can be rewritten as:

$$\frac{U_{rel}}{U} = \sqrt{\{\lambda + (1-a)\sin(\phi)\}^2 + \{(1-a)\cos(\phi)\}^2} \tag{3.198}$$

where $\lambda = \dfrac{\Omega R}{U}$ is the tip speed ratio.

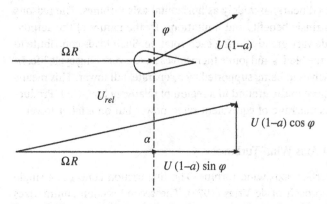

Figure 3.45 Components of wind acting on a vertical axis turbine blade

Note that at high tip-speed ratios, the second term under the square root becomes small, so that:

$$\frac{U_{rel}}{U} \approx \lambda + (1-a) \ \sin(\phi) \tag{3.199}$$

Since the chord is perpendicular to the radius of the circle, the angle of attack is:

$$\alpha = \tan^{-1}\left[\frac{(1-a)\cos(\phi)}{\lambda + (1-a)\sin(\phi)}\right] \tag{3.200}$$

At higher tip speed ratios, the second term in the denominator is small relative to the tip speed ratio and the tangent of α is approximately equal to α itself. Therefore:

$$\alpha \approx \frac{(1-a)\ \cos(\phi)}{\lambda} \tag{3.201}$$

By analogy with momentum theory for horizontal axis rotors, the forces on the blade can be related to the change in momentum in the air stream. It is assumed, as before, that in the far wake, the velocity is reduced to $U(1-2a)$, so that at the rotor it is $U(1-a)$, as indicated in the figures.

The change in velocity is:

$$\Delta U = U - U(1-2a) = 2aU \tag{3.202}$$

The force per unit height, $\tilde{F}_D$, in the wind direction is found from:

$$\tilde{F}_D = \tilde{m}\Delta U \tag{3.203}$$

where $\tilde{m}$ is the mass flow rate per unit height. This is given by:

$$\tilde{m} = \rho 2RU(1-a) \tag{3.204}$$

where R is the rotor radius, and ρ is the density of the air. The force is thus:

$$\tilde{F}_D = 4R\rho a(1-a)U^2 \tag{3.205}$$

That force must be equal to the average force on all the blades during a complete revolution. This is found using blade element theory by integrating around the circle, taking into account the contribution from each of the B blades:

$$\tilde{F}_D = \frac{B}{2\pi} \int_0^{2\pi} \frac{1}{2} \ \rho \ U_{rel}^2 \ c \ C_l \ \cos(\alpha + \phi) \ d\phi \tag{3.206}$$

where B = number of blades and C_l = lift coefficient.

By equating Equations (3.205) and (3.206), the following relation may be obtained:

$$a(1-a) = \frac{1}{8}\frac{B}{R}\frac{c}{2\pi}\frac{1}{2\pi} \int_0^{2\pi} \left[\frac{U_{rel}}{U}\right]^2 C_l \ \cos(\alpha + \phi) \ d\phi \tag{3.207}$$

This equation cannot be solved directly, so an iteration technique must be used. To do this, let $y = \lambda/(1-a)$. Dividing both sides of Equation (3.207) by $(1-a)^2$, using Equation (3.200) for the angle of attack, adding 1 to both sides and making the appropriate substitutions, one obtains the following:

$$\frac{1}{1-a} = 1 + \frac{1}{8}\frac{B}{R}\frac{c}{2\pi}\frac{1}{2\pi} \int_0^{2\pi} \{(y + \sin(\phi))^2 + \cos^2(\phi)\} \ C_l \cos\left\{\phi + \tan^{-1}\left(\frac{\cos(\phi)}{y + \sin(\phi)}\right)\right\} \ d\phi \tag{3.208}$$

For a given geometry, Bc/R is fixed. A value for y can be assumed and $(1-a)$ can be calculated. From that the corresponding tip speed ratio can be found from:

$$\lambda = y(1-a) \tag{3.209}$$

The calculation is repeated until the desired tip speed ratio is found.

The power produced by the rotor is found, as usual, from the product of the average torque and the rotational speed. The torque varies with angular position, so the expression for power is:

$$P = \Omega \frac{1}{2\pi} \int_{0}^{2\pi} Q \mathrm{d}\phi \tag{3.210}$$

The torque is the product of the radius and the tangential force. The tangential force per unit length on each blade, $\tilde{F}_T$, (which varies with ϕ) is:

$$\tilde{F}_T = \frac{1}{2}\rho U_{rel}^2 c(C_l \sin(\alpha) - C_d \cos(\alpha)) \tag{3.211}$$

The total torque, assuming B blades and a rotor of height H, is:

$$Q = BRH\tilde{F}_T \tag{3.212}$$

The average rotor power over one revolution is then:

$$P = \Omega RH \frac{Bc}{2\pi} \frac{1}{2}\rho \int_{0}^{2\pi} U_{rel}^2 (C_l \sin(\alpha) - C_d \cos(\alpha)) \ \mathrm{d}\phi \tag{3.213}$$

The power coefficient is given by the power divided by the power in the wind passing through an area defined by the projected area of the rotor, $2RH$:

$$C_P = \frac{P}{\frac{1}{2}\rho 2RHU^3} \tag{3.214}$$

Therefore:

$$C_P = \frac{\lambda}{4\pi} \frac{Bc}{R} \int_{0}^{2\pi} \left[\frac{U_{rel}}{U}\right]^2 C_l \sin(\alpha)\left[1 - \frac{C_d}{C_l \tan(\alpha)}\right] \ \mathrm{d}\phi \tag{3.215}$$

This formula can be solved numerically, but it is also of interest to consider some idealized conditions. In particular, for high tip speed ratios ($\lambda \gg 1$), the angle of attack will be relatively small. For small angles of attack (below stall), the lift coefficient is linearly related to the angle of attack and since a symmetrical airfoil is assumed, the lift coefficient may be written as:

$$C_l = C_{l,\alpha}\alpha \tag{3.216}$$

where $C_{l,\alpha}$ = slope of the lift curve and α is in radians

Thus, the angle of attack as given in Equation (3.201) is the appropriate form and, in addition, $\cos(\phi + \alpha) \approx \cos(\phi)$. Equation (3.208) can be approximated as:

$$\frac{1}{1-a} = 1 + \frac{1}{8}\frac{Bc}{R}\frac{1}{2\pi}\int_{0}^{2\pi}\{(y + \sin(\phi))^2 + \cos^2(\phi)\}C_{l,\alpha}\alpha \cos(\phi) \ \mathrm{d}\phi \tag{3.217}$$

The term in brackets can be expanded and, using Equations (3.201), (3.217) can be written as:

$$\frac{1}{1-a} = 1 + \frac{1}{16}\frac{Bc}{R}\frac{1}{\pi}\int_0^{2\pi}\left\{y^2 + 2y\sin(\phi) + \sin^2(\phi) + \cos^2(\phi)\right\} C_{l,\alpha}\frac{1}{y}\cos^2(\phi)\ d\phi \tag{3.218}$$

Taking advantage of the identity $\sin^2(\phi) + \cos^2(\phi) = 1$, Equation (3.218) yields:

$$\frac{1}{1-a} = 1 + \frac{1}{16}\frac{Bc}{R}\frac{1}{\pi}\int_0^{2\pi}\left\{y + 2\sin(\phi) + \frac{1}{y}\right\} C_{l\alpha}\cos^2(\phi)\ d\phi \tag{3.219}$$

Performing the integration, the sine term becomes zero and the cosine squared terms integrate to π. Thus:

$$\frac{1}{1-a} = 1 + \frac{1}{16}\frac{Bc}{R}C_{l,\alpha}\left(y + \frac{1}{y}\right) \approx 1 + \frac{1}{16}\frac{Bc}{R}C_{l,\alpha}\ y \tag{3.220}$$

or

$$a \approx \frac{1}{16}\frac{Bc}{R}C_{l,\alpha}\lambda \tag{3.221}$$

The expression for the power coefficient may also be simplified by using small-angle approximations and by assuming the drag coefficient to be constant. That is:

$$C_d(\alpha) \approx C_{d,0} \tag{3.222}$$

where $C_{d,0}$ is a constant drag term.

Equation (3.215) can then be integrated to give:

$$C_P = \frac{1}{4\pi}\frac{Bc}{R}C_{l,\alpha}\frac{(1-a)^4}{\lambda}(y^2+1) - \frac{1}{2}\frac{Bc}{R}C_{d,0}\lambda(1-a)^2(y^2+1) \tag{3.223}$$

Equation (3.223) can be further simplified by noting that since $y \gg 1$, it is also true that $y^2 + 1 \approx y^2$ and also that $(1-a)^2 y^2 = \lambda^2$. The expression for C_P can then be approximated as:

$$C_P \approx 4a(1-a)^2 - \frac{1}{2}\frac{Bc}{R}C_{d,0}\lambda^3 \tag{3.224}$$

When there is no drag, the optimum C_P from this equation occurs when $a = 1/3$. Thus, this vertical axis wind turbine has the same Betz limit as HAWTs. Therefore:

$$C_{P,\max} = \frac{16}{27} = 0.5926 \tag{3.225}$$

The same limit on the induction factor, a, applies as it does for HAWTs, namely, $a < 0.5$.

3.8.2.2 Multiple Stream Tube Momentum Theory
Another approach, known as the multiple stream tube theory, is used for the analysis of vertical axis wind turbines in practical situations. In this approach, it is assumed that the induction factor may vary in the direction perpendicular to the wind, but is constant in the direction of the wind. Each stream tube of constant a is parallel to the

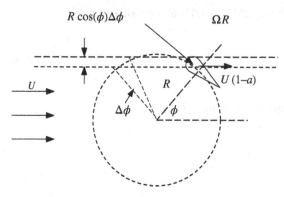

Figure 3.46 Multiple stream tube geometry

wind, as shown in Figure 3.46. The force on a stream tube of width $R\cos(\phi)\Delta\phi$ can be related to the change in momentum of the wind passing through it. The equation for force per unit height, analogous to Equation (3.203), is given by:

$$\Delta\tilde{F}_D = R\cos(\phi)\,(\Delta\phi)\,\rho\,2a\,(1-a)\,U^2 \tag{3.226}$$

During any given rotation, the blade will pass through the stream tube twice. The force on the blade is found in blade element theory. It turns out that, due to the symmetry of the airfoil, the force on the blade at both the upwind and downwind crossing position is the same. The equation for force, analogous to Equation (3.205), is accordingly given by:

$$\Delta\tilde{F}_D = B\frac{2}{2\pi}\int_{\phi}^{\phi+\Delta\phi}\frac{1}{2}\,\rho\,U_{rel}^2\,c\,\,C_l\cos(\phi+\alpha)\,\,\mathrm{d}\phi \tag{3.227}$$

The two equations above may be equated and an expression for a may be found for each stream tube. The integral in Equation (3.227) is given approximately by:

$$\int_{\phi}^{\phi+\Delta\phi}\frac{1}{2}\,\rho\,U_{rel}^2\,c\,\,C_l\cos(\phi+\alpha)\,\,\mathrm{d}\phi = \frac{1}{2}\rho\,U_{rel}^2\,c\,\,C_l\cos(\phi+\alpha)\,\Delta\phi \tag{3.228}$$

By using Equations (3.226) and (3.227), and taking advantage of Equation (3.228), it may be shown that:

$$a(1-a) = \left(\frac{1}{4\pi}\right)\left(\frac{Bc}{R}\right)\left(\frac{U_{rel}}{U}\right)^2 C_l\frac{\cos(\phi+\alpha)}{\cos(\phi)} \tag{3.229}$$

The power coefficient can be solved as before with Equation (3.223), but in this case, an iterative method must be used since a is now a function of the angle of attack. Detailed analysis of this method is beyond the scope of this text. More information may be found in de Vries (1979) and Strickland (1975).

3.8.2.3 Double Multiple Stream Tube Momentum Theory

An extension of the multiple stream tube theory is known as the "double multiple stream tube theory." The approach is similar to that of the multiple stream tube theory described above. The main difference is that the induction factors in the upwind and downwind positions need not be the same. Simulation models for vertical axis wind turbines developed by Sandia National Laboratory are based on the double multiple stream tube theory. Detailed discussion of this model is beyond the scope of this text. References such as Spera (1994) or Paraschivoiu (2002) should be consulted for further information.

3.8.3 Aerodynamics of the Darrieus Rotor

The Darrieus rotor may be analyzed with either the single or multiple stream tube methods discussed above. The main differences have to do with (1) the orientation of the blade elements, which are now different from one another, and (2) the distance of the blade elements from the axis of rotation, which is not constant over the length of the blades. Further discussion of Darrieus rotor aerodynamics is beyond the scope of this text. The reader is referred to Paraschivoiu (2002) for references that give a more in-depth coverage of this topic.

3.8.4 Savonius Rotor

The Savonius rotor, introduced in Section 3.3.6, is a vertical axis wind turbine with an S-shaped cross-section when viewed from above. A schematic is illustrated in Figure 3.47. It is primarily a drag-type device, but there may be some amount of lift contributing to the power as well.

The power from the Savonius is based on the difference in pressure across the blade retreating from the wind and advancing into the wind. This is in turn related to the difference in the drag coefficients associated with the convex side of the blade and the concave side of the blades. A detailed discussion of the aerodynamics of the Savonius rotor may be found in Paraschivoiu (2002).

Savonius rotors have, in general, fairly low efficiencies, although power coefficients close to 0.30 have been measured (see Blackwell *et al.*, 1977). The peak power coefficient for any Savonius rotor normally occurs at a tip speed ratio of less than 1.0.

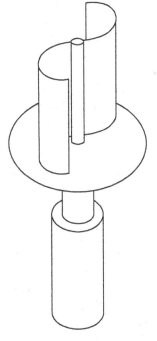

Figure 3.47 Savonius rotor (*Source: S. Kuntoff, reproduced under Creative Commons license*)

References

Abbott, I. A. and von Doenhoff, A. E. (1959) *Theory of Wing Sections*. Dover Publications, New York.

Althaus, D. (1996) *Airfoils and Experimental Results from the Laminar Wind Tunnel of the Institute for Aerodynamik and Gasdynamik of the University of Stuttgart*. University of Stuttgart.

Althaus, D. and Wortmann, F. X. (1981) *Stuttgarter Profilkatalog*. Friedr. Vieweg und Sohn, Braunschweig/Wiesbaden.

Anderson, B., Branlard, E., Vijayakumar, G. and Johnson, N. (2020) Investigation of the nacelle blockage effect for downwind wind turbines. *Journal of Physics: Conference Series*, **1618**, 062062.

Batchelor, G. K. (1967) *An Introduction to Fluid Dynamics*. Cambridge University Press.

Bertin, J. J. and Cummings, R. M. (2014) *Aerodynamics for Engineers*, 6th edition. Pearson, United Kingdom.

Betz, A. (1926) *Windenergie und Ihre Ausnutzung durch Windmüllen*. Vandenhoeck and Ruprecht, Göttingen.

Blackwell, B. F., Sheldahl, R. E. and Feltz, L. V. (1977) *Wind Tunnel Performance Data for Two- and Three-Bucket Savonius Rotors*, SAND76-0131, Sandia National Laboratories.

Bortolotti, P., Bottasso, C. L. and Croce, A. (2016) Combined preliminary-detailed design of wind turbines. *Wind Energy Science*, **1**(1), 71–88.

Bortolotti, P., Dixon, K., Gaertner, E., Rotondo, M. and Barter, G. (2020) An efficient approach to explore the solution space of a wind turbine rotor design process. *Journal of Physics: Conference Series*, **1618**, 042016.

Boorsma, K., Schepers, G., Aagard Madsen, H., Pirrung, G., Sørensen, N., Bangga, G., Imiela, M., Grinderslev, C., Meyer Forsting, A., Zhong Shen, W., Croce, A., Cacciola, S., Schaffarczyk, A. P., Lobo, B., Blondel, F., Gilbert, P., Boisard, R., Höning, L., Greco, L., Testa, C., Branlard, E., Jonkman, J. and Vijayakumar, G. (2023) Progress in validation of rotor aerodynamic codes using field data. *Wind Energy Science*, **8**, 211–230.

Branlard, E., Gaunaa, M. and Machefaux, E.2014) Investigation of a new model accounting for rotors of finite tip-speed ratio in yaw or tilt. *Journal of Physics: Conference Series*, **524**(1), 1–11. https://doi.org/10.1088/1742-6596/524/1/012124.

Branlard, E. and Gaunaa, M. (2015) Cylindrical vortex wake model: right cylinder. *Wind Energy*, **18**(11), 1973–1987.

Branlard, E. and Gaunaa, M. (2016) Superposition of vortex cylinders for steady and unsteady simulation of rotors of finite tip-speed ratio. *Wind Energy*, **19**, 1307–1323.

Branlard, E. (2017) *Wind Turbine Aerodynamics and Vorticity-Based Methods: Fundamentals and Recent Applications*. Springer International Publishing. https://doi.org/10.1007/978-3-319-55164-7.

Branlard, E., Jonkman, B., Pirrung, G., Dixon, K. and Jonkman, J. (2022) Dynamic inflow and unsteady aerodynamics models for modal and stability analyses in OpenFAST. *Journal of Physics: Conference Series*, **2265**, 032044.

Canuto, C. (1988) *Spectral Methods in Fluid Dynamics*. Springer-Verlag, Germany.

Connell, J. R. (1982) The spectrum of wind speed fluctuations encountered by a rotating blade of a wind energy conversion system. *Solar Energy*, **29**(5), 363–375.

Coleman, R. P., Feingold, A. M. and Stempin, C. W.1945) *Evaluation of the Induced-Velocity Field of an Idealized Helicopter Rotor*, NACA ARR No. L5E10, 1–28.

Cottet, G.-H. and Koumoutsakos, P. (2000) *Vortex Methods: Theory and Practice*. Cambridge University Press.

de Vries, O. (1979) *Fluid Dynamic Aspects of Wind Energy Conversion*, Advisory Group for Aerospace Research and Development, North Atlantic Treaty Organization, AGARD-AG-243.

Doubrawa, P., Quon, Eliot W., Martinez-Tossas, L. A., Shaler, K., Debnath, M., Hamilton, N., Herges, T. G., Maniaci, D., Kelley, C. L., Hsieh, A. S., Blaylock, M. L., van der Laan, P., Andersen, S. J., Krueger, S., Cathelain, M., Schlez, W., Jonkman, J., Branlard, E., Steinfeld, G., Schmidt, S., Blondel, F., Lukassen, L. J., Moriarty, P. (2020) Multimodel validation of single wakes in neutral and stratified atmospheric conditions. *Wind Energy*, **23**, 2027–2055.

Drela, M. (1989) XFOIL: An Analysis and Design System for Low Reynolds Number Airfoils, in: *Lecture Notes in Engineering*, Vol. 54, Springer-Verlag, New York, 1–12. https://link.springer.com/chapter/10.1007/978-3-642-84010-4_1.

Du, Z. and Selig, M. (1998) *A 3-D Stall-Delay Model for Horizontal Axis Wind Turbine Performance Prediction*. AIAA.

Eggleston, D. M. and Stoddard, F. S. (1987) *Wind Turbine Engineering Design*. Van Nostrand Reinhold, New York.

Ferreira, C., Yu, W., Sala, A. and Viré, A. (2022) Dynamic inflow for a floating horizontal axis wind turbine in surge motion. *Wind Energy Science*, **7**, 469–485.

Flettner, A. (1926) *The Story of the Rotor*. F. O. Wilhofft, New York.

Froude, R. E. (1889) On the part played in propulsion by differences of fluid pressure. *Transactions of the Royal Institution of Naval Architects*, **30**, 390–405.

Froude, W. (1878) On the mechanical principles of the action of propellers. *Transactions of the Royal Institution of Naval Architects*, **19**(47), 6.

Fulgsang, P., Bak, C., Gaunaa, M. and Antoniou, I. (2004) Design and verification of the Riso-B1 airfoil family for wind turbines, *Journal of Solar Energy Engineering*, **126**(4), 1002–1010.

Glauert, H. (1926) *A General Theory of the Autogyro*, NACA Reports; Memoranda No. 111.

Glauert, H. (1935) Airplane Propellers, in: W. F. Durand (Ed.), *Aerodynamic Theory*, Div. L. Chapter XI, Springer Verlag, Berlin, 169–369 (reprinted by Peter Smith (1976) Gloucester, MA).

Glauert, H. (1948) *The Elements of Aerofoil and Airscrew Theory*. Cambridge University Press, Cambridge, UK.

Goankar, G. H. and Peters, D. A. (1986) Effectiveness of current dynamic-inflow models in hover and forward flight. *Journal of the American Helicopter Society*, **31**(2), 47–57.

Gormont, R. E. (1973) *A Mathematical Model of Unsteady Aerodynamics and Radial Flow for Application to Helicopter Rotors*, US Army Air Mobility Research and Development Laboratory, Technical Report, 76–67.

Hansen, A. C. and Butterfield, C. P. (1993) Aerodynamics of horizontal axis wind turbines. *Annual Review of Fluid Mechanics*, **25**, 115–149.

Hansen, M. H., Gaunaa, M. and Madsen, H. A. (2004). *A Beddoes-Leishman Type Dynamic Stall Model in State-Space and Indicial Formulations.* Risø National Laboratory, Roskilde, Denmark.

Hansen, M. O. L. (2008) *Aerodynamics of Wind Turbines – Second Edition.* Earthscan, London, Sterling, VA.

Jonkman, J., Butterfield, S., Musial, W. and Scott, G. (2009) *Definition of a 5-MW Reference Wind Turbine for Offshore System Development*, National Renewable Energy Laboratory, TP-500-38060.

Joukowski, N. E. (1918) Travaux Du Bureau Des Calculs et Essais Aeronautiques. Ecole Superieure Technique de Moscou.

Katz, J. and Plotkin, A. (2001) *Low-Speed Aerodynamics*, 2nd edition. Cambridge Aerospace Series (No. 13). Cambridge University Press.

Lamb, H. 1916. *Hydrodynamics.* 4th edition. Cambridge University Press.

Lanchester, F. W. (1907) *Aerodynamics: Constituting the First Volume of a Complete Work on Aerial Flight.* A. Constable & Co., Ltd., London.

Leishman, J. G. and Beddoes, T. S. (1989). A semi-empirical model for dynamic stall. *Journal of the American Helicopter Society*, **34**(3), 3–17.

Lysen, E. H. (1982) *Introduction to Wind Energy.* Steering Committee Wind Energy Developing Countries, Amersfoort, NL.

Madsen, H. A., Larsen, T. J., Pirrung, G. R., Li, A. and Zahle, F. (2019) Implementation of the blade element momentum model on a polar grid and its aeroelastic load impact. *Wind Energy Science*, **5**, 1–27.

Manwell, J. F. (1990) A simplified method for predicting the performance of a horizontal axis wind turbine rotor. *Proc. of the 1990 American Wind Energy Association Conference.* Washington, DC.

Martinez-Tossas, L. A., Churchfield, M. J. and Meneveau, C. (2017) Optimal smoothing length scale for actuator line models of wind turbine blades based on gaussian body force distribution. *Wind Energy*, https://doi.org/10.1002/we.2081.

Meyer Forsting, A. R., Pirrung, G. R. and Ramos-García, N. (2019) A vortex-based tip/smearing correction for the actuator line. *Wind Energy Science*, **4**(2), 369–83. https://doi.org/10.5194/wes-4-369-2019.

Miley, S. J. (1982) *A Catalog of Low Reynolds Number Airfoil Data for Wind Turbine Applications*, Rockwell Int., Rocky Flats Plant RFP-3387, NTIS.

Moriarty, P. J. and Hansen, C. A. (2005) *AeroDyn Theory Manual.* National Renewable Energy Laboratory.

Ning, A. (2014) A simple solution method for the blade element momentum equations with guaranteed convergence. *Wind Energy*, **17**, 1327–1345.

Øye, S. (1990). A Simple Vortex Model of a Turbine Rotor. *Proc. Of the Third IEA Symposium on the Aerodynamics of Wind Turbines, ETSU, Harwell*, pp. 4.1–1.15.

Øye, S. (1991). Dynamic Stall, Simulated as a Time Lag of Separation. *Proceedings of the 4th IEA Symposium on the Aerodynamics of Wind Turbines.*

Papadakis, G. (2014). *Development of a Hybrid Compressible Vortex Particle Method and Application to External Problems Including Helicopter Flows*, PhD Thesis, National Technical University of Athens.

Paraschivoiu, I. (2002) *Wind Turbine Design with Emphasis on Darrieus Concept*, Ecole Polytechnique de Montreal.

Pitt, D. M. and Peters, D. A. (1981) Theoretical predictions of dynamic inflow derivatives. *Vertica*, **5**(1), 21–34.

Prandtl, L. (1918) Tragflügeltheorie. A. Dillmann (Ed.), *Göttinger Klassiker Der Strömungsmechanik Bd*, Universitätsverlag Göttingen, 3.

Rankine, W. J. (1865) On the mechanical principles of the action of propellers. *Transactions of the Royal Institution of Naval Architects*, **6**, 13.

Sant, T. (2007) *Improving BEM-based Aerodynamic Models in Wind Turbine Design Codes*, PhD Thesis, Delft University Wind Energy Research Institute.

Selig, M. S. and Tangler, L. T. (1995) Development and application of a multipoint inverse design method for horizontal axis wind turbines. *Wind Engineering*, **19**(2), 91–105.

Sengupta, A. and Verma, M. P. (1992) An analytical expression for the power coefficient of an ideal horizontal-axis wind turbine. *International Journal of Energy Research*, **16**, 453–456.

Simms, D., Schreck, S., Hand, M. and Fingersh, L. J. (2001) *NREL Unsteady Aerodynamics Experiment in the NASA-Ames Wind Tunnel: A Comparison of Predictions to Measurements*, NREL/TP-500-29494. National Renewable Energy Laboratory, Golden, CO.

Snel, H. and Schepers, J. G. (1991) Engineering models for dynamic inflow phenomena. *Proc. 1991 European Wind Energy Conference, Amsterdam*, pp. 390–396.

Snel, H. and Schepers, J. G. (1993) Investigation and modelling of dynamic inflow effects. *Proc. 1993 European Wind Energy Conference,* Lübeck, pp. 371–375.

Snel, H., Houwink, R. and Bosschers, J. (1994) *Sectional Prediction of Lift Coefficients on Rotating Wind Turbine Blades in Stall*, ECN Technical Report. ECN-C-93-052.

Snel, H. and Schepers, J. G. (1995) *Joint Investigation of Dynamic Inflow Effects and Implementation of an Engineering Method*, ECN-C–94-107, Energy Research Centre of the Netherlands, Petten.

Sørensen, N. N. (1995) *General Purpose Flow Solver Applied to Flow Over Hills*, PhD Thesis, Risø-DTU.

Sørensen, J. N. and van Kuik, G. A. M. (2011) General momentum theory for wind turbines at low tip speed ratios. *Wind Energy*, **14**(7), 821–39. https://doi.org/10.1002/we.423.

Sørensen, J. N. (2016) *General Momentum Theory for Horizontal Axis Wind Turbines*. Springer.

Sørensen, N., Mendez, B., Munoz, A., Sieros, G., Jost, E., Lutz, T., Papadakis, G., Voutsinas, S., Barakos, G., Colonia, S., Baldacchino, D., Baptista, C. and Ferreira, C. (2016) CFD code comparison for 2D airfoil flows. *Journal of Physics*, **753**, 082019.

Spera, D. A. (Ed.) (1994) *Wind Turbine Technology*. American Society of Mechanical Engineers, New York.

Sprague, M., Ananthan, S., Vijayakumar, G. and Robinson, M. (2020) ExaWind: a multifidelity modeling and simulation environment for wind energy. *Journal of Physics: Conference Series*, **1452**, 012071.

Strickland, J. H. (1975) *The Darrieus Turbine: A Performance Prediction Model Using Multiple Streamtubes*, Sandia National Laboratories Technical Report. SAND75-0431.

Tangler, J. L. (1987) *Status of Special Purpose Airfoil Families*, SERI/TP-217-3264, National Renewable Energy Laboratory, Golden, CO.

Theodorsen, T. (1935) *General Theory of Aerodynamic Instability and the Mechanism of Flutter*, NACA report no 496, National Advisory Committee for Aeronautics.

Timmer, W. A. and van Rooij, R. P. J. O. M. (2003) Summary of Delft University wind turbine dedicated airfoils. *Journal of Solar Energy Engineering*, **125**, 488–496.

van Garrel, A. (2003) *Development of a Wind Turbine Aerodynamics Simulation Module*, ECN-C–03-079. ECN.

van Kuik, G. A. M. and Lignarolo, L. E. M. (2015) Potential flow solutions for energy extracting actuator disc flows. *Wind Energy*, **19**(1391), 1406.

van Kuik, G. A. M. (2018) *The Fluid Dynamic Basis for Actuator Disc and Rotor Theories*. IOS Press.

Veers, P. (2019) *Wind Energy Modeling and Simulation, Volume 2: Turbine and System*. The Institution of Engineering and Technology, ISBN-13: 978-1785615238.

Versteeg, H.K. and Malalasekera, W. (1995) *An introduction to Computational Fluid Dynamics*. Longman Group Ltd, ISBN 0-582-21884-5.

Viterna, L. A. and Corrigan, R. D. (1981) Fixed pitch rotor performance of large horizontal axis wind turbines. *Proc. Workshop on Large Horizontal Axis Wind Turbines*, NASA CP-2230, DOE Publication CONF-810752, pp. 69–85, NASA Lewis Research Center, Cleveland, OH.

Wald, Q. R. (2006) The aerodynamics of propellers. *Progress in Aerospace Science* **42**, 85–128. https://doi.org/10.1016/j.paerosci.2006.04.001.

White, F. M. (2015). *Fluid Mechanics*, 8th edition. McGraw Hill Science Engineering, New York.

Wilson, R. E. and Lissaman, P. B. S. (1974) *Applied Aerodynamics of Wind Power Machine*. Oregon State University.

Wilson, R. E., Lissaman, P. B. S. and Walker, S. N. (1976) *Aerodynamic Performance of Wind Turbines*. Energy Research and Development Administration, ERDA/NSF/04014-76/1.

Winckelmans, G. S. (2004) Vortex Methods, in: E. Stein, R. de Borst, T.J.R. Hughes (Eds.), *Encyclopedia of Computational Mechanics – Vol. 1*. Chapter 5, J. Wiley & Sons, New-York, N.Y., 415–438.

Winckelmans, G. S. and Leonard, A. (1993) Contributions to vortex particle methods for the computation of 3-dimensional incompressible unsteady flows. *Journal of Computational Physics*, **109**(2), 247–273. https://doi.org/10.1006/jcph.1993.1216.

Zahle, F., Tibaldi, C., Pavese, C., McWilliam, M. K., Blasques, J. P. A. A. and Hansen, M.H. (2016) Design of an aeroelastically tailored 10 MW wind turbine rotor. *Journal of Physics Conference Series*, **753**, 062008.

4

Mechanics and Dynamics

4.1 Background

The interplay of the forces from the external environment, primarily due to the wind, and the motions of the various components of the wind turbine result not only in the desired energy production from the turbine but also in stresses in the constituent materials. For the turbine designer, these stresses are of primary concern, because they directly affect the strength of the turbine and how long it will last.

To put it succinctly, to be a viable contender for providing energy, a wind turbine must produce energy, survive, and be cost-effective. The turbine design must not only be functional in terms of extracting energy, but it must also be structurally sound so that it can withstand the loads it experiences, and the associated costs must not outweigh the value of the energy it produces.

The purpose of this chapter is to discuss the mechanical framework within which the turbine must be designed if it is to meet these three requirements. The study of the strength of the different components is discussed in Chapter 7, and the effect of loads in the design process is discussed in Chapter 8. The current chapter consists of eight sections, beyond this background. Section 4.2 provides an overview of wind turbine loads. Section 4.3 gives a summary of the fundamental principles of mechanics relevant to wind turbines. Section 4.3 provides an overview of the general concepts of wind turbine dynamics. Section 4.4 discusses methods for modeling wind turbine structural response. Section 4.5 describes the equations of motion of a wind turbine. Section 4.6 concerns wind turbine models using the assumed shape function. Section 4.7 introduces a simplified model of a wind turbine rotor known as the Linearized Hinge Spring Blade Model. Finally, Section 4.8 concerns linearization, stability, and aerodynamic damping.

Although the principles apply to all types of wind turbine design, the focus of this chapter is exclusively on horizontal axis turbines.

4.2 Wind Turbine Motions and Loads

This section provides an overview of wind turbine motions and loads: the types of loads, their sources, and their effects. As will be discussed, these loads are due primarily to aerodynamics, hydrodynamics, and the self-weight of the structure, particularly the blades as they rotate.

4.2.1 Motions

The main motions undergone by the different wind turbine components are illustrated in Figure 4.1 and further described in the subsequent paragraphs. For a continuous structure, the term deformation is preferred to refer to the continuous motion of a component. The main coordinate systems are also shown in the figure: "*b*" for the

Wind Energy Explained: On Land and Offshore, Third Edition.
James F. Manwell, Emmanuel Branlard, Jon G. McGowan, and Bonnie Ram.

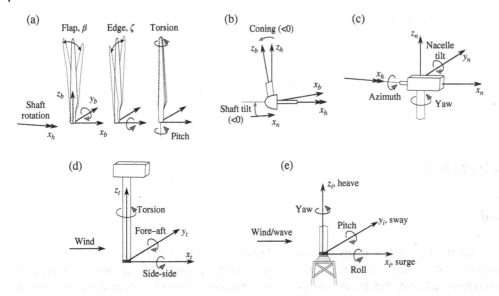

Figure 4.1 Motions of the main components of a wind turbine, main coordinate systems, and orientations. (a) Blade motions, (b) hub and blade orientations, (c) nacelle and shaft motions, (d) tower motions, (e) platform/floater motions

blade, "h" for the hub, "t" for the tower, and "i" for the inertial system. The *OpenFAST* convention (Jonkman and Buhl, 2005) is used throughout, in particular, for the shaft tilt (negative about y_n) and coning (positive about y_h).

4.2.1.1 Blade Motions
Blades deform in three directions:

1) **Flapwise**. Flapping refers to deformation parallel to the axis of rotation of the rotor, where the blade deforms upwind or downwind. The term flapwise is sometimes referred to as out-of-plane or flatwise. Forces in the flapping direction are of particular importance, since the largest stresses on the blades are associated with flapwise bending. The aerodynamic thrust is the primary driver of the flapwise motion.
2) **Edgewise**. Edgewise deformation lies in the plane of rotation. Edgewise is also known as lead–lag or in-plane motion. In leading motion, the blade is moving slightly faster than the overall rotational speed, and in lagging motion, it is moving slower. Lead–lag motions and forces are typically related to gravitational loads and fluctuations of torque in the main shaft.
3) **Torsional**. Torsional deformation refers to rotation about the long axis of the blade. For a fixed-pitch wind turbine, torsional motion is generally not of great significance. In pitch-regulated turbines, actuators located in the hub rotate the entire blade around its main axis. In this case, torsional motions and moments may be important.

4.2.1.2 Nacelle and Shaft Motions
The rotation of the main shaft defines the azimuthal position of the hub and rotor. Deformations of the shaft include torsion, and when bearings are present, bending in between the bearings (not shown in the figure). The nacelle yaws around the tower axis, usually controlled by a yaw motor. The main shaft is typically tilted by design so that the blade points further away from the tower to avoid tower strikes. Some structural models include yaw and tilt stiffnesses to account for some flexibility of the nacelle connection to the tower top.

4.2.1.3 Tower Motions
The tower deformations are like the ones of the blades, but the terminology is different. Deformations in the wind direction are referred to as fore–aft motion, and deformations in the lateral directions are referred to as side–side motion.

4.2.1.4 Offshore Support Structure/Platform Motions

The use of fixed or floating support structures for offshore wind turbines introduces more motions and greater dynamic complexity. Similar to a ship, a floating support structure may move up and down in the waves ("heave"), forward and backward ("surge"), side to side ("sway"), or rotate about its long axis ("roll"), about its transverse axis ("pitch") or its vertical axis ("yaw"). Motions are significantly greater for a floating structure, but motions are still present for fixed bottom structures such as jackets and monopiles. Monopiles are typically stiffer in torsion and heave than jackets, but softer in surge and sway bending. Fixed support structures are discussed in more detail in Chapter 10, and floating offshore wind turbines are discussed in Chapter 11.

4.2.2 Types of Loads

In this chapter, the term "load" refers to forces or moments that act upon the turbine. The loads that a turbine may experience are of primary concern in assessing the turbine's structural requirements. These loadings may be divided into five types:

1) steady loads;
2) cyclic loads;
3) transient (including impulsive);
4) stochastic;
5) resonance-induced loads.

The key characteristics of these loads and some examples of wind turbines are summarized below.

4.2.2.1 Steady Loads

Steady loads are ones that do not vary over a relatively long time period. They can be either static or rotating. Static loads, as used in this text, refer to non-time-varying loads that impinge on a nonmoving structure. For example, a steady wind blowing on a stationary wind turbine would induce static loads on the various parts of the machine. In the case of steady rotating loads, the structure may be in motion, and there could be loads that are unchanging within a rotating frame of reference. The self-weight force will also consist of a steady component about the mean position of the body. The calculation of steady aerodynamic loads has been detailed in Chapter 3.

4.2.2.2 Cyclic Loads

Cyclic loads are those which vary in a regular or periodic manner. The term applies particularly to loads that arise due to the rotation of the rotor. Cyclic loads result from such factors as the weight of the blades, wind shear, and yaw. Cyclic loads may also be associated with the vibration of the turbine structure or some of its components.

As noted in Chapter 3, a load that varies an integral number of times in relation to a complete revolution of the rotor is known as a "per revolution," or "per rev" load, and is given the symbol P. For example, a blade rotating in wind with wind shear will experience a cyclic $1P$ load. If the turbine has three blades, the turbine structure will experience a cyclic $3P$ load.

4.2.2.3 Transient Loads

Transient loads are time-varying loads that arise in response to some temporary external event. There may be some oscillations associated with the transient response, but they eventually decay. Examples of transient loads include those in the drivetrain resulting from start-up or the application of a brake.

Impulsive loads are transient time-varying loads of relatively short duration, but of perhaps significant peak magnitude. One example of an impulsive load is that experienced by a blade of a downwind rotor when it passes behind the tower (through the "tower shadow"). Another example occurs with some two-bladed rotors. Such rotors are typically "teetered" or pinned at the low-speed shaft (LSS; see Chapter 8). This allows the rotor to rock back and forth, reducing bending loads on the shaft, but necessitating the use of teeter dampers and stops. The force on a teeter damper or stops when the normal range of teeter is exceeded is an impulsive load.

4.2.2.4 Stochastic Loads

Stochastic loads are time varying, as are cyclic, transient, and impulsive loads. In this case, the loading varies in a more apparently random manner. In many cases, the mean value may be relatively constant, but there may be significant fluctuations from that mean. Examples of stochastic loads are those that arise from turbulence in the wind or ocean waves; see Chapter 6.

4.2.2.5 Resonance-Induced Loads

Resonance-induced loads are cyclic loads that result from the dynamic response of some part of the wind turbine being excited at one of its natural frequencies. They may reach high magnitudes. Resonance-induced loads are to be avoided whenever possible, but they may occur under unusual operating circumstances or due to poor design. While these loads are not truly a different type of load, they are mentioned separately because of their possibly serious consequences.

4.2.3 Sources of Loads

There are five primary sources of loads to consider in wind turbine design:

1) aerodynamics (and hydrodynamics, see Chapter 6);
2) gravity;
3) dynamic interactions and reactions;
4) mechanical control;
5) inertial loads (see Section 4.3.1.2).

A summary of the loads acting on a generic floating wind turbine is shown in Figure 4.2. The following briefly describes each of these sources.

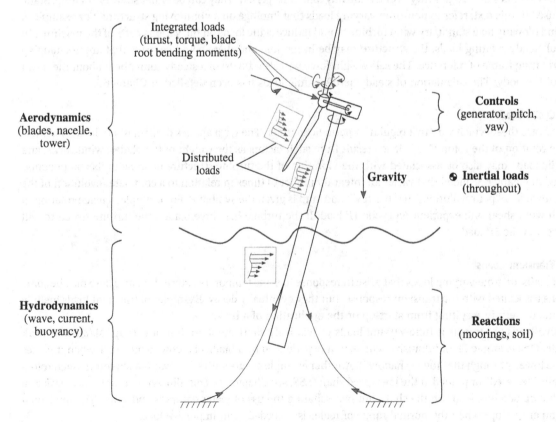

Figure 4.2 Loads acting on a floating wind turbine

4.2.3.1 Aerodynamics

The first source of wind turbine loads to consider is aerodynamics (see Chapter 3). The loadings of particular concern in the structural design are those which could arise in very high winds, or those which generate fatigue damage. When a wind turbine is stationary in high winds, drag forces are the primary consideration. When the turbine is operating, it is lift forces on the blades that create the aerodynamic loadings of concern. Secondary aerodynamic loads are present on the nacelle and tower.

4.2.3.2 Gravity

Gravity is an important source of loads on the blades of large turbines, although it is less so on smaller machines. In any case, tower top (rotor nacelle assembly) weight is significant to the tower design and to the installation of wind turbines.

4.2.3.3 Dynamic Interactions

Motion induced by aerodynamic and gravitational forces in turn induces loads in other parts of the machine. For example, virtually all horizontal-axis wind turbines allow some motion about the yaw axis. When yaw motion occurs while the rotor is turning, there will be induced gyroscopic forces. These forces can be substantial when the yaw rate is high.

4.2.3.4 Mechanical Control

Control of the wind turbine may sometimes be a source of significant loads. For example, starting a turbine that uses an induction generator or stopping the turbine by applying a brake can generate substantial loads throughout the structure.

4.2.4 Effects of Loads

The loadings experienced by a wind turbine are important in two primary areas: (1) ultimate strength and (2) fatigue. Wind turbines may occasionally experience very high loads, and they must be able to withstand them. Normal operation is accompanied by widely varying loads, due to starting and stopping, yawing, and the passage of blades through continuously changing winds. These varying loads can cause fatigue damage in machine components, such that a given component may eventually fail at much lower loads than it would have when new. More details on fatigue are given in Chapter 7. Accounting for the effect of loads in the design process is discussed in Chapter 8.

4.2.5 Expressions for the Loads Acting on a Rigid Wind Turbine

This section presents simple expressions for the loads acting on a rigid and "ideal" wind turbine, neglecting elastic motions or variations with azimuth. These expressions are useful for rapid assessments of the main loads acting on the turbine. More advanced considerations are presented in subsequent sections.

The most important rotor loads on a wind turbine are those associated with thrust on the blades, torque to drive the rotor, and hydrodynamic loads offshore. Modeling the rotor as a simple rigid, aerodynamically ideal rotor is useful to get a feeling for the steady loads on a wind turbine.

4.2.5.1 General Integration of Loads on a Component

The force (f_{root}) and moment (M_{root}) vectors at the root of a straight blade are computed as:

$$f_{root} = \int_{r_h}^{R} \tilde{f}(r)dr \tag{4.1}$$

$$M_{root} = \int_{r_h}^{R} (x_{node}(r) - x_{root}) \times \tilde{f}(r)dr \tag{4.2}$$

where the bold notation indicates vectors, $\tilde{\boldsymbol{f}}(r)$ is the force per unit length (e.g., coming from aerodynamics, gravity, inertia) acting at the given blade node located at the position vector $\boldsymbol{x}_{\mathrm{node}}$; $\boldsymbol{x}_{\mathrm{root}}$ is the position vector of the blade root; r is the radial coordinate spanning from the blade root, r_h (hub radius), to the tip of the blade, R. For a non-straight blade, the integration over r is replaced by an integration over the curvilinear coordinate following the blade mean-line. The integrated thrust force and torque from one blade are obtained as:

$$T_{\mathrm{blade}} = \boldsymbol{f}_{\mathrm{root}} \cdot \hat{\boldsymbol{x}}_h, \, Q_{\mathrm{blade}} = \boldsymbol{M}_{\mathrm{root}} \cdot \hat{\boldsymbol{x}}_h \tag{4.3}$$

where $\hat{\boldsymbol{x}}_h$ is the unit vector along the hub axis (x_h, see Figure 4.1). Other components of the integrated force and bending moment are similarly obtained by scalar products with other unit vectors. The coordinate system used for the vectors in all the expressions above is arbitrary (inertial, hub, blade, etc.), but all vectors need to be expressed in the same coordinate system (see 4.3.1.1 for a primer on coordinate systems and vectors). Also, the expressions above remain valid if the blade is moving, that is, with $\boldsymbol{x}_{\mathrm{node}}$ (and potentially $\boldsymbol{x}_{\mathrm{root}}$) changing with time.

The determination of integral loads on other components can be done in the same way. In the rest of this section, simplified expressions are provided assuming straight blades, no coning, no tilt, axisymmetric flow, and rigid components.

4.2.5.2 Aerodynamic Thrust and Torque on the Rotor

As shown in Chapter 3, the aerodynamic thrust, T, for a rigid rotor in axisymmetric flow, may be found from the following equation:

$$T = \frac{1}{2}\rho\pi R^2 U^2 \, C_T \tag{4.4}$$

where C_T is the rotor averaged thrust coefficient (ranging typically between 0.2 and 0.9, but occasionally exceeding 1.0), ρ is the density of air ($\rho = 1.225$ kg/m^3 in normal conditions), R is the radius of rotor, and U is the free stream velocity. For the ideal case, or for a quick estimate, one can use $C_T = 8/9$. In terms of this simple model, then, total thrust on a given rotor varies mostly with the square of the wind speed.

The mean torque Q is the power divided by the rotational speed. As shown in Chapter 3, the aerodynamic torque is given by:

$$Q = \frac{P}{\Omega} = \frac{1}{2}\rho\,\pi\,R^2\,\frac{U^3}{\Omega}\,C_P \tag{4.5}$$

where C_P is the power coefficient, ranging typically between 0.1 and 0.55. Torque can equivalently be expressed in terms of a torque coefficient, $C_Q = C_P/\lambda$ (the power coefficient divided by the tip-speed ratio), such that:

$$Q = \frac{1}{2}\rho\pi R^3 U^2 \, C_Q \tag{4.6}$$

For the ideal Betz rotor, $C_P = 16/27 = 0.59$, and the rotational speed varies with the wind speed, so torque varies as the square of the wind speed. Furthermore, rotors designed for higher tip-speed ratio operation have lower torque coefficients, so they experience lower torques (but not necessarily lower stresses). Again, according to this simple model, there is no variation in aerodynamic torque with blade azimuth (axisymmetric flow).

The equations above assume that the rotor-averaged thrust, torque, and power coefficients are known. Their calculation was given in Section 3.5.3 of Chapter 3. It is briefly given below. The axial and tangential loads acting on a blade ($\tilde{f}_{x_b}$, and $\tilde{f}_{y_b}$) are typical outputs of an aerodynamic solver. If a lifting-line algorithm is used, they are given by blade element theory as:

$$\tilde{f}_{x_b}(r) = \frac{dF_n}{dr} = \frac{1}{2}\rho U_{\mathrm{rel}}^2 c_n c \tag{4.7}$$

$$\tilde{f}_{y_b}(r) = \frac{dF_t}{dr} = \frac{1}{2}\rho U_{\mathrm{rel}}^2 c_t \, c \tag{4.8}$$

where c is the blade chord, U_{rel} is the relative wind velocity, and c_n and c_t are the normal and tangential aerodynamic coefficients (see Chapter 3). The loading can be integrated to get the total aerodynamic thrust using Equations (4.1) and (4.3). If the same loading is assumed on each blade, by definition of the local thrust coefficient, C_{T_r}:

$$\tilde{f}_{x_b}(r) = \frac{1}{B}\frac{dT}{dr} = \frac{1}{B}\frac{1}{2}\left(\rho\, U^2 2\pi r\right) C_{T_r}(r) \tag{4.9}$$

Combining Equations (4.7) and (4.9) leads to $C_{T_r} = \sigma\, c_n\, U_{rel}^2/U^2$, with σ the rotor solidity (see Chapter 3). Using Equations (4.1), (4.3), and (4.9), and assuming that the hub radius is zero ($r_h = 0$), the total aerodynamic thrust is:

$$T = B\int_0^R \tilde{f}_{x_b}(r)dr = \rho\, U^2 \int_0^R r\, C_{T_r}(r)\, dr = \frac{1}{2}\rho\pi R^2 U^2\, C_T \tag{4.10}$$

It is seen that Equation (4.4) is retrieved, since by definition, the local and total thrust coefficients are related by:

$$C_T = \frac{2}{R^2}\int_0^R r\, C_{T_r}(r)\, dr \tag{4.11}$$

If the thrust coefficient is constant along the blade, then it is readily seen that $C_{T_r} = C_T$. Similar integrations can be carried out to obtain the torque and power coefficient from the blade element loads.

4.2.5.3 Bending Moments and Stresses on the Rotor

As introduced in Section 4.2.1, forces that cause bending moments are designated as either flapwise or edgewise. Flapwise bending moments cause the blades to bend upwind or downwind. Edgewise moments are parallel to the rotor axis and give rise to the power-producing torque.

Axial Force and Flapwise Bending Moment

Using Equation (4.2), the root flapwise bending moment on a single rigid blade, M_β, is given by:

$$M_\beta = \boldsymbol{M}_{\text{root}}\cdot\hat{\boldsymbol{y}}_b = \int_0^R \left[(r\hat{\boldsymbol{z}}_b)\times\tilde{\boldsymbol{f}}(r)\right]\cdot\hat{\boldsymbol{y}}_b dr = \int_0^R r\tilde{f}_{x_b}\, dr \tag{4.12}$$

The symbol β is used to denote the flapping angle (see Section 4.7). Using the local thrust coefficient (Equation 4.9) gives:

$$M_\beta = \frac{1}{B}\int_0^R r\left(\frac{1}{2}\rho\, 2\pi r\, C_{T_r}(r)U^2\right)dr \tag{4.13}$$

As an approximation, it can be assumed that the thrust coefficient is constant along the blade span (i.e., $C_{T_r} = C_T$), and upon integrating and gathering terms, Equation (4.13) gives:

$$M_\beta = \frac{T}{B}\frac{2}{3}R \tag{4.14}$$

The maximum flapwise stress, $\sigma_{\beta,\max}$, due to bending at the root can be obtained based on beam theory (see Equation 4.33 for the definition of stress in beams).

The shear force, S_β, in the root of the blade is simply the thrust divided by the number of blades:

$$S_\beta = T/B \tag{4.15}$$

As a first-order approximation (straight, rigid rotor with constant loading coefficient), it is seen that bending forces and stresses vary with the square of the wind speed. Furthermore, blades on rotors designed for higher tip-speed ratio operation have smaller chords and cross-sectional area moment of inertia, so they experience higher bending stresses.

Edgewise Forces and Moments

As mentioned above, the edgewise moments give rise to the power-producing torque. In terms of blade strength, aerodynamic edgewise moments are generally of less significance than their flapwise counterparts. (It should be noted, however, that edgewise moments due to blade self-weight can be quite significant in larger turbines.)

The aerodynamic bending moment in the edgewise direction (designated by ζ) at the root of a single rigid blade, M_ζ, is simply the aerodynamic torque divided by the number of blades:

$$M_\zeta = Q/B \tag{4.16}$$

There is no correspondingly simple relation for edgewise shear force S_ζ, but it can be found by integrating the tangential force:

$$S_\zeta = \int_0^R \tilde{f}_{y_b}(r)\, dr \tag{4.17}$$

4.2.5.4 Drivetrain Loads

The main loads acting on the drivetrain are the aerodynamics torque and the generator torque. The generator torque results from the extraction of power by the generator, but it is often used to regulate the rotational speed (see Chapter 9). Additional loads may be introduced by the bearings and different elements along the drivetrain, such as the gearbox. Such loads are often omitted in simple calculations and replaced by a drivetrain efficiency (of the order of 95%) which may vary with the rotational speed. Sections 4.3.7 and 4.5.3.2 provide examples of simple models of drivetrain dynamics.

4.2.5.5 Tower Loads

The main loads along the tower come from the loads at the tower top and aerodynamic loads that are distributed along the length of the tower. The aerodynamic loads on the tower can be computed using a constant drag coefficient (typically 0.6–1). Neglecting the aerodynamic and inertial forces on the tower, and assuming a rigid tower, the force and moment at the tower bottom (TB) are:

$$F_{TB} = F_{TT}, M_{TB} = F_{TT}\, H_T + M_{TT} \tag{4.18}$$

where F_{TT} and M_{TT} are the force and moment at the tower top, and H_T is the height of the tower. This equation can be used for the forces in the fore–aft direction (leading to a fore–aft tower bending), and the force in the side–side direction (leading to a side–side bending moment). Section 4.6.2 provides an example of the dynamic motion of a tower.

4.2.5.6 Offshore Wind Turbine Substructure Loads

The substructure of an offshore turbine (i.e., the structure below the tower – see Chapters 10 and 11) undergoes hydrodynamic loading from waves and sea current and loading from the soil. A floating structure will also experience reaction loads from the stationkeeping system and buoyancy forces. The details of such loads are beyond the scope of this chapter, but for reference, the wave force per unit of length acting on a slender structure such as a monopile or a jacket brace is provided below. The horizontal hydrodynamic force on a section of height dz, moving along the x direction is given by the Morison equation (see Chapter 6):

$$dF = \rho_w \left[-C_a A\ddot{x} + A\dot{U}(1 + C_a) + \frac{1}{2}C_d D(U - \dot{x})|U - \dot{x}| \right] dz \tag{4.19}$$

where $U(z)$ is the wave velocity at the water depth z; C_a and C_d are the added mass and drag coefficients of the Morison equation (typical values around 1); A is the cross-sectional area of the structure ($A = \pi R^2$ for a cylinder); and ρ_w is the density of water. The force for a static structure is obtained with $\dot{x} = 0$ and $\ddot{x} = 0$. A simple expression of the wave velocity as function of water depth, $U(z)$, can be obtained using linear wave theory; see Chapter 6.

4.3 General Principles of Mechanics

This section presents an overview of some of the principles of basic mechanics and dynamics which are of particular concern in wind turbine design. The fundamentals of the mechanics of wind turbines are essentially the same as those of other similar structures. For that reason, the topics taught in engineering courses in statics, strength of materials, and dynamics are equally applicable here. Topics of particular relevance include Newton's Second Law, especially when applied in polar coordinates; moments of inertia; bending moments; and stresses and strains. These topics are well discussed in many physics and engineering texts, such as Hibbeler (2015), Meriam and Kraige (2011), and Meriam *et al.* (2015), and will not be pursued here in great detail, except where a particular example is especially relevant.

4.3.1 Selected Topics from Basic Mechanics

There are a few topics from basic engineering mechanics that are worth singling out because they have a particular relevance to wind energy and may not be familiar to all readers. These are summarized briefly below.

4.3.1.1 Coordinate Transformations and Rotations

Two coordinate systems can be related to each other using a coordinate transformation matrix of dimension 3×3. This is typically relevant when going from a coordinate system attached to a body (noted b) to the global coordinate system (noted g). It is recalled that a vector is an entity that exists independently of the coordinate system, but its components are different when expressed in different coordinate systems. The vector $\boldsymbol{v}$ is written $\boldsymbol{v_b} = \left(v_{x_b}, v_{y_b}, v_{z_b} \right)$ in the body coordinate system and $\boldsymbol{v_g} = \left(v_{x_g}, v_{y_g}, v_{z_g} \right)$ in the global coordinate system. The two systems are related with the 3×3 transformation matrix $\boldsymbol{R_{gb}}$ using Equation (4.20):

$$\boldsymbol{v_g} = \boldsymbol{R_{gb}} \, \boldsymbol{v_b} \tag{4.20}$$

A 3×3 matrix $\boldsymbol{M}$ is transformed from the body to a global system as:

$$\boldsymbol{M_g} = \boldsymbol{R_{gb}} \, \boldsymbol{M_b} \, \boldsymbol{R_{gb}^T} \tag{4.21}$$

where T denotes the transpose operator. In this text, we use bold typeface for matrices and vectors, and uppercase letters for matrices whenever possible.

The coordinate transformation can be constructed using successive rotations, and the transformation matrix can be obtained as a product of elementary rotation matrices. Examples of representations using three successive rotations are the Euler angles or Bryant angles representations. The use of three angles can become singular, and general-purpose code needs four parameters to represent transformations matrices, such as the Euler parameters. In this book, only simple transformations are used. If the body frame is rotated by an angle θ around the x axis compared to the global system, the coordinate transformation is given by Equation (4.22):

$$\boldsymbol{R_{gb}} = \begin{bmatrix} 1 & 0 & 0 \\ 0 & \cos\theta & -\sin\theta \\ 0 & \sin\theta & \cos\theta \end{bmatrix} \tag{4.22}$$

The matrix given in Equation (4.22) is the elementary rotation matrix about the x axis. Similar matrices are obtained about the other axes.

4.3.1.2 Inertial Forces

In most current teaching of dynamics, the forces that are considered are exclusively real forces. It is sometimes convenient, however, to describe certain accelerations in terms of fictitious "inertial forces." This is often done in the case of rotating systems, including in the analysis of wind turbine rotor dynamics. For example, the effect of the centripetal acceleration associated with the rotation of the rotor is accounted for by the inertial centrifugal force. The effect of the inertia force, as reflected in the "Principle of D'Alembert," is that the sum of all forces acting on a particle, including the inertia force, is zero. The method is most useful when dealing with larger rigid bodies, which can be considered to be made up of a large number of particles rigidly connected together. Specifically, the Principle of D'Alembert states that the internal forces in a rigid body having accelerated motion can be calculated by the methods of statics on that body under the influence of the external and inertia forces. Furthermore, a rigid body of any size will behave as a particle if the resultant of its external forces passes through its center of gravity.

4.3.1.3 Area Moment of Inertia
The area moment of inertia is important in wind energy in a variety of ways, most particularly in relation to bending, for example of a blade or a tower. It should not be confused with the mass moment of inertia presented in Section 4.3.1.4.

The area moment of inertia (expressed in m^4) is the integrated product of area times the square of the distance from an axis of interest. This is usually taken as the axis about which bending or rotation is taking place. For purposes of clarity, the axis is often indicated by a subscript, as in I_x. In equation form, the area moment of inertia with respect to the x axis is:

$$I_x = \iint_{\text{Area}} y^2 dxdy \tag{4.23}$$

where y is the perpendicular distance from the x axis to element of area $dxdy$. Similarly, the area moment of inertia with respect to the y axis is:

$$I_y = \iint_{\text{Area}} x^2 dxdy \tag{4.24}$$

The area moment of inertia I_z of a shape in the x–y plane about an axis z perpendicular to it is often called the polar moment of area and given the symbol I_p, with $I_p = I_x + I_y$. This is of interest for torsion of shafts or angular motion (e.g., vibration) of a tower.

An area moment of inertia I_{x_1} which is defined with respect to one axis x_1 can be transformed to one with respect to another parallel axis x_2, at a distance d away, by the parallel axis theorem:

$$I_{x_1} = I_{x_2} + Ad^2 \tag{4.25}$$

where A = the area of the shape.

Area moments of inertia of relevance in wind energy are those of a rectangle about axes through its center and about one side and those of an annulus (for example, of a tower in bending). Some of the most useful area moments of inertia are summarized in Table 4.1.

It is also worth noting that, in some situations where the context makes the situation clear, the subscript referring to the axis is left out of the symbol, as in I instead of I_x.

For the case of a shape which is not simple, such as an airfoil, the area moment of inertia can be obtained by numerical integration. An approximation is obtained by dividing the shape into narrow rectangles and summing the moments of inertia of each one. For example, the moment of inertia of an airfoil shape (NACA 4415 airfoil) with respect to the chord of 2 m, as illustrated in Figure 4.3, can be found to be 0.00384 m^4. Rectangles are included for illustration.

The same discretization technique is used for even more complex shapes such as a blade cross-section. A typical such cross-section, with different material properties, is illustrated in Figure 4.4.

Another useful way to consider area moments of inertia is via the radius of gyration, r_{gyr}, which is given by:

$$r_{gyr} = \sqrt{I/A} \tag{4.26}$$

For the airfoil shape here, the area is 0.411 m^2, so the radius of gyration is 0.0967 m.

Table 4.1 Basic area moments of inertia, reported at the centroid of the shape

Rectangle	
$I_{x'} = \dfrac{1}{12}bh^3 \qquad I_x = \dfrac{1}{3}bh^3$ $I_y = \dfrac{1}{12}b^3h \qquad I_y = \dfrac{1}{3}b^3h$ $I_{z'} = \dfrac{bh(b^2 + h^2)}{12}$ $Area = bh$	
Circle:	
$I_x = I_y = \dfrac{1}{4}\pi r^4$ $I_z = \dfrac{1}{2}\pi r^4$ $Area = \pi r^2$	
Annulus:	
$I_x = I_y = \dfrac{1}{4}\pi(r_2^4 - r_1^4)$ $I_z = \dfrac{1}{2}\pi(r_2^4 - r_1^4)$ $Area = \pi(r_2^2 - r_1^2)$	

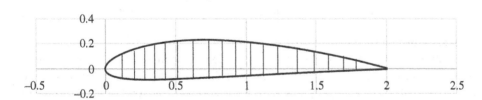

Figure 4.3 Airfoil shape for the NACA 4415 airfoil

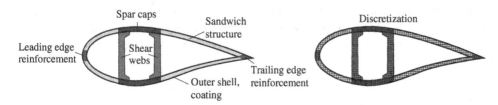

Figure 4.4 Typical blade cross-section illustrating different regions with different material properties, and its discretization for numerical integration of the inertial and stiffness properties

4.3.1.4 Mass Moment of Inertia

The mass moment of inertia is relevant to expressing the inertial loads on a rotating structure. It plays the same role in rotational dynamics as does the mass in linear dynamics. It should not be confused with the area moment of inertia, even though, the two are related in the special case where a cross section of uniform density is considered. In this text, we use the symbol I for the area moment of inertia (m^4), J for the mass moment of inertia (kg m^2), and $\tilde{J}$ for the mass moment of inertia per unit length (kg m). In the literature, the symbols for area and mass moments of inertia are sometimes used interchangeably.

The mass moment of inertia of a volume with respect to a given axis, passing through a given point, is the integral over the volume of the mass multiplied by the squared distance to the axis. For instance, the mass moment of inertia of a volume V with respect to the x axis, and passing through the origin O, is:

$$J_{xx,O} = \iiint_V \left(y^2 + z^2\right) \rho_m \, dV \tag{4.27}$$

where ρ_m is the material density (kg/m^3).

The integral can be approximated by discretizing the volume into masses m_i located at distances r_i from the axis, as follows: $J_{xx,O} \approx \sum_i m_i r_i^2$. It is noted that the mass moment of inertia is a function of the point about which distances are measured. Mass moments of inertia can be transferred to another point using the parallel axis theorem. Subscripts are often neglected in simple context, such as the polar mass moment of inertia of a shaft, where the point of reference and the axis are implicitly assumed to coincide with the shaft axis.

For a rigid three-dimensional object, six mass moments of inertia are needed to describe the dynamics of the object. These inertias are gathered into a symmetric 3×3 matrix, referred to as the inertia tensor $\boldsymbol{J}$. The subscript "xx" in Equation (4.27) indicates the term of the inertia tensor corresponding to a rotation about the x axis and generating an inertial torque in the x direction. The reader is referred to classical structural dynamics textbooks for full expressions of the inertia tensor. The tensor is diagonal when expressed with respect to the center of mass and the principal axes of the rigid body.

Applying Equation (4.27) to a cross-section of a body, one obtains the mass moment of inertia per unit length, which is typically relevant for beam structures. If the density of the cross-section is uniform, then the mass moment of inertia per unit length is $\tilde{J}_{xx} = \rho I_x$ with I_x being the area moment of inertia given in Equation (4.23). The relation between the mass and area moment of inertia only holds for the case of a uniform cross-section.

4.3.1.5 *Inertia Tensor for Three-bladed and Multibladed Rotors

In a detailed analysis of the dynamic behavior of a wind turbine, it is frequently useful to make use of the inertia tensor of the rotor. The inertia tensor of a three-bladed rotor with identical blades has the same property as the one of a uniform disk: the (mass) moment of inertia about any axis in the plane normal to the rotor axis is the same. In its most general form, the inertia tensor of one blade about the rotor center, O, is given by:

$$\boldsymbol{J}_{O,blade\,1} = \begin{bmatrix} J_{xx} & J_{xy} & J_{zx} \\ J_{xy} & J_{yy} & J_{yz} \\ J_{zx} & J_{yz} & J_{zz} \end{bmatrix} \tag{4.28}$$

where the x axis is assumed to be the rotor axis, and y and z form an arbitrary orthonormal basis with the x axis.

The inertia tensors of the other blades in this coordinate system are obtained by multiplying the matrix of blade 1 on both sides by rotation matrices of angles $2\pi/3$ and $4\pi/3$, respectively, about the x axis (see Section 4.3.1.1). The inertia of the rotor is then obtained as:

$$\boldsymbol{J}_{O,rotor} = \begin{bmatrix} 3J_{xx} & 0 & 0 \\ 0 & \frac{3}{2}\left(J_{yy} + J_{zz}\right) & 0 \\ 0 & 0 & \frac{3}{2}\left(J_{yy} + J_{zz}\right) \end{bmatrix} \tag{4.29}$$

It is observed that the moments of inertia are the same for the y and z axis, which were chosen arbitrarily. The inertia tensor has the same matricial representation for any choice of $y - z$ axes: it remains a diagonal matrix, with the same diagonal quantities, irrespective of the azimuthal position of the rotor.

The result applies to a rotor with more than three blades, replacing 3 in Equation (4.29) by the number of blades. The result does not hold for a rotor with 1 or 2 blades, for which the inertia tensor will be periodic with the azimuth and will be a function of the choice of the y and z axes.

4.3.2 Bending of Cantilevered Beams

The bending of beams is an important topic in strength of materials. Wind turbine blades, towers, and monopiles (for offshore turbines) are slender structures which can be represented as cantilevered beams. The fundamental equations for the bending of beams are given in Equations (4.30)–(4.32):

$$\frac{d^2}{dz^2}\left(EI\frac{d^2u}{dz^2}\right) = p(z) \tag{4.30}$$

$$S = -\frac{d}{dz}\left(EI\frac{d^2u}{dz^2}\right) \tag{4.31}$$

$$M = EI\frac{d^2u}{dz^2} \tag{4.32}$$

where z is the coordinate along the beam, u is the deflection of the beam in the x-direction, $p(z)$ is loading per unit length, S is the shear force, M is the moment, and E is the modulus of elasticity. The boundary conditions at the fixed end are $u(0) = 0$ and $\frac{du}{dz}\big|_{z=0} = 0$, i.e., there is no deflection and the slope is 0° (for an initially horizontal beam). Note that the signs of Equations (4.30)–(4.32) are such that the deflection is in the direction of the loading and the shear force is the reaction.

Examples of a uniform cantilever loaded in various ways are summarized in Table 4.2. More complex structures can also be considered as combinations of shells and beams, although those are generally outside the scope of this text. Table 4.2 includes the reaction force at the fixed end; $M(z)$, the moment as function of distance from the fixed end, M_0, the moment at the fixed end; $u(z)$, the deflection as function of distance from the fixed end z; and u_{max}, the maximum deflection. In this table, E = modulus of elasticity (Pa); L = total length of cantilever; I = moment of inertia, (m⁴); F = point loading (N); p = value of uniform loading (N/m); and p_0 = maximum value of variable loading (N/m). Here, I refers to I_y where y is the neutral axis; it is perpendicular to both the z axis and the direction of deflection (x axis), of the shape. In most practical cases, the neutral axis passes through the centroid of the shape. R is the reaction force at the fixed end.

Table 4.2 Deflections of a uniform cantilever subject to various loadings

Types of loading	R	$M(z)$	M_0	$u(z)$	u_{max}
Point load	F	$F(L-z)$	FL	$\frac{Fz^2}{6EI}(3L-z)$	$\frac{FL^3}{3EI}$
Uniform	pL	$\frac{p}{2}(L-z)^2$	$\frac{p}{2}L^2$	$\frac{pz^2}{24EI}(z^2-4Lz+6L^2)$	$\frac{15pL^4}{120EI}$
Decreasing from p_0 to 0	$\frac{p_0L}{2}$	$\frac{p_0(L-z)^3}{6L}$	$\frac{p_0L^2}{6}$	$\frac{p_0z^2}{120LEI}(10L^3-10zL^2+5Lz^2-z^3)$	$\frac{4p_0L^4}{120EI}$
Increasing from 0 to p_0	$\frac{p_0L}{2}$	$p_0(L-z)^2\left(\frac{1}{2}-\frac{(L-z)}{6L}\right)$	$\frac{p_0L^2}{3}$	$\frac{p_0z^2}{120LEI}(20L^3-10zL^2+z^3)$	$\frac{11p_0L^4}{120EI}$

The maximum stress in the cantilever is at the point of attachment and, in addition, it is at the maximum distance, c_b, from the neutral axis. For a wind turbine blade in bending, the neutral axis would be nearly the same as the chord line and c_b would be approximately half the airfoil thickness. (Note the subscript b has been added to distinguish of c_b is different from the chord length, c, which typically uses the same symbol!)

In equation form, the maximum stress, $\sigma_{\max}$, in a beam with area moment of inertia I would be:

$$\sigma_{\max} = \frac{M_{\max} c_b}{I}$$
(4.33)

Example 4.1 Deflections of a Cantilever Beam

Deflections for the cases above are illustrated in Figure 4.5. For these cases, the inputs are: $F = 10\,\text{N}$, $L = 20\,\text{m}$, $EI = 1.0 \times 10^6\,\text{N m}^2$, $p = 0.5\,\text{N/m}$, $p_0 = 1\,\text{N/m}$. [Deflections are shown to be down (negative), as for a weighted horizontal cantilever.] The beam with increasing loading is a simplified representation of the aerodynamic force on the blade (and the tower). The beam with a point force is a simplified representation of the tower with the aerodynamic thrust force acting at the tower's top.

4.3.2.1 Deflection of a Tapered, Hollow Cylinder

One particularly important type of cantilever is a hollow, tapered cylinder, loaded by a horizontal force F applied at the tip. A wind turbine's tower subjected to a thrust force from the rotor is the most notable example of this.

Let us begin by considering a solid constant radius cylinder of length L with radius R. Since the area moment of inertia of the circular cross-section is $I = \dfrac{\pi R^4}{4}$, from Table 4.2, the deflection is immediately:

$$u_{\max} = \frac{4FL^3}{3\pi ER^4}$$
(4.34)

Next, consider a solid tapered cylinder (i.e., a frustum) in which R_A is the radius of the top, and R_B is the radius of the base. In this case, it may be shown that the deflection is given by:

$$u_{\max} = \frac{4FL^3}{3\pi ER_A^4}\frac{1}{r^3} = \frac{4FL^3}{3\pi ER_A R_B^3}$$
(4.35)

where the taper ratio r is given by $r = \dfrac{R_B}{R_A}$.

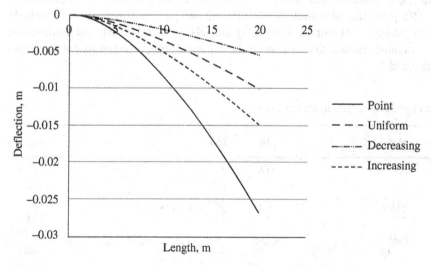

Figure 4.5 Example of deflections of cantilever subject to various loadings

For a hollow, tapered cantilevered cylinder whose wall thickness also varies, a good approximation to the deflection at the top is given by Equation (4.36).

$$u_{max} = \frac{4FL^3}{3\pi E\left(R_{A,out}R_{B,out}^3 - R_{A,in}R_{B,in}^3\right)} \tag{4.36}$$

where the subscripts *in* and *out* refer to the inner and outer radii, respectively.

Example 4.2   Deflection of the NREL 5-MW Tower

Consider the steel tower for the land-based NREL 5-MW turbine, whose properties were given in Table 1.2 of Chapter 1. The modulus of elasticity of the steel is 210 GPa. The rotor diameter is 126 m. Assuming a thrust coefficient of 0.9 at a rated wind of 11.4 m/s and an atmospheric density of 1.225 kg/m^3, the calculated thrust in this situation is 893.3 kN (see Chapter 3). The deflection can be directly predicted from Equation (4.36) to be 0.494 m. For comparison, a finite difference approach (see Section 4.3.2.3) resulted in a prediction of essentially the same value.

4.3.2.2   Mass of a Tapered, Hollow Cylinder

The mass of a tapered, hollow cylinder, m_{hc}, can be found easily as long as the density of the material is uniform. This is done by finding the mass of a solid frustum with radii corresponding to the outer dimensions and subtracting from that the mass of a solid frustum with radii corresponding to the inner dimensions. This can be combined into a single equation:

$$m_{hc} = \rho_m \frac{\pi L}{3}\left(R_{A,out}^2 + R_{A,out}R_{B,out} + R_{B,out}^2 - R_{A,in}^2 - R_{A,in}R_{B,in} - R_{B,in}^2\right) \tag{4.37}$$

where R_B is radius of base and R_A = radius of tip and *in* and *out* refer to inner and outer radii as above and ρ_m is the density of the material. For additional (inertial) properties of hollow cylinders and frustums, see Appendix D.

Example 4.3   Mass of the NREL 5-MW Tower

The tower for the NREL 5-MW turbine was described above. Assuming the density of the steel is 8500 kg/m^3, the mass of the tower, based on Equation (4.37) would be 347 374 kg.

4.3.2.3   Finite Difference Calculations for a Cantilever

The moments and deflections in Equations (4.30)–(4.32) can also be approximated by finite difference calculations in order to obtain the deflected slope for a given loading. These equations result in (4.38)–(4.42):

$$M_i = \sum_{j=i}^{N} S_j\left(z_j - i\Delta z + \frac{\Delta z}{2}\right) \tag{4.38}$$

$$u_{-1} = 0 \tag{4.39}$$

$$u_0 = 0 \tag{4.40}$$

$$u_1 = (\Delta z)^2 \frac{M_0}{EI_0} \tag{4.41}$$

$$u_i = 2u_{i-1} - u_{i-2} + (\Delta z)^2 \frac{M_i}{EI_i} \tag{4.42}$$

where $\Delta z = z_{i+1} - z_i$ for any u, M_i is the moment at the given location, and $S_i = p(z_i)\Delta z$.

This method assumes that there is no deflection at the fixed end (z_0), so $u_0 = 0$, that the slope at the fixed end is also zero and that there is a fictitious node just inside the fixed end z_{-1}. Since the slope = 0 at $z = 0$, it is also the case that $u_{-1} = 0$. The solution may be found by successive calculations.

Example 4.4 Deflection of the NREL 5-MW Tower Using Finite Differences
As already indicated, the deflection of the tower illustrated in Example 4.2 can be calculated using the finite difference approach and will give essentially the same result. Assuming 100 sections, $\Delta z = 0.876$ m. The forces S_i applied to each section are identified; in this case the force of 893.4 kN is only at the top section. The resulting moments M_i are then found for the midpoint of each section from Equation (4.38); for example, the moment at $\Delta z/2$ above the base due to the applied force at the top is 77 862 MNm. Then the displacements are found from Equations (4.38)–(4.42). For example, the displacement at the first station, Δz above the base, is 9.859×10^{-5} m. The displacement at the end (top) is 0.494 m.

4.3.3 Rigid Body Planar Rotation

4.3.3.1 Two-Dimensional Rotation

When a body, such as a wind turbine rotor, is rotating, it acquires angular momentum. Angular momentum, $\boldsymbol{H}$, is defined as:

$$\boldsymbol{H} = \boldsymbol{J}\boldsymbol{\Omega} \tag{4.43}$$

where $\boldsymbol{J}$ is the inertial tensor of the body, and $\boldsymbol{\Omega}$ is the rotational speed of the body. From Newton's law, the sum of the applied moments, $\boldsymbol{M_G}$, about the mass center is equal to the time rate of change of the angular momentum about the mass center. That is:

$$\sum \boldsymbol{M_G} = \dot{\boldsymbol{H}} \tag{4.44}$$

Most wind turbine examples provided in this chapter will only consider the planar case where the rotational speed remains along the same axis. An example can be a shaft whose support does not move. For this example, the angular momentum is $H = J\Omega$ where J is the polar moment of inertia about the rotation axis, and Ω is the rotational speed about (and directed along) this axis. In planar situations, the magnitude of the sum of the moments is:

$$\left|\sum M_G\right| = J\dot{\Omega} = J\alpha \tag{4.45}$$

where α is the angular acceleration of the inertial mass.

In the context of this text, a continuous moment applied to a rotating body is referred to as torque and denoted by "Q." The relation between applied torques and angular acceleration, α, is analogous to that between force and linear acceleration:

$$\sum Q = J\alpha \tag{4.46}$$

When rotating at a constant speed, there is no angular acceleration or deceleration. Thus, the sum of any applied torques must be zero. Accordingly, if a wind turbine rotor is turning at a constant speed in a steady wind, the driving torque from the rotor must be equal in magnitude (and opposite in sign) to the generator torque plus the loss torques in the drivetrain.

4.3.3.2 Rotational Power/Energy

A rotating body contains kinetic energy, E, given by:

$$E = \frac{1}{2}J\Omega^2 \tag{4.47}$$

The power P consumed or generated by a rotating body is given by the product of the torque and the rotational speed:

$$P = Q\Omega \tag{4.48}$$

4.3.4 Gears and Gear Ratios

Gears are frequently used to transfer power from one shaft to another, while maintaining a fixed ratio between the speeds of the shafts. While the input power in an ideal gear train remains equal to the output power, the torques and speed vary in inverse proportion to each other. In going from a smaller gear (1) to a larger one (2), the rotational speed drops, but the torque increases. In general:

$$Q_1\Omega_1 = Q_2\Omega_2 \tag{4.49}$$

The ratio between the speeds of two gears, Ω_1/Ω_2, is inversely proportional to the ratio of the number of teeth on each gear, N_1/N_2. The latter is proportional to the gear diameter. Thus:

$$\Omega_1/\Omega_2 = N_2/N_1 \tag{4.50}$$

When dealing with geared systems, consisting of shafts, inertias, and gears, as shown in Figure 4.6, it is possible to refer shaft stiffnesses (K_i) and inertias (J_i) as equivalent values on a single shaft. (Note that in the following it is assumed that the shafts themselves have no inertia.) This is done by multiplying all stiffnesses and inertias of the geared shaft by n^2 where n is the speed ratio between the two shafts. The equivalent system is shown in Figure 4.7. These relations can be derived by applying principles of kinetic energy (for inertias) and potential energy (for stiffnesses), as described in Thomson and Dahleh (1997). More information on gears and gear trains is provided in Chapter 7.

4.3.5 Gyroscopic Motion

Gyroscopic motion is of particular concern in the design of wind turbines because yawing of the turbine while the rotor is spinning may result in significant gyroscopic loads. The effects of gyroscopic motion are illustrated in the Principal Theorem of the Gyroscope, which is summarized below.

In this example, it is assumed that a rigid body, with a constant polar mass moment of inertia J, is rotating with angular momentum $J\Omega$. The Principal Theorem of the Gyroscope states that if a gyroscope of angular momentum $J\Omega$ rotates with speed ω about an axis perpendicular to Ω ("precesses"), then a moment, $J\Omega\omega$, acts on the gyroscope about an axis perpendicular to both the gyroscopic axis, Ω, and the precession axis, ω. Conversely, an applied moment that is not parallel to Ω can induce precession.

Gyroscopic motion may be considered with the help of Figure 4.8. A bicycle wheel of weight W is shown rotating in the counter-clockwise direction, supported at the end of one axle by a string, a distance a from the string. The wheel would fall down if it were not rotating. In fact, it is rotating, and rather than falling, it precesses in a horizontal plane (counter-clockwise when seen from above).

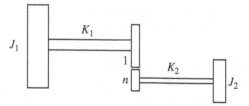

Figure 4.6 Geared system; J, inertia; K, stiffness; n, speed ratio between the shafts; subscripts 1 and 2, gear 1 and 2

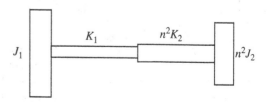

Figure 4.7 System equivalent to geared system; J, inertia; K, stiffness; n, speed ratio between the shafts; subscripts 1 and 2, gear 1 and 2

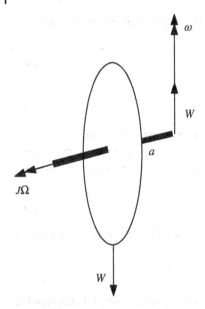

Figure 4.8 Gyroscopic motion; *a*, distance to weight; *J*, inertia; *W*, weight; *ω*, rate of precession; *Ω*, angular velocity

The moment acting on the wheel is *Wa*, so

$$Wa = J\Omega\omega \tag{4.51}$$

Thus, the rate of precession is

$$\omega = \frac{Wa}{J\Omega} \tag{4.52}$$

The relative directions of rotation are related to each other by cross products and the right-hand rule as follows:

$$\sum \boldsymbol{M} = Wa = \omega \times J\Omega \tag{4.53}$$

where $\boldsymbol{M}$, ω, and Ω are the moment, rate of precession, and angular velocity vectors, respectively.

4.3.6 Vibrations: Single Degree of Freedom

The term "vibration" refers to the limited reciprocating motion of a particle or an object in an elastic system. The motion goes to either side of an equilibrium position. Vibration is important in wind turbines because they are partially elastic structures, and they operate in an unsteady environment that tends to result in a vibratory response. The presence of vibrations can result in deflections that must be accounted for in turbine design and can also result in the premature failure due to fatigue of the materials which comprise the turbine. In addition, much of wind turbine operation can be best understood in the context of vibratory motion. The following section provides an overview of those aspects of vibrations most important to wind turbine applications.

The single degree of freedom (DOF) system is the simplest mechanical model, and it is a canonical example for the modeling of advanced structures. Subsequently, Section 4.3.8.4 will present how continuous structures can be represented using multiple DOFs, while Section 4.5 will present examples where the dynamics of the tower or blade can be reduced to a single DOF.

4.3.6.1 Undamped Vibrations

The simplest vibrating system consists of a mass *m* attached to a massless spring with spring constant *k*, as shown in Figure 4.9. When displaced a distance *x* and allowed to freely move, the mass will vibrate back and forth.

Applying Newton's Second Law, the governing equation is:

$$m\ddot{x} = -kx \tag{4.54}$$

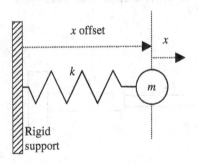

Figure 4.9 Undamped vibrating system; *k*, spring constant; *m*, mass; *x*, displacement

When $t = 0$ at $x = x_0$, the solution consists of harmonic (i.e., sine or cosine) vibrations:

$$x = x_0 \cos(\omega_n t) \tag{4.55}$$

where $\omega_n = \sqrt{k/m}$ is the natural angular frequency of the motion (rad/s). In general, when the mass is not at rest at $t = 0$, the solution contains two sinusoids as shown in Equation (4.56). The term "angular frequency" is also called "cyclic frequency" and is expressed in rad/s. The notation f is used when the "frequency" is expressed in Hz, or cycles/sec. The relation between the two definitions is given by $f = \omega/2\pi$, and the term "frequency" is used loosely depending on the context.

$$x = x_0 \cos(\omega_n t) + \frac{\dot{x}_0}{\omega_n} \sin(\omega_n t) \tag{4.56}$$

where $\dot{x}_0$ is the velocity at $t = 0$.

The solution can also be written in terms of a single sinusoid of amplitude C and phase angle ϕ. That is:

$$x = C \sin(\omega_n t + \phi) \tag{4.57}$$

The amplitude and phase angle can be expressed in terms of the other parameters as:

$$C = \sqrt{x_0^2 + \left(\frac{\dot{x}_0}{\omega_n}\right)^2} \tag{4.58}$$

$$\phi = \tan^{-1}\left[\frac{x_0 \omega_n}{\dot{x}_0}\right] \tag{4.59}$$

4.3.6.2 Damped Vibrations

Vibrations as described above will continue indefinitely. In all real vibrations, the motions will eventually die out. This effect can be modeled by including a viscous damping term. Damping involves a force, usually assumed to be proportional to the velocity, which opposes the motion. Then, the equation of motion becomes:

$$m\ddot{x} = -c\dot{x} - kx \tag{4.60}$$

where c is the damping constant and k is the spring constant. Depending on the ratio of the damping and spring constant, the solution may be oscillatory ("underdamped") or non-oscillatory ("overdamped"). The limiting case between the two is "critically damped," in which case:

$$c = c_c = 2\sqrt{km} = 2m\omega_n \tag{4.61}$$

where c_c is the critical damping coefficient. A nondimensional damping ratio ζ is used to characterize the damping. It is defined as:

$$\zeta = \frac{c}{c_c} = \frac{c}{2\sqrt{km}} \tag{4.62}$$

For $\zeta < 1$, the motion is underdamped; for $\zeta > 1$, the motion is overdamped.

The solution for underdamped oscillation is:

$$x = Ce^{-\zeta\omega_n t} \sin(\omega_d t + \phi) \tag{4.63}$$

where $\omega_d = \omega_n\sqrt{1-\zeta^2}$ is the damped frequency, characteristic of the oscillation. Note that the frequency of damped oscillation is slightly different from that of undamped vibration. The amplitude, C, and phase angle, ϕ, are determined from initial conditions.

4.3.6.3 Forced Harmonic Vibrations

Consider the mass, spring, and damper discussed above. Suppose that the system is driven by a sinusoidal force of magnitude F and angular frequency ω (which is not necessarily equal to ω_n or ω_d). The equation of motion is then:

$$m\ddot{x} + c\dot{x} + kx = F\sin(\omega t) \tag{4.64}$$

It may be shown that the steady state solution to this equation, after all transients have disappeared, is:

$$x = \frac{F}{k} \frac{\sin(\omega t - \phi)}{\sqrt{\left[1 - (\omega/\omega_n)^2\right]^2 + \left[2\zeta(\omega/\omega_n)\right]^2}} \tag{4.65}$$

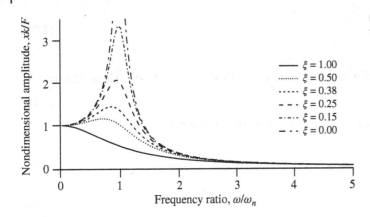

Figure 4.10 System responses to forced vibrations; ζ, nondimensional damping ratio

The output, x, oscillates at the same frequency as the input, F, but with a different amplitude and phase. The fact that the input frequency is retrieved in the output is characteristic of linear systems. The phase angle is:

$$\phi = \tan^{-1}\left[\frac{2\zeta(\omega/\omega_n)}{1-(\omega/\omega_n)^2}\right] \tag{4.66}$$

It is of particular interest to consider the nondimensional response amplitude (see Figure 4.10) which is given by:

$$\frac{xk}{F} = \frac{1}{\sqrt{\left[1-(\omega/\omega_n)^2\right]^2 + \left[2\ \zeta(\omega/\omega_n)\right]^2}} \tag{4.67}$$

As the excitation (forcing) frequency gets closer to the natural frequency, the amplitude of the response gets larger. Increasing the damping reduces the peak value and also shifts it slightly to lower frequencies.

4.3.6.4 *Equation of Motion for an Arbitrary Forcing, and Its Linearization

The study of vibrations (with no or harmonic forcing) presented above is relevant to establish the main modal characteristics of the system. In the general case, the system is under the influence of an external force, F, and the motions of the system can be obtained by solving the following differential equation:

$$m\ddot{x} + c\dot{x} + kx = F(x, \dot{x}, \ddot{x}, t) \tag{4.68}$$

Typically, the equation is recast into a first-order form (see Sections 4.5.2 and 4.5.4), and then solved using a numerical software packages (for instance the function *ode45* in MATLAB, or *solve_ivp* in Python). In simple applications, the forcing may be a function of time only, but in the general case, the forcing may be a nonlinear function of the position, speed, and even acceleration. The nonlinearities introduced by the forcing will introduce stiffness, damping, and inertia on top of the structural values and will therefore change the frequencies of the system, and affect its stability. These additional quantities can be obtained by "linearizing" (performing a first-order Taylor series expansion) the equation of the force around a point of interest, referred to as the operating point. For instance, the additional stiffness from the force is obtained using the partial derivative as:

$k_F = -\dfrac{\partial F}{\partial x}(x_0, \dot{x}_0, \ddot{x}_0, t)$, where the subscript 0 is used to denote the operating point. More details on linearization are found in Section 4.8.1.

4.3.7 Rotational Equations of Motion and Vibration

Many wind turbine components, particularly those in the drivetrain, can be modeled by a series of discs connected by shafts. In some simplified models, the discs are assumed to have inertia but are completely rigid, whereas the shafts have stiffness, but no inertia.

The case of a body with a polar mass moment of inertia J_r, attached to a rigid support via a rotational spring with rotational stiffness k, and subject to an external torque Q, is shown in Figure 4.11. The equation of motion is given by Equation (4.69).

$$J_r\ddot{\psi} + k\psi = Q \tag{4.69}$$

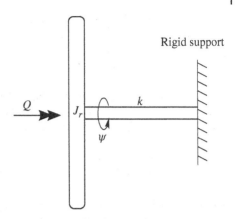

Figure 4.11 Rotational vibrating system; J_r polar mass moment of inertia; k, rotational stiffness of spring; ψ, azimuth angle

Solutions for rotational vibration are analogous to those for linear vibrations discussed previously.

Equation (4.69) can be used to model the torsion of the shaft, when the drivetrain is clamped due to the emergency break being fully active. In this case, J_r is the inertia of the rotor, and k is the torsional stiffness of the shaft. The problem shown in Figure 4.11 is said to have "fixed-free" boundary conditions. The natural frequency of the drivetrain torsion in this "fixed-free" situation is obtained as:

$$\omega_{fixed-free} = \sqrt{k/J_r} \tag{4.70}$$

In the absence of a brake (in normal operating conditions), and omitting the shaft torsion, the equation of motion of the drivetrain is:

$$J_r\ddot{\psi} = Q_a - Q_g \tag{4.71}$$

where Q_a is the aerodynamic torque, and Q_g is the generator torque. It is seen that for the rotor speed to be constant ($\ddot{\psi} = 0$), the generator torque needs to match the aerodynamic torque. The natural frequency of this "free-free" system is zero in the absence of shaft torsional stiffness. A zero natural frequency indeed corresponds to a rigid body motion of the entire structure. Similar equations are used in Section 4.5.3.1 where a two DOF model for a wind turbine is introduced, and in the linearized hinge–spring blade rotor model, which is described in Section 4.7. The drivetrain equations and "free-free" natural frequencies including shaft torsion are presented in Section 4.5.3.2.

In addition, the natural frequencies for rotating systems can be determined by Holzer's method. Here, sequence equations are used to determine each angle and torque along the shaft. The reader is referred to Thomson and Dahleh (1997) or similar texts for more information.

4.3.8 Introduction to Vibration of Cantilevered Beams

It is of interest to consider in some detail the vibration of a cantilevered beam, since many wind turbine components are slender and can therefore be modeled as cantilevered beams. These include, in particular, the tower which supports the rotor nacelle assembly and the blades themselves.

4.3.8.1 Modes and Mode Shapes

Recall that the one-degree-of-freedom system, consisting of a vibrating mass on a massless spring, oscillates with a single characteristic natural frequency. There is also only one path that the mass will take during its motion. For multiple DOFs (e.g., multiple masses), the number of natural frequencies and possible paths will increase. For continuous objects, there is an infinite number of natural frequencies and DOFs. To each natural frequency, there corresponds a characteristic mode shape of vibration. In practice, however, only the lowest few natural frequencies of a beam are usually important. In this section, the general equation of beam vibration is introduced first. Then, a uniform beam is studied since it has a closed-form solution, and it serves as a useful illustration. In Sections 4.5.5 and 4.6, the more general equation is discussed in greater detail.

4.3.8.2 General Equation for Vibration of Beams

The general case of vibration of a linear beam in two dimensions is described by the Euler–Bernoulli equation, often referred to as simply the Euler beam equation, as given in Equation (4.72).

$$\tilde{m}(z)\frac{\partial^2 u}{\partial t^2} + c(z)\frac{\partial u}{\partial t} + \frac{\partial^2}{\partial z^2}\left[EI(z)\frac{\partial^2 u}{\partial z^2}\right] = p(z,t) + \frac{\partial}{\partial z}\left[N(z)\frac{\partial u}{\partial z}\right] \tag{4.72}$$

Here, $\tilde{m}(z)$ is the mass distribution per unit length, $c(z)$ is the damping per unit length, $EI(z)$ is the bending stiffness, $p(z,t)$ is the transverse load per unit length, and $N(z)$ is the axial load along the z axis. The boundary conditions for a cantilevered (clamped-free) beam of length L are: $u(0,t) = \frac{\partial u}{\partial z}(0,t) = 0$ (at the clamped end) and $\frac{\partial^2 u}{\partial z^2}(L,t) = \frac{\partial^3 u}{\partial z^3}(L,t) = 0$ (at the free end). The area moment of inertia, $I(z)$, may also vary along the beam. The modulus of elasticity E here is assumed to be constant.

Equation (4.30), which describes the steady deflection of a beam, can be retrieved from the dynamic equation above. The reader is referred to other sources (e.g., Thomson and Dahleh, 1997) for the details of the development of the Euler–Bernoulli beam equation.

4.3.8.3 Vibration of Uniform Cantilevered Beams

For a uniform beam (i.e., constant cross-section, so that $\tilde{m}(z) = \tilde{m}$ and $I(z) = I$), without damping and applied loads, Equation (4.72) reduces to the following free vibration equation:

$$\tilde{m}\frac{\partial^2 u}{\partial t^2} + EI\frac{\partial^4 u}{\partial z^4} = 0 \tag{4.73}$$

In this case, a closed-form solution may be found (see e.g., Thomson and Dahleh, 1997). The natural frequencies ω_i(rad/s) of the uniform cantilevered beam of length L are given by:

$$\omega_i = \frac{(\beta L)_i^2}{L^2}\sqrt{\frac{EI}{\tilde{m}}} \tag{4.74}$$

for values of $(\beta L)_i$ that solve:

$$\cosh(\beta L)_i \cos(\beta L)_i + 1 = 0 \tag{4.75}$$

The dimensionless parameters $(\beta L)_i^2$ are constants, and for the case of a uniform cantilevered beam, the numerical values for the first three modes are approximately 3.52, 22.4, and 61.7. The corresponding mode shapes are given by

$$u_i = A\left\{\cosh\left(\frac{(\beta L)_i}{L}z\right) - \cos\left(\frac{(\beta L)_i}{L}z\right) - \frac{\sinh(\beta L)_i - \sin(\beta L)_i}{\cosh(\beta L)_i + \cos(\beta L)_i}\left[\sinh\left(\frac{(\beta L)_i}{L}z\right) - \sin\left(\frac{(\beta L)_i}{L}z\right)\right]\right\} \tag{4.76}$$

where u_i is the deflection as a function of distance z along the beam for the mode shape i, and where A is an arbitrary scaling constant. Since A is arbitrary, a normalized value is used for which a deflection of $u_i = 1$ at the free end of the beam is obtained.

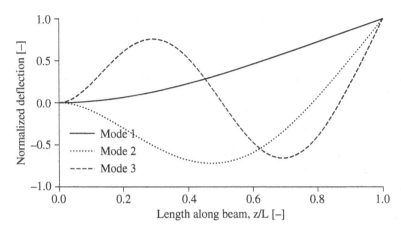

Figure 4.12 Vibration modes of uniform cantilevered beam

Figure 4.12 illustrates the shape of the first three vibration modes of a uniform cantilevered beam, based on Equation (4.76).

4.3.8.4 Vibration of Nonuniform Cantilevered Beams

Modeling the vibration of nonuniform beams is in general much more complex than that of the uniform cantilever. Some useful methods, including the Myklestad method and assumed modes shapes, are discussed in Sections 4.4–4.6. In certain situations, it is possible to simplify the problem through the use of effective stiffness and effective mass, as summarized next.

Effective Stiffness of a Cantilevered Beam

The elasticity of a cantilevered beam may be characterized by an effective stiffness, K, which corresponds to the spring constant of a single-degree-of-freedom system. One way to determine the stiffness is to use the deflection resulting from a given force, as discussed in Section 4.3.2. The stiffness is then given by:

$$K = F/u \tag{4.77}$$

Where F is force applied to the end of the beam, and u is the distance it is displaced. Other ways to find the effective stiffness of cantilevers are discussed in Section 4.6.

Effective Mass of a Cantilever with a Mass on the End

It is also often useful to characterize the combined mass of a cantilever with a mass on the end by a single effective mass, M. This effective mass is sometimes referred to as a generalized mass or modal mass. When the effective stiffness K and the first mode natural frequency are known, the effective mass can be found using:

$$M = K/\omega_n^2 \tag{4.78}$$

Where ω_n is the natural frequency in rad/s.

4.4 Methods for Modeling Wind Turbine Structural Response

It is frequently necessary to determine various aspects of wind turbine response, such as stresses within the structure, deflections, and natural frequencies. This is typically done using mathematical models. These are sets of mathematical relations that are used to describe the behavior of the real system. Different methods are used to transform the equations of motion of the complex and continuous structure into a finite set of equations that can be solved analytically or numerically. Models can differ greatly in their complexity, and so different types of models are used for different applications.

Typical applications for mathematical models in wind turbine design are the following:

- detailed model for investigating fatigue, deflection, and extreme response of the full turbine or a subsystem (tip flaps, blades, pitch linkages, gear trains, etc.);
- model for determining natural frequency of component or complete system;
- reduced-order models for control, digital-twin, educational purposes, etc.

Typically, detailed models require numerical solutions and are implemented in digital computer code. One model that is particularly comprehensive is NREL's *OpenFAST*, which is used as an example throughout this book. *OpenFAST* and some other computer models are described in Chapter 8. This section discusses some of the methods that are used in those models.

The most common methods used in the structural analysis of wind turbines are:

- lumped parameter method;
- finite element method (FEM);
- modal analysis or assumed shape function (Rayleigh–Ritz) approaches;
- multibody analysis.

The different methods are briefly introduced below. The application of the shape function approach to wind energy is presented in Section 4.6, and an example of lumped parameter method is presented in Section 4.7 with the linearized hinge-spring model.

4.4.1 Lumped Parameter Method

Lumped parameter methods are typically used to obtain simple analytical equations of motion of a system. They are typically low-fidelity approaches. A lumped parameter model is one in which a continuous system is considered to be made up of a relatively small number of simple bodies (such as lumped masses and springs). For example, the drivetrain of a wind turbine consists of a number of continuous rotating components, such as the rotor itself, shafts, gears, and the generator rotor. In modeling a drivetrain, it is common to characterize it as a few lumped inertias and stiffnesses. See Chopra (2017) for a discussion of the use of lumped masses. The Myklestad method (see Section 4.5.5) is another example of the lumped parameter method.

4.4.2 Modal Analysis Method and Shape Function Approach

Modal analysis is a method used to represent the dynamics of a system by a partial (or complete) sum of the dynamics of its mode shapes. Instead of dividing the geometry of the system into local, lumped bodies, the method uses global mode shapes, each characterizing how the full system is deforming. Prior to applying the method, the continuous system needs to be discretized into a finite number of DOFs (see Section 4.5), and an eigenvalue analysis of the discretized system is carried out to extract the mode shapes. The method then takes advantage of the fact that the individual vibration modes, for each natural frequency, are orthogonal to each other. By inserting the modes into the equations of motions, the coupled equations are transformed into uncoupled "modal" equations which can each be solved separately. The results from each of the modal equations are then added (by linear superposition) to give the complete result. Modal analysis is therefore a "linear" method and can be seen as a mid-fidelity approach. The mode shapes may be found by applying basic techniques such as the Euler or Myklestad method or more detailed techniques which use FEM. The basics of modal analysis using assumed shape functions are described in Section 4.6. More details on modal analysis may be found in Fertis (1995) or Chopra (2017).

The shape function approach, also referred to as the Rayleigh–Ritz method, can be seen as an extension of the modal analysis method. Instead of using mode shapes, the method uses arbitrary (but carefully chosen) shape functions. Since the shape functions can be different than mode shapes, they do not uncouple the equations of motion.

A few shape functions are typically selected (assuming that they are representative of the system) such that the system can be studied with a reduced number of DOFs (one DOF per shape function). Details can be found in textbooks, such as Shabana (2013), and in Section 4.6.

4.4.3 Finite Element Method

The FEM is a general method for solving partial differential equations, in particular, structural dynamics equations. The technique is based on dividing the structure into a large number of relatively small elements. Each element includes a number of "nodes." Some of these nodes may be interior to the element; others are on the boundary. Elements only interact through the nodes at the boundary. Each element is characterized by a number of parameters, such as thickness, density, stiffness, and shear modulus. Also associated with each node are displacements or DOFs. These may include translations and rotations. The FEM is most often used to study in detail individual components within a larger system. For more complex systems consisting of various components moving differently with respect to each other, other approaches such as multibody dynamics (see below) are used. A FEM model may then be used subsequently, for example, to study variations in stresses within a component. See Figure 4.13 for an illustration of blade stresses calculated on a blade using FEM. More details on the finite element method are given, e.g., by Rao (2018) and Chopra (2017).

4.4.4 Multibody Analysis

Multibody analysis refers to the modeling of the motion of a mechanical system comprising more than one component or "body." Bodies are distinguishable subdivisions of the larger structure, which are relatively uniform within themselves. Examples of bodies for a wind turbine include beams (such as the blades, tower, or shaft), shaft, couplings, generators, and gears. The bodies may be rigid or flexible and move in a variety of ways with respect to each other. Bodies are joined together by "links" (kinematic joints). The links introduce a set of constraint equations that state how the coordinates of a body are related to the ones of another body. For a wind turbine, a typical constraint is the one relating the shaft to the nacelle, which can be modeled as a hinge joint. Similar joints are also found at the yaw bearings, between the nacelle and the tower, or the pitch bearing, between the hub and the blade. Note that the description of the bodies in multibody systems is significantly more detailed than those of lumped parameter models. The multibody method involves the creation of dynamical equations involving the various bodies and their constraints. The handling of the constraints is often the difficult part in establishing the solution method. There are two main ways to impose constraints in the multibody dynamics contexts: the direct elimination method and the Lagrange multiplier method (see e.g., Cook, 2001). Constraint equations can be avoided by inherently using the kinematic relationships between bodies. This is the approach used in the ElastoDyn module of *OpenFAST* or the wind turbine elastic code *Flex* (see Jonkman, 2013; Branlard, 2019). Once the equations of motions are set, they are solved by suitable numerical techniques. More details may be found in a variety of texts, such as Shabana (2013).

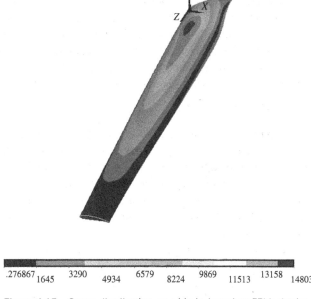

.276867 1645   3290   4934   6579   8224   9869   11513   13158   14803

Figure 4.13 Stress distribution on a blade, based on FEM; shades indicate stress level, units arbitrary

4.4.5 Multi-Physics (Aeroelastic) Wind Turbine Models

Models used to simulate wind turbines need to combine a structural analysis model (such as the ones described in the previous paragraphs) with models for aerodynamics, hydrodynamics, mooring dynamics, soil dynamics, and controls. For this reason, wind turbine models are called "multi-physics," and they require the solving of coupled equations between all models. In this chapter, we are treating the structural dynamics aspects independent of this coupling, but the reader should be aware that fully coupled models are now state of the art. Historically, mostly aerodynamic and elasticity were considered, and the term "aeroelastic" model was commonly used.

4.5 *Equations of Motion of a Wind Turbine – Discrete and Continuous systems

The different modeling methods presented in Section 4.3.8.4 lead to sets of equations, referred to as equations of motions, which are typically solved numerically to obtain the values of the DOFs of the system as functions of time. For a wind turbine, the DOFs could be the position and velocities of the rotor (discrete system), or the positions and velocities of all the individual masses forming the blades, shaft, and tower (continuous system). This section looks more closely at the notion of equations of motion for discrete and continuous systems.

4.5.1 Derivation of the Equations of Motion

There are different methods to establish the equations of motion of a given system. The common approaches mentioned in Section 4.3.8.4 (finite element method, lumped parameters, and shape function) may be grouped into two categories: direct methods, and variational methods. Direct methods include the force-balance method, e.g., Newton's Second Law. These methods are straightforward for simple systems. Variational methods include the principle of virtual work, Lagrange's equation, and Hamilton's principle. The presentation of these different methods may be found in most structural dynamics books, for example, Nayfeh and Pai (2004).

4.5.2 Equations of Motion for Multiple Degrees of Freedom System

The equation of motion of a single DOF system was presented in Section 4.3.6. The equations of motion of a multiple-DOFs system (discrete system) may be established with the methods mentioned in Section 4.5.1. They take a similar form to that of the single DOF system. Once assembled, the mechanical equations typically form a system of n, coupled, second-order, ordinary differential equations, where n is the number of DOFs of the system. The equations are written into a vectorial form by gathering the DOFs into a vector $x = [x_1, x_2, \cdots, x_n]$. The DOFs are often referred to as "generalized coordinates." The equations of motion take the form of Equation (4.79):

$$M\ddot{x} = f - Kx - C\dot{x} \tag{4.79}$$

where M is the mass matrix; C is the damping matrix; K is the stiffness matrix; and f is the vector of generalized forces. Equation (4.79) simply states that "mass times acceleration" is equal to the forcing. The damping and stiffness matrices are optional, they represent the part of the forcing vector that is linear with respect to the displacements and velocities, respectively. The matrices have dimensions $n \times n$, and the vectors have dimensions $n \times 1$. Equation (4.79) can be solved for the acceleration if the loads, velocities, and deformations are known at a given time step. From the acceleration, the velocities and deformation can be estimated for the next time step, and the integration continues.

The system given in Equation (4.79) can be rewritten as a first-order system as:

$$\begin{bmatrix} \dot{x} \\ \ddot{x} \end{bmatrix} = \begin{bmatrix} 0 & I \\ -M^{-1}K & -M^{-1}C \end{bmatrix} \begin{bmatrix} x \\ \dot{x} \end{bmatrix} + \begin{bmatrix} 0 \\ M^{-1}f \end{bmatrix} \tag{4.80}$$

The form above is referred to as the "state-space" formulation and takes the form $\dot{z} = Az + Bu$, where A is referred to as the state matrix, and $u = f$ is the input vector. Such formalism is mostly relevant when all matrices are constant and time-invariant (linear time-invariant system).

Solution methods for Equation (4.80) are discussed in Section 4.5.4. It will be shown in Section 4.5.5 that the equations of motion of continuous systems are typically discretized and result in a system of equations of the form given in Equation (4.80).

4.5.3 Examples of Rigid Body Wind Turbine Models with Few Degrees of Freedom

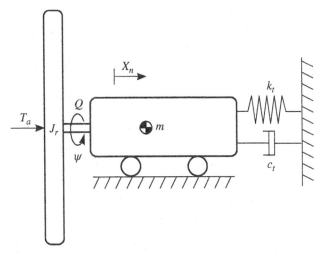

Figure 4.14 Model of a wind turbine, using tower-top displacement (x_n) and shaft rotation (ψ) as degrees of freedom

This section presents different wind turbine models to illustrate the general formalism given in Section 4.5.2. Approaches to solving the state space equations are described in Section 4.5.4.

4.5.3.1 Example: Two-DOF Model of an Onshore Wind Turbine (Shaft and Tower-top)

A simple two-DOF model of an onshore wind turbine is presented. A mix between the lumped-mass (for the rotor and drivetrain) and the assumed shape function approaches (for the tower) is used. The nacelle is assumed to translate in the fore–aft direction. The DOFs consist of the horizontal nacelle displacement, x_n, and the shaft rotation ψ. The two DOFs are gathered into the vector $x = [x_n, \psi]$. A sketch of the model is shown in Figure 4.14. The following subscripts are used to distinguish the bodies: n, nacelle; t, tower; r, rotor.

The rotor is modeled as a rigid disk, with mass moment of inertia J_r about the shaft axis. The dynamic response of the tower is modeled using a generalized spring and a generalized damper of constants k_t and c_t, respectively. The generalized mass of the system is denoted by m. It consists of the mass of the rotor nacelle assembly and the effective mass of the tower. A method to determine the generalized mass and stiffness of the tower is given in Section 4.6.2. The wind is exerting the aerodynamic thrust T_a and torque Q_a on the rotor. The generator is assumed to exert a counteracting torque on the shaft, so that the total torque on the shaft is $Q = Q_a - Q_g$. The equations of motion are readily obtained and put in the matrix form of Equation (4.81), as follows:

$$\begin{bmatrix} m & 0 \\ 0 & J_r \end{bmatrix} \begin{bmatrix} \ddot{x}_n \\ \ddot{\psi} \end{bmatrix} + \begin{bmatrix} c_t & 0 \\ 0 & 0 \end{bmatrix} \begin{bmatrix} \dot{x}_n \\ \dot{\psi} \end{bmatrix} + \begin{bmatrix} k_t & 0 \\ 0 & 0 \end{bmatrix} \begin{bmatrix} x_n \\ \psi \end{bmatrix} = \begin{bmatrix} T_a(x_n, t) \\ Q_a(x_n, t) - Q_g(x_n, t) \end{bmatrix} \tag{4.81}$$

In this situation, the system of equations is coupled via the aerodynamic loads, but the system matrices are uncoupled (they are diagonal). The aerodynamic forces can be modeled as follows:

$$T_a = \frac{1}{2} \rho \pi R^2 U_{rel}^2 C_T(\theta_p, \lambda) \qquad Q_a = \frac{1}{2} \rho \pi R^2 U_{rel}^2 C_Q(\theta_p, \lambda) \tag{4.82}$$

where $\lambda = \dot{\psi} R / U_{rel}$ is the instantaneous tip-speed ratio; θ_p is the blade pitch; $U_{rel} = U_0 - \dot{x}_n$ is the relative horizontal wind speed; and C_T and C_Q are the thrust coefficient and torque coefficients. These two functions may be obtained using an aerodynamic BEM code (Chapter 3) and stored as tabulated data. The forces are nonlinear functions of the states x_n and ψ.

4.5.3.2 Example: Two-DOF Drivetrain Model

The torsion of the main shaft of a wind turbine is an important aspect to consider when looking at the dynamics of the drivetrain (gearbox, bearings, and shaft bendings are also important but more complex to model). This example presents the equation of motions of a wind turbine drivetrain, accounting for torsion of the shaft. Equations without shaft torsion were presented in Section 4.3.7. The equations presented below are written with respect to the LSS, which is on the rotor side, as opposed to the high-speed shaft (HSS) where the generator is. The gear ratio, n_g, is introduced to relate inertias and torques from the LSS to the HSS (see Section 4.3.4). For instance, the mass moment of inertia of the generator on the LSS is related to the one on the HSS as: $J_g = J_{g,HSS}\, n_g^2$. The generator torque on the LSS is expressed with respect to the one on the HSS as: $Q_g = Q_{g,\ HSS}\, n_g$. The DOFs are the azimuthal position of the rotor (LSS), ψ, and the torsion angle of the shaft, ν. The azimuthal position of the HSS, denoted ψ_{HSS}, is related to the two DOFs as follows: $\nu = \psi - \psi_{HSS}/n_g$. The additional parameters of the model are the mass moment of inertia of the rotor, J_r, the stiffness associated with the torsion of the shaft, k, and the damping c associated with the shaft torsion. The equations of motions are obtained as:

$$\begin{bmatrix} J_r + J_g & -J_g \\ -J_g & J_g \end{bmatrix} \begin{bmatrix} \ddot{\psi} \\ \ddot{\nu} \end{bmatrix} + \begin{bmatrix} 0 & 0 \\ 0 & c \end{bmatrix} \begin{bmatrix} \dot{\psi} \\ \dot{\nu} \end{bmatrix} + \begin{bmatrix} 0 & 0 \\ 0 & k \end{bmatrix} \begin{bmatrix} \psi \\ \nu \end{bmatrix} = \begin{bmatrix} Q_a - Q_g \\ Q_g \end{bmatrix} \tag{4.83}$$

where Q_a is the aerodynamic torque acting on the LSS. Equation (4.83) is the form used in *OpenFAST*. The equations could equivalently be written using ψ_{HSS} instead of ν, but even if the two models are mathematically equivalent, Equation (4.83) is more stable numerically. The natural frequency of the drive-train torsion, in this "free-free" configuration, is given by:

$$\omega_{free-free} = \sqrt{\frac{k}{J_r} + \frac{k}{J_g n_g^2}} \tag{4.84}$$

This can be compared to the fixed-free frequency given in Equation (4.70).

4.5.3.3 Example: Two-DOF Model of a Floating Wind Turbine

The example is provided at the end of Chapter 11.

4.5.4 Solution of the System of Equations

Solutions of the equations of motion can be obtained analytically for some simple cases, but numerical integration is required for advanced cases.

An analytical treatment of the equations is possible when the system matrices are constant. The analytical methods use the eigenvectors of the second-order system, or, the left and right eigenvectors of the first-order system, to decouple the system of equations. A special, but common case for the analytical treatment, is the one where the damping matrix has the following form $\boldsymbol{C} = \alpha \boldsymbol{M} + \beta \boldsymbol{K}$, with α and β two damping coefficients. This is referred to as "Rayleigh damping." More details on analytical solution methods may be found in Nayfeh and Pai (2004).

Numerical methods are used to solve the equations of motion of general systems (with time-dependent matrices, nonlinear forces, or general damping). The common way to solve the equations of motion is to use the first-order form given in Equation (4.80). The general form of a first-order differential equation is: $\dfrac{d\boldsymbol{z}}{dt} = \boldsymbol{g}(\boldsymbol{z}, t)$. The comparison with Equation (4.80) leads to $\boldsymbol{z} = [\boldsymbol{x}, \dot{\boldsymbol{x}}]^T$, and $\boldsymbol{g}$ corresponds to the entire right-hand side of Equation (4.80). Various numerical schemes exist to obtain a solution of this equation, such as first-order forward Euler, fourth-order Runge–Kutta, Adams–Bashforth, and Adams–Moulton. These numerical schemes can be implemented in a given programming language, or preexisting integration packages can be used (e.g., the function *ode45* in MATLAB, or *solve_ivp* in Python). In the latter case, the user needs to implement a function that returns the right-hand side of Equation (4.80), as a function of the state vector $\mathbf{y}$. The Runge–Kutta–Nystrom scheme can be used to integrate the second-order system of equations directly; see (Hansen, 2008).

4.5.5 Continuous Systems

Wind turbines are continuous structures and special treatments are needed to account for the flexibility of their components. In Section 4.5.3, models using a finite number of DOFs were presented under rigid body assumptions. In this section, it is shown how the flexibility of a continuous structure can be accounted for and discretized into a finite number of DOFs so that the equation of motion can be solved. Different numerical methods are obtained based on the discretization technique used.

4.5.5.1 Equations of Motion

A continuous system consists of an infinite number of infinitesimal "particles" and accordingly requires an infinite number of DOFs and ordinary differential equations to describe it. Because the motion is continuous from one particle to its neighbor, the motion can be described using a continuous displacement field that is a function of space and time. The governing equations can be written in terms of a displacement field, leading to partial differential equations and dedicated boundary conditions. An example of the equation of motion for a continuous beam was given in Equation (4.72), which indeed involved a partial differential equation of the displacement field $u(x, t)$. Closed-form solutions of the response of a continuous system are rare. In particular cases, the use of eigenfunctions can decouple the system equations into uncoupled modal equations which can be solved. In most cases, however, the analysis of the system requires a discretization of the system, such that the equations are recast into the discrete form of Equation (4.80).

4.5.5.2 Discretization Techniques

To obtain numerical solution of the equations of motion of a continuous system, discretization techniques are needed. The most common methods, as introduced in Section 4.3.8.4, are collocation methods (lumped methods, finite differences), finite element methods, and the assumed shape function (Rayleigh–Ritz) method. The methods transform the continuous equations (requiring an infinite number of DOFs) to discrete equations with a finite number of DOFs. The equations of motion then have the form of Equation (4.79) and can readily be solved by numerical software packages. The developments of these discretization techniques are involved, so the reader is referred to books on elasticity for more details (see Nayfeh and Pai, 2004, or Bisplinghoff and Ashley, 1962). The assumed shape function method is discussed in detail in Section 4.6.

4.5.5.3 Solution

Once the discretization step is completed, the equations of motion are reduced to a set of second-order ordinary differential equations, which can be put in the matrix form presented in Section 4.5.2. Solution methods presented in Section 4.5.4 can be applied to solve the system. In the context of wind energy, the structural codes typically rely on either the finite-element method (e.g., HAWC2, HAWC2, 2021) or the Rayleigh–Ritz method (*OpenFAST*, Jonkman, 2013, *Flex*, e.g., Branlard, 2019). Analytical results, or simple models, are typically derived using lumped methods or the Rayleigh–Ritz approach. One approach is the Myklestad Method, which is described in Section 4.5.5.4.

4.5.5.4 Myklestad Method

The Myklestad method is a method to determine frequencies and mode shapes of a beam. The Myklestad method relies on a collocation technique to discretize the continuous equation of the vibration of a beam into a finite set of equations. The beam is thereby modeled with multiple lumped masses connected by massless beam elements. More details may be found in Thomson and Dahleh (1997).

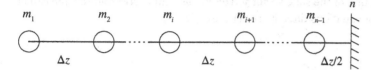

Figure 4.15 Model of cantilevered beam; m_i, mass; Δz, distance between masses; n, number of stations

The method is illustrated in Figure 4.15 for a cantilevered beam. The beam, which is assumed to lie on the x axis, is divided into $n - 1$ stations (sections) with the same number of masses, m_i. An additional station is at the point of attachment. All the distances Δz_i between masses are the same (equal to Δz). In the figure, it is assumed that the masses are located at the center of equal-sized sections, so the distance from the mass closest to the attachment is 1/2 of the others. The flexible connections have moments of inertia I_i and modulus of elasticity E_i. The beam in this example may be rotating with a speed Ω about an axis perpendicular to the beam and passing through station "n."

The Myklestad method involves solving a set of sequence equations. The sequence equations can be developed by considering the forces and moments acting on each of the masses and the point of attachment. Figure 4.16 illustrates the free-body diagram of one section of a possibly rotating beam. The diagram shows shear forces, S_i, inertial (centrifugal) forces, F_i, bending moments, M_i, deflections, u_i, distance along horizontal axis, z, natural frequency, ω_n, and angular deflections, θ_i. The beam may be rotating with an angular speed Ω.

The complete set of sequence equations for the beam follows. The centrifugal forces at distance z_j from the attachment are:

$$F_i = \Omega^2 \sum_{j=1}^{i-1} m_j z_j \tag{4.85}$$

It follows that:

$$F_{i+1} = F_i + \Omega^2 m_i z_i \tag{4.86}$$

Using small-angle approximations, the shear forces are:

$$S_{i+1} = S_i - m_i \omega_n^2 u_i - F_{i+1} \theta_i \tag{4.87}$$

It can be shown that the moments are:

$$M_{i+1} = \left[M_i - S_{i+1} \left(\Delta z - F_{i+1} \frac{\Delta z^3}{3E_i I_i} \right) + \theta_i \Delta z F_{i+1} \right] \Big/ \left(1 - F_{i+1} \frac{\Delta z^2}{2E_i I_i} \right) \tag{4.88}$$

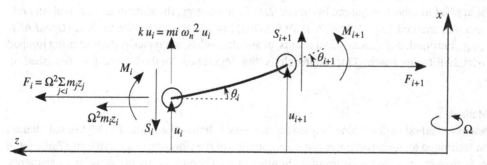

Figure 4.16 Free body diagram of a beam section for a rotating beam. For definition of variables, see text

The angles of the beam sections from horizontal are:

$$\theta_{i+1} = \theta_i + M_{i+1}\left(\frac{\Delta z}{E_i I_i}\right) + S_{i+1}\left(\frac{\Delta z^2}{2E_i I_i}\right) \tag{4.89}$$

Finally, the deflections from a horizontal line passing through the fixed end are:

$$u_{i+1} = u_i + \theta_i \Delta z + M_{i+1}\left(\frac{\Delta z^2}{2E_i I_i}\right) + S_{i+1}\left(\frac{\Delta z^3}{3E_i I_i}\right) \tag{4.90}$$

The series of equations is solved by iteratively finding natural frequencies which result in a calculated deflection of zero at the point of attachment. The process begins by assuming a natural frequency ω_n and performing a series of calculations. The calculations are repeated with a new assumption for the natural frequency until a deflection sufficiently close to zero is found. Two sets of calculations are undertaken. They start at the free end of the beam first with $u_{1,1} = 1$ and $\theta_{1,1} = 0$ and then with $u_{1,2} = 0$ and $\theta_{1,2} = 1$, where the second subscript refers to the sets of calculations. Calculations are performed sequentially for each section until section n (fixed end) is reached. The calculations yield $u_{n,1}$, $u_{n,2}$, $\theta_{n,1}$, and $\theta_{n,2}$. The desired deflection (which should approach zero) is:

$$u_n = u_{n,1} - u_{n,2}(\theta_{n,1}/\theta_{n,2}) \tag{4.91}$$

The entire process can be repeated to find additional natural frequencies. There should be as many natural frequencies as there are masses. Because inertial forces on the rotating beam effectively stiffen the beam, natural frequencies of a rotating beam will be higher than those of the same nonrotating beam.

The mode shape can be created by applying Equation (4.91) for any given natural frequency, with $u_1 = 1$ and $\theta_1 = \theta_{n,1}/\theta_{n,2}$ as initial conditions at the end of the beam.

This method is described in detail by Thomson and Dahleh (1997).

Example 4.5   Natural Frequency of a Tower

The Myklestad method can be used to estimate the natural frequency of a wind turbine tower with a rotor nacelle assembly mounted on the top. For example, consider the NREL 5-MW turbine whose tower was described previously in Section 4.3.2. The rotor nacelle assembly has a mass of 350 000 kg. With the tower subdivided into 100 equal-length sections, the first mode natural frequency was estimated to be $f_n = 0.3382$ Hz. More comprehensive assessment of the natural frequencies, including the higher order modes and mode shapes, can be found with NREL's software *BModes* (Bir, 2005). For this particular case, the result from *BModes* was 0.3362 Hz.

4.6   *Wind Turbine Models Using the Assumed Shape Function Approach

The assumed shape function approach is a convenient method to describe the motion of a flexible body. It is the underlying method used in the aeroelastic code *Flex* to model the blades, the tower and potentially the substructure, and in the ElastoDyn module of *OpenFAST* to model the blades and tower. The current section is structured with increasing levels of complexity. We start by providing the equations for an isolated beam expressed using one shape function without proof. We then use these equations in two problems highly relevant to wind energy: the tower softening as a result of its top mass, and the centrifugal stiffening of a blade as a result of its rotation. We then present a general treatment of the shape function approach. The reader may refer to Shabana (2013) for general treatment of the shape function approach and Branlard and Geisler (2022) for wind energy applications.

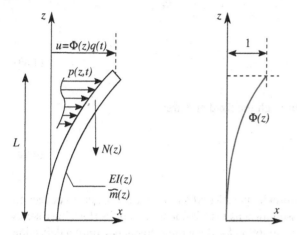

Figure 4.17 Sketch of a clamped-free beam. Its displacement is represented using one shape function

4.6.1 Equations of Motion for an Isolated Beam Described Using One Shape Function

We consider an isolated beam directed along the z axis and bending in the x direction. This model will be used in Sections 4.6.2 and 4.6.3 to model tower and blade flexibilities. This is essentially an extension and application of the cantilever discussion of Section 4.5.5. The beam properties are the following: length L, mass per unit length $\tilde{m}(z)$ and bending stiffness $EI(z)$. The external loads on the beam are assumed to consist of a distributed force $p(z)$ normal to the beam and an axial force $N(z)$. The notations are illustrated in Figure 4.17 for a clamped-free (i.e., cantilevered) beam.

The general equation of motion of the beam is a partial differential equation as given in Equation (4.72). In the shape function approach, we start by assuming a given form for the displacement field. Here, we assume that the field $u(z,t)$ in the x direction is given by a single shape function:

$$u(z, t) = \Phi(z)q(t) \tag{4.92}$$

where $\Phi(z)$ is the shape function and $q(t)$ is the generalized coordinate. We normalize the shape function such that $\Phi(L) = 1$. The shape function is typically chosen as a mode shape (determined, for example, using the Myklestad method, or NREL's *BModes* tool (Bir, 2005)), or as a function that approximates the mode shape and satisfies the geometric boundary conditions (e.g., 0 displacement at a pin or clamped location). The quality of the results will degrade if the shape function is not representative of the system. The equations of motions may be obtained by inserting Equation (4.92) into Equation (4.72), multiplying the equation by $\Phi(z)$ and integrating the equation over the length of the beam. For instance, the term $\tilde{m}(z)\dfrac{\partial^2 u}{\partial t^2}$ will become $\left[\int_0^L m(z)\Phi^2(z)dz\right]\ddot{q}(t)$, and the term $p(z, t)$ will become $\int_0^L \Phi(z)\,p(z, t)dz$. Directly inserting the shape function into the equation is not recommended because the dynamic boundary conditions may not be satisfied (in this example, the stiffness term would lead to remaining boundary terms at $z = L$ when integrating by parts). An energy approach is preferred to derive the equation of motions when the shape functions do not satisfy the dynamic boundary conditions. More details are provided in Section 4.6.4 where the equations of motions are derived. In this section, we only present the final result. The equations of motion are obtained as:

$$M\,\ddot{q}(t) + (K_{EI} + K_N)q(t) = Q(t) \tag{4.93}$$

where

$$M = \int_0^L \tilde{m}(z)\Phi^2(z)dz \qquad K_{EI} = \int_0^L EI(z)\left[\frac{d^2\Phi(z)}{dz^2}\right]^2 dz, \qquad Q(t) = \int_0^L \Phi(z)p(z, t)dz \tag{4.94}$$

and

$$K_N = -\int_0^L \Phi(z)\frac{d}{dz}\left[N(z)\frac{d\Phi}{dz}\right]dz, \qquad K_N = \int_0^L N(z)\left[\frac{d\Phi}{dz}\right]^2 dz \quad \text{(if clamped-free)} \tag{4.95}$$

The second expression for K_N was obtained using integration by parts and applying the boundary conditions of a clamped-free beam. The shape function approach leads to an ordinary second-order differential equation with constant coefficients, which we can readily be solved. It is observed that the axial load, N, acts as a stiffness term in the equations of motion. This result is important for the bending of the tower and the blades, as will be illustrated in the next sections.

4.6.2 Illustration of the Tower Softening Effect Using a Beam and a Top Mass

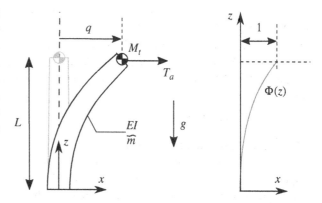

Figure 4.18 One-DOF model of a tower with top mass, using the Rayleigh–Ritz approach with one shape function

We consider the fore–aft motion of a tower with a top mass (see Branlard, 2019). We use a beam representation, and we adopt the notations of the previous paragraph. The system is illustrated in Figure 4.18. The top mass (or RNA mass) is written as M_t. The tower is assumed to be loaded using a point load T_a at the tower top, representing the thrust. In this example, we assume that the tower properties $\tilde{m}(z) = \tilde{m}$ and $EI(z) = EI$ are constant, so that the integrals can be carried out analytically.

We choose as a shape function $\Phi(z) = 1 - \cos(\pi z/2L)$ that corresponds to the first mode shape of a massless beam with a top mass. This shape function is not a mode shape for the current problem; it does not satisfy all the dynamic boundary conditions ($\frac{\partial^3 u}{\partial z^3}(L) \neq 0$), but it is a good approximate candidate. The generalized mass of the system is given by Equation (4.94) plus the extra contribution from the top mass:

$$M = M_t + \int_0^L \tilde{m}\Phi^2(z)dz = M_t + \tilde{m}L\frac{3\pi - 8}{2\pi} \tag{4.96}$$

The generalized stiffness is obtained from Equation (4.95) as follows:

$$K_{EI} = \int_0^L EI\left[\frac{d^2\Phi}{dz^2}\right]^2 dz = EI\frac{\pi^4}{32L^3} \tag{4.97}$$

The side loading from the thrust is given by $p_T(z) = \delta_L(z)T_a(t)$, where δ_L is the Dirac function, equal to 1 for $z = L$. The generalized force resulting from the thrust is obtained from Equation (4.94) as:

$$Q(t) = \int_0^L \Phi(z)\delta_L(z)T_a(t)dz = \Phi(L)T_a(t) = T_a(t) \tag{4.98}$$

The normal force induced by the top mass is $N_1(z) = -M_t g$. The stiffness contribution from this force is obtained using Equation (4.95) as:

$$K_{N_1} = -M_t g \int_0^L \left[\frac{d\Phi}{dz}\right]^2 dz = -M_t g \frac{\pi^2}{8L} \tag{4.99}$$

The normal force induced by the weight of the beam above the section z is $N_2(z) = -\int_z^L \tilde{m}g dz' = \tilde{m}g(z-L)$ leading to the following generalized stiffness:

$$K_{N_2} = \tilde{m}g\int_0^L (z-L)\left[\frac{d\Phi}{dz}\right]^2 dz = -\frac{\tilde{m}g}{16}(\pi^2 - 4) \tag{4.100}$$

Both the terms K_{N_1} and K_{N_2} are negative, which imply that they decrease the total stiffness of the system, noted: $K = K_{EI} + K_{N_1} + K_{N_2}$. This effect is referred to as "tower softening" and is mainly attributed to the presence of the top mass. The natural frequency of the system is $\omega_n = \sqrt{K/M}$. An increase in the top mass M_t increases the generalized mass M and decreases the generalized stiffness K, which then decreases the natural frequency. If the damping ratio of the system, ζ, is known, a damping term, $c = 2M\omega_n\zeta$, can be added to the equations of motion as follows:

$$M\ddot{q}(t) + c\dot{q}(t) + Kq(t) = T_a(t) \tag{4.101}$$

The results of this section can be used to simulate the motion of the tower and nacelle or estimate the natural frequency of the system. In practical applications, numerical integration is performed using the distributed beam properties of the tower, and more than one shape function is used, leading to more DOFs (see Section 4.6.4). The method presented can be used to estimate the generalized mass and stiffness of the model given in Section 4.5.3.1 for the two-DOF onshore model. It is noted that unsteady vertical loads transferred from the nacelle to the tower should also be accounted for in the axial load N for improved accuracy.

Example 4.6 Natural Frequency of a Tower

The approach described above can also be employed numerically when the mass and stiffness vary with height. For example, consider again the NREL 5-MW turbine on the land-based tower, using the $\Phi(z) = 1 - \cos(\pi z/2L)$ shape function. The tower was divided into 100 sections. The terms $\tilde{m}(z)\Phi^2(z)\Delta z$ and $EI(z)\left[\dfrac{d^2\Phi(z)}{dz^2}\right]\Delta z$ from Equation (4.94) were found for each section and then summed to approximate the integral. Note that the RNA needed to also be included, as was done in Equation (4.96). The results are $M = 409\,819$ kg, $K_{EI} = 1879$ kPa, and $f_n = 0.341$ Hz for the first mode natural frequency. The results can be improved by using a 6th-order polynomial, as employed by *OpenFAST*, such that $M = 406\,523$ kg and $K_{EI} = 1823$ kPa. The match to the frequency obtained ($f_n = 0.338$ Hz) above in Example 4.5 is even closer.

4.6.3 Illustration of Centrifugal Stiffening

In this section, we model a rotating blade using a cantilevered beam representation and one shape function. The displacement u is either in the flapping or edge direction for this example. A sketch of the system is given in Figure 4.19 for the flapwise bending. We chose the radial coordinate r instead of the coordinate z along the beam, but the equations of motion are identical to the ones presented in 4.6.1.

Once a shape function is assumed, the expressions for M and K are directly computed using Equation (4.94) based on the distributed properties of the blade ($EI(r)$ and $m(r)$). The loading $p(r)$ is either the aerodynamic load in the flapwise direction or the aerodynamic load in the edgewise direction (combined with gravitational loads). The generalized force is obtained from Equation (4.94). The important consideration for this model is the axial

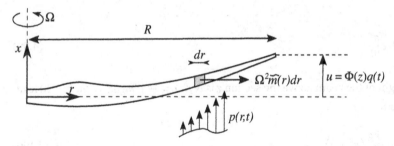

Figure 4.19 Sketch of a rotating blade, with the restoring centrifugal force

load, N. For a blade rotating with velocity Ω, the main axial load at a radial station r comes from the centrifugal force acting on all the points outboard of the current station:

$$N(r) = \int_r^R \tilde{m}(r')\,\Omega^2 r'\,dr' \tag{4.102}$$

The stiffness contribution is obtained from Equation (4.95) as:

$$K_N(\Omega) = \int_0^R N(r)\left[\frac{d\Phi}{dr}\right]^2 dr = \Omega^2 \int_0^R \int_r^R \tilde{m}(r')\,r'\,dr'\left[\frac{d\Phi}{dr}\right]^2 dr \tag{4.103}$$

The axial stiffness K_N is positive and increases with the square of the rotational speed. This restoring effect is referred to as "centrifugal stiffening." Because the mass of the blade is constant, the natural frequency of the blade will increase with the rotational speed as follows:

$$\omega_n(\Omega) = \sqrt{\frac{K_{EI} + K_N(\Omega)}{M}} = \sqrt{\omega_n^2(0) + \frac{K_N(\Omega)}{M}} = \sqrt{\omega_n^2(0) + k\Omega^2} \tag{4.104}$$

where k is a constant that depends only on the blade property and can be obtained using the expression for K_N and M. An application to the NREL 5-MW blades leads to the results presented in Figure 4.20. More details are provided in Branlard and Geisler (2022).

The variation of the blade frequency with rotational speed is something that is observed on a Campbell diagram (see Section 4.8.3) when performing stability analyses. The mode shapes of the blade will also change as function of the rotational speed. This presents a potential source of inaccuracy for the shape function approach. The shape function is typically chosen to approximate a mode shape. If the mode shape changes, then different shape functions should preferably be used for simulations at different rotational speeds. In practice, this effect is limited and alleviated by using multiple shape functions instead of one.

The general approach with multiple shape functions is presented in the following sections. The general method uses multiple shape functions in different directions (e.g., flapwise and edgewise, or fore–aft and side–side).

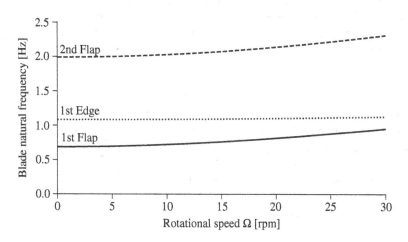

Figure 4.20 Variation of the blade natural frequencies of the NREL 5-MW turbine with the rotational speed

4.6.4   The Assumed Shape Function Approach with Multiple Shape Functions

The previous section presented some examples of the shape function approach for wind energy applications. Those examples used one shape function acting solely in one direction. The general formalism is presented in this section so that the approach can be extended to an arbitrary number of shape functions in arbitrary directions. The curious reader can extend the previous examples using the formalism of this section, for instance, to obtain the equations of motions of a tower with a top mass represented using different shape functions in the fore–aft and side–side direction or a blade using different shape functions in the flap and edge directions.

4.6.4.1   General Principle of the Assumed Shape Function Approach

The assumed shape function approach relies on carefully selected shape functions, such that only a few DOFs are necessary to represent the motion of the structure. Typically, three DOFs (and shape functions) are used to represent the blade displacements, whereas a finite element code would require hundreds of DOFs. The spatial and temporal variables are separated, and the displacement field is replaced by a linear combination of n shape functions scaled by n generalized coordinates:

$$\boldsymbol{u}(\boldsymbol{x},t) = \sum_{i=1}^{n} \boldsymbol{\Phi}_i(\boldsymbol{x}) q_i(t) \tag{4.105}$$

where q_i is referred to as generalized coordinates, and $\boldsymbol{\Phi}_i$ is referred to as associated shape functions. The shape function $\boldsymbol{\Phi}_i(\boldsymbol{x})$ provides the displacement vector of the point $\boldsymbol{x}$ from its undeflected position; it has three components ($\Phi_{i,x}, \Phi_{i,y}, \Phi_{i,z}$). It is also possible to associate a rotational field $\boldsymbol{\Psi}_i$ to each generalized coordinate, defining the orientation of the structure, $\boldsymbol{\theta} = \sum \boldsymbol{\Psi}_i q_i$. This is useful to account for torsion (see Branlard (2019)). If the shape functions satisfy both the geometrical and dynamic boundary conditions (this is the case for mode shapes), then Equation (4.105) can be directly inserted into the equations of motion. If the shape functions only satisfy the geometrical boundary conditions, then Equation (4.105) needs to be inserted either into the weak form of the equation or in the energy equations (Ritz method), or used as displacement in the principle of virtual work. In the following section, we will use Lagrange's equation to determine the equations of motion. According to Lagrange's equation, the equations of motion for each generalized DOF, q_i, are obtained as:

$$\frac{d}{dt}\left(\frac{\partial T}{\partial \dot{q}_i}\right) - \frac{\partial T}{\partial q_i} + \frac{\partial U}{\partial q_i} = Q_i \quad \text{for } i = 1..n \tag{4.106}$$

where T is the kinetic energy of the system, U is the potential energy, and Q_i is the generalized force obtained by calculating the virtual work of all the external forces for a displacement field corresponding to a virtual perturbation δq_i of the DOF q_i. The reader is referred to Bisplinghoff and Ashley (1962) for more details on the method. The final equations of motion are n coupled differential equations in the form of Equation (4.79). We provide an example of application for a beam below.

4.6.4.2   Equations for an Isolated Beam

An isolated beam directed along the z axis is considered with bending in the x and y directions. Unlike the discrete Myklestad method presented in Section 4.5.5.4, a continuous approach is used here. The beam properties are the following: length L, mass per length $\tilde{m}(z)$, and bending stiffness EI_x and EI_y. It is assumed that the displacement field is given by a simplified version of Equation (4.105), in which the shape functions are functions of z only, namely:

$$\boldsymbol{u}(z,t) = \sum_{i=1}^{N} \boldsymbol{\Phi}_i(z) q_j(t) \tag{4.107}$$

It is also assumed that the shape functions satisfy at least the geometric boundary conditions. Because the dynamic boundary conditions may not be satisfied by the Φ_i, an energy formulation is used to derive the equations of motion. The kinetic energy of the beam is $T = \dfrac{1}{2} \displaystyle\int_0^L \tilde{m}(z)(\dot{u}(z))^2 dz$, where the velocity of each point of the beam is obtained as $\dot{u}(z) = \sum_{i=1}^n \Phi_i(z)\dot{q}_i(t)$. After development, the kinetic energy is found to be $T = \dfrac{1}{2}\sum_i \sum_j M_{ij}\dot{q}_j\dot{q}_i$ where M_{ij} is obtained as the integral of the scalar product of the shape functions weighted by the mass per length:

$$M_{ij} = \int_0^L \tilde{m}(z)\Phi_i(z) \cdot \Phi_j(z)dz \quad i,j = 1..n \tag{4.108}$$

Similarly, the potential energy (strain energy) of the beam is obtained as $U = \dfrac{1}{2}\sum_i \sum_j K_{ij}q_iq_j$ where K_{ij} are the elements of the stiffness matrix, which, under the assumption of small deformations, are given by:

$$K_{ij} = \int_0^L \left[EI_x(z)\kappa_{i,x}(z)\kappa_{j,x}(z) + EI_y(z)\kappa_{i,y}(z)\kappa_{j,y}(z) \right]dz \quad i,j = 1..n \tag{4.109}$$

where $\kappa_{i,x} = \dfrac{d^2\Phi_{i,x}}{dz^2}$ is the x-curvature of the shape function i. Elongation and torsional strains (EA and GK_t) can similarly be added to the strain energy.

The external loads on the beam are assumed to consist of a distributed force vector $p(z)$. The virtual work done by the force p for each virtual displacement δq_i, provides the generalized force as: $Q_i = \int_0^L \Phi_i \cdot p \, dz$. The equations of motion are obtained using Lagrange's equation and written in matricial form as:

$$M\ddot{q} + Kq = Q \tag{4.110}$$

where $q = [q_1, \cdots, q_n]$ and the elements of the generalized mass and stiffness matrices are given in Equations (4.108) and (4.109).

Damping is typically added a posteriori to the equations, where the Rayleigh damping assumption is often used $C = \alpha M + \beta K$, where α and β are constants. If the shape functions are mode shapes, then the mass and stiffness matrices are diagonal and the stiffness values would be $K_{ii} = \omega_i^2 M_{ii}$, with ω_i the eigenfrequency of the beam mode i.

If the beam is also loaded axially by a force $N(x)$, then from Equation (4.72), it appears that this force produces a distributed load in the transverse direction equal to, $n = \dfrac{\partial}{\partial z}\left[N(z)\dfrac{\partial u}{\partial z}\right]$ with components in the y and z directions. The generalized force associated with this loading is then: $Q_{i,N} = \int_0^L \Phi_i \cdot n \, dz$. Inserting the expression of n and u (from Equation (4.107)), and after manipulation, it is found that the generalized force has the form of a stiffness term: $Q_{i,N} = -\sum_j K_{ij,N}q_j$ with

$$K_{ij,N} = -\int_0^L \Phi_i \cdot \frac{d}{dz}\left[N(z)\frac{d\Phi_j}{dz}\right]dz = \int_0^L N(z)\frac{d\Phi_i}{dz} \cdot \frac{d\Phi_j}{dz} - \left[N(z)\Phi_i \cdot \frac{d\Phi_j}{dz}\right]_0^L \tag{4.111}$$

and where integration by parts was used to obtain the second equality. The formalism presented in this section can be used to extend the examples presented in Sections 4.6.2 and 4.6.3 to include more shape functions. For instance, *OpenFAST* uses 3 shape functions per blade and 4 shape functions for the tower.

4.7 Linearized Hinge Spring Blade Model

Over the past decades, structural dynamic models of wind turbine rotors of various levels of detail have been developed. As noted previously models used in the industry at present include *OpenFAST* (Jonkman and Buhl, 2005), *Bladed* (DNV, 2021), and *HAWC2* (HAWC2, 2021). These are quite comprehensive and are particularly useful for detailed design and analysis. See Chapter 8 for details on a number of these models. On the other hand, the newer tools have become more like "black boxes" for most users. It is hard to see what the fundamental principles are except by examining the outputs. Another approach is to apply simplified linearized methods to elucidate the effect of various parameters and operating conditions. One such approach, known as the Linearized Hinge Spring Blade Model (LHSBM), is described in this section. The LHSBM was originally developed by Stoddard (1979) as one of the early structural dynamic models and then described in Eggleston and Stoddard (1987). More recently, it has been updated to be consistent with standard conventions and coordinate systems (Manwell, 2023). It is the newer form that is discussed below.

The LHSBM consists of four basic parts: (1) a model of each blade as a rigid body attached to a rigid hub by means of a hinge and a spring, (2) a linearized steady state, uniform flow aerodynamic model, (3) consideration of non-uniform flow as "perturbations," and (4) an assumed sinusoidal form for the solutions.

The model embodies a number of assumptions regarding each blade. They include:

- the blade has a uniform cross-section;
- the blade is rigid;
- the blade hinge may be offset from the axis of rotation; the offset is tuned and not necessarily equal to the hub radius;
- when rotating, the rotational speed is constant;
- the rotor axis is not tilted;
- the blades are not coned away from vertical.

4.7.1 Sources of Loads

The hinge–spring model includes an analysis of how the rotor responds to six sources of loads:

- rotor rotation;
- gravity;
- steady yaw rate;
- steady wind;
- yaw error;
- vertical wind shear.

These loads may be applied independently or in combination with the others. The LHSBM provides a general solution for the rotor response as a function of blade azimuth, i.e., the angular position of a blade. The solution will be seen to contain three terms: the first one is independent of azimuthal position, the second is a function of the cosine of the azimuth, and the third is a function of the sine of the azimuth. The development is broken into two parts: (1) "free" motion (from internal sources), and (2) externally forced motion. The free motion includes the effects of geometry, rotational speed, and blade weight. The forced motion case includes steady wind and steady yawing. Deviations from steady wind (yaw error, which results in a crosswind, and wind shear) are considered to be perturbations on the steady wind.

4.7.2 Modeling of Blade Motion

The hinge–spring model allows for three directions of blade motion and incorporates hinges and springs for all of them. The three directions of motion allowed by the hinges are: (1) flapwise, (2) lead–lag, and (3) torsion. The springs return the blade to its "normal" position on the hub. In the hinge–spring model, the blade is assumed to move rigidly in each of the DOFs.

4.7.3 Coordinate System

The coordinate system for the updated LHSBM is consistent with IEC standards (IEC, 2001) and is shown in Figure 4.21.

4.7.3.1 Flapping Equation

The assumption that the rotational speed is constant allows the solutions to be expressed in terms of the azimuth angle, rather than time. If results as a function of time are desired, they can easily be obtained by noting that azimuth angle is equal to rotational speed times time.

The LHSBM uses the hinged and offset blade to approximate a real blade. The hinge offset and spring stiffness are chosen such that the stationary and rotating hinge and spring blade have the same natural frequency and flapping inertia as the real blade.

The complete equation for flapping in the time domain is derived in Manwell (2023) and given in Equation (4.112).

$$\ddot{\beta} + \left[\Omega^2(1 + \varepsilon) - G\,\cos(\psi) + \frac{K_\beta}{J_\beta}\right]\beta = \frac{M_\beta}{J_\beta} - 2\,q\,\Omega\,\cos(\psi) \tag{4.112}$$

where β is the flapping angle, Ω is the rotational speed, ε is an offset term (expression given below), G is a gravity term (expression given below), ψ is the azimuth angle, K_β is a spring constant J_β is the blade's mass moment of inertia with respect to the flapping axis, is the aerodynamic moment, and q is the yaw rate.

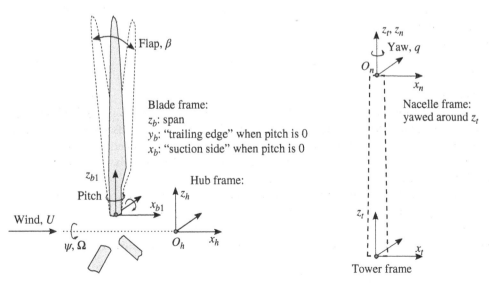

Figure 4.21 Coordinate systems for the linearized hinge–spring model, ψ = azimuth angle, β = flapping angle, q = yaw rate, Ω = rotor rotational speed; see text for other parameters

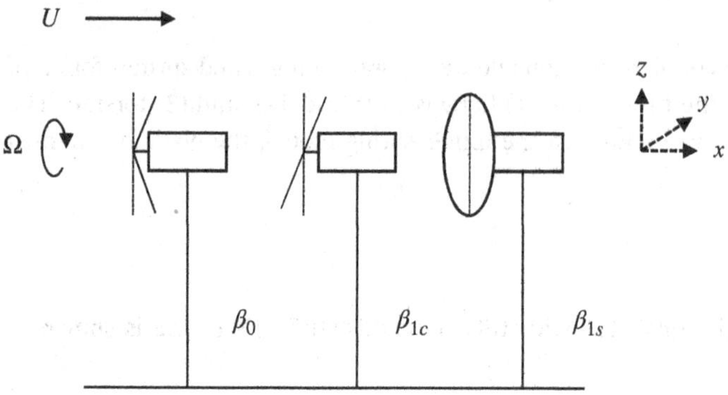

Figure 4.22 Effect of various terms in solution; β_0 collective response coefficient; β_{1c} cosine cyclic response coefficient; β_{1s} sine cyclic response coefficient

The flapping angle as a function of azimuth angle is approximated as a truncated Fourier series as shown in Equation (4.113).

$$\beta = \beta_0 + \beta_{1c}\cos(\psi) + \beta_{1s}\sin(\psi) \tag{4.113}$$

where β_0, β_{1c}, and β_{1s} are constants that are functions of the rotor parameters and the operating conditions. The terms $\beta_{1s}\sin(\psi)$ and $\beta_{1c}\cos(\psi)$ are sometimes referred to as "sine cyclics" or "cosine cyclics" as they represent sinusoidal or cosinusoidal responses which vary with the rotation cycle of the rotor. Such decomposition is also referred to as the multibladed coordinate transformation, which is required to transform the system from the rotating frame to the fixed frame of reference (see 4.8.3).

The effect is illustrated in Figure 4.22. As can be seen here, a positive β_{1c} indicates that when the blade is pointing straight up, it is pushed further downwind. When pointing downward, the blade tends to bend upwind. In either horizontal position, the cosine of the azimuth equals zero. Thus, a plane determined by a path of the blade tip would tilt about a horizontal axis, downwind at the top, and upwind at the bottom. A positive β_{1s} means that the blade which is descending tends to be bent downwind. When rising, the blade goes upwind. Overall, the plane described by the tip would tilt counterclockwise when viewed from above.

The time domain derivatives may be converted to azimuthal derivatives (see below). In that case, Equation (4.112) is expressed as Equation (4.114).

$$\beta'' + \left[1 + \varepsilon - \frac{G}{\Omega^2}\cos(\psi) + \frac{K_\beta}{\Omega^2 J_\beta}\right]\beta = \frac{M_\beta}{\Omega^2 J_\beta} - 2\,\bar{q}\,\cos(\psi) \tag{4.114}$$

where $\bar{q}$ is the dimensionless yaw rate (defined below).

The basic inputs describing the rotor are the radius, R, rotational speed, Ω (rad/s), blade mass, m_b, blade chord, c (assumed to be constant), airfoil lift coefficient slope, $C_{l,\alpha}$, and pitch angle θ_p. The rotor is offset from the yaw axis by distance, d_{yaw}. The flapwise and edgewise natural frequencies of the blade (rad/s) when stationary are $\omega_{\beta,NR}$ and $\omega_{\zeta,NR}$, and the corresponding natural frequencies when rotating are $\omega_{\beta,R}$ and $\omega_{\zeta,R}$. Note that the blade is assumed to start at the axis of rotation; the effect of the hub is ignored.

The operating conditions include the wind speed perpendicular to the rotor, U, the wind speed parallel to the rotor (i.e., crosswind), V, vertical wind shear coefficient (fractional change in wind speed from bottom to top of rotor), K_{vs}, axial induction factor, a, (assumed to be equal to 1/3) and yaw rate, q.

In order to facilitate the calculations, a number of intermediate terms are calculated. These include the following:

λ = Tip-speed ratio, $\lambda = \Omega R/U$.
e = Hinge offset.

Λ = Dimensionless inflow, $\Lambda = U(1-a)/\Omega R$.

A = First axisymmetric flow term; $A = \lambda/3 - \theta_p/4$.

A_3 = Second axisymmetric flow term; $A_3 = \lambda/2 - 2\theta_p/3$.

$\overline{V}$ = Dimensionless cross flow, $\overline{V} = V/\Omega R$.

γ = Lock number, $\gamma = \rho C_{l,\alpha} c R^4/J_\beta$.

$\overline{q}$ = Dimensionless yaw rate, $\overline{q} = q/\Omega$.

$\overline{U}$ = Dimensionless wind velocity, $\overline{U} = U/\Omega R = 1/\lambda$.

r_g = radial distance from hinge to blade center of gravity; $r_g = R(1/2 - e)$.

G = Gravity term, $g m_B r_g/J_\beta$.

B = Dimensionless gravity term, $B = G/2\Omega^2$.

ε = Flap hinge second offset term, $\varepsilon = \dfrac{3e}{2(1-e)}$.

K_{vs} = Vertical wind shear coefficient (fractional change in wind speed from bottom to top of rotor)

K = Flapping inertial natural frequency (dimensionless), $K = 1 + \varepsilon + K_\beta/J_\beta\Omega^2$.

$\overline{d}$ = Dimensionless yaw moment arm, $\overline{d} = d_{yaw}/R$.

Using the terms above, the general solution to Equation (4.114) is given in Equation (4.115):

$$
\begin{bmatrix}
K & -B & -\dfrac{\gamma\,\overline{q}\,\overline{d}}{12} \\[2mm]
-2B & K-1 & \dfrac{\gamma}{8} \\[2mm]
-\dfrac{\gamma\,\overline{V}}{6} & -\dfrac{\gamma}{8} & K-1
\end{bmatrix}
\begin{bmatrix}
\beta_0 \\[2mm] \beta_{1c} \\[2mm] \beta_{1s}
\end{bmatrix}
=
\begin{bmatrix}
\dfrac{\gamma}{2}A \\[2mm]
-2\overline{q} + \dfrac{\gamma}{2}\left[(\overline{V} - \overline{q}\,\overline{d})A_3 + \left(\dfrac{K_{vs}\overline{U}}{4}\right)\right] \\[2mm]
-\dfrac{\gamma}{8}\overline{q}
\end{bmatrix}
\tag{4.115}
$$

The natural frequencies and blade mass allow the offset, moment of inertia, and spring constant to be calculated. In the case of flapping, these terms are given by Equations (4.116)–(4.118).

$$
e = \frac{\dfrac{2}{3}\left(\dfrac{\omega_{\beta,R}^2 - \omega_{\beta,NR}^2}{\Omega^2} - 1\right)}{1 + \dfrac{2}{3}\left(\dfrac{\omega_{\beta,R}^2 - \omega_{\beta,NR}^2}{\Omega^2} - 1\right)}
\tag{4.116}
$$

$$
J_\beta = \frac{m_b R^2}{3}(1-e)^3
\tag{4.117}
$$

$$
K_\beta = \omega_{\beta,NR}^2 J_\beta
\tag{4.118}
$$

4.7.3.2 Lead–lag Equation

Lead–lag, or edgewise, loads and deflections have historically not been of as much concern as flapwise ones. As ever larger turbines are designed, however, the self-weight of the blades can become quite significant. There can also be lead-lag concerns with highly flexible blades. In this section, the lead-lag equations and their solutions are summarized. For a more in-depth discussion, the reader should consult Stoddard (1979) or Eggleston and Stoddard (1987).

Similar to flapping behavior, lead-lag behavior can be approximated by Equation (4.119).

$$
\zeta = \zeta_0 + \zeta_{1c}\cos\psi + \zeta_{1s}\sin\psi
\tag{4.119}
$$

where

ζ is the in-plane angle relative to the long axis of the blade, positive in the direction of rotation.
ζ_0 is an axisymmetric term, independent of azimuth.
ζ_{1c} is a cosine term.
ζ_{1s} is a sine term.

The sine and cosine terms can be found by solving Equation (4.120).

$$\begin{bmatrix} -2B_\zeta & (K_2-1) & 0 \\ 0 & 0 & (K_2-1) \\ K_2 & -B_\zeta & 0 \end{bmatrix} \begin{bmatrix} \zeta_0 \\ \zeta_{1c} \\ \zeta_{1s} \end{bmatrix} = \begin{bmatrix} -\dfrac{\gamma}{2}\left[K_{vs}\overline{U}A_4 + \dfrac{\theta_p\Lambda}{2}\left(\overline{V}-\overline{q}d\right)\right] \\ 2B_\zeta - \dfrac{\gamma}{2}\overline{q}A_4 \\ \dfrac{\gamma}{2}\Lambda A_2 \end{bmatrix} \tag{4.120}$$

The constants in Equation (4.120) are the same or are similar to those in Equation (4.115), except that in some cases they are based on the lead–lag offset, spring constant, and moment of inertia. These are described below. The additional or modified constants are:

A_2 = Dimensionless aerodynamic term, $A_2 = \dfrac{\lambda}{2} + \dfrac{\theta_p}{3}$.

A_4 = Dimensionless aerodynamic term, $A_4 = \dfrac{2}{3}\lambda + \dfrac{\theta_p}{4}$.

K_2 = Dimensionless spring constant, $K_2 = \varepsilon_2 + \dfrac{K_\zeta}{J_\zeta\Omega^2} = \left(\dfrac{\omega_{R,\zeta}}{\Omega}\right)^2$.

ε_2 = Second lead-lag offset term, $\varepsilon_2 = m_b e_2 x_{g,\zeta}R^2/J_\zeta = 3e_2/2(1-e_2)$, where $x_{g,\zeta}$ is the nondimensional distance of the blade center of gravity from the lead-lag hinge, as given by $x_{g,\zeta} = (1-e_2)^2/2$.

G_ζ = Gravity term, $G_\zeta = gm_b x_{g,\zeta}R/J_\zeta = 3g/(2R(1-e_2))$.

B_ζ = Dimensionless gravity term, $B_\zeta = G_\zeta/2\Omega^2$.

γ_ζ = Lock number, $\gamma_\zeta = \rho C_{l,\,\alpha}cR^4/J_\zeta$.

For edgewise motion, there is also a slight difference in the form of the offset, due primarily to the difference in the effect of centrifugal forces in "stiffening" the blade. Here, e_2 is given by Equation (4.121).

$$e_2 = \dfrac{\dfrac{2}{3}\dfrac{\omega_{\zeta,R}^2 - \omega_{\zeta,NR}^2}{\Omega^2}}{1 + \dfrac{2}{3}\dfrac{\omega_{\zeta,R}^2 - \omega_{\zeta,NR}^2}{\Omega^2}} \tag{4.121}$$

The equations for the lead-lag moment of inertia J_ζ and spring stiffness K_ζ have the same form as Equations (4.117) and (4.118) as shown in Equations (4.122) and (4.123).

$$J_\zeta = \dfrac{m_b R^2}{3}(1-e_2)^3 \tag{4.122}$$

$$K_\zeta = \omega_{\zeta,NR}^2 J_\zeta \tag{4.123}$$

4.7.4 Solutions of the Flapping Equation

The most generally useful form of the solution to the flapwise equations of motion can be found by applying Cramer's Rule to Equation (4.115). A few simplified special cases are illustrative.

i) Steady Wind and Rotation

First, consider the simplest case with rotation where gravity, crosswind, yaw rate, offset, and spring constant are all set to zero.

The only nonzero terms in Equation (4.115) are aerodynamic and centrifugal forces, which now becomes Equation (4.124):

$$
\begin{bmatrix} 1 & 0 & 0 \\ 0 & 0 & \dfrac{\gamma}{8} \\ 0 & -\dfrac{\gamma}{8} & 0 \end{bmatrix} \begin{bmatrix} \beta_0 \\ \beta_{1c} \\ \beta_{1s} \end{bmatrix} = \begin{bmatrix} \dfrac{\gamma}{2}A \\ 0 \\ 0 \end{bmatrix}
\tag{4.124}
$$

The solution to this is immediately apparent:

$$
\beta_0 = \frac{\gamma}{2}A
\tag{4.125}
$$

Thus, there is only coning in this case. It results from balance between aerodynamic thrust and centrifugal force which determines the flapping angle. There is no dependence on azimuth.

ii) Steady Wind, Rotation, Hinge–Spring, and Offset

Adding the spring and offset terms gives the same form for the solution. The flapping angle is reduced, however. The solution equation is now Equation (4.126):

$$
\begin{bmatrix} K & 0 & 0 \\ 0 & K-1 & \dfrac{\gamma}{8} \\ 0 & -\dfrac{\gamma}{8} & K-1 \end{bmatrix} \begin{bmatrix} \beta_0 \\ \beta_{1c} \\ \beta_{1s} \end{bmatrix} = \begin{bmatrix} \dfrac{\gamma}{2}A \\ 0 \\ 0 \end{bmatrix}
\tag{4.126}
$$

The solution is:

$$
\beta_0 = \gamma A/2K
\tag{4.127}
$$

The coning angle now results from a balance between the aerodynamic moments on the one hand and the centrifugal force and hinge–spring moments opposing them. As to be expected, the stiffer the spring, the smaller the coning angle.

iii) Wind Shear and Hinge–Spring

In this example, wind shear is included. The equation to be solved takes the form:

$$
\begin{bmatrix} K & 0 & 0 \\ 0 & K-1 & \dfrac{\gamma}{8} \\ 0 & -\dfrac{\gamma}{8} & K-1 \end{bmatrix} \begin{bmatrix} \beta_0 \\ \beta_{1c} \\ \beta_{1s} \end{bmatrix} = \begin{bmatrix} \dfrac{\gamma}{2}A \\ \dfrac{\gamma}{8}K_{vs}\overline{U} \\ 0 \end{bmatrix}
\tag{4.128}
$$

The determinant of the coefficient matrix is simply:

$$
D = K\left[(K-1)^2 + \left(\frac{\gamma^2}{8}\right)\right]
\tag{4.129}
$$

Applying Cramer's Rule again, one has:

$$
\beta_0 = \frac{\gamma}{2}\frac{A}{K}
\tag{4.130}
$$

$$\beta_{1c} = \frac{1}{D}\left[\frac{\gamma}{8}K_{vs}\overline{U} \quad K(K-1)\right] \tag{4.131}$$

$$\beta_{1s} = \frac{1}{D}\left[\frac{\gamma^2}{8}K_{vs}\overline{U} \quad K\right] \tag{4.132}$$

With vertical wind shear, wind speed varies only with height, that is with the cosine of the azimuth angle. It is thus considered to be a cosine input. As shown in Equations (4.131) and (4.132), a stiff rotor will have both cosine and sine responses. A teetered rotor with $K = 1$ will only have a sine response. For other examples, see Stoddard (1979) or Manwell (2023).

4.7.5 Tower Loads

Tower loads result from aerodynamic loads on the tower, the weight of the turbine and tower, and all of the forces on the machine itself, whether steady, cyclic, impulsive, etc.

4.7.5.1 Aerodynamic Tower Loads
Aerodynamic tower loads include the rotor thrust during normal operation, the moment from the rotor torque, and extreme wind loads. Extreme wind loads are those that could occur as a result of an exceptional gust when the turbine is running at rated power or due to an unusually high wind when the turbine is not operating. More details on conditions giving rise to tower loads are given in Chapter 8.

4.7.5.2 Tower Vibration
As already discussed, natural frequencies for cantilevered towers including the tower top weight can be calculated by methods described in Sections 4.5 or 4.6. Guyed towers involve methods beyond the scope of this text. The most important consideration in tower design is to avoid natural frequencies near rotor or blade-passing frequencies (1P, 2P, or 3P). A "soft-soft" tower is one whose fundamental natural frequency is below the rotor frequency, whereas a stiff tower has its dominant natural frequency above the blade passing frequency. A soft-stiff tower has its natural frequency between the rotor and blade-passing and frequencies. Further discussion of tower vibration can be found in Chapter 10.

4.7.5.3 Dynamic Tower Loads
Dynamic tower loads are loads on the tower resulting from the dynamic response of the wind turbine itself. For a rigid rotor, blade moments are the main source of dynamic tower loads. The three moments for each blade (flapwise, lead–lag, and torsion) are transferred to the tower coordinates as:

$$M_{Z'} = -M_\beta \sin(\psi) \tag{4.133}$$

$$M_{Y'} = M_\beta \cos(\psi) \tag{4.134}$$

$$M_{X'} = M_\zeta \tag{4.135}$$

where Z' refers to yawing, Y' refers to pitching backward, and X' refers to rolling of the nacelle.

For multiple blades, the contribution from each blade is summed up, adjusted by the relative azimuth angle. For three or more blades, the tower top moments are invariant with respect to azimuthal position, but not for two blades.

For a teetered rotor, the flapping moment is not transferred to the hub (or the tower) unless the teeter stops are hit, and so contributes little to dynamic tower loads under normal operation.

4.7.6 Yaw Stability

Yaw stability is an issue for free-yaw turbines. It is a complicated problem, and the LHSBM is of limited utility in its analysis. Nonetheless, it does provide insights into some of the basic physics. The key point is that various inputs contribute to cyclic responses. Any net sine response would result in a net torque about the yaw axis. (The cosine cyclic term would tend to rock the turbine up and down on its yaw bearing but would not affect yaw motion.) Conversely, for the rotor to be stable under any given conditions, the sine cyclic response term must be equal to zero.

In practical terms, the subject is more complicated than a first-order model indicates. Changing angles of attack, stall, turbulence, and unsteady aerodynamic effects all influence the yaw stability of real turbines. Additional analysis of the ability of a relatively simplified dynamic model to provide insights into yaw stability can be found in Eggleston and Stoddard (1987). Experience has shown, however, that in the matter of yaw motion in particular, a more comprehensive method of analysis is required.

4.7.7 Applicability and Limitations of the Linearized Hinge–Spring Dynamic Model

The linearized dynamic model described above can be very useful in providing insight into wind turbine dynamics. On the other hand, there are some important aspects of rotor behavior that do not appear in the model. Actual data often exhibit oscillations that are not predicted. For example, Figure 4.23 illustrates a five-second time trace of the bending moment in the flapwise (axial) direction at the blade root of a small three-blade wind turbine, using arbitrary units. The rotor in this case turns 72 rpm. As can be seen, the largest oscillations do occur at approximately that frequency, but smaller ones occur considerably more often.

The significance of this can be illustrated even more graphically in a power spectral density (psd), which provides a measure of the energy in vibrations of various frequencies. See Appendix B for details. Figure 4.24 shows a power spectrum (in arbitrary units) obtained from the data used in Figure 4.23.

As would be expected, the 1P spike in the measured data (at 72 rpm or 1.2 Hz) is very strong. In addition, spikes appear at approximately 2P and 4P. The other spikes are presumably due to turbulence in the wind and some higher mode natural frequencies of the blade. The important thing to note is that the linearized hinge–spring

Figure 4.23 Sample root flap bending moment

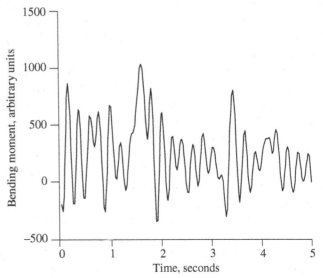

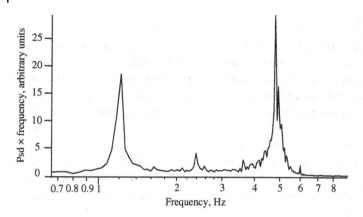

Figure 4.24 Power spectral density (psd) of root flap bending moment

model can only reproduce (to some extent) the 1*P* response. None of the higher frequency responses are predicted and as can be seen from the figure there is a significant amount of energy associated with those higher frequencies.

4.7.7.1 Comparison of LHSBM with *OpenFAST*

The following illustrates how the predictions of the LHSBM model compare to the results from the much more detailed aeroelastic wind turbine model, *OpenFAST*, which was introduced above. For this example, the NREL 5-MW reference turbine was also used, differing only in that the pitch control was turned off and an induction generator was used to keep the rotational speed nearly constant. Figure 4.25 illustrates the situation for a hub height wind speed of 11.4 m/s with a vertical power law wind shear exponent of 0.2. *OpenFAST* was run for 600 seconds and the resulting out-of-plane (flapping) and in-plane (edgewise) tip deflections were plotted vs. azimuth angle as shown in the figure.

For comparison, a simplified LHSBM model of the rotor was created, having the same diameter and blade mass, but no preconing or tilt. The chord was also set equal to a constant 4 m and the pitch angle was set to −3°. The flapping and lead-lag offsets and spring constants were calculated from the NREL 5-MW's blade natural frequencies as found from NREL's code *BModes* (Bir, 2005). The values were 0.346 and 0.243 for offsets and 1.12×10^8 and 4.68×10^8 N/rad for the spring constants, respectively. The wind shear coefficient was found from a linear fit across the top and bottom of the rotor, so as to approximately match the effect of the power law exponent. The LHSBM matrix Equations (4.115) and (4.120) were used to the find flapping and lead-lag coefficients and those were in turn used to create the curves of tip deflections as a function of azimuth angle which are also shown in Figure 4.25. The resulting constants for the hinge springs were $\beta_0 = 7.69$, $\beta_{1c} = 0.71$, $\beta_{1s} = 0.57$ and $\zeta_0 = -0.89$, $\zeta_{1c} = 0.12$, $\zeta_{1s} = -0.37$. It is apparent that the trends of the LHSBM model are very similar to those of *OpenFAST*, even though the absolute values are somewhat different (as would be expected).

When turbulent wind was used as input to *OpenFAST*, the results differ significantly from the LSHBM; this is due to the varying aerodynamic forces that result from the wind speed which varies both in time and across the plane of the rotor. This is illustrated in Figure 4.26. This figure is based on a 10 minutes average wind of 11.4 m/s with a turbulence intensity of 0.12 (IEC class C, see IEC 61400-1, 2019) with the same wind shear as used for Figure 4.25. It is interesting to note, however, that when the deflections are azimuthally averaged (lines shown in white), the trends are again very similar to Figure 4.25. Also, as might be expected, the effect of blade mass on the edgewise deflection was much greater than that of the turbulence. Blade mass dominates the lead-lag response with all larger turbines.

A final example of the difference between the LHSBM and *OpenFAST* has to do with the variability of the response. The variability of the deflection was illustrated in Figure 4.26, but that also indicates that there would

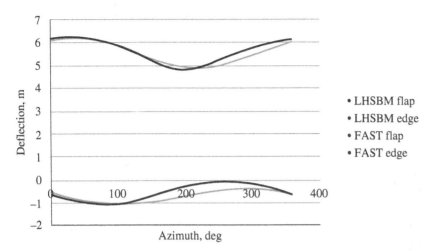

Figure 4.25 Flap and edgewise deflection vs azimuth with vertical wind shear from *OpenFAST* and the LHSBM

Figure 4.26 Flap and edgewise deflection vs. azimuth with wind shear and turbulence from *OpenFAST*

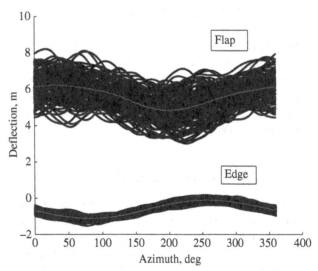

be significant variability in stress. As discussed in detail in Chapter 7, variability in stress could increase fatigue damage. Here, we consider an illustration of the variation in flapwise tip deflection. It is apparent from Figure 4.25 or Figure 4.26 that, in the absence of turbulence, the flapwise tip deflection varies by approximately 1.5 m over the course of a single rotation. For a nominal operating speed of 12 rpm, there would be approximately 120 cycles of magnitude 1.5 over a 10-minute period. By using a method known as rainflow counting (see Chapter 7), it is possible to ascertain the number of cycles of various magnitudes. Figure 4.27 illustrates the results of counting the cycles in the data used to create Figure 4.26. In this situation, there were approximately 190 cycles, with 110 of them greater than 1.5 m. This result clearly indicates that in some situations, a simplified model would be inadequate.

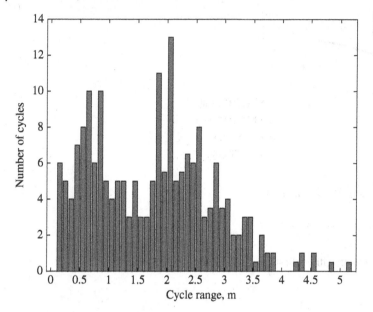

Figure 4.27 Sample blade tip deflection cycle count

4.8 Linearization and Stability

The study of structural stability consists in determining the behavior of a structure when it is subject to perturbations from a given equilibrium position. A stable condition will result in motions of finite amplitudes that decay with time and return to the equilibrium position, whereas an unstable condition will show important oscillations that increase with time. The stability of a wind turbine at different operating conditions is typically studied by linearizing the equations of motion and computing the eigenfrequencies of the system. This section describes some important aspects of this process.

4.8.1 *Linearization of Equations of Motions

In Section 4.3.6, the equations of motion of a single DOF system were introduced and the harmonic vibrations were studied. When presenting the general equations of motion (4.3.6.4), it was mentioned that the forcing on the right-hand side of the equation can provide additional stiffness, damping, or inertial contributions. These contributions are obtained by "linearizing" the system, which mathematically amounts to computing a Taylor series expansion of the equations of motion to the first order. The linearization concepts presented in Section 4.3.6 are readily extended to a system with multiple DOFs.

The equations of motion given in Equations (4.79) are recast into Equation (4.136) by assuming the stiffness and damping are absorbed into the forcing term:

$$e(x, \dot{x}, \ddot{x}, u, t) = f(x, \dot{x}, u, t) - M(x)\ddot{x} = 0 \tag{4.136}$$

where the term e represents the equations of motion with all terms put on the left-hand side, $M = -\frac{\partial e}{\partial \ddot{x}}$ is the system mass matrix and f is the forcing. The vector u represents time-dependent inputs typically related to external loads.

Linearization is done about an operating point, noted with the subscript 0, which is a solution of the nonlinear equations of motion (Equation (4.136)); that is:

$$e(x_0, \dot{x}_0, \ddot{x}_0, u_0, t) = 0 \tag{4.137}$$

The linearized equations about this operating point are obtained using a Taylor-series expansion of Equation (4.137), at $x_0 + \delta x$, $\dot{x}_0 + \delta \dot{x}$, etc, where δ indicates a small perturbation of the quantities. The expansion results in Equation (4.138).

$$M_0(x_0)\delta\ddot{x} + C_0(x_0, \dot{x}_0, u_0)\delta\dot{x} + K_0(x_0, \dot{x}_0, u_0)\delta x = Q_0(x_0, \dot{x}_0, u_0)\delta u \tag{4.138}$$

The different terms are obtained by computing the Jacobian (partial derivatives) of Equation (4.137) with respect to each variable:

$$M_0 = -\left.\frac{\partial e}{\partial \ddot{x}}\right|_0, C_0 = -\left.\frac{\partial e}{\partial \dot{x}}\right|_0, K_0 = -\left.\frac{\partial e}{\partial x}\right|_0, Q_0 = \left.\frac{\partial e}{\partial u}\right|_0 \tag{4.139}$$

M_0, C_0, and K_0 are the linear mass, damping, and stiffness matrices, Q_0 is the linear forcing vector, and $|_0$ indicates that the expressions are evaluated at the operating point. More details and examples of applications can be found in Branlard and Geisler (2022).

For a linear system with no external forces, the mass, damping, and stiffness matrices are constant matrices that are purely a function of the structural properties of the system ($e = Kx + C\dot{x} - M\ddot{x}$, with the matrices constant). In this particular case, there is no need to perform a linearization (the linear matrices are already known, i.e., $M_0 = M$, $C_0 = C$, $K_0 = K$), and an eigenvalue analysis can be performed directly. For a more advanced system (nonlinear and with nonlinear forcing), linearization is required, typically done numerically. An eigenvalue analysis of the linear matrices (M_0, C_0, and K_0), or equivalently, of the matrix A (defined in Equation (4.80)) provides the frequency, damping, and mode shapes of the system at a given operating point.

For a wind turbine, the aerodynamic and hydrodynamic (offshore) loads have important contributions to C_0, in addition to the structural damping. Gravity and inertial loads have significant stiffening/softening contributions on the tower and blades (see Sections 4.6.2 and 4.6.3). Since the structure moves the fluid around it, inertial contributions are also present. For a floating offshore wind turbine, external inertial contributions come mostly from the hydrodynamics and are referred to as added mass (see Chapter 11). Aerodynamic damping is the topic of the next section.

4.8.2 Aerodynamic Damping

Aerodynamic damping refers to the effect of the operating rotor in reducing the vibration of the wind turbine. Viewed most simply, when an element of the blade is moving downwind, the wind speed appears to be reduced slightly. This reduces the thrust, so the element tends to move upwind slightly, "pulled" by the tower, acting as a spring. Conversely, when the element moves upwind, the wind speed appears to be increased slightly. This increases the thrust correspondingly and pushes the element downwind. Note the change in velocity is due both to the motion of the nacelle and the motion of the element relative to the nacelle.

A simple model helps to explain this phenomenon further. As discussed by Stoddard (1979), among others, the aerodynamic lift per unit length $\tilde{L}$ on a blade at a distance r from the axis of rotation may be approximated by Equation (4.140).

$$\tilde{L} \approx \frac{1}{2}\rho c(r)C_{la}\left(U_n U_t - \theta_p U_t^2\right) \tag{4.140}$$

where ρ = atmospheric density, $c(r)$ = chord length at distance r, C_{la} = slope of the lift curve (assumed to be constant for high tip-speed ratio), U_n = normal (perpendicular) component of the wind, U_t = tangential component of the wind (assumed to be $U_t = \Omega r$), and θ_p = pitch angle of the blade section.

The normal component of the wind, U_n is primarily $U(1 - a)$, with a the axial induction factor, but it is also affected by the velocity of the RNA, $\dot{x}_n$. As a result, the normal component can be expressed as $U_n = U(1-a) - \dot{x}_n$. The RNA motion results in the lift force contribution $\Delta \tilde{L}$:

$$\Delta \tilde{L} \approx -\frac{1}{2}\rho c(r)C_{l\alpha}\Omega\, r\, \dot{x}_n \tag{4.141}$$

If we assume that the rotor is operating at a relatively high tip-speed ratio, then the total thrust force T is approximately equal to the lift force integrated over all B blades, so the change in thrust force ΔT due to the motion of the RNA is given by:

$$\Delta T \approx -B\frac{1}{2}\rho C_{l\alpha}\Omega\, \dot{x}_n \int\limits_{r_h}^{R} c(r)rdr \tag{4.142}$$

The integral is simply the first moment of area of one blade, which is a constant, and can be written as

$$A_{1b} = \int\limits_{r_h}^{R} c(r)rdr \tag{4.143}$$

Therefore

$$\Delta T \approx -\frac{B}{2}\rho C_{l,\alpha}\Omega A_{1b}\dot{x}_n \tag{4.144}$$

It is apparent that the terms multiplying $\dot{x}_n$ can be thought of as an aerodynamic damping term, $c_{adp} = \frac{B}{2}\rho C_{l,\alpha}\Omega A_{1b}$.

To see how this damping constant compares to the critical damping constant of the tower plus RNA, it is illustrative to express that value in terms of an effective modal mass M and spring constant K. Sections 4.5.5 and 4.6 discussed methods for determining these values.

The critical damping of the structure is, in analogy with Equation (4.61), $c_c = 2M\omega_n$. The damping ratio ζ (see Section 4.3.6) is then given by Equation (4.145):

$$\zeta = \frac{c_{adp}}{c_c} = \frac{B\rho C_{l\alpha}\Omega A_{1b}}{4M\omega_n} \tag{4.145}$$

Example 4.7 Natural Frequency of a Tower

Consider the NREL 5-MW 3-blade turbine described previously. In Section 4.6.2, the generalized mass and natural frequency were found to be $M = 406\,523$ kg and $f_n = 0.338$ Hz. The turbine's blade is 61.5 m long, and its chord averages 3.49 m. Accounting for the variation of chord length with radial position (see Jonkman *et al.*, 2009), we find A_{1b} from Equation (4.143) to be 6083 m^3. With the rated rotor speed of 1.27 rad/s (12.1 rpm), an assumed lift curve slope of $C_{l\alpha} = 2\pi$ rad^{-1}, and an atmospheric density of $\rho = 1.225$ kg/m^3, the damping ratio from Equation (4.145) becomes 0.052. This is a reasonable estimate and is also consistent with values given in Cerda Salzmann and van der Tempel (2006), where there is more discussion of aerodynamic damping.

4.8.3 Multibladed Transformation, Campbell Diagram, and Stability

The stability of a wind turbine is usually studied by linearization of the equations of motions at different operating conditions and looking at the resulting eigenfrequencies and damping for these operating conditions. An operating condition consists of a value of wind speed, rotor speed, and pitch angle. For each operating condition, different operating points and linear models are computed, corresponding to different azimuthal positions. This extra

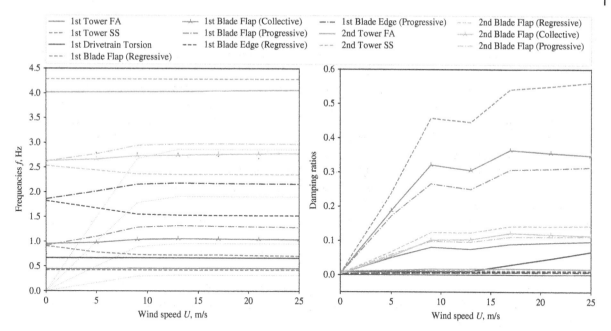

Figure 4.28 Example of a Campbell diagram of a wind turbine

complication comes from the rotation of the turbine: the steady-state equilibria of the system are periodic and not time-invariant. The multibladed coordinate transformation, or Coleman transformation (Coleman and Feingold, 1958), is applied to the linearized models at different azimuthal positions, to obtain an "averaged" linear model in the fixed frame of reference. Frequencies and damping ratios are obtained using an eigenvalue analysis of this averaged linear model. A plot of these frequencies and damping ratios as function of wind speed or rotor speed is referred to as a Campbell diagram. An example of Campbell's diagram for a wind turbine is shown in Figure 4.28.

Some of the key characteristics of a Campbell diagram are seen in the figure:

- The collective flap and edge frequencies increase with the rotational speed (and the wind speed), mostly due to centrifugal stiffening (see Section 4.6.3).
- The blade modes are separated into collective, regressive, and progressive modes. The regressive and progressive modes separate around the collective mode with an offset of +/− the rotational speed frequency. These two modes arise from the fact that not all three blades flap at the same time.
- The damping of the flap modes increases with wind speed due to the strong aerodynamic damping in the flap direction.
- Edge modes are typically lightly damped and often become negatively damped.
- Tower modes are not strongly affected by the operating conditions of the turbine.

The designer pays attention to the damping value, avoiding negative damping at all costs, which indicates instability. Change in the design, increase of stiffness, placement of center of masses, and active or passive damping devices can be used to bring back the system to a stable regime. The designer is also attentive to frequencies of the system that have low damping and that are close to a multiple of the rotational speed of the rotor because the rotation is likely to excite these modes, which can result in increased fatigue or failure. In some situations, change of control strategies are used to operate the turbine at different rotor speeds than optimal to avoid exciting certain frequencies. This is often done toward the startup region of the turbine, where the rotor speed is increased to avoid matching the tower frequency. Control strategies are discussed in Chapter 9.

References

Bir, G. (2005) *User's Guide to BModes (Software for Computing Rotating Beam Coupled Modes)*, NREL/TP-500-39133, National Renewable Energy Laboratory, Golden, CO.

Bisplinghoff, R. L. and Ashley, H. (1962) *Principles of Aeroelasticity.* Dover Books on Engineering Series. Dover Publications.

Branlard, E. (2019) Flexible multibody dynamics using joint coordinates and the Rayleigh-Ritz approximation: The general framework behind and beyond Flex. *Wind Energy*, **22**(7), 877–93, https://doi.org/10.1002/we.2327.

Branlard, E. and Geisler, J. (2022) A symbolic framework for flexible multibody systems applied to horizontal axis wind turbines. *Wind Energy Science*, **7**, 2351–2371, https://doi.org/10.5194/wes-7-2351-2022.

Cerda Salzmann, D. J. and van der Tempel, J. (2006) *Aerodynamic Damping in the Design of Support Structures for Offshore Wind Turbines.* Faculty of Civil Engineering and Geosciences, Delft University of Technology, Duwind, http://resolver.tudelft.nl/uuid:ae69666e-3190-4b22-84ed-2ed44c23e670.

Chopra, A. K. (2017) *Dynamics of Structures: Theory and Application to Earthquake Engineering*, 5th edition. Pearson Prentice Hall, Upper Saddle River, NJ.

Coleman, R. P. and Feingold, A. M. (1958) *Theory of self-excited mechanical oscillations of helicopter rotors with hinged blades*, NACA Technical Report TR 1351, 1958.

Cook, R. D. (2001) *Concepts and Applications of Finite Element Analysis.* John Wiley & Sons.

DNV GL (2021) *Wind turbine design software – Bladed*, https://www.dnv.com/services/wind-turbine-design-software-bladed-3775 Accessed 12/9/2023.

Eggleston, D. M. and Stoddard, F. S. (1987) *Wind Turbine Engineering Design.* Van Nostrand Reinhold, New York.

Fertis, D. G. (1995) *Mechanical and Structural Vibrations.* John Wiley & Sons, Inc., New York.

Hansen, M. O. L. (2008) *Aerodynamics of Wind Turbines*, 2nd edition. Earthscan, London, Sterling, VA.

HAWC2 (2021) *HAWC2 (Horizontal Axis Wind turbine simulation Code 2nd generation)*, https://www.hawc2.dk/HAWC2-info Accessed 12/9/2023.

Hibbeler, R. C. (2015), *Engineering Mechanics: Statics and Dynamics*, 14th edition. Pearson North America, New York.

IEC (2019) *Wind Turbine Generator Systems – Part 1: Design Requirements: IEC 61400-1*, 4th edition. International Electrotechnical Commission, Geneva.

IEC (2001) *Wind Turbine Generator Systems – Part 13: Measurement of Mechanical Loads: IEC/TS 61400-13*. International Electrotechnical Commission, Geneva.

Jonkman, J. (2013) *The New Modularization Framework for the FAST Wind Turbine CAE Tool*, NREL/CP-5000-57228, National Renewable Energy Laboratory, Golden, CO.

Jonkman, J., Butterfield, S., Musial, W. and Scott, G. (2009) *Definition of a 5-MW Reference Wind Turbine for Offshore System Development*, NREL/TP-500-38060, National Renewable Energy Laboratory, Golden, CO.

Jonkman, J. J. and Buhl, M. (2005) *FAST User's Guide*, NREL/EL-500-38230, National Renewable Energy Laboratory, Golden, CO.

Manwell, J. F. (2023) *The Wind Turbine Rotor Linearized Hinge Spring Blade Model Revisited*, Technical Report UMA-WEC TR 2023-1, University of Massachusetts Amherst Wind Energy Center.

Meriam, J. L., Kraige, L. G. and Bolton, J. N. (2015) *Engineering Mechanics: Dynamics*, 8th edition. John Wiley & Sons, Inc., New York.

Meriam, J. L. and Kraige, L. G. (2011) *Engineering Mechanics: Statics*, 7th edition. John Wiley & Sons, Inc., New York.

Nayfeh, A. H., and Pai, P. F. (2004) *Linear and Nonlinear Structural Mechanics.* Wiley Series in Nonlinear Science. John Wiley & Sons, Inc., New York.

Rao, S. S. (2018) *The Finite Element Method in Engineering*, 6th edition. Butterworth-Heinemann, Boston.

Shabana, A. A. (2013) *Dynamics of Multibody Systems*, 4th edition. Cambridge University Press. Cambridge.

Stoddard, F. S. (1979) *Structural Dynamics, Stability and Control of High Aspect Ratio Wind Turbine Generators*, PhD Dissertation, University of Massachusetts, Amherst, MA.

Thomson, W. T. and Dahleh, M. D. (1997) *Theory of Vibrations with Applications*, 5th edition. Prentice-Hall, Englewood Cliffs, NJ.

5

Electrical Aspects of Wind Turbines

5.1   Overview

Electrical issues are associated with many aspects of modern wind turbines. Most obviously, the primary function of the majority of wind turbines is the generation of electricity. A large number of topics in power systems engineering are thus directly relevant to wind turbines. These include generation at the turbine itself as well as power transfer from the generator, transforming to a higher voltage, interconnection with power lines, distribution, transmission, and eventual use by the consumer. Electricity is used in the operation, monitoring, and control of most wind turbines. It is also used in site assessment and data collection and analysis. For isolated or weak grids, or systems with a large amount of wind generation, storage of electricity is an issue. Finally, lightning is a naturally occurring electrical phenomenon that may be quite significant to the design, installation, and operation of wind turbines.

The principal areas in which electricity is significant to the design, installation, or operation of wind turbines are summarized in Table 5.1.

This chapter includes seven main sections. Section 5.1 provides an overview of relevant topics. Section 5.2 reviews the basic concepts of electricity. Section 5.3 discusses power transformers. Section 5.4 concerns electrical machines, particularly generators. Section 5.5 is about power converters. Section 5.6 discusses the electrical aspects of variable-speed wind turbines and Section 5.7 briefly summarizes the ancillary electrical equipment associated with typical wind turbines. The topic of the electric power system as a whole, including power transmission, is discussed in Chapter 12. Later, Chapter 16 considers power systems in which a very large fraction of the electricity comes from the wind.

5.2   Basic Concepts of Electrical Power

5.2.1   Fundamentals of Electricity

In this chapter, it is assumed that the reader has an understanding of the basic principles of electricity, including direct current (DC) circuits. For that reason, these topics will not be discussed in detail here. The reader is referred to other sources, such as Nahvi and Edminster (2003), for more information. Specific topics with which the reader is assumed to be familiar include:

- voltage;
- current;
- resistance;
- resistivity;

Wind Energy Explained: On Land and Offshore, Third Edition.
James F. Manwell, Emmanuel Branlard, Jon G. McGowan, and Bonnie Ram.
© 2024 John Wiley & Sons Ltd. Published 2024 by John Wiley & Sons Ltd.

Table 5.1 Examples of electrical issues significant to wind energy

Power generation	Generators	Storage	Batteries
	Power electronic converters		Rectifiers
			Inverters
Interconnection and distribution	Power cables	Lightning protection	Grounding
	Switch gear		Lightning rods
	Circuit breakers		Safe paths
	Transformers		
	Power quality		
Control	Sensors	End loads	Lighting
	Controller		Heating
	Yaw or pitch motors		Motors
	Solenoids		
Site monitoring	Data measurement and recording		
	Data analysis		

- conductors;
- insulators;
- DC circuits;
- Ohm's Law;
- electrical power and energy;
- Kirchhoff's laws for loops and nodes;
- capacitors;
- inductors;
- time constants of RC and RL circuits;
- series/parallel combinations of resistors.

5.2.2 Alternating Current

The form of electricity most commonly used in power systems is known as alternating current (AC) circuits. In AC circuits (at steady state), all voltages and currents vary in a sinusoidal manner. There is a complete sinusoidal cycle each period. The frequency, f, of the sine wave is the number of cycles per second (known as "Hertz" and abbreviated Hz). It is the reciprocal of the period. In the United States and much of the Western Hemisphere, the standard frequency for AC is 60 Hz. In much of the rest of the world, the standard frequency for AC is 50 Hz.

The instantaneous voltage, v, in an AC circuit may be described by Equation (5.1):

$$v = V_{max} \sin(2\pi ft + \phi) \tag{5.1}$$

where V_{max} is the maximum value of the voltage, t is time, and ϕ is the phase angle with respect to some reference angle.

The phase angle indicates the angular displacement of the sinusoid from a reference sine wave with a phase angle of zero. Phase angle is important because currents and voltages, although sinusoidal, are not necessarily in phase with each other. In the analysis of AC circuits, it is often useful to start by assuming that one of the sinusoids has zero phase, and then find the phase angles of the other sinusoids with respect to that reference.

An important summary measure of the voltage is its root mean square (rms) value V_{rms}:

$$V_{rms} = \sqrt{\int_{cycle} v^2 dt} = V_{max} \sqrt{\int_{cycle} \sin^2(2\pi ft) dt} = V_{max} \frac{\sqrt{2}}{2} \tag{5.2}$$

Note that the rms value of the voltage is $\sqrt{2}/2$, or about 70% of the maximum voltage for a pure sine wave. The rms voltage is often referred to as the magnitude of the voltage, so $|V| = V_{rms}$.

The rms current is defined similarly to that of voltage. In the rest of this chapter, the symbols V and I are used for rms values when AC currents and voltages as being referred to unless otherwise noted.

5.2.2.1 Capacitors in AC Circuits

The current through a capacitor is proportional to the derivative of the voltage across it. Thus, if the voltage across a capacitor is $v = V_{max} \sin(2\pi ft)$, then the instantaneous current, i, is:

$$i = C\frac{dv}{dt} = 2\pi f \ V_{max} C \ \sin(2\pi ft + \pi/2) \tag{5.3}$$

where C is the capacitance. Equation (5.3) can be rewritten as:

$$i = I_{max} \sin(2\pi ft + \pi/2) \tag{5.4}$$

where $I_{max} = V_{max}/X_C$ and $X_C = 1/(2\pi fC)$. The term X_C is known as capacitive reactance. Note that as the current varies in time, the current sinusoid will be displaced by $\pi/2$ radians ahead of the voltage sinusoid. For that reason, the current in a capacitive circuit is said to *lead* the voltage.

5.2.2.2 Inductors in AC Circuits

The current in an inductor is proportional to the integral of the voltage. The relation between the voltage and current in an inductor can be found from:

$$i = \frac{1}{L}\int v \, dt = \frac{V_{max}}{2\pi fL} \sin(2\pi ft - \pi/2) \tag{5.5}$$

where L is the inductance. Analogously to the equation for capacitors, Equation (5.5) can be rewritten as:

$$i = \frac{V_{max}}{X_L} \sin(2\pi ft - \pi/2) \tag{5.6}$$

where X_L = inductive reactance ($X_L = 2\pi fL$). Note that in an inductive circuit, the current *lags* the voltage. As with X_C, X_L is a positive number.

5.2.2.3 Phasor Notation

Manipulations of sines and cosines with various phase relationships can become quite complicated. Fortunately, the process can be greatly simplified, as long as the frequency is constant. This is the normal case with most AC power systems. (Note that transient behavior requires a more complicated analysis method.) The method uses phasor notation, which is summarized below.

Phasors are a means of representing sinusoids by complex numbers. For example, the voltage in Equation (5.1) can be represented by a phasor:

$$\hat{\boldsymbol{V}} = V_{rms}e^{j\theta} = a + jb = V_{rms}\angle\phi \tag{5.7}$$

The bold and circumflex are used to indicate a phasor. Here $j = \sqrt{-1}$, $\angle$ indicates the angle between the phasor and the real axis, ϕ is the phase angle, $e^{j\phi} = \cos\phi + j\sin\phi$ and, thus, $a = V_{rms}\cos(\phi)$ and $b = V_{rms}\sin(\phi)$.

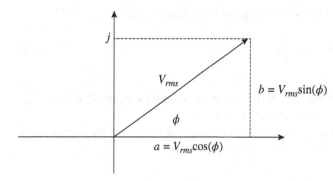

Figure 5.1 Phasor

Figure 5.1 illustrates a phasor. Note that it can be equivalently described in rectangular or polar coordinates. In phasor representation, the frequency is implicit. To recover the time series waveform, one can use Equation (5.8):

$$v = \sqrt{2}\,\mathrm{Re}\left\{V_{rms}e^{j\phi}\; e^{j2\pi ft}\right\} \tag{5.8}$$

where Re{ } signifies that only the real part is to be used.

In an analogous manner, current can be expressed as a phasor, $\hat{I}$, defined by I_{rms} and the phase angle of the current.

A few rules can be applied when using phasors. These are summarized below. Note that sometimes it is more convenient to use the rectangular form, and other times the polar form. Begin by defining two phasors $\hat{A}$ and $\hat{C}$.

$$\hat{A} = a + j\;b = A_m e^{j\phi_a} = A_m \angle \phi_a \tag{5.9}$$

$$\hat{C} = c + j\;d = C_m e^{j\phi_c} = C_m \angle \phi_c \tag{5.10}$$

where

$$\phi_a = \tan^{-1}(b/a) \tag{5.11}$$

$$\phi_c = \tan^{-1}(d/c) \tag{5.12}$$

$$A_m = \sqrt{a^2 + b^2} \tag{5.13}$$

$$C_m = \sqrt{c^2 + d^2} \tag{5.14}$$

The rules for phasor addition, multiplication, and division are:

Phasor addition

$$\hat{A} + \hat{C} = (a + c) + j(b + d) \tag{5.15}$$

Phasor multiplication

$$\hat{A}\;\hat{C} = A_m \angle \phi_a \; C_m \angle \phi_c = A_m C_m \angle (\phi_a + \phi_c) \tag{5.16}$$

Phasor division

$$\frac{\hat{A}}{\hat{C}} = \frac{A_m \angle \phi_a}{C_m \angle \phi_c} = \frac{A_m}{C_m} \angle (\phi_a - \phi_c) \tag{5.17}$$

More details on phasors can be found in most texts on AC circuits. There are also numerous Internet sites that provide useful information on phasors and other aspects of electrical circuitry. Many software packages, including Matlab, Python, and Excel, make calculations using complex numbers straightforward.

5.2.2.4 Complex Impedance

The AC equivalent of resistance is complex impedance, $\hat{Z}$, which takes into account both resistance and reactance. Impedance can be used with the phasor voltage to determine phasor current and vice versa. Impedance consists of a real part (resistance) and an imaginary part (inductive or capacitive reactance). Resistive impedance is given by $\hat{Z}_R = R$, where R is the resistance. Inductive and capacitive impedances are given by, respectively, $\hat{Z}_L = jX_L = j\,2\pi f L$ and $\hat{Z}_C = -jX_C = -j/(2\pi f C)$. Note that for a circuit that is completely resistive, the impedance is equal to the resistance. For a circuit that is completely inductive or capacitive, the magnitude of the impedance is equal to that of the reactance. Also, inductive and capacitive impedances are a function of the frequency of the AC system.

The rules relating voltage, current, and impedance in AC circuits are analogous to those of DC circuits.

Ohm's Law

$$\hat{V} = \hat{I}\,\hat{Z} \tag{5.18}$$

Impedances in series

$$\hat{Z}_S = \sum_{i=1}^{N} \hat{Z}_i \tag{5.19}$$

Impedances in parallel

$$\hat{Z}_P = 1 / \sum_{i=1}^{N} \frac{1}{\hat{Z}_i} \tag{5.20}$$

where $\hat{Z}_S$ is the effective impedance of N impedances in series and $\hat{Z}_P$ is the effective impedance of N parallel impedances. Kirchhoff's laws also apply to phasor currents and voltages in circuits with complex impedances.

5.2.2.5 Power in AC Circuits

It may be recalled that, for the case of DC, power is simply the product of the voltage and current. Calculation of power in an AC circuit is somewhat more complicated than this. There are actually three terms of interest: apparent power, S, real power, P, and reactive power, Q. The magnitude of the apparent power is simply the rms AC voltage times the rms AC current, analogous to that in a DC circuit. In general, this is larger than the real power consumed or produced in the circuit; real power is that which does work or produces light. Conversely, reactive power, which is measured in units of "volt-amperes reactive" (VAR), is an indicator of currents in the reactive components of the circuit, i.e., inductors or capacitors. For example, the currents that create the magnetic field in a generator correspond to a generator's requirement for reactive power.

These "powers" may be calculated in various ways. From the basic definition of power, and in direct analog to DC power, the instantaneous real power at any point in a cycle is given by:

$$P_{inst} = V_{\max} \sin(2\pi f t) I_{\max} \sin(2\pi f t - \phi) \tag{5.21}$$

Here, the reference angle ϕ is chosen to be that of the voltage waveform.

When integrated over a cycle, the average of the real power is:

$$P_{av} = \frac{1}{T}\int_0^T P_{inst}\,dt = \frac{1}{T}\int_0^T V_{max}\sin(2\pi ft)I_{max}\sin(2\pi ft - \phi) = \frac{V_{max}I_{max}}{T}\int_0^T \sin(2\pi ft)\sin(2\pi ft - \phi) \tag{5.22}$$

where T = length of cycle (period), s and ϕ = phase angle.

Equation (5.22) is rather cumbersome, but it can be expressed more succinctly using phasors. For example, it may be shown that the apparent power phasor in an AC circuit is given by the phasor voltage times the complex conjugate (signified by an asterisk) of the phasor current:

$$\hat{S} = \hat{V}\hat{I}^* \tag{5.23}$$

The magnitude (rms) of the real power P is then simply the real part of $\hat{S}$:

$$P = \mathrm{Re}(\hat{S}) \tag{5.24}$$

Similarly, the magnitude of the reactive power Q is the imaginary part of $\hat{S}$:

$$Q = \mathrm{Im}(\hat{S}) \tag{5.25}$$

The phase angle is given by:

$$\phi = \cos^{-1}(P/|\hat{S}|) \tag{5.26}$$

In some cases, it is convenient to express the real and reactive power in terms of the voltage, current, and cosine or sine of the phase angle. Note that the cosine of the current phase angle with respect to the voltage phase angle is used often enough that it is given its own name, power factor, and the phase angle is called the power factor angle, ϕ_{pf}. The real power in this case is:

$$P = VI\cos(\phi_{pf}) \tag{5.27}$$

The reactive power is:

$$Q = VI\sin(\phi_{pf}) \tag{5.28}$$

The power factor of a circuit or device describes the fraction of the apparent power that is real power. Thus, the power factor is simply the ratio of real to apparent power. For example, a power factor of 1 indicates that all of the power is real power.

It is important to note that the power factor angle may be either positive or negative, corresponding to whether the current sine wave is leading the voltage sine wave or vice versa, as discussed earlier. Accordingly, if ϕ_{pf} is positive, the power factor is said to be leading; if it is negative, the power factor is lagging.

An illustration of the waveforms relating voltage, current, and apparent power is shown in Figure 5.2 for a circuit with a resistor and capacitor. For this situation, the current and voltage are out of phase by 45°, so the power factor is 0.707. The current sine wave precedes the voltage wave in time, so the power factor is leading.

Example 5.1 AC Series Circuit
Consider a circuit with an AC voltage source, a resistor, an inductor, and a capacitor, all connected in series in a single loop. The resistance of the resistor is 4 Ω, the inductance of the inductor is 0.01592 H (henry), and the capacitance of the capacitor is 0.0008842 F (farad), respectively. The rms voltage is 1000 ∠30° V, and the AC frequency f is 60 Hz.

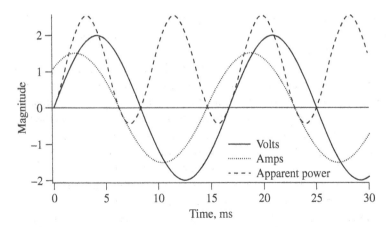

Figure 5.2 AC voltage, current, and apparent power in a circuit with a resistor and capacitor

Based on the stated assumptions, the reactance of the inductor is $X_L = 2\pi fL = 6\,\Omega$ and that of the capacitor is $X_C = 1/(2\pi fC) = 3\,\Omega$. The corresponding impedances are $\hat{Z}_L = jX_L = j6\,\Omega$ and $\hat{Z}_C = -jX_C = -j3\,\Omega$.

The total impedance, $\hat{Z}$, is $\hat{Z} = \hat{Z}_R + \hat{Z}_L + \hat{Z}_C = 4 + j6 - j3 = 4 + j3 = 5\angle 36.9°$.

The phasor form of the voltage is:

$$\hat{V} = 866.0 + j500$$

The current at the voltage source is:

$$\hat{I} = \hat{V}/\hat{Z} = (866.0 + j500)/(4 + j3) = 198.6 - j23.93 = 200\angle -6.89°$$

The apparent power is:

$$\hat{S} = \hat{V}\hat{I}^* = (866 + j500)(198.6 + j23.93) = 160\,000 + j120\,000 = 200\,000\angle 36.9°$$

The magnitude of the apparent power is:

$$|\hat{S}| = \sqrt{\hat{S}\hat{S}^*} = \sqrt{(160\,000 + j120\,000)(160\,000 - j120\,000)} = 200\,000\,\text{VA}$$

The real power is:

$$P = \text{Re}(160\,000 + j120\,000) = 160\,000\,\text{W}$$

The reactive power is:

$$Q = \text{Im}(160\,000 + j120\,000) = 120\,000\,\text{VAR}$$

The power factor angle is $\phi_{pf} = \cos^{-1}(120/200) = 36.9°$.

The real power can also be found from either $P = |\hat{I}|^2 R = 200^2(4) = 160\,000\,\text{W}$ or $P = |\hat{I}||\hat{V}|\cos\phi_{pf} = 160\,000\,\text{W}$. Note the use of the absolute value notation for the equivalent rms.

5.2.2.6 Three-phase AC Power

Power generation and large electrical loads commonly operate on a three-phase power system. A three-phase power system is one in which the voltages supplying the loads all have a fixed phase difference from each other of 120° ($2\pi/3$ radians). Individual three-phase transformers, generators, or motors all have their windings arranged in one of two ways. These are (1) Y (or wye) and (2) Δ (delta), as illustrated in Figures 5.3 and 5.4. The appearance of

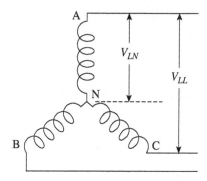

Figure 5.3 *Y*-connected coils; V_{LN} and V_{LL}, line-to-neutral and line-to-line voltage, respectively

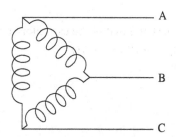

Figure 5.4 Delta-connected coils

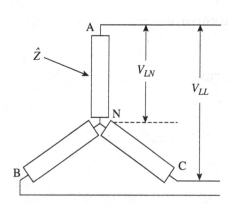

Figure 5.5 *Y*-connected loads; V_{LN} and V_{LL}, line-to-neutral and line-to-line voltage, respectively; $\hat{Z}$, impedance

the windings in circuit diagrams is responsible for the names. Note that the *Y* system has four wires (one of which is the neutral), whereas the Δ system has three wires.

Loads in a three-phase system are, ideally, balanced. That means the impedances are all equal in each phase. If that is the case, and assuming that the voltages are of equal magnitude, then the currents are equal to each other but are out of phase from one another by 120°, as are the voltages. Voltages in three-phase systems may be line-to-neutral, V_{LN}, or line-to-line voltages, V_{LL}. They may also be described as line voltages (V_{LL}) or phase voltages (voltages across loads or coils). Currents in each conductor, outside the terminals of a load, are referred to as line currents. Currents through a load are referred to as load or phase currents. In general, in a balanced *Y*-connected load, the line currents and phase currents are equal, the neutral current is zero, and the line-to-line voltage, V_{LL}, is $\sqrt{3}$ multiplied by the line-to-neutral voltage, V_{LN}. In a balanced delta-connected load, the line voltages and phase voltages are equal, whereas the line current is $\sqrt{3}$ multiplied by the phase current. Figure 5.5 illustrates *Y*-connected three-phase loads, assumed to be balanced and all of impedance $\hat{Z}$.

If a three-phase system is known to be balanced, it may be characterized by a single-phase equivalent circuit. The method assumes a *Y*-connected load, in which each impedance is equal to $\hat{Z}$. (A delta-connected load could be used by applying an appropriate $Y - \Delta$ transformation to the impedances, giving $\hat{Z}_Y = \hat{Z}_\Delta/3$.) The one-line equivalent circuit is one phase of a four-wire, three-phase *Y*-connected circuit, except that the voltage used is the line-to-neutral voltage, with an assumed initial phase angle of zero. The one-line equivalent circuit is illustrated in Figure 5.6.

More details on three-phase circuits can be found in most texts on electrical power engineering.

Power in Three-Phase Loads

It is most convenient to be able to determine power in a three-phase system in terms of easily measurable quantities. These would normally be the line-to-line voltage difference and the line currents. In a balanced load, the power in each phase is one-third of the total power. The real power in one phase, P_1, with phase current, I_P, would be:

$$P_1 = V_{LL}I_P \cos\left(\phi_{pf}\right) = \sqrt{3}V_{LN}I_P \cos\left(\phi_{pf}\right) \qquad (5.29)$$

Since the line current is given by $I_L = \sqrt{3}I_P$, and there are three phases, the total real power is:

$$P = \sqrt{3}\ V_{LL}I_L \cos\left(\phi_{pf}\right) = 3V_{LN}I_L \cos\left(\phi_{pf}\right) \qquad (5.30)$$

Similarly, the total three-phase apparent power and reactive power are:

$$S = \sqrt{3}\ V_{LL}I_L = 3V_{LN}I_L \qquad (5.31)$$

$$Q = \sqrt{3}\ V_{LL}I_L \sin\left(\phi_{pf}\right) = 3V_{LN}I_L \sin\left(\phi_{pf}\right) \tag{5.32}$$

The above relations for three-phase power also hold for balanced Y-connected loads. Calculation of power in unbalanced loads is beyond the scope of this text. The interested reader should refer to any textbook on power system engineering for more information.

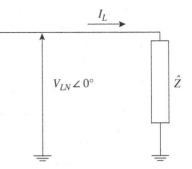

Figure 5.6 One-line equivalent circuit; I_L, line current in three-phase system; V_{LN}, line-to-neutral voltage; $\hat{Z}$, impedance

5.2.2.7 Voltage Levels
One of the major advantages of AC power is that the voltage level may be readily changed by the use of power transformers. Power may be used conveniently and safely at relatively low voltage but is normally transformed to a much higher level for transmission or distribution. To a close approximation, power is conserved during transforming, so that when the voltage is raised, currents are lowered. This lower current serves to reduce losses in transmission or distribution lines, allowing much smaller and less expensive conductors.

Wind turbines historically produced power at 480 V (in the United States) or 690 V (in Europe), but nowadays the voltage may be much higher, depending on the type of generator. Individual wind turbines are often connected to distribution lines with voltages in the range of 10–69 kV. Wind turbines in large wind power plants are connected to transmission systems of higher voltages via transformers in a substation. See Chapter 12 for more information on electrical grids and interconnection to the grid.

5.2.3 Fundamentals of Electromagnetism

The fundamental principles governing transformers and electrical machinery, in addition to those of electricity, which were summarized above, are those of the physics of electromagnetism. As with electricity, it is assumed that the reader is familiar with the basic concepts of electromagnetism. These principles are summarized below. More details can be found in most physics or electrical machinery texts. By way of a quick overview, it may be noted that the magnetic field intensity in an electromagnet is a function of the current. The forces due to magnetic fields are a function of the magnetic flux density. This depends on the magnetic permeability of the materials within the magnetic field as well as the intensity of the magnetic field.

5.2.3.1 Ampere's Law
Current flowing in a conductor induces a magnetic field of intensity H (A/m) in the vicinity of the conductor. This is described by Ampere's Law:

$$\oint H \cdot d\ell = I \tag{5.33}$$

which relates the current in the conductor, I, to the line integral of the magnetic field intensity along a path, ℓ, around the conductor.

5.2.3.2 Flux Density and Magnetic Flux
The magnetic flux density, B (Tesla, T, or Wb/m^2), is related to the magnetic field intensity by the permeability, μ, of the material in which the field is occurring:

$$B = \mu\ H \tag{5.34}$$

where $\mu = \mu_0\mu_r$ is the permeability (Wb/A-m) which can be expressed as the product of two terms: μ_0, the permeability of free space, $4\pi \times 10^{-7}$ Wb/A-m, and μ_r, the dimensionless relative permeability of the material. Note that Wb is the symbol for weber, the unit of magnetic flux.

The permeability of nonmagnetic materials is close to that of free space and thus the relative permeability, μ_r, is close to 1.0. The relative permeability of ferromagnetic materials is very high, in the range of 10^3 to 10^5. Consequently, ferromagnetic materials are used in the cores of windings in transformers and electrical machinery in order to create strong magnetic fields (i.e., with high flux density).

These relations can be used to characterize the magnetic field in a coil of wire. Current flowing in a wire coil will create a magnetic field whose strength is proportional to the current and the number of turns, N, in the coil. The simplest case is a solenoid, which is a long wire wound in a close-packed helix. The direction of the field is parallel to the axis of the solenoid. Using Ampere's Law, the magnitude of the flux density inside a solenoid of length L can be determined:

$$B = \mu I \frac{N}{L} \tag{5.35}$$

The flux density is relatively constant across the cross-section of the interior of the coil.

Magnetic flux Φ (Wb) is the integral of the product of the magnetic field flux density, and the cross-sectional area A through which it is directed:

$$\Phi = \int B \cdot dA \tag{5.36}$$

Note that the integral takes into account, via the dot product, the directions of the area through which the flux density is directed as well as that of the flux density itself. For example, magnetic flux inside a coil is proportional to the magnetic field strength and the cross-sectional area, A, of the coil:

$$\Phi = B \, A \tag{5.37}$$

5.2.3.3 Faraday's Law

A changing magnetic field will induce an electromotive force (EMF, or voltage) E in a conductor within the field. This is described by Faraday's Law of Induction:

$$E = -\frac{d\Phi}{dt} \tag{5.38}$$

Note the minus sign in Equation (5.38). This reflects the observation that the induced current flows in a direction such that it opposes the change that produced it (Lenz's Law). Note also that, in this text, the symbol E is used to indicate induced voltages, while V is used for voltages at the terminals of a device.

As a result of Faraday's Law, a coil in a changing magnetic field will have a voltage induced in it that is proportional to the number of turns:

$$E = -\frac{d(N\,\Phi)}{dt} \tag{5.39}$$

The term $\lambda = N\Phi$ is often referred to as the flux linkages in the device. Coils are particularly relevant since the magnetic poles in many generators are based on coils of wire wrapped around a ferromagnetic core.

5.2.3.4 Induced Force

A current flowing in a conductor in the presence of a magnetic field will result in an induced force acting on the conductor. This is the fundamental property of motors. Correspondingly, a conductor that is forced to move through a magnetic field will have a current induced in it. This is the fundamental property of generators. In either case, the force $d\boldsymbol{F}$ in a conductor of incremental length $d\boldsymbol{\ell}$ (a vector), the current, and the magnetic field $d\boldsymbol{B}$ are related as shown in Equation (5.40):

$$dF = I\,d\ell \times dB \tag{5.40}$$

Note the cross product ($\times$) in Equation (5.40). This indicates that the conductor is moving at right angles to the field when the force is greatest. The force is also in a direction perpendicular to both the field and the conductor.

5.2.3.5 Reluctance

A concept that can be useful in understanding some electrical conversion devices such as transformers and rotating machinery is that of "reluctance." First of all, in electromagnetic circuits in such devices, the quantity that drives the flux is the product of the number of turns in a coil, N, and the current, I. This product, NI, is sometimes referred to as the magnetomotive force (MMF). The reluctance is the ratio of MMF to flux and can be thought of as resistance to the generation of magnetic flux by the MMF. In general, the reluctance is proportional to the distance through which the flux travels and is inversely proportional to the permeability of the material, which is much higher than that of air (see Section 5.2.3.2). In electrical machines, magnetic flux flows across air gaps from stationary to rotating parts of the system. Normally, it is important to keep these gaps as narrow as possible because of their reluctance. In some cases, the geometry of a rotor causes the reluctance to vary during rotation. This variation in reluctance can be taken advantage of in certain types of generators (see Section 5.4.6).

5.2.3.6 Additional Considerations

In practical electromagnetic devices, additional considerations affect machine performance, including leakage and eddy current losses, and the nonlinear effects of saturation and hysteresis.

Magnetic fields can never be restricted to exactly the regions where they can do useful work. Because of this, there are invariably losses. These include leakage losses in transformers and electrical machinery. The effect of the leakage losses is to decrease the magnetic field from what would be expected from the current in the ideal case, or conversely to require additional current to obtain a given magnetic field. Eddy currents are secondary and generally undesired currents induced in parts of a circuit experiencing an alternating flux. These contribute to energy losses.

Ferromagnetic materials, such as are used in electrical machinery, often have nonlinear properties. For example, the magnetic flux density, $\boldsymbol{B}$, is not always proportional to field intensity, $\boldsymbol{H}$, especially at higher intensities. At some point, $\boldsymbol{B}$ ceases to increase even though $\boldsymbol{H}$ is increasing. This is called saturation. The properties of magnetic materials are typically shown in magnetization curves. An example of such a curve is shown in Figure 5.7.

Figure 5.7 Sample magnetization curve; μ, permeability

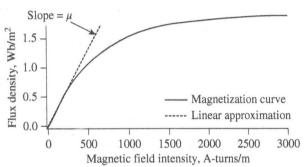

Another nonlinear phenomenon that affects electrical machine design is known as hysteresis. This describes a common situation in which the material becomes partially magnetized, so **B** does not vary when **H** is decreasing in the same way that it does when **H** is increasing.

5.3 Power Transformers

Power transformers are important components in any AC power system. Most wind turbine installations include at least one transformer for converting the generated power to the voltage of the electrical network to which the turbine is connected. In addition, other transformers may be used to obtain voltages of the appropriate level for various ancillary pieces of equipment at the site (lights, monitoring and control systems, tools, compressors, etc.). Transformers are rated in terms of their apparent power (kVA). Distribution transformers are typically in the 5–50 kVA range and may well be larger, depending on the application. Substation transformers are typically between 1000 and 60 000 kVA.

A transformer is a device that has two or more coils, coupled by a mutual magnetic flux. Transformers are usually comprised of multiple turns of wire, wrapped around a laminated metal core. In the most common situation, the transformer has two windings, one known as the primary, and the other as the secondary. The wire is normally of copper and is sized so there will be minimal resistance. The core consists of laminated sheets of metal, separated by insulation so that there will be a minimum of eddy currents circulating in the core.

The operating principles of transformers are based on Faraday's Law of Induction (Section 5.2.3.3). An ideal transformer is one that has: (1) no losses in the windings, (2) no losses in the core, and (3) no flux leakage. An electrical circuit diagram of an ideal transformer is illustrated in Figure 5.8.

Assume that E_1 is applied to the primary of an ideal transformer with N_1 coils on the primary and N_2 coils on the secondary. The ratio between the voltages across the primary and the secondary is equal to the ratio of the number of turns:

$$E_1/E_2 = N_1/N_2 = a \tag{5.41}$$

The parameter a is known as the "turns ratio" of the transformer.

The primary and secondary currents are inversely proportional to the number of turns (as they must be to keep the power or the product IV constant):

$$I_2/I_1 = N_1/N_2 = a \tag{5.42}$$

Real, or nonideal, transformers do have losses in the core and windings, as well as leakage of flux. A nonideal transformer can be represented by an equivalent circuit as shown in Figure 5.9.

In Figure 5.9, R refers to resistances, X refers to reactances, 1 and 2 refer to the primary and secondary coils, respectively. R_1 and R_2 represent the resistance of the primary and secondary windings. X_1 and X_2 represent the leakage inductances of the two windings. The subscript M refers to the mutual (also called magnetizing) inductance and the subscript c to core resistance. V refers to terminal voltages and E_1 and E_2 are the induced voltages at the primary and secondary whose ratio is the turns ratio.

Parameters on either side of the coils may be referred to (or viewed from) one side by using a^2. Figure 5.10 illustrates the equivalent circuit of the transformer when referred to the primary side.

A transformer will draw some current whether or not there is a load on it. There will be losses associated with the current, and the power factor will invariably be lagging. The magnitude of the losses and the power factor can be

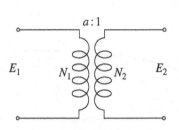

Figure 5.8 Ideal transformer; *a*, turns ratio; *E*, induced voltage; *N*, number of turns; subscripts 1 and 2 refer to the primary and secondary windings, respectively

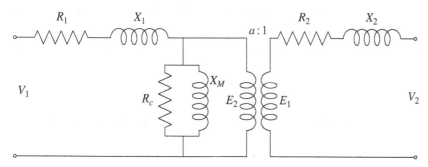

Figure 5.9 Nonideal transformer; for the notation, see text

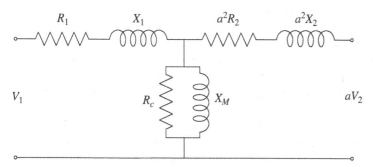

Figure 5.10 Nonideal transformer, referred to primary winding; for notation, see text

estimated if the resistances and reactances in Figure 5.10 are known. These parameters can be calculated by the use of two tests: (1) measurement of the voltage, current, and power at no load (open circuit on one of the coils) and (2) measurement of voltage, current, and power with one of the coils short-circuited. The latter test will be at a reduced voltage to prevent burning out the transformer. Most texts on electrical machinery describe these tests in more detail. For example, see Nasar and Unnewehr (1984).

It is worth noting here that the equivalent circuit of the transformer is similar in many ways to that of induction machines, which are discussed in Section 5.4.4 and are used as generators in many wind turbines.

5.4 Electrical Machines

Generators are devices that convert mechanical power to electrical power; motors convert electrical power to mechanical power. Both generators and motors are frequently referred to as electrical machines because they can usually be run as one or the other. The electrical machines most commonly encountered in wind turbines are those acting as generators.

The two fundamental types of generators are induction (IG) and synchronous (SG). Both of these two types may be further divided into two subtypes. Induction generators are classified according to their rotor design: squirrel cage (SCIG) and wound rotor (WRIG) . Synchronous generators are classified according to how their magnetic field is produced, which is either through electrical excitation (EESG) or permanent magnets (PMSG). A common method of using the WRIG is known as a doubly fed induction generator; these are referred to by the acronym DFIG. A further classification relates to the generator's rotational speed, which may be considered high speed

(HS), medium speed (MS), or direct drive (DD). Typically, induction generators are HS machines, EESGs are DD and PMSGs may be either MS or direct drive.

The following sections present the principles of electrical machines in general and then discuss these generators in more detail.

5.4.1 Basic Electrical Machines

Many of the important characteristics of most electrical machines are evident in the operation of the simplest electrical machine, as shown in Figure 5.11.

In this simple electrical machine, the two magnetic poles (i.e., a pair of poles) create a field. The loop of wire is the armature. The armature can rotate, and it is assumed that there are brushes and slip rings to connect the armature to a stationary circuit. If a current is flowing in the armature, a force acts on the wire. The force on the left side is down, and on the right side is up. The forces then create a torque, causing the machine to act as a motor. In this machine, the torque will be a maximum when the armature loop is horizontal, and a minimum (of zero) when the loop is vertical.

Conversely, if there is initially no current in the wire, but if the armature loop is rotated through the field, a voltage will be generated in accordance with Faraday's Law and current will then flow into the stationary circuit. In this case, the machine is acting as a generator. In general, the directions of current or voltage, velocity, field direction, and force are specified by cross-product relations and conveniently summarized in Fleming's right-hand rule.

When slip rings are used, there are two metal rings mounted to the shaft of the armature with one ring connected to one end of the armature coil and the other ring connected to the other coil. Brushes on the slip rings allow the current to be directed to a load. As the armature rotates, the direction of the voltage will depend on the position of the wire in the magnetic field. In fact, the voltage will vary sinusoidally if the armature rotates at a fixed speed. In this mode, this simple machine acts as an AC generator. Similarly in the motoring mode, the force (and thus torque) reverses itself sinusoidally during a revolution.

This simple machine could be converted to a DC machine with a commutator. A simple commutator in this case would have two segments, each spanning 180° on the armature. Brushes would contact one segment at a time, but segment–brush pairing would reverse itself once during each revolution. The induced voltage would then consist of a sequence of half-sine waves, all of the same sign. In the motoring mode, the torque would always be in the same direction. The commutator principle is the basis of conventional DC motors and generators.

Real electrical machines are similar in many ways to this simple one, but there are also some major differences:

- Except in machines with permanent magnets, the fields are produced electrically.
- The fields are most often on the rotating part of the machine (the rotor), while the armature is on the stationary part (the stator).
- There is also a magnetic field produced by the armature which interacts with the rotor's field. The resultant magnetic field is often of primary concern in analyzing the performance of an electrical machine.

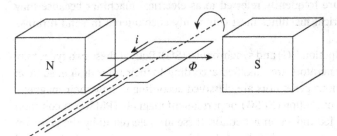

Figure 5.11 Simple electrical machine; *i*, current; Φ, magnetic flux; N, North magnetic pole; S, South magnetic pole

5.4.2 Rotating Magnetic Fields

By suitable arrangement of windings in an electrical machine, it is possible to establish a rotating magnetic field, even if the windings are stationary. This property forms an important basis of the design of most AC electrical machines. In particular, it is the interaction of the stator's rotating magnetic field with the rotor's magnetic field that determines the operating characteristics of the machine.

The principle of rotating fields in a 3-phase machine can be developed in a number of ways, but the key points to note are that: (1) the coils in the stator are 120° apart, (2) the magnitude of each field varies sinusoidally, with the current in each phase differing from the others by 120°, and (3) the windings are such that the distribution of each field is sinusoidal. The resultant magnetic field, H, expressed in phasor form in terms of the three individual magnetic fields H_i is:

$$H = H_1 \angle 0 + H_2 \angle 2\pi/3 + H_3 \angle 4\pi/3 \tag{5.43}$$

Substituting in sinusoids for the currents, and introducing an arbitrary constant C to signify that the field results from the currents, we have:

$$H = C[\cos(2\pi f\, t)\angle 0 + \cos(2\pi f\, t + 2\pi/3)\angle 2\pi/3 + \cos(2\pi f\, t + 4\pi/3)\angle 4\pi/3] \tag{5.44}$$

After performing the algebra, we obtain the interesting result that the magnitude of H is constant, and its angular position is $2\pi ft$ radians. The latter result implies immediately that the field is rotating at a constant speed of f revolutions per second, which is the same as the electrical system frequency.

The above discussion implicitly involved a pair of magnetic poles per phase. It is quite possible to arrange windings so as to develop an arbitrary number of pole pairs per phase. By increasing the number of poles, the resultant rotating magnetic field will rotate more slowly. At no load, the rotor of an electrical machine will rotate at the same speed as the rotating magnetic field, called the synchronous speed. In general, the synchronous speed is:

$$n = \frac{60\ f}{P/2} \tag{5.45}$$

where n is the synchronous speed in rpm, and P is the number of poles per phase.

Equation (5.45) implies, for example, that any two-pole AC machine connected to a 60 Hz electrical network would turn with no load at 3600 rpm, a four-pole machine would turn at 1800 rpm, a six-pole machine at 1200 rpm, etc. It is worth noting here that historically many wind turbine generators were of the four-pole type, thus having a synchronous speed of 1800 rpm when connected to a 60 Hz power system. In a 50 Hz system, such generators would turn at 1500 rpm. Today, however, most generators for wind turbines have more poles than that.

5.4.3 Electrically Excited Synchronous Generators (EESG)

5.4.3.1 Overview of Synchronous Machines

Synchronous machines are used as generators in large central station power plants. In wind turbine applications, they may be used in conjunction with power electronic converters in variable-speed wind turbines (see Section 5.6). Synchronous machines may also be used as a means of voltage control and a source of reactive power in autonomous AC networks. In this case, they are known as synchronous condensers.

This section focuses on conventional synchronous machines, in which the magnetic field is produced by electromagnets (in contrast to permanent magnets, see Section 5.4.4). The poles are created by wrapping coils of wire around a ferromagnetic core. The EESG consists of (1) electromagnetic poles on the rotor and (2) a stationary armature containing multiple windings. The field is produced by a DC current (referred to as excitation) in the field windings. The field current is normally provided by a small AC generator mounted on the rotor shaft of the machine. This small generator is known as the "exciter", since it provides excitation to the field. The exciter has its field stationary, and its output is on the rotor. The output of the exciter is rectified to DC on the rotor and fed directly into the field windings. Alternatively, the field current may be conveyed to the synchronous machine's rotor via slip rings and brushes. In either case, the synchronous machine's rotor field current is controlled externally, and the generator is said to be electrically excited, hence the acronym EESG.

A simple view of a synchronous machine can give some insight into how it works. Assume, as in Section 5.4.2, that a rotating magnetic field has been established in the stationary windings. Assume that there is also a second field on the rotor. These two fields generate a resultant field that is the sum of the two fields. If the rotor is rotating at synchronous speed, then there is no relative motion between any of these rotating fields. If the fields are aligned, then there is no force acting upon them that could change the alignment. Next, suppose that the rotor's field is displaced somewhat from that of the stator, causing a force, and hence an electrical torque, which tends to align the fields. If an external torque is continually applied to the rotor, it could balance the electrical torque. There would then be a constant angle between the fields of the stator and the rotor. There would also be a constant angle between the rotor field and the resultant field, which is known as the power angle, and it is given the symbol delta (δ). It can be thought of as a spring, since, other things being equal, the power angle increases with torque. It is important to note that as long as $\delta > 0$, the machine is a generator. If input torque drops, the power angle may become negative and the machine will act as a motor. A detailed discussion of these relations is outside the scope of this text, however, and is not needed to understand the operation of a synchronous machine for the purposes of interest here. A full development is given in most electrical machinery texts. See, for example, Masters (2013) or Brown and Hamilton (1984).

5.4.3.2 Theory of Synchronous Machine Operation

The following presents an overview of the operation of synchronous machines, based on an equivalent circuit that can be derived for it and two electrical angles. The latter are the power factor angle, ϕ_{pf}, and the power angle, δ.

The current in the synchronous machine's field windings, I_f, induces a magnetic flux. The flux, Φ, depends on the material and dimensions of the core and the number of turns in the windings, as explained in Section 5.2.3, but, to a first approximation, the flux is proportional to the current. The magnetic field in turn induces a voltage in the stationary armature, E, which is proportional to (1) the magnetic flux and (2) the speed of rotation, n, so:

$$E = k_1\, n\, \Phi \tag{5.46}$$

where k_1 is a constant of proportionality.

The generator armature winding is an inductor, which can be represented by a reactance, called the synchronous reactance, X_s, and a small resistance, R_s. Recall that $X_s = 2\pi f L$ where $L =$ inductance. The synchronous impedance $\hat{Z}_s$ is then:

$$\hat{Z}_s = R_s + j\, X_s \tag{5.47}$$

The resistance is usually small compared to the reactance, so the impedance is often approximated by considering only the reactance.

An equivalent circuit, as shown in Figure 5.12, can be used to facilitate analysis of the machine's operation. For completeness, the resistance, R_s, is included, even though it is often ignored in analyses.

The equivalent circuit can be used to develop phasor relations, such as are shown in Figure 5.13. This figure illustrates the phasor relations between the field-induced voltage, E, the terminal voltage, V, and the armature current, I_a, for a synchronous machine with a lagging power factor. The figure may be derived by first assuming a terminal voltage, with reference angle of zero. With a known apparent power and power factor (lagging or leading), the magnitude and angle of the current may

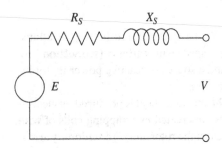

Figure 5.12 Equivalent circuit of a synchronous machine; *E*, field-induced voltage; *R_s*, resistance; *V*, terminal voltage; *X_s*, synchronous reactance

be found. Using the equivalent circuit, field-induced voltage may be determined. The equation corresponding to the equivalent circuit, ignoring the resistance, is:

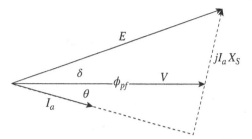

$$\hat{E} = \hat{V} + j\, X_s \hat{I}_a \qquad (5.48)$$

The power angle, δ, shown in Figure 5.13 is, by definition, the angle between the field-induced voltage and the terminal voltage. As described above, it is also the angle between the rotor and the resultant fields. In a simple circuit, $\hat{I}_a$ is also the load current, which is given by $\hat{I}_a = \hat{V}/\hat{Z}_L$ where $\hat{Z}_L$ is the load impedance.

Figure 5.13 Phasor diagram for synchronous generator, lagging power factor; E, field-induced voltage; I_a, armature current; j, $\sqrt{-1}$; V, terminal voltage; X_s, synchronous reactance; δ, power angle; ϕ_{pf}, power factor angle

The things to note are (1) the armature current in Figure 5.13 is lagging the terminal voltage, indicating a lagging power factor and (2) the field-induced voltage leads the terminal voltage, giving a positive power angle, as it should with the machine in a generating mode.

By performing the phasor operations with reference to Figure 5.13, as explained in Section 5.2.2.5, one can find that the real power, P, is:

$$P = \mathrm{Re}\left(\hat{V}\hat{I}_a^*\right) = \frac{|\hat{E}||\hat{V}|}{X_S}\sin(\delta) \qquad (5.49)$$

Similarly, the reactive power, Q, is:

$$Q = \mathrm{Im}\left(\hat{V}\hat{I}_a^*\right) = \frac{|\hat{E}||\hat{V}|\cos(\delta) - |\hat{V}|^2}{X_S} \qquad (5.50)$$

Example 5.2   Power from a Synchronous Generator

Consider a synchronous generator in a 60 Hz system supplying a load consisting of resistance and inductance in parallel, with $R_L = 0.6\,\Omega$ and $X_L = 0.8\,\Omega$. The reactance of the generator is $X_s = 0.03\,\Omega$, and its terminal voltage is $\hat{V} = 480$ V. Based on the assumptions, the impedance of the load is $\hat{Z}_L = 0.384 + j0.278\,\Omega$. The total impedance is $\hat{Z}_T = 0.384 + j0.318\,\Omega$. The current is $\hat{I}_a = \hat{V}/\hat{Z}_L = 800 - j600$ A so $\hat{E} = I_a \hat{Z}_T = 498 + j24$ V. The apparent power is $\hat{S} = \hat{V}\hat{I}_a^* = 384\,000 + j288\,000$ VA. The real power is then 384 kW, the apparent power is 288 kVAR, the power factor angle is -36.87, and power factor is 0.8. The same results for real power and reactive power are found from the right-hand expressions in Equations (5.49) and (5.50).

It is worth emphasizing that, in a grid-connected application with a constant terminal voltage (controlled by other generators), a synchronous machine may serve as a source of reactive power, which may be required by loads on the system. Changing the field current will change the field-induced voltage, E, while the power stays constant. For any given power level, a plot of armature current vs. field current will have a minimum at unity power factor. An example of this is shown in Figure 5.14. This example is for a generator with a line-to-neutral terminal voltage of 2.4 kV. In the generator mode, higher field current (and therefore higher field voltage) will result in a lagging power factor; lower field current (lower field voltage) will result in a leading power factor. In practice, the field current is generally used to regulate the generator's terminal voltage. A voltage regulator connected to the synchronous generator automatically adjusts the field current so as to keep the terminal voltage constant.

An additional consideration relates to the details of the synchronous machine's windings. Most synchronous generators have salient poles, though there are some with round rotors. Salient poles are poles that project out

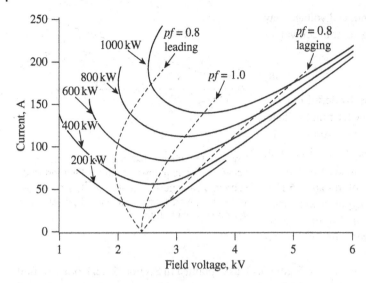

Figure 5.14 Synchronous machine armature current vs. field current in generator mode

from the rest of the rotor and that are individually wound. The terms direct-axis and quadrature-axis are associated with salient pole machines. These terms are discussed in detail in electrical machinery texts such as Kirtley (2011) or Chapman (2001).

5.4.3.3 Starting Synchronous Machines

Synchronous machines are not intrinsically self-starting. In some applications, the machine is brought up to speed by an external prime mover and then synchronized to the electrical network. For other applications, a self-starting capability is required. In this case, the rotor is built with "damper bars" embedded in it. These bars allow the machine to start like an induction machine does (as described in the next section). During operation, the damper bars also help to damp oscillations in the machine's rotor.

Regardless of how a synchronous machine is brought up to operating speed, particular attention must be given to synchronizing the generator with the network to which it is to be connected. A very precise match is required between the angular position of the rotor and the electrical angle of the AC power at the instant of connection. Historically, synchronization was done manually with the help of flashing lights, but it is now done with electronic controls.

Wind turbines with synchronous generators are normally started by the wind (unlike many turbines with induction generators, which can be motored up to speed). When the turbine is to be connected to an AC network that is already energized, active speed control of the turbine may be needed as part of the synchronizing process. In some isolated electrical grids, the AC power is supplied by a synchronous generator on either a diesel generator or a wind turbine, but not both. This obviates the need for a synchronizer.

5.4.3.4 Pole Numbers and Operational Speed

As previously indicated, the speed of a synchronous generator producing power of a given frequency is directly proportional to the number of poles, so generators with only a few poles would turn much faster than a typical wind turbine rotor. For that reason, gearboxes are used to increase the input speed to the generator; see Chapter 7. By increasing the number of poles in the generator, the rotational speed will be reduced. In some cases, the generator speed will be such that a gearbox will still be needed, but that it would have a significantly lower speed up ratio than it otherwise would have. Such generators are referred to as medium speed (MS). In other cases, there will be a sufficient number of poles to enable the generator rotor to turn at the same speed as the wind turbine rotor. This eliminates the need for a gearbox; the generator can be directly driven (DD) from the turbine's rotor. Because of the large number of poles, the diameter of the generator is relatively large, and thus overall weight will be quite high. The benefit of eliminating the gearbox, however, can be substantial.

5.4.4 Permanent Magnet Synchronous Generators (PMSGs)

A type of electrical machine that is being used more frequently in wind turbine applications is the permanent magnet synchronous generator (PMSG). This is now the generator of choice in most small wind turbine generators, up to at least 10 kW, and they are widely used in larger wind turbines as well. In these generators, permanent magnets provide the magnetic field, so there is no need for field windings or supply of current to the field. In one example, the magnets are integrated directly into a cylindrical cast aluminum rotor. The power is taken from a stationary armature, so there is no need for commutators, slip rings, or brushes. Because the construction of the machine is so simple, the permanent magnet generator is quite rugged.

The operating principles of PMSGs are similar to those of EESGs. The main difference is that the magnetic field is provided by permanent magnets instead of electromagnets. The permanent magnets are fabricated from materials that are intrinsically magnetic, such as alloys of neodymium, iron, and boron. Neodymium is a rare earth element, however, and there are some issues in securing an adequate supply. For that reason, the option for producing synthetic permanent magnets is now an area of active research; see, for example, Cui *et al.* (2022). One implication of using permanent magnets, however, is that adjusting the voltage by varying the field current is not applicable. Electronic power converters must take over that function.

PMSGs are usually run asynchronously. That is, they are not generally connected directly to the AC network. The power produced by the generator is initially variable voltage and frequency AC, but this AC is rectified almost immediately to DC. The DC power is then directed to battery storage (with smaller turbines), or more commonly it is inverted to AC with a fixed frequency and voltage. See Section 5.5 for a discussion of power electronic converters that can do this conversion. With such converters, PMSGs are an option for variable-speed wind turbines, as discussed in Section 5.6.1. Indeed, PMSGs have become common among wind turbines designed for offshore use. An example of a 10 MW direct-drive PMSG for offshore use is shown in Figure 5.15. PMSGs for wind turbine applications are discussed by Sethuraman and Dykes (2017), Sethuraman *et al.* (2017), and Gaertner *et al.* (2020).

5.4.5 Squirrel Cage Induction Generators (SCIG)

5.4.5.1 Overview of Induction Machines
Induction machines (also known as asynchronous machines) are commonly used for motors in most industrial and commercial applications. It had long been known that induction machines may be used as generators, but they were seldom employed that way until the advent of distributed generation in the mid-1970s. Induction machines became a common type of generator on wind turbines, and they are now used for other distributed generation applications (e.g., hydrokinetic) as well. As noted previously, induction machines are of two basic types, those with squirrel cage rotors and those with wound rotors. This section focuses on the squirrel cage type (SCIG).

SCIGs are popular because (1) they have a simple, rugged construction, (2) they are relatively inexpensive, and (3) they may be connected and disconnected from the grid relatively simply.

The stator on the SCIG consists of multiple windings, similar to that of a synchronous machine. The rotor, however, has no windings. Rather it has conducting bars, embedded in a solid, laminated core. The bars make the rotor resemble a squirrel cage, hence the name.

Induction machines require an external source of reactive power. They also require an external constant frequency source to control the speed of rotation. For these reasons, they are most commonly connected to a larger electrical network. In these networks, synchronous generators connected to prime movers with speed governors ultimately set the grid frequency and supply the required reactive power.

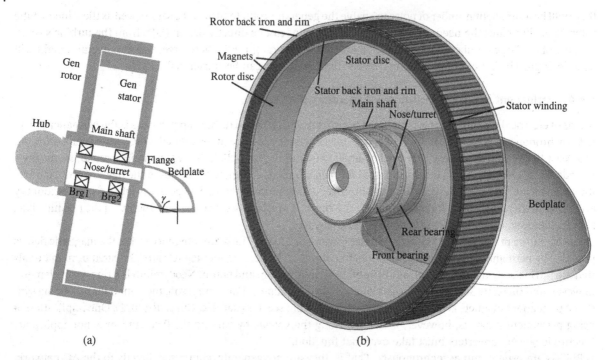

Figure 5.15 An example of a 15 MW direct-drive PMSG. (a) Side view cross section of PMSG, (b) Oblique front view of PMSG. Here, the Gen stator is the generator stator, which is attached to the stationary bedplate (main frame) via the Nose/turret. The Gen rotor is the generator rotor which is attached to the turbine rotor's hub. The main shaft of the generator rotor is supported on the Nose/turret via a front bearing (Brg1) and a rear bearing (Brg2) (*Source:* figure from Gaertner *et al.*, 2020)

When operated as a generator, an induction machine can be connected to the network and brought up to operating speed as a motor, or it can be accelerated by its prime mover (e.g., a wind turbine rotor), and then connected to the network. There are issues to be considered in either case. Some of these are discussed later in this section.

Unlike EESGs, induction machines are not able to control their voltage; thus, they operate with a less than unity power factor. To improve the power factor, capacitors are frequently connected to the machine at or near the point of connection to the electrical network. Care must be taken in sizing the capacitors when the machine is operated as a generator. In particular, it must not be possible for the generator to be "self-excited" if connection to the grid is lost due to a fault.

SCIGs can be used autonomously in small electrical networks or even in isolated applications. In these cases, special measures must sometimes be taken for them to operate properly. The measures involve providing reactive power, maintaining frequency stability, and bringing a stationary machine up to operating speed.

5.4.5.2 Theory of Squirrel Cage Induction Machine Operation
The theory of operation of a squirrel cage induction machine can be summarized as follows:

- The stator has windings arranged such that the phase-displaced currents produce a rotating magnetic field in the stator (as explained in Section 5.4.2).
- The rotating field rotates at exactly synchronous speed (e.g., 1800 rpm for a four-pole generator in a 60 Hz electrical network).
- The rotor turns at a speed slightly different from synchronous speed (so that there is relative motion between the rotor and the field on the stator).

- The rotating magnetic field induces currents and hence a magnetic field in the rotor due to the difference in speed of the rotor and the magnetic field.
- The interaction of the rotor's induced field and the stator's field causes elevated voltage at the terminals (in the generator mode) and current to flow from the machine.

One parameter of particular importance in characterizing induction machines is the slip, s. Slip is the ratio of the difference between synchronous speed n_s and rotor operating speed n, and synchronous speed:

$$s = \frac{n_s - n}{n_s} \tag{5.51}$$

When slip is positive, the machine is a motor; when negative it is a generator. Slip is often expressed as a percentage. Typical values of slip at rated conditions are on the order of $\pm 2\%$.

Operation of the SCIG may be illustrated with the help of the equivalent circuit shown in Figure 5.16. The equivalent circuit and the relations used with it are derived in most electrical machinery texts. In Figure 5.16, V is the terminal voltage, I is the stator current, I_M is the magnetizing current, I_R is the rotor current, X_S is the stator leakage inductive reactance, R_S is the stator resistance, X_R is the rotor leakage inductive reactance, R_R is the rotor resistance, X_M is the mutual (magnetizing) reactance, and R_M is the resistance in parallel with the mutual reactance. Note that the rotor parameters are referred to the stator, by accounting for the turns ratio between the stator and the rotor; see Section 5.3.

A few items of particular note are:

- X_M is always much larger than X_S and X_R.
- The term $R_R(1 - s)/s$ is essentially a variable resistance. For a motor it is positive; for a generator it is negative.
- R_M is a large resistance and is often ignored.

The values of the various resistances and reactances can be derived from tests. These tests are:

- stator resistance measurement (with an ohmmeter capable of measuring small resistances);
- locked rotor test (current, voltage, and power measured at approximately rated current and reduced voltage);
- no-load current and voltage test (current, voltage, and power measured at no load);
- mechanical tests to quantify windage losses and friction losses (described below).

Measurements of induction machine parameters are described in many books on electrical machinery. In the no-load test, the slip is very close to zero, so the resistance $\frac{1-s}{s}R_R$ in the right-hand loop of the circuit shown in Figure 5.16 is very large, and currents in that loop can be ignored. From measurements of voltage, current, and power, it is straightforward to determine X_M (assumed to be much larger than X_S) and R_M. In the locked rotor test, the rotor is held fast; the slip is therefore equal to 1 so the resistance in the rotor circuit is just R_R. A reduced voltage is applied to the stator terminals so that the rated current will flow into the machine. Most of this current

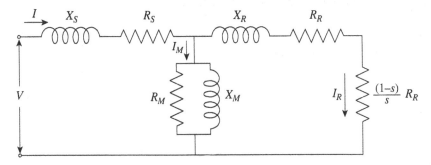

Figure 5.16 Induction machine equivalent circuit; for notation, see text

will flow through the right-hand side of the circuit in Figure 5.16, with little going through the mutual inductance. Once the voltage, current, and power are measured, the values of $R_S + R_R$ and $X_S + X_R$ may be determined. R_R may be found by subtracting the previously measured value of R_S. The stator and rotor leakage reactances are often assumed to be the same.

Not all the power converted in an induction machine is useful power. There are some mechanical losses, $P_{loss,mech}$ as well as electrical losses. The primary losses are (1) mechanical losses due to windage and friction, (2) resistive and magnetic losses in the rotor, and (3) resistive and magnetic losses in the stator. Windage losses are those associated with drag on the rotor from air friction. Friction losses are primarily in the bearings. More information on losses may be found in most texts on electrical machinery.

In the generator mode, mechanical power input to the machine, P_{in}, that is available to produce electricity is reduced by mechanical losses. The mechanical power available to be converted at the generator's rotor, P_m, is:

$$P_m = -(P_{in} - P_{loss,mech}) \tag{5.52}$$

Note: The minus sign is consistent with the convention of generated power as negative. (Elsewhere in this text, however, the sign may be omitted when the context is clear.)

It is most straightforward to find the power from the generator when the slip is given. If the power from the wind turbine's rotor is known rather than the slip, then an iterative process may be used to find the corresponding slip. The power may be found by first determining the overall impedance of the generator, which is based on the generator parameters and the slip.

From the equivalent circuit in Figure 5.16, the impedance of the stator is:

$$\hat{Z}_S = R_S + jX_S \tag{5.53}$$

Similarly, the impedance of the rotor is:

$$\hat{Z}_R = R_R + R_R(1-s)/s + jX_R \tag{5.54}$$

The rotor is in parallel with the mutual inductance (ignoring R_M), and the impedance of the stator is in series with the combination. Accordingly, the total impedance of the induction machine is:

$$\hat{Z}_{IM} = \hat{Z}_S + \frac{1}{1/\hat{Z}_M + 1/\hat{Z}_R} \tag{5.55}$$

where $\hat{Z}_M = jX_M$.

Assuming that the voltage at the terminals of the induction machine is $\hat{V}$ then the current flowing to or from the machine is:

$$\hat{I} = \hat{V}/\hat{Z}_{IM} \tag{5.56}$$

The apparent power delivered or consumed is found as described in Section 5.2.2.5:

$$\hat{S} = \hat{V}\hat{I}^* \tag{5.57}$$

Similarly, the real and reactive power and power factor are:

$$P = \mathrm{Re}(\hat{S}) \tag{5.58}$$

$$Q = \mathrm{Im}(\hat{S}) \tag{5.59}$$

$$pf = \frac{P}{|\hat{S}|} \tag{5.60}$$

The electrical efficiency of the machine is the ratio of the power in to the power out. The power in, P_m, may be expressed in terms of the rotor current, $\hat{I}_R$, rotor resistance, and slip:

$$P_m = \frac{1-s}{s} R_R \hat{I}_R \hat{I}_R^*$$ (5.61)

where again referring to Figure 5.16:

$$\hat{I}_R = \hat{V}_R / \hat{Z}_R$$ (5.62)

$$\hat{V}_R = \hat{V} - \hat{I}\hat{Z}_S$$ (5.63)

The electrical efficiency is the power out divided by the power in:

$$\eta = P / P_m$$ (5.64)

It is worth noting that one of the reasons that efficiency is less than 100% is due to I^2R losses in the rotor resistance. The power is lost as heat. The rotor loss is sometimes referred to as slip power loss. More will be said about this in connection with WRIGs.

Example 5.3   Power from a Squirrel Cage Induction Generator

Consider a wind turbine with a 3-phase SCIG connected to an electrical grid with a line-neutral system voltage, $\hat{V} = 389.25\angle 0°$. The generator has the following (per phase) resistances and reactances of $X_S = X_R = 0.3\,\Omega$, $R_S = 0.028\,\Omega$, $R_R = 0.0272\,\Omega$ and $X_M = 10\,\Omega$. It is operating with a slip of $s = -0.025$. Windage and friction losses are ignored here for simplicity.

Performing the algebraic calculations with Equations (5.53)–(5.57), the results (per phase) are as follows: $\hat{Z}_{IM} = -0.9863 + j0.6984\,\Omega$, $\hat{I} = -262.87 - j186.15A$, $\hat{S} = -102\,320 + j72\,458\,VA$. From Equations (5.58)–(5.64), the real power (per phase) delivered to the grid is 102.3 kW, the reactive power (per phase) is 72.46 kVAR, the power factor is 0.82, and the electrical efficiency is 0.949. The total real power and reactive power for the three phases are 307.0 kW and 217.4 kVAR, respectively. The input power from the wind turbine's rotor is 323.6 kW.

The total mechanical torque, Q_m, applied to a three-phase induction machine (as a generator) is the total input power divided by the generator's rotational speed, n (rpm):

$$Q_m = P_{in} / (2\pi\, n / 60)$$ (5.65)

Note that in this chapter Q without a subscript refers to reactive power. Q with a subscript is used for torque to maintain consistency with other chapters and conventional engineering nomenclature. The mechanical torque is, to a very good approximation, linearly related to slip over the range of slips that are generally encountered in practice.

Figures 5.17 and 5.18 illustrate the results of applying the equivalent circuit to a typical three-phase squirrel cage induction machine rated at 100 kW. Figure 5.17 shows power, current, and torque for operating speeds ranging from 0 to twice synchronous speed. Between standstill and 1800 rpm, the machine is motoring; above that it is generating. As a generator in a wind turbine, the machine would never operate above approximately 3% above synchronous speed, but it would operate at speeds under 1800 while starting. Note that the peak current during startup is over 730 A, more than five times the rated value of 140 A. Peak torque is about two and half times rated (504 Nm). The zero-speed starting torque is approximately equal to the rated value. Peak terminal power is approximately three times the rated value (100 kW).

Figure 5.18 shows efficiency and power factor from startup to 2000 rpm. The machine has roughly the same efficiency and power factor when motoring during normal operation as when generating, but both decrease to zero at no load.

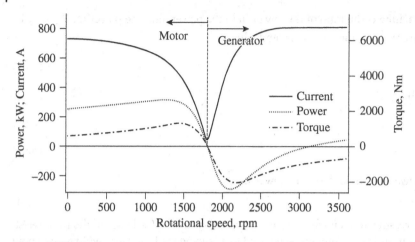

Figure 5.17 Power, current, and torque of a typical squirrel cage induction machine

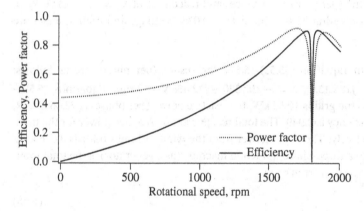

Figure 5.18 Efficiency and power factor of an induction machine

Matrix Form of Squirrel Cage Induction Machine Equations

It is often convenient to express the relations between currents, voltages, and machine parameters in matrix form. The result is a compact equation that can be conveniently solved using matrix methods. These methods are available, for example, in MATLAB® or can be programmed from procedures in Press *et al.* (1992). It can be shown that the matrix equation corresponding to the circuit in Figure 5.16 (assuming R_M can be ignored) is:

$$\begin{bmatrix} R_S + j(X_S + X_M) & -(0 + jX_M) \\ (0 - jX_M) & (R_R/s + j(X_R + X_M)) \end{bmatrix} \begin{bmatrix} I_{S,r} + jI_{S,i} \\ I_{R,r} + jI_{R,i} \end{bmatrix} = \begin{bmatrix} V \\ 0 \end{bmatrix} \tag{5.66}$$

Note that the subscripts r and i refer to the real and imaginary components of the current. The matrix form of the equation will also be useful in describing more complex circuits such as those involving wound rotor induction generators (see Section 5.6.3). It is assumed here that the stator voltage, V, has a phase angle of zero.

5.4.5.3 Starting Wind Turbines with Squirrel Cage Induction Generators

There are two basic methods of starting a wind turbine with a squirrel cage induction generator:

1) Using the wind turbine rotor to bring the generator rotor up to operating speed, and then connecting the generator to the grid.
2) Connecting the generator to the grid and using it as a motor to bring the wind turbine rotor up to normal operating speed.

When the first method is used, the wind turbine rotor must obviously be self-starting. This method is common with pitch-controlled wind turbines, which are normally self-starting. Monitoring of the generator speed is required so that it may be connected when the speed is as close to synchronous speed as possible.

The second method is commonly used with stall-controlled wind turbines (see Chapter 9). In this case, the control system must monitor the wind speed and decide when the wind is in the appropriate range for running the turbine. The generator may then be connected directly "across the line" to the electrical grid, and it will start as a motor. In practice, however, across the line is not a desirable method of starting. It is preferable to use some method of voltage reduction or current limiting during starting. Options for doing this are discussed in Section 5.7.1. As the speed of the wind turbine rotor increases, the aerodynamics become more favorable. Wind-induced torque will impel the generator rotor to run at a speed slightly greater than synchronous, as determined by the torque–slip relation described in Section 5.4.5.2.

5.4.5.4 Squirrel Cage Induction Machine Dynamic Analysis

When a constant torque is applied to the rotor of a SCIG, it will operate at a fixed slip. If the applied torque is varying, then the speed of the rotor will vary as well. The relationship is:

$$J \frac{d\omega_r}{dt} = Q_e - Q_r \tag{5.67}$$

where J is the combined moment of inertia of all mechanically connected components of the drivetrain, ω_r is the angular speed of the generator rotor (rad/s), Q_e is the electrical torque, and Q_r is the mechanical torque applied to the combined inertia. The combined inertia includes the inertia of the generator rotor, the gears (if any), and the aerodynamic rotor (blades, hub, and shafts.) All of these inertias must be referred to rotational speed of the generator (see Chapter 7).

When the applied torque varies slowly relative to the electrical grid frequency, a quasi-steady state approach may be taken for the analysis. That is, the electrical torque may be assumed to be a function of slip as described in Equation (5.65) and preceding equations. The quasi-steady state approach may normally be used in assessing wind turbine dynamics. This is because the frequency of fluctuations in the wind-induced torque and those of mechanical oscillations are generally much less than the grid frequency.

It is worth noting that SCIGs are "softer" in their dynamic response to changing conditions than are EESGs when the latter are directly connected. This is because the former undergo a small, but significant, speed change (slip) as the torque in or out changes. EESGs, as indicated previously, operate at constant speed, with only the power angle changing as the torque varies. SCIGs thus have a much less "stiff" response to fluctuating conditions than do directly connected EESGs.

5.4.5.5 Off-design Operation of Squirrel Cage Induction Generators

SCIGs are designed for a particular operating point. This operating point is normally the rated power at a given frequency and voltage. In wind turbines, there are many situations when the machine may run at off-design conditions. Four of these situations are:

1) starting (mentioned above);
2) operation below rated power;

3) variable-speed operation;
4) operation in the presence of harmonics.

Operation below rated power, but at rated frequency and voltage, is a very common occurrence. It generally presents few problems. The efficiency and power factor are normally both lower under such conditions, however. Operating behavior below rated power may be examined through application of the induction machine equivalent circuit and the associated equations.

There are numerous benefits of running a wind turbine rotor at variable speed. A wind turbine with a SCIG can be run at variable speed if an electronic power converter of appropriate design is included in the system between the generator and the rest of the electrical network. Such converters work by varying the frequency of the AC supply at the terminals of the generator. These converters must also vary the applied voltage. This is because an SCIG performs best when the ratio between the frequency and voltage ("Volts to Hertz ratio") of the supply is constant (or nearly so). When that ratio deviates from the design value, a number of problems can occur. Currents may be higher, for example, resulting in higher losses and possible damage to the generator windings. This is illustrated in Figure 5.19, which shows currents for a squirrel cage induction machine operated at its rated frequency and half that value, but both at the same voltage. The machine in this case is the same one used in the previous example (Figure 5.17). See Section 5.6.2 for more details.

Operation in the presence of harmonics can occur if there is a power electronic converter of significant size on the system to which the machine is connected. This would be the case for a variable-speed wind turbine and could also be the case in isolated electrical networks (see Chapter 16). Harmonics are AC voltages or currents whose frequency is an integer multiple of the fundamental grid frequency. The source of harmonics is discussed in Section 5.5.4. Harmonics may cause bearing and electrical insulation damage and may interfere with electrical control and data signals as well.

5.4.6 Wound Rotor Induction Generators (WRIG)

In contrast to SCIGs, some induction generators do have windings on the rotor. These are known as wound rotor generators (WRIG). These machines are often used in variable-speed wind turbines (see Section 5.6.3). They are more expensive and less rugged than SCIGs, but they offer a number of advantages for variable-speed operation. Depending on how these machines are used, they may also be referred to as doubly fed. This is because power may be sent to or taken from the rotor, as well as from the stator. The acronym DFIG is commonly used for these generators.

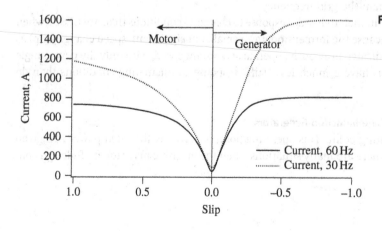

Figure 5.19 Current at different frequencies

WRIGs are similar in many ways to SCIGs, except in the design of their rotors, which have windings of copper wire rather than metal bars. (The stator has windings just as with the SCIG.) The ends of the rotor windings are accessible, normally via brushes and slip rings. Power can either be extracted or injected into the rotor via these brushes and slip rings. WRIGs have many of the features of SCIGs, in that they are compact and fairly rugged (except for the brushes and slip rings). They are also less expensive than synchronous generators, although they are more expensive than SCIGs. The main advantage of the WRIG is that it is possible to have variable-speed operation of the wind turbine while using power converters of approximately 1/3 of the capacity that would be required if all the power were to go through the converters. The converters will not only be smaller, but they will also be less expensive than those that would be used with either SGs or SCIGs of the same rating. Section 5.6.3 provides a worked example of a WRIG.

5.4.7 Other Types of Generators

There are at least two other types of generators that have been used or may be considered for wind turbine applications: (1) shunt wound DC generators and (2) switched reluctance generators.

5.4.7.1 Shunt Wound DC Generator

A historically important type of wind turbine generator is the shunt wound DC generator. These were once common in smaller, battery-charging wind turbines. In these generators, the field is on the stator, and the armature is on the rotor. The electric field is created by currents passing through the field winding which is in parallel ("shunt") with the armature windings. A commutator on the rotor in effect rectifies the generated power to DC. The full generated current must be passed out through the commutator and brushes.

In these generators, the field current, and hence the magnetic field (up to a point), increases with operating speed. The armature voltage and electrical torque also increase with speed. The actual speed of the turbine is determined by a balance between the torque from the turbine rotor and the electrical torque.

DC generators of this type are seldom used today because of high costs and maintenance requirements. The latter are associated particularly with the brushes. More details on generators of this type may be found in Johnson (1985).

5.4.7.2 Switched Reluctance Generators

Switched reluctance generators employ a rotor with salient poles (without windings). As the rotor turns, the reluctance of the magnetic circuit linking the stator and rotor changes. The changing reluctance varies the resultant magnetic field and induces currents in the armature. A switched reluctance generator thus does not require field excitation. The switched reluctance generators currently being considered are intended for use with power electronic converters. Switched reluctance generators need little maintenance due to their simple construction. There are no switched reluctance generators currently being used in commercial wind turbines, but that could change in the future. See Hansen *et al.* (2001) or Arifin *et al.* (2012) for more information.

5.4.8 Generator Mechanical Design

There are numerous issues to consider regarding the mechanical design of a generator. The rotor shaft and main bearings are designed according to the basic principles to be discussed in Chapters 7 and 8. The stator housing of the generator is normally made of steel. Commercial generator housings come in standard frame sizes. Windings of the armature (and field when applicable) are of copper wire, laid into slots. The wire is not only insulated, but also additional insulation is added to protect the windings from the environment and to stabilize them. Different types of insulation may be specified depending on the application.

The exterior of the generator is intended to protect the interior from condensation, rain, dust, blowing sand, etc. Two designs are commonly used: (1) open drip-proof and (2) totally enclosed, fan-cooled (TEFC). The open

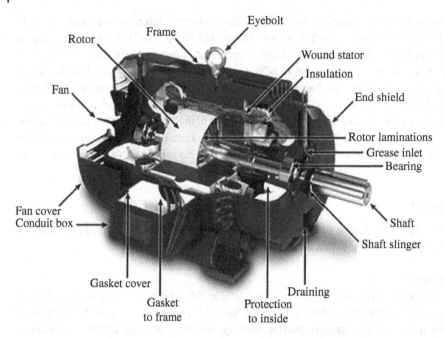

Figure 5.20 Construction of typical three-phase squirrel cage induction machine (*Source:* Rockwell International Corp.)

drip-proof design has been used on many wind turbines, because it is less expensive than other options, and it has been assumed that the nacelle is sufficient to protect the generator from the environment. In many situations, however, it appears that the additional protection provided by a TEFC design may be worth the cost.

A schematic of a typical squirrel cage induction machine is illustrated in Figure 5.20.

5.4.9 Generator Specification

Wind turbine designers are not, in general, generator designers. They either select commercially available electrical machines, with perhaps some minor modifications, or they specify the general requirements of the machine to be specially designed. The basic characteristics of the important generator types have been discussed previously. The following is a summary list of the key considerations from the point of view of the wind turbine designer:

- operating speed;
- efficiency at full load and part load;
- power factor and source of reactive power (induction machines);
- voltage regulation (synchronous machines);
- ability to run at variable speed
- method of starting;
- starting current (squirrel cage induction machines);
- synchronizing (synchronous machines);
- frame size and generator weight;
- type of insulation;
- protection from environment;
- ability to withstand fluctuating torques;
- heat removal;
- operation with high electrical noise on conductors.

5.5 Power Converters

5.5.1 Overview of Power Converters

Power converters are devices used to change electrical power from one form to another, as in AC to DC, DC to AC, one voltage to another, or one frequency to another. Power converters have many applications in wind energy systems. They are being used more often as the technology develops and as costs drop. For example, power converters are used in generator starters, variable-speed wind turbines, and isolated networks. This section provides an overview of many of the types of power converters that are used with wind turbines. For more details see Masters (2013), Tawfiq *et al.* (2019), Baroudi *et al.* (2007), Bhadra *et al.* (2005), or Thorborg (1988).

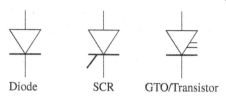

Figure 5.21 Converter circuit elements; SCR, silicon-controlled rectifier; GTO, gate turn off thyristor

Modern converters are power electronic devices. Basically, these consist of an electronic control system turning on and off electronic switches. Some of the key circuit elements used in the inverters include diodes, silicon-controlled rectifiers (SCRs, also known as thyristors), gate turn off thyristors (GTOs), and power transistors. Diodes behave like one-way valves, allowing current to go only one way. SCRs are essentially diodes that can be turned on by an external pulse (at the "gate") but are turned off only by the voltage across them reversing. GTOs are SCRs that may be turned off as well as on. Transistors require the gate signal to be continuously applied to stay on. The overall function of power transistors is similar to GTOs, but the firing circuitry is simpler. The term "power transistor," as used here, includes Darlingtons, power MOSFETS, and insulated gate bipolar transistors (IGBTs). The present trend is toward increasing the use of IGBTs. Figure 5.21 shows the symbols used in this chapter for the most important power converter circuit elements.

5.5.2 Rectifiers

Rectifiers are devices that convert AC into DC. They may be used in: (1) battery-charging wind systems or (2) as part of a variable-speed wind power system.

The simplest type of rectifier utilizes a diode bridge circuit to convert the AC to fluctuating DC. An example of such a rectifier is shown in Figure 5.22. In this rectifier, the input is three-phase AC power; the output is DC.

The average DC voltage, V_{DC}, resulting from rectifying a three-phase voltage V_{LL}:

$$\overline{V}_{DC} = \frac{3\sqrt{2}}{\pi} V_{LL} \tag{5.68}$$

Filtering may be done to remove some of the fluctuations; see Section 5.5.4.

Figure 5.22 Diode bridge rectifier using three-phase supply

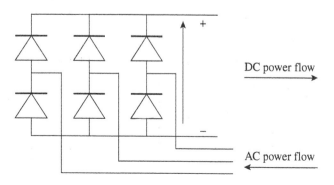

Example 5.4   DC Voltage from a 3 Phase Diode Rectifier

Figure 5.23 illustrates the DC voltage that would be produced from a three-phase, 480 V (line–line) supply using the type of rectifier shown in Figure 5.22. The upper line is the DC voltage; the others are the phase voltages. $\overline{V}_{DC}$ based on Equation (5.68) would be 648 V.

5.5.2.1   Controlled Rectifiers

In some cases, it is useful to be able to vary the output voltage of a rectifier. This may be done by using a controlled rectifier. Here, the primary elements in the bridge circuit are SCRs rather than diodes, as illustrated in Figure 5.24. The SCRs are held off for a certain fraction through the cycle, corresponding to the firing delay angle, α, and then they are turned on. $\overline{V}_{DC}$ is reduced in accordance with the cosine of the firing delay angle, as shown in Equation (5.69).

$$\overline{V}_{DC} = \frac{3\sqrt{2}}{\pi} V_{LL} \cos(\alpha) \tag{5.69}$$

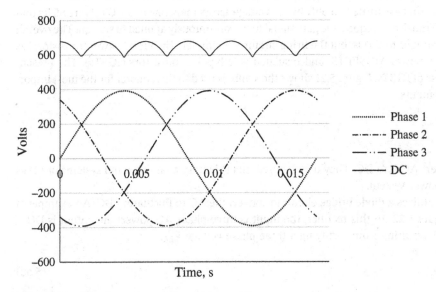

Figure 5.23   DC voltage from a three-phase rectifier

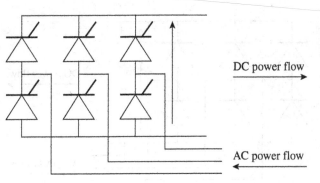

Figure 5.24   Controlled bridge rectifier using three-phase supply

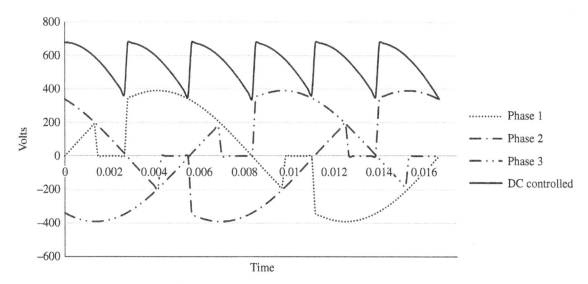

Figure 5.25 DC voltage from phase-controlled rectifier

Example 5.5   DC Voltage from a 3 Phase Controlled Rectifier

A typical controlled rectifier waveform (with a firing delay angle of 30°) is illustrated in Figure 5.25. In this case, by Equation (5.69) the average voltage would be approximately 561 V.

5.5.3   Inverters

5.5.3.1   Overview of Inverters

In order to convert DC to AC, as from a battery or rectified AC in a variable-speed wind turbine, an inverter is used. Historically, motor generator (MG) sets produced AC from DC. These are AC generators driven by DC motors. This method is very reliable but is also expensive and inefficient. Because of their reliability, however, MGs are still used in some demanding situations.

At the present time, most inverters are electronic. An electronic inverter consists of circuit elements that switch high currents and control circuitry that coordinates the switching of those elements. The control circuitry determines many aspects of the successful operation of the inverter. There are two basic types of inverters: line-commutated and self-commutated. The term commutation refers to the switching of current flow from one part of a circuit to another.

Inverters that are connected to an AC grid and that take their switching signal from the grid are known as line-commutated inverters. Figure 5.26 illustrates an SCR bridge circuit, such as is used in a simple three-phase line-commutated inverter. The circuit is similar to the three-phase bridge rectifier shown above, but in this case, the timing of the switching of the circuit elements is externally controlled such that the current flows from the DC supply to the three-phase AC lines.

Self-commutated inverters do not need to be connected to an AC grid. Thus, they can be used for autonomous applications. They tend to be more expensive than line-commutated inverters.

The actual circuitry of inverters may be of a variety of designs, but inverters fall into one of two main categories: (1) voltage source inverters and (2) current source inverters. In current source inverters, the current from the DC source is held constant regardless of the load. They are typically used to supply high power factor loads where the impedance is constant or decreasing at harmonic frequencies. Overall efficiencies are good (around 96%), but the control circuitry is relatively complex. Voltage source inverters operate from a constant voltage DC power source.

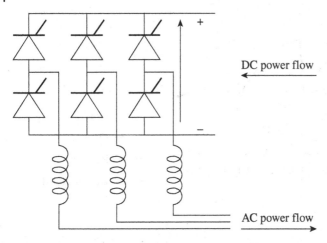

Figure 5.26 Line-commutated silicon-controlled rectifier (SCR) inverter

They are the type most commonly used to date in wind energy applications. (Note that most of the devices described here can operate as rectifiers or inverters, so the term converter is also appropriate.)

5.5.3.2 Voltage Source Inverters
Within the voltage source inverter category are two main types of interest: (1) six-pulse inverters and (2) pulse width modulation (PWM) inverters.

The simplest self-commutated voltage source inverter, referred to here as the "six-pulse" inverter, involves the switching on and off of a DC source through different elements at specific time intervals. The switched elements are normally GTOs or power transistors, but SCRs with turn-off circuitry could be used as well. The circuit combines the resulting pulse into a staircase-like signal, which approximates a sinusoid. Figure 5.27 illustrates the main elements of such an inverter. Once again, the circuit has six sets of switching elements, a common feature of three-phase inverter and rectifier circuits, but in this case both switching on and off of the elements can be externally controlled.

If the elements are switched on one-sixth of a cycle apart, in sequence according to the numbers shown in the figure, and if they are allowed to remain on for one-third of a cycle, an output voltage of a step-like form will appear

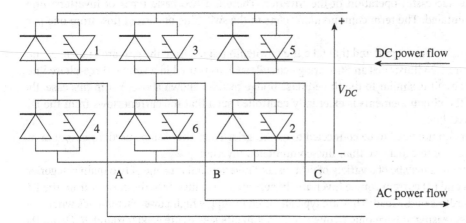

Figure 5.27 Voltage source inverter circuit

between any two phases of the three-phase terminals, A, B, and C. A few cycles of such a waveform (with a 60 Hz fundamental) are illustrated in Figure 5.28.

It is apparent that the voltage is periodic, but that it also differs significantly from a pure sine wave. The difference can be described by the presence of harmonic frequencies resulting from the switching scheme. These harmonics arise because of the nature of the switching. Some type of filtering is normally needed to reduce the effect of these harmonics; see Section 5.5.4.

In PWM, an AC signal is synthesized by switching the supply voltage on and off at high frequency to create pulses of a fixed height. The duration ("width") of the pulse may vary. Many pulses will be used in each half-wave of the desired output. Switching frequencies on the order of 8–20 kHz may be used. The rate of switching is limited by the losses that occur during the switching process. Even with such losses, inverter efficiencies can be 94%. PWM inverters normally use power transistors (IGBTs) or GTOs as the switching elements.

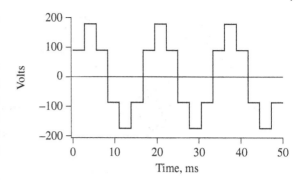

Figure 5.28 Self-commutated six-pulse inverter voltage waveform

Figure 5.29 illustrates the principle behind one method of obtaining pulses of the appropriate width. In this method, a reference sine wave of the desired frequency is compared with a high-frequency offset triangle wave. Whenever the triangle wave becomes less than the sine wave, the transistor is turned on. When the triangle next becomes greater than the sine wave, the transistor is turned off. An equivalent approach is taken during the second half of the cycle. Figure 5.29 includes two complete reference sine waves, but note that the triangle wave is of a much lower frequency than would be used in a real application. The pulse train corresponding to Figure 5.29 is shown in Figure 5.30. It is apparent that the pulses (in the absolute sense) are widest near the peaks, so the average magnitudes of the voltages are greatest there, as they should be. The voltage wave is still not a pure sine wave, but it does contain a few low-frequency harmonics, so filtering is easier. Other methods are also used for generating pulses of the appropriate width. Some of these are discussed in Thorborg (1988) and Bradley (1987).

5.5.4 Harmonics

Harmonics are AC voltages or currents whose frequency is an integer multiple of the fundamental grid frequency. Harmonic distortion refers to the effect on the fundamental waveform of non-sinusoidal or higher frequency voltage or current waveforms resulting from the operation of electrical equipment using solid-state switches. Harmonic distortion is caused primarily by inverters, industrial motor drives, electronic appliances, light dimmers,

Figure 5.29 Pulse width modulation (PWM) control waves

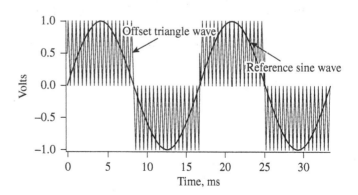

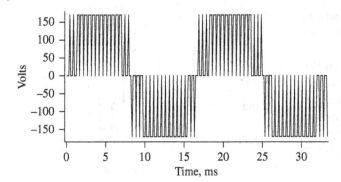

Figure 5.30 Pulse width modulation (PWM) voltage pulse train

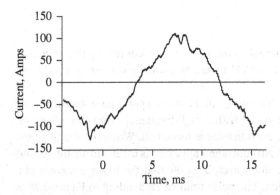

Figure 5.31 Example of a waveform with harmonic distortion

fluorescent light ballasts, and personal computers. It can cause overheating of transformers and motor windings, resulting in premature failure of the winding insulation. The heating caused by resistance in the windings and eddy currents in the magnetic cores is a function of the square of the current. Thus, small increases in current can have a large effect on the operating temperature of a motor or transformer. A typical example of harmonic distortion of a current waveform is illustrated in Figure 5.31.

The presence of harmonics is often characterized by total harmonic distortion (THD), which is the ratio of the total energy in the waveform at all of the harmonic frequencies to the energy in the waveform at the fundamental frequency. The higher the THD, the worse the waveform.

Any periodic waveform $v(t)$ can be expressed as a Fourier series of sines and cosines:

$$v(t) = \frac{a_0}{2} + \sum_{n=1}^{\infty}\left\{a_n \cos\left(\frac{2n\pi t}{T}\right) + b_n \sin\left(\frac{2n\pi t}{T}\right)\right\} \tag{5.70}$$

where n is the harmonic number, T is the period of the fundamental frequency, and:

$$a_n = \frac{2}{T}\int_0^{T/2} v(t) \cos\left(\frac{2n\pi t}{T}\right) dt \tag{5.71}$$

$$b_n = \frac{2}{T}\int_0^{T/2} v(t) \sin\left(\frac{2n\pi t}{T}\right) dt \tag{5.72}$$

The fundamental frequency corresponds to $n = 1$. Higher-order harmonic frequencies are those for which $n > 1$. For normal AC voltages and currents, there is no DC component, so in general $a_0 = 0$. See also Appendix B for more on Fourier series and the related transforms.

Figure 5.32 shows an approximation to the converter voltage output shown originally in Figure 5.28. The approximation uses 15 frequencies in the Fourier series. Note that if more terms were added, the mathematical approximation would be more accurate, and the ripples would decrease. Since the voltage as shown is an odd function, that is $f(-x) = -f(x)$, only sine terms are nonzero.

The relative magnitudes of the coefficients of the various harmonic voltages are shown in Figure 5.33. (It can be shown, in general, for this particular wave that all higher harmonics are zero except those for which $n = 6k \pm 1$ where $k = 0, 1, 2, 3$, etc. and $n > 0$.)

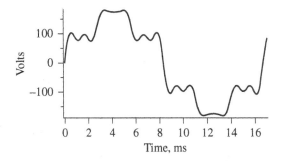

Figure 5.32 Fourier series of voltage waveform from a six-pulse inverter; fifteen terms

5.5.4.1 Quantifying Harmonic Distortion

The harmonic distortion caused by the nth harmonic of the fundamental frequency, HD_n, is defined as the ratio of the rms value of the harmonic voltage of order n over some time T (an integral number of periods of the fundamental) to the rms value of the fundamental voltage, v_F, over the same time T:

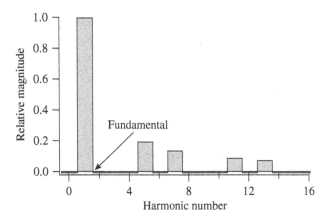

Figure 5.33 Relative harmonic content of voltage

$$HD_n = \frac{\sqrt{\frac{1}{T}\int_0^T v_n^2 dt}}{\sqrt{\frac{1}{T}\int_0^T v_F^2 dt}} \tag{5.73}$$

Similarly, the THD can be expressed as:

$$THD = \frac{\sqrt{\sum_{n=2}^{\infty}\frac{1}{T}\int_0^T v_n^2 dt}}{\sqrt{\frac{1}{T}\int_0^T v_F^2 dt}} = \sqrt{\sum_{n=2}^{\infty}(HD_n)^2} \tag{5.74}$$

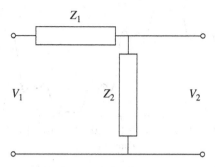

Figure 5.34 AC voltage filter; V_1. AC voltage at filter input; V_2. AC voltage at filter output; Z_1 resonant filter series impedance; Z_2 resonant filter parallel impedance

5.5.4.2 Harmonic Filters

Since the voltage and current waveforms from power electronic converters are never pure sine waves, electrical filters are frequently used. These improve the waveform and reduce the adverse effects of harmonics. Harmonic filters of a variety of types may be employed, depending on the situation. The general form of an AC voltage filter includes a series impedance and a parallel impedance, as illustrated in Figure 5.34.

In Figure 5.34 the input voltage is V_1, and the output voltage is V_2. The ideal voltage filter results in a low reduction in the fundamental voltage and a high reduction in all the harmonics.

One type of AC voltage filter is the series–parallel resonance filter. It consists of one inductor and capacitor in series with the input voltage and another inductor and capacitor in parallel, as shown in Figure 5.35.

As discussed in most basic electrical engineering texts, a resonance condition exists when inductors and capacitors have a particular relation to each other. The effect of resonance in a capacitor and inductor in series is, for example, to greatly increase the voltage across the capacitor relative to the total voltage across the two. Resonance in a filter helps to make the higher-order harmonics small relative to the fundamental. The resonance condition requires, for the capacitor and inductor in series, that:

$$L_1 2\pi f = \frac{1}{C_1 2\pi f} (= X') \tag{5.75}$$

where X' is the reactance of either the capacitor or the inductor and f is the frequency of the fundamental (Hz). For the capacitor and inductor in parallel, resonance implies that:

$$C_2 2\pi f = \frac{1}{L_2 2\pi f} (= Y'') \tag{5.76}$$

where $Y'' =$ the reciprocal of the reactance ("admittance") of either the capacitor or the inductor.

For this particular filter, the output voltage harmonics will be reduced relative to the input by the following scale factor:

$$f_{scale}(n) = \left| \frac{V_n}{V_1} \right| = \left| \frac{1}{1 - [n - (1/n)]^2 X'Y'} \right| \tag{5.77}$$

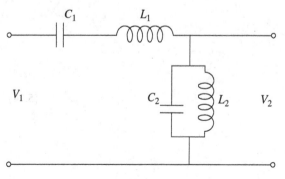

Figure 5.35 Series–parallel resonance filter; C_1 and L_1 series capacitance and inductance, respectively; C_2 and L_2 parallel capacitance and inductance, respectively; V_1 and V_2 input and output voltages, respectively

From Equation (5.77), $f_{scale}(1) = 1$ and $f_{scale}(n)$ approaches $1/n^2$ for high values of n. For certain values of n, the denominator of Equation (5.77), goes to zero, indicating a resonant frequency. For example, there is a resonant frequency for a value of $n > 1$ at:

$$n = \frac{1}{2}\left[\sqrt{\frac{1}{X'Y''}} + \sqrt{\left(\frac{1}{X'Y''} + 4\right)}\right] \tag{5.78}$$

Near this resonant frequency, input harmonics are amplified. Thus, this resonant frequency should be chosen lower than the lowest occurring input harmonic voltage.

Further discussion of filters is outside the scope of this text. It is important to realize, however, that sizing of the inductors and capacitors in a filter is related to the harmonics to be filtered. Higher frequency harmonics can be filtered with smaller components. Accordingly, when considering filtering it is preferrable to have the switching rate in PWM inverters be as high as possible. This is because higher switching rates can reduce the lower frequency harmonics but increase higher ones. More information on filters is given in Thorborg (1988) and Masters (2013).

Example 5.6   Filtering a Square Wave to Produce a Sine Wave
A square wave of height V and frequency f may be generated by the following series:

$$sq(t_i) = \frac{4V}{\pi}\sum_{n=1}^{\infty}\frac{\sin[2\pi(2n-1)t_if]}{2n-1} \tag{5.79}$$

where $t_i = i\Delta t$ is the time of the ith time step and Δt is the time step interval.

Suppose a 60 Hz square voltage wave is produced by reversing the voltage from a 480 VDC source 120 times per second. The square wave is then filtered by the series–parallel resonant circuit such as described above. The inductor and capacitor in series have values $L_1 = 0.01$ H and $C_1 = 0.0007036$ F; those in parallel have values $L_2 = 0.0007036$ H and $C_2 = 0.01$ F. Accordingly, $X' = 3.77\ \Omega$ and $Y' = 3.77\ \Omega$ (Note that in this example X' and Y' have the same values, but in general that need not be the case). The first few useful scaling terms are $f_{scale}(3) = 0.01$ and $f_{scale}(5) = 0.0031$. Figure 5.36 illustrates the original and a filtered square wave in which 100 terms were used.

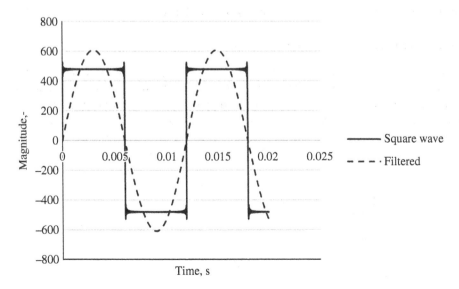

Figure 5.36   Example of filtered square wave

5.6 Electrical Aspects of Variable-Speed Wind Turbines

Variable-speed operation of wind turbines is often desirable for two reasons: (1) below rated wind speed, the wind turbine rotor can extract the most energy if the tip speed ratio can be kept constant, requiring that the rotor speed varies with the wind speed, and (2) variable-speed operation of the turbine's rotor can result in reduced fluctuating stresses, and hence reduced fatigue, of the blades and components of the drivetrain. While variable-speed operation of the turbine rotor may be desirable, such operation complicates the generation of AC electricity at a constant frequency, as discussed below.

There are, at least in principle, a variety of ways to allow variable-speed operation of the turbine rotor, while keeping the generating frequency constant. These may be either mechanical or electrical, as described by Manwell *et al.* (1991). Nearly all the approaches to variable-speed operation of wind turbines in use today, however, are electrical. Variable-speed operation is possible with any of the main types of generators discussed previously.

5.6.1 Variable-speed Operation of Synchronous Generators

As explained in Section 5.4, there are basically two types of synchronous generators: (1) those whose fields are electrically excited, and (2) those whose fields are provided by permanent magnets. In either case, the output frequency is a direct function of the speed of the generator and the number of poles that it has. This was quantified in Equation (5.45). For a synchronous generator to be used in a variable-speed wind turbine, the output of the generator must first be rectified to DC and then converted back to AC. An arrangement that would allow this to take place is shown in Figure 5.37. In this figure, the wind turbine's rotor is shown on the left-hand side. Moving progressively toward the right are the main shaft, gearbox if any (GB), generator (SG), rectifier (AC/DC), DC link, inverter (DC/AC), and grid. The rectifier may be either a diode rectifier or a controlled rectifier, depending on the situation. The inverter may either be an SCR inverter or a PWM inverter. In the case of an electrically excited synchronous generator, there may be voltage control on the generator itself. If it is a PMSG, voltage control must take place somewhere in the converter circuit. As noted previously, some wind turbines use a multiple-pole synchronous generator with a sufficient number of poles that the generator may be directly connected to the main shaft, with no need for a gearbox and so is called "direct drive." Accordingly, the gearbox in Figure 5.37 is shown with a dotted line, indicating that it may or may not be included. It should be recalled that direct drive, multipole generators are physically much larger than are generators commonly used with gearboxes.

5.6.2 Variable-speed Operation of Squirrel Cage Induction Generators

Conventional, squirrel cage induction generators (SCIG) may be used in variable-speed wind turbines, although the method of doing so is not as conceptually straightforward as that of synchronous generators. In particular, all induction generators require a source of reactive power, which must be supplied by a power electronic converter. These converters are expensive, and they also introduce additional losses into the system. The losses are often of the same order of magnitude as the gains in aerodynamic efficiency, so the net gain in energy production may be

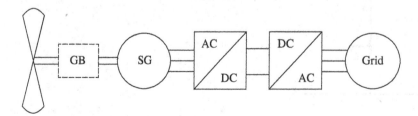

Figure 5.37 Variable-speed wind turbine with synchronous generator

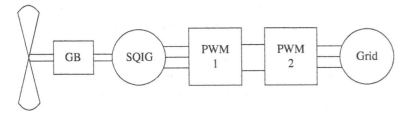

Figure 5.38 Variable-speed wind turbine with squirrel cage induction generator

relatively small. In this case, the main benefit of variable-speed operation is likely to be in the reduction of fatigue damage to the rest of the turbine.

A typical variable-speed SCIG configuration is illustrated in Figure 5.38. This figure shows the wind turbine rotor, gearbox, SCIG, two PWM power converters, a DC link between them, and the electrical grid. In this configuration, the generator side converter (PWM 1) provides reactive power to the generator stator as well as accepting real power from it. It controls the frequency at the generator stator, and hence the speed of both the generator rotor and wind turbine rotor (the latter via the gearbox). PWM 1 also converts the stator power to DC. The grid side converter (PWM 2) converts the DC power to AC at the appropriate voltage and frequency. Note that there may be other components in the circuit, including capacitors, inductors, or transformers.

5.6.3 Variable-speed Operation with Wound Rotor Induction Generators

Another approach to variable-speed turbine operation is through the use of a wound rotor induction generator (WRIG). These generators can be used in a variety of ways to facilitate variable-speed operation. Possibilities include (1) high slip operation, (2) slip power recovery, and (3) true variable-speed operation. The following section provides an overview of some of these. Table 5.2 summarizes the key features of these topologies.

The first thing to observe is that the slip for a given power output of a conventional SCIG could be increased by increasing the resistance of the rotor. This would increase the range of possible operating speeds. Accordingly, one could, in principle, create a variable-speed induction generator by simply increasing the rotor resistance. On the other hand, the losses in the rotor also increase with the resistance, and so the overall efficiency decreases. (Note that this method would not be actually used in practice.)

5.6.3.1 High Slip Operation

When a wound rotor is used rather than a squirrel cage rotor, there are more options available because the rotor windings are accessible. For example, with access to the rotor circuit from the outside, it is also possible to introduce resistance into the circuit at certain times and remove it at other times. Furthermore, the resistance may be changed depending on the situation. A schematic of this arrangement is shown in Figure 5.39. In this figure, the

Table 5.2 Topology options for variable-speed wind turbine using wound rotor induction generators

Topology	Rotor side converter	Line side converter	Speed range
High slip	Adjustable resistor	N/A	Limited super-synchronous
Slip power recovery	Rectifier	Inverter	Super-synchronous only
	Phase-controlled rectifier		Super-synchronous; limited subsynchronous
True variable speed	PWM converter	PWM converter	Super or subsynchronous

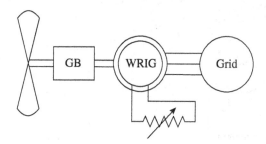

Figure 5.39 Wind turbine with WRIG and external rotor resistance

outer circle in the WRIG element corresponds to the stator, and the inner circle corresponds to the rotor. The external resistor is shown to be variable. The key observation is that when the rotor circuit resistance is made higher by whatever means, the operating speed range is increased. On the other hand, this comes at the cost of greater losses, unless the rotor power loss (known as "slip power") can be recovered.

The equivalent circuit for one phase of this configuration is illustrated in Figure 5.40. Note that this circuit is very similar to Figure 5.16. The main differences are: (1) the mutual resistance R_M is assumed to be very large, and hence does not appear in the circuit diagram, (2) an external resistance, R_x. has been added, and (3) use has been made of the following:

$$R_R + (1 - s)R_R/s = R_R/s$$

For many calculations, Equations (5.53)–(5.63) can be used directly, except that $R_R + R_X$, must be used instead of R_R. The matrix form of the equation corresponding to the equivalent circuit in Figure 5.40 may also be useful; it is shown in Equation (5.80). Note the similarity to Equation (5.66), the only difference being that R_R has been replaced by $R_R + R_X$.

$$\begin{bmatrix} R_S + j(X_S + X_M) & -(0 + jX_M) \\ (0 - jX_M) & ((R_R + R_X)/s + j(X_R + X_M)) \end{bmatrix} \begin{bmatrix} I_{S,r} + jI_{S,i} \\ I_{R,r} + jI_{R,i} \end{bmatrix} = \begin{bmatrix} V \\ 0 \end{bmatrix} \tag{5.80}$$

The capability of adjusting the external resistance in the rotor circuit can be put to good use. In order to do this safely, however, the stator current must be kept at or below its rated level. This can be done by varying the external rotor resistance, R_X, according to the following relation:

$$(R_R + R_X)/s = \text{constant} \tag{5.81}$$

When the resistance is varied according to Equation (5.81), it is possible to allow the speed of the rotor to vary when desired but to keep losses low under other conditions. The most straightforward way to vary the external rotor resistance is to have no resistance when the generator output is less than rated, but then to gradually increase the rotor resistance at higher input powers.

When higher winds result in greater shaft input power, stator power can be held constant (with currents kept at rated) by varying the external rotor circuit resistance in accordance with Equation (5.80). The excess shaft power can be dissipated in the external resistance. This provides a convenient way to quickly shed power, for example from a sudden gust of wind. Otherwise, this excess power would have to be accommodated by changing the blade pitch (which might be too slow) or allowing it to overload the generator. This type of control has been applied by Vestas® Wind Systems A/S under the name of Opti-Slip®. This latter implementation is also of particular interest,

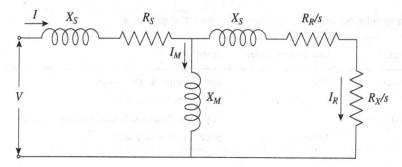

Figure 5.40 Equivalent circuit for WRIG with external rotor resistance

in that the external resistors remain on the rotor, so no slip rings or brushes are required. Communication is done with fiber optics.

5.6.3.2 Slip Power Recovery

In the method of using a WRIG described above, all of the slip power loss is dissipated immediately and serves no useful purpose. There are, however, a number of ways to recover the slip power. A few of these are described here. First of all, it must be noted that the power coming off the rotor is AC, but its frequency is that of the slip multiplied by the line frequency. Accordingly, the slip power cannot be used directly by conventional AC devices, nor can it just be fed directly into the grid. The slip power can be made useful ("recovered"), however, by first rectifying it to DC and then inverting the DC to grid frequency AC power.

The conceptually simplest arrangement for doing this is illustrated in Figure 5.41. This method changes the effective resistance of the rotor but captures the power otherwise lost in the resistance and allows for a wider range of slip. In this arrangement, the power can flow in only one direction, that is to say, out of the rotor. This takes place when the rotor is driven above its synchronous speed. The converter options are the same as those described previously.

The matrix form of the equation for the circuit corresponding to the topology shown in Figure 5.41 is:

$$\begin{bmatrix} R_S + j(X_S + X_M) & -(0 + jX_M) \\ (0 - jX_M) & \left(\dfrac{R_R}{s} + j(X_R + X_M)\right) \end{bmatrix} \begin{bmatrix} I_{S,r} + jI_{S,i} \\ I_{R,r} + jI_{R,i} \end{bmatrix} = \begin{bmatrix} V \\ \dfrac{V_{RT,r}}{s} + j\dfrac{V_{RT,i}}{s} \end{bmatrix} \tag{5.82}$$

where $V_{RT,r}$ is the real component of the rotor terminal voltage, and $V_{RT,i}$ is the imaginary component (both referred to the stator). Note that the difference between Equations (5.82) and (5.80) is that the effect of the voltage at the rotor terminals is now being considered rather than an external resistance in the rotor circuit.

The question arises as to what the voltage across the rotor is. One of the prime requirements is that the stator currents be kept below the current for which the machine is rated. That condition can be maintained as long as $(R_R + R_X)/s \geq$ constant. The rectifier/inverter is not really a resistor. Nonetheless, it could be viewed as a resistive load in which $R_x = V_{RT}/I_R$, where V_{RT} = rotor voltage and I_R = rotor current. The limiting condition for the rotor voltage can be derived from Equation (5.80) above. That will ensure that the stator current does not exceed its allowed value.

We may assume that the rotor current is also limited at some rated value, $I_{R,rated}$. The value of the constant can be determined from the rated conditions, which we assume to occur when the rotor circuit is short-circuited:

$$\text{constant} = R_{R,rated}/s_{rated} \tag{5.83}$$

Accordingly, for any given slip the magnitude of the rotor voltage is:

$$V_{RT} = I_{R,rated}R_{R,rated}(s/s_{rated} - 1) \tag{5.84}$$

Figure 5.41 Variable-speed wind turbine with WRIG and slip power recovery

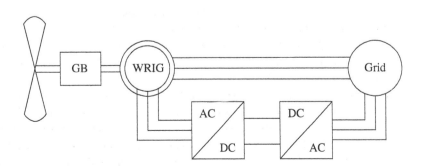

The resistance R_X can also be found:

$$R_X = R_{R,rated}(s/s_{rated} - 1) \tag{5.85}$$

The power that is dissipated or converted externally is:

$$P_{ext} = R_X \hat{I}_R \hat{I}_R^* \tag{5.86}$$

The rotor side rectifier can be of the phase-controlled type. In this case, its output is adjustable downwards to zero from some maximum, which is a function of the line voltage and the turns ratio between the rotor and the stator. The actual rotor voltage is then a function of the phase angle of the rectifier. It should also be noted that the rotor voltage and current are out of phase by 180° (indicating that the power is going out of the rotor) in this arrangement.

The control of the rectifier and inverter in the rotor circuit is of great interest, but it is beyond the scope of this text. More details may be found in Petersson (2005) or Bhadra *et al.* (2005).

Example 5.7   Stator Power and Slip Power in a WRIG

Consider a WRIG with the following parameters: $R_s = 0.005\,\Omega, X_s = 0.01\,\Omega, R_R = 0.004\,\Omega, X_R = 0.008\,\Omega, X_m = 0.46\,\Omega$. The line-neutral voltage is 277.13 V. The generator has a nominal rated slip of $s = -0.025$. The first task is to find the rated stator and rotor currents, the output power, and the efficiency when the rotor is short-circuited, i.e., when the WRIG acts like a SCIG. Next, suppose that an external resistor is added to the rotor circuit to dissipate power and that the generator is to be allowed to run with a slip of $s = -0.4$. The question is how much power can be delivered to the external resistor, assuming that the original current values cannot be exceeded.

Analogously to Example 5.3, we can use Equations (5.53)–(5.64) to find that the stator and rotor currents are $I_S = 1867$ A and $I_R = 1736$ A. The generated power is 1.394 MW, the input power is 1.482 MW, and the efficiency is 0.94. Using Equation (5.85), together with the rotor resistance of 0.004 Ω and the slip of −0.4, we find the external resistance R_X to be 0.06 Ω. We then use $R_R + R_X$ in place of R_R in Equation (5.54) to find the new rotor circuit impedance, $\hat{Z}_R = -0.16 + j0.008\,\Omega$, Equation (5.62) for the rotor current $\hat{I}_R = -1723 - j216.5$, and then Equation (5.86) to find the power dissipated in the external resistor, which is $P_{ext} = 0.54$ MW. If this power were recovered and supplied to the grid, the effective rating of the generator would now be 1.937 MW, with an efficiency of 0.956.

5.6.3.3   True Variable-Speed Operation

In the various types of slip power recovery discussed above, speed during generation would vary from synchronous speed to above synchronous (super-synchronous). This type of operation is always accompanied by conversion of power coming out of the rotor. It is also possible, however, to feed power into the rotor. This arrangement allows operation at sub-synchronous speeds, which is a feature of what may be called true variable-speed operation. There are various ways that this can be accomplished, but the essential characteristic of these arrangements is that there is a rotor-side converter that is capable of interacting directly with the AC of the rotor. The most common arrangement in wind turbine applications is illustrated in Figure 5.42.

In this arrangement, there are two PWM power converters, separated by a DC link. These power converters are both bidirectional. In super-synchronous operation, converter PWM 1 operates as a rectifier and PWM 2 operates as an inverter; power flows out of the rotor as well as out of the stator. In sub-synchronous operation, rotor power goes the other way; PWM 1 operates as an inverter and PWM 2 operates as a rectifier. The complete WRIG generator/converter system often goes under the name of "doubly fed induction generator (DFIG)." The term emphasizes the ability of this arrangement to transfer power into or out of the rotor, as well as out of the stator.

In the simplest view of the DFIG, power transfer into or out of the rotor takes place with both the voltage and the current in phase. Thus, the rotor load looks like either a resistor or a "negative resistor." In reality, however, the PWM converters have the capability of providing current and voltage at different phase angles. This capability has

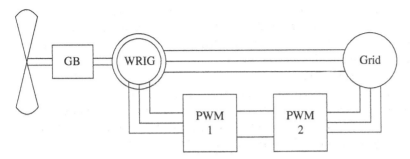

Figure 5.42 Variable-speed operation with turbine with doubly fed induction generator (DFIG)

many advantages. In particular, it allows the power factor of the DFIG to be adjusted at will, and even to supply reactive power to the network if so required. The presence of PWM converters in the circuit allows a wide range of control options. These are discussed in various sources, including Petersson (2005) and Masters (2013), but they are outside the scope of this text.

The DFIG arrangement allows a range of operating speeds of the induction generator, from approximately 50% below synchronous speed to 50% above. DFIGs are used in many modern wind turbines.

5.7 Ancillary Electrical Equipment

There is a variety of ancillary electrical equipment associated with a wind turbine installation. It normally includes both high-voltage (generator voltage) and low-voltage items. Figure 5.43 illustrates the main high-voltage components for a typical installation. Dotted lines indicate items that are often not included. These items are discussed briefly below.

5.7.1 Power Cables

Power must be transferred from the generator, which is at that the top of the tower, to the electrical switchgear at the base. This is done via power cables. Three-phase generators have four conductors, including the ground or neutral. Conductors are normally made of copper, and they are sized to minimize voltage drop and power losses.

In most larger wind turbines, the conductors are continuous from the generator down the tower to the main contactor. In order that the cables not be wrapped up and damaged as the turbine yaws, a substantial amount of slack is left in them so that they "droop" as they hang down the tower. The power cables are thus often referred to as droop cables. The slack is taken up as the turbine yaws and then released as it yaws back the other way. With sufficient slack, the cables seldom or never wrap up tight in most sites. When they do wrap up too far, however, they must be unwrapped. This is normally done by using the yaw drive.

5.7.2 Slip Rings

Some turbines, particularly smaller ones, use discontinuous cables. One set of cables is connected to the generator. Another set goes down the tower. Slip

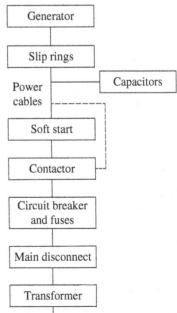

Figure 5.43 Wind turbine high-voltage equipment

rings and brushes are used to transfer power from one set to the other. In a typical application, the slip rings are mounted on a cylinder attached to the bottom of the main frame of the turbine. The axis of the cylinder lies on the yaw axis, so the cylinder rotates as the turbine yaws. The brushes are mounted on the tower in such a way that they contact the slip rings regardless of the orientation of the turbine.

Slip rings are not commonly used on larger wind turbines to carry power down the tower, since they become quite expensive as the current-carrying capacity increases. In addition, maintenance is required as the brushes wear. As previously noted, however, slip rings and brushes may be used to carry the slip power out of the rotor of a WRIG into the stationary electrical system of the nacelle. Slip rings have also been proposed for use with multi-rotor wind turbines, for which traditional yaw control and droop cables would be difficult.

5.7.3 Soft Start

As indicated in Section 5.4.5.3, induction generators will draw much more current while starting across the line than they produce when running. Starting in this way has numerous disadvantages. High currents can result in early failure of the generator windings and can result in voltage drops for loads nearby on the electrical network. Rapid acceleration of the entire wind turbine drivetrain can result in fatigue damage. In isolated grids with a limited supply of reactive power, it may not be possible to start a large induction machine at all.

Due to the high currents that accompany across-the-line starting of induction machines, most stall-regulated wind turbines with induction generators employ some form of soft start device. These can take a variety of forms. In general, they are a type of power electronic converter that, at the very least, provides a reduced current to the generator during start up.

5.7.4 Contactors

The main contactor is a switch that connects the generator cables to the rest of the electrical network. When a soft start is employed, the main contactor may be integrated with the soft start or it may be a separate item. In the latter case, power may be directed through the main contactor only after the generator has been brought up to operating speed. At this point, the soft start is simultaneously switched out of the circuit.

5.7.5 Circuit Breakers and Fuses

Somewhere in the circuit between the generator and the electrical grid are circuit breakers or fuses. These are intended to open the circuit if the current gets too high, presumably due to a fault or short circuit. Circuit breakers can be reset after the fault is corrected. Fuses need to be replaced.

5.7.6 Main Disconnect

A main disconnect switch is usually provided between the electrical grid and the entire wind turbine electrical system. This switch is normally left closed but can be opened if any work is being done on the electrical equipment of the turbine. The main disconnect would need to be open if any work were to be done on the main contactor and would, in any case, provide an additional measure of safety during any electrical servicing.

5.7.7 Power Factor Correction Capacitors

Power factor correction capacitors are frequently employed to improve the power factor of a SCIG when viewed from the utility. These are connected as close to the generator as is convenient, but typically they are at the base of individual turbines or in banks associated with multiple turbines.

5.7.8 Turbine Electrical Loads

There may be a number of electrical loads associated with the operation of wind turbines. These could include actuators, hydraulic motors, pitch motors, yaw drives, air compressors, and control computers. Such loads typically require 120 or 240 V. Since the generator voltage is normally higher than that, a low-voltage supply needs to be provided by the utility, or step-down transformers need to be obtained.

References

Arifin, A., Al-Bahadly, I. and Mukhopadhyay, S. C. (2012) State of the art of switched reluctance generator. *Energy and Power Engineering*, **4**, 447–458, https://doi.org/10.4236/epe.2012.46059.

Baroudi, J. A, Dinavahi, V. and Knight, A. M. (2007) A review of power converter topologies for wind generators. *Renewable Energy*, **32**(14), 2369–2385.

Bhadra, S. N., Kastha, D. and Banerjee, S. (2005) *Wind Electrical Systems*. Oxford University Press, Delhi.

Bradley, D. A. (1987) *Power Electronics*. Van Nostrand Reinhold, Wokingam, UK.

Brown, D. R. and Hamilton, E. P. (1984) *Electromechanical Energy Conversion*. Macmillan Publishing, New York.

Chapman, S. J. (2001) *Electric Machinery and Power System Fundamentals*. McGraw-Hill Companies, Columbus.

Cui, J., Ormerod, J., Parker, D., Ott, R., Palasyuk, A., Mccall, S., Paranthaman, M. P., Kesler, M. S., McGuire, M. A., Nlebedim, I. C. and Pan, C. (2022) Manufacturing processes for permanent magnets: Part I—Sintering and casting. *JOM*, **74**, 1279–1295, https://doi.org/10.1007/s11837-022-05156-9.

Gaertner, E., Rinker, J., Sethuraman, L., Zahle, F., Anderson, B., Barter, G., Abbas, N., Meng, F., Bortolotti, P., Skrzypinski, W., Scott, G., Feil, R., Bredmose, H., Dykes, K., Shields, M., Allen, C., and Viselli, A. (2020) *IEA Wind TCP Task 37: Definition of the IEA 15-Megawatt Offshore Reference Wind Turbine*. NREL/TP-5000-75698. National Renewable Energy Lab. (NREL), Golden, CO.

Hansen, L. H., Helle, L., Blaabjerg, F., Ritchie, E., Munk-Nielsen, S., Bindner, H., Sørensen, P. and Bak-Jensen, B. (2001) *Conceptual Survey of Generators and Power Electronics for Wind Turbines*, Risø-R-1205 (EN). Risø National Laboratory, Roskilde, DK.

Johnson, G. L. (1985) *Wind Energy Systems*. Prentice-Hall, Inc., Englewood Cliffs, NJ. https://www.rpc.com.au/pdf/contents.pdf.

Kirtley, J. (2011), *Electric Power Principles: Sources, Conversion, Distribution and Use*, John Wiley & Sons, New York.

Manwell, J. F., McGowan, J. G. and Bailey, B. (1991) Electrical/mechanical options for variable speed wind turbines. *Solar Energy*, **41**(1), 41–51.

Masters, G. M. (2013) *Renewable And Efficient Electric Power Systems*, 2nd edition. John Wiley & Sons.

Nahvi, M. and Edminster, J. A. (2003) *Electric Circuits*, 4th edition, Schaum's Outline Series in Engineering, McGraw Hill Book Co., New York.

Nasar, S. A. and Unnewehr, L. E. (1984) *Electromechanics and Electric Machines*, 2nd edition. John Wiley & Sons, Inc., New York.

Petersson, A. (2005) *Analysis, Modeling and Control of Doubly-Fed Induction Generators for Wind Turbines*, PhD Dissertation, Chalmers University of Technology, Göteborg.

Press, W. H., Flannery, B. P., Teukolsky, S. A. and Vetterling, W. T. (1992) *Numerical Recipes in FORTRAN: The Art of Scientific Computing*. Cambridge University Press, Port Chester.

Sethuraman, L. and Dykes, K. (2017). *Generator SE: A Sizing Tool for Variable-Speed Wind Turbine Generators*. NREL/CP-5000-64625, National Renewable Energy Lab. (NREL) Golden CO.

Sethuraman, L., Maness, M., and Dykes, K. (2017). Optimized generator designs for the DTU 10-MW offshore wind turbine using GeneratorSE. NREL/CP-5000-67444, National Renewable Energy Lab. (NREL) Golden CO.

Tawfiq, K. B., Mansour, A. S., Ramadan, H. S., Becherif, M., El-kholy, E. E. (2019) Wind energy conversion system topologies and converters: comparative review, *Energy Procedia*, **162**, 38–47.

Thorborg, K. (1988) *Power Electronics*. Prentice-Hall, Englewood Cliffs, NJ.

6

Environmental External Design Conditions

6.1 Overview of External Design Conditions

As is obvious from the source of their energy and their names, wind turbines are dependent on and are affected by the wind. The wind is both the resource (or "fuel") and the source of most of the structural loads on land-based turbines. Chapter 2 discussed the wind primarily in relation to the resource aspect. This chapter considers those aspects that affect the structural loads. In much of the literature on wind turbines, those topics (wind as resource and wind as source of loads) are considered together. The emergence of offshore turbines has changed the situation because the structural design of such turbines is strongly affected by other phenomena in their environment. Ocean waves, which originate primarily due to the wind, are particularly significant, but currents, tides, salinity, and floating ice are important as well and so they are all grouped together under the general heading of meteorological oceanographic, or "metocean" conditions. This full set of metocean conditions is considered when evaluating the design and performance of offshore wind turbines. The wind, in so far as it affects design of land-based turbines and metocean conditions for offshore turbines, may be considered to constitute external design conditions. Other environmental factors can also be important for design, such as temperature, possibility of earthquakes or tsunamis, and lightning.

This chapter discusses external design conditions that affect any turbine, both land-based and offshore. Section 6.2 discusses the wind as an external design condition; Section 6.3 does the same for waves; Section 6.4 concerns the forces from waves; Section 6.5 is about the combined effects of wind and waves; Section 6.6 is about currents; Section 6.7 has to do with floating sea and lake ice; Section 6.8 provides an overview of exceptional conditions, such as hurricanes; Section 6.9 is about other marine conditions; Section 6.10 concerns offshore metocean data collection; and Section 6.11 discusses the role of external conditions in wind turbine design standards.

6.1.1 Wind Turbine Design Basis and External Conditions

Wind turbines themselves and wind energy projects in which they are used are designed with respect to a "design basis," which is the set of factors that need to be considered in the turbine's design and utilization. There are many factors to be considered in the design basis, but the most critical ones are associated with the specific external design conditions for which the turbine is being designed. Most of this chapter focuses on the science and modeling of all of the external conditions experienced by a wind turbine. The specific choices of external design conditions that have been chosen for characterizing wind turbine design bases in international design standards are discussed in Section 6.11. The design basis is discussed in more detail in Chapter 8. For this chapter, it is important to note that the fundamental wind turbine requirement is that the turbine be able to withstand ultimate loads and fatigue loads. Ultimate loads are the highest load that the turbine may be expected to encounter over its lifetime. Such loads could occur when the turbine is operating or when it is parked, depending on the design. These loads may result from extremes in the environment or from a combination of factors. Fatigue loads are associated with

Wind Energy Explained: On Land and Offshore, Third Edition.
James F. Manwell, Emmanuel Branlard, Jon G. McGowan, and Bonnie Ram.
© 2024 John Wiley & Sons Ltd. Published 2024 by John Wiley & Sons Ltd.

varying conditions in the environment itself (e.g., turbulent wind) and the turbine's operation (such as the rotation of the rotor). For the purpose of the design basis, the external conditions are codified into "design load cases" (DLCs) in international design standards. These DLCs are idealized descriptions of the situations that could occur in the physical environment. They are discussed in Section 6.11, while the underlying atmospheric and meteorological oceanographic conditions are described in Sections 6.2 through 6.10.

6.2 Wind as an External Design Condition

The wind is one of the major sources of structural loads and is of critical concern to the design of the wind turbine, whether land based or offshore. Loads due to the wind must be considered in the design of all parts of the wind turbine. These parts include the rotor/nacelle assembly (RNA) and the various components of the support structure, which for land-based turbines consists of a tower and foundation. The support structure of offshore turbines includes an additional substructure, which is largely below the surface of the water. Floating turbines also have a station keeping system to keep them in place. Although the loads on the substructures of offshore wind turbines arise primarily from ocean waves, the wind is a significant source as well.

The manner in which the wind produces loads on a wind turbine is quite complex and nonlinear and is a function of the turbine's design and its control system as well as its operating state. Predicting those loads is discussed in other chapters, but they arise due to some combination of the following:

- Variations in speed from one point in time or location to another;
- Variations in wind direction;
- Wind gusts;
- Turbulence;
- Extreme wind speeds.

Wind turbines are designed with consideration of a suite of characteristics which are categorized into classes, Class I being the most severe and Class III the least severe (see IEC, 2019a). These classes are representative of the wind climate that is expected at the site of prospective installations. Accordingly, the turbine designer will design a turbine that should be suitable for a site with a particular wind climate. A wind energy project developer will need to assess the wind climate of a site in sufficient detail to ensure that turbines of the appropriate class are selected. Offshore turbines require information about their eventual location before the detailed design can begin.

The mean wind climate at a prospective site is amongst the first quantities to be evaluated. This includes the mean wind speed at hub height, the probability density function of the 10-minute averaged wind (Weibull parameters), and distribution of wind speed with wind direction (wind rose). These topics were all discussed in Chapter 2.

In addition, there are other conditions at a site that need to be assessed, mostly associated with turbulence and the other parameters summarized above. These are discussed below.

There are many ways to consider the wind, but in this text, we will follow the approach used in the international design standard for wind turbines (IEC, 2019a), in which wind conditions are divided into two main categories: normal and extreme.

6.2.1 Normal Wind Conditions

6.2.1.1 Fundamental Considerations

As discussed in Chapter 2, one of the fundamental features of the wind is its variability. For the design process as well as for performance estimates, occurrences of wind speed are averaged into 10-minute intervals with mean U. The amount of time that the wind is blowing at various speeds is relevant to the design process for assessing the likelihood of extreme loads as well as the accumulation of fatigue damage. The relative frequency of occurrence of 10-minute wind speeds generally can be described by a Weibull distribution, using a scale parameter, c, and a shape parameter, k, which are in turn related to the long-term standard deviation, σ_U. Variations within those intervals are characterized first of all by the short-term standard deviation, σ_u, but the variability is most commonly reported as the turbulence intensity, $TI = \sigma_u/U$.

6.2.2 Wind Profile (Wind Shear)

As also discussed in Chapter 2, vertical wind shear refers to the variation of wind speed with height. Wind shear is relevant to design because cyclic loads are increased as the rotor rotates through a wind field which varies with height; see Chapter 4. In the mean, wind shear is typically modeled with a power law or log law. Over short time periods, wind shear can differ significantly from its mean value, resulting in an even greater range of cyclic loads.

6.2.3 Stability

Stability of the atmosphere refers to the tendency of the air to move upwards or downwards due to a vertical temperature gradient. Stability can have a significant effect on wind shear as well as turbulence. Stability was discussed in Chapter 2.

6.2.4 Turbulence

Turbulence in the wind refers to short-term variations in wind speed and direction. It is present to some degree in all naturally occurring wind. It is generally either (1) roughness (mechanically) generated or (2) thermally generated, and the underlying processes are quite complex. Turbulence eventually results in the dissipation of the wind's kinetic energy into thermal energy via the creation and destruction of progressively smaller eddies (or gusts). Turbulent wind may have a relatively constant mean over time periods of an hour or more, but over shorter times (minutes or less), it may be quite variable. The wind's variability superficially appears to be quite random, but actually it has distinct features. These features are characterized by a number of properties:

- Gust speeds;
- Turbulence intensity;
- Probability density function;
- Autocorrelation;
- Crosscorrelation;
- Integral time scale/length scale;
- Power spectral density function.

Turbulence is relevant to wind turbine design because it can result in component fatigue and may also be a factor in extreme and possibly ultimate loads. Understanding turbulence is important in preparing inputs to aeroelastic simulation models such as NREL's *OpenFAST* (Jonkman, 2013). These sets of input data are prepared by ancillary codes such as NREL's *TurbSim* (Kelly and Jonkman, 2007).

Turbulent wind consists of longitudinal, lateral, and vertical components which vary with time t and height z. The longitudinal component, in the prevailing wind direction, is designated u. The lateral component (perpendicular to u) is v, and the vertical component is w. Each component is frequently conceived of as consisting of a short-term mean wind, for example, U, with a superimposed fluctuating wind of zero mean, $\tilde{u}$, added to it, thus:

$$u = U + \tilde{u} \tag{6.1}$$

where u = instantaneous longitudinal wind speed.

The lateral and vertical components can be decomposed into a mean and a fluctuating component in a similar manner.

Note that the short-term mean wind speed, U, refers to the mean longitudinal wind speed averaged over some (short) time period, Δt, longer than the time scale of the fluctuations in the turbulence. This time period is usually taken to be 10 minutes, but it can be as long as an hour. In equation form:

$$U = \frac{1}{\Delta t} \int_0^{\Delta t} u \, dt \tag{6.2}$$

Instantaneous turbulent wind is not actually observed continuously; it is sampled at some relatively high rate (typically 1 Hz or higher). Assuming that the sample interval is δt, such that $\Delta t = N_s \delta t$ where N_s = number of samples during each short-term interval, then turbulent wind can be expressed as a sequence, u_i. The short-term mean wind speed can then be expressed in sampled form as:

$$U = \frac{1}{N_s} \sum_{i=1}^{N_s} u_i \tag{6.3}$$

The short-term average longitudinal wind speed, U, is the one most often used in time series observations. This is the form introduced in Chapter 2.

6.2.4.1 Gusts

Gusts are relatively coherent, short-term changes in wind speed. In fact, turbulent wind may be considered to consist of multiple gusts superimposed upon a constant mean. Gusts are characterized by their magnitude relative to the short-term mean and their duration. For wind energy applications, gusts are important particularly in that they may result in extreme loads and must be accounted for in control system design. Synthetically generated gusts used to assess these situations are discussed in Section 6.11.1.

6.2.4.2 Turbulence Intensity

The most basic measure of turbulence is the turbulence intensity. It is defined as the ratio of the standard deviation of the wind speed to the short-term mean wind speed. In this calculation, both the mean and standard deviation are calculated over a time period longer than that of the turbulent fluctuations, but shorter than periods associated with other types of wind speed variations (such as diurnal effects). The length of this time period is normally no more than an hour, and by convention in wind energy engineering, it is usually equal to 10 minutes. The turbulence intensity, TI, is defined by:

$$TI = \frac{\sigma_u}{U} \tag{6.4}$$

where σ_u is the standard deviation, given in sampled form by Equation (6.5). Note the use of the lower case u to distinguish σ_u from σ_U (the long-term standard deviation of 10 minute averages).

$$\sigma_u = \sqrt{\frac{1}{N_s - 1} \sum_{i=1}^{N_s} (u_i - U)^2} \tag{6.5}$$

Turbulence intensity is frequently in the range of 0.1 to 0.4. In general, the highest turbulence intensities occur at the lowest wind speeds, but the lower limiting value at a given location will depend on the specific terrain features and surface conditions at the site. The rougher the terrain, the higher the turbulence intensity. Offshore, turbulence intensity levels are usually lower than onshore, but are highest at low wind speeds, reaching a minimum for medium speeds and then increasing; see Section 6.2.8.4. Figure 6.1 illustrates a plot of a 10-minute segment of wind

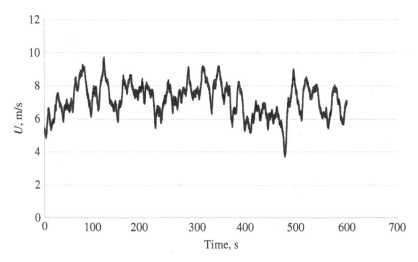

Figure 6.1 Sample turbulent wind data

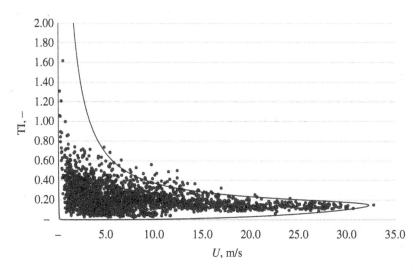

Figure 6.2 Typical turbulence intensity vs. wind speed on land

data sampled at 20 Hz. The data has a mean of 7.16 m/s and a standard deviation of 0.97 m/s. Thus, the turbulence intensity, over the 10-minute period, is 0.14.

Figure 6.2 illustrates typical turbulence intensity vs. wind speed. This plot was based on one month of 10-minute data from a 135 m tower at NREL in Colorado. The smooth line in the figure is an environmental contour (EC); see Section 6.2.7.1.

6.2.4.3 Turbulent Wind Probability Density Function

As discussed elsewhere, occurrences of multiple, 10-minute average wind speeds are best modeled with a Weibull distribution. The distribution of turbulent wind within any 10-minute average is different, however. Experience has shown that the turbulent wind speed is nearly as likely to be below the mean as above it. The probability

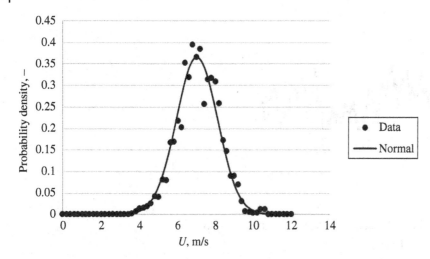

Figure 6.3 Normal (Gaussian) probability density function and histogram of wind data

density function that best describes this type of behavior for turbulence is the normal, or Gaussian, distribution (see Appendix C). The normal probability density function for continuous data in terms of the variables used here is given by:

$$f(u) = \frac{1}{\sigma_u \sqrt{2\pi}} \exp\left[-\frac{(u-U)^2}{2\sigma_u^2}\right]$$

(6.6)

Figure 6.3 shows a normalized histogram (with a bin width of 0.2 m/s) of the wind data illustrated in Figure 6.1. A probability density function that represents the data is superimposed on the histogram for comparison.

6.2.4.4 Autocorrelation

The autocorrelation function provides a measure of how much the wind speed at one location varies from one moment in time to another. This is relevant to wind turbine design in relation to the dynamic response of the rotor in the presence of time-varying turbulent winds. The normalized autocorrelation function (see Appendix B) for sampled turbulent wind speed data is given by Equation (6.7):

$$R(r\delta t) = \frac{1}{\sigma_u^2(N_s - r)} \sum_{i=1}^{N_s - r} \tilde{u}_i \tilde{u}_{i+r}$$

(6.7)

where r = lag number, δt is the sampling period, N_s is the number of points in the sample population, and σ_u is the standard deviation based on the population. Figure 6.4 shows a plot of the autocorrelation function of the data illustrated above in Figure 6.1.

6.2.4.5 Integral Time Scale/Length Scale

Integral time scales and length scales provide a convenient measure of the duration and magnitude of turbulent fluctuations. They can be estimated from the autocorrelation function as described below.

The autocorrelation function will, if any trends are removed before starting the process, decay from a value of 1.0 at a lag of zero to a value of zero and will then tend to take on small positive or negative values as the lag increases. The integral time scale is found by integrating the autocorrelation from zero lag to the first zero crossing. While typical values are on the order of ten seconds, the integral time scale is a function of the site, atmospheric stability,

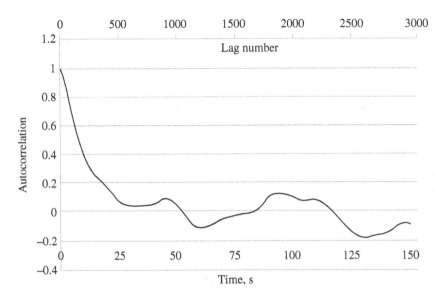

Figure 6.4 Autocorrelation function of sample wind data

and other factors and may be significantly greater than 10 seconds. Gusts have characteristic times (i.e., increase and decrease) of the same order as the integral time scale.

Multiplying the integral time scale by the mean wind velocity gives the integral length scale. The integral length scale tends to be more constant over a range of wind speeds than is the integral time scale, and thus is somewhat more representative of a site.

Based on the autocorrelation function illustrated above, the integral time scale is 12.1 seconds. Since the mean wind velocity is 7.16 m/s, the characteristic size of the turbulent eddies in the mean flow, or the integral length scale, is on the order of 86.6 m.

6.2.4.6 Crosscorrelation

The crosscorrelation function provides a measure of how similar the wind speed is at one location relative to that at another location. This is relevant to wind turbine design since it is an indicator of how the wind varies across the rotor and so contributes to unsteady loading. Crosscorrelation $R_{xy}(\ )$ is defined in Equation (6.8). The crosscorrelation of the turbulent wind across the rotor plane is one of the major considerations in NREL's code *TurbSim* (Kelly and Jonkman, 2007).

$$R_{xy}(r\ \delta t) = \frac{1}{\sigma_x \sigma_y (N_s - r)} \sum_{i=1}^{N_s - r} \tilde{x}_i\ \tilde{y}_{i+r} \tag{6.8}$$

where $\tilde{x}_i, \tilde{y}_i$ are velocities at different locations (with means subtracted), and σ_x, σ_y are the corresponding standard deviations based on the population.

6.2.4.7 Power Spectral Density Function

The fluctuations in the wind can be thought of as resulting from a composite of sinusoidally varying winds superimposed on the mean steady wind. These sinusoidal variations will have a variety of frequencies, amplitudes, and phases. The term "spectrum" is used to describe functions of frequency. The function that characterizes turbulence as a function of frequency is known as a "spectral density" function. Since the average value of any sinusoid is zero, the amplitudes are characterized in terms of their mean square values. This type of analysis originated in electrical power applications, where the square of the voltage or current is proportional to the power. The complete name for the function describing the relation between frequency and amplitudes of sinusoidally varying waves making up the fluctuating wind speed is therefore "power spectral density" or psd. The units of the psd function as used in the characterization of turbulence are wind speed variance per Hertz, $(m/s)^2/Hz$. See Appendix B for more information on spectral analysis. Among other things, Appendix B explains how to determine a psd from time series data.

There are two points of particular importance to note regarding power spectral densities. The first is that the variance in the turbulence over a range of frequencies may be found by integrating the psd between the two frequencies. Second, the integral over all frequencies is equal to the total variance in the data being analyzed.

Power spectral densities are often used in dynamic analyses of wind turbines. Analytical models of psds are used when an empirical one is not available or convenient. One of these was developed by von Karman for turbulence in wind tunnels (Freris, 1990). It is given by Equation (6.9). This is referred to as the von Karman psd elsewhere in this text.

$$S(f) = \frac{\sigma_u^2 4(L/U)}{\left[1 + 70.8(f\,L/U)^2\right]^{5/6}}$$

(6.9)

where f is the frequency (Hz), L is the integral length scale, and U is the 10-minute mean wind speed at the height of interest.

Other psds are also used for turbulent wind. These include the Kaimal and the Mann spectra (see IEC, 2019a). The choice of turbulence models can be significant. Somoano *et al.* (2021) provide examples of the effect of different models in the dynamic response of floating offshore wind turbines.

The Kaimal spectrum is given in Equation (6.10):

$$S(f) = \frac{\sigma_k^2 4f\,L_k/U_{hub}}{(1 + 6fL_k/U_{hub})^{5/3}}$$

(6.10)

The subscript k indicates the component of the turbulence, with u being longitudinal, v lateral, and w upward. A scale parameter Λ_1 is used on which length scales are based:

$$\Lambda_1 = \begin{cases} 0.7z \text{ if } z \le 60\,\text{m} \\ 42 \ \text{ if } z > 60\,\text{m} \end{cases}, z \text{ is the height above the surface}$$

(6.11)

For example, the longitudinal length scale L_u is given by $L_u = 8.1\Lambda_1$.

The Kaimal spectrum has the advantage that in its full description it is more comprehensive than the von Karman; see IEC (2019a).

The empirical psd of the wind data from Figure 6.1 is illustrated in Figure 6.5. The graph also includes the von Karman and Kaimal psds for comparison. As can be seen, the latter two are very similar for this situation. It may also be noted that the magnitude of the empirical psd is lower than from either model at high frequencies, particularly above 0.1 Hz. This is likely due to the inability of the anemometer to respond to rapid fluctuations.

6.2.4.8 Coherence

Coherence of wind turbulence is a measure of how similar the frequency components of turbulence are from one location to another. This is important because the response of the turbine's structure, particularly the blades, will differ depending on how similar the wind is at various points. The coherence is the frequency domain counterpart of the crosscorrelation. Coherence between two-time series is defined by Equation (6.12):

$$\gamma_{xy}^2(f) = \frac{|S_{xy}(f)|^2}{S_{xx}(f)S_{yy}(f)}$$

(6.12)

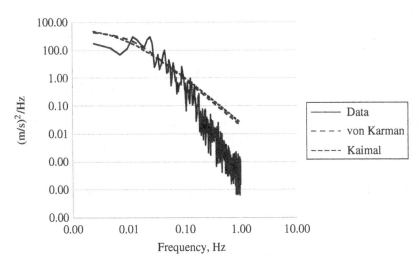

Figure 6.5 Sample wind data power spectral density functions

where $S_{xy}(f)$ is known as the cross-spectral density and $S_{xx}(f)$ and $S_{yy}(f)$ are the power spectral densities at the two locations. The subscripts x and y are added to emphasize that the relation is between two data sets.

For turbulent wind, the magnitudes of the power spectral densities are to a first approximation independent of location, but the phase angles of the different frequency components may differ. In particular, higher frequencies are less well correlated than lower frequencies, and the phase angles are correspondingly different. Thus, coherence decreases at higher frequencies.

A commonly used model to represent the coherence of turbulent wind is an exponential function of frequency as given in Equation (6.13); see Beyer *et al.* (1989).

$$\gamma_{xy}^2(f) = e^{-a\frac{\Delta_{xy}}{U}f} \tag{6.13}$$

where Δ_{xy} is the distance between the two locations, a is the coherence decay constant, taken to be 50, U is the 10-minute wind speed, and f is the frequency.

Note that Equation (6.13) reflects the observation that the coherence decreases with distance and at higher frequencies and increases with higher wind speeds.

Coherence is relevant to generating time series of turbulent wind which are needed in detailed wind turbine simulation models. This topic is discussed below.

6.2.5 Turbulent Data Synthesis

Real wind turbines operate in a wind velocity field that varies over many time scales and dimensions. This variability can have a significant effect on the instantaneous power production, forces, moments, stresses, and deflections and must be accounted for in the design process. To obtain real wind data varying in time and space across a large rotor would be prohibitively expensive. Fortunately, one can synthesize realistic turbulent wind data. The methods commonly used to synthesize such data rely on their ability to construct time series that have the appropriate correlation and spectral properties as summarized above. The methods have been incorporated in NREL's code *TurbSim*, which has been mentioned previously. The purpose of *TurbSim* is to produce a velocity field, i.e., a grid of multiple wind speed time series, which are assumed to pass through the plane of the wind turbine rotor in an aeroelastic simulation model such as NREL's *OpenFAST*. For example, for a turbine with a rotor diameter of 100 m *TurbSim* could produce inflow time series for a grid 100 m high and 100 m wide. With grid points spaced 10 m apart, there would be 100 time series. Typically, one data point with three components (*u*, *v*, *w*) would be generated for each time series every 0.05 seconds. A single *OpenFAST* simulation might model turbine behavior for a 10-minute period. For such a simulation, *TurbSim* would generate 1.2 million data points.

The fundamental principle of data synthesis is the use of the wind spectrum to create a time series that would have the same spectral characteristics as the wind speed, that is the same range of frequencies with the same amplitudes. The spectrum $S(f)$ is first divided into n segments, each of width Δf, centered at f_i. A realistic time series of the fluctuating component of the turbulence $\tilde{u}(t_j)$ can then be created by summing up the products of amplitudes, A_i, times the sinusoids for the number of time steps needed, as shown in Equations (6.14) and (6.15).

$$A_i = \sqrt{2\, S(f_i)\Delta f} \tag{6.14}$$

Each sinusoid must be associated with a random phase angle for this process to work, as shown in Equation (6.15).

$$\tilde{u}(t_j) = \sum_{i=1}^{n} A_i \sin\left(2\pi f_i t_j + \phi_i\right) \tag{6.15}$$

where j is the index of time series, and ϕ_i is a random phase angle from a uniform distribution

The method described here is sometimes referred to as the Shinozuka Method (Shinozuka and Jan, 1972). In the case of turbulence at various locations in the rotor plane, a coherence function, such as Equation (6.13), is used to ensure that the time series is suitably correlated, depending on the spacing. More details on the Shinozuka Method and some of the other principles underlying *TurbSim* are given in Appendix B. Other methods are also available, such as the Mann wind field model (Mann, 1998).

6.2.6 Extreme Wind Speeds

The anticipated extreme wind speed is one of the prime considerations in wind turbine design. This is the highest wind speed expected over some relatively long period of time. Extreme wind speeds are of particular concern in the design process since the turbine must be able to withstand those possibly damaging but infrequent conditions.

6.2.6.1 Return Period

Extreme winds are normally described in terms of return period (or recurrence period), T, as measured in years. Specifically, an extreme wind U_e is that wind speed, averaged over some appropriate time interval, which will be exceeded with a probability $Pr(U > U_e)$ as defined by Equation (6.16):

$$Pr(U > U_e) = \frac{1}{T \times I_{yr}} \tag{6.16}$$

where I_{yr} = intervals/yr. Also, note that $Pr(U > U_e) = 1 - F(U_e)$ where $F(U_e)$ is the CDF of U_e. For example, the highest 10 minute average wind within a 50-year recurrence period would have a probability of exceedance of $1/(50 \times 365.25 \times 24 \times 6) = 3.81 \times 10^{-7}$.

Note that an event with a given return period may well occur before the end of that period. For example, if a turbine has a design life of T_L, the probability that the wind speed would exceed U_e for at least one interval is given by:

$$Pr_{T_L}(U > U_e) = 1 - \left(1 - \frac{1}{T \times I_{yr}}\right)^{T_L I_{yr}} \tag{6.17}$$

Thus, if $T_L = 25$ years and $T = 50$ years, then $Pr_{T_L}(U > U_e) = 0.393$.

6.2.6.2 Models for Extreme Events

Determination of extreme wind speeds by actual measurement is difficult since it would require measurements over a long period of time. It is possible to estimate extreme wind speeds, however, by using extremes over some shorter periods, together with a suitable statistical model. Various models may be used for estimating extreme wind speeds, but the most common is the Gumbel distribution. The CDF for the Gumbel distribution is shown in Equation (6.18).

$$F(U_e) = \exp\left(-\exp\left(\frac{-(U_e - \mu)}{\beta}\right)\right) \tag{6.18}$$

where $\beta = (\sigma_e\sqrt{6})/\pi$, $\mu = \overline{U}_e - 0.577\beta$, $\overline{U}_e$ = the mean of a set of measured extreme values, and σ_e = the standard deviation of that set; see also Appendix C. The set of extremes is typically based on relatively long intervals, typically many years.

Example: The extreme one-minute wind speeds for each of 17 years at the 80 m height of tower M2 at the National Renewable Energy Laboratory were identified (https://midcdmz.nrel.gov/apps/day.pl?NWTC). The mean of these extremes was 36.1 m/s and their standard deviation was 3.14 m/s. From these, β and μ were found to be 2.45 m/s and 34.7 m/s, respectively. Using Equation (6.18) the CDF from the Gumbel was found and illustrated in Figure 6.6. The cumulative occurrence from the data is included as well for comparison. More specifically, the 17-year extreme was predicted to be 41.6 m/s, which is very close to the highest measured extreme of 42.6 m/s. The 50-year extreme value was found to be 44.3 m/s.

6.2.7 Joint Occurrence of Environmental Conditions

It is frequently the case that designers must take into account multiple conditions in the environment that take place at the same time, for example, high wind speeds and high turbulence. Fortunately, the worst of those conditions seldom occur simultaneously, and there are methods available to estimate the likelihood of combinations of events. The following discussion provides the basis for considering two events and uses the case of wind speed and

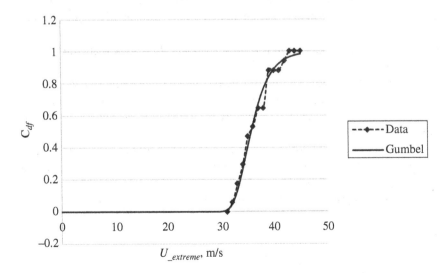

Figure 6.6 Illustration of Gumbel cumulative distribution function

turbulence intensity (as indicated by standard deviation) as an illustration. Later in this chapter, other simultaneous events are also discussed, and the method is essentially the same.

The likelihood of occurrence of multiple events is typically characterized by joint probability distributions (see also Appendix C). An example of such a joint probability distribution is shown by the density function in Equation (6.19).

$$f(U, \sigma_u) = f(U)f(\sigma_u|U) \tag{6.19}$$

where $U = 10$ minute average wind speed and $\sigma_u = $ standard deviation.

The joint probability density function of wind speed and standard deviation is given by the product of the marginal probability density function of wind speed and the conditional probability density function of the standard deviation, given the wind speed. Assuming appropriate functions can be found for the above density functions on the right side of Equation (6.19), then an expression for the joint probability density function of wind speed and standard deviation is at hand. Note that the probability of an occurrence of wind speed between U_1 and U_2 and standard deviation between $\sigma_{u,1}$ and $\sigma_{u,2}$ is given by the integral in Equation (6.20).

$$Pr(U_1 < U \leq U_2, \sigma_{u,1} < \sigma_u \leq \sigma_{u,2}) = \int_{U_1}^{U_2} \int_{\sigma_{u,1}}^{\sigma_{u,2}} f(U, \sigma_u)d\sigma_u dU \tag{6.20}$$

In the case that U and σ_u are uncorrelated, then Equation (6.19) simplifies to:

$$f(U, \sigma_u) = f(U)f(\sigma_u) \tag{6.21}$$

In general, however, the wind speed and standard deviation are partially correlated. That is, the standard deviation, on average, increases with wind speed. One approach to working with variables that are partially correlated is to divide them into intervals, or bins, and then assume that the variables are uncorrelated within those intervals. In that case, Equation (6.21) is applicable for those intervals. The environmental contour method, summarized in the next section, is useful for dealing with this situation.

6.2.7.1 *Environmental Contours

The effect of partial correlation of environmental variables may be conveniently accounted for by the method of environmental contours (EC); see Haver and Winterstein (2009). This method produces pairs of events that are all equally probable. Methods described in Chapter 8 can then be used to determine which pairs result in an extreme response.

The environmental contour method is an application of the Inverse First Order Reliability Method (IFORM), which is also discussed in more detail in Chapter 8.

In the discussion here, we assume that there are two partially correlated physical variables, x_1 and x_2, although the method can be extended to more. In the environmental contour method, the partial correlation is accounted for by first separating the available variables into N intervals according to the value of x_1 and then assuming that the values of x_2 are uncorrelated with x_1 within the intervals. (The variables may be subscripted $i = 1$ to N, but the subscripts are omitted from most of the equations below for clarity.) A transformation allows pairs of values with any given probability of occurrence to be found easily. It involves converting the distributions to standard normal form (see Appendix C) and then back again as described below.

Two important and useful relations regarding the standard normal distribution are given in Equations (6.22) and (6.23).

$$\Phi(x) = \frac{1}{2}\left[1 + \text{erf}\left(\frac{x}{\sqrt{2}}\right)\right] \tag{6.22}$$

$$\Phi^{-1}(p) = \sqrt{2}\text{erf}^{-1}(2p - 1) \tag{6.23}$$

where $\Phi()$ is the CDF of the standard normal distribution, $\Phi^{-1}()$ is its inverse, $erf()$ is the error function, $erf^{-1}()$ is its inverse, and x and p are arbitrary nondimensional variables.

Another useful variable is the reliability index, β. It is an indicator of the probability of occurrence of a rare event or combination of events. It derives its name for its use in structural reliability, where it is used to characterize the probability of failure. It is discussed in that context in Chapter 8. The reliability index is defined by Equation (6.24).

$$\beta = -\Phi^{-1}(1/T\,I_{yr}) = \Phi^{-1}(1 - 1/T\,I_{yr}) \tag{6.24}$$

where T is the return period in years and I_{yr} is the number of sample intervals in a year.

There will be two transformed variables, u_1 and u_2, which are defined such that the transformed variables and the original physical variables have the same cumulative distribution function, as shown in Equations (6.25) and (6.26).

$$\Phi(u_1) = F(x_1) \tag{6.25}$$

$$\Phi(u_2) = F(x_2|x_1) \tag{6.26}$$

where $F(x_1)$ is the CDF of the marginal distribution and $F(x_2|x_1)$ is the CDF of the conditional distribution. The inverses of Equations (6.27) and (6.28) are also useful:

$$u_1 = \Phi^{-1}[F(x_1)] \tag{6.27}$$

$$u_2 = \Phi^{-1}[F(x_2|x_1)] \tag{6.28}$$

In standard normal space u_1 and u_2 are uncorrelated, and since they are uncorrelated, it may be shown that:

$$u_2 = \pm\sqrt{\beta^2 - u_1^2} \tag{6.29}$$

Where a positive sign for u_2 will result in a corresponding physical value above the mean and a negative sign will result in the opposite. Equation (6.29) provides the link between β, u_1, and u_2. Given known values for the target joint probability and for the variable x_1, the probability for the co-occurring variable x_2 can be determined and its value found.

Since x_1 and x_2 are assumed to be uncorrelated within each bin of the physical variables, it is also the case that Equation (6.21) applies.

Weibull Distribution in Environmental Contours

For many situations both $f(x_1)$ and $f(x_2|x_1)$ and their corresponding cumulative distribution functions can be modeled by Weibull distributions. Thus, for example, the CDF in Equation (6.25) would be:

$$F(x_1) = 1 - e^{-(x_1/c)^k} \tag{6.30}$$

where c is a Weibull scale parameter, and k is a Weibull shape parameter, as discussed elsewhere. The values of the parameters depend on the situation.

Equation (6.30) can be rearranged to express x_1 in terms of $F(x_1)$:

$$x_1 = c(-\ln(1 - F(x_1)))^{1/k} \tag{6.31}$$

Substituting Equation (6.25) into Equation (6.31) yields Equation (6.32), which allows x_1 to be found from u_1 where appropriate.

$$x_1 = c(-\ln(1 - \Phi(u_1)))^{1/k} \tag{6.32}$$

Equation (6.32) may be used more conveniently by substituting Equation (6.22) to yield:

$$x_1 = c\left(-\ln\left(1 - \frac{1}{2}\left[1 + \text{erf}\left(\frac{u_1}{\sqrt{2}}\right)\right]\right)\right)^{1/k} \tag{6.33}$$

A similar approach allows x_2 to be found from u_2 where u_2 comes from Equation (6.29).

$$x_2 = c\left(-\ln\left(1 - \frac{1}{2}\left[1 + \text{erf}\left(\frac{u_2}{\sqrt{2}}\right)\right]\right)\right)^{1/k} \tag{6.34}$$

The environmental contour process starts with finding β, which is done from the target return period T of the joint occurrence of x_1 and x_2 from Equation (6.24). Then a value of the physical variable x_1 is selected, and u_1 is found from Equation (6.27). When a Weibull distribution is assumed, Equation (6.27) becomes Equation (6.35):

$$u_1 = \sqrt{2}\text{erf}^{-1}\left(1 - 2e^{-(x_1/c)^k}\right) \tag{6.35}$$

Conversely, given u_1 the physical variable x_1 can be found from Equation (6.32).

Similarly, with u_2 found from Equation (6.29), x_2 conditioned on x_1 may be determined from Equation (6.34).

Example: The following is an example of the EC method for 10-minute wind speed means and standard deviations. It is based on the guidelines for turbulent wind speeds given in the IEC 61400-1 design standard (IEC, 2019a). The example provides an illustration of how an EC is created for combinations expected to occur in one 10-minute interval every 50 years. It assumes that a Weibull distribution can model the wind speeds U using Weibull scale and shape parameters, c_U and k_U. These could be found in the usual way; see Chapter 2. Here, $x_1 = U$. Within each ith interval of mean 10-minute wind speed, the standard deviation of the wind, $\sigma_{u,i}$, is also assumed to have a Weibull distribution with shape and scale parameters as expressed by:

$$k_{\sigma_{u,i}} = 0.27U_i + 1.4 \tag{6.36}$$

$$c_{\sigma_{u,i}} = I_{ref}(0.75U_i + 3.3) \tag{6.37}$$

where I_{ref} is the reference turbulence intensity, assumed here to be 0.18. For this example, we use the average and standard deviation of the data on which Figure 6.2 was based, $\overline{U} = 7.84$ m/s and $\sigma_U = 6.48$ m/s. From those $k_U = 1.23$ and $c_U = 8.38$ m/s.

With the return period $T = 50$ years and $I_{yr} = 52\,596$, the reliability index β is found from Equation (6.24) to be 4.945. Consider the bin with $U_i = 15$ m/s. From Equation (6.35), the transformed wind speed variable u_1 is 1.742. The transformed standard deviation variable is found from Equation (6.29) to be $u_2 = 4.628$. From Equations (6.36) and (6.37), the Weibull parameters for $\sigma_{u,i}$ are $k_{\sigma_{u,i}} = 5.45$ and $c_{\sigma_{u,i}} = 2.62$. From Equation (6.34), the physical extreme value of the standard deviation is $\sigma_{u,i} = 4.2$ m/s.

We can repeat the process for a range of wind speeds to find corresponding standard deviations such that the joint events all have the same combined probability of occurrence. Alternatively, the process can start from a range of transformed wind speed variables, u_1. In either case, the standard deviations can be converted into turbulence intensities by dividing by the corresponding mean. For example, for $U_i = 15$ m/s, $TI = 0.28$. The results are included in Figure 6.2. In that figure, the upper line of the contour corresponds to unusually high values of TI and lower line to unusually low values.

Log Normal Distribution in Environmental Contours

It is worth noting that the log-normal distribution is often used in the EC method. For example, when x_2 is log normally distributed with mean μ_{x_2} and standard deviation σ_{x_2}, the equation analogous to Equation (6.34) is Equation (6.38). See also Appendix C.

$$x_2 = \exp\left(\mu_{\ln x_2} + \sigma_{\ln x_2} u_2\right) \qquad (6.38)$$

where

$$\mu_{\ln x_2} = \ln \mu_{X_2} - \ln \sqrt{1 + \left[\sigma_{x_2}/\mu_{x_2}\right]^2}$$

$$\sigma_{\ln x_2} = \sqrt{\ln\left(1 + \left[\sigma_{x_2}/\mu_{x_2}\right]^2\right)}$$

6.2.8 Design Wind Conditions Offshore

Design wind conditions offshore have, of course, many similarities to those on land, but there are some significant differences as well. These are a result primarily of the heat capacity of the seawater, and hence stability, and the effect of the waves on the turbulence and wind shear. Because wave height is affected by fetch, wind shear, and turbulence may vary depending on direction. (Fetch is a measure of the distance over open water across which the wind blows before reaching a particular location.) In addition, since the sea surface is nearly flat (compared to the land), the effect of terrain is minimal. One significant difference between land and offshore is the possibility of hurricanes, also known as tropical cyclones, in some locations. Hurricanes can affect land as well as the offshore, but the origin of hurricanes and the storms of greatest intensity occur offshore. Some of the differences between wind conditions on land and offshore were summarized in Chapter 2. In the following sections, we discuss those aspects that are particularly relevant to the design of offshore wind turbines. See Section 6.8 for more discussion of hurricanes.

6.2.8.1 Stability Offshore
As discussed in Chapter 2, there can be significant differences in atmospheric stability offshore from that on land. This is due primarily to the difference in the heat capacity of water compared to the land: daily and seasonal water temperatures vary more slowly than land temperatures. Warm seawater will tend to decrease stability, and increase mixing, in the presence of cold air. Cold seawater will have the opposite effect, making the atmosphere more stable than it otherwise might be.

6.2.8.2 Offshore Wind Shear
As discussed in Chapter 2, vertical wind shear is the variation of wind speed with height above the surface, and it is strongly affected by the roughness of the surface. Because the roughness of the sea surface varies with the sea state, the log law is commonly used to characterize the offshore wind shear. It is repeated here as Equation (6.39):

$$U(z)/U(z_r) = \ln\left(\frac{z}{z_0}\right) / \ln\left(\frac{z_r}{z_0}\right) \qquad (6.39)$$

where z is the height of interest, z_r is a reference height, and z_0 is the surface roughness. A major difference from land is that surface roughness will change with wind speed through its effect on waves. Surface roughness offshore is discussed in more detail in the next Section 6.2.8.3.

6.2.8.3 Surface Roughness Offshore
The surface roughness offshore is typically lower than that on land. For example, based on observations at the Vindeby offshore wind farm, Barthelmie *et al.* (1996) concluded that a value of $z_0 = 0.0002$ m (0.2 mm) could be used most of the time. With this assumption of $z_0 = 0.0002$ m, the wind speed at a height of 48 m could be predicted from data at a height of 7 m to within +/− 5%. In fact, however, the surface roughness of the sea depends on wave height, which is in turn a function of the wind speed and fetch. In the absence of data, a number of models can be used to estimate offshore surface roughness.

Surface Roughness: Wind Speed and Fetch

At offshore sites close to land, fetch is direction dependent: winds that originate from the land will have different characteristics than those from the open sea. Since fetch is significant in the creation of ocean waves, it also indirectly affects the wind itself, particularly the surface roughness, wind shear, and turbulence. Lange and Højstrup (1999) showed that sea surface roughness length increases with wind speed over the range of wind speeds typically used for power production. See also Lange *et al.* (2001) for a fetch-dependent model of sea surface roughness.

The Charnock model, Equation (6.40), is often used for modeling the change in sea surface roughness length as a function of wind speed:

$$z_0 = A_C \frac{(u^*)^2}{g} \tag{6.40}$$

where g is the gravitational constant, u^* is the friction velocity, and A_C is the Charnock constant. $A_C = 0.011$ is recommended for open sea, while $A_C = 0.034$ may be used for near-coastal locations (IEC, 2019b).

Another approach, the Johnson model (Equation (6.41); see Lange and Højstrup, 1999), uses an implicit equation for z_0 as a function of u^* and fetch, x (m).

$$z_0 = 0.64 \frac{(u^*)^3}{x^{\frac{1}{2}} g^{\frac{3}{2}} \kappa} \ln\left(\frac{10}{z_0}\right) \tag{6.41}$$

where κ is von Kármán's constant.

Comparisons of each of these models with data from locations in the Baltic Sea show that the Johnson model does a better job of modeling z_0 than the Charnock equation for fetches from 10 to 20 km. At longer distances (>30 km), the Charnock model appears to be better (Lange and Højstrup, 1999).

Calculation of friction velocity may be challenging, but it is often approximated by:

$$u^* = \sqrt{C_{D,10} U_{10}^2} \tag{6.42}$$

where $C_{D,10}$ is the effective surface drag coefficient based on the wind velocity, and U_{10} is the wind speed measured at an elevation of 10 m.

Measurements of $C_{D,10}$ offshore indicate that it may vary between 0.001 in low winds and 0.003 in high winds (Garratt, 1994).

The result of the low surface roughness over the ocean is the increase in low-level wind speeds at offshore sites compared to nearby onshore sites. Data from Vindeby shows that the annual mean wind speed at a 38 m height is about 4% greater at the wind farm, 1.4–1.6 km from shore, than at the nearby shoreline (Barthelmie *et al.*, 1996).

Surface Roughness and Wave Height

Alternatively, surface roughness can be estimated from wave heights and wave periods, since surface roughness offshore is strongly affected by both of those. As will be discussed in more detail subsequently (Section 6.3.3), real (random) waves are characterized predominantly by the significant wave height, H_s, which is approximately the average of the highest third of waves and the wave peak period, T_p, which is typically in the range of 5–20 seconds. Wave period is closely related to the peak wavelength, L_p.

To a first approximation, surface roughness, z_0, increases with wave height. Upon a closer look, the ratio z_0/H_s is not constant. It is, however, typically between 6.0×10^{-6} and 5.0×10^{-3}. Observation has shown that z_0/H_s depends on both wave height and wavelength, but most particularly on their ratio, H_s/L_p. In general, the significant wave height and the wavelength both increase with mean wind speed (although they are both affected by fetch and duration of the wind), so in general the ratio H_s/L_p is between 0.02 and 0.06. Based on observations, Taylor and Yelland (2001) concluded that the ratio, z_0/H_s, is given approximately by Equation (6.43):

$$\frac{z_0}{H_s} = 1200 \left(\frac{H_s}{L_p}\right)^{4.5} \tag{6.43}$$

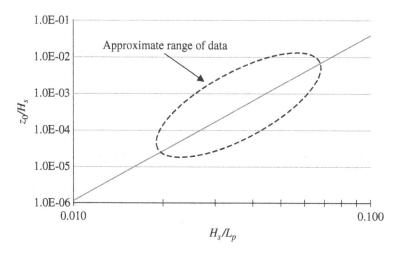

Figure 6.7 Roughness height z_0; significant wave height, H_S; and wave length, L_p

The implications of Equation (6.43) are illustrated in Figure 6.7. This figure also includes a dotted oval showing the approximate range of the observed values.

Example: Suppose $H_s = 2.5$ m and $L_p = 76.5$ m. Then $H_s/L_p = 0.033$ so $z_0/H_s = 0.000247$ and $z_0 = 0.000619$ m.

6.2.8.4 Turbulence Offshore

As with wind turbines on land, turbulence offshore can be significant to both operation and design. In the ocean, however, surface roughness is much less than on land, so roughness effects on turbulence are less (in absence of other turbines). Thermal effects are thus more important. Roughness effects become more significant with higher winds and result in greater wave heights. The effect is to increase the turbulence intensity at higher wind speeds. This is illustrated in Figure 6.8.

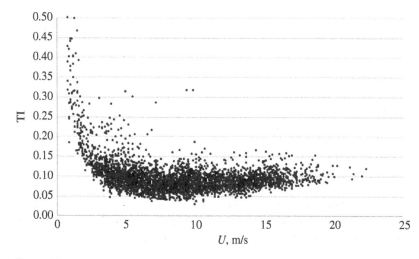

Figure 6.8 Typical turbulence intensity offshore

6.3 Waves as External Design Condition

Consideration of waves is critical to the design of offshore wind turbines, particularly their support structures. As applied to offshore wind energy, a wave may be thought of as a line or ridge that moves across the surface of the water. As the wave travels, it transmits energy, sometimes over long distance although the water particles themselves do not actually move very far. The result is forces on structures that the wave encounters. These forces may be substantial.

Waves may take a variety of forms, but for purposes here they may be divided into four main categories: (1) regular waves, (2) random waves, (3) breaking waves, and (4) other types of waves. The easiest waves to describe are regular waves. These are also known as first-order, small-amplitude, or Airy waves. Their behavior was first described by Airy (1845). Waves of relevance to offshore wind energy are primarily wind-driven waves, although some are driven by other processes such as tsunamis. The following sections focus on wind-driven waves. More information on waves may be found in such sources as the Coastal Engineering Manual (USACE, 2002), Ochi (1998), or Krogstad and Arntsen (2000).

6.3.1 Overview of Wind-Driven Waves

Ocean waves are the result of the shear force of the wind as it passes over the surface of the sea, causing some of the water to move. This motion ultimately becomes periodic and manifests itself as waves. In deep water and under the influence of relatively low wind speeds, the wave motion is quite regular. In the presence of higher wind speeds, the motion can become very complex. At any given location, the waves are a composite of waves originating at various distances and directions, but in any case, there is a substantial correlation between the waves and the adjacent wind.

Some terminology is useful here. The term "sea" describes a situation in which waves are being formed; "sea state" refers to characteristics of the sea in terms of wave height, period, and direction; "fully developed sea state" indicates that power absorbed from the wind equals power being dissipated in breaking waves.

6.3.2 Regular (Airy) Waves

Airy's wave model is most applicable to deep water swell, but it provides a useful and elegant starting point for discussion of more complex wave behavior as well. The assumptions of the Airy wave model are the following:

- water is homogeneous and incompressible;
- surface tension is neglected;
- Coriolis effects are neglected;
- the pressure at the free surface is constant;
- viscosity is neglected;
- waves do not interact with other water motions;
- the flow is irrotational;
- there is no velocity at the seabed;
- the amplitude is small and the waveform is invariant in time and space;
- the wave is long-crested (two-dimensional).

The wave equations resulting from these assumptions are given in the following discussion. The wave is assumed to propagate along the x direction, and the vertical direction is along the z axis. The mean water level is located at $z = 0$, while the seabed is located at $z = -d$, where d is the water depth to mean water level.

The main parameters used to define a regular wave are the wavelength, L, which is the distance between wave crests, and the frequency, f. Related variables include the wave number, k, and the wave period, T, as follows:

$$k = \frac{2\pi}{L} \qquad T = 1/f \tag{6.44}$$

The wave celerity, c, is the speed at which the wave crest moves and is given by:

$$c = L/T \tag{6.45}$$

In the Airy model, the surface elevation $\eta(x,t)$ is sinusoidal and takes the following form:

$$\eta(x, t) = A \cos(2\pi f t - kx) \tag{6.46}$$

where A is the amplitude of the waves, which is half the wave height H, so $A = H/2$.

6.3.2.1 Dispersion Relation

The free surface condition imposes the following relation between the wave number and the frequency, referred to as the dispersion relation:

$$f^2 = \frac{gk}{4\pi^2} \tanh(kd) \tag{6.47}$$

where g is the gravitational constant.

Expressed using the wavelength and wave period, the dispersion relation is given by:

$$L = \frac{gT^2}{2\pi} \tanh\left(\frac{2\pi d}{L}\right) \tag{6.48}$$

For water of moderate depth (50 m) or more and periods of 10 seconds or less, the wavelength may be approximated as:

$$L = \frac{gT^2}{2\pi} \sqrt{\tanh\left(\frac{4\pi^2 d}{T^2 g}\right)} \tag{6.49}$$

Equation (6.49) is convenient because the wavelength only appears on one side. In even deeper water (>100 m), the yet simpler Equation (6.50) may be used, since the hyperbolic tangent of a large number is approximately equal to 1.0:

$$L = \frac{gT^2}{2\pi} \approx 1.56 \ T^2 \tag{6.50}$$

The approximations above apply to longer periods when the water is deeper.

6.3.2.2 Wave Velocity and Acceleration

Under the Airy wave assumptions, the wave motion can be described in terms of a velocity potential, Φ, which is the solution of a Laplace equation within the fluid. The equation is solved by applying the proper boundary conditions at the free surface and at the seabed located at $z = -d$.

The velocity potential may be shown to be the following:

$$\Phi(x, z, t) = -A \frac{2\pi f}{k} \frac{\cosh(k(z + d))}{\sinh(kd)} \sin(2\pi f t - kx) \tag{6.51}$$

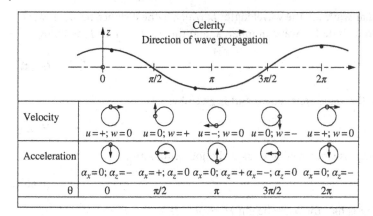

Figure 6.9 Illustration of Airy waves. Top: surface elevation as function of $\theta = 2\pi ft - kx$; Bottom plots: particle motion; u and w and α_x and α_z are velocities and accelerations in the x and z direction (*Source:* USACE (2002)/U.S. Army/Public Domain)

The velocity and acceleration of the water, U_w and $\dot{U}_w$, respectively, are obtained from Equation (6.51) as follows:

$$U_w = \frac{\partial \Phi(x,z,t)}{\partial x} = A(2\pi f)\frac{\cosh(k(z+d))}{\sinh(kd)}\cos(2\pi f\, t - kx) \tag{6.52}$$

$$\dot{U}_w = \frac{\partial U_w}{\partial t} = -A(2\pi f)^2\frac{\cosh(k(z+d))}{\sinh(kd)}\sin(2\pi f\, t - kx) \tag{6.53}$$

Velocity and acceleration are particularly useful for the calculation of the hydrodynamic forces on offshore structures. It can also be shown that the water particle trajectories are circular orbits (in deep water) with radius $= Ae^{kz}$. The motion of the water in the Airy model is illustrated in Figure 6.9. Note that the actual motion of the water particles also changes according to depth, whereas in deep water the path is circular, the path becomes more elliptical when the depth is shallower.

6.3.3 Random Waves

Real ocean waves are only approximately described by the Airy model. First of all, they are irregular in that they do not have a single wavelength and uniform height. In addition, their heights and wavelengths tend to vary in a non-repeating and apparently random way. Such waves are referred to as random waves. They arise due to the superposition of waves from various sources and directions, as illustrated schematically in Figure 6.10. Fortunately, these waves can be characterized probabilistically.

As introduced earlier, the fundamental characterization of a sea is the significant wave height, H_s. Historically, this was estimated by sailors as the average of the highest third of the waves that could be seen from their ship. Nowadays, it is taken to be four times the standard deviation of the sea surface elevation. It can be shown that the mean energy density in waves is proportional to the square of this standard deviation. The standard deviation can also be related to the spectral description of the waves, as is discussed below.

Analysis of the random sea is carried out using statistical and spectral methods (see also Appendix B). Random seas are analyzed over intervals during which the statistics of the waves are on average relatively constant, or "stationary." These intervals are the "sea states"; by convention, sea states are considered to last three hours. A sea state consists of multiple waves of different heights, periods, and directions. (Recall that wind conditions are normally evaluated over 10-minute intervals, so there is a disconnect between the two types of intervals. For offshore wind energy, there are various approaches to bridging this disconnect – one of them is to evaluate both wind and waves over hourly intervals.)

Figure 6.10 Formation of random waves (*Source:* USACE (2002)/U.S. Army/Public Domain)

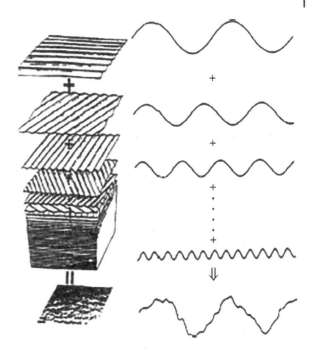

The important considerations in characterizing a random sea state are the following:

- Significant wave height, H_s;
- Peak period, T_p;
- Frequency spectrum, $S_w(f)$.

The details of these concepts are presented in Section 6.3.4.

6.3.4 Wave Spectra

A wave spectrum, $S_w(f)$, is the psd of the sea surface elevation. Analogously to atmospheric turbulence, the wave spectrum provides the wave variance density in terms of the wave frequency. It is also convenient to use the wave spectrum as a characterization of a given sea state. The spectrum can be computed from a time series of sea surface elevation or can be assumed to follow a probabilistic model, such as the Pierson–Moskowitz or the JONSWAP models, which are described below. It is also possible to use a directional spectrum, $S_w(f, \theta)$ which is a function of the frequency and wave direction, θ, to capture more information on the directivity of the waves at a given site. In such a case, the wave spectrum is related to the directional spectrum as $S(f) = \int_0^{2\pi} S_w(f, \theta) \, d\theta$; this approach will not be elaborated here.

6.3.4.1 Spectral Parameters
Several parameters are defined from the knowledge of the wave spectrum:

Variance
By the definition of a spectrum, the integral over all frequencies is equal to the variance of the data series. In this case, the variance of the sea surface elevation, σ_w^2, is related to the wave spectrum by Equation (6.54).

$$\sigma_w^2 = \int_0^\infty S_w(f)\, df \tag{6.54}$$

Note that σ_w is not the same as the standard deviation of the wave heights.

Significant Wave Height

Significant wave height is defined by Equation (6.55):

$$H_s = 4\sigma_w = 4\sqrt{\int_0^\infty S_w(f)df} \tag{6.55}$$

The significant wave height increases with wind speed, with values typically ranging from 0 to 12 m, and a polynomial of order 2 or 3 may be used to fit the experimental data to a model for H_s vs. U as necessary.

Peak Period

The peak period, T_p, is the wave period associated with the most energetic waves. Strictly speaking, it is the reciprocal of the peak frequency, f_p, at which the wave spectrum reaches its maximum value. The peak wavelength, L_p, is the wavelength associated with the peak period. The peak period typically increases with H_s, with values ranging up to approximately 30 seconds. It is usually assumed that the peak period follows the following relationship with respect to the significant wave height:

$$T_p \cong a + b\sqrt{H_s/g} \tag{6.56}$$

where a and b are constants determined from measurements of H_s and T_p (obtained from measured spectra) for different sea states at a given site.

6.3.4.2 Pierson–Moskowitz Spectrum

The most common wave spectrum model used in offshore wind is known as the Pierson–Moskowitz (PM) spectrum. It is most applicable to deep water, but it is often used for other situations as well. The PM spectrum is given by Equation (6.57):

$$S_{PM}(f) = 0.3125 \cdot H_s^2 \cdot f_p^4 \cdot f^{-5} \cdot \exp\left(-1.25\left(\frac{f_p}{f}\right)^4\right) \tag{6.57}$$

Figure 6.11 illustrates the Pierson–Moskowitz spectrum for a sea state with H_s equal to 8.0 m and f_p equal to 0.1 Hz.

6.3.4.3 JONSWAP Spectrum

Another commonly used model is the Joint North Sea Wave Project (JONSWAP) spectrum. This spectrum is a modification of the PM. It is more applicable to a developing sea in a fetch-limited situation. It is also more versatile in that it includes a "peak factor" and a "normalizing factor" and so can be tuned to adapt to various depths and fetches. In particular, the spectrum can account for a higher peak and a narrower spectrum for the same total energy in a storm situation as compared with the PM spectrum. The JONSWAP spectrum is often used for extreme event analysis but is also commonly used for characterization of sea states in the European North Sea. It is represented by Equation (6.58).

$$S_{JS}(f) = C(\gamma) \cdot S_{PM}(f) \cdot \gamma^\alpha \tag{6.58}$$

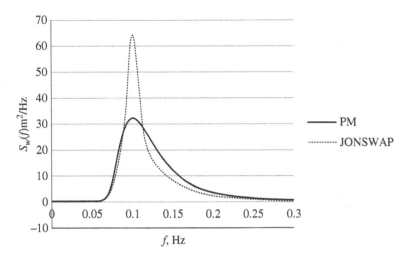

Figure 6.11 Illustration of Pierson–Moskowitz and JONSWAP spectra

where γ is known as the peak enhancement factor, which is given by:

$$\gamma = \begin{cases} 5 & \text{for} \quad \dfrac{T_P}{\sqrt{H_s}} \leq 3.6 \\[2mm] \exp\left(5.75 - 1.15\dfrac{T_P}{\sqrt{H_s}}\right) & \text{for} \quad 3.6 \leq \dfrac{T_P}{\sqrt{H_s}} \leq 5 \\[2mm] 1 & \text{for} \quad \dfrac{T_P}{\sqrt{H_s}} > 5 \end{cases} \tag{6.59}$$

α is an exponent given by:

$$\alpha = \exp\left(-\frac{\left(f - f_p\right)^2}{2\sigma^2 f_p^2}\right) \tag{6.60}$$

where $\sigma = 0.07$ *for* $f \leq f_p$; $\sigma = 0.09$ *for* $f > f_p$.
And $C(\gamma)$ is a normalizing factor, given by:

$$C(\gamma) = 1 - 0.287 \cdot \ln\gamma \tag{6.61}$$

Figure 6.11 also includes the JONSWAP spectrum for the same conditions as the PM.

6.3.4.4 Wind and Wave Spectra

For offshore wind turbines, both the spectrum of the turbulent wind and the wave spectrum can be important. Both the wind and the waves can affect the response of the wind turbine (although in different ways). It is worth noting that the peak frequency of waves is considerably higher than that of wind. Figure 6.12 illustrates two typical spectra on the same plot. The wind is modeled with the Kaimal spectrum (note the linear scales here as opposed to the log scales of Figure 6.5). The wave spectrum is the same as the PM spectrum in Figure 6.11. A related and important consideration is how these spectra relate to the natural frequencies of the wind turbine. This topic is discussed in Chapter 10 with respect to fixed offshore wind turbines and in Chapter 11 for floating turbines.

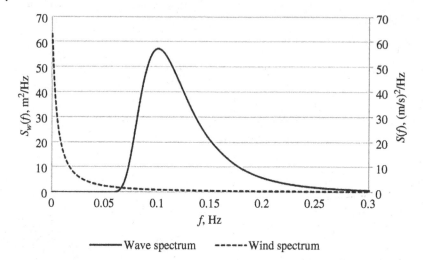

Figure 6.12 Typical wind and wave spectra

6.3.4.5 Occurrences of Wave Heights in a Sea State

In any given random sea state, there will be waves of various heights. One way to determine the number of them is by means of a cycle counting algorithm. One such algorithm is known as rainflow cycle counting, which is commonly used for assessing material fatigue (Downing and Socie, 1982). This technique is summarized in Chapter 7 and described in more detail in Appendix B. In ocean wave analysis, a rainflow counter can also be used to facilitate estimates of the average wave height, significant wave height, and maximum wave height.

It has also been shown (Longuet-Higgins, 1952) that the occurrence of wave heights in a sea state can be described by the Rayleigh distribution, analogous to the Rayleigh's use with wind speeds. In this case, the PDF and CDF of wave heights are given in Equations (6.62) and (6.63).

$$f(H) = \frac{\pi H}{2H_{av}^2} \exp\left[-\frac{\pi}{4}\left(\frac{H}{H_{av}}\right)^2\right] = \frac{4H}{H_s^2} \exp\left[-2\left(\frac{H}{H_s}\right)^2\right] \tag{6.62}$$

$$F(H) = 1 - \exp\left[-\frac{\pi}{4}\left(\frac{H}{H_{av}}\right)^2\right] = 1 - \exp\left[-2\left(\frac{H}{H_s}\right)^2\right] \tag{6.63}$$

where H_{av} is the average wave height.

Note that when the waves are Rayleigh distributed then the average wave height and significant wave height are related by Equation (6.64):

$$H_{av} = \frac{\sqrt{\pi/2}}{2} H_s \cong 0.627 H_s \tag{6.64}$$

More information on the Rayleigh distribution and waves may be found in Krogstad and Arntsen (2000) or USNA (2023).

6.3.4.6 Maximum Wave in a Sea State

The maximum wave in a sea state will, of course, be considerably greater than the significant wave height. When complete data is available, a rainflow counter will find that height. When only the significant wave height is known and the waves can be assumed to be Rayleigh distributed, the expected value of the maximum, $E(H_{\max})$ over a given period, is given approximately by Equation (6.65) where N_w is the number of waves over that time period, based on the average wave period; see Krogstad and Arntsen (2000) for details.

$$E(H_{\max}) \cong H_s\left[\sqrt{\ln(N_w)/2} + \frac{0.57}{\sqrt{8\ln(N_w)}}\right] \tag{6.65}$$

Example: Consider a three-hour sea state in which the average wave period is 10 seconds and the significant wave height is 10 m. There would be 1080 waves over the three-hour period; by Equation (6.65) the expected maximum of these would be 19.5 m.

6.3.4.7 Overall Maximum Wave Height

Sea states will vary over the course of the year and longer periods, so the maximum waves will vary as well. The overall maximum, or extreme, wave height will be the highest wave observed over some extended period. Such waves are often characterized by their return period, analogous to the return period of extreme wind speeds. As may be surmised, high waves are likely to be associated with high wind speeds. There is not a perfect correlation, however, so the relation is more complex than it otherwise might appear. This topic is discussed in more detail in Section 6.5.

For longer periods, such as 50 years, analogously to extreme wind speeds (Section 6.2.6), the extreme wave height can be estimated by applying the Gumbel distribution to the occurrences of highest waves over multiple, shorter periods (such as years) and applying a similar process.

6.3.4.8 Illustrative Example of Modelling Real Waves

A data set of sea surface elevation, sampled at 2 Hz, came from the SHOWEX data collection program (Zhang et al., 2009). This example considers a three-hour period, the first 60 seconds of which are illustrated in Figure 6.13.

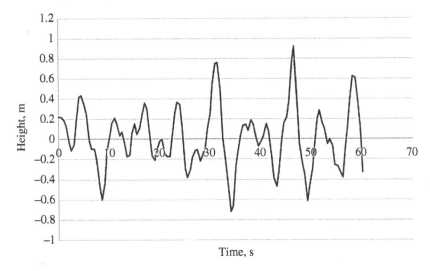

Figure 6.13 Sixty seconds of sample SHOWEX wave data

A basic zero upcrossing count (based on when the height increases above mean sea level) indicated that there were 2193 waves during the three-hour period. Accordingly, the average wave period was 4.9 seconds. A spectral analysis of the data yielded the psd which is illustrated in Figure 6.14. From the psd, H_s was found to be 1.44 m and the peak frequency was 0.142 Hz. From those two results, a PM spectrum was also created and superimposed on that of the data in Figure 6.14.

A rainflow counting algorithm was also applied to the data. The normalized results are shown in Figure 6.15. From the output, it was possible to determine the total number of waves, which was 2194.5, nearly identical to that of the upcrossing count. With the number of waves of different heights, it was also possible to find the average of the highest third of the waves. That value was 1.44 m, which was the same value from the psd, as expected. From the rainflow counting, the standard deviation of the wave heights was found to be 0.513 m, while that of the sea surface elevation (see Equation (6.55)) was $H_s/4 = 0.361$ m.

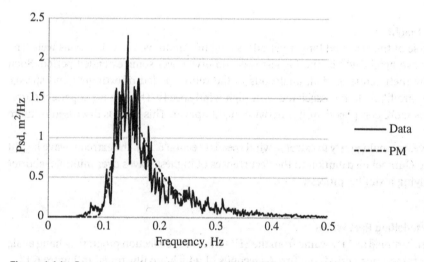

Figure 6.14 Power spectral density of sample SHOWEX data

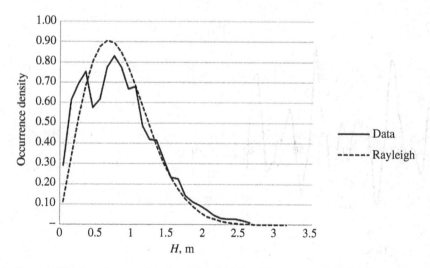

Figure 6.15 Sample wave data normalized rainflow histogram and Rayleigh approximation

A Rayleigh distribution, using the mean wave height as obtained from Equation (6.64), was also created and superimposed on the normalized rainflow histogram for comparison; see Figure 6.15.

The maximum height as found from the rainflow counting of the data was 3.15 m. The maximum value predicted from Equation (6.65) was 2.9 m, so the values were similar although not identical.

6.3.5 Breaking Waves

So far, we have only considered non-breaking waves. For some applications, particularly in shallow water or in very high winds, breaking waves are likely. A breaking wave is one whose base can no longer support its top, causing it to collapse. Waves will break in shallow water when the height of the wave is greater than 0.78 times the water depth. Accordingly, they are more common in shallow water, but they do still need to be considered in deeper water, since they may occur there during storms. See Barltrop and Adams (1991) for more detail on this. The forces resulting from a breaking wave can be considerably higher than from a non-breaking wave of equivalent height. This phenomenon is examined in detail by Wienke (2001). Additional discussion regarding breaking waves in relation to offshore wind turbines is provided by Lackner *et al.* (2018).

6.3.6 Other Types of Waves

There are other types of waves and wave models as well. In some cases, waves appear to be nearly regular Airy waves, as described previously, but they do not have a sinusoidal cross-section. Approaches to modeling them include Stokes waves (of various orders), Cnoidal waves, and Boussinesq waves. Detailed discussion of the other wave models is outside the scope of this text.

One distinction worth noting is that regarding water depth. Waves in deep water are fundamentally similar to waves in shallow water, except that since the wind speeds are generally higher in deeper water (which typically is farther from shore) waves tend to be higher as well. In addition, since there is no depth constraint, waves in deep water tend to be closer to the Airy model shape at low wind speeds than they are in shallow water.

Finally, some other waves worth being aware of are solitons, which are isolated, unusually high waves, and tsunamis, which are waves produced by subsea earthquakes.

6.3.7 Synthetic Waves

As mentioned previously, a random sea state can be thought of as a superposition of regular waves. The continuous superposition can be approximated by summing a discretized set of waves at frequencies separated by Δf. The amplitudes of a given wave at frequency f_i is then related to the spectral amplitude at this frequency in a matter analogous to that of atmospheric turbulence (Equation 6.15), such that for a wave spectrum $S_w(f)$:

$$A_i = \sqrt{2 S_w(f_i)\Delta f} \tag{6.66}$$

Where A_i is the wave amplitude.

The water surface elevation as a function of time and location is then obtained as:

$$\eta(t_j,x) = \sum_{i=1}^{n} A_i \sin(2\pi f_i t_j - k_i x + \phi_i) \tag{6.67}$$

where k_i is the wave number associated with frequency f_i, obtained from the dispersion relation, and ϕ_i is the wave phase. Equation (6.67) can be used to generate times series of sea surface elevation (i.e., waves) from a given spectrum. Figure 6.16 shows an example of a time series of synthetic waves with a PM psd over a 10-minute period. In this case, $H_s = 2$ m and $T_p = 15$ seconds (f_p 0.067 Hz).

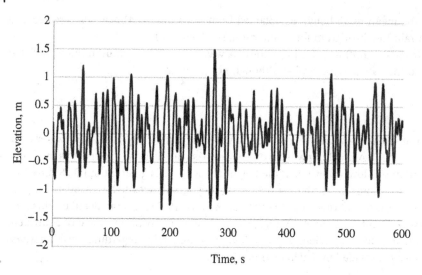

Figure 6.16 Example of synthetic waves

6.4 Forces Due to Waves

When waves impinge on a structure, they produce both viscous drag and inertia forces on that structure. Morison's equation is commonly used for the calculation of those forces.

Morison's equation for a cylindrical member is:

$$\tilde{F} = \frac{1}{2}C_d\rho_w D \mid U_w \mid U_w + C_m\rho_w A_c \dot{U}_w \tag{6.68}$$

where

$\tilde{F}$ = force per unit length of the member
C_d = drag coefficient
C_m = inertia coefficient
ρ_w = density of water, kg/m³
D = member diameter, m
A_c = cross-sectional area of the member, m²
U_w = velocity of the water, resolved normal to the member, m/s
$\dot{U}_w$ = acceleration of the water, resolved normal to the member, m/s²

Typical values for a monopile are $C_d = 1.0$ and $C_m = 2.0$. These values need to be determined based on the geometry of the structure, possible appurtenances attached to the structures, marine growth attached to the structure, and any secondary effects such as diffraction.

If the structure is moving with a velocity U_b with respect to the water (normal to the member), then the equation is modified as follows:

$$\tilde{F} = \frac{1}{2}C_d\rho_w D \mid U_w - U_b \mid (U_w - U_b) + C_m\rho_w A_c \dot{U}_w - C_a\rho_w A_c \dot{U}_b \tag{6.69}$$

where C_a is the added mass coefficient which is related to the inertia coefficient with $C_m = 1 + C_a$.

6.4.1 Wave Loads on a Monopile

It is of interest to consider the forces and moments due to Airy waves on a slender monopile, because closed-form equations may be developed to describe them. In general, calculations of forces and moments are not so simple.

The loads on a slender monopile of constant diameter are obtained by applying the Morison equation and integrating over its length from the seabed to the still water level (average water level). The wave velocity and acceleration at each height are determined from the Airy wave theory. Integration leads to analytical expressions of the loads as function of time. The inertial and drag forces due to waves on a monopile of diameter D in water of depth d are as follows (see van der Tempel, 2006):

$$F_I = -\rho_w g \frac{C_m \pi D^2}{4} A \tanh(kd) \sin(2\pi f t - kx) \tag{6.70}$$

$$F_D = \rho_w g \frac{C_d D}{2} A^2 \left[\frac{1}{2} + \frac{kd}{\sinh(2kd)} \right] \cos(2\pi f t - kx) \tag{6.71}$$

Similarly, the moments at the sea floor due to the inertial and drag forces are:

$$M_I = -\rho_w g \frac{C_m \pi D^2}{4} A d \left[\tanh(kd) + \frac{1}{kd} \left(\frac{1}{\cosh(kd)} - 1 \right) \right] \sin(2\pi f t - kx) \tag{6.72}$$

$$M_D = \rho_w g \frac{C_d D}{2} A^2 \left[\frac{d}{2} + \frac{2(kd)^2 + 1 - \cosh(2kd)}{4k \, \sinh(2kd)} \right] \cos(2\pi f t - kx) \tag{6.73}$$

Example: Consider the situation of a monopile of diameter 5 m in water depth of 30 m. The wave height equals 4 m, the period is 5 seconds, the drag coefficient $C_d = 1.0$, the inertia $C_m = 2.0$ and the density of the seawater is 1025 kg/m³. From Equations (6.70) and (6.71), $F_{D,max} = 50.3$ kN and $F_{I,max} = 789.6$ kN. Figure 6.17 illustrates the forces over one complete period, and Figure 6.18 illustrates the maximum forces per unit length for the still water level to the sea bed. Note that in this example, the inertia force is dominant and that most of the wave force acts on the upper 10 m of the pile. The graphs were created numerically using Equations (6.70) and (6.71). A similar example can be created for the moments.

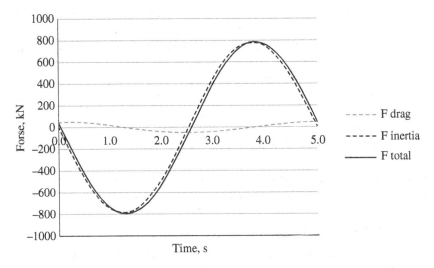

Figure 6.17 Wave forces over one period

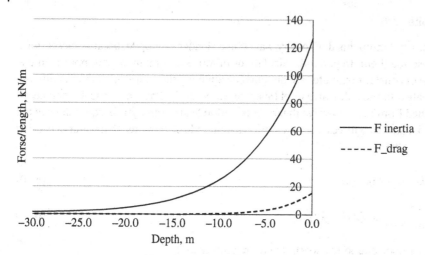

Figure 6.18 Maximum wave forces per unit length vs. depth

6.5 Wind and Waves: Combined Effects

There is a significant but not perfect correlation between wind speed and waves. An understanding of this correlation is important for turbine design. At the same time, there is not a direct relation between the sea state and the loads on an offshore wind turbines:

- The structural loads on the rotor of an operating turbine are normally highest when the wind is at or close to the rated wind speed.
- An offshore wind turbine's structural response is nonlinear, so loads may not follow wind/wave distributions.
- An aero/hydro/elastic simulation model with appropriate control algorithms, such as *OpenFAST* is needed to determine just when and where the highest loads occur.

In some cases, it is of interest to know about the likelihood of more than one type of event occurring at the same time, such as high winds and high waves. While it may well be the case that high waves tend to occur most often when the winds are also high, it is less likely that the highest wind and the highest wave will occur at exactly the same time. See, for example, Figure 6.19, which is a plot of significant wave height vs. wind speed for one year off the coast of Alabama (NDBC buoy 42040, 2020). The figure also includes estimated joint occurrences of significant wave height and wind speed with a 100-year return period as discussed below in Section 6.5.1.

It is accordingly not necessary to design a structure to withstand an extreme wind event and an extreme wave event at the same time. Savings in material and overall cost can be achieved by designing the structure for a combined extreme event.

The occurrences of multiple events in a sea state are typically characterized by joint probability distributions in a manner analogous to that of turbulent wind (Section 6.2.7.) The use of environmental contours for this situation is discussed in the next section.

Another method for considering partial correlation is an approach suggested by Nerzic and Prevosto (2000). This is based on a method developed by Plackett (1965). This approach is discussed in Appendix C.

6.5.1 *Environmental Contours of Sea States

It is often of use to be able to characterize combinations of relatively infrequent sea states in terms of their joint probability of occurrence. The method of environmental contours is frequently used for this purpose, and it is very

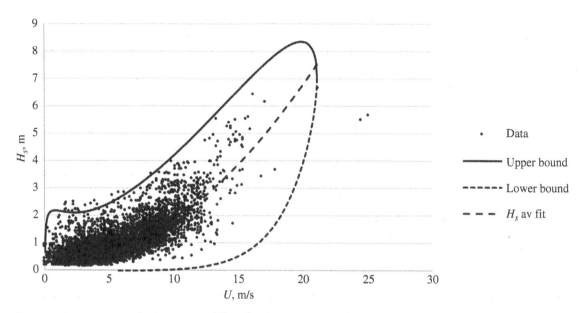

Figure 6.19 Plot of significant wave height vs. wind speed, NDBC buoy 42040, 2020

similar to that discussed previously for combinations of wind speed and turbulent standard deviation. Here, the idea is to identify combinations of events such as wind speed, wave height, and wave period that have equal probability. For example, a designer may be concerned that a structure is able to survive a combination of events that occurs once in 100 years. The are many possible combinations of wind speed, wave height, and wave period that would have a return period of 100 years. The designer would identify a number of those combinations and find the one that would cause the highest structural load, and make sure that the design was adequate. The example below provides an illustration.

Example: This example illustrates the use of IFORM (Section 6.2.7.1) for finding the environmental contours for the joint occurrence of wind speeds and significant waves for the data illustrated in Figure 6.19. This example was inspired by one given by Cheng *et al.* (2003). It follows the same method as that of the one in Section 6.2.7.1. In this case, we are interested in the extreme three-hour sea state that may occur every 100 years, which corresponds to a probability $1/[8760 \times 100/3] = 3.425 \times 10^{-6}$. Using Equation (6.24), the reliability index $\beta = 4.498$. The Weibull parameters for the wind data were found to be $c = 6.425$ m/s and $k = 2.134$. Mean and standard deviation of significant waves were determined for each 1 m/s wind speed bin and from those Weibull parameters for significant wave heights conditioned on the wind speeds were estimated with curve fits to be:

$$k_{H_{s,i}} = 0.001U_i^2 + 0.1732U_i + 1.3606 \tag{6.74}$$

$$c_{H_{s,i}} = 0.0169U_i^2 - 0.2116U_i + 0.4821 \tag{6.75}$$

For any given wind speed, U_i, the transformed variable u_1 can be found from Equation (6.35); u_2 is found from Equation (6.29), and then $H_{s,i}$ from Equations (6.34), (6.74), and (6.75). The resulting contours were included in Figure 6.19. It may be seen in Figure 6.19 that the highest predicted significant wave (8.3 m) occurs at a somewhat lower wind speed (20 m/s) than the highest wind speed, and vice versa.

Also note that if the effect of joint probabilities had not been considered, then the predicted extreme wave height with a 100-year return period could have been significantly higher. For example, the most conservative prediction could be obtained first by estimating the 100-year wind speed from the Weibull CDF, which would be 22.1 m/s. Then, $k_{H_{s,l}}$ and $c_{H_{s,l}}$ could have been found at that wind speed from Equations (6.74) and (6.75). Finally, Equation (6.34) with $u_2 = \beta$ would have yielded 12.0 m. In summary, it is recommended to use joint probabilities in predictions of this type.

6.6 Currents

Current is the flow of water past a fixed location. It is usually described in terms of its speed and direction.

The following are the main types of currents of relevance:

- Subsurface currents generated by tides, storm surge atmospheric pressure variations, etc.
- Wind-generated, near-surface currents.
- Near shore, breaking wave-induced surf currents running parallel to the coast.

In most cases, currents will not have a significant impact on the design of the rotor nacelle assembly of an offshore wind turbine. They could be relevant to the support structure or, even more likely, to the need for and type of scour protection (Chapter 10). For example, tidal currents can occur at many locations and the magnitude of the tidal current is based on the constructive or destructive addition of tidal contributions. Regardless, currents should be evaluated. In particular, the 1-year and 50-year recurrence values of the currents may be relevant to the design process. When estimating the effect of currents, principles of fluid mechanics can be applied. For example, the magnitude of the force per unit length $\tilde{F}_c$ on a cylinder due to a current of velocity u_c is given by:

$$\tilde{F}_c = \frac{1}{2} \rho_w C_d D \, u_c^2 \tag{6.76}$$

where D is the diameter and C_d is the drag coefficient, which may be estimated in the usual way (by consideration of the Reynolds number).

6.6.1 Subsurface Currents

The vertical velocity profile of the subsurface currents, U_{SS}, may be modeled by a power law (similar to that which is used with wind):

$$U_{SS}(z) = U_{SS}(0) \left[\frac{(z+d)}{d} \right]^{1/7} \tag{6.77}$$

where z is the vertical distance from water surface (positive up) and d is the water depth.

6.6.2 Wind Generated, Near-Surface Currents

The velocity profile of wind-generated current $U_w(z)$ is often assumed to be linearly related to the surface velocity of the water $U_w(0)$ down to 20 m below the surface or the maximum depth by:

$$
\begin{aligned}
U_w(z) &= U_w(0)(1 + z/20) \text{ for } z \geq -20 \text{ m} \\
&= 0 \text{ for } z < -20 \text{ m}
\end{aligned}
\tag{6.78}
$$

The wind-generated surface velocity is assumed to be 1% of the one hour mean wind speed at 10 m above the water level.

6.6.3 Breaking Wave-Induced Surf Currents

Breaking waves near the shore may induce surf currents which will run parallel to the coast. The velocity of these currents U_{bw} can be in many cases approximated by:

$$U_{bw} = 2s\sqrt{gH_B}$$ (6.79)

where

H_B is the breaking wave height.
s is the sea floor slope.
g is the acceleration due to gravity.

6.7 Floating Sea/Lake Ice

Floating sea or lake ice can be a significant external condition influencing the design of an offshore wind turbine, particularly the support structure, and so needs to be considered in cold climates. Floating ice can either be stationary or it can be moving. Stationary ice is referred to as "fast." Moving ice sheets are known as "floes" and can take a variety of forms as the ice breaks up, piles on top of other chunks of ice, and refreezes. In that case, it may form ridges or hummocks. Fast ice exerts force on offshore structures as it expands and contracts with changes in temperature or moves up and down with the tides. Moving ice floes exert force which is a function of the local ice pressure and the area upon which the ice acts. The force of the moving ice may also be increased due to movement caused by wind or currents.

The formation of ice and its thickness are closely related to the temperature at which seawater freezes, the sea temperature and the amount of time that the sea temperature has been below the freezing point of the seawater.

It is well known that presence of salt and other dissolved material reduces the freezing point of the water. Equation (6.80) gives the relation as a function of salinity.

$$T_f = -0.05411\,S_w$$ (6.80)

where T_f = freezing point (°C), S_w = salinity of the water, in parts per thousand (ppt).
Factors influencing the nature and thickness of the ice include:

- Winter temperature;
- Duration of cold periods;
- Salinity of the water;

Examples of ice loads that may need to be considered include:

- Horizontal loads due to temperature fluctuation in a fast ice cover (thermal ice pressure).
- Horizontal load from moving ice floes.
- Pressure from hummocked ice and ice ridges.
- Wind and current-induced loads
- Dynamic loads
- Vertical force from fast ice covers subject to water level fluctuations.

An estimate of the thickness of ice, h (m) after a cold period is:

$$h = 0.032\sqrt{0.9K_{max} - K_0}$$ (6.81)

where

$K_{\max}$ is the absolute value of the total of those 24 hours mean temperatures T_{mean} that are less than the freezing point of water in a period of cold weather (degree-days). $K_{\max}$ is defined in Equation (6.82).

$$K_{\max} = \sum_{days}\left(T_f - T_{mean}\right)^+ \text{ where } T_{mean} < T_f \tag{6.82}$$

K_0 is a constant that depends on the region. In Northern Europe, a value of $K_0 = 50$ is used. The superscript $+$ indicates that only consecutive days in which T_{mean} is less than the freezing point of water are to be considered (see IEC, 2019b).

6.7.1 Forces Due to Floating Ice

The forces that ice will exert on a structure will depend on the nature of the ice and the situation. Empirical equations are available that express the crushing strength of the ice as a function of the thickness of the ice, dimensions of the structure, the contact area between the ice and the structure, and the crushing strength of the ice. As a matter of reference, the crushing strength of ice ranges from approximately 1.0 MPa for deteriorated ice near the melting point to 3.0 MPa at the coldest time of the year. Details are beyond the scope of this text. The reader is referred to Annex D of IEC 61400-3-1 (IEC, 2019b) for more information.

6.8 Exceptional Conditions

Exceptional events are those that occur beyond the normal expectation of extreme events. They can be of great significance to the design and successful operation of offshore wind turbines. Some of the major sources of exceptional events are tropical cyclones, hurricanes, and extratropical cyclones. Some of the key points are listed in Sections 6.8.1 and 6.8.2.

6.8.1 Hurricanes/Tropical Cyclones

A hurricane is defined as an intense tropical weather system with a well-defined circulation and maximum sustained winds of 38.2 m/s or higher. In the western Pacific, hurricanes are called "typhoons," and similar storms in the Indian Ocean are called "cyclones." Hurricanes are most likely to form in the tropics, but they can travel farther away from the equator as well. A typical hurricane is illustrated in Figure 6.20.

Hurricanes are characterized by a distinct low-pressure core, or "eye," which is relatively small (30–60 km) and calm. Immediately outside the core, the wind speed is much higher, gradually decreasing with distance from the eye. The overall diameter of the storm may be on the order of 1000 km. Rainfall is typically quite intense, up to 150 mm/h with totals considerably more than that. Hurricanes are strongest over water, from which they derive most of their energy. As coherent atmospheric structures, hurricanes follow distinct paths from their formation until they eventually dissipate. Their forward velocity can reach 11 m/s. Wind speeds within the storm may exceed 70 m/s. Details on hurricanes in the United States may be found at https://www.nhc.noaa.gov/. General information on the physics and history of hurricanes is given by Emanuel (2005).

The winds associated with hurricanes invariably generate large waves. In addition, a zone of "extended fetch" can result in particularly large waves. Extended fetch refers to the effect of the forward velocity of the hurricane augmenting the creation of waves ahead of the storm; see Young (2003) for details.

Figure 6.20   Hurricane Humberto, as captured by a NOAA satellite on September 15, 2019 (*Source:* NOAA Satellites)

6.8.2   Extratropical Cyclones

Extratropical cyclones are a type of severe storm that originates outside the tropics. They occur most often in late fall or winter, and they generally have a cold core. They may have a very large diameter, even exceeding 1000 km. These storms are often accompanied by very high waves. Northeast storms, also known as "nor'easters," which occur fairly often off the northeast coast of the United States, are prime examples.

For illustration, the significant wave heights for an eight-year period off the coast of Massachusetts are shown in Figure 6.21. The highest significant wave heights there are associated with northeast storms.

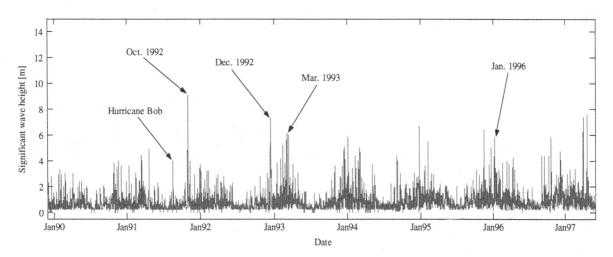

Figure 6.21   Typical northeast storm events off Boston, MA

6.9 Other Marine Conditions

6.9.1 Tides

Tides can have a significant effect on the design of an offshore wind turbine. For example, they can affect the height at which waves impinge on the support structure. They can also affect access to the turbine. Tides are caused primarily by the gravitational effect of the moon as it rotates around the earth and to a lesser extent by the gravitational pull of the sun. These changes in water level are known as astronomical tides. In most locations, there are on average two high tides and two low tides every day. The tidal range is the difference between high tide and low tide. Mean sea level is halfway between high tide and low tide. Tidal range can be very little or quite large. For example, the Bay of Fundy sometimes has a tidal range of 16 m. The tidal range is also affected by the seasons. Figure 6.22 illustrates the nomenclature related to tidal range. The water level can also be increased or decreased by storm surges.

6.9.2 Salinity

Salinity has a significant effect on the design of offshore wind turbines. Salt tends to increase the rate of corrosion of steel and other metals. It also affects the sea ice that forms in cold climates. Salinity is defined as the amount of salt (grams) found in 1000 grams of water. It is measured in parts per thousand. Accordingly, if there is 1 gram of salt in 1000 grams of water, the salinity is 1 part per thousand, or 1 ppt. The average salinity of the earth's oceans is approximately 35 ppt, but ranges between 31 and 39 ppt, depending on location. The density of seawater is also a function of salinity. For example, whereas freshwater has a density of 1000 kg/m³, the density of seawater with a salt content of 35 ppt is approximately 1027 kg/m³.

6.9.3 Marine Growth

Marine growth includes mollusks and other organisms that can attach themselves to the structure, mostly at depths less than 100 m. The effect is to increase drag on the structure in the presence of waves or currents. The accumulation may be sufficient to increase the effective mass of the structure as well. ISO 19901-1 (ISO, 2015) provides more information on this topic for some of the world's seas.

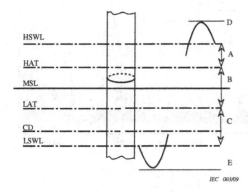

Figure 6.22 Definition of water levels (*Source:* From IEC 61400-3-1:2019, used by permission, Copyright © 2019 IEC Geneva, Switzerland. www.iec). HSWL, highest still water level; HAT, highest astronomical tide; MSL, mean sea level; LAT, lowest astronomical tide; CD, chart datum (often equal to LAT); A, positive storm surge; B, tidal range; C, negative storm surge; D, maximum crest elevation; E, minimum trough elevation; LSWL, lowest still water level.

6.9.4 Additional Conditions

A variety of other conditions may also need to be considered: range of atmospheric temperature, humidity, possibility of lightning, icing, earthquakes, tsunamis, etc.

6.10 Offshore Metocean Data Collection

The gathering of wind, wave, and other metocean data offshore is an extensive topic unto itself, and it is beyond the scope of this text. The following is a brief overview. For more on some of the relevant issues, see Manwell *et al.* (2007) or DNV GL (2018).

6.10.1 Platform and Tower Options

Platforms and tower options for collecting metocean data in general are similar to those for collecting wind data in particular (see Chapter 2). They include the following:

- Offshore structures;
- Buoys;
- Ships;
- Remote sensing from land, sea, or space.

6.10.2 Instrument Options

Instrumentation for collecting wind data offshore was discussed in Chapter 2. Instrumentation options for wave or current data include the following:

- Wave staffs;
- Wave gauges;
- Accelerometers;
- Acoustic Doppler current profilers (ADCP).

A wave staff is a capacitive type water level sensor. It is intended for use when it can be attached to a fixed structure. A processing unit is installed above normal water level and the staff projects downwards into the water up to 20 m.

A wave gauge is a submersible pressure sensor. It is also intended to be attached to a fixed structure. Typical wave gauges can operate ranges from sea level to a depth 100 m.

An accelerometer may be installed within a buoy. The accelerometer registers the rate at which the buoy is rising or falling as it follows the pattern of waves. By integrating against time, the acceleration signal can be converted to vertical displacement. The displacement values are relayed to a recording station on the shore. The Ocean Data Acquisition System (ODAS) buoys, such as are used by the US National Oceanic and Atmospheric Administration (NOAA), incorporate accelerometers of this type.

Acoustic Doppler current profilers (ADCPs) are devices that operate in a manner similar to that of sodars (see Chapter 2). Sound is sent from the device and the returned signal is analyzed to determine water speed and direction at various distances. Wave height, period, and direction as well as any submarine currents may be measured with ADCPs. These devices may be installed on a fixed structure, buoys, or the sea bed.

More details on commercially available instrumentation may be found in product literature.

6.10.3 Available Offshore Data in the United States

Similar to wind, existing wave data are available from a variety of sources. These include:

- NOAA buoys;
- Voluntary Observation Ship (VOS) program;
- Comprehensive Ocean Atmospheric Data Set (COADS).

See Manwell *et al.* (2007) for a summary of these sources.

6.10.4 Buoy Measurement Issues

There are issues measuring wind speed and waves with conventional data buoys: the presence of waves can affect wind measurement. The most significant of these is that the anemometer height may be so low that the data is not representative of the wind speed at hub height. Other factors are that when the waves are high, they may partially block the wind and indicate lower speeds than are actually occurring. Finally, the motion of the buoy (pitch, roll, and heave) may also introduce errors.

It should be noted, however, that recently developed lidar buoys (introduced in Chapter 2) differ from traditional data buoys in a number of ways. They can extract sea surface measurements from motion, provide motion-corrected wind data at rotor heights, and can also be connected to an ADCP so as to gather wave and wind data simultaneously. On the other hand, it is not yet possible for the lidars to accurately measure turbulence intensity.

6.10.5 Wave Data Estimation

In some cases, wave data can be estimated. The method is referred to as "hindcasting." This method makes use of numerical modelling such as was introduced in Chapter 2 with reference to wind. See, for example, SWAN (2013). Data for hindcasting is available from a range of commercial sources. The procedure for wave hindcasting is summarized below.

- Wind speed and direction are estimated from atmospheric pressure data by a numerical wind model.
- The sea state is estimated using oceanographic information and the estimated wind.
- Values are calculated at each point on a geographic grid. Calculations are based on momentum transfer.

6.11 External Conditions in Wind Turbine Design Standards

External design conditions are among the most important considerations in wind turbine design. Design standards connect the quantification, science, and modeling of external conditions discussed previously in this chapter back to the design basis (Section 6.1.1). The International Electrotechnical Commission (IEC) has developed a suite of standards to facilitate that process. These standards are relevant in three fundamental ways. For onshore turbines, the turbine designer will use the conditions defined in the standard in the design basis for turbines of the intended class. Subsequently, a project developer will choose turbines based on the conditions of the site: the conditions must be no worse than those corresponding to the class. Offshore turbines are designed somewhat differently. The support structure design requires site-specific information. The RNA design will be designed preliminarily based on expected class conditions, but it may be adjusted based on the specifics of the intended site. The IEC design standards are discussed in detail in Chapter 8. Here, we summarize the wind conditions as presented in the fundamental design standard, IEC 61400-1 (IEC, 2019a), and the metocean (wind and marine) conditions in the standard for fixed offshore wind turbines, IEC 61400-3-1 (IEC, 2019b).

Table 6.1 IEC wind classes

Wind turbine class		I	II	III	S
Average wind speed, U (m/s)		10	8.5	7.5	Values to be specified by designer
Reference wind speed, U_{ref} (m/s)	Normal	50	42.5	37.5	
	Tropical cyclone	57	57	57	
A+	I_{ref} (−)		0.18		
A	I_{ref} (−)		0.16		
B	I_{ref} (−)		0.14		
C	I_{ref} (−)		0.12		

6.11.1 IEC Design Wind Conditions

The design wind conditions comprise the most important part of the design basis for all wind turbines. IEC 61400-1 defines three classes of wind conditions that a wind turbine might reasonably be expected experience – I, II, and III – ranging from those with the highest winds (50 m/s reference wind speed) to the lowest (37.5 m/s reference wind speed). Within those classes, four ranges of turbulence are defined. They are denoted A+, A, B, and C, corresponding to high, medium, and low turbulence. These classes are summarized in Table 6.1. The important thing to note is that each class is characterized by a reference speed and a turbulence intensity. Other conditions of interest are referenced to the basic characterizations. To cover special cases, a fourth class, S, is also provided where the specific parameters are specified by the designer. These wind conditions are fundamental to the design load cases (DLCs), which were introduced in Section 6.1.1.

6.11.1.1 Normal Wind Conditions
Under normal wind conditions, the frequency of occurrences of wind speeds are assumed to be described by the Rayleigh distribution.

Normal Wind Profile (NWP)
The wind profile, $U(z)$, is the variation in wind speed with the height, z, above the ground. For the purposes of the IEC requirements, the variation in wind speed with height is assumed to follow a power law model, with an exponent of 0.2. It is then known as the normal wind profile (NWP).

Normal Turbulence Model (NTM)
The standard deviation of the turbulence in the longitudinal direction (direction of the mean wind), σ_x, is assumed to be given by:

$$\sigma_x = I_{ref}(0.75\,U_{hub} + 5.6) \tag{6.83}$$

where I_{ref} = turbulence intensity at 15 m/s and U_{hub} = wind speed at hub height.
   The psd of the turbulence can be modeled with the Mann or Kaimal spectrum; see Section 6.2.4.7.

6.11.1.2 Extreme Wind Conditions
There are five extreme wind conditions to be used in determining extreme loads under the IEC standards: (1) extreme wind speed (EWM), (2) extreme operating gust (EOG), (3) extreme coherent gust (ECG), (4) extreme coherent gust with change in direction (ECD), and (5) extreme wind shear (EWS).

Extreme Wind Speed (EWM)

Extreme wind speeds are very high, sustained winds which are likely to occur, but only rarely. Two extreme wind speeds are defined by the frequency with which they are expected to recur: the 50-year extreme wind (U_{e50}) and the 1-year extreme wind (U_{e1}). They are based on the reference wind (see Table 6.1). The 50-year wind is approximately 40% higher than the reference wind, while the 1-year wind is 30% higher than the reference wind.

Extreme Operating Gust (EOG)

An extreme operating gust is a sharp increase and then decrease in wind speed which occurs over a short period of time while the turbine is operating. The magnitude of the 50-year extreme operating gust (U_{gust50}) is related to the turbulence intensity, the scale of the turbulence, and the rotor diameter of the turbine. The gust is also assumed to rise and fall over a period of 10.5 seconds. An illustration of an extreme operating gust is shown in Figure 6.23. The situation here is a turbine with a diameter of 84 m operating in Class 1A conditions at an initial wind speed of $U = 25$ m/s. More details on the EOG can be found in the standard (IEC, 2019a).

Extreme Direction Change (EDC)

Extreme direction changes are defined in an analogous manner to extreme gusts. In a typical example, the wind direction may change by 64° over six seconds.

Extreme Coherent Gust (ECG)

A coherent gust is a rapid rise in wind speed across the rotor. The IEC extreme coherent gust is assumed to have an amplitude of 15 m/s and to be superimposed on the mean wind. The wind rises sinusoidally to a new level over a period of 10 seconds.

Extreme Coherent Gust with Change in Direction (ECD)

In the extreme coherent gust with a change in direction, a rise in wind speed is assumed to occur simultaneously with a direction change. Details are provided in the standard (IEC, 2019a).

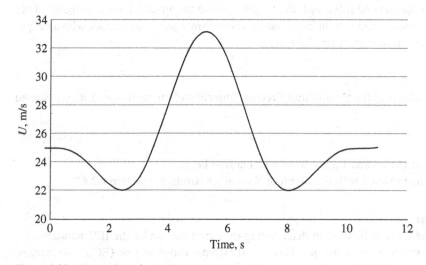

Figure 6.23 Illustration of extreme operating gust

Extreme Wind Shear (EWS)

Two transient wind shear conditions are also defined, one for horizontal shear and the other for vertical wind shear. Transient wind shears will be much larger than in the normal wind profile.

Random Wind Velocity Field

Analysis of turbulent wind DLCs relies on a synthetic random wind velocity field as input to an aeroelastic model of the wind turbine. These are created by application of methods such as the one described in Section 6.2.5.

6.11.2 IEC Design Conditions Offshore

External design conditions offshore, as specified in IEC 61400-3-1, include, of course, wind conditions, but in addition they include waves, currents, and sea ice, as well as a number of others. These are referred to as marine conditions.

6.11.2.1 IEC Wind Conditions Offshore

Design wind conditions offshore are similar to those on land, but there are some differences as well. The primary difference is in the turbulence model, as described immediately below.

Offshore Normal Turbulence Model

As previously noted, (Section 6.2.8.4) turbulence intensity offshore increases at higher wind speeds, when the waves are higher. Wang *et al.* (2014) have developed an Offshore Normal Turbulence Model (ONTM) which accounts for that effect. This model is referenced in IEC 61400-3-1. The results of this model are illustrated in Figure 6.24 for the IEC 61400-1 turbulence classes, A, B, and C.

6.11.2.2 Marine Conditions

The marine conditions defined in IEC 61400-3-1 are the metocean conditions other than the wind itself. They consist primarily of the waves and joint occurrence of wind and waves. In addition, currents, sea ice, and exceptional conditions, such as hurricanes, may be critical in some locations. Other marine conditions including water level range and marine growth also need to be considered.

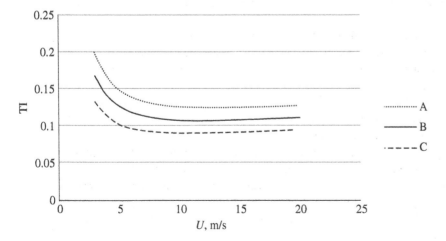

Figure 6.24 Turbulence intensity vs. wind speed offshore, as predicted by the Offshore Normal Turbulence Model

Waves

Waves are to be categorized by a wave spectrum, significant wave height, peak spectral period, and mean wave direction. Spectra include (by reference to ISO, 2015) the Pierson–Moskowitz and JONSWAP models; see Section 6.3.4.

A stochastic wave model is to be used to synthesize a wave train for input to an aero/hydro/elastic model of an offshore wind turbine; see Section 6.3.7.

Joint Occurrences of Wind and Waves

The joint occurrence of wind and waves should consider significant wave height, peak spectral period, and wind speed over an appropriate time period. Because significant wave heights are typically reported over three hours and wind speed over 10 minutes, an intermediate period of one hour is advised. In that case, both the wind speed and wave heights must be adjusted. Combinations of wind and waves are to be accounted for using "severe sea states." The use of environmental contours is recommended for this purpose; see Section 6.5.1.

6.11.2.3 Other Conditions

Guidance on currents in IEC 61400-3-1 is substantially as discussed in Section 6.6. For sea ice, see Section 6.7. For hurricanes and tropical cyclones see Section 6.8. For tides, salinity, marine growth, etc. see Section 6.8. More details on all of these may be found in the standard itself.

References

Airy, G. B. (1845) On tides and waves. *Encyclopedia Metropolitana*, **5**, 241–396.

Barltrop, N. D. P. and Adams, A. J. (1991) *Dynamics of Fixed Marine Structures*. Butterworth Heinemann, Oxford.

Barthelmie, R. J., Courtney, M. S., Højstrup, J. and Larsen, S. E. (1996) Meteorological aspects of offshore wind energy: observations from the Vindeby wind farm. *Journal of Wind Engineering and Industrial Aerodynamics*, **62** (2–3), 191–211.

Beyer, H. G., Luther, J. and Steinberger-Willms, R. (1989) Power fluctuations from geographically diverse, grid coupled wind energy conversion systems. *Proc. of the 1989 European Wind Energy Conference*, Glasgow, pp. 306–310.

Cheng, P. W., van Bussel, G. J. W., van Kuik G. A. M. and Vugts, J. H. (2003) Reliability-based design methods to determine the extreme response distribution of offshore wind turbines, *Wind Energy*, **6**, 1–22, https://doi.org/10.1002/we.80.

Downing, S. D. and Socie, D. F. (1982) Simple rainflow counting algorithms. *International Journal of Fatigue*, **4**(1), 31–40.

DNV GL (2018) *Metocean Characterization Recommended Practices for U.S. Offshore Wind Energy*, https://www.boem.gov/sites/default/files/environmental-stewardship/Environmental-Studies/Renewable-Energy/Metocean-Recommended-Practices.pdf (Accessed 6/23/2023).

Emanuel, K. (2005) *Devine Wind: The History and Science of Hurricanes*, Oxford University Press, Oxford, UK.

Freris, L. (1990) *Wind Energy Conversion Systems*, Prentice Hall, Hoboken.

Garratt, J. R. (1994) The atmospheric boundary layer. *Earth-Science Reviews*, **37**(1–2), 89–134. Elsevier.

Haver, S. and Winterstein, S. R. (2009) Environmental contour lines: a method for estimating long term extremes by a short term analysis. *Transactions – Society of Naval Architects and Marine Engineers*, 116–127.

IEC (2019a) *Wind Turbines, Part 3: Design Requirements for Wind Turbines, 61400-1*, 4th edition. International Electrotechnical Commission, Geneva.

IEC (2019b) *Wind Turbines, Part 3: Design Requirements for Offshore Wind Turbines, 61400-3-1*. International Electrotechnical Commission, Geneva.

ISO (2015) *Petroleum and Natural Gas Industries-Specific Requirements for Offshore Structures — Part 1: Metocean Design and Operating Considerations, ISO 19901:2015(E)*, 2nd edition. International Standards Organization, Geneva.

Jonkman, J. M. (2013) *The new modularization framework for the FAST wind turbine CAE tool*, NREL/CP-5000-57228. National Renewable Energy Laboratory, Golden, CO

Kelly, N. D. and Jonkman, B. J. (2007) *Overview of the TurbSim stochastic inflow turbulence simulator Version 1.21 (Revised)*, NREL/TP-500-41137, National Renewable Energy Laboratory, Golden, CO.

Krogstad, H. and Arntsen, O. (2000) *Linear wave theory*. Dept. of Structural Engineering, Technical University of Norway, Trondheim, https://folk.ntnu.no/oivarn/hercules_ntnu/LWTcourse/ (Accessed 12/14/2021).

Lackner, M. A., Schmidt, D. P., Arwade, S. R., Myers, A. T. and Robertson, A. N. (2018) *Simulating breaking waves and estimating loads on offshore wind turbines using computational fluid dynamic models*, https://www.bsee.gov/sites/bsee.gov/files/research-reports//776aa.pdf (Accessed 12/14/2021).

Lange, B. and Højstrup, J. (1999) The influence of waves on the offshore wind resource. *Proc. 1991 European Wind Energy Conference*, Nice, pp. 191–194.

Lange, B., Højstrup, J., Larsen, S. E. and Barthelmie, R. J. (2001) A fetch dependent model of sea surface roughness for offshore wind power utilization, Wind Energy for the new millennium. *Proceedings of the European Wind Energy Conference* (Copenhagen 2001), WIP, Munich and ETA, Florenz, pp. 830–833.

Longuet-Higgins, M. S. (1952) On the statistical distribution of the heights of sea waves. *Journal of Marine Research*, **11**(3), 245–266.

Mann, J. (1998) Wind field simulation. *Probabilistic Engineering Mechanics*, **13**(4), 269–282.

Manwell, J. F., Elkinton, C., Rogers, A. L. and McGowan, J. G. (2007) Review of design conditions applicable to offshore wind energy in the United States. *Renewable and Sustainable Energy Reviews*, **11** (2), 183–364, Elsevier, Amsterdam.

NDBC (2020) *Buoy 42040, historical data download, 2020*, https://www.ndbc.noaa.gov/download_data.php?filename=42040h2020.txt.gz&dir=data/historical/stdmet/ (Accessed 12/13/2021).

Nerzic, R. and Prevosto, M. (2000) Modelling of wind and wave joint occurrence probability and persistence duration from satellite observation data. *Proceedings of the Tenth (2000) International Offshore and Polar Engineering Conference*, Seattle, USA, May 28–June 2, 2000, https://onepetro.org/ISOPEIOPEC/proceedings-abstract/ISOPE00/All-ISOPE00/ISOPE-I-00-238/7016 (Accessed 12/14/2021).

Ochi, M. K. (1998) *Ocean Waves - The Stochastic Approach*, Cambridge University Press, Cambridge.

Plackett, R. L. (1965) A class of bivariate distributions. *Journal of the American Statistical Association*, **60**(310) 516–522, Taylor & Francis, Ltd. on behalf of the American Statistical Association, https://www.jstor.org/stable/2282685 Accessed 8/15/2018.

Shinozuka, M. and Jan, C. M. (1972) Digital simulation of random processes and its application *Journal of Sound and Vibration*, **25**(1), 111–128.

Somoano, M., Battistella, T., Rodríguez-Luis, A., Fernández-Ruano, S. and Guanche, R. (2021) Influence of turbulence models on the dynamic response of a semi-submersible floating offshore wind platform, *Ocean Engineering*, **237**, 109629, Elsevier.

SWAN (2013) *SWAN Cycle III version 40.49A. User manual*, https://swanmodel.sourceforge.io/download/zip/swanuse.pdf (Accessed 6/23/2023).

Taylor P. K. and Yelland, M. (2001) The dependence of sea surface roughness on the height and steepness of the waves. *Journal of Physical Oceanography*, **31**(2), 572–590.

USACE (2002) *Coastal Engineering Manual*, CEM M 1110-2-1100; US Army Corps of Engineers (USACE), Washington, https://www.publications.usace.army.mil/USACE-Publications/Engineer-Manuals/u43544q/636F617374616C20656E67696E656572696E67206D616E75616C/ (Accessed 6/23/2023).

USNA (2023) Rayleigh probability distribution applied to random wave heights https://www.usna.edu/NAOE/_files/documents/Courses/EN330/Rayleigh-Probability-Distribution-Applied-to-Random-Wave-Heights.pdf (Accessed 3/10/2023).

van der Tempel, J. (2006) *Design of Support Structures for Offshore Wind Turbines*, PhD Dissertation, Delft University of Technology, Delft.

Wang, H., Barthelmie, R. J., Pryor, S. C. and Kim, H. G. (2014) A new turbulence model for offshore wind turbine standards. *Wind Energy*, **17**(10), 1587–1604.

Wienke, J. (2001) *Druckschlagbelastung auf Schlanke Zylindrische Bauwerke durch Brechende Wellen*, PhD Dissertation, TU Carolo-Wilhelmina zu Braunschweig, Braunschweig, Germany.

Young, I. R. (2003) A review of the sea state generated by hurricanes. *Marine Structures*, **16**, 201–218.

Zhang, F. W., Drennan, W. M., Haus, B. K., Graber, H. C. (2009) On wind-wave-current interactions during the Shoaling Waves Experiment, *Journal of Geophysical Research*: Oceans. Available at https://agupubs.onlinelibrary.wiley.com/doi/full/10.1029/2008JC004998 Accessed 4/2/2024.

7

Wind Turbine Materials and Components of the Rotor Nacelle Assembly

7.1 Overview

This chapter considers the materials and components that are commonly used in wind turbines. Since fatigue is so significant to wind turbines, this topic is discussed at some length early in the chapter (Section 7.2). The focus is then on materials that are of particular concern in wind turbines, particularly composites. This is the subject of Section 7.3. Components are discrete items that are used to make up a larger structure. Some components that are widely available from a variety of sources and are common to a wide range of machinery are often called elements. Some of the machine elements most commonly used in wind turbines are discussed in Section 7.4. As indicated previously, an entire wind turbine is comprised of a rotor nacelle assembly (RNA) and support structure. This chapter focuses on the RNA; fixed support structures are discussed in Chapter 10, and floating support structures are discussed in Chapter 11. The various larger components in wind turbines are discussed in Section 7.5. How these components are then combined to form a complete turbine is the subject of Chapter 8.

7.1.1 Review of Basic Mechanical Properties of Materials

In this text, it is assumed that the reader has a familiarity with the fundamental concepts of material properties, as well as with the most common materials themselves. The following is a list of some of the essential concepts [for more details, see a text on mechanical design, such as Shigley's Mechanical Engineering Design (Budynas and Nisbett, 2015)]:

- Hooke's Law;
- modulus of elasticity;
- yield strength, breaking strength;
- ductility and brittleness;
- hardness and machinability;
- failure by yielding or fracture.

7.2 Material Fatigue

Many materials, which can withstand a load when applied once, will not survive if that load is applied, removed, and then applied again ("cycled") a number of times. This increasing inability to withstand loads applied multiple times is called fatigue damage. The underlying causes of fatigue damage are complex, but they can be most simply

Wind Energy Explained: On Land and Offshore, Third Edition.
James F. Manwell, Emmanuel Branlard, Jon G. McGowan, and Bonnie Ram.
© 2024 John Wiley & Sons Ltd. Published 2024 by John Wiley & Sons Ltd.

conceived of as deriving from the growth of tiny cracks. With each cycle, the cracks grow a little more, until the material fails. This simple view is also consistent with another observation about fatigue: the lower the magnitude of the load cycle, the greater the number of cycles that the material can withstand.

7.2.1 Fatigue in Wind Turbines

Wind turbines are subject to a great number of cyclic loads. The lower bound on the number of many of the fatigue-producing stress cycles in turbine components is proportional to the number of blade revolutions over the turbine's lifetime. The total cycles, n_L, over a turbine's lifetime would be:

$$n_L = 60\,k\,n_{rotor}H_{op}\,Y \tag{7.1}$$

where k is the number of cyclic events per revolution, n_{rotor} is the rotational speed of rotor (rpm), H_{op} is the operating hours per year, and Y is the years of operation.

For blade root stress cycles, k would be at least equal to 1, while, for the drivetrain or tower, k would be at least equal to the number of blades. A turbine with rotational speed of 15 rpm operating 6000 hours per year would experience more than 10^8 cycles over a 20-year lifetime. This may be compared to many other manufactured items, which would be unlikely to experience more than 10^6 cycles over their lifetime. In fact, the number of cyclic events on a blade can be much more than once per revolution of the rotor, due to such causes as wind shear and structural vibrations. Simulation models such as *OpenFAST*, which is discussed elsewhere in this text, can help to quantify those events.

7.2.2 Assessment of Fatigue

Procedures have been developed to estimate fatigue damage. Many of these techniques were developed for analysis of metal fatigue, but they have been extended to other materials, such as composites, as well. The following sections summarize fatigue analysis methods most commonly used in wind turbine design. More detail on fatigue in general may be found in Ansell (1987) and Budynas and Nisbett (2015). Information on fatigue in wind turbines is given in Sutherland (2000) and Veers (2011). The software *MLife* and accompanying theory manual (Hayman, 2012) are useful for assessing wind turbine fatigue damage. For fatigue of offshore structures, see ABS (2018) and Mendes *et al.* (2021).

The estimation of the fatigue life of a wind turbine component requires three things: (1) appropriate fatigue life properties of the material in question, (2) a model or theory that can be used to determine material damage and component lifetime from the loading and material properties, and (3) a method to characterize the loading incurred by the component during the turbine lifetime. Each of these topics is described in Sections 7.2.2.1–7.2.2.3.

7.2.2.1 Fatigue Life Characterization

Fatigue resistance of materials is traditionally tested by subjecting successive samples to a sinusoidally applied load until failure. One type of test is a rotating beam test. A sample is mounted on a test machine and loaded to a given stress with a side load. The sample is then rotated in the test machine so that the stress reverses every cycle. The test is continued until the sample fails. The first load is somewhat less than the ultimate load (the maximum load that the sample will withstand). The number of cycles and the load are recorded. The load is then reduced on another sample and the process is repeated. The data is summarized in a fatigue life, or "*S–N*," curve, in which *S* refers to the stress range (despite the customary use of σ to indicate stress) and *N* refers to the number of cycles to failure. With tests of this type, it is important to note that (1) the mean stress is zero and (2) the stresses are fully reversing (for example, an object may be first bent one way and then the other way an equal amount). The stress range is defined as the difference between the maximum stress and the minimum stress (more details are given below). Another important parameter is damage per cycle, *d*, which is the reciprocal of *N*.

A commonly used *S–N* relationship is given by:

$$\sigma = \left(\frac{A}{N}\right)^{1/m} \qquad (7.2)$$

where σ is used for stress range (MPa), N is the number of cycles to failure, A is known as the fatigue strength coefficient, and m is the Wöhler, or fatigue, exponent, which depends on the material used. Typical values for steel (e.g., a wind turbine tower) and for fiber glass (e.g., wind turbine blades) are $m = 3$ and $m = 10$, respectively. Values of A for steel range from 10^{10} to 10^{15}. Figure 7.1 illustrates a typical *S–N* curve (labeled "S"). In this curve, the solid line represents a relation of the form shown in Equation (7.2). The dotted line ("S2") corresponds to the second segment of a two-segment *S–N* curve (with different fatigue strength coefficient and exponent), which is also widely used, especially for offshore steel structures – see (ABS, 2018) and Section 7.2.3.2.

The straight lines in Figure 7.1 are convenient, but they somewhat obscure how much more damaging higher stress cycles are than lower stress cycles. Another way to depict the relationship is to plot damage/cycle (i.e., $1/N$) vs. stress range. This is illustrated in Figure 7.2 for the single slope line *S* in Figure 7.1.

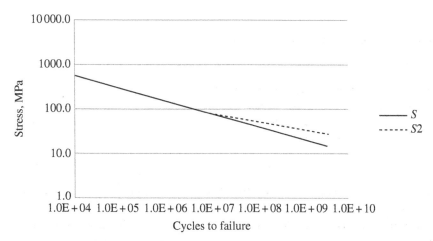

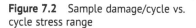

Figure 7.1 Typical *S–N* curves with $A = 4.23 \times 10^{13}$ and $m = 3.5$. Logarithmic scales are used for both axes, resulting in a slope of $-1/m$ for *S*

Figure 7.2 Sample damage/cycle vs. cycle stress range

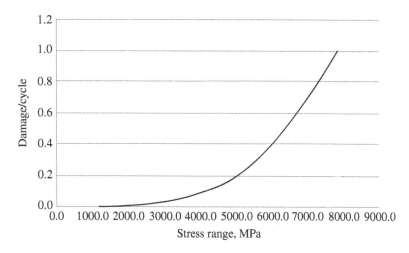

Most manufactured items of commercial interest, with the exception of wind turbines, do not experience more than 10 million cycles in their lifetimes. Thus, if a sample does not fail after 10^7 cycles at a particular stress (as the loads are progressively reduced), the stress is referred to as the endurance limit, σ_{el}. Endurance limits typically reported are in the range of 20–50% of the ultimate stress, σ_u, for many materials. (Recall that the ultimate stress is the maximum stress that a material can withstand.) In reality, the material may not actually have a true endurance limit, and it may be inappropriate to use that assumption in wind turbine design.

Alternating Stresses with Nonzero Mean

Alternating stresses typically do not have a zero mean. In this case, they are characterized by the mean stress, σ_m, the maximum stress, σ_{max}, and the minimum stress σ_{min}, as shown in Figure 7.3.

As noted above and shown in Equation 7.3, the stress range, σ, is defined as the difference between the maximum and the minimum.

$$\sigma = \sigma_{max} - \sigma_{min} \tag{7.3}$$

The stress amplitude, σ_a, is half the range:

$$\sigma_a = (\sigma_{max} - \sigma_{min})/2 \tag{7.4}$$

The stress ratio, R, is the ratio between the maximum and the minimum. That is:

$$R = \sigma_{min}/\sigma_{max} = (\sigma_m - \sigma_a)/(\sigma_m + \sigma_a) \tag{7.5}$$

For alternating stress with zero mean ("completely reversing"), $R = -1$.

An example of alternating stresses with non-zero mean which is more realistic than the idealized one of Figure 7.3 is shown in Figure 7.4. This is from an *OpenFast* simulation of the NREL 5 MW wind turbine. Characterizing stresses such as these is discussed subsequently.

Goodman Diagram

Material fatigue tests may also be performed at a variety of mean stress levels and amplitudes (or R values). The resulting data may be presented in an *S–N* diagram with multiple curves for the various test conditions or in a Goodman diagram. A Goodman diagram depicts constant-life curves on a plot of stress amplitude vs. mean stress. Often, the stress amplitude and mean stress are normalized by the ultimate strength of the material. An example is shown in Figure 7.5.

In Goodman diagrams, all points with a constant R value are on a straight line with its origin at zero mean and zero amplitude. Fully reversed bending with a zero mean stress level ($R = -1$) is on the vertical axis. The ultimate

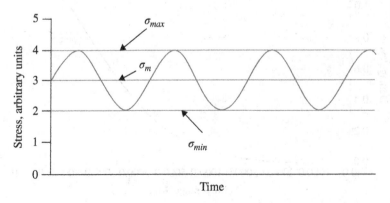

Figure 7.3 Idealized alternating stress with nonzero mean

Figure 7.4 Example of alternating stress in a typical of wind turbine

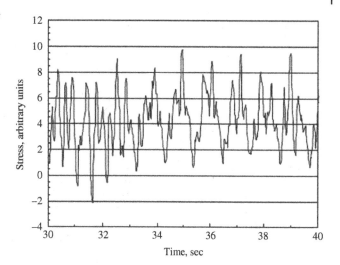

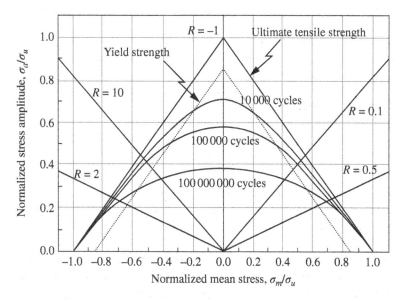

Figure 7.5 Sample Goodman diagram (*Source:* Sutherland, 1999/Sandia National Laboratories)

compressive strength bounds the fatigue life data on the compressive (left) side of the diagram and the ultimate tensile strength bounds the curves on the tensile (right) side of the diagram. Constant-life curves are shown for 10^5, 10^6, and 10^7 cycles. This example is of a symmetric material for which the left and right sides of the diagram are mirror images of each other.

Goodman diagrams for fiberglass materials are typically not symmetric. For example, a Goodman diagram (based on strain rather than stress) for one sample fiberglass composite is shown in Figure 7.6.

As mentioned above, the fatigue life for a given alternating stress depends on R (i.e., the mean). In fact, the allowed alternating stress for a given fatigue life will decrease as the mean increases. This relationship is often approximated by Goodman's Rule:

$$\sigma_{al} = \sigma_e (1 - \sigma_m/\sigma_u)^c \qquad (7.6)$$

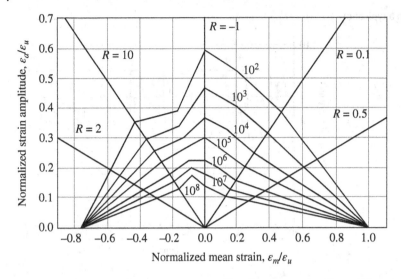

Figure 7.6 Goodman diagram for a sample fiberglass composite (*Source:* Sutherland, 1999/Sandia National Laboratories)

where σ_{al} is the allowable stress range for a given fatigue life, σ_e is the equivalent zero mean (i.e., $R = -1$) alternating stress for the desired fatigue life, σ_u is the ultimate stress, and c is an exponent that depends on the material, but which is often assumed to be equal to one.

Goodman's Rule can be inverted and used with nonzero mean alternating stress data to find the equivalent zero mean alternating stress (assuming $c = 1$) [see, for example, Ansell (1987)]:

$$\sigma_e = \sigma_{al}/(1 - \sigma_m/\sigma_u) \tag{7.7}$$

7.2.2.2 Fatigue Damage Model

The effect of fatigue on a component is characterized by a damage model. Although mean stress is a factor in damage as indicated above, stress cycle ranges are the most important factor. In the fatigue model described here, three assumptions are made: (1) damage is a function of stress range, (2) damage is cumulative, and (3) the order in which stress cycles are applied is immaterial.

According to the damage model, if a component undergoes fewer load cycles than would cause it to fail, it will still suffer damage. The component might fail at a later time, after additional load cycles are applied. To quantify this, a damage term, d_n, is defined. It is the ratio of the number of cycles applied, n, to the number of cycles to failure, N, at the given amplitude:

$$d_n = n/N \tag{7.8}$$

Cumulative Damage and Miner's Rule

A component may experience multiple load cycles of different ranges, as illustrated in Figure 7.7. In this case, the cumulative damage, D, is defined by Miner's Rule as the sum of the damages due to each of the cycles at each range. The component is deemed to have failed when the total damage is equal to 1.0. In the general case of load cycles of M different ranges, the cumulative damage would be given by:

$$D = \sum_{i=1}^{M} n_i/N_i \leq 1 \tag{7.9}$$

where n_i is the number of cycles at the ith range and N_i is the number of cycles to failure at the ith range.

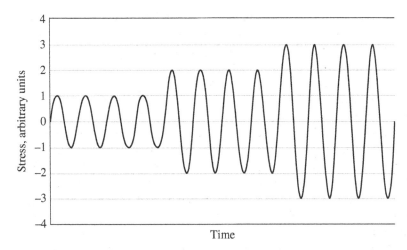

Figure 7.7   Load cycles of different range but same frequencies

To illustrate this, suppose that a material undergoes four cycles each of magnitude 1, 2, and 3 stress units (as in Figure 7.7) and that the number of cycles to failure, N, at those stress units is 20, 16, and 10 cycles, respectively. The cumulative damage due to the 12 cycles would be:

$$D = (4/20) + (4/16) + (4/10) = 0.85$$

Randomly Applied Load Cycles

When loads are not applied in blocks (as was the case in Figure 7.7) but rather occur more randomly, it is difficult to identify individual load cycles. Figure 7.8 illustrates such a situation.

A technique known as rainflow cycle counting (Downing and Socie, 1982) has been developed to identify alternating stress cycles and mean stresses from time series of randomly applied loads. Once the mean and alternating stress data have been found, they can be converted to zero mean alternating stresses and the total damage estimated using Miner's Rule, as discussed above.

Figure 7.8   Random load cycles

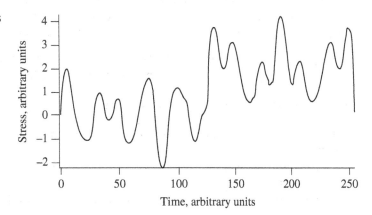

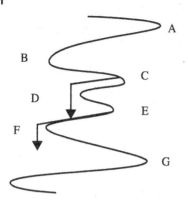

Figure 7.9 Example of rainflow cycle counting

The rainflow counting method is most appropriately applied to strain data, rather than stress data, and can deal with inelastic as well as elastic regions of the material. For most wind turbine applications, the material is in the elastic region, so either stress or strain data can be used. When strain is used, results are ultimately converted to stresses.

The essence of the rainflow counting technique for cycle ranges is summarized next, and an algorithm for performing the counting is given in Appendix B. Local highs and lows in the data are identified as "peaks" or "valleys." The differences between every peak and valley and between every valley and every peak are all considered to be "half cycles." The algorithm then pairs the half cycles to find complete cycles to the extent possible. The entire range of the data is divided into bins, and the output of the algorithm consists of the number of cycles that fall into each of the bins. The algorithm may also be extended to output cycle means as well.

The term "rainflow" derives from an aspect of the method, in which the completion of a cycle resembles rainwater dripping from a roof (a peak) and meeting water flowing along another roof (from a valley below). In this view of the method, the peak–valley history is imagined to be oriented vertically so that the "rain" descends with increasing time. An example of a complete cycle is shown in Figure 7.9, which is a subset of Figure 7.8. Here rain (indicated by heavy arrows) running down roof C–D encounters rain running down roof E–F. The magnitude of the peak C is retained for use in identifying subsequent cycles, and the counter for the corresponding bin will be increased by one. The cycle C–D–E is no longer needed by the algorithm and is eliminated from further consideration. In this particular case, point F is slightly to the left of point B, so the next complete cycle is B–C–F. The algorithm returns to the starting point, and the process is repeated until all cycles have been accounted for.

For another example of rainflow cycle counting, consider a sinusoidally varying stress of range σ (i.e., of amplitude $\sigma/2$) and frequency f, for a time series containing n periods. Suppose the total range is divided into $N_{B,r} = 20$ bins of width $\sigma/20$. The algorithm would return a count of n for the 20th bin, i.e., the bin with endpoints σ and $\sigma - \sigma/20$, and zero for all the other bins.

Effect of Operating Conditions

The loads that contribute to fatigue of a wind turbine originate from a variety of sources, which are closely connected to the operating conditions (see Chapter 4). The loads include steady loads from high winds; periodic loads from wind shear, yaw error, yaw motion, and gravity; stochastic loads from turbulence; transient loads from such events as gusts, starts, and stops, etc.; and resonance-induced loads from vibration of the structure. The magnitude of the contribution from each of these load sources to component fatigue will depend on the operational state of the wind turbine. Thus, the task of characterizing the loading on a component, during its lifetime, requires the cumulative counting of damage over all operating states and all stress levels that are encountered during its lifetime. The damage to a component over the course of a year, D_y, can be expressed as:

$$D_y = \sum_{\text{All conditions}} \sum_{\text{All stress levels}} \frac{\text{number of stress cycles during year}}{\text{number of cycles to failure}} \tag{7.10}$$

The conditions that need to be considered include:

1) Normal power production where the fatigue is assumed to be a function of wind speed (between cut-in and cut-out wind speeds).
2) Nonoperating states when the turbine is being buffeted by the wind (typically at wind speeds greater than the cut-out wind speed).
3) Normal or emergency start/stop operations or other operations that cause transient loading.

Typically, the stress levels are determined using measurements on turbine prototypes or computational models that include a detailed description of the turbulent inflow to the turbine and the structural dynamics of the turbine. The measurements or model outputs describe the strains, forces, or moments at specific points in the turbine structure. These must be converted to stress levels in specific components using stress concentration factors, based on the location of the measurements and the geometry of the turbine, and partial safety factors (see Chapter 8). Data are collected for multiple ten-minute periods at discrete operating states in order to characterize the stress histogram (the number of cycles of various ranges) at each operating state. A total of five hours of data at each operating state may be needed to ensure a representative data set. Equations for the damage that occurs during each of the three types of operating states follow. See Sutherland (1999) for more details.

The cumulative damage over the lifetime of turbine, D_T, is a function of the stress levels at each wind speed, the frequency of occurrence of each wind speed, and the operating state. For a given operating state (x), the damage in the jth wind speed bin is given by Equation 7.11. The cumulative damage over all wind speeds in that state is then given by Equation (7.12). Recall that fatigue damage is generally determined in accordance with the stress cycle ranges, but mean level can also have an effect. When analysis is to include mean stress level, Equation (7.12) can be extended with an additional summation.

$$d_j\big|_x = \sum_{i}^{N_{B,r}} \frac{n_{i,j}}{N_i} \tag{7.11}$$

$$D_{T,x} = N_T \sum_{j=1}^{N_{B,U}} F_{U_j} d_j\big|_x \tag{7.12}$$

where

$x = O$ for normal operation, $x = P$ for parked, $x = T$ for transient.
N_T = Total number of time intervals over the lifetime.
$N_{B,U}$ = Number of wind speed bins.
$N_{B,r}$ = Number of stress range bins.
i = index for stress ranges (covering the range of all stress ranges encountered).
j = index for wind speeds that may cause damage, including winds beyond cut-out.
Δt = the length of the time interval during which fatigue cycles are evaluated (typically 10 min).
U_j = the free stream wind speed.
σ_i = i^{th} stress range.
$n_{i,j}$ = the number of cycles of ith stress range at j^{th} wind speed during interval of length Δt; these will come from the rainflow counter.
N_i = the number of cycles to failure at i^{th} stress range; these will come from the S–N diagram.
F_{U_j} = the fraction of time that the wind speed, U, is between U_{j-1} and U_j:

$$F_{U_j} = \int_{U_{j-1}}^{U_j} f(U)dU \tag{7.13}$$

where $f(U)$ = the wind speed probability density function.

The total cumulative damage from all conditions is then expressed as:

$$D_T = \sum_x D_{T,x} = D_{T,O} + D_{T,P} + D_{T,T} \tag{7.14}$$

For many situations, most fatigue occurs during operation. When the *S–N* relation is of the form given in Equation (7.2), it can be rewritten for each ith cycle range bin as Equation (7.15):

$$\frac{1}{N_i} = \frac{\sigma_i^m}{A} \tag{7.15}$$

Equation (7.14) becomes:

$$D_T = N_T \sum_{j=1}^{N_{B,U}} F_{U_j} d_j \Big|_{x=0} = N_T \sum_{j=1}^{N_{B,U}} F_{U_j} \sum_i^{N_{B,r}} \frac{n_{i,j} \sigma_i^m}{A} \tag{7.16}$$

7.2.2.3 Damage Equivalent Loads

A common method used to quantify fatigue damage is by means of the "damage equivalent load" or "damage equivalent stress" abbreviated σ_{del}. This is a varying load (or stress) with a constant range that would result in the same damage as that from Equation (7.16) or Equation (7.25) (q.v.), given the same number of cycles. This method may be applied for the case of a single slope *S–N* curve as given in Equation (7.2). It is also convenient in that the fatigue constant *A* in that equation is not needed until the time comes to actually compute the damage. The method involves finding the damage contribution for the ith cycle range bin in the jth wind speed bin, $d_{del,i,j}$, according to Equation 7.17:

$$d_{del,i,j} = \frac{n_{i,j}}{N_{ref,tot}} \sigma_i^m \tag{7.17}$$

where $N_{ref,tot}$ is a reference number of total cycles in the wind speed interval.

The damage equivalent load is found by summing all the damage contributions over all the stress bins (indexed by *i*) and then raising the result to the $1/m$ power, as shown in Equation 7.18.

$$\sigma_{del,j} = \left[\sum_i \frac{n_{i,j}}{N_{ref,tot}} \sigma_i^m \right]^{1/m} \tag{7.18}$$

The total damage over the jth wind speed interval is found simply by Equation 7.19:

$$d_j = N_{ref,tot} \frac{\sigma_{del,j}^m}{A} \tag{7.19}$$

Note that in determining the damage equivalent load, the actual value of the reference total number of cycles ($N_{ref,tot}$) is arbitrary when it comes to finding the total damage. It only needs to be the case that the same value is used in both Equations 7.18 and 7.19.

7.2.3 Statistical Methods in Fatigue Estimation

The sum over the stress ranges in Equation (7.16) can be replaced by an integral, by observing that the cycle ranges are approximately Weibull distributed for each jth wind interval. First, the rainflow algorithm is run. Then, the mean $\mu_{c,j}$ and standard deviation, $\sigma_{c,j}$, of the cycle ranges are computed (for each interval *j*), and a Weibull distribution is fitted based on these values. The rest of the approach again involves using the damage per stress cycle *d* from Equation (7.2), so:

$$d = \frac{1}{N} = \frac{\sigma^m}{A} \tag{7.20}$$

Since the stress cycles are assumed to be Weibull distributed, the average damage over a wind interval j (see Appendix C) is:

$$\bar{d}_j = \int_0^\infty \frac{\sigma^m}{A} f(\sigma) d\sigma = \int_0^\infty \frac{\sigma^m}{A} \frac{k_j}{c_j} \left(\frac{\sigma}{c_j}\right)^{k_j-1} e^{-(\sigma/c_j)^{k_j}} d\sigma \qquad (7.21)$$

where $f(\sigma)$ is the Weibull probability density function and k_j and c_j are the shape and scale parameters for the wind interval j. As shown in Appendix C, there are direct relationships between those parameters and the mean $\mu_{c,j}$ and standard deviation $\sigma_{c,j}$ of the Weibull distribution (the subscript c is added to avoid confusion with σ as stress range). As long as $1 \leq k \leq 10$, which is usually the case, they are related as:

$$k_j \approx \left(\sigma_{c,j}/\mu_{c,j}\right)^{-1.086} \qquad (7.22)$$

$$c_j = \frac{\mu_{c,j}}{\Gamma\left(1 + 1/k_j\right)} \qquad (7.23)$$

Equation 7.21 can be integrated analytically (see Equation C.21 in Appendix C), and the total damage d_j in each wind speed interval is given by:

$$d_j = n_j \bar{d}_j = \frac{n_j}{A} c_j^m \Gamma\left(1 + \frac{m}{k_j}\right) \qquad (7.24)$$

where n_j is the number of cycles in the jth wind speed interval.

If we focus on operating conditions, then the total damage (Equation (7.16)) becomes:

$$D_T = N_T \sum_{j=1}^{N_{B,U}} F_{U_j} \frac{n_j}{A} c_j^m \Gamma\left(1 + \frac{m}{k_j}\right) \qquad (7.25)$$

There are two other points worth noting in this discussion. First, the Weibull parameters c_j and k_j here have no relation to the parameters c and k used in characterizing the wind resource, which were discussed in Chapter 2. Second, when a rainflow algorithm is used to count cycles and a statistical model is applied, the results may be improved if small cycle ranges are not included in the count. This is because most damage is due to higher range values but including the numerous small values in the count may change the Weibull parameters adversely. A better match with results from actual cycle counts may be obtained by filtering out the small cycles, such as by only considering cycles that cross the mean of the data.

7.2.3.1 Damage Equivalent Load Using the Weibull Distribution

When the stress (or moment) cycles are Weibull distributed, it may be shown that Equation (7.26) provides a good and very compact estimate of damage equivalent load in a given wind speed interval j.

$$\sigma_{eq,j} = c_j \left[\Gamma\left(1 + \frac{m}{k_j}\right)\right]^{1/m} \qquad (7.26)$$

where c_j and k_j are the Weibull parameters as above. In analogy with Equation 7.19, the damage over the interval may also be expressed as:

$$d_j = n_j \frac{\sigma_{eq,j}^m}{A} \qquad (7.27)$$

Example: In this example, the notion of wind speed bins is omitted. This is equivalent to setting: $N_{B,U} = 1, j = 1$, $F_{U_j} = 1$, $N_T = 1$ and omitting the index j. Consider a situation in which the NREL 5-MW turbine is operated for a

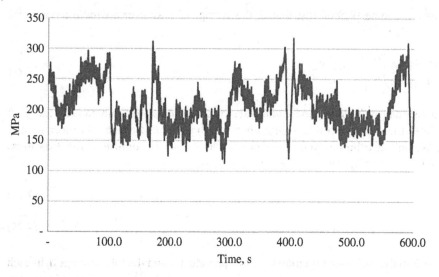

Figure 7.10 Example of maximum tower base stress

10-minute interval during which the average wind speed is just over 10 m/s. An *OpenFAST* simulation yielded a time series of tower base bending moments, which were converted to bending stresses as illustrated in Figure 7.10. (This conversion used the classic bending stress equation of $\sigma_B = Mc/I$ where M = moment, I = area moment of inertia of the tower base, and c = tower base radius. Axial loading was ignored.)

The cycles were counted with a rainflow cycle counter. The maximum range was $\sigma_{max} - \sigma_{min}$ = 210 MPa, therefore, using a bin width of 10, resulted in $N_{B,r}$ = 21 bins. The damage was estimated by using $A = 1.0 \times 10^{14}$ and $m = 3$. The results are illustrated in Figure 7.11. There was a total of 1337.5 cycles, most of which were quite small (between 0 and 10 MPa). Summing up the damage from each bin, the total damage was 4.649×10^{-7}. The damage equivalent load, using Equation 7.18 and the total cycles as reference, was 32.6 MPa.

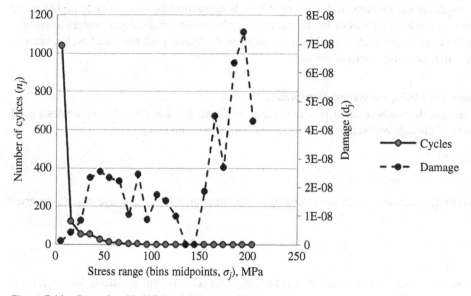

Figure 7.11 Example of Rainflow cycle counts and damage

The mean and standard deviation of the cycle ranges were $\mu_c = 10.93$ MPa and $\sigma_c = 16.51$ MPa, respectively. From these, the Weibull c and k values were calculated to be 7.85 MPa and 0.64. Using Equation (7.26), the damage equivalent load was 32.6 MPa and from Equation (7.27) the damage was 4.652×10^{-7}. In this example, the total damage and damage equivalent loads during the interval were virtually the same with either the method of Section 7.2.2.1 or the Weibull. In general, that will not necessarily be the case.

7.2.3.2 *S–N curve with Two Segments

In some cases, an S–N curve with two segments provides a better fit to test data. In that situation, we must use the incomplete gamma function to properly account for the two ranges. Here, the line at the higher number of cycles to failure is defined by $C = N\sigma^r$ and the transition between this line and the line defined by Equation (7.2) occurs at

$\sigma_Q = \exp\left(\dfrac{\ln(A/C)}{(m-r)}\right)$; see also Figure 7.1.

As will be recalled, the gamma function is defined as $\Gamma(x) = \int_0^\infty t^{x-1}e^{-t}dt$.

The incomplete gamma function can take one of two forms. The upper incomplete gamma function is defined as:

$$\Gamma(a,z) = \int_z^\infty t^{a-1}e^{-t}dt \tag{7.28}$$

The lower incomplete gamma function is:

$$\Gamma_0(a,z) = \int_0^z t^{a-1}e^{-t}dt \tag{7.29}$$

In these equations' formulation, the transition point z is given by:

$$z = \left(\frac{\sigma_Q}{c_i}\right)^{k_i} \tag{7.30}$$

It may be shown that the damage in the jth interval in the two-segment case is:

$$d_j = \frac{n_j c_j^m}{A}\Gamma\left(\frac{m}{k_j}+1,z\right) + \frac{n_j c_j^r}{C}\Gamma_0\left(\frac{r}{k_j}+1,z\right) \tag{7.31}$$

The total lifetime damage over all wind speeds in this case is given by:

$$D_T = N_T \sum_{j=1}^{N_{bins}} F_{U_j}\left[\frac{n_j c_j^m}{A}\Gamma\left(\frac{m}{k_j}+1,z\right) + \frac{n_j c_j^r}{C}\Gamma_0\left(\frac{r}{k_j}+1,z\right)\right] \tag{7.32}$$

7.3 Wind Turbine Materials

Many types of materials are used in wind turbines, as summarized in Table 7.1. Two of the most important of these are steel and composites. Composites typically comprise fiberglass and often carbon fibers with a matrix of polyester or epoxy. Wood has also been used in composite blades, but that is presently seldom the case. Other common materials include copper and concrete. The following provides an overview of some of the aspects of materials most relevant to wind turbine applications.

Table 7.1 Materials used in wind turbines

Subsystems or components	Material category	Material subcategory
Blades	Composites	Glass fibers, carbon fibers, wood laminates, polyester resins, epoxies
Hub	Steel	
Gearbox	Steel	Various alloys, lubricants
Generator	Steel, copper	Rare earth-based permanent magnets
Mechanical equipment	Steel	
Nacelle cover	Composites	Fiberglass
Tower	Steel	
Foundation	Steel, concrete	
Electrical and control system	Copper, silicon	

7.3.1 Steel

Steel is one of the most widely used materials in wind turbine fabrication. Steel is used for many structural components including the tower, hub, main frame, shafts, gears and gear cases, fasteners, and the reinforcing in concrete. Information on steel properties can be found in Budynas and Nisbett (2015), Sadegh and Worek (2018), and data sheets from steel suppliers.

7.3.2 Composites

Composites are described in more detail in this text than are most other materials, because it is assumed that they may be less familiar to many readers than are more traditional materials. They are also the primary material used in blade construction. Composites are materials comprising at least two dissimilar materials, most commonly fibers held in place by a binder matrix. Judicious choice of the fibers and binder allows tailoring of the composite properties to fit the application. Composites used in wind turbine applications include those based on fiberglass, carbon fiber, and wood. Binders include polyester, epoxy, and vinyl ester. The most common composite is fiberglass-reinforced plastic, known as GRP. In wind turbines, composites are most prominently used in blade manufacture, but they are also used in other parts of the machine, such as the nacelle cover. The main advantages of composites are: (1) ease of fabrication into the desired aerodynamic shape, (2) high strength, and (3) high stiffness-to-weight ratio. They are also corrosion resistant, are electrical insulators, are resistant to environmental degradation, and lend themselves to a variety of fabrication methods.

7.3.2.1 Glass Fibers

Glass fibers are formed by spinning glass into long threads. The most common glass fiber is known as E-glass, a calcium aluminosilicate glass. It is a low-cost material, with reasonably good tensile strength. Another common fiber is S-glass, a calcium-free aluminosilicate glass. It has approximately 25–30% higher tensile strength than E-glass, but it is significantly more expensive (more than twice as much). There is also a glass fiber, HiPER-tex, from Owens–Corning, which claims the strength of S-glass with a cost near E-glass.

Fibers are sometimes used directly but are most commonly first combined into other forms (known as "preforms"). Fibers may be woven or knitted into cloth, formed into continuous strand or chopped strand mat, or prepared as chopped fibers. Where high strength is required, unidirectional bundles of fibers known as "tows" are used. Some fiberglass preforms are illustrated in Figure 7.12. More information is presented in McGowan *et al.* (2007) and Mishnaevsky *et al.* (2017).

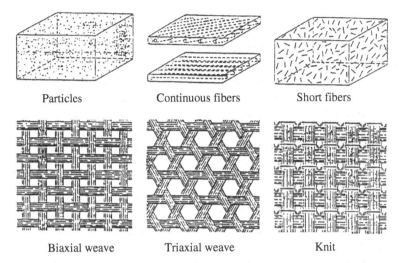

Particles   Continuous fibers   Short fibers

Biaxial weave   Triaxial weave   Knit

Figure 7.12 Fiberglass preforms (*Source:* National Research Council, 1991/National Academy Press)

7.3.2.2 Matrix (Binder)

There are three types of resins commonly used in matrices of composites. They are (1) unsaturated polyesters, (2) epoxies, and (3) vinyl esters. These resins all have the general property that they are used in the liquid form during the layup of the composite, but when they are cured they are solid. As solids, all of the resins tend to be somewhat brittle. The choice of resin affects the overall properties of the composite.

Polyesters have been used most frequently in the wind industry because they have a short cure time and low cost. Cure time is from a few hours to overnight at room temperature, but with the addition of an initiator, curing can be done at elevated temperatures in a few minutes. Shrinkage upon curing is relatively high, however.

Epoxies are stronger, have better chemical resistance, good adhesion, and low shrinkage upon curing, but they are also more expensive (approximately 50% more expensive than polyesters) and have a longer cure time than polyesters.

Vinyl esters are epoxy-based resins that have become more widely used over recent years. These resins have similar properties to epoxies but are somewhat lower in cost and have a shorter cure time. They have good environmental stability and are widely used in marine applications.

7.3.2.3 Carbon Fiber Reinforcing

Carbon fibers are more expensive than glass fibers (by approximately a factor of 5–10), but they are stronger and stiffer. One way to take advantage of carbon fibers, without paying the full cost, is to use some carbon fibers along with glass in the overall composite.

7.3.2.4 Wood–Epoxy Laminates

Wood has been used instead of synthetic fibers in some composites. In this case, the wood is preformed into laminates (sheets) rather than fibers, or fiber-based cloth. The use of wood together with an epoxy binder was developed for wind turbine applications based on previous experience from the high-performance boat-building industry. A technique known as the wood–epoxy saturation technique (WEST) is used in this process. These laminates have good fatigue characteristics (National Research Council, 1991). Wood epoxy laminates are not common blade materials at present but may be reconsidered in the future due to possible advantages in disposal of blades at the end of their design life.

7.3.2.5 Fatigue Damage in Composites

Fatigue damage occurs in composites as it does in many other materials, but it does not necessarily occur by the same mechanism. The following sequence of events is typical. First, the matrix cracks, then cracks begin to combine and there is debonding between the matrix and the fibers. Then there is debonding and separation (delamination) over a wider area. This is followed by breaking of the individual fibers, and finally by complete fracture.

The same type of analysis techniques that are used for other materials (explained in Section 7.1.1) are used for predicting fatigue in composites. That is, rainflow cycle counting is used to determine the range and mean of stress cycles, and Miner's Rule is used to calculate the damage from the cycles and the composite's *S–N* curve. *S–N* curves for composites are modeled by an equation that has a somewhat different shape than that used in metals, however:

$$\sigma = \sigma_u(1 - B \log N). \tag{7.33}$$

where σ is the cyclic stress range, σ_u is the ultimate strength, B is a constant, and N is the number of cycles to failure.

The parameter B is approximately equal to 0.1 for a wide range of E-glass composites when the reversing stress ratio $R = 0.1$. This is tension–tension fatigue. Life is reduced under fully reversed tension–compression fatigue ($R = -1$) and compression–compression fatigue.

Fatigue strength of glass fibers is only moderate. The ratio of maximum stress to static strength is 0.3 at 10 million cycles. Carbon fibers are much more fatigue resistant than are glass fibers: the ratio of maximum stress to static strength is 0.75 at 10 million cycles, two and half times that of glass. Fatigue life characteristics of E-glass, carbon fiber, and some other common fibers are shown in Figure 7.13.

Owing to the complexity of the failure mode of composites and the lack of complete test data on all composites of interest, it is, in practice, still difficult to predict fatigue life accurately. Equation 7.33 and as well as others have been developed for this purpose. It is common practice, however, to apply Equation (7.2) with a Wöhler slope of $m = 10$. In addition, summary information on the degradation of composites used in wind turbines may be found in McGowan *et al.* (2007). More details on the fatigue life of composites in wind turbine blades may be found in Hu *et al.* (2016).

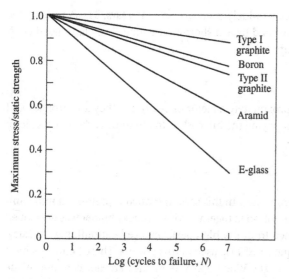

Figure 7.13 Fatigue life of some composite fibers (*Source: National Research Council, 1991/National Academy Press*)

7.3.3 Copper

Copper has excellent electrical conductivity, and for that reason, it is used in nearly all electrical equipment on a wind turbine, including the power conductors. The mechanical properties of copper are, in general, of much less interest than the conductivity. The weight, however, can be significant. A substantial part of the weight of the electrical generator is due to the copper windings, and the weight of the main power conductors may also be of importance. Information on copper relative to its use in electrical applications can be found in many sources, including Sadegh and Worek (2018).

7.3.4 Concrete

Reinforced concrete is frequently used for the foundations of wind turbines. It has sometimes been used for the construction of towers as well. Discussion of reinforced concrete, however, is outside the scope of this text.

7.3.5 Materials for Offshore Applications

Materials for offshore applications are an important topic. One of the most significant issues is corrosion, particularly of steel since it is used throughout the structure. Other metals may corrode as well. The saltwater environment can result in accelerated corrosion in general and can also reduce the fatigue life of the structure.

Components that are intermittently immersed in the water corrode more rapidly than those that are completely submerged. This problem is therefore most notable near the water surface, whose elevation will change with the tides and wave action. Spray due to combined wind and waves may reach well above sea level and hasten corrosion higher up as well. Junctions and contacts between different materials may also experience higher rates of corrosion due to electrolytic action. In this process, electrons move from the anode to the cathode, resulting in corrosion of the anode. An example is the contact of aluminum (an anode) with stainless steel (the cathode). The aluminum will deteriorate much more rapidly than it would if the stainless steel were not present. Note, however, that stainless steel will actually be protected by this process.

Materials are also an important topic for blades offshore; see Nijssen and de Winkel (2016).

7.3.5.1 Corrosion Prevention

To the extent possible, corrosion-resistant materials are used in offshore wind turbines in saltwater environments. These may include steels such as Cor-Ten or stainless steel. These are relatively expensive, so there is a trade-off between the effect and rate of the corrosion and the cost to reduce its impact. Measures may be taken to slow the rate of corrosion. These include painting exposed surfaces and cathodic protection.

Cathodic Protection of Steel Structures

Sometimes cathodic protection is used to reduce corrosion. It is common to use anodes that are more active than steel, such as aluminum or zinc. See, for example, Singh (2014). Some cathodic protection is passive, and some is active. In either case, a sacrificial metal is applied to the surface to be protected. The sacrificial metal is consumed during the process, rather than the desired material, i.e., the steel. With passive protection an electrical reaction takes place due to the proximity of dissimilar metals; this produces electrons which facilitate the protection, as described below. With active protection, an outside source of electricity is used to increase the rate of production of electrons. The process is summarized below. More details may be found in other sources, such as Singh (2014).

The corrosion of steel results in the formation of ferrous oxide first and then ferric oxide, i.e., rust. The process is illustrated in Figure 7.14, Equations (7.34)–(7.37) describe the reactions. Initially, ferrous ions go into solution at the anode, releasing some electrons which then travel to the anode:

$$2Fe \rightarrow 2Fe^{++} + 4e^{-} \tag{7.34}$$

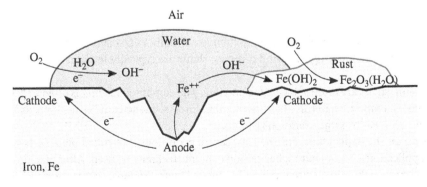

Figure 7.14 Schematic of corrosion of steel

Oxygen combines with water at the cathode and the electrons to form hydroxyl ions:

$$O_2 + 2H_2O + 4e^- \rightarrow 4OH^- \tag{7.35}$$

The overall reaction results in the formation of ferrous oxide at the cathode:

$$2Fe + O_2 + 2H_2O \rightarrow 2Fe(OH)_2 \tag{7.36}$$

In the presence of more oxygen from the air, the ferrous oxide is converted to hydrated ferric oxide, i.e., rust, so the overall reaction is then

$$4Fe + 3O_2 + 2H_2O \rightarrow 2Fe_2O_3(H_2O) \tag{7.37}$$

Consider now the presence of an aluminum anode as a cathodic protection system. The following reaction takes place:

$$4Al \rightarrow 4Al^{++} + 12e^- \tag{7.38}$$

Hydroxyl ions are created as summarized in Equation (7.35). In this process, the aluminum anode is consumed, but it does produce sufficient electrons to drastically slow the corrosion of the steel. With sufficient electrons relative to the amount of oxygen, corrosion is prevented.

7.3.5.2 Floating Systems – Lightweight Designs

Towers and substructures are normally constructed of steel. In some situations, particularly for floating offshore wind turbines, there are advantages to having a lighter structure. For such cases, composite material may be attractive. See Young *et al.* (2017) for some more information.

7.4 Machine Elements

Many of the principal components of a wind turbine are composed, at least partially, of machine elements. These are manufactured items that typically have a much wider applicability, and with which there has been a great deal of experience. Many of these elements are commercially available and are fabricated according to recognized standards. This section presents a brief overview of some machine elements that are often found in wind turbine applications. Specifically, those discussed here include shafts, couplings, bearings, gears, and dampers. Some other commonly used machine elements include clutches, brakes, springs, screws, and wire rope. For details on any of these, the reader should consult a text on machine design, such as Budynas and Nisbett (2015).

7.4.1 Shafts

Shafts are cylindrical elements designed to rotate. Their primary function is normally to transmit torque, and so they carry or are attached to gears, pulleys, or couplings. In wind turbines, shafts are typically found in the rotor, gearboxes, generators, and linkages.

In addition to being loaded in torsion, shafts are often subjected to bending. The combined loading is often time-varying, so fatigue is an important consideration. Shafts also have resonant natural frequencies at "critical speeds." Operation near such speeds is to be avoided, or large vibrations can occur.

Materials used for shafting depend on the application. For the least severe conditions, hot-rolled plain carbon steel is used. For greater strength applications, somewhat higher carbon content steel may be used. After machining, shafts are often heat treated to improve their yield strength and hardness. Under the most severe conditions, alloy steels are used for shafts.

7.4.2 Couplings

Couplings are elements used for connecting two shafts together for the purpose of transmitting torque between them. A typical use of couplings in wind turbines is the connection between the generator and the high-speed shaft of the gearbox.

Couplings consist of two major pieces, one of which is attached to each shaft. They are often kept from rotating relative to the shaft by a key or by interference fit. The two pieces are, in turn, connected to each other by bolts. In a solid coupling, the two halves are bolted together directly. In a typical flexible coupling, teeth are provided to carry the torque, and rubber bumpers are included between the teeth to minimize the effects of impact. Shafts to be connected should ideally be collinear, but flexible couplings are designed to allow some slight misalignment. An example of a solid coupling is shown in Figure 7.15.

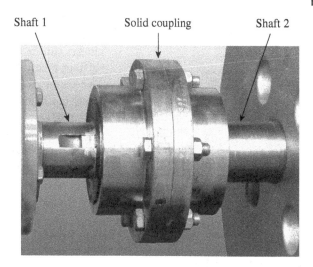

Figure 7.15 Typical solid coupling

It is worth noting that high-speed shaft couplings, such as between the gearbox and the generator, may be designed for both static misalignment and loading situations. Static misalignment may occur as a result of assembly. During loading, the structures and isolation elastomers will deflect and move. These couplings can also be used to provide damping through the use of suitable material or momentary slipping (during overtorque) by including a clutch.

7.4.3 Bearings

Bearings are used to reduce frictional resistance between two surfaces undergoing relative motion. In the most common situations, the motion in question is rotational. There are many bearing applications in wind turbines. They are found in main shaft mountings, gearboxes, generators, yaw systems, blade pitch systems, and teetering mechanisms.

Bearings come in a variety of forms, and they are made from a variety of materials. Rolling element bearings are used for many applications in a wind turbine. These include ball bearings, cylindrical roller bearings, spherical roller bearings, and tapered roller bearings. These bearings are typically made from steel. In special situations, ceramic rolling elements have been used in generator shaft bearings to provide electrical isolation. In multimegawatt turbines, journal bearings (those without rolling elements) are being increasingly used in lower-speed applications. These bearings may be plain or hydrostatically pressurized and they are coated or fabricated with special alloys that can tolerate operation with momentary loss of lubrication.

Rolling element bearings consist of four types of parts: an inner ring, an outer ring, the rolling elements, and the cage. The rolling elements run in curvilinear grooves machined specific to the arrangement into the rings. In most arrangements, the inner ring is fit to the supported shaft and the outer ring to a gear, the bearing support frame (pillow block), or the gearbox housing. The cage holds the rolling elements in place and keeps them spaced properly to evenly support the radial and axial forces. Bearings of this type are commonly used in wind turbines in applications such as gearboxes and main shaft support. A typical cylindrical rolling element bearing is shown in Figure 7.16.

Ball bearings are made in a range of types. They may be designed to take radial loads or axial thrust loads. Radial ball bearings can also withstand some axial thrust. For instance, blade pitch systems commonly utilize a preloaded

Figure 7.16 Cutaway view of typical cylindrical roller bearing (*Source:* The Timken Company/https://science. howstuffworks.com/transport/engines-equipment/bearing2.htm/printable/last accessed August 31, 2023)

double-row ball bearing arranged with four-point angular contact to accommodate rotor blade moments and forces, while the axial forces change sinusoidally during rotation.

Other types of bearings also have applications in wind turbine design. Two examples are the sleeve bearings and thrust bearings used in the teetering mechanism of some two-bladed wind turbines.

Generally speaking, the most important considerations in the design of a bearing are the load it experiences and the number of revolutions it is expected to survive. Detailed information on all types of bearings may be found in manufacturers' engineering manuals and online selection guides; see, for example, Timken (2023). Bearings used in wind turbine gearboxes are discussed in the IEC gearbox standard (IEC, 2012).

7.4.4 Gears

Gears are elements used in transferring torque from one shaft to another. There are numerous applications for gears in wind turbines. The most prominent of these is the drivetrain gearbox. Other examples include yaw drives, pitch linkages, and erection winches. Common types of gears include spur gears, helical gears, worm gears, and internal gears. Some common types of gears are illustrated in Figure 7.17. All gears have teeth. Spur gears have teeth whose axes are parallel to the rotational axis of the gear. The teeth in helical gears are inclined at an angle relative to the gear's rotational axis. Worm gears have helical teeth, which facilitate transfer of torque between shafts at right angles to each other. An internal gear is one which has teeth on the inside of an annulus.

Gears may be made from a wide variety of materials, but the most common material in wind turbine gears is low-carbon steel. It is easier to shape the complex geometries of gears with low or medium carbon steel than it is with high carbon steel. High contact durability and surface hardness in steel gear teeth is then obtained by carburizing or inductive hardening.

Gears may be grouped together in gear trains. Typical gear trains used in wind turbine applications are discussed in Section 7.5.

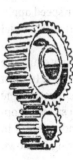

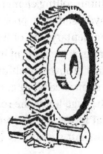

Spur Helical Herringbone

Figure 7.17 Common gear types

7.4.4.1 Gear Terminology

The most basic, and most common, gear is the spur gear. Figure 7.18 illustrates the most important characteristics. The pitch circle is the circumference of a hypothetical smooth gear (or one with infinitesimally small teeth). Two smooth gears would roll around each other with no sliding motion at the point of contact. The diameter of the pitch circle is known as the pitch diameter, d. With teeth of finite size, some of each tooth will extend beyond the pitch circle, some of it below the pitch circle. The face of the tooth is the location that meets the corresponding face of the mating gear tooth. The width of the face, b, is the dimension parallel to the gear's axis of rotation. The circular pitch, p, of the gear is the distance from one face on one tooth to the face on the same side of the next tooth around the pitch circle. Thus, $p = \pi d/N$ where N is the number of teeth.

Ideally, the thickness of a tooth, measured on the pitch circle, is exactly one-half of the circular pitch (i.e., the width of the teeth and the space between teeth are the same on the pitch circle). In practice, the teeth are cut a little smaller. Thus, when the teeth mesh, there is some free space between them. This is known as backlash. Excessive backlash can contribute to accelerated wear, so it is kept to a minimum. Backlash is shown in Figure 7.19.

Figure 7.18 Principal parts of a gear

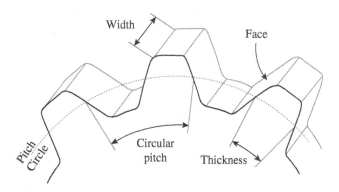

Figure 7.19 Backlash between gears

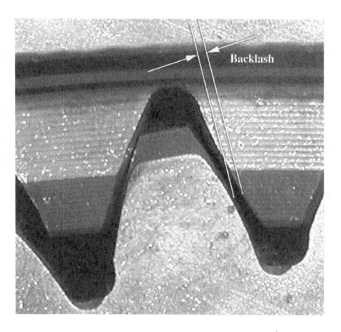

7.4.4.2 Gear Speed Relations

When two meshing gears, 1 and 2, are of different diameters, they will turn at different speeds. The relation between their rotational speeds n_1 and n_2 is inversely proportional to their pitch diameters d_1 and d_2 (or number of teeth). That is:

$$n_1/n_2 = d_2/d_1 \tag{7.39}$$

7.4.4.3 Gear Loading

Loading on a gear tooth is determined by the power being transmitted and the speed of the tooth. In terms of power, P, and pitch circle velocity, $V_{pitch} = \pi\, dn$, the tangential force, F_t, on a tooth is:

$$F_t = P/V_{pitch} \tag{7.40}$$

As the gear turns, individual teeth will be subjected to loading and unloading. At least one pair of teeth is always in contact, but, at any given time, more than one pair is likely to be in contact. For example, one pair may be unloading while another is taking a greater fraction of the load.

The bending stress, σ_b, on a gear tooth of width b and height h is calculated by application of the bending equation for a cantilevered beam:

$$\sigma_b = \frac{6M}{bh^2} \tag{7.41}$$

The moment, M, is based on a load F_b (which is closely related to F_t) applied at a distance L to the weakest point on the tooth. The result is:

$$\sigma_b = \frac{F_b}{b}\frac{6L}{h^2} \tag{7.42}$$

The factor $h^2/6L$ is a property of the size and shape of the gear and is frequently expressed in terms of the pitch diameter as the form factor (or Lewis factor), $y = h^2/6pL$. In this case, Equation (7.42) can be expressed as:

$$\sigma_b = \frac{F_b}{ypb} \tag{7.43}$$

The form factor is available in tables for commonly encountered numbers of teeth and pressure angles. Typical values for spur gears range from 0.056 for 10 teeth/gear to 0.170 for 300 teeth/gear.

7.4.4.4 Gear Dynamic Loading

Dynamic loading can induce stresses that are also significant to the design of a gear. Dynamic effects occur because of imperfections in the cutting of gears. The mass and spring constant of the contacting teeth and the loading and unloading of the teeth as the gear rotates are also contributing factors. Dynamic effects can result in increased bending stresses and can exacerbate deterioration and wear of the tooth surfaces.

The effective spring constant, k_g, of two meshing gear teeth can be important in the dynamic response (natural frequency) of a wind turbine drivetrain. Equation (7.44) gives an approximation to that spring constant. This equation accounts for gears (1 and 2) of different materials. Assuming moduli of elasticity E_1 and E_2, k_g is given by:

$$k_g = \frac{b}{9}\frac{E_1 E_2}{E_1 + E_2} \tag{7.44}$$

Dynamic effects and wear are very significant to the design of gears for wind turbine gearboxes. Information on gear tooth wear in general can be found in Budynas and Nisbett (2015). Gears in wind turbine gearboxes in particular are discussed in McNiff *et al.* (1990) and IEC 61400-4 (2012).

7.4.5 Dampers

Wind turbines are subject to dynamic events, with potentially adverse effects. These effects may be decreased by the use of appropriate dampers. There are at least three types of devices that act as dampers and that have been used on wind turbines: (1) fluid couplings, (2) hydraulic pumping circuits, and (3) linear viscous fluid dampers.

Fluid couplings are sometimes used between a gearbox and a generator to reduce torque fluctuations. They are used most commonly in conjunction with synchronous generators, which are inherently stiff. Hydraulic pumping circuits consist of a hydraulic pump and a closed hydraulic loop with a controllable orifice. Such circuits may be used for damping yaw motion. Linear viscous fluid dampers are essentially hydraulic cylinders with internal orifices. They may be used as teeter dampers on one- or two-bladed rotors.

More information on the general topic of hydraulics, on which many damper designs are based, can be found in Hunt and Vauchan (1996).

7.4.6 Fastening and Joining

Fastening and joining is an important concern in wind turbine design. The most important fasteners are bolts and screws. Their function is to hold parts together, but in a way that can be undone if necessary. Bolts and screws are tightened so as to exert a clamping force on the parts of interest. This is often accomplished by tightening the bolt to a specified torque level. There is a direct relation between the torque on a bolt and its elongation. Thus, a tightened bolt acts like a spring as it clamps. Fatigue can be an important factor in specifying bolts. The effects of fatigue can often be reduced by prestressing the bolts.

Bolts and screws on wind turbines are frequently subjected to vibration, and sometimes to shock. These tend to loosen them. To prevent loosening, a number of methods are used. These include washers, locknuts, lock wire, and chemical locking agents.

There is also a variety of other fasteners, and the use of ancillary items, such as washers and retainers, may be critical in many situations. Joining by means that are not readily disassembled, such as welding, riveting, soldering, or bonding with adhesives, is also frequently applied in wind turbine design. More details on fastening and joining may be found in Parmley (1997).

7.5 Principal Components of the Rotor Nacelle Assembly (RNA)

The principal component groups in the RNA are the rotor, the drivetrain, the main frame, and the yaw system. The rotor includes the blades, hub, and aerodynamic control surfaces. The drivetrain includes the gearbox (if any), the generator, the mechanical brake, and shafts and couplings connecting them. The yaw system components depend on whether the turbine uses free yaw or driven yaw. The type of yaw system is usually determined by the orientation of the rotor (upwind or downwind of the tower). Yaw system components include a yaw bearing and may include a yaw ring gear, one or more yaw drives (geared motor drive), yaw brake, and yaw damper. The main frame provides support for mounting the other components and a means for protecting them from the elements (the nacelle cover). See Figure 7.3 in Chapter 1 and Figure 7.3 in Chapter 9 for illustrations.

The following sections discuss each of the component groups. Unless specifically noted, it is assumed that the turbine has a horizontal axis.

7.5.1 Rotor

The rotor is unique among the component groups. Other types of machinery have drivetrains, brakes, and towers, but only wind turbines have rotors designed for the purpose of extracting significant power from the wind and converting it to rotary motion. As discussed elsewhere, wind turbine rotors are also unusual in that they must operate under conditions that include steady as well as periodically and stochastically varying loads. These varying loads occur over a very large number of cycles, so fatigue is a major consideration. The designer must strive to keep the cyclic stresses as low as possible and to use material that can withstand those stresses as long as possible. The rotor is also a source of cyclic loadings for the rest of the turbine, in particular the drivetrain.

Sections 7.5.1.1–7.5.1.3 focus on the topics of primary interest in the rotor: (1) blades, (2) aerodynamic control surfaces, and (3) hub.

7.5.1.1 Blades

The most fundamental components of the rotor are the blades. They are the devices that convert the force of the wind into the torque needed to generate useful power.

Design Considerations

There are many things to consider in designing blades, but most of them fall into one of two categories: (1) aerodynamic performance (Chapter 3) and (2) structural strength. Underlying all of these, of course, is the need to minimize the life cycle cost of energy, which means that not only should the cost of the turbine itself be kept low, but the operation and maintenance costs should be kept low as well. There are other important design considerations as well; they are summarized in the following list.

- Aerodynamic performance;
- Structural strength;
- Blade materials;
- Recyclability;
- Blade manufacturing;
- Worker health and safety;
- Noise reduction;
- Condition/health monitoring;
- Blade roots and hub attachment;
- Passive control or smart blade options;
- Costs.

The first nine of these considerations are discussed immediately below. Passive control options are discussed in Section 7.5.1.2. Cost is related to all the topics here, but it is outside the scope of this chapter. See Chapter 13, however, for more information on wind turbine costs.

There are a number of sources of information on wind turbine blades. See Veers *et al.* (2003) for a historical overview and the archived presentations of the Sandia National Laboratory blade workshops (Sandia, 2023).

Blade Shape Overview

The basic shape and dimensions of the blades are determined primarily by the overall topology of the turbine (see Chapter 8) and aerodynamic considerations, which were discussed in Chapter 3. Details in the shape, particularly near the root, are also influenced by structural considerations. For example, the planform of most real wind turbines differs significantly from the optimum shape, because the expense of blade manufacture would otherwise be too high. Figure 7.20 illustrates a typical modern blade planform. Material characteristics and available methods of fabrication are also particularly important in deciding upon the exact shape of the blades.

Figure 7.20 Typical modern blade planform (*Source:* Reproduced by permission of Siemens Gamesa, Inc.)

Aerodynamic Performance

The primary aerodynamic factors affecting blade design are:

- design rated power and rated wind speed;
- design tip speed ratio;
- solidity;
- airfoil;
- number of blades;
- rotor power control (stall or variable pitch);
- rotor orientation (upwind or downwind of the tower).

The overall size of the rotor swept area, and hence the length of the blades, is directly related to design rated power and rated wind speed. Other things being equal, it is usually advantageous to have a high design tip speed ratio. A high tip speed ratio results in a low solidity, which in turn results in less total blade area. This, in turn, should result in lighter, less expensive blades. The accompanying higher rotational speed is also of benefit to the rest of the drivetrain. On the other hand, high tip speed ratios result in more aerodynamic noise from the turbine. Because the blades are thinner, the flapwise stresses tend to be higher. Thinner blades are also more flexible. This can sometimes be an advantage, but thinner blades may also experience vibration problems, and extreme deflections can result in blade–tower impacts. The tip speed ratio also has a direct effect on the chord and twist distribution of the blade.

As design tip speed ratios increase, the selection of the proper airfoil becomes progressively more important. In particular, it is necessary to keep the lift-to-drag ratio high if the rotor is to have a high power coefficient. It is also of note that the lift coefficient will have an effect on the rotor solidity and hence the blade's chord: the higher the lift coefficient, the smaller the chord. In addition, the choice of airfoil is, to a significant extent, affected by the method of aerodynamic control used on the rotor. For example, an airfoil suitable for a pitch-regulated rotor may not be appropriate for a stall-controlled turbine. One concern is fouling (or "soiling"): certain airfoils, particularly on stall-regulated turbines, are quite susceptible to fouling (due, for example, to a build up of insects on the leading edge). This can result in a substantial decrease in power production. Selection of an airfoil can be done with the help of databases such as those developed by Selig (2023).

Wind turbine blades typically do not have just one airfoil shape along the entire length. See, for example, Figure 7.21. Sometimes the airfoils are all from the same family, but the relative thickness varies. Thicker cross-sections near the root provide greater strength. Thicker airfoils provide reduced aerodynamic performance but are often more resilient to soiling.

With present manufacturing techniques, it is generally advantageous to have as few blades as possible. This is primarily because of the fixed costs in fabricating the blades. In addition, when there are more blades (for a given solidity), they will be less stiff and may have higher stresses at the roots. At the present time, all commercial wind turbines have either two or three blades, and that will be assumed to be the case here as well. Two-bladed wind turbines have historically had a lower solidity than three-bladed machines. This keeps the blade cost low, which is one of the presumed advantages of two blades over three blades.

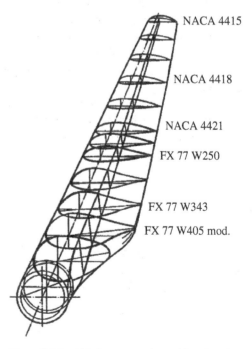

Figure 7.21 Airfoil cross-sections with radius (from Gasch, 1996) (*Source:* Reproduced by permission of B. G. Teubner GmbH)

The method of power control (stall or variable pitch) has a significant effect on the design of the blades, particularly in regard to the choice of the airfoil. A stall-controlled turbine depends on the loss of lift that accompanies stall to reduce the power output in high winds. It is highly desirable that the blades have good stall characteristics. They should stall gradually as the wind speed increases, and they should be relatively free of transient effects, such as are caused by dynamic stall. In pitch-controlled turbines, stall characteristics are generally much less important. On the other hand, it is important to know that the blades perform acceptably when being pitched in high winds. It is also worth noting that blades can be pitched toward either feather (decreasing angle of attack) or stall (increasing angle of attack). Pitching to feather is more common, however.

The rotor orientation with respect to the tower has some effect on the geometry of the blades, but mostly in a secondary manner related to the preconing of the blades. This preconing is a tilting of the blades away from a plane of rotation as defined by the blade roots. Many downwind turbines have historically operated with free yaw. The blades then must be coned away from the plane of rotation to enable the rotors to track the wind and maintain some yaw stability. Some upwind rotors also have preconed blades. In this case, the preconing is in the upwind direction, and the purpose is to keep the blades from hitting the tower.

Blade design often involves a number of iterations to properly account for both aerodynamic and structural requirements. In each iteration, a tentative design is developed and then analyzed. One approach to expedite this process, known as an inverse design method, was developed by Selig and Tangler (1995). It involves the use of a computer code (PROPID) to propose designs that will meet certain requirements. For example, it is possible to specify overall dimensions, an airfoil series, a peak power, and blade lift coefficient along the span, and then use the code to determine the chord and twist distribution of the blade. More comprehensive analysis that includes structural details is available in codes such as NREL's *OpenFAST* (see Chapter 8).

Structural Strength

The exterior shape of wind turbine blades is based on aerodynamics, but the interior architecture is determined primarily by considerations of strength. The blade structure must be sufficiently strong both to withstand extreme loads and to survive many fatigue cycles. The blade must also not deflect more than a specified amount when under load. Figure 7.22 illustrates the cross-section of a typical blade. As shown in the figure, an interior spar provides the interior strength, primarily via forward and aft shear webs. Spar caps transfer loads from the outer skin to the shear webs.

In order to provide sufficient strength, particularly near the root of the blade, inboard sections are relatively thick. As blades have become larger and larger, chords near the root have become larger as well. This has caused some problems, including during transportation. One solution has been to develop "flatback" airfoils, such as the one shown in Figure 7.23.

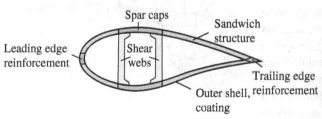

Figure 7.22 Typical wind turbine blade structural architecture (*Source:* Right: Roach *et al.*, 2014/Sandia National Laboratories)

Blade Materials

Historically, wind turbine blades were made from wood, sometimes covered with cloth. Until the middle of the 20th century,
blades for larger wind turbines were made from steel. Examples
include both the Smith–Putnam turbine (1940s) and the Danish
Gedser turbine (1950s).

Since the 1970s, most blades for horizontal-axis wind turbines
have been made from composites. The most common composites
consist of fiberglass in a polyester resin, but vinyl ester and epoxy-
based laminates have also been used. More recently, carbon fibers
have become widely used in blade construction, not necessarily as
a replacement for fiberglass, but to augment it. Typical composites used for wind turbine blades were described in more detail in Section 7.2.3.

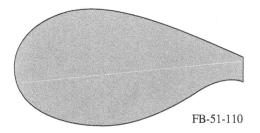

FB-51-110

Figure 7.23 Example of a flatback airfoil (*Source:
Veers et al.*, 2003/John Wiley & Sons)

Some wind turbines have used aluminum for blade construction. Aluminum has been a popular choice for vertical-axis wind turbines. VAWT blades normally have a constant chord with no twist, so lend themselves to formation by aluminum pultrusion (see blade manufacturing discussion below.) A few horizontal-axis wind turbines
have used aluminum blades, but aluminum is not commonly used for HAWTs at this time.

Recyclability

As more and more turbines are built, and as older turbines are replaced by newer ones, the issue of blade disposal
and its impact on the environment, and, of securing raw materials for new blades will become more significant.
One way to deal with both issues, at least partly, is by making the blades of recyclable materials. This is a relatively
new topic of study, but it is expected to become more significant over time. For example, see Jensen and Skelton
(2018) and Liu et al. (2022).

Blade Manufacturing

There are a number of options for manufacturing composite wind turbine blades. The most important of these are
summarized as follows:

Wet Lay-up Wet lay-up of fiberglass involves placing multiple layers (plies) of fiberglass cloth into a mold. Each
layer is soaked in a binder (polyester or epoxy resin plus a hardener). The lay-up is typically done by hand, with the
resin being forced into the glass with rollers or squeegees. In this method, there are two parts to mold: one for the
upper surface and one for the lower surface. When the two halves of the blade are completed, they are removed
from the molds. They are then bonded together, with the spar (separately fabricated) in between. Hand lay-up is
very labor intensive, and it is hard to ensure consistency and avoid defects in the product. Another difficulty with
this approach is that fumes are released into the air, making the process hazardous to the workers (Cairns and
Skramstad, 2000).

Pre-preg In the pre-preg method, the fiberglass cloth has been previously impregnated with a resin which
remains firm at room temperature. After all the cloth has been placed in the mold, the entire mold is heated
so that the resin will flow and then harden permanently. The heating is done in a low-temperature oven or an
autoclave. This method has the advantage that the resin-to-glass ratio is consistent, and it is easier to achieve good
results. On the other hand, the material cost is higher, and it is both difficult and expensive to provide heating for
large blades.

Resin Infusion In resin infusion, dry fiberglass cloth is laid into the mold and then resin is impelled by some
means to flow into the glass. One method is known as vacuum-assisted resin transfer molding (VARTM). In this

Figure 7.24 Laying fiberglass cloth into blade molds (*Source:* Reproduced by permission from TPI)

method, air can be removed from one side of the mold via a vacuum pump. A vacuum bag is placed over the cloth and arranged in such a way that resin from an external reservoir will be pushed through the cloth by atmospheric pressure into the space evacuated by the vacuum pump. This method avoids many of the quality issues associated with wet lay-up. An additional advantage is that toxic byproducts are not released into the air. A variant of the VARTM method is the SCRIMP™ method, which has been developed by TPI, Inc.

An illustration of one step of the process of fabricating a blade using this method is shown in Figure 7.24. Here, the laying of the fiberglass cloth into the mold is underway.

Compression Molding In compression molding, the fiberglass and resin are placed in a two-part mold which is then closed. Heat and pressure are applied until the product is ready. This method has a number of advantages, but so far has not proven to be adaptable to large blades. It may be useful for smaller blades, however.

Pultrusion Pultrusion is a fabrication method in which material is pulled through a die to form an object of a specific cross-sectional shape but indefinite length. The object is cut to the desired length afterward. The material can be either metal or a composite. If a composite is used, provision is also made for adding resin and heating the mold to assist in curing. Pultrusion is well suited to making blades of constant cross-section such as blades for vertical axis wind turbines. So far, however, pultrusion has not proven attractive for twisted and tapered blades of horizontal axis turbines.

Worker Health and Safety
As noted previously, some methods of blade fabrication, particularly wet lay-up, can result in toxic gases being released into the air. These gases can be hazardous for the workers who are building the blades. It may therefore be expected that there will be a continuing trend to replace wet lay-up with other fabrication methods, such as VARTM.

Noise Reduction

A wind turbine's rotor can be a source of unwanted sound (i.e., noise), as will be discussed in Chapter 14. One way to reduce the noise is through the selection of suitable airfoils. Another way is to design the rotor to operate best at a relatively low tip speed ratio.

Condition/Health Monitoring

Wind turbine blades can suffer various types of damage over time. Ideally, they are inspected periodically. In practice, however, it is difficult to do a thorough inspection, especially of the blades' interior. One way to deal with this issue is via condition or health monitoring. In this technique, sensors are embedded in the blades during manufacture. The sensors are afterward integrated into a monitoring system, which can be used to alert operators that repairs are necessary. See Hyers *et al.* (2006) and Do and Söffker (2021) as well as Chapter 15 for more information on this topic.

Blade Roots and Hub Attachment

Blades are attached to the hub via the blade root, which is the end of the blade nearest to the hub. The root experiences the highest loads, and it is also the location that must provide for the connection to the hub. To reduce stresses, the root is generally made as thick as is practical in the flapwise direction. The connection between the root and the hub has often proven to be difficult. This is largely due to dissimilarities in material properties and stiffnesses between the blades, the hub, and the fasteners. Highly variable loads also contribute to the problem.

One type of root is known as the Hütter design, named after its inventor, the German wind energy pioneer Ulrich Hütter. In this method, long fiberglass strands are bonded into the lower part of the blade. Circular metal flanges are provided at the base of the blade, and attached to these flanges are circular hollow spacers. The fiberglass strands are wrapped around the spacers and brought back into the rest of the blade. Resin keeps all the strands and the flanges in place. The blades are eventually attached to the hub via bolts through the flanges and spacers. As described here, this root design is most applicable to fixed-pitch rotors. A variant of this design was widely used in the 1970s and 1980s but has largely been supplanted by other methods, largely because of fatigue problems.

The most common method of attachment today is to use studs or threaded inserts bonded directly into the blades. The use of studs was originally developed in conjunction with wood–epoxy blades, but it has proven applicable to fiberglass blades as well. A blade root with studs is illustrated in Figure 7.25.

Figure 7.25 Close-up of blade root with studs (*Source:* J. F. Manwell (Author))

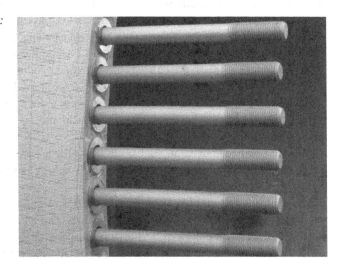

Fixed-pitch wind turbine blades normally are attached directly to the hub with bolts or studs which are aligned radially, and perpendicular to the bottom of the blade root. These fasteners must withstand all the loads arising from the blades.

The attachment of a variable-pitch blade root has some differences from that of a fixed-pitch blade. In particular, the root–hub connection must incorporate bearings so that the blade can be rotated (see Figure 7.29). These bearings must be able to withstand the bending moments and shear forces imposed by the rest of the blade. In addition, these, or other, bearings must take the radial load resulting from the rotor's rotation.

The blade attachment methods discussed above are most common on medium size or larger turbines. Blades on small turbines normally employ different attachment techniques. In one method, the root is thickened, and bolts are placed through the root and a matching part on the hub. The bolts are perpendicular to both the long axis and chord of the blade.

Blade Properties

Properties of the overall blade, such as total weight, stiffness, and mass distributions, and moments of inertia are needed in the structural analysis of the rotor. Important concerns are the blade's strength, its tendency to deflect under load, its natural vibration frequencies, and its resistance to fatigue. These were all discussed previously. Some of the blade properties can be difficult to obtain due to the complex geometry of the blade, which varies from root to tip. The normal method used is to divide the blade into sections, in a manner similar to that for aerodynamic analysis. Properties for each section are found, based on the dimensions and material distribution, and then combined to find values for the entire blade. Sandia National Laboratory has also developed software known as NuMAD (Berg and Resor, 2012) to facilitate this process.

7.5.1.2 Aerodynamic Control Options

There are a number of ways to modify the aerodynamic performance of a blade. These include pitch control, control surfaces, and passive control, as discussed below. More details on wind turbine control in general are provided in Chapter 9.

Control Surfaces

An aerodynamic control surface is a device that can be moved to change the aerodynamic characteristics of a rotor. A variety of types of aerodynamic control surfaces can be incorporated into wind turbine blades. They must be designed in conjunction with the rest of the rotor, especially the blades. The selection of aerodynamic control surfaces is strongly related to the overall control philosophy. Stall-regulated wind turbines usually incorporate some type of aerodynamic brake. These can be tip brakes, flaps, or spoilers.

Turbines that are not stall-controlled usually have much more extensive aerodynamic control. In conventional pitch-controlled turbines, the entire blade can rotate about its long axis. Thus, the entire blade forms a control surface. Some turbine designs use partial span pitch control. In this case, the inner part of the blade is fixed relative to the hub. The outer part is mounted on bearings and can be rotated about the radial axis of the blade. The advantage of partial span pitch control is that the pitching mechanism need not be as massive as it must be for full span pitch control.

Another type of aerodynamic control surface is the aileron. This is a movable flap, located at the trailing edge of the blade. The aileron may be approximately 1/3 as long as the entire blade, and extend approximately 1/4 of the way toward the leading edge.

Any control surface is used in conjunction with a mechanism that allows or causes it to move as required. This mechanism may include bearings, hinges, springs, and linkages. Aerodynamic brakes often include electromagnets to hold the surface in place during normal operation, but to release the surface when required. Mechanisms for active pitch or aileron control include motors for operating them.

Figure 7.26 Illustration of swept blade

Passive Control

One way to passively limit the loads on a wind turbine blade is via pitch-twist coupling. In this concept, thrust forces create a moment about the pitch axis of the blade, causing it to twist. The twisting changes the pitch angle along the blade, decreasing the lift force and thus limiting the load. Blades that employ pitch-twist coupling typically incorporate carbon fibers, laid in asymmetrically with respect to the glass. They sometimes have a swept platform, such as the one shown in Figure 7.26.

Another means of passive control utilizing the blade architecture is pre-bending. In this concept, the blade is fabricated with a built-in bend. The prebend is such that the unloaded tip will project upwind from the plane of rotation when the blade is affixed to the hub (on an upwind rotor). Under load, the tip will bend back into the plane of rotation, rather than bending downwind of the rotor plane and possibly into the tower. The disadvantage of this concept is that the molds themselves must be larger than they otherwise would be so as to incorporate the prebend. Shipping large bent blades can also be problematic.

Smart Blades

The term "smart blades" refers to the aerodynamic control of wind turbine blades via distributed and embedded intelligence and actuators. Smart blades may incorporate flaps, micro-tabs, boundary layer suction or blowing, piezo-electric elements, and shape memory alloys. See Alejandro Franco *et al.* (2015) and Barlas and van Kuik (2010) for a discussion of this topic.

7.5.1.3 Hub
Function

The hub of the wind turbine is the component that connects the blades to the main shaft and ultimately to the rest of the drivetrain. The hub transmits and must withstand all the loads generated by the blades. Hubs are generally made of steel, either welded or cast. Details of hubs differ considerably depending on the overall design philosophy of the turbine.

Types

There are three basic types of hub design that have been applied in horizontal axis wind turbines: (1) rigid hubs, (2) teetering hubs, and (3) hubs for hinged blades. A rigid hub, as the name implies, has all major parts fixed relative to the main shaft. This is the most common design, and it is nearly universal for machines with three blades. A teetering hub allows relative motion between the part that connects to the blades and that which connects to the main shaft. Like a child's teeter-totter (seesaw), when one blade moves one way, the other blade moves the other way. Teetering hubs are commonly used for two- and one-bladed wind turbines. Hubs for hinged blades allow independent flapping motion relative to the plane of rotation. Such hubs are not presently used on any commercial wind turbines, but they have been employed on some historically important turbines (Smith–Putnam) and have recently received renewed attention. Some of the common types of hubs are illustrated in Figure 7.27.

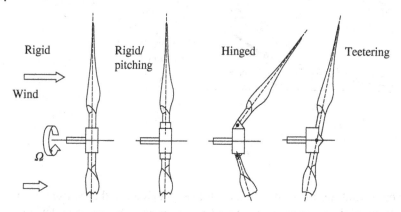

Figure 7.27 Hub options (Gasch, 1996) (*Source:* Reproduced by permission of B. G. Teubner GmbH)

Rigid Hub As indicated above, a rigid hub is designed to keep all major parts in a fixed position relative to the main shaft. The term rigid hub does, however, include those hubs in which the blade pitch can be varied, but in which no other blade motion is allowed.

The main body of a rigid hub is a casting or weldment to which the blades are attached, and which can be fastened to the main shaft. If the blades are to be preconed relative to the main shaft, provision for that is made in the hub geometry. A rigid hub must be strong enough to withstand all the loads that can arise from the aerodynamic forces on the blades, as well as dynamically induced loads, such as those due to rotation and yawing. These loads are discussed in Chapters 4 and 8. A typical rigid hub for a pitch-controlled turbine is shown in Figure 7.28.

A hub on a pitch-controlled turbine must provide bearings at the blade roots, a means for securing the blades against all motion except pitching, and a pitching mechanism. Pitching can either be collective (all the blades move together) or individual. A collective pitching mechanism may use a pitch rod passing through the main shaft, together with a linkage on the hub. This linkage is, in turn, connected to the roots of the blades. The pitch rod is driven by a motor mounted on the main (nonrotating) part of the turbine. An alternative method is to mount electric gear motors on the hub and have them pitch the blades directly. This approach is used with individual pitch control. In this case, power still needs to be provided to the motors. This can be done via slip rings or a rotary transformer. Regardless of the design philosophy of the pitching mechanism, it should be fail-safe. In the event

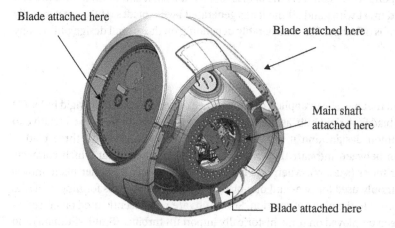

Figure 7.28 Typical wind turbine hub (*Source:* Reproduced by permission from Siemens Gamesa, Inc.)

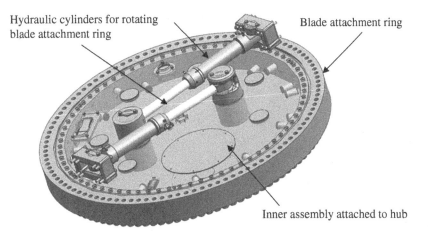

Hydraulic cylinders for rotating blade attachment ring

Blade attachment ring

Inner assembly attached to hub

Figure 7.29 Blade pitching mechanism (*Source:* Reproduced by permission of Siemens Gamesa, Inc.)

of a power outage, for example, the blades should pitch themselves into a no-power position. An example of a blade pitching mechanism is shown in Figure 7.29.

The hub must be attached to the main shaft in such a way that it will not slip or spin on the shaft. Smaller turbines frequently employ keys, with keyways on the shaft and the hub. The shaft is also threaded, and the mating surfaces are machined (and perhaps tapered) for a tight fit. The hub can then be held on with a nut. Such a method of attachment is less desirable on a larger machine, however. First of all, a keyway weakens the shaft, and machining threads on a large shaft can be inconvenient. A common method of hub attachment involves the use of a permanent flange on the end of the main shaft. The hub is attached to the flange by bolts. An example of this can be seen in Figure 7.30. The other end of the main shaft may be connected to the low-speed shaft of the gearbox – this is shown in Figure 7.33. The main shaft itself is discussed in more detail below.

Teetering Hub   Teetering hubs are used on nearly all two-bladed wind turbines. This is because a teetering hub can reduce loads due to aerodynamic imbalances or loads due to dynamic effects from rotation of the rotor or yawing of the turbine (three-bladed turbines are less affected due to an overall averaging of these effects over the blades).

Figure 7.30   Example of main shaft with bolt flange for hub attachment (*Source:* Reproduced by permission of Siemens-Gamesa, Inc.)

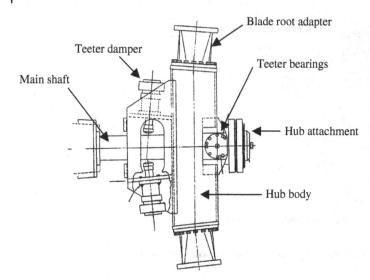

Figure 7.31 Example of a teetering hub (ESI-80)

Teetering hubs are considerably more complex than are rigid hubs. They consist of at least two main parts (the main hub body and a pair of trunnion pins) as well as bearings and dampers. A typical teetering hub is illustrated in Figure 7.31. The main hub body is a steel weldment. At either end are the attachment points for the blades. This hub has blades that are preconed downwind from the plane of rotation, so the planes of attachment are not perpendicular to the long axis of the hub. On either side of the hub body are teeter bearings. They are held in place by removable bearing blocks. The arrangement is such that the bearings lie on an axis perpendicular to the main shaft, and equidistant from the blade tips. The teeter bearings carry all of the loads passing between the hub body and the trunnion pin. The trunnion pin is connected rigidly to the main shaft.

Most teetering hubs have been built for fixed-pitch turbines, but they can be used on variable-pitch turbines as well. Design of the pitching system is more complex since the pitching mechanism is on the part of the hub which moves relative to the main shaft. An example of a pitching/teetering hub design is described by Bergami *et al.* (2014).

Teetering hubs require two types of bearing. One type is a cylindrical, radially loaded bearing; the other is a thrust bearing. There is one bearing of each type on each pin. The cylindrical bearings carry the full load when the pin axis is horizontal. When the pin axis is not horizontal, there is an axial component due primarily to the weight of the rotor. One of the thrust bearings will carry that part of the load. Teeter bearings are typically made of special-purpose composites.

During normal operation, a teetering hub will move only a few degrees forwards and backwards. During high winds, starts and stops, or high yaw rates, greater teeter excursions can occur. To prevent impact damage under these conditions, teeter dampers and compliant stops are provided. In the hub shown in Figure 7.31 (which has a maximum allowed range of ±7.0°), the dampers are on the side of the hub opposite the bearings.

The options for attaching a teetering hub to the main shaft are the same as for rigid hubs.

Hinged Hub   A hinged hub is, in some ways, a cross between a rigid hub and a teetering hub. It is basically a rigid hub with "hinges" for the blades. The hinge assembly adds some complexity, however. As with a teetering hub, there must be bearings at the hinges. Teetering hubs have the advantage that the two blades tend to balance each other, so lack of centrifugal stiffening during low rpm operation is not a major problem. There is no such counter-balancing on a hinged blade, however, so some mechanism must be provided to keep the blades from flopping over during low rotational speed. This could include springs. It would almost certainly include dampers as well.

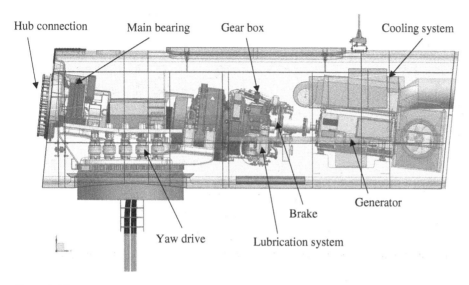

Figure 7.32 Typical drivetrain and associated components (*Source:* Reproduced by permission of Siemens Gamesa, Inc.)

7.5.2 Drivetrain

A complete wind turbine drivetrain consists of all the rotating components: rotor, main shaft, couplings, gearbox, brakes, and generator. With the exception of the rotor components, which were considered above, all of these are discussed in the following sections. In addition, there will be lubrication and cooling systems associated with the drivetrain. These systems will include pumps, filters, piping, and monitoring. Figure 7.32 illustrates a typical drivetrain.

7.5.2.1 Main Shaft

Every wind turbine has a main shaft, sometimes referred to as the low-speed or rotor shaft. The main shaft is the principal rotating element, providing for the transfer of torque from the rotor to the rest of the drivetrain. It also supports the weight of the rotor. The main shaft is supported, in turn, by bearings, which transfer reaction loads to the main frame of the turbine. The type and arrangement of the main bearings and gearbox support are key elements of the drivetrain to ensure that gears only transfer torque. There are several options currently in use as shown in Figure 7.33.

Three-point Suspension

In this arrangement, a single main bearing (usually dual rolling elements in the same housing) supports the main shaft in a separate housing close to the hub. The gearbox is mounted near its center through trunnion mounts integral to the housing bolted down to either side of the main frame with elastomer for isolation of noise and vibration. Sometimes the support is a circular frame mounted to the housing along the ring gear housing (also with elastomers). Internal gearbox bearings, such as the first stage planet carrier support bearings, transfer some of the reaction to the non-torque rotor loads and must be designed to tolerate that. The generator is normally foot mounted directly to the main frame (also with isolation) and connected to the gearbox high-speed shaft with a coupling that can tolerate the motions.

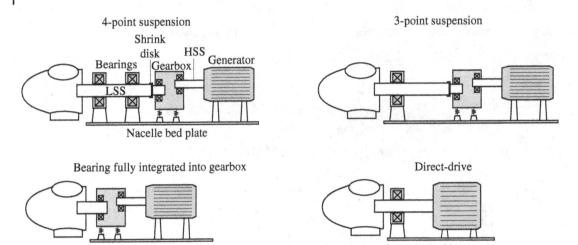

Figure 7.33 Main drivetrain options and transfer of loads to the nacelle bed plate. LSS: low-speed shaft. HSS: high-speed shaft

Four-point Suspension

In this arrangement, two separate main bearings mounted to the main frame support the main shaft with enough spacing to allow for effective load sharing. Depending on the types of bearings used, the load share may be in the form of one taking thrust and a small share of rotor moments. The bearings can be sized accordingly and independent of the gearbox design. The gearbox is then subject to only torque and can be attached to the end of the main shaft and suspended to similar trunnion clamps as three points.

Integrated Main Bearings

Depending on the design of the gearbox, the shaft and the bearings may be integrated into the gearbox housing. The housing in this case would need to transfer non-torque loads to the mainframe or be integrated in some form with the mainframe. Turbines with direct drive generators (Section 7.5.2.4) will also have integrated bearings.

Main Shaft Considerations

The main shaft is sized in accordance with methods described in Chapter 8, taking into account the combined loads of torque and bending. The main shafts are normally made of steel. Methods of connecting the main shaft to the rotor were discussed in Section 7.5.1.

7.5.2.2 Couplings

Function

Couplings, as discussed in Section 7.4.2, are used to connect shafts together. There are two main locations where large couplings are likely to be used in wind turbines: (1) between the main shaft and the gearbox and (2) between the gearbox output shaft and the generator. In addition, there may be spline couplings within gearboxes, such as between planetary and parallel stages.

The primary function of the coupling is to transmit torque between two shafts, but it may have another function as well. Sometimes it is advantageous to dampen torque fluctuations in the main shaft before the power is converted to electricity. A coupling of appropriate design can serve this role. A fluid coupling may be used for this purpose. Couplings were described in Section 7.4.2.

7.5.2.3 Gearbox

Function

With the exception of turbines with direct drive generators (see Section 7.5.2.5), drivetrains include a gearbox to increase the speed of the input shaft to the generator. An increase in speed is needed because wind turbine rotors, and hence main shafts, turn at a much lower speed than is required by most electrical generators. Small wind turbine rotors turn at speeds on the order of a few hundred rpm. Larger wind turbines turn more slowly. In any case, most generators turn considerably faster than the rotor does. Their speed, when directly connected to the grid, depends on the number of poles they have; see Chapter 5. Generator speed in variable speed wind turbines is to some degree independent of the number of poles, but except in direct drive applications, will still be higher than the rotor rpm.

Some gearboxes also perform functions other than increasing speed, such as supporting the main shaft bearings. These are secondary to the basic purpose of the gearbox, however.

The gearbox is one of the heaviest and most expensive components in a wind turbine. Gearboxes are normally designed and supplied by a different manufacturer to the one actually constructing the wind turbine. Since the operating conditions experienced by a wind turbine gearbox are significantly different than those in most other applications, it is imperative that the turbine designer understand gearboxes, and that the gearbox designer understand wind turbines. Experience has shown that underdesigned gearboxes are a major source of wind turbine unreliability due to operational problems.

Types

All gearboxes have some similarities: they consist of torque transmitting parts, such as shafts and gears, machine elements such as bearings and seals, and structural components, such as the case. In most cases, there is a single input shaft and a single output shaft, although in some cases there have been multiple output shafts connected to multiple generators. Beyond that there are two basic types of gear arrangements used in wind turbine applications: (1) parallel-shaft and (2) planetary arrangements. The ratio of the speed of the output (high-speed) shaft to the input (low-speed) shaft of a wind turbine gearbox is known as the gear ratio or speed-up ratio For example, the gear ratio of the NREL 5-MW is 97:1.

In parallel-shaft gearboxes, gears are carried on two or more parallel shafts. These shafts are supported by bearings mounted in the case. In a single-stage gearbox, there are two shafts: a low-speed shaft and a high-speed shaft. Both of these shafts pass out through the case. One of them is connected to the main shaft or rotor and the other to the generator. There are also two gears, one on each shaft. The two gears are of different sizes, with the one on the low-speed shaft being the larger of the two. The smaller of two gears meshing together is commonly referred to as a pinion. The ratio of the pitch diameter of the gears is inversely proportional to the ratio of the rotational speeds (as described in Section 7.4).

There is a practical limit to the size ratio of the two gears that can be used in a single-stage parallel-shaft gearbox. For this reason, gearboxes with large speed-up ratios use multiple stages of shafts and gears. These gears then constitute a gear train. A two-stage gearbox, for example, would have three shafts: an input (low-speed) shaft, an output (high-speed) shaft, and an intermediate shaft. There would be gears on the intermediate shaft, the smaller one driven by the low-speed shaft. The larger of these gears would drive the gear on the high-speed shaft. A typical parallel-shaft gearbox is shown in Figure 7.34.

Planetary gearboxes have a number of significant differences from parallel-shaft gearboxes. Most notably, the input and output shafts are coaxial. In addition, there are multiple pairs of gear teeth meshing at any time, so the loads on each gear are reduced. This makes planetary gearboxes relatively light and compact. A typical planetary gearbox is illustrated in Figure 7.35.

In typical planetary gearboxes, a low-speed shaft, supported by bearings in the case, is rigidly connected to a planet carrier. The carrier holds three (or more) identical small gears, known as planets. These gears are mounted on short shafts and bearings and are free to turn. These planets mesh with a large-diameter internal or ring gear

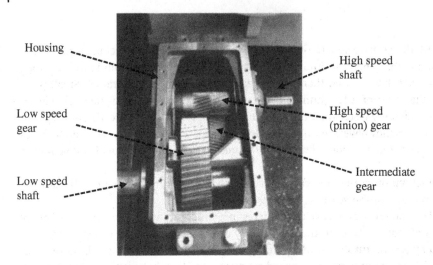

Figure 7.34 Parallel-shaft gearbox (*Source:* Hau, 1996/Reproduced with permission from Springer Nature)

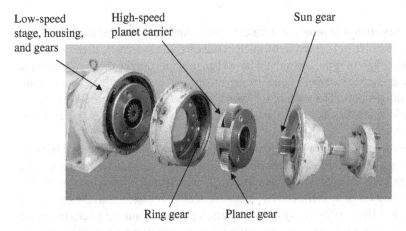

Figure 7.35 Exploded view of two-stage planetary gearbox

and a small-diameter sun gear. When the low-speed shaft and carrier rotate, meshing of the planets in the ring gear forces the planets to rotate, and to do so at a speed higher than the speed of the carrier. The meshing of the planets with the sun gear causes it to rotate as well. The sun gear then drives the high-speed shaft, to which it is rigidly connected. The high-speed shaft is supported by bearings mounted in the case. Figure 7.36 illustrates the relation between the gears and the angles made during a small angle of rotation. Note that before the rotation the sun and planet gear mesh at point B, while the planet and ring gear mesh at point A. After the rotation, the corresponding meshing points are $B1$ and $A1$. The centers of the sun and the planet are at O and OP, respectively.

The gear ratio (speed-up ratio), n_g, for the configuration shown in Figure 7.36 (with the ring gear stationary) is:

$$n_g = \frac{n_{HSS}}{n_{LSS}} = 1 + \frac{D_{Ring}}{D_{Sun}} \tag{7.45}$$

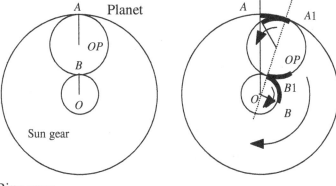

Ring gear

Figure 7.36 Relations between gears in a planetary gearbox

where n_{HSS} is the speed of the high-speed shaft, n_{LSS} is the speed of the low-speed shaft, D_{Ring} is the diameter of the ring gear, and D_{Sun} is the diameter of the sun gear.

As with the parallel-shaft gearbox, there is a limit to the speed-up ratio that can be achieved by a single-stage planetary gear set. Generally, any one stage will not provide a speed-up of more than 6 : 1. To achieve a higher speed-up ratio, multiple stages are placed in series. When there are multiple stages in series, the overall speed-up is the product of the speed-up of the individual stages. For example, one could gain a speed-up of 30 : 1 by having two stages of 5 : 1 and 6 : 1 in series.

As wind turbines matured above the 1 MW size, it was found that the low-speed gear in a parallel shaft gearbox was challenging to fabricate consistently and to be of consistent quality at the required pitch diameter. Gear designers instead opted for hybrid gearboxes with a planetary as the first stage. One to two stages of planetary and one or more stages of parallel shafts have since become the norm in large turbines (except when a direct drive generator is used).

Gears in wind turbine gearboxes are spur or helical gears. Helical gears are quieter and more tolerant of deflection and misalignment although they cause resultant axial forces that need to be tolerated. Bearings are ball bearings, cylindrical roller bearings, or tapered roller bearings, depending on the loads and support stiffness. Gears and bearings were discussed in more detail in Section 7.4.

Design Considerations

There are a great many issues to consider in the design and selection of a gearbox. These include:

- basic type (parallel-shaft, planetary, or hybrid), as discussed above;
- separate gearbox and main shaft bearings, or an integrated gearbox;
- speed-up ratio;
- number of stages;
- gearbox weight and cost;
- gearbox loads;
- lubrication system design including filtering, cooling, and distribution;
- effects of intermittent operation;
- noise;
- reliability.

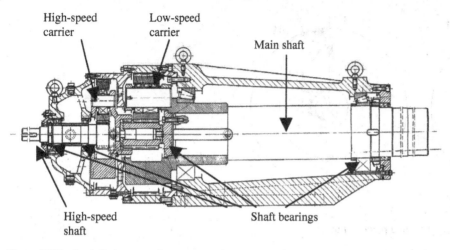

High-speed carrier   Low-speed carrier   Main shaft

High-speed shaft   Shaft bearings

Figure 7.37   Partially integrated, two-stage planetary gearbox

Wind turbine gearboxes are either separate components, or they are combined with other components. In the latter case, they are known as integrated or partially integrated gearboxes. For example, in a number of turbines with a partially integrated gearbox, the main shaft and main shaft bearings are integrated into the rest of the gearbox. A fully integrated gearbox is one in which the gearbox case is really the main frame of the wind turbine. The rotor is attached to its low-speed shaft. The generator is coupled to the high-speed shaft and is also bolted directly to the case. Part of the yaw system is integrated into the bottom of the case. Figure 7.37 illustrates a partially integrated planetary gearbox.

The speed-up ratio of a gearbox is directly related to the desired rotational speed of the rotor and the speed of the generator. As previously indicated, the rotor speed is determined primarily by aerodynamic considerations. For example, a wind turbine with a rotor designed to operate at 15 rpm and a 450 rpm generator would need a gearbox with a 30 : 1 speed-up ratio.

The number of stages in a gearbox is generally of secondary concern to the wind turbine designer. It is important primarily because it affects the complexity, size, weight, and cost of the gearbox. With more stages there are also more internal components, such as gears, bearings, and shafts.

The weight of a gearbox increases dramatically with increasing power rating of the turbine. In fact, the gearbox weight will scale approximately with the cube of the radius, as does the weight of the rotor. Since planetary gearboxes are lighter than parallel-shaft boxes, there is a weight advantage to be gained by using them. On the other hand, due to their greater complexity, they also cost more than would be indicated by their reduced weight.

The loads that the gearbox must withstand are due primarily to those imposed by the rotor. These will include at least the main shaft torque and may include the weight of the rotor and various dynamic loads, depending on the degree of integration of the gearbox with the main shaft and bearings. Loads are also imposed by the generator, both during normal operation and while starting, and by any mechanical brake located on the high-speed side of the gearbox. Over an extended period of time, the gearbox, like the rotor, will experience some loads that are relatively steady, other loads that vary periodically or randomly, and still others that are transient. All of these contribute to fatigue damage and wear on the gear teeth, bearings, and seals.

Lubrication is a significant issue in gearbox operation. Oils must be selected to minimize wear on the gear teeth and bearings and to function properly under the external environmental conditions in which the turbine will operate. In all large turbines, an active lubrication system is used to distribute clean, cool oil directly to all bearings and gears. This involves pumps, filters, coolers, internal and external manifolds, hoses, piping, and jets.

Online oil condition monitoring is also common, but, periodic oil samples should be taken to assess the state of the oil, as well as to check for signs of internal wear.

Intermittent operation, a common situation with wind turbines, can have a significant impact on the life of a gearbox. When the turbine is not running, oil may drain away from the gears and bearings, resulting in insufficient lubrication when the turbine starts. For this reason, active lubrication is facilitated by both a PTO-driven mechanical pump and an electrical pump to assure oil flow during operation.

In cold weather, the oil may have too high a viscosity until the gearbox has warmed up. Turbines in such environments may benefit by having gearbox oil heaters. Condensation of moisture may occur in various situations and conditions and absorbent filter media and/or rechargeable desiccant breathers are commonly installed to avoid the risk of corrosion. When the rotor is parked (depending on the nature and location of a shaft brake), the gear teeth may move slightly back and forth. The movement is limited by the backlash, but it may be enough to result in so-called fretting damage and premature tooth wear.

Gearboxes may be a source of noise. The amount of noise is a function of, among other things, the type of gearbox, the materials from which the gears are made, and how they are cut. Designing gearboxes for a minimum of noise production is presently an area of significant interest.

Reliability is a key consideration in the design of a gearbox. A detailed and carefully planned process is required to ensure that the design is sufficiently reliable. It requires close coordination between the turbine designer and the gearbox designer. Similar to the turbine as a whole, the gearbox design requires the following steps:

- specification of design load cases (see IEC, 2019);
- calculation of design loads at pertinent locations;
- preliminary detailed design;
- fabrication of a prototype;
- workshop testing of the prototype;
- installation of prototype gearbox in a wind turbine;
- field testing of the gearbox;
- final design of gearbox.

More details on wind turbine gearboxes, relating particularly to design, are given in design guidelines from AGMA (1997) and IEC (2012).

7.5.2.4 Generator

The generator converts the mechanical power from the rotor into electrical power. Generator options were described in detail in Chapter 5 and will not be discussed here. One of the important things to recall is that, until recently, most grid-connected generators have turned at constant or nearly constant speed. This resulted in the rotors also turning at constant or nearly constant speed. At present, however, most large wind turbines operate at variable speeds, such that both the rotor and the generator speeds vary by roughly a factor of two during normal operation.

7.5.2.5 Direct-Drive Generators

Many turbines these days have direct-drive generators (see Figure 7.33 and Chapter 5). That means that there is no gearbox, and the main shaft of the rotor connects directly to or is part of the generator. The generator has multiple poles. These poles may use permanent magnets, although that is not necessarily the case. For generators that are directly connected to the grid at above-rated wind speed, the rotational speed is proportional to the number of poles. In the case of variable speed, direct-drive generators with full-rated power converters, rotational speed is controlled electronically.

7.5.2.6 Brakes

Function

Nearly all wind turbines employ a mechanical brake somewhere on the drivetrain. Such a brake is normally included in addition to an aerodynamic brake. In fact, some design standards (Germanischer Lloyd, 1993) require two independent braking systems, one of which is usually aerodynamic and the other of which is on the drivetrain. With a stall-controlled rotor, the mechanical brake is capable of stopping the turbine. In other cases, the mechanical brake is used only for parking. That is, it keeps the rotor from turning when the turbine is not operating.

Types

Two types of mechanical brakes have been used on wind turbines: disc brakes and clutch brakes. The disc brake operates in a manner similar to that of an automobile. A steel disc is rigidly affixed to the shaft to be braked. During braking a hydraulic-applied caliper pushes brake pads against the disc. The resulting force creates a torque opposing the motion of the disc, thus slowing the rotor. An example of a disc brake is shown in Figure 7.38.

Clutch-type brakes consist of at least one pressure plate and at least one friction disc. Actuation is normally done via springs, so they are fail-safe by design. These brakes are released by compressed air or hydraulic fluid. Brakes of this type are no longer common on wind turbines, however.

Location

Mechanical brakes can be located at any of a variety of locations on the drivetrain. For example, they may be on either the low-speed or high-speed side of the gearbox. If on the high-speed side, they may be on either side of the generator.

It is important to note that a brake on the low-speed side of the gearbox must be able to exert a much higher torque than would be the case with one on the high-speed side. It will thus be relatively massive. On the other hand, if the brake is on the high-speed side, it will necessarily act through the gearbox, possibly increasing the gearbox wear. Furthermore, in the event of an internal failure in the gearbox, a brake on the high-speed side might be unable to slow the rotor.

Activation

Brake activation depends on the type of brake used. Disc brakes require hydraulic pressure. This is normally supplied by a hydraulic pump, sometimes in conjunction with an accumulator. There are also designs in which springs apply brake pressure, and the hydraulic system is used to release the brakes.

Disc

Caliper

Figure 7.38 Disc brake (*Source:* Reproduced by permission of Svendborg Brakes A/S)

Clutch-type brakes are normally spring-applied. Either a pneumatic system or hydraulic system is used to release the brake. In the case of pneumatics, an air compressor and storage tank must be provided, as well as appropriate plumbing and controls.

Performance

Three important considerations in the selection of a brake include:

1) maximum torque;
2) length of time required to apply;
3) energy absorption.

A brake intended to stop a wind turbine must be able to exert a torque in excess of what could plausibly be expected to originate from the rotor. Recommended standards indicate that a brake design torque should be equal to the maximum design torque of the wind turbine (Germanischer Lloyd, 1993).

A brake intended to stop a turbine should begin to apply almost immediately, and it should ramp up to full torque within a few seconds. The ramp-up time selected is a balance between instantaneous (which would apply a very high transient load to the drivetrain) and so slow that acceleration of the rotor and heating of the brake during deceleration could be concerns. Normally the entire braking event, from initiation until the rotor is stopped, is less than five seconds.

Energy absorption capability of the brake is an important consideration. First of all, the brake must absorb all the kinetic energy in the rotor when turning at its maximum possible speed. It must also be able to absorb any additional energy that the rotor could acquire during the stopping period.

7.5.3 Yaw System

7.5.3.1 Function

All modern horizontal-axis wind turbines must be able to yaw so as to orient themselves in line with the wind direction. Some turbines also use active yaw as a means of power regulation or wake mitigation. In any case, a mechanism must be provided to enable the yawing to take place, and to do so at a slow enough rate that large gyroscopic forces are avoided.

7.5.3.2 Types

There are two basic types of yaw systems: active yaw and free yaw. Active yaw relies on yaw drives (motors) to actively align the turbine with the wind direction. Passive yaw relies on the aerodynamic thrust of the rotor (for downwind machines) or the aerodynamics of a tail fin (for upwind machines) to align the turbine. Large-scale machines typically rely on active yaw.

7.5.3.3 Description

Regardless of the type of yaw system, a yaw bearing must be present. This bearing must carry the weight of the main part of the turbine, as well as transmitting thrust loads to the tower.

In a turbine with active yaw, the yaw bearing includes gear teeth around its circumference, and the bearing industry refers to this as a slewing ring. The gear can be on the inner ring or the outer ring. A pinion gear on each yaw drive engages with these teeth, so that it can be driven in either direction.

The yaw drive normally consists of an electric motor, speed reduction gears (multistage planetaries), and a pinion gear. The speed must be reduced so that the yaw rate is slow, and so that adequate torque can be supplied from a small motor.

One problem encountered with active yaw has been rapid wear or breaking of the yaw drive due to continuous small yaw movements of the turbine. This is possible because of backlash between the yaw drive pinion and the

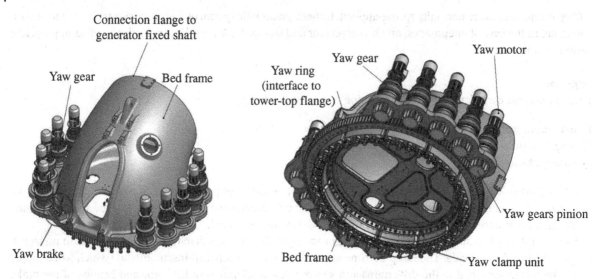

Figure 7.39 Typical yaw drive (*Source:* Reproduced with permission from Siemens Gamesa, Inc.)

slewing ring gear. The motion results in many shock load cycles between those gears. In order to reduce these cycles, a yaw brake is now used in active yaw systems. This brake is engaged whenever the turbine is not yawing. It is released just before yawing begins. A typical yaw drive with a brake is shown in Figure 7.39.

Large turbines currently use 4 to 10 yaw drives with inverter control to provide this braking and limit backlash. However, installing and balancing these systems require careful timing methodologies.

The yaw motion in an active yaw system is controlled using yaw error as an input. Yaw error is monitored by means of a wind vane mounted on the turbine. When the yaw error is outside the allowed range for some period of time, the drive system is activated, and the turbine is moved in the appropriate direction.

In turbines with free yaw, the yaw system is normally simpler. Often there is nothing more than the yaw bearing. Some turbines, however, include a yaw damper. Yaw dampers are used to slow the yaw rate, helping to reduce gyroscopic loads. They are most useful for machines that have a relatively small polar moment of inertia about the yaw axis.

7.5.4 Main Frame and Nacelle

The nacelle is the housing for the principal components of the RNA (with the exception of the rotor). It includes the main frame and the nacelle housing (cover).

7.5.4.1 Main Frame
Function
The main frame is the structural component to which the gearbox, generator, and brake are attached. It provides a rigid structure to maintain the proper alignment among those other components. It also provides a point of attachment for the yaw bearing, which in turn is bolted to the top of the tower.

Description
The main frame is a rigid steel casting or weldment. Threaded holes or other attachment points are provided in appropriate locations for bolting on the other components.

Figure 7.40 Exterior and interior of the nacelle of a 5 MW wind turbine (*Source:* J. F. Manwell (Author))

Loads

The main frame must transmit all the loads from the rotor and reaction loads from the generator and brake to the tower. It must also be rigid enough that it allows no relative movement between the rotor support bearings, gearbox, generator, and brake.

7.5.4.2 Nacelle Housing

The nacelle housing provides weather protection for the components which are located inside the nacelle. These include, in particular, electrical and mechanical components that could be affected by sunlight, rain, ice, or snow.

Nacelle housings are normally made from a lightweight material, such as fiberglass. On larger machines, the nacelle housing is of sufficient size that it can be entered by personnel for inspecting or maintaining the internal components. On small and medium-size turbines, a separate nacelle housing is normally attached to the main frame in such a way that it can be readily opened for access to items inside. The exterior and interior of the nacelle of a 5 MW turbine are shown in Figure 7.40. A component that most turbines have, and which is closely related to the nacelle cover, is the spinner or nose cone. This is the cover for the hub. One of these can also be seen in Figure 7.40.

References

ABS (2018) *Fatigue Assessment of Offshore Structures*, American Bureau of Shipping, Houston, TX, https://safety4sea.com/wp-content/uploads/2016/06/Fatigue_Guide_e-Mar18.pdf Accessed 6/25/2023.

AGMA (1997) *Recommended Practices for Design and Specification of Gearboxes for Wind Turbine Generator Systems*, AGMA Information Sheet. AGMA/AWEA 921-A97. Available from https://www.nrel.gov/docs/legosti/old/7076.pdf. Accessed 6/23/2023.

Alejandro Franco, J., Carlos Jauregui, J. and Toledano-Ayala, M. (2015) Optimizing wind turbine efficiency by deformable structures in smart blades. *Journal of Energy Resources Technology*, **137**(5), 051206.

Ansell, M. P. (1987) Laymans Guide to Fatigue, *Wind Energy Conversion*, Mechanical Engineering Publications Ltd., 39–54.

Barlas, T. K. and van Kuik, G. A. (2010) Review of state of the art in smart rotor control research for wind turbines. *Progress in Aerospace Sciences*, **46**(1), 1–27.

Berg, J. and Resor, B. R. (2012) *Numerical Manufacturing and Design Tool (NuMAD V. 2.0) for Wind Turbine Blades*, SAND2012-7028, Sandia National Laboratory, Albuquerque, NM.

Bergami, L., Madsen, H. A. and Rasmussen, F. (2014) *A Two-Bladed Teetering Hub configuration for the DTU 10 MW RWT: loads considerations*, European Wind Energy Association (EWEA), 1–8.

Budynas, R. G and Nisbett, J. K., (2015) *Shigley's Mechanical Engineering Design*, 10th edition. McGraw Hill Education, New York.

Cairns, D. and Skramstad, J. (2000) *Resin Transfer Molding and Wind Turbine Blade Construction, SAND99-3047*. Sandia National Laboratory, Albuquerque, NM.

Do, M. H. and Söffker, D. (2021) State-of-the-art in integrated prognostics and health management control for utility-scale wind turbines. *Renewable and Sustainable Energy Reviews*, **145**, 111102.

Downing, S. D. and Socie, D. F. (1982) Simple rainflow counting algorithms. *International Journal of Fatigue* **4**(1), 31–40.

Gasch, R. (Ed.) (1996) *Windkraftanlagen (Windpower Plants)*. B. G. Teubner, Stuttgart.

Germanischer, Lloyd (1993) *Regulation of the Certification of Wind Energy Conversion Systems*. Germanischer Lloyd, Hamburg.

Hau, E. (1996) *Windkraftanlagen (Windpower Plants)*. Springer, Berlin.

Hayman, G. J. (2012) *MLife Theory Manual for Version 1.00*. National Renewable Energy Laboratory, Denver.

Hu, W., Choi, K.K., Zhupanska, O. and Buchholz, J.H.J. (2016) Integrating variable wind load, aerodynamic, and structural analyses towards accurate fatigue life prediction in composite wind turbine blades. *Structural and Multidisciplinary Optimization*, **53**, 375–294.

Hunt, T. and Vauchan, N. (1996) *Hydraulic Handbook*, 9th edition. Elsevier, Oxford.

Hyers, R. W. McGowan, J. G. Sullivan, J. F. and Syret, B. C. (2006) Condition monitoring and prognosis of utility scale wind turbines. *Energy Materials*, **1**(3), 187–203.

IEC (2012) *Design Requirements for Wind Turbine Gearboxes*, IEC 61400-4:2012. International Electrotechnical Commission, Geneva.

IEC (2019) *Wind Turbines Part 1: Design Requirements*, 61400-1, 4th edition. International Electrotechnical Commission, Geneva.

Jensen, J. P. and Skelton, K. (2018) Wind turbine blade recycling: experiences, challenges and possibilities in a circular economy. *Renewable and Sustainable Energy Reviews*, **97**, 165–176.

Liu, P., Meng, F. and Barlow, C. Y. (2022) Wind turbine blade end-of-life options: an economic comparison. *Resources, Conservation and Recycling*, **180**, 106202.

McGowan, J. G., Hyers, R. W., Sullivan, K. L., Manwell, J., Nair, S. V., McNiff, B. and Syrett, B. C. (2007) A review of materials degradation in utility scale wind turbines. *Energy Materials*, **2**(1), 41–64.

McNiff, B. P., Musial, W. D. and Erichello, R. (1990) Variations in gear fatigue life for different braking strategies. *Proc. 1990 American Wind Energy Association Conference*, Washington, DC. Available from https://www.osti.gov/servlets/purl/5828926. Accessed 6/25/2023.

Mendes, P., Correia, J. A., De Jesus, A. M., Ávila, B., Carvalho, H. and Berto, F. (2021) A brief review of fatigue design criteria on offshore wind turbine support structures. *Frattura ed Integrità Strutturale*, **15**(55), 302–315.

Mishnaevsky, L., Branner, K., Petersen, H. N., Beauson, J., McGugan, M. and Sørensen, B. F. (2017) Materials for wind turbine blades: an overview. *Materials*, **10**(11), 1285–1309.

National Research Council (1991) *Assessment of Research Needs for Wind Turbine Rotor Materials Technology*. National Academy Press, Washington, DC.

Nijssen, R. and de Winkel, G. D. (2016) Developments in Materials for Offshore Wind Turbine Blades, in *Offshore Wind Farms: Technologies, Design and Operation*, Woodhead Publishing, 85–104.

Parmley, R. D. (1997) *Standard Handbook of Fastening and Joining*, 3rd edition. McGraw Hill, New York.

Roach, D., Neidigk, S., Rice, T. Duvall, R. and Paquette, J. (2014) *Blade Reliability Collaborative: Development and Evaluation of Nondestructive Inspection Methods for Wind Turbine Blades*, Sandia Report SAND2014-16965, Sandia National Laboratories, Albuquerque, NM.

Sadegh, A. M. and Worek, W. M., eds. (2018) *Marks' Standard Handbook for Mechanical Engineers*. 12th edition. McGraw-Hill Education, New York.

Sandia (2023) *Archived Workshops*, https://energy.sandia.gov/programs/renewable-energy/wind-power/workshops/archived-workshops/ Accessed 6/23/2023.

Selig, M. and Tangler, J. L. (1995) Development of a multipoint inverse design method for horizontal axis wind turbines. *Wind Engineering*, **19**(2), 91–105.

Selig, M. (2023) UIUC *Airfoil Coordinates Data Base*, https://m-selig.ae.illinois.edu/ads/coord_database.html. Accessed 6/25/2023.

Singh, R. (2014) *Corrosion control for offshore structures: cathodic protection and high-efficiency Coating*. Gulf Professional Publishing.

Sutherland, H. J. (1999) *On the Fatigue Analysis of Wind Turbines*, SAND99-0089, Sandia National Laboratories, Albuquerque, NM.

Sutherland, H. J. (2000) A summary of the fatigue properties of wind turbine materials. *Wind Energy*, **3**, 1–34.

Timken (2023) *Timken Engineering Manual*, https://www.timken.com/resources/timken-engineering-manual. Accessed 6/25/2023.

Veers, P. (2011) Chapter 5 – Fatigue Loading of Wind Turbines, in: J. D. Sørensen, J. N. Sørensen (Eds.), *Wind Energy Systems*, Woodhead Publishing Series in Energy, Woodhead Publishing, 130–158.

Veers, P. Ashwill, T. D. Sutherland, H. J. Laird, D. L. Lobitz, D. W. Griffin, D. A. Mandell, J. F. Musial, W. D. Jackson, K. Zuteck, M. Miravete, A. Tsai, S. W. and Richmond, J. L. (2003) Trends in the design, manufacture and evaluation of wind turbine blades. *Wind Energy*, **6**(3), 245–259.

Young, A. C., Goupee, A. J., Dagher, H. J. and Viselli, A. M. (2017) Methodology for optimizing composite towers for use on floating wind turbines. *Journal of Renewable and Sustainable Energy*, **9**(3), 033305.

8

Wind Turbine Design and Testing

8.1 Overview

The design process of a wind turbine includes conceptual design, detailed design of components to withstand the loads they will be expected to experience, and then testing of full wind turbines and components to ensure that they have indeed met those design goals. This chapter covers all of these aspects of the design process. The substance of the chapter begins in Section 8.2 with an introduction to the concept of wind turbine design basis. Section 8.3 presents an overview of the design process and continues with a more in-depth examination of the various steps involved. This is then followed in Section 8.4 by a review of the basic wind turbine topologies. Section 8.5 gives an overview of design standards related to wind turbines. Section 8.6 examines the types of loads that a wind turbine experiences, particularly with reference to the key international design standard IEC 61400-1. Section 8.7 is about assessing the effects of loads and presents an introduction to probabilistic design in so far as it relates to wind turbines. After that, Section 8.8 provides an overview of scaling relationships for loads and natural frequencies. These can be used for the initial starting point of a new wind turbine design. This is followed in Section 8.9 by an overview of some of the analysis tools that are available to assist in the development of a new design. Section 8.10 describes how to predict a turbine's power curve once the basic aspects of the design have been chosen. After a new design has been developed, it must be evaluated. This is the topic of Section 8.11. Section 8.12 provides a review of wind turbine testing methods and issues and aspects of component and certification testing. Finally, Section 8.13 focuses specifically on offshore wind turbines.

8.1.1 Overview of Design Issues

The process of designing a wind turbine involves the conceptual assembling of a large number of mechanical and electrical components into a machine that can convert the fluctuating power in the wind into a useful form. This process is subject to a number of constraints, but the fundamental ones involve the potential economic viability of the design. Ideally, the wind turbine should be able to produce power at a cost lower than its competitors, which are typically petroleum-derived fuels, natural gas, nuclear power, or other renewables. In many situations, wind energy is the most cost-effective, but otherwise, incentives may be provided by governments to make up the difference. Regardless of whether there are such incentives, it is a fundamental design goal to keep the cost of energy lower than it would be from a turbine of a different design.

The cost of energy from a wind turbine is a function of many factors, but the primary ones are the cost of the turbine itself and its annual energy productivity. In addition to the first cost of the turbine, other costs (as will be discussed in more detail in Chapter 13) including installation, operation, and maintenance, need to be accounted for in the design process. Operation and maintenance costs can amount to 1/3 of wind farm life cycle expenditures for offshore projects. The productivity of the turbine is a function of both the turbine's design and the wind resource. The designer cannot control the resource but must consider how best to utilize it. Other factors that affect

Wind Energy Explained: On Land and Offshore, Third Edition.
James F. Manwell, Emmanuel Branlard, Jon G. McGowan, and Bonnie Ram.

the cost of energy, such as loan interest rates and discount rates, tend to be of secondary importance and are largely outside the purview of the designer.

The constraint of minimizing the cost of energy has far-reaching implications. It impels the designer to minimize the cost of the individual components and thus to use the most cost-effective materials. There is also an impetus to keep the weight of the components as low as possible to minimize their cost. On the other hand, the turbine design must be strong enough to survive any probable extreme events and to operate reliably and with a minimum of maintenance or replacements over a long period of time. Finally, wind turbine components tend to experience relatively high stresses, as the components are designed to be as light as possible. By the nature of the turbine's operation, the stresses also tend to be highly variable (see Chapter 4). Varying stresses result in fatigue damage (see Chapter 7). This may eventually lead to either failure of the component or the need for replacement.

The need to balance the initial cost of the wind turbine with the requirement that the turbine have a long, safe, and fatigue-resistant life should be the fundamental concern of the designer.

8.2 Design Basis for Wind Turbines

As introduced in Chapter 6, the design basis for a wind turbine is the set of conditions or requirements that must be considered in the turbine's design and utilization. The design basis for a wind turbine may be fairly broad, considering safety, reliability, performance, maintainability, environmental factors, serviceability, public acceptance, etc. One of the core requirements of a wind turbine be that it be able to withstand the forces resulting from the external conditions. A key consideration in this regard is whether the design of the turbine and its support structure are to be class-based or site-based, as discussed next.

8.2.1 Wind Turbine Design Philosophy

Two types of design philosophies may be applicable to wind turbines: class-based design and site-based design. In class-based design, the turbine is designed so that it can meet the minimal requirements of a particular "class." The class is defined by a specified set of external conditions (e.g., mean and extreme wind speeds. turbulence intensity, etc.). If the conditions that a particular turbine is likely to encounter are not more extreme than those that define the class, the design is deemed adequate. Land-based wind turbines and the rotor nacelle assemblies (RNAs) of offshore wind turbines are typically designed according to class-based criteria. In site-based design, the turbine is designed with respect to external conditions that are known to exist at the site where the turbine is to be installed. Support structures for offshore wind turbines are normally designed according to site-based design philosophy. In either class-based design or site-based design, the external environmental conditions are relevant and must be considered. The difference is that in class-based design, the site must be assessed to ensure that a design of a given class will be suitable. For site-based design, the site must be assessed before the design can be completed (or sometimes before it can be started).

It is worth noting that there is a distinction to be made between the turbine designer and the project developer. The designer can design an onshore turbine for a given class without needing to worry about the actual site. The project developer, on the other hand, will need to assess the site to know which class of turbine to use. In the offshore case, there will be communication between the turbine manufacturer (and so the designer) and the project developer during the planning process (see as well, Chapter 14).

8.2.2 Guiding Considerations for Wind Turbine Design

The guiding considerations in the design of a wind turbine of any sort are:

- The turbine must produce an adequate amount of electricity.
- The turbine must withstand occasional extreme loads which may arise from external sources.

- The turbine must be able to withstand fatigue damage which will arise due to the operation of the turbine itself and the varying conditions of the environment, during its installation and operation (including standstill and idling conditions).

8.2.2.1 Operating Conditions

Operating conditions are also important to consider in the design of the wind turbine. The turbine may be operating for long periods of time at rated power, or it may be running below rated power such that the power output varies with the wind speed. The wind may be too low to run at all (below cut-in) or so high it has been stopped (above cut-out). At any given time, it may be transitioning from one operating condition to another or undergoing various control actions. The weather may also affect operations, such as periods of very cold or very hot conditions, or extreme events, such as hurricanes, when the turbine will be shut down but still needs to be protected. All these situations need to be accounted for in the design basis. For offshore turbines, both the atmospheric and marine conditions (metocean conditions, which were discussed in Chapter 6) are considered in the design basis. In large wind power plants, it is necessary to consider the presence of the other turbines in the plant (see Chapter 12) and neighboring plants.

8.3 Design Process

There are a number of approaches that can be taken toward wind turbine design, and there are many issues that must be considered. This section outlines the steps in one approach. The subsequent sections provide more details on those steps.

The key design steps include the following:

1) Determine application;
2) Review previous experience;
3) Select topology;
4) Estimate preliminary loads;
5) Develop tentative design;
6) Predict performance;
7) Evaluate design;
8) Estimate costs and cost of energy;
9) Refine design;
10) Build prototype;
11) Test prototype;
12) Design production machine.

Steps 1 through 7 and 11 are the subjects of this chapter. Turbine cost and cost of energy estimates (Step 8) can be made using methods discussed in Chapter 13. Steps 9, 10, and 12 are beyond the scope of this text, but they are based on the principles outlined here. The integration of the engineering design process within the entire project development workflow (including permitting, environmental considerations, financing, etc.) is discussed in Chapter 14.

8.3.1 Determine Application

The first step in designing a wind turbine is to determine the application. Wind turbines for producing bulk power for supply to large utility networks, for example, will have a different design than will turbines intended for operation in remote communities.

The application will be a major factor in choosing the size of the turbine, the type of generator it has, the method of control, and how it is to be installed and operated. For example, wind turbines for utility power will tend to be as large as practical. At the present time, onshore turbines typically have power ratings in the range of 1–5 MW with rotor diameters in the range of 50–150 m. Offshore turbines are even larger, with power ratings approaching 15 MW and diameters up to 250 m. Such machines are usually installed in clusters or wind power plants (a.k.a. wind farms) and may be able to utilize well-developed infrastructure for installation, operation, and maintenance.

Turbines for use by utility customers, or for use in remote communities, tend to be smaller, typically in the 10–500 kW range. Ease of installation and maintenance and simplicity in construction are important design considerations for these turbines.

8.3.2 Review Previous Experience

The next step in the design process should be a review of previous experiences. This review should consider wind turbines built for similar applications. A wide variety of wind turbines has been conceptualized. Many have been built and tested, at least to some degree. Lessons learned from those experiences should help guide the designer and narrow down the options.

A general lesson that has been learned from every successful project is that the turbine must be designed in such a way that operation, maintenance, and servicing can be done in a safe and straightforward way.

8.3.3 Select Topology

There is a wide variety of basic configurations or "topologies" for a wind turbine. Most of these relate to the rotor. The most important choices are listed below. These choices are discussed in more detail in Section 8.4.

- Rotor axis orientation. Horizontal or vertical;
- Power control. Stall, variable pitch, controllable aerodynamic surfaces, or yaw control;
- Rotor position. Upwind of tower or downwind of tower;
- Yaw control. Driven yaw, free yaw, or fixed yaw;
- Rotor speed. Constant or variable;
- Design tip-speed ratio and solidity;
- Type of hub. Rigid, teetering, hinged blades, or gimballed;
- Number of blades;
- Generator speed. Synchronous speed, multiple synchronous speeds, or variable speed;
- Tower structure.

8.3.4 Preliminary Loads Estimate

Early in the design process, it is necessary to make a preliminary estimate of the loads that the turbine must be able to withstand. These loads will serve as inputs to the design of the individual components. Estimation of loads at this stage may involve the use of scaling of loads from turbines of similar design, "rules of thumb," or simple computer analysis tools. These estimates are improved throughout the design phase as the details of the design are specified. At this stage, it is important to keep in mind all the loads that the final turbine will need to be able to withstand. This process can be facilitated by referring to recommended design standards. It is normal in this stage to prepare a design basis for the turbine, as described in Section 8.2. Note that throughout this chapter, the term "load effect" is frequently used to indicate the effect of the load in terms of stress or deflection.

8.3.5 Develop Tentative Design

Once the preliminary layout has been chosen and the loads approximated, a preliminary design will be developed. The design may be considered to consist of a number of subsystems. The primary subsystems are the rotor/nacelle assembly (RNA) and the support structure. The principal components of the RNA are listed as follows. These were discussed in more detail in Chapter 7.

- rotor (blades, hub, aerodynamic control surfaces);
- drivetrain (shafts, couplings, gearbox, mechanical brakes, generator);
- nacelle and main frame;
- yaw system.

The support structure for onshore turbines consists of the tower and foundation. With offshore turbines, there is an additional substructure, the specifics of which depend on the situation, as discussed in Chapters 10 (fixed) and 11 (floating).

There are also a number of general considerations, which may apply to the entire turbine. Some of these include:

- fabrication methods;
- ease of maintenance;
- aesthetics;
- noise;
- other environmental conditions.

8.3.6 Predict Performance

Early in the design process, it is also necessary to predict the performance (power curve) of the turbine. This will be primarily a function of the rotor design, but it will also be affected by the type of generator, the efficiency of the drivetrain, the method of operation (constant speed or variable speed), and choices made in the control system design; see Section 8.10.

8.3.7 Evaluate Design

The preliminary design must be evaluated for its ability to withstand the loading the turbine may reasonably be expected to encounter. The wind turbine must be able to easily withstand any loads likely to be encountered during normal operation. In addition, the turbine must be able to withstand extreme loads that may only occur infrequently, as well as holding up to cumulative, fatigue-induced damage, as discussed in Chapter 7.

The categories of loads the wind turbine must withstand (e.g., steady, cyclic, and impulse) are described in Chapter 4. The turbine must be able to withstand these loads under all plausible conditions, including during normal operation, idling, standstill, and extreme events. These conditions will be discussed in more detail in Section 8.6.

The loads of primary concern are those in the rotor, especially at the blade roots, but any loads at the rotor also propagate through the rest of the structure. Therefore, the loading at each component must also be carefully assessed.

Analysis of wind turbine loads and their effects is typically carried out with the use of computer-based analysis codes (see Section 8.9). In doing so, reference is normally made to accepted practices or design standards (see Sections 8.5). The principles underlying the analysis of wind turbine loads were discussed in detail in Chapter 4. A more in-depth discussion of wind turbine loads as related to design is given in Sections 8.5, 8.6, and 8.7.

8.3.8   Estimate Costs and Levelized Cost of Energy

An important part of the design process is the estimation of the cost of energy from the wind turbine. The key factors in the cost of energy are the cost of the turbine itself and its productivity. It is therefore necessary to be able to predict the cost of the machine, both in the prototype stage and, most importantly, in production. This topic is discussed in detail in Chapter 13.

8.3.9   Refine Design

When the preliminary design has been analyzed for its ability to withstand loads, its performance capability has been predicted, and the eventual cost of energy has been estimated, it is normal that some areas for refinement will have been identified. At this point, another iteration of the design is made. The revised design is analyzed in a similar manner to the process summarized in Sections 8.3.4–8.3.8. This design, or perhaps a subsequent one if there are more iterations, will be used in the construction of a prototype.

8.3.10   Build Prototype

Once the prototype design has been completed, a prototype should be constructed. The prototype may be used to verify the assumptions in the design, test any new concepts, and ensure that the turbine can be fabricated, installed, and operated as expected. Normally, the turbine will be very similar to the expected production version, although there may be provision for testing and instrumentation options that the production machine will not need.

8.3.11   Test Prototype

After the prototype has been built and installed, it is subjected to a wide range of field tests. Power is measured and a power curve is developed to verify and certify the performance predictions. Strain gauges are applied to critical components. Actual loads are measured and compared to the predicted values.

8.3.12   Design Production Turbine

The final step is the design of the production turbine. The design of this turbine will likely be very close to the prototype. It may have some differences, however. Some of these may be improvements, the need for which was identified during testing of the prototype. Others may have to do with lowering the cost of mass production. For example, a weldment may be appropriate in the prototype stage, but a casting may be a better choice for mass production.

8.4   Wind Turbine Topologies

This section provides a summary of some of the key issues relating to the most commonly encountered choices in the overall topology of modern wind turbines. The purpose of this section is not to advocate a particular design philosophy but to provide an overview of what must be considered. It should be noted that there are, in the wind energy community, strong proponents of particular aspects of design, such as rotor orientation and number of blades. A good overview of some issues of design philosophy is given by Dörner (2008).

One of the general topics of great interest at the present time is how light a turbine can be and still survive for the desired amount of time. Some of the issues in this regard are discussed by Johnson *et al.* (2021).

8.4.1 Rotor Axis Orientation: Horizontal or Vertical

The most fundamental decision in the design of a wind turbine is probably the orientation of the rotor axis. In most modern wind turbines, the rotor axis is horizontal (parallel to the ground), or nearly so. The turbine is then referred to as a "horizontal-axis wind turbine" (HAWT), as discussed in Chapter 1. There are a number of reasons for that trend; some are more obvious than others. Two of the main advantages of horizontal axis rotors are the following:

1) HAWT rotors do not undergo significant cyclical aerodynamic variation, e.g., dynamic stall and large angle of attack variations per cycle.
2) The rotor solidity of a HAWT, and hence total blade mass relative to swept area, is lower when the rotor axis is horizontal (at a given design tip-speed ratio); see Dörner (2008). This tends to keep costs lower on a per kW basis.

The aerodynamics of vertical-axis wind turbines (VAWTs) were discussed in Chapter 3. As also noted there, a major advantage of VAWTs is that they are omnidirectional, i.e., the rotor can accept wind from any direction. This means a yaw system is not needed, simplifying mechanical design. Another advantage is that in most VAWTs the blades can have a constant chord and no twist. These characteristics should enable the blades to be manufactured relatively simply (e.g., by aluminum extrusion or composite pultrusion) and thus inexpensively on a cost-per-length basis. This has the potential to offset the added blade length required when compared to a HAWT with the same swept area. The third advantage is that much of the drivetrain (gearbox, generator, and brake) can be located on a stationary support structure and can be relatively close to the ground (when desirable). One more potential advantage is that at very large scales, things might start to look different: HAWTs start to have massive oscillating blade root moments due to gravitational loading. Very large VAWTs do not suffer from this. Last, it has been argued that the land footprint necessary for the same energy capture could be lower for VAWTs due to more rapid wake recovery and constructive turbine interactions. This would be advantageous as fewer sites on land become available for wind energy utilization.

Despite some promising advantages of the vertical axis rotor, the design has not met with widespread acceptance. Due to the cyclical aerodynamics of VAWTs, these machines go through many more aerodynamically driven fatigue cycles than do equivalently sized HAWTs. Additionally, dynamic stall at low tip-speed ratios is difficult to model. Because of these challenges, many machines that were built in the 1970s and 1980s suffered fatigue damage of the blades, especially at connection points to the rest of the rotor. This was an outcome of the cyclic aerodynamics and the fatigue properties of the aluminum from which the blades were commonly fabricated. These challenges may be overcome using modern composite materials and advances in aeroelastic modeling.

VAWT designs can also suffer from a lack of options for aerodynamic control. This can play an important role in the design as it limits the options for braking or stopping the turbine, possibly resulting in more expensive mechanical braking solutions for VAWTs. In perspective though, the braking system would still not be as expensive as a blade pitch system.

In summary, a horizontal axis has so far proved to be preferable. There are enough advantages, however, to the vertical axis rotor that it may be worth considering for some applications. In this case, however, the designer should have a clear understanding of what the limitations are and should also have some plausible options in mind for addressing those limitations.

Because of the predominance of HAWTs presently in use or under development, the remainder of this chapter, unless otherwise specified, applies primarily to wind turbines of that type.

8.4.2 Rotor Power Control: Stall, Pitch, Yaw, or Aerodynamic Surfaces

There are a number of options for controlling power aerodynamically. The selection of which of these is used will influence the overall design in a variety of ways. The following is a summary of the options, focusing on those aspects that affect the overall design of the turbine. Details on control issues are discussed in Chapter 9.

Stall control takes advantage of reduced aerodynamic lift at high angles of attack to reduce torque at high wind speeds. For stall to function, the rotor speed must be separately controlled, most commonly by a squirrel cage induction generator (see Chapter 5) connected directly to the electrical grid. Blades in stall-controlled turbines are fastened rigidly to the rest of the hub, resulting in a simple connection. The nature of stall control, however, is such that maximum power is reached at a relatively high wind speed. The drivetrain must be designed to accommodate the torques encountered under those conditions, even though such winds may be relatively infrequent. Stall-controlled machines invariably incorporate separate braking systems to ensure that the turbine can be shut down under all eventualities. The trend at present is away from stall control.

Variable-pitch turbines have blades that can be rotated about their long axis, changing the blade's pitch angle. Changing pitch also changes the angle of attack of the relative wind and the amount of torque produced. Variable pitch provides more control options than does stall control. On the other hand, the hub is more complicated because pitch bearings need to be incorporated, and some form of pitch actuation system must also be included. In some wind turbines, only the outer part of the blades may be pitched. This is known as partial span pitch control.

Some wind turbines utilize aerodynamic surfaces on the blades to control or modify power. These surfaces can take a variety of forms, but in any case, the blades must be designed to hold them, and means must be provided to operate them. In most cases aerodynamic surfaces are used for braking the turbine. In some cases, specifically when ailerons are used (see Chapter 7), the surfaces may also provide a fine-tuning effect.

Another option for controlling power is yaw control. In this arrangement, the rotor is turned away from the wind, reducing power. This method of control requires a robust yaw system. The hub must be able to withstand gyroscopic loads due to yawing motion but can otherwise be relatively simple.

8.4.3 Rotor Position: Upwind of Tower or Downwind of Tower

The rotor in a HAWT may be either upwind or downwind of the tower. A downwind rotor in principle allows the turbine to have free yaw, which is simpler to implement than active yaw. (In reality, free yaw is not necessarily desirable, however; see Section 8.4.4). Another advantage of the downwind configuration is that it is easier to take advantage of centrifugal forces to reduce the blade root flap bending moments. This is because the blades are normally coned downwind, so centrifugal moments tend to counteract moments due to thrust. The third advantage of downwind rotors is that during operation, in contrast to upwind rotors, the blades bend away from the tower. This minimizes the chance of tower strikes and simplifies their stiffness requirements. Finally, downwind rotors are tilted in the opposite direction of upwind rotors, which drives the wake toward the ground and can potentially reduce wake effects in wind plants (see Chapter 12).

On the other hand, since the tower produces a wake in the downwind direction, the blades must pass through that wake every revolution. This wake is a source of periodic loads, which may result in increased fatigue damage to the blades and impose a ripple on the electrical power produced. Blade passage through the wake is also a source of noise. The effects of the wake (known as "tower shadow") may, to some extent, be reduced by utilizing a tower design that provides minimal obstruction to the flow. The main disadvantage of downwind rotors is that the rotor swept area is decreased as the blade deflects, thereby reducing the energy that can be extracted.

8.4.4 Yaw Control: Free or Active

All HAWTs must provide some means to orient the machine as the wind direction changes. In downwind machines, yaw motion has historically been free of constraints. The turbine follows the wind like a weathervane. For free yaw to work effectively, the blades are typically coned a few degrees in the downwind direction. Free yaw machines sometimes incorporate yaw dampers to limit the yaw rate and thus gyroscopic loads in the blades.

Small-scale upwind machines may include a passive yaw mechanism by means of a tail vane. Large-scale upwind turbines rely on active yaw control. This usually includes a yaw motor, gears, and a brake to keep the turbine stationary in yaw when it is properly aligned (see Chapter 7). Support structures must be capable of resisting the torsional loads that will result from use of the yaw system.

8.4.5 Rotor Speed: Constant or Variable

Historically, most rotors on grid-connected wind turbines have operated at a nearly constant rotational speed, determined by the electrical generator (squirrel cage induction) and the gearbox. In many turbines today, however, the rotor speed is allowed to vary. Variable-speed rotors can be operated at the optimum tip-speed ratio to maximize power conversion in low wind and at lower tip-speed ratios in high winds to reduce loads in the drivetrain. On the other hand, variable-speed rotors require more complicated and expensive power conversion equipment than would otherwise be the case (see Chapter 5 for discussion of the electrical aspects of variable-speed wind turbines and Chapter 9 for rotor speed control).

8.4.6 Design Tip-Speed Ratio and Solidity

The design tip-speed ratio of a rotor is the tip-speed ratio at which the power coefficient is a maximum. Selection of this value will have a major impact on the design of the entire turbine. First of all, there is a direct relation between the design tip-speed ratio and the rotor's solidity, as was discussed in Chapter 3. A high tip-speed ratio rotor will have lower solidity than a slower rotor. For a constant number of blades, the chord and thickness will decrease as the solidity decreases. Owing to structural limitations, there is a lower limit to how thin the blades may be. Thus, as the solidity decreases, the number of blades may decrease as well.

There are a number of incentives for using higher tip-speed ratios. First of all, reducing the number of blades or their weight reduces the cost. Second, higher rotational speeds imply lower torques for a given power level. This should allow the balance of the drivetrain to be relatively light. On the other hand, there are some drawbacks to high tip speed ratios as well. For one thing, high tip-speed rotors tend to be noisier than slower ones because the aerodynamic noise increases with the relative wind speed perceived by the blade sections (see Chapter 14).

8.4.7 Hub: Rigid, Teetering, Hinged Blades, Gimballed

The hub design of a HAWT is an important constituent of the overall layout. The main options are rigid, teetering, or hinged. Most wind turbines employ rigid rotors. This means that the blades cannot move freely in the flapwise and edgewise directions. The term "rigid rotor" does include those with variable-pitch blades, however.

The rotors in two-bladed turbines are usually teetering. That means that a portion of the hub is mounted on bearings and can teeter back and forth, in and out of the plane of rotation. The blades in turn are rigidly connected to the teetering portion of the hub, so during teetering one blade moves in the upwind direction, while the other moves downwind. See Chapter 7 for more discussion of teetering. An advantage of teetering rotors is that the flapwise bending moments on the hub can be very low during normal operation.

Some two-bladed wind turbines have used hinges on the hub (e.g., the Smith–Putnam turbine – see Chapter 1). The hinges allow the blades to move into and out of the plane of rotation independently of each other. Since the blade weights do not balance each other, however, other provisions must be made to keep them in the proper position when the turbine is not running or is being stopped or started.

One design variant is known as a "gimballed turbine." It uses a rigid hub, but the entire RNA is mounted on horizontal bearings so that the machine can tilt up or down horizontally. This motion can help to relieve imbalances in aerodynamic forces. Such design has only been applied to small-scale wind turbines.

8.4.8 Rigidity: Flexible or Stiff

Turbines with lower design tip-speed ratios and higher solidities tend to be relatively stiff. Lighter, faster turbines are more flexible. Larger turbines are also more flexible than smaller turbines of a similar design. Flexibility may have some advantages in relieving stresses and reducing mass, but blade motions may also be more unpredictable. Most obviously, a flexible blade in an upwind rotor may be far from the tower when unloaded but could conceivably hit it in high winds. Flexible components such as blades or towers may have natural frequencies near the operating speed of the turbine. This is something to be avoided. Flexible blades may also experience "flutter" motion, which is a form of unstable and undesirable operation.

8.4.9 Number of Blades

Most large-scale wind turbines have three blades, although some have two or even one. Three blades have the particular advantage that the polar moment of inertia of the rotor with respect to yawing is constant and is independent of the azimuthal position of the rotor. This characteristic contributes to relatively smooth operation even while yawing. A two-bladed rotor, however, has a lower moment of inertia when the blades are vertical than when they are horizontal. This "imbalance" is one of the reasons that most two-bladed wind turbines use a teetering rotor. Using more than three blades could also result in a rotor with a moment of inertia independent of position, and slightly increase the power output, but more than three blades are seldom used, at least on electricity-generating turbines. This is primarily because of the higher costs but minimal benefit that would be associated with the additional blades. It is worth noting, however, that mechanical water-pumping windmills do have many more than three blades and also higher solidity. This facilitates their production of relatively high torque, which is beneficial to their function. See Chapter 16 for more details.

A few single-blade turbines have been built in the last 30 years. The presumed advantage is that the turbine can run at a relatively high tip-speed ratio, and that the cost should be lower because of the need for only one blade. On the other hand, a counterweight must be provided to balance the weight of the single blade. The aesthetic factor of the appearance of imbalance is another consideration.

8.4.10 Generator Speed

Generator speed choices include having one synchronous speed, multiple synchronous speeds, or a range of continuously variable speeds. Directly connected squirrel cage induction and synchronous generators, as a function of their design, have one synchronous speed. Another, now less common, option is a generator with two sets of windings, resulting in a generator with two operational speeds, depending on which winding is energized. Most large wind turbines today can operate at variable speeds. In any case, the nominal operating speed of the generator will be related to the rotor size and the speed-up ratio of the gearbox (if there is one).

The choice of generator speed has a significant effect on the design and weight of other components, particularly the gearbox: the higher the generator speed, the heavier the gearbox.

8.4.11 Support Structure

The support structure of land-based turbines consists of the tower and the foundation. Offshore wind turbines may be fixed or floating. Fixed offshore turbines have an additional substructure, founded on the seabed. Floating wind turbines have a floating substructure with a stationkeeping system.

The tower of a wind turbine serves to elevate the RNA up into the air to reach higher wind speeds and to provide clearance from the ground or ocean surface. Typically, tower heights are between 1.2 and 1.6 times the blade length for modern turbines. Winds are nearly always much stronger as elevation above ground increases, and they are less turbulent. All other things being equal, the tower should be as high as practical. Choice of tower height is based on an economic trade-off of increased energy capture vs. increased cost. Towers, foundations, and fixed offshore substructures are discussed in more detail in Chapter 10. Substructures of floating turbines are discussed in Chapter 11.

8.4.12 Design Constraints

There are inevitably numerous other factors that will influence the general design of a wind turbine. These include climatic factors, site-specific factors, and environmental factors, as summarized in Sections 8.4.12.1–8.4.12.3.

8.4.12.1 Climatic Factors Affecting Design

Turbines designed for more energetic or turbulent sites need to be stronger than those in more conventional sites. Expected conditions at such sites must be considered if the turbines are to meet international standards. This topic is discussed in more detail in Section 8.5.

General climate can affect turbine design in a number of ways. For example, turbines for use in hot climates may need provisions for extra cooling, whereas turbines for cold climates may require heaters, special lubricants, or even different structural materials. Turbines intended for use in marine climates need protection from salt and should be fabricated of corrosion-resistant materials wherever possible.

8.4.12.2 Site-specific Factors Affecting Design

Turbines that are intended for relatively inaccessible sites have their designs constrained in a number of ways. For example, they might need to be self-erecting. Difficulty in transport could also limit the size or weight of any one component. The maximum chord and maximum radius of the tower of modern onshore wind turbines are usually limited by transportation constraints, such as heights of bridges or width of highways; see Chapter 15 for more details.

Limited availability of expertise and equipment for installation and operation would be of particular importance for machines intended to operate singly or in small groups. This would be particularly important for applications in remote areas or developing countries. In this case, it would be especially important to keep the machine simple, modular, and designed to require only commonly available mechanical skills, tools, and equipment.

Offshore turbines, even if close to shore, may be inaccessible in high seas for both personnel and especially the vessels that would be needed for servicing. These and other site-specific constraints on the design of offshore wind turbines are discussed in Chapters 10, 11, and 15.

8.4.12.3 Social and Environmental Factors Affecting Design

Wind energy proponents inevitably emphasize the environmental benefits that accrue to society through the use of wind-generated electricity. On the other hand, there will always be some impacts on the immediate environment where the turbine may be installed, and not all of these may be appreciated by the neighbors. There is also an impact in terms of CO_2 production, electricity use, and materials needed to manufacture the turbine. Careful design, however, can minimize many of the adverse effects. Three of the most commonly noted environmental impacts of wind turbines are noise, visual appearance, and effects on birds. Some of the key issues affecting overall wind turbine design are summarized in the following paragraphs.

Wind turbines invariably produce some sound when they are operating. The main source of unwanted sound (noise) is blade aerodynamics, in particular toward the tip, so that the value of the tip velocity often becomes a design constraint. Excessive sound can be minimized through careful design, however. In general, upwind rotors are quieter than downwind rotors, and lower tip-speed ratio rotors are quieter than those with higher tip-speed ratios. Selection of airfoils, fabrication details of the blades, and design of tip brakes (if any) will also affect noise. Gearbox noise can be reduced by including sound proofing in the nacelle or eliminated by using a direct drive generator. Variable-speed turbines tend to make less noise at lower wind speeds, since the rotor speed is reduced under those conditions.

Visual appearance is very subjective, but there are reports that people prefer the sight of three blades to two, slow rotors to faster ones, and solid towers to lattice ones. A neutral color is often preferred as well.

In general, it appears that turbines with lower tip speeds and towers with few perching opportunities (such as tubular ones) are the least likely to adversely affect birds.

See Chapter 14 for more details on visual impact, noise, birds and other social and environmental factors.

8.4.13 Unconventional Wind Turbines

A variety of unconventional wind turbines have been proposed, and prototypes constructed and tested. These include those with ducted rotors, turbines with multiple rotors, and airborne turbines. They are beyond the scope of this text, but examples can be found in Nelson (2019).

8.5 Wind Turbine Design Standards, Technical Specifications, and Certification

Standards and guidelines provide a useful adjunct to the design process. There are many such documents relevant to the design, testing, and operation of wind turbines currently available. This section provides a summary of wind turbine-related standards and discusses how they are applied through the certification process. Section 8.6 then gives more details on the application of the key design standard (IEC 61400-1). Closely related to standards are technical specifications. These are similar to standards but are considered to be more recommendations than requirements. We do not make the distinction here, however.

8.5.1 Wind Turbine Standards and Technical Specifications

Until recently, standards and technical specifications were developed country by country, or by classification societies with relevant expertise. The lead in wind turbine standards is now being taken by the International Electrotechnical Commission (IEC). The most important IEC standards that are in use are listed in Table 8.1. Some other useful, but non-IEC, standards and guidelines are listed in Table 8.2.

All of the standards referred to in Tables 8.1 and 8.2 have some relevance to wind turbine design, although some are more directly relevant than others. It should be noted that the IEC standards are in a continuous process of revision and updating. The most important IEC standard is IEC 61400-1 (IEC, 2019a). It deals explicitly with design requirements. It applies particularly to larger, land-based turbines, but it is relevant to smaller turbines and

Table 8.1 Wind turbine-related International Electrotechnical Commission (IEC) standards

Source/Number	Title
IEC 61400-1	Wind Turbines – Part 1: Design Requirements
IEC 61400-2	Wind Turbines – Part 2: Safety Requirements for Small Wind Turbines
IEC 61400-3-1	Wind Turbines – Part 3-1: Design Requirements for Fixed Offshore Wind Turbines
IEC 61400-3-2	Wind Turbines – Part 3-2: Design Requirements for Floating Offshore Wind Turbines
IEC 61400-4	Wind Turbines – Part 4: Design Requirements for Wind Turbine Gearboxes
IEC 61400-5	Wind Turbines – Part 5: Wind Turbine Blades
IEC 61400-6	Wind Turbines – Part 6: Tower and Foundation Design Requirements
IEC 61400-11	Wind Turbines – Part 11: Acoustic Emission Measurement Techniques
IEC 61400-12	Wind Turbines – Part 12: Power Performance Measurements
IEC 61400-13	Wind Turbines – Part 13: Measurement of Mechanical Loads
IEC 61400-14	Wind Turbines – Part 14: Apparent Sound Power Levels and Tonality
IEC 61400-21	Wind Turbines – Part 21: Power Quality Measurements
IEC 61400-23	Wind Turbines – Part 23: Full-scale Structural Testing of Rotor Blades
IEC 61400-24	Wind Turbines – Part 24: Lightning Protection
IEC 61400-25	Wind Turbines – Part 25: Communications for Monitoring and Control

Table 8.2 Other wind turbine-related standards and guidelines

Source/Number	Title
DNV GL	Certification of Wind Turbines
DNV/Risø	Guidelines for the Design of Wind Turbines, 2nd edition
DNV-OS-J101	Design of Offshore Wind Turbine Structures
DNV-OS-J102	Design and Manufacture of Wind Turbine Blades

offshore turbines as well. IEC 61400-1 will be discussed in more detail in Section 8.6.2. IEC 61400-2 (IEC, 2013) is, as its name implies, concerned with small turbines. IEC 61400-3-1 (IEC, 2019b) applies to fixed offshore wind turbines, and IEC 61400-3-2 (IEC, 2019c) is for floating offshore wind turbines. They are both consistent with IEC 61400-1 and focus particularly on those design issues of relevance (waves, ocean currents, and substructures, etc.) that are not considered in IEC 61400-1. IEC 61400-4 (IEC, 2012a) is concerned specifically with the design of wind turbine gearboxes. IEC 61400-5 (IEC, 2020a) provides guidance on the design of wind turbine blades. IEC 61400-6 (IEC, 2020b) is used for the design of wind towers and foundations, particularly onshore.

A number of the standards apply to the evaluation and testing of aspect of a turbine design. IEC 61400-11 (IEC, 2012b) deals with the measurement of acoustic emissions from wind turbines, and IEC 61400-14 (IEC, 2005) is concerned with the suitable characterization of the sounds emitted from a wind turbine. IEC 61400-12 (IEC, 2002) is concerned with power performance measurements. IEC 61400-13 (IEC, 2015) concerns the proper measurement of mechanical loads. IEC 61400-21 (IEC, 2019d) is concerned with power quality measurements. This standard has the most relevance for the design of electrical and electronic components of a wind turbine. IEC 61400-23 (IEC, 2014) concerns testing of wind turbine blades. IEC 61400-24 (IEC, 2019e) concerns lightning protection. It is relevant to certain details of the wind turbine's design, specifically those that are particularly likely to be adversely affected by lightning. Finally, IEC 61400-25 (IEC, 2017) is concerned with monitoring and control of wind turbines and is relevant to the associated aspects of the design. Each of these might be used to evaluate the operation of a prototype turbine design and to provide information for subsequent modification of the design to, for example, reduce noise emissions, help identify areas where structural design changes might be needed or could possibly result in modifications to the blades' structural design.

The first document included in Table 8.2 is particularly concerned with the certification of wind turbines (see Section 8.5.2) but has indirect implications for the turbine design. The second document (DNV/Risø, 2002) is directly relevant to the design process. The third document listed, DNV-OS-J101 (DNV, 2014), provides guidelines for the design of offshore wind turbine support structures. The fourth document listed, DNV-OS-J102 (DNV, 2006), is concerned with the design of wind turbine blades.

It may be noted that there are many other standards that are relevant to the design of wind turbines. These standards are typically referred to where applicable. For example, IEC 61400-1 includes references to a number of other IEC standards that are of broader scope than wind turbines, as well as International Organization for Standardization (ISO) standards. One example of another IEC standard is IEC 60204-1:1997: Safety of Machinery – Electrical Equipment of Machines – Part 1: General requirements. Similarly, an example of an ISO standard is ISO 4354:2009: Wind Actions on Structures (ISO, 2009). Also note that the development of the wind turbine gearbclassification societies with relevant expertise.ox standard (IEC, 2014) is a joint ISO/IEC activity. For offshore wind turbine design, ISO 19901-1 (ISO, 2015b) is particularly relevant.

8.5.2 Certification

Certification is a process used to ensure that the wind turbines are actually built and installed in accordance with the standards. Certification in general is the confirmation of compliance of a product or a service with defined requirements (e.g., guidelines, codes, and standards). The certification of activities associated with wind turbines is undertaken by an approved, independent entity such as a national laboratory. These entities are referred to as "Certified Verification Agents" (CVA). Certification typically involves application of computer models and tests, undertaken in accordance with design standards, to verify that the design is consistent with the standard. Certification may apply to the following: (1) a specific design or type, (2) turbines, their support structures, and associated equipment or activities within a project, (3) components, or (4) prototypes. The certification process for wind turbines is under the auspices of the IEC System for Certification to Standards Relating to Equipment for Use in Renewable Energy Applications (IECRE). Details are given in IECRE (2017) and IECRE (2018).

8.6 Wind Turbine Design Loads

8.6.1 Overview

Once the basic layout of the turbine is selected, the next step in the design process is to consider in detail the loads that the turbine must be able to withstand. As is commonly used in mechanics, the loads are the externally applied forces or moments to the entire turbine or any of the components considered separately. The "load effect," as indicated previously, is the effect of the load in terms of stress or deflection. Knowledge of dimensions and often other properties of the component is required to determine the effect.

Wind turbine components are designed for two types of loads: (1) ultimate loads and (2) fatigue loads. Ultimate loads refer to likely maximum loads, multiplied by a safety factor. Fatigue loads refer to the component's ability to withstand an expected number of cycles of possibly varying magnitude. As discussed in Chapter 4, wind turbine loads can be considered to fall into five categories: (1) steady (here including static loads), (2) cyclic, (3) stochastic, (4) transient (here including impulsive loads), and (5) resonance-induced loads.

Vibrations of wind turbine components were also discussed in Chapter 4. It was noted there that operation of the turbine in a manner which excites blade and support structure natural frequencies should be avoided and that one way to identify points of correspondence between natural frequencies and excitation from the rotor is to use a Campbell diagram.

Sometimes operation at or near a natural frequency cannot be completely avoided. This may occur during start-up or shutdown, or at some rotor speeds of a variable-speed wind turbine. Effects of operation under such conditions must be considered. Wind turbine design standards developed by DNV (DNV GL, 2016) offer some guidance in this area.

8.6.2 Wind Turbine Design Loads and the IEC 61400-1 Design Standard

Many manufactured items are designed with reference to a particular "design point." This corresponds to an operating condition such that, if the item can meet that condition, it will perform at least adequately at any other realistic set of conditions.

A single design point is not adequate for wind turbine design. Rather, the wind turbine must be designed for a range of conditions. Some of these will correspond to normal operation, where most of the energy will be produced. Others are extreme or unusual conditions that the turbine must be able to withstand with no significant damage. The most important considerations are (1) expected events during normal operation, (2) extreme events, and (3) fatigue.

Enough experience has been gained with wind turbines that it has been possible to define a set of design conditions under which a turbine should be able to perform. These have been codified in IEC 61400-1 (IEC, 2019a; see also Section 8.5). The designer needs to be familiar with this (and other applicable standards), since a turbine's ability to meet these conditions must be demonstrated if it is intended for use in any country which enforces those standards.

8.6.2.1 The IEC 61400-1 Design Standard

The following sections provide a summary of the IEC 61400-1 design standard as it applies to the design process. It should be noted that a complete assessment of a turbine's ability to meet these requirements is not possible until a full design has been completed and analyzed. Information in this standard, however, provides a target for the design, so it should be considered early in the design process.

As previously mentioned, IEC 61400-1 may be considered to be the fundamental wind turbine design standard. Its purpose is to specify "design requirements to ensure the engineering integrity of wind turbine [and] ... to provide an appropriate level of protection against damage from all hazards during the planned lifetime." As such, it does not contain all the information that is needed to design a wind turbine and there are issues to consider that are outside the scope of the standard.

In accordance with IEC 61400-1, the process of incorporating loads into the design process consists of the following:

- determine a range of design environmental conditions;
- specify design load cases of interest, including operating and extreme wind conditions;
- calculate the loads for the load cases;
- evaluate load effects (stresses or deflections) to verify that they are acceptable.

8.6.3 Design Environmental Conditions

The design environmental conditions consist primarily of design wind conditions and for offshore wind turbines the other metocean conditions as well. These were all discussed in Chapter 6 and will not be repeated here. It is important to note the widely used acronyms, such as NWP (normal wind profile), NWS (normal wind shear), and EOG (extreme operating gust).

8.6.4 Design Load Cases

The next step is defining the design load cases (DLC). The load cases are based on the turbine's various operating states, as they are affected by the wind conditions and possible electrical or control system faults. Many of the situations have more than one load case. Most of the cases deal primarily with ultimate loads (U) but also include one fatigue load (F) case. These are summarized in the following paragraphs.

DLC 1 Power Production
"Power production" has five load cases which cover the full range of design wind conditions.

DLC 2 Power Production Plus Fault
This has three load cases which assume normal wind conditions and one with an extreme operating gust. Each of the load cases considers some type of fault, either with the turbine or the electrical network.

DLC 3 Start-up
"Start-up" load cases include a one-year extreme operating gust and a one-year extreme wind direction change for ultimate loading, as well as normal wind conditions (resulting in multiple starts) for fatigue.

DLC 4 Normal Shutdown
"Shutdown" includes a one-year extreme operating gust for ultimate loading and normal wind conditions (resulting in multiple stops) for fatigue.

DLC 5 Emergency Shutdown

"Emergency shutdown" includes one case in which normal winds are assumed. More extreme wind conditions are not considered here since the emergency shutdown event is the focus of the loading being evaluated.

DLC 6 Parked

The "parked" case considers extreme wind speed together with a loss of electrical connection (to make sure that the machine will not start-up) and normal turbulence for fatigue. Note that "parked" can refer to either standstill or idling.

DLC 7 Parked Plus Fault

"Parked plus fault" considers extreme wind speed, together with a possible fault (other than loss of electrical connection).

DLC 8 Transport, Assembly, Maintenance, and Repair

The eighth category, as the name implies, considers transport, assembly, maintenance, and repair. These cases are to be specified by the manufacturer.

Load Calculations

Calculation of the loads can be a very complex process. The principles that affect the loads were discussed in Chapter 4. Simplified methods can be used in the early stages of design for rough sizing of the components. The estimates from simple methods can also be improved if there are some data available from similar machines with which to "calibrate" the predictions. When such data are available, scaling methods, discussed in Section 8.8, can help to facilitate the calibrations. For final design in accordance with the IEC 61400-1 process, load calculations are carried out with an aeroelastic model such as *OpenFAST*, *Bladed*, or *HAWC2*, which are discussed in Section 8.9.

Evaluation of Load Effects

The design load cases are used to guide the analysis of critical components to ensure that they are adequate. Four types of analyses are undertaken:

1) Analysis of maximum strength;
2) Analysis of fatigue failure;
3) Stability analysis (e.g., buckling).
4) Deflection analysis (e.g., preventing blades from striking tower).

Fundamentally, the analyses first involve calculation of the expected loads for the various operating wind conditions as discussed above. From the loads and the dimensions of the components, the maximum load effects (stresses or deflections) are then found. Those load effects are then compared with the design stresses (or allowed deflections) of the material from which the component is constructed to make sure that they are low enough. Methods to do this are the subject of the next section.

8.7 Design Values, Safety Factors, and Probabilistic Design

In order to ensure that a design is suitable, it must be verified that all the components are adequate for the loads that they are expected to experience. The most common concern is that of strength, but other criteria are possible, such as deflection. Ensuring adequacy is the subject of this section, with a focus on strength.

8.7.1 Conventional Strength Assessment

In the traditional way to determine whether a component is strong enough, the designer evaluates the load and expresses the load effect as stress, S (Pa). That stress is compared to the resistance, R (i.e., strength), of the material, which is also measured in units of Pa. As long as the load effect is sufficiently less than the resistance, then the design is deemed to be adequate. Safety factors are used to account for uncertainties in the load or the material properties. In this case, S times the safety factor, γ, should be less than R. Another way to express the safety factor is as shown in Equation (8.1) in which $\gamma > 1$ for the design to be adequate.

$$\gamma = \frac{R}{S} \tag{8.1}$$

Consider the case when the applied stress S is 200, and the resistance R is 350 (units arbitrary). The safety factor γ in this case would be $350/200 = 1.75$. This method is simple to apply but has some serious limitations. In particular, it does not take into account the uncertainties either in the load or the material properties in a quantifiable way. Nowadays, other methods are applied, as described in the as described in Sections 8.7.2–8.7.5.

It is worth recalling at this point that the output of analysis codes such as *OpenFAST* are forces or moments, which need to be converted to stresses. A simple example is finding the stress at the base of a tower due to a bending moment. The stress in this case can be found readily from the base radius, c, and wall thickness, t from:

$$S = Mc/I \tag{8.2}$$

where M is the moment and I is the area moment of inertia of the tower base, which is given by:

$$I = \pi\left(c^4 - (c - t)^4\right)/4 \tag{8.3}$$

In general, more sophisticated tools, such as finite element methods (FEM), are used.

8.7.2 Varying Loads and Material Properties

In real situations for wind turbines, the load effect will vary significantly, depending on the conditions, so S is inherently quite variable and is best characterized by a distribution of values, rather than by a single value. These distributions can be quantified by collecting data over a long period of time. Alternatively, an aeroelastic code may be run many times with different inputs and the peak loads from all of them can be used to create the distribution. Similarly, manufacturing entails variability in the strength of the components, particularly in the blades but in other parts of the turbine as well, so R is also best characterized by a distribution. A series of tests of the material can be used to develop the distribution. In more specific terms, S could represent the distribution of 10-minute maximum stresses at a particular location, such as a blade root, over the lifetime of a turbine and R could represent the strength determined from hundreds of tests of the material used at that location.

Accordingly, the situation looks more like that shown in Figure 8.1. As illustrated here both S and R are random variables representing stress (MPa) modeled by probability distributions. The third curve, M, is for the safety margin, which is described in Section 8.7.2.1. In Figure 8.1, the curves are shown as normal distributions. In general, that is not the case, but that assumption simplifies the discussion. A description of how these situations are analyzed is provided in Sections 8.7.3–8.7.5. There are basically two approaches: (1) characteristic values and partial safety factors and (2) probabilistic design (reliability analysis) focusing on probability of failure.

8.7.2.1 Accounting for Uncertainties

In accounting for variations, or uncertainties, in load effects or material properties, what matters most at any given time is the difference between the load effect and the resistance: if the resistance is greater than the load effect, then there is no problem. If the converse is true, the component fails. As a result, the assessment is facilitated by the use

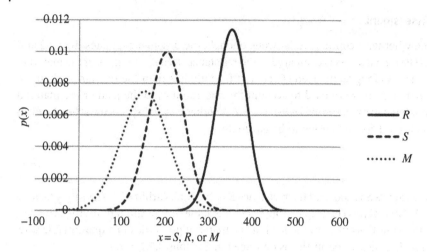

Figure 8.1 Example of load (*S*), resistance (*R*), and safety margin (*M*) distributions

of a safety margin, *M* which is defined as the difference between *R* and *S* as given in Equation (8.4). The resulting curve was also included in Figure 8.1. A component is considered strong enough if *R* is greater than *S*, which is equivalent to *M* > 0.

$$M = R - S \tag{8.4}$$

The following discussion summarizes the key aspects of approaches to account for these uncertainties beginning with the simplest method possible. In this case, there are two random variables of interest, the load *S* and the resistance *R* and they are normally distributed and uncorrelated. Note the convention is followed that random variables are denoted by upper case letters and specific realizations thereof are denoted by lower case letters; see Appendix C. Also, the symbols are bold to denote vectors, as in **x**, where applicable.

8.7.3 Design Values, Characteristic Values, and Partial Safety Factors

From the designer's point of view, the topics of particular relevance are the specific values that can be used in the design with the expectation that component will not fail except under improbable conditions. There are two values of special interest: the design load effect, which is the maximum acceptable value of the load effect, and the design resistance, which is the minimum acceptable value of the resistance. In particular, the design load effect should not be greater than the design resistance under any plausible situation.

It is apparent that as long as *M* > 0, as given in Equation (8.1), the resistance is greater than the load effect, so failure has not occurred. The border between the region of failing and not failing is expressed by the limit state function:

$$g(x) = r - s \tag{8.5}$$

The probability of failure is then the integral of the distribution of *M* over values less than zero. In other words:

$$p_F = \int_{-\infty}^{0} g(x)\,dx \tag{8.6}$$

Since we are assuming that both R and S are normally distributed and uncorrelated, the mean of the safety margin μ_M and its standard deviation σ_M are given in Equations (8.7) and (8.8) as:

$$\mu_M = \mu_R - \mu_S \tag{8.7}$$

$$\sigma_M = \sqrt{\sigma_R^2 + \sigma_S^2} \tag{8.8}$$

It is apparent from Equations (C.7) and (C.10) in Appendix C that the probability of failure as given in Equation (8.6) can be expressed as the cumulative distribution function (CDF) of the standard normal distribution, Φ, by Equation (8.9):

$$p_F = \Phi\left(\frac{0 - \mu_M}{\sigma_M}\right) = \Phi\left(-\frac{\mu_M}{\sigma_M}\right) \tag{8.9}$$

The ratio μ_M/σ_M is the argument of the CDF and, in this context, is known as the reliability index and given the symbol β. Therefore, the probability of failure is related to the reliability index by Equation (8.10). Also recall that the reliability index was introduced in Chapter 6 and used in a similar manner there.

$$p_F = \Phi(-\beta) = \frac{1}{\sqrt{2\pi}} \int_{-\infty}^{-\beta} e^{-\frac{u^2}{2}} du = \frac{1}{2}\left[1 - \operatorname{erf}\left(\frac{\beta}{\sqrt{2}}\right)\right] \tag{8.10}$$

Some typical values of β and corresponding probabilities of failure, p_F are shown in Table 8.3.

8.7.3.1 Consequence of Failure

An important aspect in the design process is the consequence of failure of a component. The most relevant factors that must be considered are risk to human life or injury, potential economic losses, and the cost to prevent failure. In accordance with these factors, consequence classes have been developed to assist in the design process; see ISO 2394 (ISO, 2015a), for example. An illustration of the relation between consequence, class of failure, and the corresponding reliability indices and probabilities of failure is given in Table 8.4. These consequences underlie the development of consequence of failure safety factors, which are discussed in Section 8.7.4.

Table 8.3 Typical values of reliability index and corresponding probabilities of failure assuming a normal distribution

p_F	10^{-1}	10^{-2}	10^{-3}	10^{-4}	10^{-5}	10^{-6}	10^{-7}
β	1.28	2.33	3.09	3.72	4.26	4.75	5.2

Table 8.4 Values of acceptable failure, annual reliability indices, and probabilities of failure

Class of failure	Consequence of failure	
	Less serious	Serious
I – redundant structure	$\beta = 3.09\,(p_F \approx 10^{-3})$	$\beta = 3.71\,(p_F \approx 10^{-4})$
II – significant warning before failure; nonredundant structure	$\beta = 3.71\,(p_F \approx 10^{-4})$	$\beta = 4.26\,(p_F \approx 10^{-5})$
III – No warning before failure; nonredundant structure	$\beta = 4.26\,(p_F \approx 10^{-5})$	$\beta = 4.75\,(p_F \approx 10^{-6})$

(*Source:* Adapted from DNV, 1992)

8.7.3.2 Characteristic Values

In analogy with Equation (8.1), one could define a "central" safety factor γ_c as Equation (8.11):

$$\gamma_c = \frac{\mu_R}{\mu_S} \tag{8.11}$$

In fact, however, the central safety factor is not too useful, since by conventional usage, a safety factor is used to compare a high value of a load effect to a low value of resistance. One method that is useful, however, is to apply "characteristic" values in conjunction with "partial" safety factors to determine design load effects and resistances. This is known as the Method of Partial Safety Factors and is discussed in Section 8.7.4. In this method, the characteristic value of the load effect is some relatively high value, which is exceeded only rarely (commonly, 2% of the time, i.e., $q = 0.98$ in Equation (8.12a)), and the characteristic value of the resistance is exceeded almost all of the time (commonly, 95% of the time, i.e., $p = 0.05$ in Equation (8.12b)).

The characteristic values are found by reference to the probability with which they are associated. Let us assume that the characteristic value of the load effect s_c is that value which is *greater* than all other loads with a probability of q, and that the characteristic value of the resistance r_c is that value which is *less* than all other resistances with a probability of $1 - p$. Since the normal distribution is assumed, then the characteristic values are:

$$s_c = \mu_S + k_q \sigma_S \tag{8.12a}$$

where

$$k_q = \Phi^{-1}(q)$$
$$r_c = \mu_R - k_{1-p}\sigma_R \tag{8.12b}$$

where

$$k_{1-p} = \Phi^{-1}(1-p)$$

For the common situation in which $p = 0.05$ and $q = 0.98$, the following values may be found:

$$k_q = 2.05 \text{ and } k_{1-p} = 1.64$$

8.7.4 Method of Partial Safety Factors

The method of partial safety factors is used by the IEC and many other organizations to define design values through characteristic values and specified safety factors. The method consists of three parts:

1) Determining design properties for materials by de-rating their characteristic (or published) properties using a material safety factor.
2) Determining design load effects by applying a load safety factor to expected peak loads.
3) Applying a consequence of failure safety factor.

The general requirement for ultimate loading is that the expected load effect, s_d, multiplied by a consequence of failure safety factor, γ_n, must be equal to or less than the design resistance, r_d. As a consequence, the design load effect is the highest allowable value of the expected load effect, and the design resistance is the minimum allowable resistance value. The requirement can be expressed as:

$$\gamma_n \ s_d \leq r_d \tag{8.13}$$

The design values for load effects in Equation (8.11) are found from the characteristic values of the load effect, s_c, by applying a partial safety factor for loads, γ_f:

$$s_d = \gamma_f \ s_c \tag{8.14}$$

The design values for the resistance in Equation (8.11) are found from the characteristic values of the materials, r_c, by applying a partial safety factor for materials, γ_m:

$$r_d = (1/\gamma_m)r_c \tag{8.15}$$

Partial safety factors are typically greater than 1.0, with certain exceptions. Normally, partial safety factors for loads range from 1.0 to 1.5 and the recommended value for many situations is 1.35. Partial safety factors for materials are usually at least 1.1, and partial safety factors for consequence of failure are 0.9, 1.0, and 1.2, depending on the situation. More discussion can be found in IEC (2019a). Partial safety factors for materials can also be found in many sources.

Example 8.1 Method of Partial Safety Factors

This example and the two subsequent examples in Section 8.7.5 were based on a hypothetical case of the tower base bending stress of the NREL 5-MW on a slightly modified tower. The tower diameter was assumed to be 6 m and the wall thickness 0.027 m such that the area moment of inertia was 2.259 m^4. The average bending moment at the base was taken to be 90 MNm, resulting in a mean stress (load effect) of $\mu_S = 120$ MPa. The standard deviation of the stress was assumed to be $\sigma_S = 20$ MPa. The tower was presumed to be of steel with a mean yield stress of $\mu_R = 250$ MPa and a standard deviation of $\sigma_R = 15$ MPa. For this example 8.1, it was also assumed that k_p and k_q were as given in Section 8.7.3, i.e., $k_p = 2.05$ and $k_{1-p} = 1.64$, and that the partial safety factors for loads, material and consequence were $\gamma_s = 1.25$, $\gamma_m = 1.1$, and $\gamma_n = 1.0$. The results are summarized in Table 8.5.

In this example, the design value of the load effect is less than that of the resistance, so failure is unlikely.

8.7.5 *Probabilistic Design and the Probability of Failure

The Method of Partial Safety Factors works well in many situations, particularly when there is one random variable for the load effect and one for the resistance, and both of those are normally distributed. In other situations, the method may be difficult to apply. This may be the case when there are more than two random variables of significance or when there are nonlinear relations between the variables. In these cases, a mathematically more comprehensive approach may be more appropriate. This is known as probabilistic design, and it deals more directly with the probability of failure. Partial safety factors are not necessarily used. (It should be noted that probabilistic design is closely related to methods of structural reliability and the terms are sometimes used interchangeably.)

Table 8.5 Results from Example 8.1

Variable	Symbol	Equation	Value
Safety margin, mean	μ_M	$\mu_M = \mu_R - \mu_S$	130
Safety margin, standard deviation.	σ_M	$\sigma_M = \sqrt{\mu_R^2 + \mu_S^2}$	25.0
Reliability index	β	$\beta = \mu_M/\sigma_M$	5.2
Probability of failure	p_f	$p_f = \Phi(-\beta)$	1.0×10^{-7}
Characteristic value of the load effect	s_c	$s_c = \mu_S + k_q\sigma_S$	161.0
Characteristic value of the resistance	r_c	$r_c = \mu_R - k_p\sigma_R$	225.4
Design load effect	s_d	$s_d = \gamma_s s_c$	201.3
Design resistance	r_d	$r_d = r_c/\gamma_m$	204.9

8.7.5.1 Multiple Linearly Related Normally Distributed Random Variables

When there are more than two random variables, the two-value approach summarized above may be readily extended as long as the random variables are normally distributed, and the limit state function is linear. For example, we may be investigating the probability of failure at the base of a tower, in a situation that depends linearly on the diameter, wall thickness, height, and material. Then, the limit state function is:

$$g(x) = a_0 + \sum_{i=1}^{n} a_i x_i \tag{8.16}$$

where the constants $a_0...a_n$ define the linear relations. The safety margin will also be normally distributed with mean value and variance given by Equations (8.17) and (8.18):

$$\mu_M = a_0 + \sum_{i=1}^{n} a_i \mu_{X_i} \tag{8.17}$$

$$\sigma_M^2 = \sum_{i=1}^{n} a_i^2 \sigma_{X_i}^2 \tag{8.18}$$

In the case that the variables are correlated, i.e., not independent, then Equation (8.18) is replaced by Equation (8.19):

$$\sigma_M^2 = \sum_{i=1}^{n} a_i^2 \sigma_{X_i}^2 + \sum_{i=1}^{n} \sum_{j=1, j \neq i}^{n} \rho_{ij} a_i a_j \sigma_i \sigma_j \tag{8.19}$$

where

ρ_{ij} is the correlation coefficient between X_i and X_j (ρ_{ij} is zero for uncorrelated variables).

In both cases, the reliability index, as before, is

$$\beta = \mu_M / \sigma_M \tag{8.20}$$

As given in Equation (8.10), the probability of failure is thus:

$$p_F = \Phi(-\beta) \tag{8.21}$$

8.7.5.2 Reliability and the Standard Normal Distribution

In many situations involving probabilistic design, the analysis is facilitated by transforming the original "basic" dimensioned variables (such as stress) into standard normal form (see Appendix C). After the calculations are complete, the standard normal variables can be transformed back into dimensioned form. The transformation into standard normal form is illustrated below for the case of two normally distributed basic variables (resistance and load effect) R and S with means and standard deviations μ_R, μ_S, σ_R, σ_S.

As before, the limit state function is:

$$g(r, s) = r - s = 0 \tag{8.22}$$

These can be written in terms of standard normally distributed random variables U_R, U_S such that:

$$U_R = \frac{R - \mu_R}{\sigma_R}, U_S = \frac{S - \mu_S}{\sigma_S} \tag{8.23}$$

The limit state function can be written in terms of u_R and u_S:

$$g(u_R, u_S) = r - s = (u_R \sigma_R + \mu_R) - (u_S \sigma_S + \mu_S) = 0 \tag{8.24}$$

The limit state function in terms of the safety margin M is then:

$$g(u_M) = (u_M \sigma_M + \mu_M) = 0 \tag{8.25}$$

where

$$U_M = \frac{M - \mu_M}{\sigma_M}$$

8.7.5.3 First Order Reliability Method (FORM)

The previous sections dealt with situations in which the random variables were linearly related. For more complicated situations, that method is insufficient. One commonly used method for these situations is the First Order Reliability Method (FORM).

In this method, the probability of failure is related to a "failure event" F which is given by the Equation (8.26).

$$F = \{g(\mathbf{x}) \leq 0\} \tag{8.26}$$

where $g(\mathbf{x})$ is the limit state function and $\mathbf{x}$ is a random vector representing the contributing factors to the failure. The brackets indicate that F is the set of all events for which a failure occurs.

In an extension of Equation (8.6), the probability of failure is given by the fraction of all the events for which $F \leq 0$. In other words:

$$p_f = \int_{g(\mathbf{X}) \leq 0} f_x(\mathbf{x}) d\mathbf{x} \tag{8.27}$$

where $f_x(\mathbf{x})$ is the joint probability density function of $\mathbf{x}$.

The following is an illustration of the method. This is simplified in that it assumes normally distributed variables and a linear limit state equation, so the same result could have been found in Section 8.7.5.1. Applying this method to a nonlinear limit state equation is discussed in Section 8.7.5.4.

We begin by assuming that there are two basic variables X_1 and X_2, which have been transformed to standard normal variables, U_1 and U_2, as illustrated in Figure 8.2.

When converted to standard normal form the limit state equation is given by $g(u_1, u_2) = 0$. We want to find the minimum value of the combination of u_1 and u_2. This method seeks to find the minimum of $\sqrt{u_1^2 + u_2^2}$ such that:

$$\beta = \min \sqrt{u_1^2 + u_2^2} \tag{8.28}$$

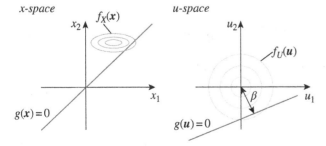

Figure 8.2 Transformation to standard normal variables

With the assumption that the limit state equation is linear, we use:

$$g(u_1, u_2) = a_0 + a_1 u_1 + a_2 u_2 = 0 \qquad (8.29)$$

It may be shown (e.g., by using Lagrange multipliers) that, for the minimum reliability index:

$$u_1 = \frac{-a_0 a_1}{a_1^2 + a_2^2} \qquad (8.30)$$

$$u_2 = \frac{-a_0 a_2}{a_1^2 + a_2^2} \qquad (8.31)$$

The minimum reliability index is then:

$$\beta = \sqrt{u_1^2 + u_2^2} = \sqrt{\left(\frac{-a_0 a_1}{a_1^2 + a_2^2}\right)^2 + \left(\frac{-a_0 a_2}{a_1^2 + a_2^2}\right)^2} = \frac{a_0}{\sqrt{a_1^2 + a_2^2}} \qquad (8.32)$$

The probability failure is found from Equation (8.10).

Example 8.2   First Order Reliability Method (Linear)

Suppose that the resistance R and load effect S have the same means and standard deviations as in Example 8.1 with limit state equation $g(r, s) = r - s = 0$. Converting to standard normal form and using the format of Equation (8.29), the limit equation becomes: $g(u_1, u_2) = 130 + 50u_1 - 20u_2 = 0$

In this case, $a_0 = 130$, $a_1 = 50$, and $a_2 = -20$. Equations (8.30) and (8.31) yield u_1 and u_2:

$$u_1 = \frac{-a_0 a_1}{a_1^2 + a_2^2} = -3.120 \text{ and } u_2 = \frac{-a_0 a_2}{a_1^2 + a_2^2} = 4.160$$

The reliability index from Equation (8.28) is $\beta = \sqrt{u_1^2 + u_2^2} = 5.2$, and the probability of failure, from Equation (8.10), is 1.0×10^{-7}.

Finally, since the limit state equation is linear, we can check the results directly by applying the method of Section 8.7.5.1. Recalling that since u_1 and u_2 are standard normal variables, their means and standard deviations are 0 and 1, respectively. Therefore, from Equations (8.17), (8.18), and (8.20) we have:

$$\mu_M = 130 + (-3.120)(0) + (4.160)(0) = 130 \text{ and } \sigma_M = \sqrt{(15)^2 + (-20)^2} = 25, \text{ and so again } \beta = 130/25 = 5.2 \text{ and}$$
$p_f = 1.0 \times 10^{-7}$, as expected.

8.7.5.4   First Order Reliability Method (FORM) for Nonlinear Limit State Equations

When the limit state equation is nonlinear, there are a number of approaches that can be used. One of them is to linearize the limit state equation using the first terms of a Taylor series and then apply the methods described in Section 8.7.5.3. For an equation which is a function of two variables, the Taylor series approximation about the initial points $u_{1,0}$, $u_{2,0}$ is:

$$g(u_1, u_2) \cong g(u_{1,0}, u_{2,0}) + \frac{\partial g(u_1, u_2)}{\partial u_1}(u_1 - u_{1,0}) + \frac{\partial g(u_1, u_2)}{\partial u_2}(u_2 - u_{2,0}) = 0 \qquad (8.33)$$

Similar to Equation (8.29), Equation (8.33) can be written more conveniently as:

$$g(u_1, u_2) \cong a_0 + a_1 u_1 + a_2 u_2 = 0 \qquad (8.34)$$

where

$$a_1 = \frac{\partial g(u_1, u_2)}{\partial u_1} \qquad (8.35)$$

$$a_2 = \frac{\partial g(u_1, u_2)}{\partial u_2} \tag{8.36}$$

$$a_0 = g(u_{1,0}, u_{2,0}) - a_1 u_{1,0} - a_2 u_{2,0} \tag{8.37}$$

The values of the constants a_0, a_1, and a_2 are found by evaluating the partial derivatives at the initial points $u_{1,0}$ and $u_{2,0}$. These will result in predictions for u_1 and u_2, as given in Equations (8.38) and (8.39). Note that in these equations an asterisk is added to distinguish the predicted values from the initial ones.

$$u_1^* = \frac{-a_0 a_1}{a_1^2 + a_2^2} \tag{8.38}$$

$$u_2^* = \frac{-a_0 a_2}{a_1^2 + a_2^2} \tag{8.39}$$

The solution is found by iteration, beginning by guessing initial values $u_1 = u_{1,\,0}$ and $u_2 = u_{2,\,0}$, predicting new values, substituting them for the initial guesses, and repeating until convergence. The probability of failure is given by Equation (8.32), and the design values are found by converting the standard normal variables back to dimensioned ones.

In some cases, it may be preferable to start with a target probability of failure or reliability index β (see Table 8.4, for example) and seek the combination of random variables that results in the desired failure probability. For example, it may be desirable to change the parameters of a component to meet a specific requirement. This can be done by extending the iteration process, as illustrated in the second part of the following example. Finally, Equations (8.33) through (8.39) can be readily extended when there are more variables involved.

Example 8.3   First Order Reliability Method (Nonlinear)

Consider the situation of the same material (steel) as in the previous examples, now with variable name x_1. In this case, tower height is the random variable, called x_2. The load effect is proportional (by the constant A) to the square of the height, so the limit state equation is $g(x_1, x_2) = x_1 - A x_2^2 = 0$. The basic variables have means and standard deviations μ_1, μ_2, σ_1, σ_2. Assume, for the sake of illustration, that the numeric values are those shown in Table 8.6 and that A equals 0.022. The first part of the example involves finding design values for the resistance, tower height, load effect, and the probability of failure. In the second part of the example, new design values are sought when a probability of failure of 10^{-6} is desired, and the standard deviation of the tower height is allowed to change. Resistance and load effect are assumed to have units of MPa.

The basic variables can be converted to standard normal form:

$$u_1 = \frac{x_1 - \mu_1}{\sigma_1}, u_2 = \frac{x_2 - \mu_2}{\sigma_2}$$

Substituting the standard normal variables into the limit state equation yields:

$$g(u_1, u_2) = \sigma_1 u_1 + \mu_1 - A(\sigma_2 u_2 + \mu_2)^2$$

From Equation (8.33), the approximation to the limit state equation is: now:

$$g(u_1, u_2) \cong g(u_{1,0}, u_{2,0}) + \frac{\partial}{\partial u_1}\left[\sigma_1 u_1 + \mu_1 - A(\sigma_2 u_2 + \mu_2)^2\right](u_1 - u_{1,0})$$

$$+ \frac{\partial}{\partial u_2}\left[\sigma_1 u_1 + \mu_1 - A(\sigma_2 u_2 + \mu_2)^2\right](u_2 - u_{2,0}) = 0$$

After taking the derivatives indicated, evaluating them at the initial point, and applying some algebra, the limit state equation can be put into the form of Equation (8.29):

$$g(u_1, u_2) \cong a_0 + a_1 u_1 + a_2 u_2 = 0$$

Table 8.6 Constants for nonlinear FORM example

Resistance		Height	
μ_1	σ_1	μ_2	σ_2
250	15	90	2.0

Table 8.7 Summary of standard normal calculations from nonlinear FORM example

a_1	a_2	a_0	u_1	u_2
15	−8.28	−72.16	−3.67	2.10

Table 8.8 Final results from nonlinear FORM example

Design resistance, $R_d = x_{1,d}$	Design height, $x_{2,d}$	Design load effect, $S_d = x_{2,d}$	β	p_f
194.7	94.07	194.7	4.21	1.27×10^{-5}

where

$$a_1 = -\sigma_1 \qquad a_2 = [2A(\sigma_2 u_2 + \mu_2)\sigma_2] \qquad a_0 = g(u_{1,0}, u_{2,0}) - (a_1 u_{1,0} + a_2 u_{2,0})$$

New values for the normalized variables u_1^* and u_2^* can be found from Equations (8.38) and (8.39). When the constants from Table 8.6 are substituted, the results from the standard normal form calculations are shown in Table 8.7. The iteration here began with $u_{1,0} = u_{2,0} = 1$.

The final results, including the design resistance $R_d = x_{1,d}$, design tower height, $x_{2, d}$, design load effect $S_d = A x_{2,d}^2$, reliability index β, and probability of failure p_f, are shown in Table 8.8.

Suppose a probability of failure of 10^{-6} is desired and the method used is to reduce the variability of the tower height. New values of standard deviation are considered, and the steps above are repeated until convergence. Note that this is a double iteration. The results are shown in Tables 8.9 and 8.10. When the iterations are complete, the standard deviation of the tower height is 0.45, and its mean is 90.25. The design load effect R_d has decreased.

Table 8.9 Summary of standard normal calculations from nonlinear FORM example, $p_f = 10^{-6}$

a_1	a_2	a_0	u_1	u_2
25 994	−6000	128 386	−4.634	1.056

Table 8.10 Final results from nonlinear FORM example, $p_f = 10^{-6}$

Design resistance, $R_d = x_{1,d}$	Design height, $x_{2,d}$	Design load effect, $S_d = x_{2,d}$	β	p_f
179.20	90.25	179.20	4.75	10^{-6}

8.7.5.5 Inverse First Order Reliability Method (IFORM)

The methods discussed above work well when the distributions are normal, but they are less convenient otherwise. In these situations, the Inverse First Order Reliability Method (IFORM) can be used. IFORM was first proposed for use with offshore structures by Winterstein *et al.* (1993) and makes use of the Rosenblatt transformation (Rosenblatt, 1952). This facilitates the transformation of arbitrary distributions into standard normal form and back. As above, certain calculations are much easier in standard normal form.

The most common use of IFORM in wind turbine design is the creation of environmental contours. The process to do this for two variables was summarized in Chapter 6 for wind speed and turbulent standard deviation and then for wind speed and significant wave height. The method can also be extended to more environmental variables. In terms of design evaluation, a structural dynamics model, such as *OpenFAST* (see Section 8.9), is then used to evaluate the loads at a range of the situations identified on the environmental contour. The highest loads are then used as the basis of the design.

In fact, the basic environmental contour method identifies the mean "worst" loads, whereas those loads are actually random variables themselves. Accordingly, the method may be extended to examine the distribution of those loads and take into account possible extremes beyond the mean. This approach has been considered by Agarwal (2008).

In addition, the IFORM method can be used when one of the variables represents the material properties, e.g., allowed stress, and the other variables are those that contribute to the load effect. In this case, the limit state equation divides the difference into "safe" and "unsafe" regions. In general, the limit state equation is not linear, and one of various methods, such as were discussed in the previous section, can be used to deal with the nonlinearity. Further discussion of IFORM is beyond the scope of this text. More information may be found in Melchers and Beck (2017), Haver and Winterstein (2009), and Saranyasoontorn (2006).

8.7.5.6 Monte Carlo Methods

In many situations, there are numerous random variables that are related in highly nonlinear ways. In such conditions, it may be difficult or impossible to use the methods discussed previously to ascertain the probability of failure. For these cases, Monte Carlo methods may be used. These methods involve running simulations many times with randomly selected inputs representing conditions with plausible distributions. The results are examined for rare events to see whether they are acceptable. Discussion of these methods is beyond the scope of this text. The reader should see Melchers and Beck (2017) or Sørensen (2011) for more information.

8.7.5.7 Probabilistic Load Extrapolation

Estimating extreme loads is a necessary but notoriously difficult and uncertain process. The fundamental wind turbine design standard, IEC 61400-1, recommends that multiple simulations be carried out over a range of wind speeds and the results of those simulations be extrapolated to estimate the highest loads that can be reasonably expected. In particular, the wind speed range is first divided into bins no less than 2 m/s wide. Then a minimum of fifteen simulations, 10 minutes long, are to be carried out for each wind speed bin from just below rated wind speed to cut-out, and six simulations for each wind speed below rated wind speed. Acceptable methods of extrapolation are described in detail in an annex to the standard. Discussion of various methods and some possible methods for dealing with uncertainties are provided by Moriarty *et al.* (2004), Fogle *et al.* (2008), Freudenreich and Argyriadis (2007), Agarwal and Manuel (2009), and van Eijk *et al.* (2017).

8.8 Scaling Relations

Sometimes design information is available about one turbine, and one wishes to design another turbine that is similar, but of a different size. In this case, one can take advantage of some scaling relations for the rotor in laying out the preliminary design.

Table 8.11 Summary of scaling relations

Quantity	Symbol	Relation	Scale dependence
Power, forces, and moments			
Power	P	$P_1/P_2 = (R_1/R_2)^2$	$\sim R^2$
Torque	Q	$Q_1/Q_2 = (R_1/R_2)^3$	$\sim R^3$
Thrust	T	$T_1/T_2 = (R_1/R_2)^2$	$\sim R^2$
Rotational speed	Ω	$\Omega_1/\Omega_2 = (R_1/R_2)^1$	$\sim R^{-1}$
Weight	W	$W_1/W_2 = (R_1/R_2)^3$	$\sim R^3$
Aerodynamic moments	M_A	$M_{A,1}/M_{A,2} = (R_1/R_2)^3$	$\sim R^3$
Centrifugal forces	F_c	$F_{c,1}/F_{c,2} = (R_1/R_2)^2$	$\sim R^2$
Stresses			
Gravitational	σ_g	$\sigma_{g,1}/\sigma_{g,2} = (R_1/R_2)^1$	$\sim R^1$
Aerodynamic	σ_A	$\sigma_{A,1}/\sigma_{A,2} = (R_1/R_2)^0 = 1$	$\sim R^0$
Centrifugal	σ_c	$\sigma_{c,1}/\sigma_{c,2} = (R_1/R_2)^0 = 1$	$\sim R^0$
Resonances			
Natural frequency	ω	$\omega_{n,1}/\omega_{n,2} = (R_1/R_2)^{-1}$	$\sim R^{-1}$
Excitation	Ω/ω	$(\Omega_1/\omega_{n,1})/(\Omega_2/\omega_{n,2}) = (R_1/R_2)^0 = 1$	$\sim R^0$

Note: R, rotor radius

8.8.1 Load Scaling Relations

Scaling relations for wind turbines start with the following assumptions:

- the tip-speed ratio remains constant;
- the number of blades, airfoil, and blade material are the same;
- geometric similarity is maintained to the extent possible.

The scaling relations for a number of important turbine characteristics are described in this section, first when the radius is doubled and then for the general case. They are also summarized in Table 8.11.

It should be noted that these scaling relations are idealized and, in fact, large, modern turbines are not as heavy as these relations suggest they should be. This is because of advances in the technology over the past several decades. For example, recent developments of larger machines indicate an increase of rotor mass at a rate significantly less than the "square–cube law" (power and mass vs. radius) predicts. More discussion on this topic is provided by Veers *et al.* (2019). The scaling relations remain useful as reference points, however.

8.8.2 Power

Power, as discussed previously, is proportional to the swept area of the rotor, so doubling the radius will quadruple the power. In general, power is proportional to the square of the radius.

8.8.3 Rotor Speed

With the tip-speed ratio held constant, the rotor speed will be halved when the radius is doubled. In general, rotor speed will be inversely proportional to the radius.

8.8.4 Torque

As noted above, when the radius is doubled, the power is quadrupled. Since the rotor speed will drop by half, the torque will be increased by a factor of 8. In general, the rotor torque will be proportional to the cube of the radius.

8.8.5 Aerodynamic Moments

The forces in the blades go up as the square of the radius, and the moments are given by the forces multiplied by the distances along the blade. When the radius is doubled, the aerodynamic moments will increase by a factor of 8. In general, aerodynamic moments will be proportional to the cube of the radius.

8.8.6 Rotor Weight

By the assumption of geometric similarity, as the turbine size gets larger, all dimensions will increase. Therefore, if the radius doubles, the volume of each blade goes up by a factor of 8. Since the material remains the same, the weight must also increase by a factor of 8. In general, rotor weight will be proportional to the cube of the radius. Note that the fact that the weight goes up as the cube of the dimension, whereas the power output goes up as its square gives rise to the famous "square–cube law" of wind turbine design. Some people have argued that it is this "law" that may eventually limit the ultimate size that turbines may reach. Whether or not that is the case, or if some other factor is most important, remains to be seen.

8.8.7 Maximum Stresses

Maximum bending stresses, σ_b, in the blade root due to flapwise moments applied to the blade, M, are related to the thickness of the root, t, and its area moment of inertia, I, by $\sigma_b = M(t/2)/I = M/(2I/t)$, as should be clear from discussions in Chapter 4. (The root is assumed to be solid here).

To determine how bending stresses change with rotor size, scaling laws for the area moment of inertia need to be determined. For simplicity, consider the blade root to be approximated by a rectangular cross-section of width c (corresponding to the chord) and thickness t. The area moment of inertia about the flapping axis is $I = ct^3/12$. If the radius is doubled, then the moment of inertia goes up by a factor of 16, and the thickness by a factor of 2. The ratio $2I/t$, which is given by $2I/t = ct^2/6$, is then increased by a factor of 8, just like the aerodynamic moments. In general, the blade root area moment of inertia scales as R^3.

Maximum stresses due to aerodynamic moments, blade weight, and centrifugal force are a function of the area moment of inertia and the applied moments. They are discussed in more detail in the next several paragraphs.

8.8.7.1 Stresses Due to Aerodynamic Moments

Aerodynamically induced stresses, σ_A, are unchanged with scaling. This is true for both the flapwise and lead–lag directions, as should be apparent from the discussion above. This can be demonstrated simply as follows: Under simplifying assumptions, the force varies with square of the rotor diameter and the moment M at the root with the cube. The dimensions of the root vary in proportion to the rotor diameter and the moment of inertia of the root varies as the 4th power, so Mc/I stays constant.

8.8.7.2 Stresses Due to Blade Weight

In-plane (edgewise) stresses due to blade weight, unlike most other stresses in the rotor, are not independent of size. In fact, they increase in proportion to the radius. Allowance for that difference must be made during the design process.

Consider a horizontal blade of weight, W, and center of gravity distance, r_{cg}, from the hub. The maximum moment due to gravity, M_g, is:

$$M_g = Wr_{cg} \tag{8.40}$$

The maximum stress due to gravity, σ_g, in the edgewise direction for a rectangular blade root (here with $I = tc^3/12$) is therefore:

$$\sigma_g = \left(Wr_{cg}\right)(c/2)/I = Wr_{cg}/\left(tc^2/6\right) \tag{8.41}$$

Since weight scales as R^3 and the other dimensions scale as R, the stress due to weight also scales as R. The general relation is: then:

$$\sigma_{g,1}/\sigma_{g,2} = (R_1/R_2)^1 \tag{8.42}$$

8.8.7.3 Stresses Due to Centrifugal Force

Stresses due to centrifugal force are unchanged by scaling. This can be illustrated as follows. The tensile stress, σ_c, due to centrifugal force, F_c, applied across area A_c is given by:

$$\sigma_c = F_c/A_c \tag{8.43}$$

Centrifugal force itself is found from:

$$F_c = \frac{W}{g}\, r_{cg}\, \Omega^2 \tag{8.44}$$

where Ω is the rotor rotational speed and g is the gravitational constant. Blade weight scales as R^3, r_{cg} scales as R^1, and Ω scales as R^{-1}. Thus $F_c \sim R^2$. It is also the case that $A_c \sim R^2$, so σ_c is independent of R. In general

$$\sigma_{c,1}/\sigma_{c,2} = (R_1/R_2)^0 = 1 \tag{8.45}$$

8.8.8 Blade Natural Frequencies

Blade natural frequencies decrease in proportion to the radius. This can be seen by modeling a blade as a rectangular cantilevered beam of dimensions c wide, t thick, and R long. As shown in Chapter 4, the angular natural frequency ω_n of a simple cantilevered beam is given by:

$$\omega_n = \frac{(\beta R)_i^2}{R^2}\sqrt{\frac{E\,I}{\tilde{\rho}}} \tag{8.46}$$

where E is the modulus of elasticity, I is the area moment of inertia, $\tilde{\rho}$ is the mass per unit length, and $(\beta R)_i^2$ is a series of constants such that $(\beta\,R)_i^2 = (3.52, 22.4, 61.7, \ldots)$.

For example, $I = ct^3/12$ and $\tilde{\rho} = \rho_b c\, t$ (where ρ_b = mass density of blade). In this case:

$$\omega_n = \frac{(\beta R)_i^2}{R^2}\sqrt{\frac{E\,c\,t^3}{12\,\rho_b\,c\,t}} = \frac{(\beta R)_i^2}{R^2}t\sqrt{\frac{E}{12\,\rho_b}} \tag{8.47}$$

Blade thickness is proportional to the radius. Therefore, it is apparent that $\omega_n \sim R^{-1}$. In general, the relation of natural frequencies between two blades (1 and 2) is:

$$\omega_{n,1}/\omega_{n,2} = (R_1/R_2)^{-1} \tag{8.48}$$

Since rotor rotational speed also decreases with radius, the propensity of the rotor to excite a particular resonance condition is independent of radius.

8.9 Computer Codes for Wind Turbine Design

8.9.1 Overview of the Use of Computer Codes in the Wind Industry

Computer codes play a very large role in the overall design process. They are used to specify and evaluate the design of components and of the complete wind turbine itself. Testing of components and prototype wind turbines provides a significant database for the evaluation of a design, but no testing program can duplicate all of the operating conditions that a wind turbine might experience in its lifetime without prohibitively long and costly testing. Computer codes may be used to model many of those operating conditions and to provide assurances that the wind turbine will operate as desired and last as long as desired. These computer codes may be developed by private companies or may be publicly available codes developed by government research laboratories.

Testing of a wind turbine design requires models for (1) the turbulent inflow, the sea state (for offshore turbines), and other environmental conditions that might be important, (2) the generation of aerodynamic torques and forces at the surfaces of the blades and the structure, (3) the dynamics of the drivetrain and structure (including the foundation), and (4) the conversion of the loads into stresses, fatigue, and any other variables that are of interest. It is important to ensure that the models that are used are designed to simulate those aspects of the system operation that are of most importance to the end-user. Modeling results should also be compared with test results to validate the models.

8.9.2 Categories of Models Used in the Wind Industry

Figure 8.3 categorizes the codes used in onshore wind turbine design into five groups. The turbulent wind modeling codes take input parameters such as turbulence intensity, shear, mean hub height, wind speed, and various statistical parameters that describe the turbulent flow field (turbulent length scale, characterization of the frequency spectrum of the turbulence, etc.) and generate wind speeds across the rotor that represent a turbulent wind field (see Appendix C for more information on the synthesis of wind data). Many different wind speed cases are often generated, including special cases of extreme conditions specified in the IEC 61400-1 design standard. Similar sea state modeling codes are used for offshore applications. The component design codes shown in Figure 8.3 might be provided by suppliers of turbine components to simulate the performance of their products. The aerodynamic design codes are those codes that take the input aerodynamic conditions and the current rotor configuration and motions to determine the aerodynamic loading on the blade surface. These are closely coupled with the turbine design codes which model the rest of the wind turbine structural dynamics and control system to generate a description of the loads in the system. Those loads are used to determine the ultimate strength and fatigue life of the various components of the wind turbine. All of the load and

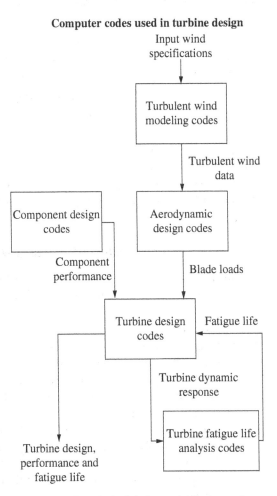

Computer codes used in turbine design

Figure 8.3 Overview of design codes for an onshore wind turbine

Table 8.12 Some computer codes used in wind turbine design.

Code	Source	Purpose
OpenFAST	NREL	Detailed structural dynamics model
TurbSim	NREL	Wind simulator; preprocessor
HydroDyn	NREL	Sea states simulator, hydrodynamics loads calculation; preprocessor or dynamic module.
IECWind	NREL	Wind simulator; preprocessor
AirfoilPrep	NREL	Generates airfoil data; aerodynamics preprocessor
PROPID	Univ. of Illinois	Rotor design code
PreComp	NREL	Computes coupled section properties of composite blades; structural response preprocessor
BModes	NREL	Computes mode shapes and frequencies of blades, towers, and monopiles; structural response preprocessor
AeroDyn	NREL	Generates aerodynamic forces; input to structural models
MoorDyn	NREL	Mooring structural dynamics model.
NuMAD	Sandia	Blade design; preprocessor for finite element model, ANSYS
Bladed	DNV	Detailed structural dynamics model
HAWC2	DTU	Time domain aeroelastic analysis
ROSCO	NREL	Control system design
MLife	NREL	Fatigue

fatigue life information is then used to evaluate and improve the wind turbine design until the performance and cost are acceptable. The final design specifications are then used to evaluate the performance of the turbines at the intended installation site. The codes may be separate and free-standing, or multiple capabilities may be included in a single code.

8.9.3 Examples of Computer Codes Used in Wind Turbine Design

Over the past 30 years, a many computer codes for wind turbine design have been developed and refined. Some of them are listed in Table 8.12 and then described in more detail below.

8.9.3.1 OpenFAST

OpenFAST is a medium-complexity aero-servo-hydro-elastic open source code for multi-physics (aerodynamics, hydrodynamics, structural dynamics, generator dynamics, and control) analysis of HAWTs with one to three blades and a teetering or rigid hub. The code can be used for time domain or linearization analyses. The code models the wind turbine as a combination of rigid and flexible bodies and uses a combination of lumped parameter, shape-functions, and finite-elements approaches. In its current form, *OpenFAST* incorporates multiple modules (some described below): *AeroDyn* to calculate the aerodynamic forces on the blades, *HydroDyn* to compute the hydrodynamics, *SoilDyn* to compute the soil response, and *ServoDyn* for the control system. In its baseline onshore formulation, 14 degrees of freedom are used, including 4 tower bending modes, three blade-bending modes, yaw, teeter, and drivetrain torsion. The flexibility of fixed-bottom and floating platforms is modeled using the module *SubDyn*, which is a linear finite element software that uses a Craig-Bampton reduction to limit the number of degrees of freedom. Mooring dynamics are accounted for using the *MoorDyn* module. Blades with high flexibility and torsion can be modeled using the geometrically exact *BeamDyn* module. *BeamDyn*, *MoorDyn*, and *SubDyn* add degrees of freedom to the system. *OpenFAST* has been approved for calculation of both onshore and offshore wind turbine loads for design and certification. See https://openfast.readthedocs.io/ for details.

8.9.3.2 TurbSim
TurbSim is used to synthesize time series of turbulent wind data, including coherent structures, for input to design codes such as *YawDyn* and *OpenFAST*. It is based on techniques such as are described in Appendix B and Veers (1988). See https://www.nrel.gov/wind/nwtc/turbsim.html for details.

8.9.3.3 HydroDyn
HydroDyn is used to synthesize time series of wave and sea current data to input to the design code *OpenFAST*. *HydroDyn* is also used as part of *OpenFAST* to compute the hydrodynamic loads on the structure. See https://www.nrel.gov/wind/nwtc/hydrodyn.html for details.

8.9.3.4 IECWind
IECWind may be used to synthesize wind data for input to design codes. The synthesized data is of the form required by the IEC 61400-1 standard. See Section 8.6 and https://www.nrel.gov/wind/nwtc/iecwind.html.

8.9.3.5 AirfoilPrep
AirfoilPrep generates airfoil data for use by other codes. It adjusts two-dimensional data to account for rotational effects, using methods from Selig for stall delay and Eggers for drag. It extrapolates to high angles of attack using the Viterna method or flat plate theory. It also computes dynamic stall characteristics. See https://www.nrel.gov/wind/nwtc/airfoil-prep.html for details.

8.9.3.6 PropID
PropID was developed to facilitate the design of wind turbine blades, considering aerodynamics, structure, cost, and noise. It can also be used to analyze a given rotor. See https://m-selig.ae.illinois.edu/propid.html.

8.9.3.7 PreComp
This code is a preprocessor that calculates section properties of composite blades. It finds section properties such as inertia and stiffnesses. Inputs are external blade shape and internal lay-up of composite layers. See https://www.nrel.gov/wind/nwtc/precomp.html for details.

8.9.3.8 BModes
BModes is a preprocessor that calculates mode shapes and frequencies of cantilevered beams, particularly blades and towers (and monopiles for offshore). The beam is divided into a number of finite elements. Each element has three internal and two boundary nodes and is characterized by 15 degrees of freedom. Inputs for a blade include geometric and structural properties, rotational speed, and pitch angle. More details may be found at https://www.nrel.gov/wind/nwtc/bmodes.html.

8.9.3.9 Aerodyn
Aerodyn is a code that calculates aerodynamic loads for HAWTs and VAWTs. It is typically used in conjunction with a structural model such as *OpenFAST*. *AeroDyn* implements multiple lifting-line algorithms, including an unsteady blade element momentum (BEM) code (see Chapter 3), a vortex-lattice method, and a double multiple stream tube method (for VAWTs). Its BEM code includes the following features: skewed wake, unsteady airfoil aerodynamics (e.g., Beddoes–Leishman dynamic stall approach), tower shadow, and unsteady inflow (turbulence). More details may be found at https://www.nrel.gov/wind/nwtc/aerodyn.html.

8.9.3.10 MoorDyn

MoorDyn is used to compute the dynamic response of mooring lines as part of *OpenFAST*. See https://www.nrel.gov/wind/nwtc/moordyn.html for details.

8.9.3.11 NuMAD

NuMAD (Numerical Manufacturing And Design) is a preprocessor and post-processor for ANSYS, which is commercial finite element analysis software. It is intended for use in analyzing wind turbine blade structures. It may also be used to create blade design inputs for *OpenFAST*. This code is available from Sandia National Laboratory at https://energy.sandia.gov/tag/numad/; see Berg *et al.* (2012) for more details.

8.9.3.12 Bladed

Bladed is a medium-complexity aero-servo-hydro-elastic commercial software for wind turbine performance and load calculations. It is used for preliminary or final design and wind turbine certification. Its features are similar to *OpenFAST*. Details may be found in Bossanyi (2006) and https://www.dnv.com/services/wind-turbine-design-software-bladed-3775.

8.9.3.13 HAWC2

HAWC2 is a medium-complexity aero-servo-hydro-elastic licensed software intended for calculating the response of wind turbines in the time domain. Its features are similar to *OpenFAST,* but the structural formulation of *HAWC2* uses a multi-body co-rotational finite element beam formulation throughout. It was developed by the Danish Technical University (DTU). More details are found at http://www.hawc2.dk/.

8.9.3.14 ROSCO

ROSCO (Reference Open-Source Controller) was developed by NREL to facilitate modeling a variable speed wind turbine controller for use in *OpenFAST*. Details may be found in Abbas *et al.* (2021)

8.9.3.15 MLife

MLife computes statistical information and fatigue estimates for time-series data, such as would be output from *OpenFAST* or similar codes. See Chapter 7 for more information.

8.9.4 Verification and Validation of Computer Codes

The computer codes described above have all been verified or validated to some degree. Verification typically refers to making comparisons to other codes or analytical results, whereas validation is generally assumed to involve comparison to test data. These distinctions are not always made consistently, however, so the user must pay close attention to each situation. Some examples of code verifications are presented in Buhl and Manjock (2006) and Jonkman and Buhl (2007). An example of model comparison to test data is given in Schepers *et al.* (2002). More recently, the International Energy Agency has facilitated code verification via a number of working groups, such as the Offshore Code Comparison Collaboration (OC3) of IEA Task 23 (see Jonkman and Musial, 2010).

8.10 Power Curve Prediction

Prediction of a wind turbine's power curve is an important step in the design process. It involves consideration of the rotor, gearbox, generator, and control system. In general, the power curve is obtained by running an aeroelastic tool at different wind speed, for multiple shear conditions and a range of turbulence intensities.

Variable-speed wind turbines, as discussed in Chapter 5, are connected to the grid indirectly. They typically have either a synchronous generator or a wound rotor induction generator, together with power electronic converters, as also explained in Chapter 5. Power curves of variable-speed, variable-pitch wind turbines are discussed in Chapter 9. A simplified method for directly connected, fixed-pitch turbines is summarized below.

Subsequent to the prediction of the power curve the wind turbine manufacturer will need to certify their results by field tests. The IEC 61400-12 standard, which was introduced in Section 8.5, will be used for this. Note that data from such tests will typically exhibit a lot of scatter. The power is generally averaged over wind speed bins to create a useful curve.

8.10.1 Simplified Power Curve Prediction of Directly Connected, Fixed Pitch Wind Turbines

Predicting the power curve for a directly connected, fixed-pitch wind turbine involves matching the power output from the rotor as a function of wind speed and rotational speed to the power produced by the generator, also as a function of rotational speed. The effects of component efficiencies are also considered where appropriate. In this discussion, it is assumed that all drivetrain efficiencies are accounted for by adjusting the rotor power. The process may be done either graphically or in a more automated fashion. The graphical method best illustrates the concept and will be described here.

Rotor power as a function of rotational speed is predicted for a series of wind speeds by applying estimates for the power coefficient, C_P. The power coefficient as a function of tip-speed ratio, and hence rotational speed, may be obtained from a BEM code as described in Chapter 3. Preferably, the power coefficient is computed using an aeroelastic software for a set of wind speeds, and rotor speeds (and pitch angles), as the deflection of the structure will affect the aerodynamic performances. The rotor power, P_{rotor}, is then:

$$P_{rotor} = C_P \eta \frac{1}{2} \rho \pi R^2 U^3 \tag{8.49}$$

where η = drivetrain efficiency, ρ = air density, R = rotor radius, and U = wind speed. The rotor speed, n_{rotor}, in rpm is found from the tip-speed ratio, λ:

$$n_{rotor} = \frac{30}{\pi} \lambda \frac{U}{R} \tag{8.50}$$

A power vs. rpm relation is found for the generator and referred to the low-speed side of the gearbox by dividing the generator speed by the gearbox speed-up ratio. This relation is superimposed on a series of plots (for a range of wind speeds) for rotor power vs. rotor rpm. Every point where a generator line crosses a rotor line defines a pair of power and wind speed points on the power curve. These points also define the operating speed of the rotor.

As was explained in Chapter 5, generators that are directly connected to the grid are usually of the squirrel cage induction type (SCIG). These generators turn at a nearly fixed speed, determined primarily by the number of poles and grid frequency but also by the power level, in accordance with their slip. The relation may also be expressed as:

$$P_{generator} = \frac{g\,n_{rotor} - n_{sync}}{n_{rated} - n_{sync}} P_{rated} \tag{8.51}$$

where $P_{generator}$ is the generator power, g is the gearbox ratio, P_{rated} is the rated generator power, n_{sync} is the synchronous speed of the generator, and n_{rated} is the speed of the generator at rated power.

Example: The following example illustrates the process of estimating the power curve for a hypothetical wind turbine with a SCIG. The turbine has a rotor of 20 m diameter with a power coefficient vs. tip-speed ratio relation illustrated in Figure 8.4.

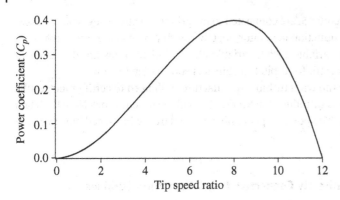

Figure 8.4 Rotor power coefficient vs. tip-speed ratio

The overall mechanical and electrical efficiency is assumed to be 0.9. Two possible pairs of gear ratios and generator ratings are considered. Six wind speeds are used, ranging from 6 to 16 m/s. It is assumed that power will be regulated at above-rated wind speed (16 m/s), so only the part of the power curve at or below 16 m/s is shown. Gearbox 1 has a speed-up ratio of 36 : 1, whereas gearbox 2 has a speed-up ratio of 24 : 1. The rated power of generator 1 is 150 kW and that of generator 2 is 225 kW. Both generators are of the squirrel cage induction type. They have a synchronous speed of 1800 rpm and a speed of 1854 rpm at rated power. Figure 8.5 illustrates the power vs. rotational speed curves for the six wind speeds and two generator/gearbox combinations.

The power curves that can be derived from Figure 8.5 are shown in Figure 8.6. For comparison, an ideal variable-speed power curve is shown for the same wind speed range. The ideal curve was obtained by assuming a constant power coefficient of 0.4 over all wind speeds. As can be seen from the figure, gearbox/generator combination 1 would produce more power than combination 2 at winds less than about 8.5 m/s, but less than combination 2 at higher winds.

Curves of the type developed in this example can be useful in selecting the generator size and gearbox ratio. By combining the power curves with characterizations of prospective wind regimes (as described in Chapter 2), the effect on annual energy production can be estimated. Generally speaking, as illustrated in this example, a smaller generator and slower rotor speed (larger gearbox ratio) will be beneficial when the wind speeds are lower. Conversely, a larger generator and faster rotor speed are more effective in higher winds.

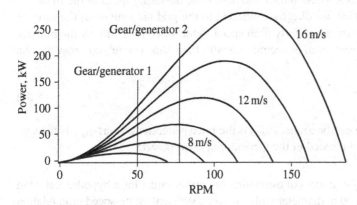

Figure 8.5 Rotor and generator power vs. rotor speed for turbine with SCIG

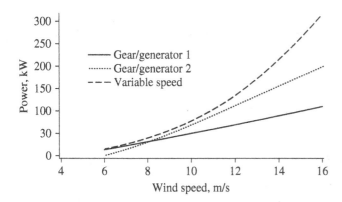

Figure 8.6 Power curves

8.11 Design Evaluation

Once a detailed design for the wind turbine has been developed, its ability to meet basic design requirements, such as those discussed in Section 8.8, must be assessed. This design evaluation should use the appropriate analytical tools. Where possible, validated computer codes, such as those described in Section 8.9, should be used. When necessary, models specific to the application may need to be developed.

There are five steps that need to be taken in performing a detailed design evaluation:

1) Prepare the wind (and metocean for offshore) input;
2) Model the turbine;
3) Perform a simulation to obtain loads;
4) Convert predicted loads to stresses;
5) Assess damage;

Each of these steps is summarized in the following subsections. An extensive discussion of detailed design evaluations for a number of turbine types is given in Laino (1997).

8.11.1 Wind and Metocean Inputs

Wind and metocean input data that will correspond to the design input conditions need to be generated. For extreme winds and discrete gusts, specifying the wind input is relatively straightforward, given the guidelines summarized in Section 8.5. Converting that wind input to time series inputs can also be done fairly simply. Synthetic turbulent wind can be produced by using public domain computer codes such as *TurbSim*, which was described in Section 8.9. A knowledge of the sea state at a given site can be used to generate wave data relevant for the design evaluation and provide the model with corresponding sea state conditions (wave directions, current, sea level, etc.)

8.11.2 Model of Turbine

The next step is developing a detailed model of the wind turbine. This should include aerodynamics, structural dynamics, generator dynamics, control systems, and hydrodynamics (for offshore applications). This can be done from basics, using the methods discussed in Chapters 3 and 4, but, when possible, it is preferable to use models that are already available. Some of the presently available models that may be appropriate include *OpenFAST*, *Bladed*, and *HAWC2*. These codes and others were discussed in more detail in Section 8.9.

Once the model has been selected or developed, inputs describing the specific turbine need to be assembled. These generally include mass and stiffness distributions, dimensions, aerodynamic properties, and hydrodynamic parameters.

8.11.3 Simulation

The simulation is the actual running of the computer model to generate predictions. Multiple runs may have to be made to study the full range of design conditions.

8.11.4 Converting Simulation Outputs to Load Effects

Outputs from simulation codes are frequently in the form of time series loads; that is, forces, bending moments, and torques at function of time. In that case, they must be converted to load effects (e.g., stresses or deflections). This can be done with the help of computer codes of varying complexity, which use the loads together with geometric properties of the components of interest. Laino (1997) describes one approach to this task.

8.11.5 Damage Assessment

As discussed above, there are two basic aspects of design evaluation: (1) ultimate loads and (2) fatigue loads. If the maximum stresses are low enough during the extreme load design cases, then the turbine passes the ultimate loads test.

The fatigue case is more complicated. For one thing, the total amount of damage that is generated over an extended period of time will depend on the damage arising as a result of particular wind conditions and the fraction of time that those various conditions occur. Thus, the distribution of the wind speed is an important factor that needs to be taken into account. In order to expedite fatigue damage estimates, it is advantageous to use such codes as *MLife* (see https://www.nrel.gov/wind/nwtc/mlife.html) to carry out the assessment. See also Chapter 7.

8.12 Wind Turbine and Component Testing

Testing of complete wind turbines and wind turbine components is an integral part of the design process. Testing is critical to ensure a safely operating turbine, confirm design calculations, and identify improvements that may be needed in the design process or future test procedures. Testing occurs during research in order to refine our understanding of scientific issues, during the design process as part of the development of novel designs, in prototype testing to ensure that the intended design will meet the design criteria, and, finally, as part of the turbine certification process. Each of these tests is performed according to quality control standards appropriate to the goal of the testing. Testing may also be done as part of the certification process, as discussed in Section 8.5.

8.12.1 A Wind Turbine Test Program

8.12.1.1 Overview

The successful testing of a wind turbine or wind turbine components requires:

- **Testing goals.** This goal of testing might be testing the validity of the results of a computer model of a wind turbine, testing the hypothesis that a newly designed component will withstand certain forces, or testing the hypothesis that the failure of a part is due to fatigue failure due to resonance. The test goal will determine what is measured and under what conditions and against what measure the results will be evaluated.
- **A test plan.** The test plan should clearly lay out the purposes of the tests, the important characteristics of the sensors and data acquisition systems to be used, the method for producing or identifying the presence of the

desired test conditions, the duration of the tests, how the results are to be interpreted, and what documentation is required.

- **Testing facilities.** Testing facilities need to be able to safely reproduce the desired test conditions with accommodation for staff and data acquisition equipment. If the tests are for certification, the test facilities will need to be an accredited testing laboratory. Safety concerns and quality control are very important.
- **Appropriate data acquisition systems.** The data acquisition systems will need to include appropriate sensors, signal conditioning, and measurement and recording systems. The sensors should be able to respond with acceptable accuracy and with adequate response time and should survive the environmental conditions. The signal conditioning filters the signals and may transform them for measurement. Finally, the data measurement and recording system needs to have adequate data storage space and needs to have adequate data acquisition rates.
- **Procedures for testing, data analysis, and reporting**. The details of the test set-up, instrument calibration, the progress of the tests, and the analysis of the test results all need to be well documented to ensure that the results are accurate and useful. The results of the analysis need to be presented in a complete and detailed report.

8.12.2 Full-scale Wind Turbine Testing

Tests of full-scale wind turbines may have many different purposes:

- prototype testing to support the design and development of a new wind turbine design;
- testing according to international standards for the purpose of obtaining design certification;
- testing to identify the source of problems or to check a solution for problems that have been identified during operation;
- testing for research purposes to advance the state of the art.

These tests might focus on any aspect of wind turbine operation, including power production, component loads, system dynamics, operation of any one of a number of components, control system performance, power quality, sound emissions, and aerodynamic performance. Such full-scale tests generally require a fairly large set of sensors to ensure that the inflow and operating conditions can be correlated with whatever output variable is being measured.

8.12.2.1 Sensors for Wind Turbine Testing

The sensors that might be needed for full-scale testing of a wind turbine include, or might include, sensors to measure:

- **Inflow**. These include wind speed and direction, temperature (overall or differential temperatures across various heights of the boundary layer), and atmospheric pressure. Wind speed and direction sensors that may be used include cup anemometers, sonic anemometers, direction vanes, and possibly sodar or lidar.
- **Strain**. The strain of components is used to determine stresses and loads in the system. Strain is most usually measured with strain gauges. Strain gauges are often employed at various locations on the blades, tower, shafts, and bed plate. Networks of strain gauges can be set up to measure either linear strain or torsion. More recently, fiber-optic Bragg grating (FBG) strain gauges have been used. FBG strain gauges use an optical grating that consists of a series of changes in refractive index in a glass fiber. As the length of the glass fiber changes, the frequency of the light reflected by the grating changes. FBG strain gauges may be used in fiber-optic cables embedded in wind turbine blades.
- **Forces and torques**. Forces and torques may also be measured directly with load cells. A variety of technologies may be used to measure forces or torques, but many rely on the strain of the instrument and use strain gauges of some sort.

- **Position or motions**. The positions or speeds and accelerations of the blade, blade pitch angle, yaw angle, rotor azimuth, and shafts are all important for evaluating the turbine operation and controlling the operation of the turbine. Position and motion may be measured with linear variable differential transformers (LVDTs), inductive, magnetic, or optical proximity sensors or encoders.
- **Pressures**. Pressures in hydraulic or pneumatic systems are measured with a variety of possible technologies that usually provide an output voltage or current directly proportional to pressure.
- **Voltages, currents, power, or power factor**. Voltages, currents, power, and power factor measurements are generally derived from a sensor that measures voltage or a voltage that is proportional to current. Signal conditioning units then convert the instantaneous voltage and current measurements into measures of real and reactive power, power factor, etc.
- **Sound**. Wind turbine sound emissions are measured with calibrated microphones. Care must be taken to account for background noise. Sound measurements may also be used to measure the emissions of individual components like a gearbox to help identify changes in operation. This type of measurement might be done with a small piezoelectric microphone.
- **Vibration**. There are many components in a wind turbine that vibrate, and the technology for measuring those vibrations will depend on the nature of the vibrations to be measured. Vibrations are typically measured with an accelerometer but may be measured with strain gauges or other motion sensors or derived from sound measurements.

8.12.2.2 Modal Testing

Modal testing refers to a technique for determining the modal frequencies and mode shapes of a structure. There are a number of methods of carrying out modal testing on wind turbines, but fundamentally they involve exciting the wind turbine, measuring the response, and analyzing the data acquired. More details may be found in Lauffer *et al.* (1988). One of these methods, known as the Natural Excitation Technique (NExT), is especially applicable to an operating wind turbine (James *et al.*, 1993).

8.12.2.3 Example of Full-scale Wind Turbine Testing

An example of the data that can be gleaned from full-scale testing is illustrated in Figure 8.7. This figure shows data obtained from the Long-term Inflow and Structural Test turbine (LIST) (Sutherland, 2002). The LIST turbine was located in Bushland, TX. It was a 115 kW, three-bladed, upwind, stall-regulated wind turbine. The turbine was part of a long-term research study of factors contributing to fatigue. An array of sensors measured the inflow to the wind turbine. Figure 8.7a illustrates the differences in the wind speeds at the bottom and the top of the rotor for a 20-second period. Figure 8.7b shows the fluctuations in electrical power. Figure 8.7c shows the changing rotor RPM as the slip of this squirrel cage induction generator changes with power level. Finally, Figure 8.7d shows the edgewise bending moment at the root of one of the blades.

8.12.3 Component Testing

8.12.3.1 Material Tests

The properties of all of the materials that comprise a wind turbine need to be understood in order to design the system. Typically, these are determined by testing material coupons (smaller samples of materials) in tension, compression, and torsion, either with constant loads or fluctuating loads to determine fatigue properties.

The testing of composite materials has presented a particular challenge. Composite structures come in many different combinations of matrices, fibers, layout patterns, geometries, and manufacturing methods. Each of these affects the strength of the structure. Testing of entire composite structures or important substructures is important for confirming the endurance limits of the item but may involve significant expense and effort (see Section 8.12.3.3 on blade testing). Nevertheless, the many possible structural failure modes in composites can be reduced to tests on

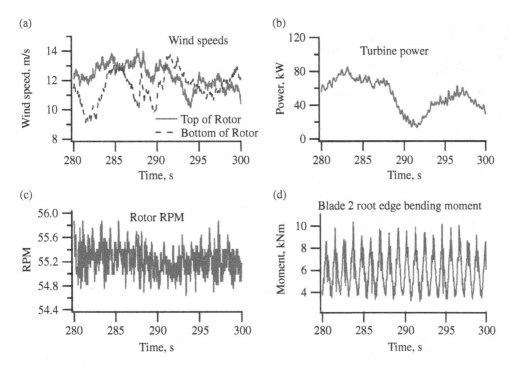

Figure 8.7 Sample test results from the LIST turbine (a) wind speeds at the turbine rotor; (b) electrical power; (c) rotor RPM; (d) blade root edge bending moment (*Source:* Sutherland (2002) / Sandia National Laboratories)

coupons for determining material properties. The test results can then be used to evaluate the strength of a full structure. Coupon tests of composite materials need to be carefully designed to avoid issues with geometry-dependent failure modes, test results influenced by the type of connection between the coupon and the test stand, the anisotropy of the material, or by failure due to heating of the coupon as the shear energy is dissipated in the material.

8.12.3.2 Testing of Drivetrain Components

One of the more problem-prone subsystems in wind turbines has been the drivetrain, including the gearbox, shafts, and generator. Gearbox and generator manufacturers test their products, in consultation with the wind turbine manufacturers, prior to shipping them to the manufacturers. In addition to this, manufacturers may test the drivetrain as a unit. For example, the National Renewable Energy Laboratory (NREL) in the United States maintains three wind turbine dynamometer test beds for this specific purpose, with capacities of 225 kW, 2.5 MW, and 5 MW.

The NREL dynamometers are designed specifically to test wind turbine drivetrains. Wind turbine drivetrains are unique in that they have very high input torques at very low shaft speeds. They are also subject to shaft bending and thrust loads that vary significantly over time.

The mid-size NREL dynamometer is driven by a 4169 V motor coupled to a three-stage epicyclic gearbox and can deliver 2.5 MW to the drivetrain to be tested. The whole motor/gearbox assembly that provides the torque to the test drivetrain can be raised as much as 2.4 m and tilted up to 6° to accommodate drivetrains with a shaft tilt. The motor is driven by a variable-speed drive whose output can be controlled to provide either a user-defined torque profile or a user-defined speed. In addition, the control system of the variable-speed drive can be programmed to mimic the inertial characteristics of the rotor that would be connected to the drivetrain of an actual wind turbine. The dynamometer is pictured in Figure 8.8.

Figure 8.8   NREL's 2.5 MW wind turbine drivetrain dynamometer test bed (Musial and McNiff, 2000) (*Source:* NREL)

Shaft bending loads can be reproduced using a hydraulic actuator system that provides a side load of up to 489 kN. The side loading can be controlled and can be synchronized with the rotation of the low-speed shaft to provide the periodic per-rev shaft loading that would be seen in actual turbine operation.

The tests that can be performed include:

- **Drivetrain endurance testing**. Drivetrain endurance testing involves prolonged operation at high loads combined with the additional application of side loads, and also transient load cases, to reflect the operating conditions that would be experienced in the field. Loads as much as 1.7 times the rated capacity of the drivetrain are used to generate, in three to six months of testing, the expected wear experienced by a drivetrain over its lifetime.
- **Turbulent wind conditions**. Turbulent wind testing is used to demonstrate the proper operation of the drivetrain and control system under normal operating conditions and a variety of conditions near the operational limits of the system.
- **Load event testing**. Load event testing can be used to test the operation of the full drivetrain and control system under extreme operating conditions. In this manner, the brake system and the control system response to generator failures and grid faults can be tested.
- **Component testing**. These tests focus on the operation of individual components within the system, such as the gearbox, the generator, or, more specifically, gears, shafts, housing, bearings, isolated control systems, or power electronics, with or without the full drivetrain assembly.

See also Musial and McNiff (2000) and https://www.nrel.gov/docs/fy11osti/45649.pdf for more information on this and other dynamometers.

8.12.3.3   Full-scale Testing of Blades

Full-scale testing of wind turbine blades is done to confirm the ultimate fatigue strength of the blades. Full-scale blade structural testing is covered by IEC standard IEC 61400-23 (IEC, 2014). The goal of tests is to load the blade or a section of the blade in a manner that will approximate ultimate or fatigue design loads and to demonstrate that the blade will, indeed, survive these loads. The standard defines three types of tests (static testing, fatigue testing, and selected load cases), the partial safety factors to use for the tests, methods with which to document that the desired load or the desired fatigue damage cycle has indeed been achieved, what constitutes structural failure, and, finally, reporting requirements. Full-scale blade tests provide confidence that the blade design and production have no large errors that would lead to unsafe operation. They may also provide direct measurements of blade strength and information with which to evaluate and improve design computations. Thus, they provide critical feedback in the design process but do not replace the detailed design calculations required to evaluate the strength, stiffness, fatigue life, and resistance to multiple modes of failure.

The static tests are used to ensure that the blade successfully survives the ultimate design loads. Tests might apply loads in a variety of directions or with a variety of load distributions to represent different extreme load cases. These are not tests to failure but tests to ensure that the blade will survive the extreme case design loads. Additionally, tests to failure may be conducted to determine the static strength design margin. Failure is generally indicated by a reduction in stiffness at some point along the blade. Figure 8.9 illustrates a blade set-up for static flapwise loading.

The fatigue test cyclically loads the blade in one or more directions to confirm estimates of the blade fatigue life and to identify the mode and location of initial fatigue failure. Again, failure is defined by a reduction in stiffness at some point along the blade. Fatigue tests generally use higher loads than the wind turbine is expected to experience to reduce the length of the tests

Figure 8.9 Test stand and blade set-up for static flapwise loading (*Source:* Reproduced by permission of TPI Composites, Inc.)

(Freebury and Musial, 2000). In the fatigue tests, it may be difficult to duplicate the exact distribution of loading expected along the blade and the load history expected over the life of the blade. In addition, time and cost usually limit the number of samples that can be tested. Nevertheless, full-scale blade testing provides a test of many aspects of the design analysis, including basic blade properties, fatigue life, and the nature and location of the most likely point of failure.

Fatigue Test Blade Loading

Ideally, full-scale fatigue tests of blades would subject the blade to the full range of blade loading that it would experience in operation, in terms of magnitudes, directions, load distribution, and time histories. Wind turbine blades are typically subjected to more than 10^8 loading cycles during their lifetime. These loads include axial, in-plane, out-of-plane, and torsional loads. Typically, the out-of-plane loads have a nonzero mean and, depending on operating conditions, may reverse direction. For smaller blades, the out-of-plane loading is the primary direction of loading and fatigue damage. As wind turbine blades have increased in size, the in-plane bending loads, due to the increased blade weight, have become an important factor. In-plane loading tends to have a negative R value (see Chapter 7) with reversing loading as the rotor rotates (White, 2004). Typically, loading is performed in only one direction at a time and with higher-than-expected loads to approximate, in a reasonable amount of time, the total fatigue damage that a blade would experience. On the other hand, two-axis loading may become standard for large blades. With two-axis loading, the location of the greatest stresses in the blade is a function of the vector sum of the two loads, or equivalently, of the phase angle between the two. Analysis has shown that, in the field, the

Figure 8.10 NREL's dual-axis forced displacement test set-up (White, 2004) (*Source:* NREL)

phase angle between these two components varies substantially over time, but that the mean phase angle is linearly related to the wind speed (White and Musial, 2004).

Blade Fatigue Testing Approaches

There are only a few laboratories in the world that can do full-scale blade testing. They generally use variations of two approaches to apply the loads for the fatigue tests: (1) forced displacement and (2) resonant excitation.

The forced displacement method uses hydraulic actuators to load the blade in one or two different directions simultaneously. Figure 8.10 shows a two-axis forced displacement testing system. When performing two-axis testing, the primary advantage of the system is that loading in two axes with a desired phase angle can be achieved, thus duplicating with more fidelity the loading that the blade experiences. The disadvantages of the method are the size and power requirements of the equipment required and the resulting costs and the duration of fatigue testing with the equipment. Cycling rates are typically below the natural frequencies of the blades, resulting in very long tests to failure.

The resonance method excites the blade using an eccentric mass on a rotating shaft or a reciprocating mass driven by hydraulic actuators to excite the blade at its natural frequency. The force amplification near resonance results in much lower power requirements than with the forced displacement method. Weights can be added to the blade to get the desired load distribution and, finally, the resonant frequency is typically much higher than the excitation frequency of the forced displacement systems, resulting in shorter testing periods. On the other hand, these systems can only test the blade in one axis at a time. For dual-axis testing, a hybrid approach has been developed by the NREL (White *et al.*, 2004). It uses the resonance method to excite vibrations in one axis while using a hydraulic actuator to load the blade in the perpendicular direction. Figure 8.11 shows a hybrid resonance/forced displacement testing system. The equipment requirements for the forced displacements in the edgewise direction are lower than in the flap direction, so the actuator can excite the edgewise motions at the same frequency as the resonance in the flap direction. The result is a system that can provide two-axis tests with shorter durations of the resonance method.

Resonance induction system

Bell Crank

Figure 8.11 NREL's hybrid resonance/forced displacement blade test system (White *et al.*, 2004) (*Source:* NREL)

Nondestructive Blade Testing

In many of the cases discussed above, the component or sample is tested to failure or until some damage occurs, and then the damage is assessed by cutting up the item prior to its examination. In many situations, it is desirable to undertake a test without destroying the item. This is known as nondestructive testing. There are numerous approaches that can be undertaken. Three of them are briefly described below.

In one technique, an infrared camera is used to measure dynamic temperature distribution (due to thermoelastic effects) on the surface of the sample that is being stressed. The temperature distribution gives a measure of the sum of the principal stresses at each point on the surface and can be used to determine mode shapes and also to find cracks. See Beattie and Rumsey (1998).

Another means of nondestructive testing is acoustic emissions testing. In general, a sample that is being stressed will emit low energy sound. In acoustic emissions testing, this sound is monitored and then analyzed to determine fatigue damage. See Beattie (1997).

A third method of nondestructive testing is the acousto-ultrasonic technique (Gieske and Rumsey, 1997). In one variant of this technique, which has been successfully applied to a wind turbine blade joint, an ultrasonic pulse is sent through a sample. The pulse is then monitored on the other side of the sample, and the data is analyzed. The results of the analysis can be used to identify any material damage inside the blade.

8.12.4 Turbine Testing for Certification

Once a wind turbine has been designed and prototyped, testing for certification requires numerous types of test programs, some of which can be pursued in parallel. A number of these testing programs are described below, and the requirements for each are described in the relevant IEC standard (see Section 8.5).

8.12.4.1 Load Measurements

One aspect of the certification process is a description of the design loads and the presentation of evidence that that turbine will survive IEC design load cases. In order to demonstrate that the calculated loads are indeed correctly calculated, the manufacturer needs to either (1) demonstrate that the codes used to calculate the loads provide correct values at various measured load conditions or (2) measure the loads under a variety of design load conditions. The IEC standard for the measurement of mechanical loads, IEC 61400-13, provides guidance for the measurement of mechanical loads in the wind turbine.

8.12.4.2 Power Performance Testing

The most critical measure of performance of a wind turbine is the power curve. IEC 61400-12 specifies the testing requirements for measuring power curves of grid-connected wind turbines. These tests require anemometer measurements upwind of the turbine and electrical power measurements. The standard specifies the location of the met tower, the calibration procedures, specifications for the anemometers and other sensors used in the test, and the data analysis procedures. The location of the wind speed measurement, uneven flow effects at the site, and issues with anemometer measurement uncertainties play an important role in the standard. To ensure that the data are representative, there is also a requirement that each 0.5 m/s wind speed bin has a minimum of 30 minutes of data.

8.12.4.3 Sound Emission Tests

Wind turbine sound emission levels are very important when communities are considering new wind turbine projects. IEC 61400-11 and IEC 61400-14 specify just how the sound emission levels are to be determined (for more details on wind turbine sound emissions, see Chapter 14). Because sounds are emitted from the machinery in the nacelle, from the blade surfaces and tips, and may also be radiated by the tower, it is impractical to measure and track the sound emissions from all locations on a wind turbine. Thus, sound emissions are measured a distance away from the wind turbine and at multiple locations around the wind turbine, and the effective wind turbine sound emission levels are calculated from these data. Simultaneous wind speed measurements are also made, as the sound emission levels will depend on the wind speed and the operation of the wind turbine, which will be a function of wind speed. The calculations use additional background sound measurements (without the turbine operating) at all the wind speeds at which turbine measurements are done to correct for the additional noise generated at the test site from wind or other activities. As with other standards, the IEC standards specify the location and specifications of the wind speed and sound measuring equipment and the procedures for analyzing the data. Typically, sound levels averaged over 10 minutes are used to characterize the wind turbine operation.

8.13 Design of Offshore Wind Turbines

Offshore wind turbines have many similarities with land-based turbines, but there are some significant differences as well. This section focuses on the differences. The overarching differences between land-based and offshore wind turbines are listed in Table 8.13 and discussed below.

Table 8.13 Differences of offshore from land-based wind turbines

Factor	Implication
Support structure	Significant marine substructure required
Scale	Offshore turbines are generally larger than land-based turbines
Metocean conditions	Waves, currents, etc. are major factors in design
Marine environment	Implications for permitting and support structure design
Infrastructure	Range of implications
Distance from habitation	Implications for rotor design, and more
Accessibility	Relatively difficult access militates for high reliability

Support structure. First of all, it will be recalled that all wind turbines may be considered to be composed of two major parts, the RNA and the support structure. In the case of offshore turbines, the support structure is quite significant. For fixed offshore wind turbines, the support structure consists of the tower, the substructure (which is below the tower, and primarily in the water) and the foundation, which is based on the seabed and supports the rest of the turbine. These are all discussed in Chapter 10. For floating turbines, the support structure consists of the tower, a floating substructure (or hull), and the stationkeeping system. The substructure provides buoyancy and stability and the stationkeeping system keeps the turbine in place. The floating substructure and the stationkeeping system are discussed in Chapter 11.

Scale. Offshore wind turbines are almost invariably large, and in general larger than land-based turbines. At the time of this writing, there are already offshore turbines with rated capacities greater than 14 MW, with rotor diameters in excess of 220 m. The size has a number of design implications, including for the turbine itself as well as for the infrastructure required to fabricate, assemble, install, and service the machines.

Metocean design conditions. Meteorological oceanographic (metocean) conditions, as discussed in Chapter 6, have a major effect on the design of the turbine, particularly the support structure. Offshore turbines by definition exist in a high humidity environment. Most offshore turbines are also installed in salt water. Thus, humidity and salt can affect the design and selection of components and materials.

Marine environment. In addition to design conditions, other aspects of the marine environment have implications for the design of the turbine itself as well as the entire wind power plant, including permitting.

Infrastructure. Offshore turbines by their nature require a significant maritime infrastructure, including ports, harbors, and a variety of vessels. The turbine design will need to take into account the requirements and attributes of such infrastructure.

Distance from habitation. Offshore wind turbines are normally far enough from shore that they are not visually intrusive nor can they be heard during operation

Accessibility. It is invariably more difficult and more expensive to access offshore turbines than land-based ones, particularly when major components need to be replaced. Thus, designs need to be as robust as possible and repairs as easy and modular as possible. Special features such as helicopter pads, boat landings, and integrated cranes need to be incorporated in the design.

8.13.1 Important Considerations for the Design of the Offshore RNA

Other than the support structure, the main part of the offshore wind turbine is the RNA. As with land-based turbines, the RNA is composed of a number of subsystems and components. The most important of these are discussed below and the key differences with land-based turbines are highlighted.

8.13.1.1 Rotor – Offshore

As discussed in Section 8.8, with classical scaling, the blade (and thus rotor) weight increases by the cube of the dimensions, while the power goes as the square. Actual weight increase is less than that (see Veers *et al.*, 2019) but still by more than the square, so the RNA weight increases by more than the power as the rotors get larger, at least for the case of a constant tip-speed ratio.

On the other hand, the greater distance from inhabited areas of most offshore wind turbines offers alternatives in the rotor design. In particular, downwind turbines, which have a number of advantages, but also tend to generate more noise, may be more acceptable. The major advantage is that blades do not tend to bend in toward the tower under load, so they do not necessarily need to be as stiff, nor does the rotor axis need to be tilted. Similarly, two-blade turbines also become more interesting. The two-blade rotor often can be designed for a higher tip-speed ratio than a three-blade rotor, resulting in less rotor weight and a lower gearbox ratio, and thus lower gearbox weight. Finally, even without changing the number of blades or the orientation of the rotor with respect to the tower, the rotor is still likely to be designed for a higher tip-speed ratio, since operational noise offshore is not an important consideration. Also, transportation constraints are not as restrictive and the maximum chord or the maximum tower radius can take larger values.

8.13.1.2 Drivetrain – Offshore

The most obvious difference in the drivetrain components of an offshore turbine compared to a land based one is that they will be larger. In addition, it is likely that most offshore turbines will be direct drive, so in that case, there will be no gearbox. In certain cases, however, a "hybrid" system may be used, which would incorporate a multiple pole, medium-speed generator with a gearbox with a reduced speed-up ratio. Other components of the drivetrain will be comparable to direct drive, land-based turbines.

8.13.1.3 Generator – Offshore

As noted above, offshore wind turbines may well be direct drive. That implies that the generators would have multiple poles, as discussed in Chapter 5, the number depending on the rotor speed, the grid frequency, and details of the power electronics. Other than that, generators for offshore turbines are likely to be similar to those on land. For a number of reasons, variable speed generators are most likely.

8.13.1.4 Power Control System – Offshore

Control for offshore wind turbines is most likely to include blade pitch control, probably individual pitch control, for wind speeds above rated as well as variable speed generator control for wind speeds below rated. Active dampers may be used to limit the excitations from the waves, and specific control strategies are needed to avoid instabilities for floating platforms. Such control systems are discussed in Stewart and Lackner (2013).

8.13.1.5 Yaw System – Offshore

Offshore wind turbines as presently designed are all of horizontal axis type, and their substructures, whether fixed or floating, do not rotate. Accordingly, their yaw systems are similar to land-based turbines, that is to say located on the tower top. On the other hand, floating offshore wind turbines could conceivably employ sea-level yaw systems, similar to the turret moorings systems sometimes used by the oil and gas industry. Such mooring systems would actually allow a redesign of other parts of the structure as well. For example, a braced tower rather than a cantilever could be relatively light, and multiple rotors on a single substructure would be possible.

8.13.1.6 Nacelle – Offshore

The remainder of the RNA, which includes the mainframe and cover as well as the components inside, would likely be similar to those of large land-based turbines. Differences would reflect the difficulty of access and would facilitate servicing and repairing the turbine on site. Climate control inside the nacelle would also be used to minimize corrosion.

8.13.2 Trends in the Design of Offshore Wind Turbines

It is, of course, difficult to predict what future offshore wind turbines will look like, but based on recent experience, one can surmise that they will continue to become larger, be designed for deeper water, and be simpler and easier to maintain. The support structures will continue to evolve, with increased standardization and industrialization of the fabrication process.

The most radical changes may occur with floating offshore turbines, where designers will attempt to take advantage of the maritime aspects of the floating substructures. For example, turbines that can be assembled close to shore and towed out to the wind plant are attractive options. Floaters with multiple rotors (as mentioned above) also become interesting. These may build on the designs of the early visionaries such as Heronemus (1972) or more recent investigators such as Henderson and Patel (2003).

8.13.3 Additional Considerations for the Design of Offshore Wind Turbines

In addition to what has been discussed above, there are a number of other considerations for the design of offshore wind turbines. First of all, they experience additional structural loads due to the marine environment. The metocean conditions that are the source of these loads were discussed in Chapter 6. Support structures for fixed offshore wind turbines are discussed in Chapter 10, and those for floating offshore wind turbines are in Chapter 11. The design standards for fixed and floating offshore wind turbines (IEC, 2019b, c) provide guidance for assessment of these loads as well as other relevant information. The American Petroleum Institute (API) also has created documents that are particularly applicable to the design of the substructures. The most notable of these is API (2014), which is a recommended practice for planning, designing, and constructing fixed offshore platforms. Many countries have their own documents as well. In the United States, for example, the American Clean Power Association has created (ACP, 2022) and the Bureau of Ocean Energy Management has produced (BOEM, 2011), among other documents.

References

Abbas, N. J., Zalkind, D. S., Pao, L., Wright, A. (2021) A reference open-source controller for fixed and floating offshore wind turbines, *Wind Energy Science*, https://doi.org/10.5194/wes-2021-19.

ACP (2022) *ACP Offshore Compliance Recommended Practices (OCRP)*, 2nd edition. American Clean Power Association, Washington, DC.

Agarwal, P. (2008) *Structural Reliability of Offshore Wind Turbines*, Ph.D. Dissertation, University of Texas at Austin, https://citeseerx.ist.psu.edu/viewdoc/download?doi=10.1.1.390.6468&rep=rep1&type=pdf Accessed 12/2/2023.

Agarwal, P. and Manuel, L. (2009) Simulation of offshore wind turbine response for long-term extreme load prediction. *Engineering Structures*, **31**, 2236–2246, https://doi.org/10.1016/j.engstruct.2009.04.00.

API (2014) *API RP 2A-WSD, Planning, Designing and Constructing Fixed Offshore Platforms – Working Stress Design*, 22nd edition. American Petroleum Institute, Washington, DC.

Beattie, A. G. (1997) Acoustic emission monitoring of a wind turbine blade during a fatigue test. *Proc. of the 1997 AIAA Aerospace Sciences Meeting*, pp. 239–248.

Beattie, A. G. and Rumsey, M. (1998) *Non-destructive Evaluation of Wind Turbine Blades Using an Infrared Camera*, SAND98-2824C, Sandia National Laboratory, Albuquerque, NM.

Berg, J., Resor, B. R., Owens, B. C. and Laird, D. (2012) *Numerical Manufacturing and Design Tool (NuMAD) for Wind Turbine Blades*, v2.0. Sandia National Laboratory. Albuquerque, NM.

BOEM (2011) *Structural Integrity of Offshore Wind Turbines: Oversight of Design, Fabrication, and Installation - Special Report 305*, Bureau of Ocean Energy Management, Washington, DC.

Bossanyi, E. A. (2006) *GH Bladed Theory Manual 282/BR/009*, Garrad-Hassan, Brighton, UK.

Buhl, M. and Manjock, A. (2006) A comparison of wind turbine aeroelastic codes used for certification. *Proc. of the AIAA Aerospace Sciences Meeting*, Reno.

DNV (1992) *Structural Reliability Analysis of Marine Structures*, Classification Note 30.6, Det Norske Veritas, Oslo.

DNV (2006) *Design and Manufacture of Wind Turbine Blades, Offshore and Onshore Wind Turbines*, DNV-JS-102, Det Norske Veritas, Oslo.

DNV (2014) *Design of Offshore Wind Turbine Structures*, DNV-JS-101, Det Norske Veritas, Oslo.

DNV GL (2016) *Type and component certification of wind turbines, DNVGL-SE-0441*, https://www.dnv.com/energy/standards-guidelines/dnv-se-0441-type-and-component-certification-of-wind-turbines.html Accessed 12/2/23.

DNV/Risø (2002) *Guidelines for the Design of Wind Turbines*, 2nd edition. Det Norske Veritas, Copenhagen.

Dörner, H. (2008) *Philosophy of the wind power plant designer a posteriori*, http://www.heiner-doerner-windenergie.de/edesignphil.html (Accessed 3/12/2023).

Fogle, J., Agarwal, P. and Manuel, L. (2008) Towards an improved understanding of statistical extrapolation for wind turbine extreme loads, *Wind Energy*, **11**, 613–635, https://doi.org/10.1002/we.303.

Freebury, G. and Musial, W. (2000) *Determining Equivalent Damage Loading for Full-Scale Wind Turbine Blade Fatigue Tests*, NREL/CP-500-27510, National Renewable Energy Laboratory, Golden, CO.

Freudenreich, K. and Argyriadis, K. (2007) The load level of modern wind turbines according to IEC 61400-1, *Journal of Physics: Conference Series*, **75**, 12075, https://doi.org/10.1088/1742-6596/75/1/012075.

Gieske, J. H. and Rumsey, M. A. (1997) Non-destructive evaluation (NDE) of composite/metal bond interface of a wind turbine blade using an acousto-ultrasonic technique. *Proc. of the 1997 AIAA Aerospace Sciences Meeting*, pp. 249–254.

Haver, S. and Winterstein, S. R. (2009) Environmental contour lines: a method for estimating long term extremes by a short term analysis. *Transactions – Society of Naval Architects and Marine Engineers*, 116–127.

Henderson, A. R. and Patel, M. H. (2003) On the modelling of a floating offshore wind turbine. *Wind Energy*, **6**, 53–86, https://doi.org/10.1002/we.83.

Heronemus, W. E. (1972) Pollution-free energy from offshore winds. *Proceedings of 8th Annual Conference and Exposition*, Marine Technology Society, Washington, DC.

IEC (2002) *Wind Turbines – Part 12: Power Performance Measurements of Grid Connected Wind Turbines, IEC 61400-12-1*, 3rd edition, International Electrotechnical Commission, Geneva.

IEC (2005) *Wind Turbines – Part 14: Declaration of Apparent Sound Power Level and Tonality Values of Wind Turbines, IEC 61400-14*, 1st edition, International Electrotechnical Commission, Geneva.

IEC (2012a) *Wind Turbines – Part 4: Design Requirements for Wind Turbine Gearboxes, IEC 61400-4*, International Electrotechnical Commission, Geneva.

IEC (2012b) *Wind Turbines – Part 11: Acoustic Noise Measurement Techniques, IEC 61400-11*, 3rd edition, International Electrotechnical Commission, Geneva.

IEC (2013) *Wind Turbines – Part 2: Design Requirement for Small Wind Turbines, IEC 61400-2*, 3rd edition, (FDIS) International Electrotechnical Commission, Geneva.

IEC (2014) *Wind Turbines – Part 23: Full-scale Structural Testing of Rotor Blades. IEC 61400-23*, 1st edition, International Electrotechnical Commission, Geneva.

IEC (2015) *Wind Turbines – Part 13: Measurement of Mechanical Loads, IEC 61400-13:2015(B)*, 1st edition. International Electrotechnical Commission, Geneva.

IEC (2017) *Wind Turbines – Part 25: Communications for Monitoring and Control of Wind Power Plants 61400-25* (in 4 documents). International Electrotechnical Commission, Geneva.

IEC (2019a) *Wind Turbines – Part 1: Design Requirements, IEC 61400-1*, 4th edition, International Electrotechnical Commission, Geneva.

IEC (2019b) *Wind Turbines – Part 3-1: Design of Offshore Wind Turbines, IEC 61400-3-1*, International Electrotechnical Commission, Geneva.

IEC (2019c) *Wind Turbines – Part 3-2: Design Requirements for Floating Offshore Wind Turbines, IEC 61400-3-2*, International Electrotechnical Commission, Geneva.

IEC (2019d) *Wind Turbines – Part 21: Measurement and Assessment of Power Quality Characteristics of Grid Connected Wind Turbines, IEC 61400-21-1*. International Electrotechnical Commission, Geneva.

IEC (2019e) *Wind Turbines – Part 24: Lightning Protection, IEC 61400-24:2019*, 2nd edition, International Electrotechnical Commission, Geneva.

IEC (2020a) *Wind Turbines – Part 5: Wind Turbine Blades, IEC 61400-5*, International Electrotechnical Commission, Geneva.

IEC (2020b) *Wind Turbines – Part 6: Tower and foundation design requirements, IEC 61400-6*, International Electrotechnical Commission, Geneva.

IECRE (2017) *Conformity Assessment and Certification of Loads by RECB's: OD-501-4* IECRE System, International Electrotechnical Commission, Geneva.

IECRE (2018) *Type and Component Certification Scheme: ECRE OD-501*, IECRE System, International Electrotechnical Commission, Geneva.

ISO (2009) *Wind Actions on Structure*, ISO 4354:2009, International Organization for Standardization, Geneva.

ISO (2015a) *General Principles on Reliability for Structures*, International Organization for Standardization, Geneva.

ISO (2015b) *Petroleum and Natural Gas Industries — Specific Requirements for Offshore Structures — Part 1: Metocean Design and Operating Considerations*, International Organization for Standardization, Geneva.

James, G. H. III, Carrie, T. J. and Lauffer, J. P. (1993) *The Natura, l Excitation Technique (NExT) for Modal Parameter Extraction From Operating Wind Turbines*, SAND92–1666 UC-261, Sandia National Laboratory, Albuquerque, NM.

Johnson, N., Paquette, J., Bortolotti, P., Mendoza, N., Bolinger, M., Camarena, E., Anderson, E. and Ennis, B. (2021) *Big Adaptive Rotor Phase I Final Report*, NREL/TP-5000-79855, National Renewable Energy Laboratory, Golden, CO.

Jonkman, B. J. and Buhl, M. L. (2007) Development and verification of a fully coupled simulator for offshore wind turbines. *Proc. of the 45th AIAA Aerospace Sciences Meeting and Exhibit, Wind Energy Symposium,* Reno, NV.

Jonkman, J. and Musial, W. (2010) *Offshore Code Comparison Collaboration (OC3) for IEA Task 23 Offshore Wind Technology and Deployment*, NREL/TP-5000-48191, National Renewable Energy Laboratory, Golden, CO.

Laino, D. J. (1997) *Evaluating Sources of Wind Turbine Fatigue Damage*. Ph.D. Dissertation, University of Utah, Salt Lake City.

Lauffer, J. P., Carrie, T. G. and Ashwill, T. D. (1988) *Modal Testing in the Design Evaluation of Wind Turbines*. SAND87–2461 UC–60. Sandia National Laboratory, Albuquerque, NM.

Melchers, R. E and Beck, A. T. (2017) *Structural Reliability Analysis and Prediction*, 3rd edition. John Wiley & Sons, Chichester.

Moriarty, P. J., Holley, W. E. and Butterfield, S. P. (2004) *Extrapolation of extreme and fatigue loads using probabilistic methods*, NREL/TP-500-34421, National Renewable Energy Laboratory, Golden, CO.

Musial, W. and McNiff, B. (2000) Wind turbine testing in the NREL dynamometer test bed. *Proc. 2000 American Wind Energy Association Conference*. American Wind Energy Association, Washington, DC.

Nelson, V. (2019) *Innovative Wind Turbines: An Illustrated Guidebook*. CRC Press.

Rosenblatt, M. (1952) Remarks on a multivariate transformation, *The Annals of Mathematical Statistics*, **23**(3), 470–472, http://www.jstor.org/stable/2236692 (Accessed 12/13/2021).

Saranyasoontorn, K. (2006) *A Simulation-based Procedure for Reliability Analysis of Wind Turbines*, Ph.D. Dissertation, University of Texas at Austin.

Schepers, J. G., Heijdra, J., Foussekis, D., Ralinson-Smith, R., Belessis, M., Thomsen, K., Larsen, T., Kraan, I., Ganander, H. and Drost, L. (2002) *Verification of European Wind Turbine Design Codes, VEWTDC*. Final Report, ECN-C-01-055, ECN, Petten.

Sørensen, J.D. (2011) *Notes in 'Structural Reliability Theory - and Risk Analysis*. Aalborg University.

Stewart, G. and Lackner, M. (2013) Offshore wind turbine load reduction employing optimal passive tuned mass damping systems. *IEEE Transactions on Control Systems Technology*, **21**(4), 1090–1104.

Sutherland, H. J. (2002) Analysis of the structural and inflow data from the LIST turbine. *ASME Journal of Solar Energy Engineering*, **124**(44), 432–445.

van Eijk, S. F., Bos, R. and Wim A. A. M. Bierbooms, W.A.A.M (2017) The risks of extreme load extrapolation *Wind Energy Science*, **2**, 377–386, https://doi.org/10.5194/wes-2-377-2017.

Veers, P. S. (1988) *Three-dimensional Wind Simulation*, SAND88-0152 UV-261. Sandia National Laboratory, Albuquerque, NM.

Veers, P., Dykes, K., Lantz, E., Barth, S., Bottasso, C. L., Carlson, O., Clifton, A., Green, J., Green, P., Holttinen, H., Laird, D., Lehtomäki, V., Lundquist, J. K., Manwell, J., Marquis, M., Meneveau, C., Moriarty, P., Munduate, X., Muskulus, M., Naughton, J., Pao, L., Paquette, J., Peinke, J., Robertson, A., Rodrigo, J. S., Sempreviva, A. M., Smith, J. C., Tuohy, A. and Wiser, R.: (2019) Grand challenges in the science of wind energy, *Science*, **366**(6464), 1–17, https://doi.org/10.1126/science.aau2027.

White, D. (2004) *New Method for Dual-Axis Fatigue Testing of Large Wind Turbine Blades Using Resonance Excitation and Spectral Loading*, NREL/TP-500-35268, National Renewable Energy Laboratory, Golden, CO.

White, D. and Musial, W. (2004) The effect of load phase angle on wind turbine blade fatigue damage. *Proc. of the 42nd AIAA Aerospace Sciences Meeting and Exhibit,* Reno, NV.

White, D., Musial, W. and Engberg, S. (2004) *Evaluation of the New B-REX Fatigue Testing System for Multi-Megawatt Wind Turbine Blades,* NREL/CP-500-37075, National Renewable Energy Laboratory, Golden, CO.

Winterstein, S., Ude, T.C., Cornell, C.A. and Haver, S. (1993) Environmental parameters for extreme response: inverse FORM with omission factors, *Proc. of Intl. Conf. on Structural Safety and Reliability (ICOSSAR93),* in: G.I. Schuëller, M. Shinozuka, J.T.P. Yao (Eds.), *6th International Conference on Structural Safety and Reliability,* Innsbruck, Austria, 9–13 August 1993.

9

Wind Turbine Control

The previous chapters have discussed the numerous components of wind turbines and their operation. To successfully generate power from these various components, wind turbines need a control system that ties the operation of all the subsystems together and manages the safe, automatic operation of the turbine. This reduces operating costs, provides consistent dynamic response and improved power quality, and helps to ensure safety. For example, a control system might monitor the wind speed, check the health of system components, release the parking brake, and connect the wind turbine to the grid. Control systems may dynamically adjust the generator torque on variable-speed wind turbines to maximize power output and adjust the blade pitch settings to limit power in high winds. Without some form of control system, a wind turbine cannot successfully and safely produce power. In addition, a wind turbine is subject to complex external loading from turbulent wind and waves (for offshore turbines), and these loadings will induce vibrations in the different components of the turbines and fluctuations in the power output. Control systems may also be used to dampen some of these effects.

The main goal of the control system is to ensure maximum and stable power production while maintaining the structural integrity of the turbine. Often, these two objectives are in conflict. For instance: high power production is accompanied by higher loads; stable power production often requires aggressive control strategies which may increase loads on the actuators or the turbine components. With the increased penetration of wind into the energy portfolio and the increased scale of wind power plants, the role of the control system is progressively shifting from an optimization at the turbine level to an optimization of production and cost of electricity at the wind plant level, accounting for the energy market.

This chapter is intended to provide the reader with an overview of the important aspects of control systems that are specifically relevant to wind turbine control. Section 9.1 gives an overview of wind turbine control. In Section 9.2, the key aspects of dynamic control are presented. In Section 9.3, details of the dynamic control in each operating region are described. Advanced control features are explained in Section 9.4. Controller design and challenges are discussed in Section 9.5. Supervisory control is discussed in Section 9.6.

Control system design and modeling for wind turbines is a complex topic area. There are several specialized books on the control of wind turbines, for example Gambier (2022), Luo *et al.* (2014), Munteanu *et al.* (2008), and Bianchi *et al.* (2007). Various aspects of wind turbine control systems can also be found in Andersson et al. (2021), Garcia-Sanz and Houpis (2012), Pao and Johnson (2011), and Gasch and Twele (2002).

9.1 Wind Turbine Control Overview

9.1.1 Levels of Control Systems in Wind Turbines

Wind turbine control systems can be divided into four main levels (from high to low level): (1) a wind farm controller (2) a supervisory controller for each individual turbine, (3) an operational controller that communicates

Wind Energy Explained: On Land and Offshore, Third Edition.
James F. Manwell, Emmanuel Branlard, Jon G. McGowan, and Bonnie Ram.
© 2024 John Wiley & Sons Ltd. Published 2024 by John Wiley & Sons Ltd.

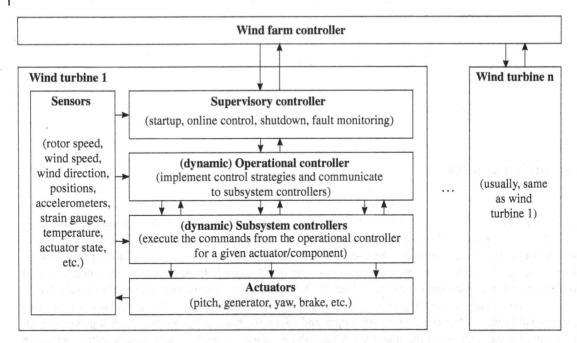

Figure 9.1 Four levels of control system: wind farm, wind turbine supervisory controller, operational controller, and subsystem/component controllers

between the supervisory and subsystem controllers, and (4) the subsystem controllers for the various turbine subsystems (components or actuators) in each turbine. These levels of control operate hierarchically with interlocking control loops (see Figure 9.1).

The wind farm controller, often called a supervisory control and data acquisition (SCADA) system, can initiate and shutdown turbine operation, coordinate the operation of multiple wind turbines, and store measurement signals from them all. Recent developments in wind farm controllers include implementing strategies such as derating or wake steering to optimize the power output at the wind farm level. See Chapter 15 for more information on SCADA systems.

The supervisory controller manages and monitors turbine operation in response to medium- and long-term changes in environmental and operating conditions, such as supervising the sequence of actions needed to startup or shutdown the turbine, monitoring extreme events (storms, extreme winds, or loads), monitoring the turbine health, and performing fault diagnostics. See Section 9.6 for more details.

The operational controller is responsible for the implementation of a given control strategy for the operation of the turbine. It will typically take the form of a computer program. It sends commands to the subsystem controllers, which are dedicated controllers that operate individual actuators or components. The subsystem controllers typically involve power electronics to communicate with the actuators. Most controllers are "dynamic": they use dynamic inputs and dynamic commands to achieve a given strategy. Examples of dynamic operational control include controlling the power production and loads, changing the blade pitch in response to turbulent winds, aligning the wind turbine with the main wind direction, using tuned mass dampers (TMDs) to avoid resonance at certain frequencies, and controlling power flow using power electronic converters. An example of dynamic subsystem control would consist of ensuring that the pitch angle requested by the operational controller is indeed achieved by the actuator, irrespective of the external loads applied on the pitch axis. Since turbines spend most of their lifetime producing power and using dynamic controllers, most of the chapter focuses on dynamic control during power production (Sections 9.2 and 9.3). In the majority of the chapter, the subsystem controllers are considered to be "black boxes" that can respond to the commands of the operational controller.

9.1.2   Active and Passive Control

Control can be achieved in two ways: passively and actively. Passive control refers to techniques that do not require external source of electricity, while active control includes at least one computer that uses sensors and sends commands to various actuators. For example, active control techniques rely on actuators to pitch the blades, employ motors to yaw the nacelle, and use the generator torque to control the turbine rotational speed or power. Most modern turbines utilize active control and therefore need a backup system to be able to operate even if the power to the grid is lost.

Historically, passive control strategies were used. One example is stall-regulated (fixed pitch) turbines with directly connected squirrel cage induction generators. The generator in these turbines keeps the rotor speed nearly constant (subject to slip, as described in Chapter 5). Power production is passively limited by aerodynamic stall: as the wind speed increases, so too does the angle of relative wind. Since the pitch is fixed, the angle of attack along the blade also increases until it reaches the stalling point of the airfoils. Rotor torque then starts to drop, resulting in decreased aerodynamic power.

Another example of a passive strategy is the aerodynamic tip-brake. This is a mechanical device placed at the tip of the blade, designed to deploy when the rotational speed exceeds a predetermined value. When the tip brakes are deployed, they significantly decrease the rotor torque, preventing a potentially catastrophic overspeed. These devices have been used with fixed-pitch turbines for situations when there is a grid fault or the generator is otherwise disconnected from the grid.

A third example of a passive strategy used to align small-scale wind turbines with the wind is the use of an aerodynamic tail fin, similar to a weathervane.

Most of this chapter focuses on active strategies used by modern wind turbine controllers. The term controller is then used to denote an onboard computer that interfaces with the different electrical circuits, sensors, and electrical and mechanical actuators.

9.1.3   Power Curve and Power Regions

Power curves were introduced in Chapter 1 and then discussed in some detail for fixed pitch, directly connected turbines in Chapter 8. Power curves for pitch-regulated, variable-speed turbines are a key result of how the controller functions. The power curve of the NREL 5-MW wind turbine is shown on the left of Figure 9.2. The power curve was obtained using *OpenFAST* and the *ROSCO* controller (see Chapter 8 and Abbas *et al.*, 2022), which differs slightly from the original NREL 5-MW DISCON controller.

There are four important operating regions in this curve, as summarized below. In between these regions, transition regions (e.g., 1½, 2½, 3½) are present to smoothly switch between the different modes of operation. More details are given in Section 9.3.

- **Region 1**. The turbine is not producing power. This region extends to the cut-in wind speed.
- **Region 2**. The turbine is operating and the power increases with the wind speed.
- **Region 3**. The turbine operates at a fixed (rated) power.
- **Region 4**. The turbine stops producing power after a cut-out wind speed to avoid excessive loads, which could occur during unusually high winds.

Power curves of modern wind turbines typically follow the same regions, which is generally accepted as the optimal, cost-effective way to operate a wind turbine. For reasons that will appear clear subsequently in this chapter, active control is key to obtain such a performance curve, because the rotational speed and pitch angle of the wind turbine need to be continuously adjusted.

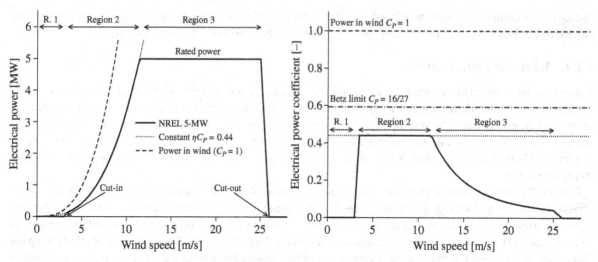

Figure 9.2 Example of power curves (left) and corresponding power coefficients (right)

The dashed line in the left of Figure 9.2 shows the total power available in the wind. As discussed in detail in Chapter 3, the wind turbine rotor only extracts a portion of the power available in the wind. The power coefficient expresses the ratios of the rotor power to the total power available in the wind. The notation C_P is used for the aerodynamic power coefficient. It is also possible to define the electrical (generator) power coefficient, equal to ηC_P, where η is the drivetrain and generator efficiency ($\eta \leq 1$), accounting for any kind of losses between the aerodynamic power and the electrical power. Values of the electrical power coefficient for the NREL 5-MW are shown at the right of Figure 9.2.

It is observed that the turbine operates at a constant power coefficient with value $\eta C_P = 0.44$ in Region 2, and this is the region when the rotor efficiency is at maximum. In Region 3, the power coefficient is lower. The typical maximum power coefficient of modern turbines is around 0.4 to 0.5. These values are lower than the theoretical Betz limit of 16/27, as discussed in Chapter 3. If one were able to maintain a constant power coefficient throughout the range of wind speeds, one would obtain the power curves labeled "Constant $\eta C_P = 0.44$." One could harvest more power at higher wind speed but at the expense of requiring a more powerful generator and ensuring that the turbines can withstand the loads which also increase with wind speed. It is recalled from Chapter 3 that the power in the wind increases with the cube of the speed while the forces increase with the square. To reach a compromise between cost (generator size, structural ability to withstand loads) and power generated, the power is capped at a design value (the rated power) in Region 3. The controller ensures that the power production is in line with the designed power curve: starting and shutting down the turbine, adjusting the rotor speed in Region 2, and limiting the power in Region 3.

9.2 Key Aspects of Dynamic Control

This section presents the key aspects of dynamic control. Dynamic control consists of controlling the wind turbine (actuators, generators, and motors) based on a set of inputs (from sensors) in order to achieve a given strategy (such as maximizing power or limiting loads). The time scales of dynamic control are of the order of milliseconds to seconds. The focus of this section is on dynamic control during normal operation. Normal operation refers to cases where the wind speed is between the cut-in and cut-out wind speeds, and the turbine is "available," that is, not in a faulty state, or under maintenance, and the turbine is connected to the electrical grid. For a windy site, this

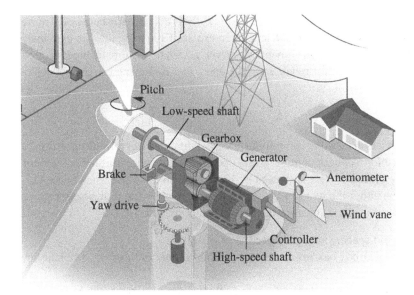

Figure 9.3 Sketch of the main sensors and actuators used in wind turbine active dynamic control (*Source:* Based on a figure provided with U.S. Department of Energy / Public Domain)

represents most of the lifetime of the wind turbine. During normal operation, the controller uses sensors and actuators to satisfy given objectives. The main components of the rotor nacelle assembly of a modern wind turbine are shown in Figure 9.3, together with some of the key control actuators (pitch, yaw, and generator torque) and examples of sensors (wind vane and anemometer) that will be discussed in this section. Background information on dynamic control system design can be found in textbooks such as Nise (2019).

9.2.1 Control Objectives

During normal operations, different control objectives may be considered:

- **Power regulation**. The objective is to produce power according to the power curve and maximize the annual energy production within the speed and load constraints of the turbine components.
- **Power quality**. Providing acceptable power quality at the point of connection to the grid.
- **Speed regulation**. The rotational speed is limited to reduce noise emission and to be within the range of operation of the generator. Aerodynamic noise is an important concern for wind energy, and it typically increases with the relative wind speed, which is greatest at the tip of the blade and close to the tip-speed ΩR; see also Chapter 14.
- **Safety limits and load mitigation**. The objective is to ensure that the turbine keeps its structural integrity. The controller should ensure safe turbine operation, prevent extreme loads from occurring, including excessive transient and resonance-induced loads, and minimize fatigue damage. The loads and vibrations need to remain within a certain level to ensure that the turbine will last as long as its expected lifetime (wind turbines are designed to last between 20 and 25 years). Control strategies may be used if a problem is observed in the field that might lower the expected lifetime of the turbine.

In most cases, all the control objectives above need to be considered. The control objectives are often in conflict, requiring compromises and tradeoffs.

As wind turbines become larger, and as offshore operation continues to grow, these objectives become even more challenging to meet. Modern offshore wind turbines have blade lengths in excess of 100 m, which are extremely long and flexible structures that operate in a complex environment. Moreover, access to offshore turbines is often limited during much of the year, increasing the need for reliable turbine operation. Achieving the control objectives therefore requires careful design of the control system.

9.2.2 Wind Turbine Sensors

On a large modern wind turbine, many sensors are used to communicate important aspects of turbine operation to the control system. The main variable used for feedback control during operation of the wind turbine is the rotor speed. Therefore, rotor speed sensors are placed on the low-speed or high-speed shaft. Anemometers are also placed on the nacelle to obtain information about the wind speed, which may be used by the supervisory control to determine whether to start the wind turbine. Other measured variables might include:

- speeds (generator speed, rotor speed, wind speed, yaw rate, direction of rotation);
- temperatures (gearbox oil, hydraulic oil, gearbox bearing, generator bearing, generator winding, ambient air, electronic temperatures);
- position (blade pitch, teeter angle, aileron position, blade azimuth, yaw position, yaw error, tilt angle, wind direction);
- electrical characteristics (grid power, current, power factor, voltage, grid frequency, ground faults, converter operation);
- fluid flow parameters (hydraulic or pneumatic pressures, hydraulic oil level, hydraulic oil flow);
- motion, stresses, and strain (tower top acceleration, tower strain, shaft torque, gearbox vibration, blade root bending moment);
- environmental conditions (turbine or sensor icing, humidity, lightning).

Reviewing the different technologies used to measure these measured variables is beyond the scope of this book. It is important to note, however, that all sensors will have limitations, such as (1) a limited measuring range and frequency band in which the sensor can operate and is reliable; (2) possible drift, noise, and biases introduced with time and therefore requiring recalibration; and (3) the sensors may only provide a partial estimate of the quantity of interest due to physical limitations. For instance, the wind sensors located on the nacelle suffer from their placement in the wake of the rotor, and therefore the data they produce does not correspond to the free stream wind speed. An important part of control engineering consists of filtering the data coming from the sensors and applying transfer function and estimation techniques to attempt to obtain a cleaner value of the quantity of interest. For instance, most control systems use a wind speed estimator that estimates a rotor-averaged wind speed based on sensors different from the nacelle anemometer. The most common type of estimator is a Kalman filter, which is used to obtain an "optimal" estimate of a quantity by combining different measurements with a physical model of the system (see e.g., Zarchan and Musoff, 2015). In this chapter, it will be assumed that the filtering and estimation of physical quantities have already been performed before the sensor data is used by the controller.

9.2.3 Control Actuators

The control system performs actions on the wind turbine via actuators. The main control actuation options of a modern wind turbine are the pitch and yaw motors and generator torque. The pitch actuator is mostly active in Region 3 and transition regions, whereas the generator torque controller is active throughout the operation. Details on these control strategies will be given in Section 9.3, but as an introduction is important to note the following:

- Changing the pitch of the blade will change the angle of attack throughout the blade, thereby changing the aerodynamic thrust and torque.
- Changing the generator torque will change the rotational speed (and the rotor torque).

Both the pitch and torque actuations can affect the speed and torque on the shaft, and since power is the product of the two, both can be used to control the power.

Actuators in a wind turbine may be of the following kind:

- **Electromechanical devices.** Electromechanical devices include DC motors, stepper motors, AC motors with solid-state controllers, linear actuators, magnets, and solid-state switching components.
- **Hydraulic pistons.** Hydraulic pistons are often used in positioning systems that need high power and speed such as the pitching system.
- **Resistance heaters and fans.** Resistance heaters and fans are used to control temperature.

Actuator systems may include gears, linkages, and other machine elements that modify the actuating force or direction. When the control signal from the controller is not powerful enough to power the actuator, an amplifier is needed between the controller and the actuator. Typical power amplifiers in a wind turbine include electrical amplifiers, hydraulic pumps, and various power electronic switches.

9.2.3.1 Generator Torque Control

Generator torque control can be used to control rotor speed, tip-speed ratio, and power. This control is active during the entire operation of the wind turbine. The generator torque has a fast response time, at least an order of magnitude faster than the time constant of the rotor speed, which makes it an effective actuator for controlling the rotor speed (Pao and Johnson, 2011). The generator torque may be regulated by the design characteristics of the generator or independently controlled with the use of power electronic converters:

- Directly connected generators operate over a very small or no speed range and provide whatever torque is required to maintain operation at or near synchronous speed (see Chapter 5).
  - Directly connected synchronous generators have no speed variations and, thus, any imposed torque results in an almost instantaneous compensating torque. This can result in high torque and power spikes under some conditions. As a result, very few wind turbines today use directly connected synchronous generators.
  - Directly connected conventional (squirrel cage) induction generators change speed by as much as a few percent of the synchronous speed (this is referred to as slip). This results in a softer response and lower torque spikes than with a synchronous generator.
- Alternatively, the generator can be connected to the grid through a power electronic converter. This allows the generator torque to be very rapidly set to almost any desired value. The converter determines the frequency, phase, and voltage of the current flowing from the generator, thus controlling generator torque. This is the approach used in most modern turbines. Either wound rotor induction generators or synchronous generators may be employed in this way.

9.2.3.2 Pitch Control

Hydraulic pistons or electro-mechanical actuators are typically used to pitch the blades. The pitching system has a physical limit on the pitch rate (i.e., the rate at which the angle may be varied). The pitch rate is typically between 2 and 10 deg/s on modern turbines. Depending on the system, some inertia, stiffness, and damping are present in the actuator, leading to a dynamic response such that the actuator torque on the blade has to balance the aerodynamic torque and structural twisting moment of the blade. Consequently, there is often a lag between the commanded pitch and the actual pitch of the blade. Until recently, turbines used a collective signal to pitch all the blades simultaneously. With recent advancements in controls and actuators, more and more control systems are adopting individual pitch control (IPC), whereby each blade can be operated at a different pitch angle. IPC makes a lot of sense as rotors get larger since the blades experience very different wind as they sweep through the atmospheric boundary layer (for instance due to shear). Another advantage of IPC is that it gives stronger control authority to reduce unwanted motions of the turbine. A strong control authority means that the controller can easily change the value it is trying to control. More details on IPC are given in Section 9.4.

9.2.3.3 Yaw Orientation Control

An active yaw system includes one or more motors at the nacelle/tower interface (see Chapter 7). The control system ensures that the nacelle remains aligned with the main wind direction. Gyroscopic loads need to be considered in the design of a yaw power regulation system. If gyroscopic loads are a concern, yaw rate can be limited (typically with a maximum of 1 deg/s for modern wind turbines), but limiting yaw rate may affect the ability to regulate power output. It is noted that the yawing occurs at a lower frequency than the pitch or generator torque control.

Both the yaw and pitch system typically involve gears. To avoid wear and tear of the gears when the gear system remains at a fixed position, the controller may apply small step changes of yaw and pitch angles every few minutes.

9.2.4 Turbine Power Production and Efficiency

The turbine power production depends upon a combination of aerodynamic performance and control system operation, which can be understood by examining the C_P-λ curve, where C_P is the power coefficient and λ is the tip-speed ratio. Both terms were defined in Chapter 3, but their definitions are recalled below. The tip-speed ratio is defined as:

$$\lambda = \frac{\Omega R}{U} \tag{9.1}$$

where Ω is the rotor speed, R is the rotor radius, and U is a characteristic wind speed (under turbulent wind conditions, this might be taken as the average over the rotor area). The power coefficient represents the fraction of the power in the wind that is extracted by the rotor:

$$C_p = \frac{P}{\rho U^3 A/2} = \frac{\text{Rotor power}}{\text{Power in the wind}} \tag{9.2}$$

where P is the power extracted by the rotor, A is the area swept by the blades (e.g., πR^2), and ρ is the air density.

An example C_P-λ curve is shown in Figure 9.4 for the NREL 5-MW turbine. The maximum power coefficient is obtained for a pitch angle of 0° for this turbine. In this case, the aerodynamic efficiency peaks at a tip-speed ratio of approximately 7.5. The optimal tip-speed ratio, optimal pitch angle, and associated maximum C_P are design features of any rotor and can only be adjusted by changing the aerodynamic design (see Chapter 3). To maximize power production, the control system should attempt to control the rotor speed such that the tip-speed ratio is held constant at the optimal value. Therefore, as wind speed inevitably changes over time, the rotor speed should be controlled proportionally to preserve a constant value of λ and therefore maximize aerodynamic efficiency. Rotor speed control is therefore critical for power production control. A commonly employed practical control algorithm for achieving optimal tip speed control is presented in Section 9.3.3.

The impact of blade pitch on the C_P-λ curve is also shown in Figure 9.4. The performance of the rotor is greatly reduced as the pitch angle is increased. This explains why blade pitching is used to lower the power production at high wind speed, or even stop the rotation when the turbine is placed in parked position. Three-dimensional contours of the power and thrust coefficients as a function of tip-speed ratio are often used to devise a control strategy. An example of such 3D surface is given in Figure 9.5. It is seen from the figure that a maximum power coefficient is obtained for a given value of the tip-speed ratio and pitch angle, referred to as the optimal values. For values other than optimal, the same power coefficient can be obtained for different values of pitch and tip-speed ratio, but at these operating points, the thrust coefficient will be different, which can lead to different loads. Therefore, different controller strategies can be implemented based on the compromise between thrust and power.

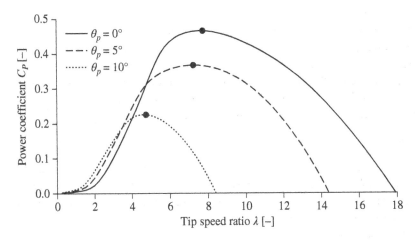

Figure 9.4 Example of C_P-λ curve for the NREL 5-MW wind turbine. The pitch angle θ_p is kept fixed while the tip-speed ratio is changed

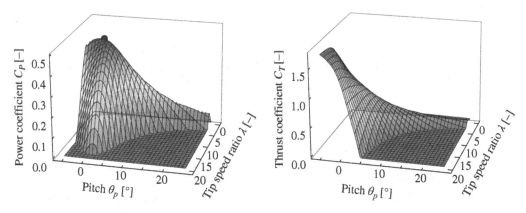

Figure 9.5 Example of 3D surface plot of the power and thrust coefficients for the NREL 5-MW. The black dot on the left figure indicates the maximum power coefficient, which is obtained for the optimal pitch angle and optimal tip-speed ratio

9.2.5 Control Theory with Application to a Pitch Actuator System

This section presents some of the key aspects of control theory used in dynamic control: (1) establishing a model (typically, a set of differential equations) from a physical system; (2) linearizing the system; (3) designing a controller using an open-loop or closed-loop system; (4) and studying the system using tools such as transfer functions or state-space models. Throughout this section, these different steps and notions are illustrated using a model of a pitch mechanism driven by an AC servo motor.

9.2.5.1 The Physical System

A simple pitch control mechanism is considered, which uses an electric motor to rotate the blade about its pitch axis. To rotate the blade, a specific amount of current needs to be applied to the motor for a predetermined amount of time. The motor needs to balance the pitching moment applied on the blade from aerodynamic and inertial loads (disturbances to the system). The role of the controller is to ensure that the blade pitches by the right amount without being affected by system disturbances and accounting for the possible variability in the system.

9.2.5.2 Model Development and Linearization

Before designing a controller, a mathematical model of the system should be developed first. Here, the combined blade and pitch mechanism is modeled using a one-degree-of-freedom mass-spring-damper system. The torque from the AC servo motor is modeled as a linear combination of terms that are a function of motor speed and applied voltage (see Golnaraghi and Kuo, 2017). In the model presented here, the motor torque and speed are referred to the blade side of the pitch mechanism. The differential equation for this system has the system dynamics terms on the left and the external torques from the motor and disturbance on the right:

$$J\ddot{\theta}_p + C\dot{\theta}_p + K\theta_p = kv(t) + c\dot{\theta}_p + Q_p \tag{9.3}$$

where θ_p is the angular position of the motor (the blade pitch angle), J is the total inertia of the blade and motor, C is the pitch system coefficient of viscous friction, K is the pitch system spring constant, k is the slope of the torque–voltage curve for motor/pitch mechanism combination, $v(t)$ is the voltage applied to the motor terminals, c is the slope of the torque–speed curve for motor/pitch mechanism combination, and Q_p is a pitching moment due to dynamic and aerodynamic forces that act as a disturbance in the system.

If the system is at steady state, then the derivatives of the pitch angle are zero, and $v(t)$ is a constant value, v. In this case, the differential equation reduces to:

$$\theta_p = \frac{k}{K}v + \frac{Q_p}{K} \tag{9.4}$$

Most of the tools used in control theory manipulate linear differential equations. If the dynamic equations are nonlinear, they will need to be linearized (see Chapter 4). In this example, Equation (9.3) is already linear except for the pitching moment which may be a nonlinear function of the pitch angle. In this example, the pitching moment is the input of the system, and its linearization is not required.

9.2.5.3 Open-loop and Closed-loop Response to Disturbances

There are two categories of controllers: open-loop and closed-loop controllers. The basic differences between the two systems are illustrated in Figure 9.6. The relationships between the various dynamic elements in a controlled system are often illustrated with the use of block diagrams, such as the one shown in Figure 9.6. In an open-loop system, the controller actions are based on the desired system state, with no reference to the actual state of the process (i.e., no measurements). In closed-loop feedback systems, the controller is designed to use the difference between the desired and the actual system output to determine its actions. Closed-loop controllers offer superior control authority; they are generally preferred and are most common. For this reason, the term "open-loop system" is often used to refer to a system without a controller.

9.2.5.4 Example of Open-loop system

Assuming that the airfoil designer has tried to minimize any pitching moment, one could design a control system that applied a specific voltage to the motor for a desired "reference" pitch angle, $\theta_{p,ref}$. When the disturbances are zero, Equation (9.5) then gives a relationship between the voltage and the desired pitch angle:

$$v = \frac{K}{k}\theta_{p,ref} \tag{9.5}$$

In this case, the differential equation for the open-loop system is:

$$J\ddot{\theta}_p + (C - c)\dot{\theta}_p + K\theta_p = K\theta_{p,ref} + Q_p \tag{9.6}$$

Here, the pitch angle is the output, and the system has two inputs, the voltage to the pitch motor (v) and a disturbance torque from the aerodynamic and dynamic pitching moments on the blade (Q_p). If one assumes that all derivatives and the pitching disturbance torque are zero, it can be seen that the steady-state response of the system

Figure 9.6 Open-loop and closed-loop response to disturbances

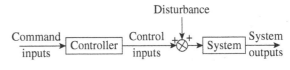

Open-loop control system

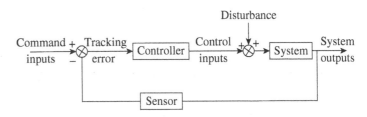

Closed-loop control system

to a desired pitch angle command is, indeed, that desired pitch angle. A block diagram for the open-loop system (without damping C) is shown in Figure 9.7.

9.2.5.5 Equation Solution Using Laplace Transform – Transfer Function

Laplace transforms can be used to solve differential equations like Equation (9.6). The Laplace transform of the system response, about steady state operation, to an impulse input is referred to in classical control theory as the system transfer function. The transfer function provides the relationship between the inputs and outputs of the system in the Laplace-transformed space. It is often used in control system design to characterize system dynamics.

We briefly present the relevant notions of Laplace transforms in this paragraph. The Laplace transform, noted $\mathcal{L}$, can be applied to linear time-invariant systems. It transforms a function, f, of the time variable, t, into a function of the complex variable, s, as follows:

$$\mathcal{L}(f(t)) = \int_0^\infty f(t)e^{-st}\,dt = F(s) \tag{9.7}$$

By convention, the transformed function is written with an uppercase. We note the following relationships:

$$\mathcal{L}(af(t)) = a\,F(s), \quad \mathcal{L}\big(\dot{f}(t)\big) = s\,F(s), \quad \mathcal{L}\big(\ddot{f}(t)\big) = s^2 F(s) \tag{9.8}$$

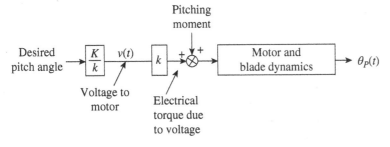

Figure 9.7 Open-loop pitch control mechanism; for notation, see text

where a is a scalar. For more background information, the reader is referred to Golnaraghi and Kuo (2017) or Nise (2019).

The system transfer function can be found by taking the Laplace transform of the open-loop differential equation, solving for the pitch angle, and assuming that initial conditions are all zero. Applying the rules given in Equation (9.8) to Equation (9.6), the transfer function of this system is:

$$\Theta_p(s) = \frac{K\Theta_{p,ref}}{Js^2 + (C-c)s + K} + \frac{Q_p(s)}{Js^2 + (C-c)s + K} \tag{9.9}$$

where $\Theta_p(s)$, $\Theta_{p,ref}$, and $Q_p(s)$ are the Laplace transforms of θ_p, $\theta_{p,ref}$, and $Q_p(t)$. The factor in front of $\Theta_{p,ref}$ in Equation (9.9) is the transfer function from the voltage input to the pitch angle output. The factor in front of Q_p is the transfer function from the disturbance (the pitching moment, Q_p) to the dynamic pitch angle ($\Theta_{p,dyn}(s)$):

$$\frac{\Theta_{p,dyn}(s)}{Q_p(s)} = \frac{1}{Js^2 + (C-c)s + K} \tag{9.10}$$

The time responses can be obtained by taking the inverse Laplace transforms of the transfer functions and assuming that $\Theta_{p,ref}(s)$ and $Q_p(s)$ are known. For instance, a unit step input would be given by $Q_p(s) = 1/s$. Responses to unit steps and impulse steps are readily obtained for transfer functions consisting of polynomials on the numerator and denominators, using analytical tools, or software packages (such as Wolfram, 2023, MATLAB®, or Sympy). If, for example, the motor and blade parameters are $J = \frac{1}{41}$, $(C-c) = \frac{1}{13}$, and $K = 1$, then the corresponding response to a step disturbance is shown in Figure 9.8.

In Figure 9.8, the deviation from ideal operation should be zero, but the pitch disturbance results in a steady pitch angle error (this error is arbitrarily scaled to 1 in the figure). From Equation (9.4), the steady-state positioning error due to the pitching moment is:

$$\theta_p - \theta_{p,ref} = \frac{Q_p}{K} \tag{9.11}$$

In practice, open-loop control systems often have deficiencies that significantly affect system operation. Manufacturing variability, wear, changes in operation with temperature and time, and outside disturbances can adversely affect open-loop system performance. In this example, changes in the wind speed, rotor speed, blade icing, or anything else that might affect the pitching moment would change the pitch. If the effects of these changes in operation become a problem, closed-loop control systems are used to improve system performance without significantly complicating the control system.

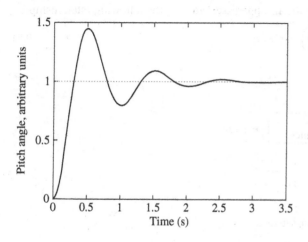

Figure 9.8 Sample pitch system response to a step change of the disturbance torque

9.2.5.6 Example of Closed-loop System Using a PID-controller

In a closed-loop control system, a measurement of the pitch angle is incorporated into the input to the pitch control system, and corrections can be made for errors in the position of the blade. Typically, the controller provides a control output that is a function of the "tracking error," which is the difference between the desired system output and the measured output. The block diagram of such a closed-loop system (without the damping C) is shown in Figure 9.9.

The controller can be designed with any dynamic properties that might help the system achieve the desired operation. There are, however, some standard approaches to control systems that are often implemented and are used as references in considering control system design. Some of these standard approaches include *proportional* (P), *derivative* (D), and *integral* (I) control. Often these approaches are combined to yield a proportional–integral (PI) controller or a proportional–integral–derivative (PID) controller.

The differential equation of the PID controller for the voltage $v(t)$ of the pitch system motor is:

$$v(t) = K_p e(t) + K_I \int e(t)dt + K_D \dot{e}(t) \tag{9.12}$$

where the constants of proportionality are K_P, K_I, and K_D and where $e(t) = \theta_{p,ref}(t) - \theta_p(t)$ is the error, the difference between the desired and the measured pitch angle. Equation (9.12) is a general equation describing a PID controller where a control signal (on the left-hand side) is given as a function of a signal error. The constants of proportionality, K_P, K_I, and K_D, are referred to as "gains" in the control engineering literature. The proportional gain, K_P, acts on the present value of the error and is effective as long as the error is non-zero. The integral gain, K_I, uses "past" values of the error and ensures that the steady-state error is zero. The derivative gain acts on the predicted ("future") value of the error because the time derivative (or "slope") of the error indicates where the error is likely to be in the next time steps.

In our example, a PI controller will be used ($K_D = 0$). The closed-loop equation is obtained by substituting the definition of the controller, Equation (9.12) without the derivative term, into Equation (9.3), and using the definition of $e(t)$ from above. After differentiating the complete equation and rearranging, Equation (9.13) is obtained. The resulting controlled system is now a third-order system with two controller constants, K_P and K_I:

$$J\dddot{\theta}_p + (C-c)\ddot{\theta}_p + (K + kK_P)\dot{\theta}_p + kK_I\theta_p = kK_P\dot{\theta}_{p,ref} + kK_I\theta_{p,ref} + \dot{Q}_p \tag{9.13}$$

If we assume here that $\theta_{p,ref}$ is zero and does not change, then the transfer function for the closed-loop controller corresponding to Equation (9.12) is:

$$\frac{\Theta_p(s)}{Q_p(s)} = \frac{s}{Js^3 + (C-c)s^2 + (K + kK_P)s + kK_I} \tag{9.14}$$

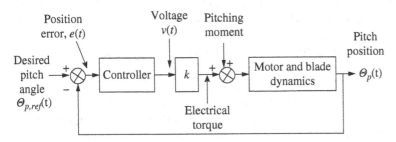

Figure 9.9 Closed-loop pitch mechanism example

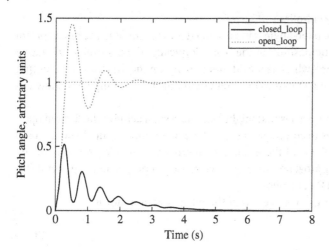

Figure 9.10 Sample closed-loop and open-loop pitch system response to a step disturbance

Once again, the dynamic response of the closed-loop system to a unit step disturbance can be found by using inverse Laplace transforms. For example, if it is assumed that the motor and blade dynamics remain the same and that $K_P = K_I = 2$ and that $k = 1$, then the response to a step disturbance is shown in Figure 9.10. The figure includes the open-loop response for comparison. The disturbance clearly affects the closed-loop system less than the open-loop system.

While further improvements might be made to improve dynamic response of the system, a PI controller would correctly position the blade under a variety of wind and operating conditions. Thus, without adding too much complexity, the blade can be positioned at any desired position in high or low winds, under icing conditions, and with a sticking bearing on the pitch mechanism. This is a significant improvement over the open-loop controller.

9.2.5.7 Typical Control Issues – Resonances

In the typical design process of a controller, it is important to ensure that the controller provides a stable response when the system is excited by the environmental loads (which excites the system with a broad range of frequencies), and that the controller does not excite the system at critical natural frequencies of the system. The response of the control system to step and sinusoidal disturbances is useful to study the stability of the closed-loop system.

The pitch controller presented in Section 9.2.5 provides reasonable disturbance rejection of a step input. This indicates that step changes and less severe changes in operating conditions can be compensated for by the pitch control system. Analysis of the differential equation for the closed-loop system (Equation 9.13) or of the system transfer function shows that the closed-loop system has a natural frequency of 6.76 rad/s (1.0 Hz), and a damping ratio of 0.24 (see Chapter 4 for information on natural frequencies and damping ratios). If this pitch control system experienced a sinusoidal disturbance, as it might with wind shear, with a frequency near the natural frequency of the closed-loop system, then the system response might be significantly magnified over that due to other disturbances. For example, the response to a sinusoidal disturbance with a magnitude of 1 and a frequency of 1.04 Hz is shown in Figure 9.11.

The closed-loop system clearly does a much poorer job of managing disturbances near the system's natural frequency than it does managing step disturbances. The magnitude of the response depends on the frequency of the disturbance and the damping in the system. Not only could the pitch fluctuations induced by the disturbance wear out the pitch mechanism, but oscillating shaft torque from the aerodynamics caused by the oscillating blade pitch could introduce resonances in other feedback paths. This sample control system was designed specifically to

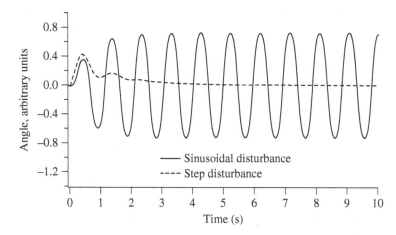

Figure 9.11 Closed-loop pitch system response to unit sinusoidal and step disturbances

illustrate these issues. A well-designed control system would avoid exciting system natural frequencies and might even be able to provide additional damping at those frequencies; see Section 9.5.

9.3 Main Regions of Dynamic Control

While maximizing power output is one goal of the dynamic control system, it clearly cannot be the only goal. In high wind speeds, the allowable power output is limited by the size of the generator, and thus the dynamic controller must limit the power production as wind speed increases. Thus, typical power curves are broken up into three primary operating regions as introduced in Section 9.1.3 and illustrated in Figure 9.2. This section presents some of the control strategies implemented to produce such curves.

9.3.1 Power Curve – Power vs. Wind Speed

To understand the different control regions, one needs to look at the operating conditions of the turbines (wind speed, rotor speed, pitch, and tip-speed ratio) and the aerodynamic performance (power and thrust). These quantities are shown for the NREL 5-MW in Figure 9.12.

The three main control regions and "intermediate" regions are defined as follows:

- **Region 1** is defined as the region where the turbine does not produce power, and the mean wind speed is below the cut-in wind speed. In this region, the generator is disconnected (no generator torque and therefore no power), the turbine is allowed to freewheel, and the rotor speed will increase progressively. For the NREL 5-MW the cut-in wind speed is 3 m/s, which is typical of modern turbines, though some designs with large rotors can start at even lower wind speeds.
- **Region 1½.** When the mean wind speed exceeds the cut-in wind speed, the generator is connected, and the turbine begins to produce power. This region is typically labeled as Region 1½, because the turbine is not yet operating at its optimal tip-speed ratio; instead, because there is a minimum generator speed, the tip-speed ratio exceeds the optimal value. In some control system designs, the pitching system is active in this region to maximize the power output. For this example of the NREL 5-MW, the rotor speed is kept constant to a specified minimum value in this region.

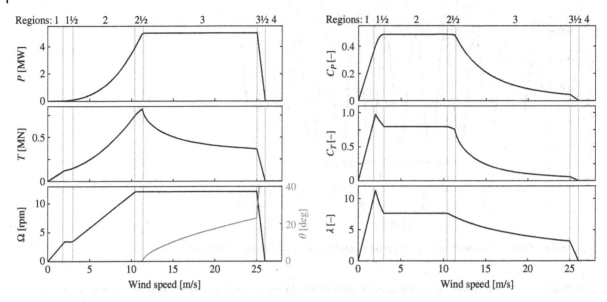

Figure 9.12 Operating conditions and performance curves for the NREL 5-MW turbine (onshore): Power (*P*), Thrust (*T*), Rotor speed (*Ω*), Power coefficient (*C_P*), Thrust coefficient (*C_T*), and tip-speed ratio (*λ*)

- **Region 2.** As the mean wind speed increases, the turbine enters Region 2, or optimal tip-speed ratio operation. In this region, the variable speed generator control is used to track the optimal tip-speed ratio, maximizing aerodynamic efficiency (as discussed in Section 9.2.4). The tip-speed ratio and pitch are held constant at their optimal values, resulting in a constant power coefficient. Because the tip-speed ratio is constant, the rotational speed must change when the wind speed changes. The generator torque increases with the square of the rotational speed in Region 2. See Section 9.3.3 for a detailed discussion of optimal tip-speed ratio control algorithms.
- **Region 2½.** As the wind speed continues to increase, and thus the rotor speed, a maximum (rated) rotor speed is reached, typically before the rated power production is reached. This is Region 2½. At this point, the rotor speed is held constant while the generator torque and therefore power output increases until the rated power is reached at the rated wind speed. The rotor speed is often regulated using the blade pitch control system, discussed next for Region 3. The maximum rotor speed for the NREL 5-MW is 12.1 rpm. The maximum thrust values often occur in Region 2½.
- **Region 3.** In Region 3, the turbine operates at its rated power. When the wind speed exceeds the rated wind speed, the dynamic control aims to keep the rotor speed and the power output constant. For the NREL 5-MW, the rated power is 5 MW, and the rated wind speed is approximatively 11.4 m/s. There are many factors influencing the rated values, such as the size of the generator, the size of the rotor, the wind speed at hub height, and the design of the turbine. In Region 3, the generator power is constant, but the thrust decreases because the blades are pitching to feather, decreasing the angle of attack and reducing the aerodynamic forces on the blades. The rotor speed is primarily controlled using the blade pitch system. The blades are usually pitched "collectively" for this purpose, meaning they all have the same pitch angle. IPC for load reduction can be superimposed upon this collective pitch angle (see Section 9.4.4). The blade pitch is adjusted to minimize the difference between the actual measured rotor speed and the rated rotor speed. Modern wind turbines are "pitch-to-feather": as the wind speed increases, the blade pitch angle also increases, which reduces the angle of attack on the blades, and therefore reduces the lift coefficient. Another approach is "pitch-to-stall" which increases the angle of attack until the airfoil stalls, which decreases the loads but can introduce stability issues. The dynamic pitch control system often employs a PI controller to minimize the rotor speed error (see Section 9.2.5). The values of the PI gains are

determined by the control system designer and typically vary with the mean wind speed. The use of control system gains that vary according to system operating conditions is referred to as "gain scheduling." To control the power output in Region 3, the generator torque is controlled in concert with the blade pitch. There are two strategies. The generator torque may be held constant, which in combination with a constant generator speed, will produce constant power. Alternatively, the generator torque can be controlled to be inversely proportional to the generator speed, again producing constant power. More details on Region 3 control are given in Section 9.3.4.

- **Region 3½.** To avoid excessive loads on the turbine, the turbine's power is progressively reduced after a specified wind speed typically around 25 m/s. This is achieved by increasing the pitch angle.
- **Region 4.** The blades are pitched even more, and the turbine stops producing power at the cut-out wind speed. As seen in Figure 9.12, the cut-out wind speed is around 27 m/s, which is typical of modern wind turbines. The original NREL 5-MW turbine model did not have a Region 3½ and therefore the cut-out wind speed of 25 m/s is often used in the literature.

The overall closed-loop dynamic control loop is shown in Figure 9.13. The rotor speed is measured using a speed sensor and is used as the input to the pitch and torque controller.

9.3.2 Equation of Motion of the Drivetrain

An equation of motion of the drivetrain is needed for the rotational speed control of the turbine. The equation was introduced in Chapter 4 for a rigid and flexible shaft. The rigid shaft equation written on the low-speed shaft (LSS) is given by:

$$J\dot{\Omega} = Q_a\left(U, \Omega, \theta_p\right) - Q_g(\Omega) \tag{9.15}$$

where J is the effective drivetrain inertia (rotor plus drivetrain), U is the wind speed, Ω is the rotor speed, $\dot{\Omega}$ is the rotor acceleration, θ_p is the pitch angle, Q_a is the aerodynamic torque, and Q_g is the generator torque referred to the LSS. The effective inertia is obtained by combining the inertias of all the components of the drivetrain onto the LSS. Neglecting the gearbox and shaft inertias (unless these inertias are already included in the rotor inertia), the effective inertia may be obtained as:

$$J = J_r + J_g = J_r + n_g^2 J_{g,HSS} \tag{9.16}$$

where J_r the rotor inertia, J_g is the generator inertia on the LSS, $J_{g,HSS}$ is the generator inertia on the high-speed shaft (HSS), and n_g is the gear ratio of the gearbox. The generator torque on the LSS is obtained from the generator torque on the HSS as follows:

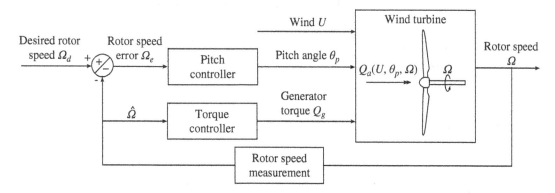

Figure 9.13 Illustration of dynamic control of pitch and generator torque based on the rotor speed

$$Q_g = \frac{n_g}{\eta} Q_{g,HSS} \tag{9.17}$$

where η is the efficiency of the drivetrain, and $Q_{g,HSS}$ is the generator torque on the HSS. Equation (9.15) will be used in Section 9.3.3 to obtain closed-loop equations that include the controller and the turbine dynamics.

9.3.3 Region 2 Control

9.3.3.1 Overview

In Region 2, the objective of the controller is to maximize power output. This is obtained by keeping the pitch angle and the tip-speed ratio to their optimum values. To achieve maximum power, a simple strategy consists of adjusting the generator torque to control the rotor speed and maintaining an optimal tip-speed ratio as the wind speed changes. Modern controller may use PI controllers (see Section 9.3.4), or more advanced strategies such as adaptive controllers. In the simple strategy for Region 2, the generator torque is set to be proportional to the square of the rotational speed of the controller:

$$Q_{g,HSS} = K_g \Omega^2 \tag{9.18}$$

where $Q_{g,HSS}$ is the generator torque on the high-speed shaft (HSS), and K_g is the proportional constant derived in a subsequent paragraph. An example of a generator torque curve as function of the rotor speed is given in Figure 9.14.

In Region 2, an increase of wind speed leads to an increase of aerodynamic torque, rotational speed, generator torque, and power. An example of generator torque response in Region 2 is shown in Figure 9.15 for two cases: a wind step, and a gust. For the wind step, it is seen that the tip-speed ratio is indeed the same at the beginning and the end of the simulation despite the change of wind speed, which confirms that the generator torque response is such as to maintain a constant tip-speed ratio in Region 2. The generator torque response is observed to follow closely the rotor speed response. Indeed, the controller is implemented so as to follow Equation (9.18) quasi-instantaneously (a low pass filter is often employed). Despite the fast response of the generator torque to changes in rotor speed, there are transient regions where the tip-speed ratio is not constant, due to: (1) the drivetrain inertia which prevents the drivetrain from being in equilibrium instantaneously (see Equation 9.15), and (2) the delayed response of the aerodynamics due to dynamic inflow (see Chapter 3) which implies that the aerodynamic torque

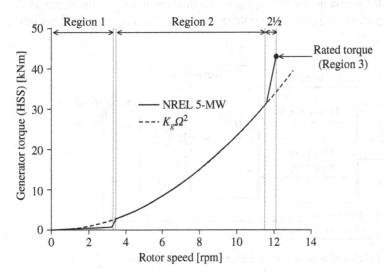

Figure 9.14 Generator torque vs. rotor speed for the NREL 5-MW wind turbine with focus on Region 2 where the generator torque is a quadratic function of the rotor speed

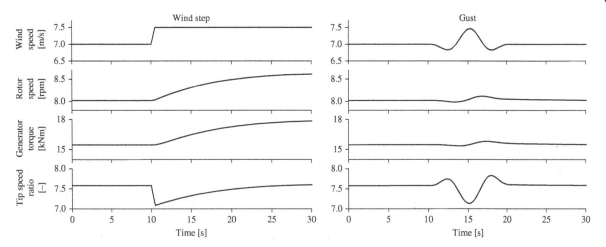

Figure 9.15 Rotor speed and generator torque responses in Region 2 to a wind step (left) and gust (right) for the NREL 5-MW wind turbine. The wind speed and tip-speed ratio are also shown on the graph

requires time before reaching a new equilibrium value. In this example, the large inertia of the rotor is the main contributor to the transient response. The aerodynamic torque increases abruptly after the wind speed increases, but the large inertia leads to a progressive increase of the rotor speed, until a new equilibrium is reached on the drivetrain, following Equations (9.15) and (9.18).

Similar conclusions are drawn from the gust case. For a realistic wind field with turbulence, the tip-speed ratio in Region 2 will be close to the design values, but variations will be expected due to the delayed inertial and aerodynamic responses.

9.3.3.2 Derivation of the Proportionality Constant in Region 2

The optimum pitch, $\theta_{p,opt}$, and the optimal tip-speed ratio, λ_{opt}, are usually chosen as the ones providing the maximum power coefficient. They are therefore defined as:

$$C_{p,\max} = C_p\left(\theta_{p,opt}, \lambda_{opt}\right) \tag{9.19}$$

where C_P is the power coefficient. In Region 2, the generator torque is used to regulate the rotational speed so that λ stays at λ_{opt} over the range of wind speeds of this region.

If the controller successfully achieves the condition $\lambda \equiv \lambda_{opt}$, the aerodynamic power, P_a, is:

$$P_{a,opt} = \frac{1}{2}\rho A U^3 C_P\left(\theta_{opt}, \lambda_{opt}\right) \tag{9.20}$$

The power is seen to be directly proportional to the cube of the wind speed. Similarly, the aerodynamic torque is proportional to the square of the rotational speed:

$$Q_{a,opt} = \frac{P_{a,opt}}{\Omega} = \left[\frac{1}{2}\rho A \frac{R^3}{\lambda_{opt}^3} C_P\left(\theta_{opt}, \lambda_{opt}\right)\right]\Omega^2 = K_a\Omega^2 \tag{9.21}$$

where K_a is the aerodynamic torque constant with dimension in $\dfrac{\text{Nm}}{\left(\dfrac{\text{rad}}{\text{s}}\right)^2}$. If the rotor's speed is given in RPM, the

constant is $K_{a,RPM} = K_a\left(\dfrac{2\pi}{60}\right)^2$.

To achieve a torque balance on the shaft, the generator torque, Q_g, and the aerodynamic torque need to be in equilibrium on the LSS ($Q_g = Q_{a,\,opt}$), therefore:

$$Q_g = K_a \Omega^2 \tag{9.22}$$

Using Equation (9.17), the generator torque needed on the HSS is then:

$$Q_{g,HSS} = \frac{\eta}{n_g} Q_g = \frac{\eta}{n_g} K_a \Omega^2 = K_g \Omega^2 \tag{9.23}$$

where K_g is the generator torque constant.

The controller strategy in this region consists of setting the generator torque according to the above equation using the measured rotor speed as input. The torque command is usually filtered to avoid excessively rapid changes in the rotor speed.

The control diagram corresponding to the generator torque controller is illustrated in Figure 9.16. It is noted that Ω is the rotational speed of the LSS. If the rotational speed is the rotational speed of the HSS, then the proportionality constant needs to be divided by n_g^2.

The closed-loop equation of the controller in Region 2 is therefore obtained by inserting the expression for the generator torque, Equation (9.22), into the equation of motion of the drivetrain, Equation (9.15).

$$J\dot{\Omega} = \frac{1}{2} \rho A R^3 \left[\frac{1}{\lambda^3} C_P(\theta, \lambda) - \frac{1}{\lambda_{opt}^3} C_P(\theta_{opt}, \lambda_{opt}) \right] \Omega^2 \tag{9.24}$$

Example: For the NREL 5-MW, $\lambda_{opt} = 7.65$, $C_{P,\,opt} = 0.48$, $\rho = 1.225$, we obtain $K_a = 0.002R^5$ and $K_g \approx$ $21\,300 \, \dfrac{\text{Nm}}{\left(\dfrac{\text{rad}}{\text{s}}\right)^2} = 233.6 \, \dfrac{\text{Nm}}{(\text{RPM})^2}$.

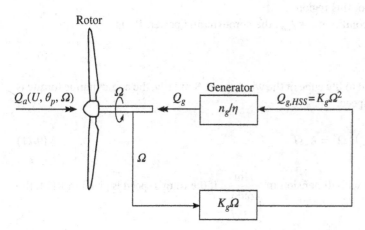

Figure 9.16 Control diagram in Region 2 for the regulation of the rotor speed using the generator torque

9.3.4 Region 3 Control

9.3.4.1 Overview

In Region 3, the controller strategy consists of using the pitch mechanism to maintain the rotor speed constant and use the generator torque to regulate the power. The reason for limiting the power output in this region is to limit the loads and limit the generator size (and therefore limit the cost). As a rule of thumb, the power is equal to the product of the thrust force times the wind speed. Therefore, a constant power will lead to decreasing thrust values for increasing wind speeds and a reduction of loads at higher wind speeds, which can be translated into cost reduction (less material needed to withstand high loads). Generator sizes and costs scale with the maximum power output supported by the generator. Because the probability of high wind speeds is relatively low, diminishing financial returns will be obtained by increasing the generator rating. High wind speeds are not frequent enough to justify a machine strong enough to harvest as much wind energy as possible. This justifies the choice of constant power in Region 3. Keeping the rotational speed constant in Region 3 helps the control of average power (and avoid sudden dips in power).

The generator torque control in Region 3 is described in Section 9.3.4.2; it is achieved either using a constant power strategy or a constant torque strategy.

A common way to implement the pitching strategy for rotor speed regulation of Region 3 is to use a PI controller (see Section 9.2.5), which takes as input the error of the rotor speed and returns the desired pitch angle to reach the objective of constant rotational speed:

$$\Delta\theta_p = K_P \Delta\Omega + K_I \int_0^t \Delta\Omega \, dt \tag{9.25}$$

where $\Delta\Omega = \Omega - \Omega_0$, with Ω_0 the rated rotational speed, and where K_P and K_I are the proportional and integral gains, which can be obtained using the developments of Section 9.3.4.3. The PI controller is used to drive the rotor speed error to zero, and therefore achieve the constant rotor speed objective. A control diagram of the PI pitch controller expressed in Equation (9.25) is shown in Figure 9.17.

As previously discussed, most wind turbines use a "pitch to feather" strategy, meaning that the pitch angle is set to increase when the wind speed increases. The main reason is that pitching to feather reduces the angle of attack, and therefore this strategy avoids stall. Stall can induce vibrations on the blades, and sudden drops of loading, which are best to avoid. If the wind speed increases while the pitch is held constant, the angle of attack will increase, leading to increased aerodynamic torque, and therefore an increase in rotational speed. Pitching the blade to feather will reduce the angle of attack, thereby reducing the aerodynamic torque and so the rotational speed. The value K_P is therefore positive in a pitch-to-feather strategy. We note that some authors use an opposite convention for $\Delta\Omega$, or, introduce negative signs in front of K_P and K_I in Equation (9.25) (this is the case for the ROSCO controller, see e.g., Abbas *et al.*, 2022), in which case the gains have opposite signs.

In Section 9.3.4.3, the closed-loop equations (equations combining the controller and drivetrain response) are derived and expressions for the gains of the controller are obtained. It will be seen that the gains need to be adjusted ("scheduled"), so that they decrease as the wind speed increases. Indeed, the aerodynamic forces are proportional to the square of the relative wind velocity, so the aerodynamic forces become more sensitive to changes in pitch as the wind speed increases. Said differently, smaller changes in pitch angles are required to get the same changes in torque. It will also be seen that the closed-loop system behaves as a second-order system. Therefore, a frequency and damping can be associated with the system response, which is a function of the gains of the controller. The tuning of the gains can lead to different responses (slow, fast, highly damped, or poorly damped). This behavior is illustrated in Figure 9.18 where the pitch and rotor speed responses to a wind step are plotted for the NREL 5-MW. Changing K_P or K_I influences the damping and frequencies of the responses (the relationships between damping, frequencies, and the gains are found in Section 9.3.4.3). See also Section 9.2.5.

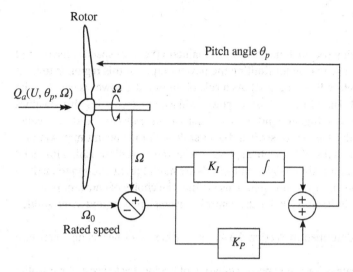

Rotor

Figure 9.17 Control diagram for a PI-controller used in Region 3 to regulate the rotor speed using the pitch actuator

9.3.4.2 Generator Torque Control Strategies in Region 3

The limitation of power output in Region 3 is achieved using a constant power strategy or a constant torque strategy. In the constant power strategy, the generator torque (referred to the LSS) is obtained as:

$$Q_g = \frac{n_g}{\eta} Q_{g,HSS} = \frac{n_g}{\eta} \frac{P_O}{n_g \Omega} = \frac{1}{\eta} \frac{P_O}{\Omega} \tag{9.26}$$

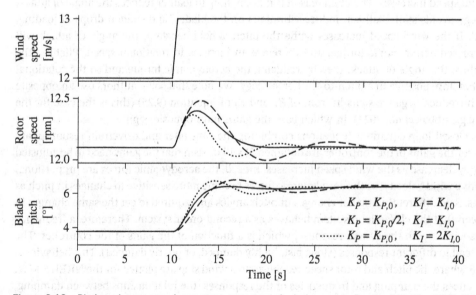

Figure 9.18 Pitch and rotor speed responses to a step of wind speed in Region 3 for the NREL 5-MW wind turbine, and for different values of the proportional and integral gains (K_P and K_I) The generator torque is kept constant here

where P_O is the operating point (rated) power, η the efficiency of the drivetrain, and n_g is the gear ratio of the gearbox. From Equation (9.26), it is seen that if the rotor speed increases, the generator torque will drop, thereby ensuring a constant power value.

In the constant torque strategy, as indicated by its name, the generator torque is kept constant to its rated value:

$$Q_g = Q_{g,O} = \frac{1}{\eta} \frac{P_O}{\Omega_O} \tag{9.27}$$

where Ω_O is the operating point (rated) rotational speed of the LSS.

The constant power strategy is effective at keeping a constant power because the power electronics of the generator can rapidly change the generator torque and respond rapidly to changes of rotational speed. The second strategy can lead to large power fluctuations. More advanced control strategies may use a PI controller. The control diagrams of three different control strategies are illustrated in Figure 9.19.

9.3.4.3 *Closed-loop Equations for the Pitch Controller
To obtain the closed-loop equations of the pitch controller, first, the equation of motion of the system is linearized, then, the equation for the PI-controller is inserted into the linearized equation.

9.3.4.4 *Open-loop Linearized Drivetrain Equation of Motion
The controller takes as input the error in rotational speed, denoted $\Delta\Omega$, with:

$$\Delta\Omega = \Omega - \Omega_0 \tag{9.28}$$

where Ω_0 is the rated rotational speed. To introduce the error into the equation of motion, one needs to linearize the drivetrain equation (Equation 9.15). A simple way to achieve this consists of performing a Taylor expansion of all the terms. The variables are expressed as function of the operating point values (e.g., the rated values) and small variations: $\Omega = \Omega_0 + \Delta\Omega$, $U = U_0 + \Delta U$, $\theta = \theta_{p,0} + \Delta\theta_p$. The terms of Equation (9.3) are then expressed using a first-order Taylor expansion:

$$Q_a(U,\Omega,\theta_p) = Q_a\big|_0 + \frac{\partial Q_a}{\partial U}\bigg|_0 \Delta U + \frac{\partial Q_a}{\partial \Omega}\bigg|_0 \Delta\Omega + \frac{\partial Q_a}{\partial \theta_p}\bigg|_0 \Delta\theta_p$$
$$Q_g(\Omega) = Q_g\big|_0 + \frac{\partial Q_g}{\partial \Omega}\bigg|_0 \Delta\Omega \tag{9.29}$$

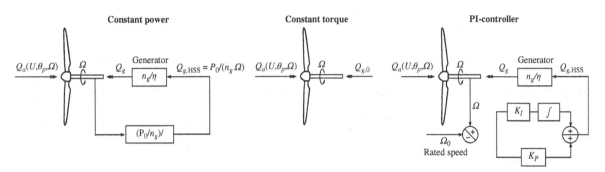

Figure 9.19 Control diagram in Region 3 for three generator torque control strategies. Left: constant power. Center: constant torque. Right: PI controller

At the operating point, the drivetrain is assumed to be in equilibrium (i.e., at a constant rotor speed, $\dot{\Omega}_0 = 0$). Therefore, the zero-order terms satisfy the following equation of motion:

$$J\dot{\Omega}_0 = Q_a|_0 - Q_g|_0 = 0 \tag{9.30}$$

which states that the aerodynamic and generator torque are in balance. The notation $|_0$ will be further dropped for conciseness. Inserting the linear expansion of each term into the drivetrain equation of motion, and using $\dot{\Omega} = \dot{\Omega}_0 + \Delta\dot{\Omega} = \Delta\dot{\Omega}$, leads to:

$$J\Delta\dot{\Omega} + \left[-\frac{\partial Q_a}{\partial \Omega} + \frac{\partial Q_g}{\partial \Omega} \right]\Delta\Omega - \frac{\partial Q_a}{\partial U}\Delta U - \frac{\partial Q_a}{\partial \theta_p} = 0 \tag{9.31}$$

The partial derivatives present in this equation can be obtained numerically or analytically, as will be discussed later in this section.

9.3.4.5 *Closed-loop Linearized Drivetrain Equation of Motion

A PI controller is used to obtain the pitch angle based on the error in rotor speed, therefore:

$$\Delta\theta_p = K_P\Delta\Omega + K_I \int_0^t \Delta\Omega\, dt = K_P\dot{\varphi} + K_I\varphi \tag{9.32}$$

where the intermediate variable φ is introduced for convenience, defined as:

$$\varphi = \int_0^t (\Omega - \Omega_0)dt, \quad \dot{\varphi} = \Delta\Omega, \quad \ddot{\varphi} = \Delta\dot{\Omega} \tag{9.33}$$

The closed-loop equation of motion, containing the action of the controller, is obtained by inserting Equation (9.32) into Equation (9.31), leading to:

$$J\ddot{\varphi} + \left[-\frac{\partial Q_a}{\partial \Omega} - \frac{\partial Q_a}{\partial \theta_p}K_P + \frac{\partial Q_g}{\partial \Omega} \right]\dot{\varphi} - \frac{\partial Q_a}{\partial \theta_p}K_I\varphi - \frac{\partial Q_a}{\partial U}\Delta U = 0 \tag{9.34}$$

It is common to assume that the variation of aerodynamic torque with wind speed and rotor speed, close to the operating point, can be neglected. Therefore, Equation (9.34) reduces to:

$$J\ddot{\varphi} + \left[-\frac{\partial Q_a}{\partial \theta_p}K_P + \frac{\partial Q_g}{\partial \Omega} \right]\dot{\varphi} - \frac{\partial Q_a}{\partial \theta_p}K_I\varphi = 0 \tag{9.35}$$

In Region 3, $\frac{\partial Q_a}{\partial \theta_p} < 0$, therefore, if $K_P > 0$ (pitch to feather), then the term $-\frac{\partial Q_a}{\partial \theta_p}K_P$ adds damping to the drivetrain. The term $\frac{\partial Q_g}{\partial \Omega}$ is either negative or zero, which means that it can potentially introduce negative damping which needs to be balanced by the proportional term to make the system stable. If $k_I > 0$, the term $-\frac{\partial Q_a}{\partial \theta_p}K_I$ provides restoring stiffness to the system. Equation (9.35) is a second-order differential equation, which can be rewritten as:

$$M_\varphi\ddot{\varphi} + C_\varphi\dot{\varphi} + K_\varphi\varphi = 0 \tag{9.36}$$

where the inertial, damping, and stiffness terms are directly identified from Equation (9.36). The natural frequency and damping ratio associated with this equation can be computed as:

$$\omega_\varphi = \sqrt{\frac{K_\varphi}{M_\varphi}}, \qquad \zeta_\varphi = \frac{C_\varphi}{2\sqrt{M_\varphi K_\varphi}} \tag{9.37}$$

Hansen *et al.* (2005) suggest using the following values to obtain acceptable responses:

$$\omega_\varphi \approx 0.6 \, \text{rad/s}, \qquad \zeta_\varphi = 0.6 - 0.7 \tag{9.38}$$

From an *a priori* knowledge of the frequency and damping coefficient, it is possible to obtain the values of the gains K_P and K_I, as will be detailed in the next paragraph. As mentioned earlier, the fact use of gains that vary with the operating conditions is referred to as gain scheduling.

9.3.4.6 *Gain Scheduling of the Pitch Controller

The gains of the pitch controller can be identified from the frequency and damping using Equation (9.37) and the definitions of M_φ, K_φ and C_φ as follows:

$$K_P = \frac{2\zeta_\varphi \omega_\varphi J - \frac{\partial Q_g}{\partial \Omega}\Big|_0}{\frac{\partial Q_a}{\partial \theta_p}\Big|_0} \tag{9.39}$$

$$K_I = \frac{\omega_\varphi^2 J}{-\frac{\partial Q_a}{\partial \theta_p}\Big|_0} \tag{9.40}$$

where the $|_0$ has been reintroduced. The term $\dfrac{\partial Q_a}{\partial \Omega}\Big|_0$ varies with wind speed, rotor speed, and pitch; therefore, the controller needs to adapt the value of K_P and K_I continuously, a process referred to as gain scheduling.

Gain scheduling is required because a wind turbine is a nonlinear system, which responds differently at different operating conditions, whereas a PI controller is based on a linear model that is only applicable within a certain operational range (see Bianchi *et al.*, 2007 for more details).

The term $-\dfrac{\partial Q_g}{\partial \Omega}\Big|_0$ is obtained as follows, depending on the control strategy:

$$\frac{\partial Q_g}{\partial \Omega}\Big|_0 = \begin{cases} 0, & \text{Constant generator torque strategy} \\ -\dfrac{1}{\eta}\dfrac{P_0}{\Omega_0^2}, & \text{Constant power strategy} \end{cases} \tag{9.41}$$

For the constant power strategy, the generator torque (on the LSS) is given by Equation (9.26). Therefore, the Taylor series expansion of the generator torque is:

$$Q_g = \frac{1}{\eta}\frac{P_O}{\Omega_O} - \frac{1}{\eta}\frac{P_O}{\Omega_O^2}\Delta\Omega \tag{9.42}$$

For the constant torque strategy, the torque is $Q_g = \dfrac{1}{\eta}\dfrac{P_0}{\Omega_0}$, and the derivative with respect to Ω is zero.

The sensitivity of the aerodynamic torque with respect to the pitch, $\dfrac{\partial Q_a}{\partial \theta_p}\Big|_0$, can be computed numerically using an unsteady BEM code (see Chapter 3). For each wind speed above rated, one needs to find the pitch angle such that the turbine produces the desired rated power, while the rotational speed is kept constant at the rated value. An iterative procedure may be used to find this pitch angle. Once the reference pitch angle is found for each wind speed, the rate of change $\left.\dfrac{\partial Q_a}{\partial \theta_p}\right|_0$ can be determined in two ways: either by performing linearization using a dedicated tool (such as the *OpenFAST* linearization functionality) or by computing the slope numerically when running an unsteady BEM simulation where a sudden pitch change is introduced. When computing the rate of change using a BEM code and finite differences (for instance computing the torque at two pitch angles surrounding the operating point value), it is important that the induction values from the BEM algorithm do not "jump" between two steady-state values but instead, that the induced velocities are kept constant and equal to the values at the operating point. This is referred to as the "Frozen Wake" assumption. Alternatively, if a dynamic inflow model is available, such a model should be used, as it will ensure that the inductions close to the operating point vary smoothly.

By plotting $\left.\dfrac{\partial Q_a}{\partial \theta_p}\right|_0$ as function of θ_p, a quasi-linear relationship is obtained. The torque rate is then modeled as follows:

$$\left.\frac{\partial Q_a}{\partial \theta_p}\right|_0 (\theta) = \frac{\partial Q_a}{\partial \theta_p}(\theta_p = 0)\left[1 + \frac{\theta_p}{\theta_K}\right] \tag{9.43}$$

where θ_k is a parameter that can be obtained by performing a linear fit to the data. Using this model, and neglecting $-\left.\dfrac{\partial Q_g}{\partial \Omega}\right|_0$, the gain scheduling (continuous adaptation of the gains) can be obtained as follows:

$$K_P = \frac{2\zeta_\varphi \omega_\varphi J}{\left.\dfrac{\partial Q_a}{\partial \theta_p}\right|_0 (\theta_p = 0)} GK(\theta_p) \tag{9.44}$$

$$K_I = \frac{\omega_\varphi^2 J}{-\left.\dfrac{\partial Q_a}{\partial \theta_p}\right|_0 (\theta_p = 0)} GK(\theta_p) \tag{9.45}$$

where $GK(\theta_p)$ is the gain-correction factor, defined as:

$$GK(\theta_p) = \left[1 + \frac{\theta_p}{\theta_k}\right] \tag{9.46}$$

In the literature, some authors use the rate of change of the aerodynamic power with respect to the pitch angle, $\left.\dfrac{\partial P_a}{\partial \theta_p}\right|_0$. Considering only the variation in pitch, the Taylor expansion of the aerodynamic torque can be obtained as:

$$Q_a = Q_a|_0 + \frac{\partial Q_a}{\partial \theta_p}\Delta \theta_p = \frac{P_{a,0}}{\Omega_0} + \frac{1}{\Omega_0}\left.\frac{\partial P_a}{\partial \theta_p}\right|_0 \Delta \theta_p \tag{9.47}$$

The rates are then directly related by the rated rotor speed:

$$\left.\frac{\partial Q_a}{\partial \theta_p}\right|_0 = \frac{1}{\Omega_0}\left.\frac{\partial P_a}{\partial \theta_p}\right|_0 \tag{9.48}$$

The constant θ_k is the same whether the aerodynamic power or torque is used to determine the gain scheduling. We recall that if an opposite convention is used for $\Delta\Omega$, then the gains will have an opposite sign (this is the case for the ROSCO controller). An example of gain scheduling for the NREL 5-MW is obtained using the ROSCO toolbox as shown in Figure 9.20.

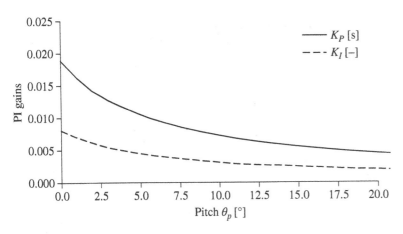

Figure 9.20 Gain scheduling (functions $K_P(\theta)$ and $K_I(\theta)$) for the pitch PI-controller of the NREL 5-MW wind turbine in Region 3

9.3.5 Region Matching

As discussed in Section 9.3.1, intermediate regions are present between the main control regions. Region matching refers to the design of a control system that smoothly transitions from one main region to the next. Of the different transition regions, Region 2½ is the one that requires the most attention. As shown in Figure 9.14, the generator torque increases rapidly in this region and the generator torque is often directly proportional to the rotor speed. The linear slope is sometimes called generator slip in reference to how conventional (squirrel cage) induction genera-tors work (see Chapter 5). Because of the steep slope of the generator torque, a small drop in rotor speed will result in a large drop of generator torque, which can lead to a significant drop in power. This is mainly an issue when the turbine is operating in Region 3 and a drop of rotor speed occurs due to a gust or turbulence. This region is also the region where the thrust (and other loads) are the largest; therefore maximum structural damage is likely to occur during this transition. Different algorithms are implemented in Region 3 to prevent sudden switch from Region 3 to Region 2. A simple option is to keep the generator torque at its maximum value as long as a positive pitch angle is present (indicating that the turbine is in Region 3), even if the rotor speed drops below its rated value.

9.3.6 Example Behaviors

Examples of the steady-state response of the NREL 5-MW wind turbine were shown previously in Figure 9.12. The results of a dynamic simulation of the NREL 5-MW turbine using *OpenFAST* in turbulent winds in different regions are given in Figure 9.21 through Figure 9.23.

Results for Region 2, with a mean wind speed of 7 m/s are shown in Figure 9.21. In this region, the rotor speed is seen to respond slowly to the changes of wind speed, only the generator torque control is active, and it follows the rotational speed.

Results corresponding to Regions 2½ and 3 are shown in Figure 9.22, for a mean wind speed of 11 m/s. It can be seen that active pitch is used to regulate the rotational speed when the wind speed is sufficiently large (Region 3). At times, the rotor speed and power drops, and the pitch control is no longer active (Region 2½ and 2).

Results for Region 3 only are shown in Figure 9.23, for a mean wind speed of 13 m/s. As can be seen in this simulation, the generator torque and blade pitch are controlled to regulate the generator power around its rated value of 5 MW and the rotational speed around the rated value of 12.1 rpm.

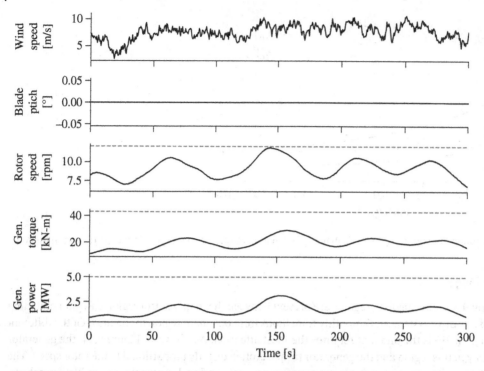

Figure 9.21 Example of dynamic response of the NREL 5-MW in Region 2 for a turbulent wind simulation with mean speed 7 m/s

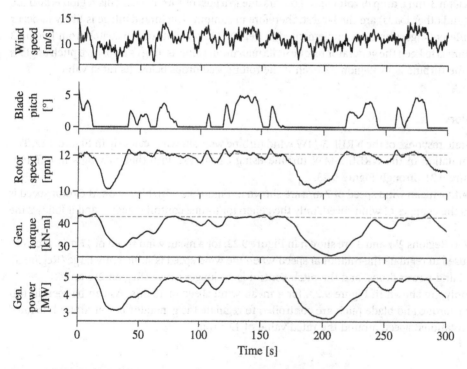

Figure 9.22 Example of dynamic response of the NREL 5-MW in Region 2.5 and 3 for a turbulent wind simulation with mean speed 11 m/s. Rated values are indicated with horizontal-dashed lines

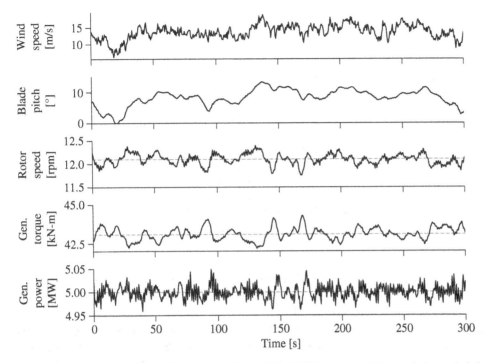

Figure 9.23 Example of dynamic response of the NREL 5-MW in Region 3 for a turbulent wind simulation with mean speed 13 m/s. Rated values are indicated with horizontal-dashed lines

9.4 Advanced Control Strategies

The methods presented in Section 9.3 are basic control strategies used to regulate the power and rotational speed of a wind turbine. Nowadays, manufacturers rely on more advanced strategies to improve the power quality and employ a wide range of methods to control and reduce the loads on the wind turbine. The underlying implementation of controllers is highly protected by manufacturers and typically inaccessible to the community. Therefore, only the main features and strategies are presented in this section, without detailing the implementation.

9.4.1 Thrust Peak Shaving

Peak shaving of thrust (or thrust limiting) is a control strategy used in Region 2½, which is the region where the aerodynamic thrust is at its largest and the change of thrust with wind speed is large. Without special care, major loads may occur in this region, in particular at the tower base. Pitching can be introduced in this region to reduce the thrust force and transition progressively to Region 3, at the expense of a slight power loss. An example of this strategy consists of introducing a minimum blade pitch angle in this region (see Abbas *et al.*, 2022). An illustration of the pitch-shaving strategy is given in Figure 9.24.

9.4.2 Speed Constraints in Region 2

In Region 2, the controller may need to introduce constraints to avoid certain rotational speeds. This can be the case if the cut-in speed of the generator requires a minimum speed that is above the optimal speed at low wind speed. In this case, pitch scheduling may be used to achieve maximum power extraction at low wind speed and at

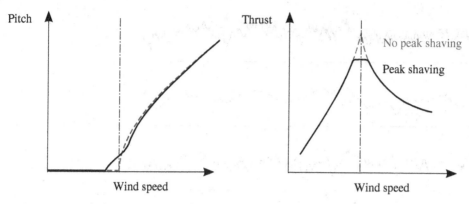

Figure 9.24 Illustration of thrust shaving using adaptative pitch scheduling around the rated wind speed to minimize the thrust loads

sub-optimal tip-speed ratios (due to the rotational speed constraint). This situation is illustrated on the left of Figure 9.25.

In Region 2, the optimal rotational speed may coincide with natural frequencies of the wind turbine, which can lead to resonance issues during a certain range of wind speed. This is often the case for the side-side mode of the tower which has typically low damping and can be excited by small mass or pitch imbalances of the rotor. To avoid resonance, which can lower the structural lifespan of the turbine (see Chapter 8), the controller may adopt a strategy that avoids the rotational speeds that correspond to the tower frequencies. This is illustrated on the right of Figure 9.25.

9.4.3 Controlling Structural Vibrations

As discussed in Chapter 4, wind turbines consist of long lightly damped members that vibrate and operate in an environment in which turbulence, gusts, wind shear, tower shadow, and dynamic effects from other turbine components can excite vibrations. Structural vibrations of the turbine can have a drastic impact on the lifetime of the system. Structural damping is typically low and will only provide limited attenuation. The external loads may

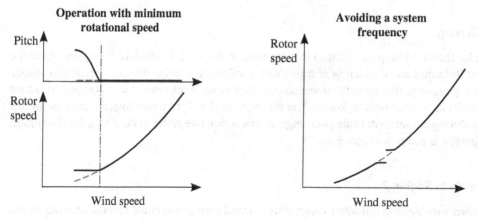

Figure 9.25 Strategies to mitigate the potential constraints on the rotational speed in Region 2. Left: minimum rotational speed. Right: avoiding a system frequency

provide some damping (such as the aerodynamic and hydrodynamic forces), but this damping is often limited to some direction of motions (e.g., fore-aft or flap motion), and only present when the external forces are present (e.g., aerodynamic damping becomes negligible at standstill). Therefore, dedicated control strategies of additional devices are often needed to avoid excessive vibrations of the structure.

9.4.3.1 Controlling Tower Vibrations

Changes in blade pitch have a significant effect on the rotor thrust, which excites tower vibrations. Tower vibrations (and therefore translations of the rotor back and forth) affect the relative wind speed seen by the blades and, thereby, the rotor torque. This feedback path between tower and pitch system is strong enough that careful controller design is needed to avoid exacerbating tower vibrations or making a pitch control system that is completely unstable (Bossanyi, 2003). One approach is to use a filter tuned to the tower vibration frequency that shifts the phase of any control actions at that frequency and effectively damps the tower vibrations. Another approach uses the signal from a tower accelerometer as an input to the control system and to the tower damping algorithm. Alternatively, active or passive tuned-mass-dampers system may be installed in the tower to damp the tower vibrations (see Section 9.4.3.3). Such approaches have been shown to be very effective at damping the excitation of the first tower vibration mode. When implementing this approach, however, care must be taken to ensure that the two control goals of the pitch control system, control of aerodynamic torque and reduction of loads, do not conflict, thereby degrading the efforts to achieve both goals.

9.4.3.2 Controlling Drivetrain Vibrations

In constant-speed wind turbines with conventional induction generators, the generator slip provides a significant amount of damping of drivetrain oscillations. In variable-speed wind turbines, when the torque of the generator is fixed, there is very little damping in the drivetrain, which acts like two large masses (the generator and the rotor) on a flexible shaft. The first collective rotor torsional vibration mode may act like an additional inertial mass connected to the hub by a torsional spring. Finally, if the second side-to-side tower vibration mode is excited (which causes angular displacements at the top of the tower), this can add additional rotational oscillations to the system. If left undamped, these can lead to large torque oscillations in the gearbox (Bossanyi, 2003). One approach to controlling drivetrain torsional vibrations is the use of bandpass filters that effectively add small torque ripples to the drivetrain of a magnitude and phase that cancel out the vibrations excited by the other turbine dynamics.

9.4.3.3 Additional Structural Dampers

Active or passive structure dampers are often placed in the tower, nacelle, and floating platform and can potentially be placed in the blades to reduce the vibrations of these components. Out of the many technologies available, TMDs and tuned liquid column dampers (TLCD) are among the most common. The structural dampers are usually tuned to damp a given frequency of the system, typically the first fore-aft mode of the turbine. A characteristic of structural dampers is that they typically introduce a frequency split that will introduce two frequencies on both sides of the target frequency. The deployment of structural dampers can drastically reduce vibrations, and they are particularly relevant to reduce the fatigue occurring during standstill in offshore wind plants (Stewart and Lackner, 2013; Brodersen *et al.*, 2017).

9.4.4 Individual Pitch Control

The control systems described in the previous sections assume that the blades are pitched similarly, either collectively or cyclically (as a function of azimuth angle). IPC consists of pitching each blade independently, allowing the turbine to respond to asymmetric loads caused by wind shear across the rotor, turbulence, soiling of a blade, etc. (for example, see Geyler and Caselitz, 2007; Bossanyi, 2003). It offers additional possibilities for load reduction but requires more complicated control algorithms, faster processors, and some measure of asymmetrical rotor loads.

Loads that might be useful as control inputs are blade root bending moments, shaft bending moments, or yaw-bearing moments, but typical strain gauges have not been robust enough for this type of application. Recent advances in optical and solid-state strain gauges make IPC more feasible. Some of the most effective approaches to IPC transform the asymmetric loading into a mean load and loads in two orthogonal directions. Fairly simple control algorithms can then be used to determine control actions in this new coordinate system. These are then transformed back into control commands for each of the blade pitch actuators. This approach has been shown to provide significant additional load reduction over collective pitch control.

As wind turbines increase in size, the changes in wind speed and direction across the rotor increase proportionally. Additionally, large wind turbines operate within regions of the upper atmospheric boundary layer where organized, coherent turbulent structures and low-level jets may occur. These conditions may increase fatigue damage significantly (Hand *et al.*, 2006). An example of the use of individual blade pitch actuation for load mitigation due to large, coherent vortices in the atmosphere is described in Hand and Balas (2007). The authors showed that, with adequate sensors and models for the structure of atmospheric vortices, blade root bending moments could be reduced by 30%.

9.4.5 Smart Rotors

Smart rotors refer to technologies applied to the blades to control the aerodynamical forces on the blades to provide fast response to load changes. Indeed, the pitch control has a slow response rate, and the generator torque control (which is fast) acts on the entire rotor and not on individual blades. Smart rotor technologies are also called distributed control devices because they are placed at key locations along the blades. Aerodynamic control devices change the aerodynamic properties of the blades. Examples of concepts (illustrated in Figure 9.26) are:

- **Flaps**. Usually located at the trailing edge of the airfoil, the flaps affect the shape of the airfoil, and therefore the pressure distribution and the lift and drag force. The flaps can either be rigid (similar to airplane flaps), or deformable.
- **Microtabs**. Microtabs are small devices that can deploy continuously in the direction normal to the surface of the airfoil and are typically placed toward the trailing edge. The device affects the flow near the trailing edge of the airfoil and effectively changes the camber of the airfoil. Decrease of lift is achieved by deploying microtabs on the suction side of the airfoil and increase of lift is achieved using microtabs on the pressure side.

Other technologies include camber control (where the shape of the airfoil is changed) and boundary layer control (using suction methods, jets, active vortex generators, or plasma actuators). In general, distributed aerodynamic controls have high response rate but low control authority as their effect is localized. To effectively integrate such technologies, sensing is required to be able to control the device and have the desired load mitigation behavior. Measurements on the blades would be the best for such purposes, but they are more difficult to implement than in the nacelle and can be more expensive. Reliability of the actuator is also critical because any replacement would have significant cost as they are difficult to access. Active smart rotor technologies require power to function and therefore, a compromise between load mitigation and power consumption needs to be reached. The reader is referred to Barlas and Van Kuik (2007) for a review of smart rotor technologies.

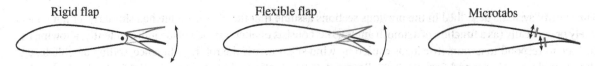

Rigid flap   Flexible flap   Microtabs

Figure 9.26 Examples of smart rotor concepts used to locally control the aerodynamics of the blade

9.4.6 Frequency Control and Grid Integration

In order to provide electricity that is of sufficient quality to be used in the grid (see Chapter 12), the wind turbine controller uses advanced power electronics to regulate the frequency and amplitude of the power output. As more and more wind turbines are integrated on the grid, a new branch of control is emerging in which the wind turbines are used to answer the needs of the utility grid. The wind turbines can be used to actively control the overall power output of a wind plant in order to meet the power needs of the grid and contribute to the frequency regulation of the grid. Grid operators indeed need to constantly balance the generation and loads on the grid to ensure that the grid frequency remains stable. Wind turbines can provide a fast response to the grid need, and in contrast to conventional power plants, they can switch from 0% to 100% capacity with low inertia and a fast response time (as long as the wind resource allows it). Wind plants can therefore provide a contingency reserve. For instance, they can be downregulated (set to produce less than the nominal power), so that, when needed, they can provide a quick response to an increase of load on the grid by increasing their power production to nominal. Similarly, wind turbines can potentially provide fast-frequency response, detecting changes of frequency in the grid automatically and adapting their response without a need for an operator. The topic of frequency control and grid integration is a topic of active research (see Andersson *et al.*, 2021).

9.4.7 Signal Processing and Filtering

An important aspect of modern control consists of filtering and synchronizing the signals from the various sensors and actuators on the turbine. Different sensors may have various levels of noise, biases or may become unavailable for indeterminate periods of time. A typical signal processing tool uses low-pass or high-pass filters to eliminate some range of frequencies. Excessive filtering may nevertheless introduce lags in the signals, which can be counterproductive and prevent the controller from responding in a timely manner. Signal processing is required to ensure a continuous and reliable response from the controller. Signal processing techniques are beyond the scope of this text, but they are an important aspect of the design of modern wind turbine controllers.

9.4.8 Lidar Feedforward

Long-range lidar systems can be used to forecast the production of a wind turbine, which can potentially be used to accommodate the energy market and plan a strategy that is economically advantageous (for instance, providing a contingency reserve to the grid). Short-range lidars can measure the incoming wind seconds before its arrival at the turbine and adjust the control strategy accordingly. For instance, the system may detect an incoming gust and start pitching the blades in anticipation to reduce the load response. Lidar technologies remain currently too expensive to be deployed systematically, but more and more research is being undertaken to use lidar measurements as part of the control system (see e.g., Scholbrock *et al.*, 2016).

9.4.9 Control of Floating Systems

Floating wind turbines (see Chapter 11) require different control systems than onshore or fixed-bottom turbines. A common issue observed for spars and semisubmersibles is the negative damping of the platform pitch motion which occurs above rated conditions. This issue can be explained as follows: when the platform pitches upwind, the relative wind speed increases, which increases the rotor loads and therefore the rotor speed. This will lead the blade pitch controller to increase the blade pitch angle so as to maintain a constant rotational speed. This results in a decrease of the thrust force, and therefore, the platform will pitch upwind even more. The opposite occurs if the platform pitches downwind. In that case, the controller is seen to introduce negative damping into the system. The pitching frequency of the platform, which is usually relatively slow, is therefore an integral part of the design of a floating turbine's controller. Wind and wave loading can cause sizable platform motions for a floating system, and dedicated control strategies are required to mitigate the motions and loads. The reader is referred to Ha *et al.* (2021) for an example control strategy.

9.4.10 Wind Farm Control

Control of wind farms (see Chapter 12) consists of implementing wind-farm-level control strategies to achieve an overall optimal performance of the entire plant. Typically, the strategies employed attempt to achieve maximum annual energy production (AEP), but more advanced objectives can include load mitigation or real-time adaptation to the energy market. To achieve such objectives, the wind farm controller (or "supercontroller") instructs the controller of each turbine to follow a given strategy and operate at a given operating point. The supercontroller will typically ask some turbines to "derate," which means that they will operate at a suboptimal point. These turbines will produce less power, operate at lower axial induction, and undergo lower thrust and lower loads. The benefit of such strategy is that a turbine downstream of a derated turbine can produce more energy since the wake deficit from the upstream turbine is not as pronounced (due to the reduced thrust). Another tool available consists of changing the yaw angle of a given turbine. This will deflect its wake away from the downstream turbine and potentially increase the overall power production of the plant. An example of optimization of a wind farm layout in conjunction with yaw-based control is given in the study of Gebraad *et al.* (2017). In this study, AEP gains of the order of 5% are obtained by optimizing the layout of a wind farm and implementing a yaw-steering strategy.

9.5 Design, Implementation, and Challenges of Dynamic Control

This section elaborates on the concept of dynamic control introduced in Sections 9.2 and 9.3. The section presents the dynamic control system design process, examples of dynamic control system design issues specific to wind turbine applications, and how the dynamic control systems are implemented.

9.5.1 Design of a Dynamic Control System

9.5.1.1 Classic Control System Design Methodology

The classic control system design process, as described in Grimble *et al.* (1990) and De LaSalle *et al.* (1990), involves the development of a mathematical model of the machine and a linearization of the model. The controller is then designed iteratively based on the linear model such that it meets the desired specifications which are tested with various simulation studies. The different steps are illustrated in Figure 9.27.

9.5.1.2 Other Control Design Approaches

Other control design approaches that build on or deviate from classical linear system design include adaptive control, optimal control, search algorithms, and quantitative feedback robust design (see De LaSalle *et al.*, 1990; Di Steffano *et al.*, 1967; Horowitz, 1993; Munteanu *et al.*, 2008, Gambier, 2022).

Adaptive Control

The dynamic behavior of a wind turbine is highly dependent on wind speed due to the nonlinear relationship between wind speed, turbine torque, and pitch angle. System parameter variations can be accommodated by designing a controller for minimum sensitivity to changes in these parameters. Adaptive control schemes continuously measure the value of system parameters and then change the control system, adapting to the environment and potential changes in the system. As an example, an adaptive controller may vary the optimal tip-speed ratio and power coefficient instead of assuming these values to be constant as presented in Section 9.3.3. This is the approach for instance presented in Johnson (2004).

Figure 9.27 Control system design methodology (Grimble *et al.*, 1990) (*Source:* Reproduced by permission of the Institution of Mechanical Engineers)

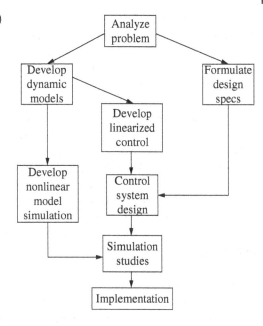

Optimal Control

Optimal design is a time domain approach in which variances in the system output (for example, loads) are balanced against variances in the input signal (for example, pitch action). Optimal control theory formulates the control problem in terms of a performance index. The performance index is often a function of the error between the commanded and actual system responses. Sophisticated mathematical techniques are then used to determine values of design parameters to maximize or minimize the value of the performance index. Optimal control algorithms often need a measurement of the various system state variables or a "state estimator," an algorithm to estimate unmeasured variables based on a machine model. A typical state estimator is a Kalman filter algorithm (see, e.g., Zarchan and Musoff, 2015).

Search Algorithms

Search algorithms can also be used to control wind turbines. These algorithms might constantly change the rotor speed in an attempt to maximize measured rotor power. If a speed reduction resulted in decreased power then the controller would slightly increase the speed. In this manner, the rotor speed could be kept near the maximum power coefficient as the wind speed changed. The controller would not need to use a machine model and would, thus, be immune to changes in operation due to dirty blades, local airflow effects, or incorrectly pitched blades.

Quantitative Feedback Robust Control

Quantitative feedback theory (Horowitz, 1993) is a frequency domain approach for designing robust controllers for systems with uncertain dynamics. It provides methods for designing a closed-loop system with specific performance and stability criteria for systems within a defined range of operating parameters. It can ensure that one controller design will perform well for a system with a range of dynamic properties.

9.5.1.3 Wind Turbine System Models

System models used for control system design are often mathematical models based on the principles of physics. This was the approach illustrated in Section 9.2.5. When such models cannot be developed, an experimental approach termed "system identification" can be used. The two approaches are presented below.

Models Based on Physical Principles

As discussed previously, dynamic models are used to understand, analyze, and characterize system dynamics for control system design. The models consist of one or more differential equations describing system operation. These differential equations are usually written in one of two forms: the transfer function representation (for linear models) or a state-space representation (linear or nonlinear models). The transfer function representation involves the use of Laplace transforms (as in Section 9.2.5.5) and characterizes the system in the frequency domain, while the state-space representation characterizes the system in the time domain. For a linear system, these representations are interchangeable and the choice of approach depends on the degree of complexity of the system and the analytical tools available to the system designer.

The results of control system design are only as good as the model used to describe the system. A control system based on a model that ignores critical dynamics of any part of the machine can result in catastrophic failure. However, an excessively detailed model adds complexity and cost to the analysis and may require certain input parameters that are unknown. If a nonlinear system has been linearized for control system design, simulations of system behavior over a wide range of nonlinear operation should be carried out to check operation of the full controlled system.

System Identification

In cases where models for disturbances or for complicated systems cannot easily be determined, the experimental approach of system identification can be used (see Ljung, 1999, Franklin *et al.*, 2022). System identification involves measuring the system output, given a controlled input signal or measurements of the system inputs. Simpler approaches to system identification require sinusoidal or impulsive inputs. Other approaches can use any input but may require increased computational capability to determine the system model. In order to correctly identify the dynamics of a system, the input signal should be designed to excite all the modes of the system sufficiently to provide measurable outputs.

9.5.2 Summary of Control Issues in Wind Turbine Design

Wind turbines present a number of unique challenges for a control system designer (see De LaSalle *et al.*, 1990; Bianchi *et al.*, 2007; Munteanu *et al.*, 2008, Gambier, 2022). Conventional power generation plants have an easily controllable source of energy and are subject to only small disturbances from the grid. In contrast, the energy source for a wind turbine is subject to large rapid fluctuations, resulting in large transient loads in the system. The consequences of these large fluctuations and other aspects of wind turbines introduce unique issues in control system design for wind turbines:

- The penalty for a poorly regulated wind turbine is higher structural loads and reduced lifetime of the components.
- As mentioned in 9.4.3, wind turbine system consists of numerous lightly damped structures that are excited by the external environment and by forcing functions at the frequency of the rotor rotation and its harmonics. The control system needs to be designed to avoid resonance conditions (see Section 9.2.5) and avoid exciting the system with its control action.
- The aerodynamics are highly nonlinear. This results in significantly different dynamic descriptions of turbine behavior at different operating conditions. These differences may require the use of nonlinear controllers or different control laws for different wind regimes.
- Transitions under dynamic operation from one control law or algorithm to another require careful design (see Section 9.3.5).
- The control goal is not only the reduction of transient loads but also of fatigue loads, caused by the load fluctuations about the mean load.
- Reliable torque and rotor speed measurements for feedback are often difficult to obtain.

- Adequate reduced-order models are often difficult to obtain or establish. Nonlinearities in the system can result in behavior that deviates from that of a simple model (see Novak *et al.*, 1995).

9.5.3 Dynamic Control System Implementation

Dynamic control can be implemented as mechanical systems, analog electrical circuits, digital computers, or in combinations of these.

9.5.3.1 Mechanical Control Systems

Hardware dynamic control systems use linkages, springs, and weights to actuate system outputs in response to some input. Two examples of hardware control systems are tail vanes to orient wind turbines into the wind and pitch mechanisms (for small wind turbines) that vary blade pitch on the basis of aerodynamic forces or rotor speed.

9.5.3.2 Analog Electrical Circuit Control Systems

Analog electrical circuits are often used as distributed controllers in a larger control network. Once a control algorithm has been developed and tested it can be hardwired into circuit boards that are robust and easy to manufacture. One disadvantage of using analog circuits is that changes in the control algorithm can only be made by changing the hardware.

9.5.3.3 Digital Control Systems

Modern dynamic control systems are implemented in digital controllers (e.g., computers). Digital control algorithms are relatively easy to upgrade and allow for easy implementation of nonlinear control approaches.

Digital control systems, as illustrated in Figure 9.28, may need to communicate with analog sensors or actuators via analog-to-digital (A/D) converters and digital-to-analog (D/A) converters (see Astrom and Wittenmark, 1996).

Digital control systems are not continuous but rather are sampled. Sampling gives rise to a number of issues specific to digital control systems. The sampling rate, controlled by the system clock, affects (1) the frequency content of processed information, (2) the design of control system components, and (3) system stability.

The effect can be illustrated by considering a sinusoidal signal, $\sin(2\pi f t)$ where f is the frequency of the sinusoid and t is the time. If this signal is sampled at a frequency f, then the nth sample is sampled at time $t_n = n/f + t_0$ for some starting time t_0 and for integer, n, the value of each sample, s_n, is then:

$$s_n = \sin[2\pi f(n/f + t_0)] = \sin(2\pi n + 2\pi f\, t_0) = \sin(2\pi f t_0) \tag{9.49}$$

But $\sin(2\pi f t_0)$ is a constant. Thus, sampling the signal at the frequency of the signal yields no information at all about fluctuations at that frequency. In fact, in a system that samples information at a frequency f, there will be no useful information at frequencies above $f_{Ny} = f/2$, referred to as the Nyquist frequency. Furthermore, unless the input signal is filtered with a cut-off frequency below f_{Ny}, then the high-frequency information in the signal will distort the desired lower-frequency information.

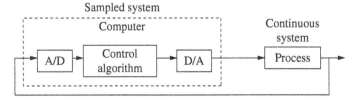

Figure 9.28 Schematic of computer-controlled system; A/D, analog-to-digital converter; D/A, digital-to-analog converter

Sampling rate also affects the design of the control system. Digital control system dynamics are a function of the sampling rate. Thus, the sampling rate affects the subsequent control system design and operation, including the determination of the values of constants in the controller and the final system damping ratio, system natural frequency, etc. Because of this, changes in sampling rate can also turn a stable system into an unstable one. Stability can be a complicated issue, but, in general, a closed-loop digital control system will become unstable if the sampling rate is slowed down too much.

9.6 Supervisory Control

This section provides an introduction to the wind turbine supervisory control system. This system manages safe, automatic operation of the turbine, identifying problems, and activating safety systems. The supervisory control system switches between the various operating states of the turbine (ready for operation, freewheeling, startup, power production, shutdown, etc.), perform continuous fault diagnosis and safely shuts down the turbine when needed.

9.6.1 Operating States

The main operating states of a wind turbine are illustrated in Figure 9.29. The turbine may remain in some operating states for long periods of time, depending on the wind and operating conditions. These states are designated as stationary states. Other operating states may only be transitional if they are entered during changes from one stationary operating state to another.

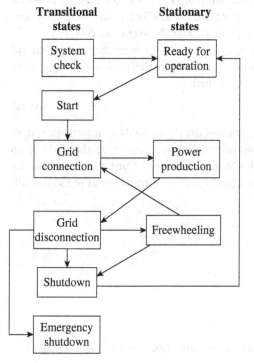

Figure 9.29 Typical turbine operating states

9.6.2 Fault and Component Health Diagnosis

The continuous fault diagnosis capabilities of the supervisory controller must include monitoring for component failures, including sensor failures, operation beyond safe operating limits, grid failure or grid problems, and other undesirable operating conditions.

Component failures may be detected directly or indirectly. For example, the failure of a coupling between the generator and the gearbox could be directly detected if the generator and rotor speeds are known and do not correspond to each other. Such a failure could also be indirectly detected by noting that the rotor speed is increasing or is too high.

While the most robust and accurate sensors that can be afforded should be chosen for a wind turbine, sensor failures can also occur.

Fault diagnosis and component health diagnosis using digital twins technologies (combining multiple sensors, physics-based models, machine learning, and virtual sensing) is an active area of research as it can enable condition-based maintenance and lead to significant savings on operation and maintenance costs. See e.g., Branlard *et al.* (2020).

9.6.3 Fail-safe and Fault-tolerant Systems

Wind turbine control systems rely on (1) power, (2) control logic, and (3) sensors and actuators. The control system must include fail-safe systems in case any of these elements fail. These fail-safe systems must safely shut down the turbine in cases of grid loss, rotor over-speed, excessive vibration, and other emergency situations. On the other hand, as wind turbines become more complex, shutdown due to failures of nonessential sensors, or nondestructive damage of a component, can result in unnecessary downtimes, and therefore, loss of income. Fault-tolerant control system that allows the turbine to continue to operate (potentially at lower rating) until maintenance is performed is therefore an active area of research.

References

Andersson, L. E., Anaya-Lara, O., Tande, J. O., Merz, K. O. and Imsland, L. (2021) Wind farm control-Part I: A review on control system concepts and structures. *IET Renewable Power Generation*, **15**(10), 2085–2108.

Astrom, K. J. and Wittenmark, B. (1996) *Computer Controlled Systems*, 3rd edition. Prentice-Hall, Englewood Cliffs, NJ.

Abbas, N. J., Zalkind, D. S., Pao, L. and Wright, A. (2022) A reference open-source controller for fixed and floating offshore wind turbines. *Wind Energy Science*, **7**, 53–73, https://doi.org/10.5194/wes-7-53-2022.

Barlas, T. and van Kuik, G. (2007) State of the art and prospectives of smart rotor control for wind turbines. *Journal of Physics: Conference Series*, https://doi.org/10.1088/1742-6596/75/1/012080.

Bianchi, F. D., de Battista, H. and Mantz, R. J. (2007) *Wind Turbine Control Systems*. Spring-Verlag London Limited, London.

Bossanyi, E. A. (2003) Wind turbine control for load reduction. *Wind Energy*, **6**(3), 229–244, Wiley Interscience, Hoboken, NJ.

Branlard, E. Giardina, D. and Brown, C. S. D. (2020) Augmented Kalman filter with a reduced mechanical model to estimate tower loads on a land-based wind turbine: a step towards digital-twin simulations. *Wind Energy Science*, **5**, 3. https://doi.org/10.5194/wes-5-1155-2020.

Brodersen, M. L., Bjørke, A.-S. and Høgsberg, J. (2017) Active tuned mass damper for damping of offshore wind turbine vibrations. *Wind Energy*, **20**, 783–796, https://doi.org/10.1002/we.2063.

De LaSalle, S. A., Reardon, D., Leithead, W. E. and Grimble, M. J. (1990) Review of wind turbine control. *International Journal of Control*, **52**(6), 1295–1310.

Di Steffano, J. J., Stubberud, A. R. and Williams, I. J. (1967) *Theory and Problems of Feedback and Control Systems*. McGraw-Hill, New York.

Franklin, G., Powell, J. D. and Workman, M. L. (2022) *Digital Control of Dynamic Systems-Third Edition*. Ellis-Kagle Press. ISBN: 0-9791226-3-5.

Gambier, A. (2022) *Control of Large Wind Energy Systems*. Springer, Switzerland.

Garcia-Sanz, M. and Houpis, C. H. (2012) *Wind Energy Systems: Control Engineering Design. Taylor & Francis*, United Kingdom.

Gasch, R. and Twele, J. (2002) *Wind Power Plants*. James and James, London.

Gebraad, P., Thomas, J. J., Ning, A., Fleming, P. and Dykes, K. (2017) Maximization of the annual energy production of wind power plants by optimization of layout and yaw-based wake control. *Wind Energy*. https://doi.org/10.1002/we.1993.

Geyler, M. and Caselitz, P. (2007) Individual blade pitch control design for load reduction on large wind turbines, *Scientific Proceedings, 2007 European Wind Energy Conference*, pp. 82–86. Milan.

Golnaraghi, F. and Kuo, B. C. (2017) *Automatic Control Systems*, 10th Edition, McGraw-Hill Education.

Grimble, M. J., De LaSalle, S. A., Reardon, D. and Leithead, W. E. (1990) A lay guide to control systems and their application to wind turbines. *Proc. of the 12th British Wind Energy Assoc. Conference*, pp. 69–76, Mechanical Engineering Publications, London.

Ha, K., Truong, H. V. A., Dang, T. D. and Ahn, K. K. (2021) Recent control technologies for floating offshore wind energy system: a review. *International Journal of Precision Engineering and Manufacturing-Green Technology*, **8**, 281–230. https://doi.org/10.1007/s40684-020-00269-5.

Hand, M. M. and Balas, M. J. (2007) Blade load mitigation control design for a wind turbine operating in the path of vortices. *Wind Energy*, **10**(4), 339–355. Wiley Interscience, Hoboken, NJ.

Hand, M. M. Robinson, M. C. and Balas, M. J. (2006) Wind turbine response to parameter variation of analytic inflow vortices. *Wind Energy*, **9**(3), 267–280, Wiley Interscience, Hoboken, NJ.

Hansen, M. H., Hansen, A., Larsen, T. J., Oye, S., Sorensen, F. and Fuglsand, P. (2005) Control design for a pitch-regulated variable speed wind turbine, Riso Report Ris0-R-1500(EN).

Horowitz, I. M. (1993) *Quantitative Feedback Design Theory*, **1**. QFT Publications, Boulder, CO.

Johnson, K. (2004) *Adaptive torque control of variable speed wind turbines*. NREL TP 500-36265.

Luo, N., Vidal, Y. and Acho, L. (2014) *Wind Turbine Control and Monitoring*. Springer, Switzerland.

Ljung, L. (1999) *System Identification – Theory for the User*. Prentice-Hall, Upper Saddle River NJ.

Munteanu, I., Bratcu, A. I., Cutululis, N. and Changa, E. (2008) *Optimal Control of Wind Energy Systems*. Spring-Verlag London Limited, London.

Nise N. S. (2019) *Control System Engineering*, 8th edition. John Wiley & Sons, Inc., New York.

Novak, P., Ekelund, T., Jovik, I. and Schmidtbauer, B. (1995) Modeling and control of variable-speed wind-turbine drive-system dynamics. *IEEE Control Systems*, **15**(4), 28–38.

Pao, L. E. and Johnson, K. E. (2011) Control of wind turbines: approaches, challenges and recent developments. *IEEE Control Systems Magazine*, **31**(2), 44–62, https://doi.org/10.1109/MCS.2010.939962.ind Turbine Control.

Scholbrock, A., Fleming, P., Schlipf, D., Wright, A., Johnson, K. and Wang, N. Lidar-enhanced wind turbine control: Past, present, and future, *2016 American Control Conference (ACC)*, Boston, MA, USA, 2016, pp. 1399–1406, 10.1109/ACC.2016.7525113.

Stewart, G. and Lackner, M. (2013) Offshore wind turbine load reduction employing optimal passive tuned mass damping systems. *IEEE Transactions on Control Systems Technology*, **21**(4), 1090–1104.

Wolfram (2023) *Laplace Transform*, https://www.wolframalpha.com/input?i=laplace+transform Accessed 12/9/23.

Zarchan, P. and Musoff, H. (2015) Fundamentals of Kalman filtering: a practical approach, Fourth Edition, AIAA, Progress in astronautics and 732 aeronautics.

10

Soils, Foundations, and Fixed Support Structures

10.1 Overview

As discussed previously in this text, all wind turbines are comprised of rotor nacelle assemblies (RNAs) which are mounted on some type of support structure. According to accepted definitions (IEC, 2019a, b), for land-based or fixed offshore turbines, the support structure consists of at least a tower and a foundation. In addition, fixed offshore turbines (i.e., those founded on the seabed) include a substructure (see Figure 10.1). The function of the substructure is to connect the tower, whose base should be well above still water level, to the foundation, which is on the seabed. Most fixed offshore substructures consist of a transition piece and a primary steel structure (e.g., a monopile or a jacket). The transition piece contains many secondary steel structures (boat landing, ladders, maintenance platforms, tower flange, etc.); therefore, installing the transition piece after the main structure is installed avoids damaging these components. Most onshore foundations are of the gravity type; that is, they rely on the weight of the foundation to provide stability. Most offshore foundations utilize piles, either multiple piles or a single pile, known as a monopile, but more recently suction buckets are being introduced. In this chapter, we discuss all these topics. Since the foundation is ultimately supported by the soil, we provide an overview of relevant soil properties as well.

Floating offshore wind turbines have buoyant substructures that are significantly different than land-based or fixed offshore turbines. These are discussed in Chapter 11, although the soil discussion here is relevant for the anchors of floating turbines as well.

10.2 Soil

For the purpose of this section, we will consider the soil to consist of all-natural material on or in the vicinity of the earth's surface. This material could range from something as soft as fine silt to solid rock. In general, the soil occurs in layers of a variety of types of material. It may include boulders, gravel, sand, silt or clay, and varying amounts of organic material.

The soil is important for the design of wind turbines because it provides for the ultimate connection of the wind turbine to the earth. The soil must prevent the turbine from sinking into the ground, either straight down or tilting over (like the leaning tower of Pisa!). In the case of floating offshore wind turbines, the soil must hold the anchors stationary. In any case, the soil must react to the vertical and horizontal loads that are transferred from the tower, substructure, or stationkeeping system.

Wind Energy Explained: On Land and Offshore, Third Edition.
James F. Manwell, Emmanuel Branlard, Jon G. McGowan, and Bonnie Ram.
© 2024 John Wiley & Sons Ltd. Published 2024 by John Wiley & Sons Ltd.

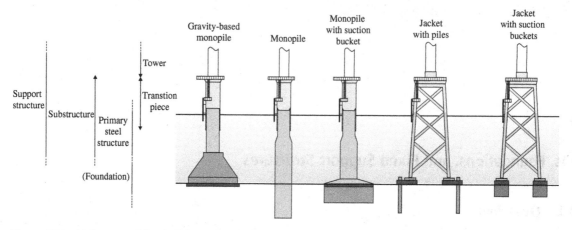

Figure 10.1 Main types of fixed offshore support structures and their components. Monopiles do not have separate foundations. The transition piece may be integrated as part of the jackets or monopile

Soil may be required to resist forces in at least three directions:

- Downward (as with the conventional gravity foundations)
- Laterally (as with pilings or anchors)
- Vertically (as with certain types of anchors or jacket piles)

To a first approximation, soil may be conceived of as consisting of myriad of very small particles. When an object applies a relatively small force to the particles they remain in place; when an object applies a higher force, the particles may be pushed sideways. The object would then sink into the soil. Resistance to lateral forces is similar in this simplified model. Conversely, if an object is embedded in the soil and a lifting force is applied, then the particles will tend to resist upward motion due to friction between the particles and the object.

Real soils resist motion through several mechanisms. The behavior of soils underwater is further affected by the presence of the water itself, often resulting in a slurry or colloidal suspension of solid particles and water. The soil is characterized by:

- Compressive strength
- Shear strength
- Tensile strength

Compressive strength is a measure of the applied stress that a soil can withstand before it yields. Compressive strength is often called the "unconstrained compressive strength" to emphasize that there is no external constraint (such as a wall) keeping the soil in place.

Shear strength refers to the ability of the soil to withstand a shear force, such as could result from a lateral load. The tensile strength refers to the ability to resist uplift forces.

Strength of soil and its ability to withstand forces in various directions is related to the composition of the soil as discussed in the next section.

10.2.1 Soil Properties

The soil has a number of properties that are relevant to the foundation of a wind turbine. This section provides a summary description of the most important of them. These include:

- Specific weight, γ (kN/m^3)
- Unconfined compressive strength, q_{ucs} (kPa)

- Undrained shear strength, s_u (kPa)
- Friction angle $\phi(°)$
- Soil cohesion, c (kPa)
- Modulus of elasticity, E_s (MPa)

10.2.1.1 Soil Types and Soil Grain Size

Soil is characterized most commonly by reference to the size of the grains of material of which it consists. For example, gravel is coarser than sand, which is in turn coarser than silt or clay. Size ranges for various particles are shown in Table 10.1.

10.2.1.2 Types of Soil Based on Moisture Content

Soil may have different properties depending on its moisture content. For this purpose, it may be categorized as dry, saturated, or partially saturated. Soil which supports land-based turbines in normally dry. Most soil of interest in offshore wind energy would be considered saturated.

10.2.1.3 Specific Weight of Soil

The specific weight of soil, γ, is its weight per unit volume. Typical values for sand, gravel, and clay vary by more than a factor of 2, ranging from 11.2 to 24.5 kN/m^3.

Table 10.1 Grain size range of various particles

Name			Size range, mm
Very coarse soil		Large boulder	>630
		Boulder	200–630
		Cobble	63–200
Coarse soil	Gravel	Coarse gravel	20–63
		Medium gravel	6.3–20
		Fine gravel	2.0–6.3
	Sand	Coarse sand	0.63–2.0
		Medium sand	0.2–0.63
		Fine sand	0.063–0.2
Fine soil	Silt	Coarse silt	0.02–0.063
		Medium silt	0.0063–0.02
		Fine silt	0.002–0.0063
	Clay		<0.002

(*Source:* Based on data from ISO, 2017)

Table 10.2 Undrained shear strength of clay

Consistency of clay	Undrained shear strength, s_u (kPa)
Very soft	0–13
Soft	13–25
Firm	25–50
Stiff	50–100
Very stiff	100–200
Hard	200–400
Very hard	>400

(*Source:* Based on data from ASTM D-2488 (ASTM, 2018))

10.2.1.4 Unconfined Compressive Strength

The unconfined compressive strength, q_{ucs}, is the compressive stress at which an unconfined cylindrical specimen of soil will fail in a simple compression test. Failure here means it allows unacceptable movement. This strength is one of the most important in evaluating the bearing capacity of the soil under a gravity foundation. The unconfined compressive strength is also often used to estimate the undrained shear strength (see below.)

The compressive strength of soils varies over a wide range. Typical values are 100 kPa for clay and 400 kPa for gravel. The strength of rock is in excess of 1000 kPa.

10.2.1.5 Undrained Shear Strength

The undrained shear strength, s_u, is used to characterize the shear strength of the soil. Table 10.2 shows typical values of this parameter for clays of various consistencies. As may be intuitively expected, a very hard clay has a much higher undrained shear strength than soft clay.

Under most circumstances, the undrained shear strength is taken to be one-half the unconfined compressive strength, as in Equation (10.1):

$$s_u = \frac{q_{ucs}}{2}$$

$$(10.1)$$

10.2.1.6 Shear Strength Values from Test Data

The shear strength of soil, and hence the compressive strength, is most commonly inferred from soil tests. Two of these are the "standard penetration test" (SPT) and the "cone penetrometer test" (CPT).

Standard penetration test The SPT is a field test that measures the resistance of the soil to the penetration of a certain sampling device according to a specified procedure. The output of the test is the number of hammer blows (N) that are required for the sampler to travel the prescribed distance.

Cone penetrometer test In the CPT, a metal cone is pushed into the soil at a specified rate. The force to do this is recorded and provides a measure of the characteristics of the soil as a function of depth. More information on these tests may be found in the State Materials Office (2022).

10.2.1.7 Friction Angle

One particularly important soil property is the friction angle, ϕ. The friction angle is a measure of the soil's shear strength and is used in estimating the bearing capacity of the soil. It can be determined in the laboratory based on soil samples or it can be estimated from the SPT as shown in Table 10.3 for granular soils.

Table 10.3 Friction angle vs. SPT test values for granular soils

SPT, N	ϕ (°)
0	25–30
4	27–32
10	30–35
30	35–40
50	38–43

(*Source:* Adapted from Bowles, 1996)

10.2.1.8 Cohesion of Soil

Cohesion is the force that holds together molecules or particles within a soil, so cohesive soils tend to stick together. Cohesive soils, such as clays, have a higher bearing capacity than those that are not cohesive, such as a gravel. Typical values range essentially from 0 to 100 kPa. For conservatism, the cohesion parameter, c, is sometimes ignored.

10.2.1.9 Modulus of Elasticity of Soil

The modulus of elasticity of the soil, E_s, is used to estimate settlement of a foundation due to static loads. It is also used in assessing the effect of lateral loads on pile foundations; see Section 10.3.3.1.

The modulus of elasticity can vary drastically, by more than a factor of 300. Some estimates summarized from USACE (1990) are presented below in Table 10.4.

10.2.2 Assessment of Soil Conditions

Prior to the installation of a wind turbine, a thorough assessment of the soil conditions should be undertaken to ensure that the foundation is properly designed. This will include:

- geological survey of the site (i.e., the basic characteristics of the earth, rock chemistry, etc. at the site soil and/or rock)
- geophysical investigation (having to do with the physics of the soil)
- geotechnical investigations (having to do with the behavior of the soil as it relates to construction, a subdiscipline of civil engineering) consisting of in-situ testing and laboratory tests.

Table 10.4 Typical soil elastic moduli

Soil	E_s (MPa)
Very soft clay	0.5–4.8
Soft clay	4.8–19.2
Medium clay	19.2–47.9
Stiff clay, silty clay	47.9–95.8
Sandy clay	23.9–191.5
Clay shale	95.8–191.5
Loose sand	9.6–23.9
Dense sand	23.9–95.8
Dense sand and gravel	95.8–191.5
Silty sand	23.9–191.5

(*Source:* Adapted from USACE, 1990)

For offshore projects, the following will also be required:

- a bathymetric survey of the sea floor including measurement of water depth, registration of boulders, sand waves, items of archaeological interest, or obstructions. This is an extensive process, which could take two years to complete.

In general, soil borings are required, but they may be augmented by CPT tests. Specifics will be based on the general conditions at the site, such as uniformity (or lack thereof!) of the soil and the number of turbines. Offshore borings may go 50 m or more into the soil.

The soil investigation should provide the following data as the basis of the foundation design:

- data for soil classification and description of the soil;
- shear strength parameters;
- deformation properties, including consolidation parameters;
- permeability; and
- stiffness and damping parameters.

More details for offshore sites in particular may be found in IEC (2019b) and Fugro Marine (2017)

10.3 Foundations and Soil Reaction

The purpose of a foundation is to provide a secure connection between the rest of the wind turbine's support structure and the ground. It must be able to keep the turbine upright and stable under the most extreme design conditions. For a land-based turbine, the tower will be connected directly to the foundation. For a fixed offshore turbine, a substructure will be connected to the foundation at one end and to the tower at the other end (often via a transition piece). Monopiles do not have a separate foundation as they are directly driven into the soil. As will be discussed below, there are four main types of foundations: gravity foundations, gravity foundations above rock, pile foundations, and suction caissons.

10.3.1 Bearing Capacity and Foundations

A prime concern in the design of any foundation is the bearing capacity of the soil. This is the ability of the underlying soil to support the foundation loads without shear failure. Bearing capacity is particularly important for wind turbines installed on gravity-based support structures (vertical capacity) and monopile structures (lateral capacity), which are discussed subsequently, but it is relevant for other types of foundations as well.

Ultimate bearing capacity, q_u, is the theoretical maximum pressure that the soil can support without failure. This is closely related to the soil compressive strength discussed previously, but it also takes into account the effect of the embedment depth of the foundation and the cohesion of the soil. In practice, it is estimated as described below.

Allowable bearing capacity, q_a, as used in geotechnical design is the pressure that should be used in design and takes into account uncertainties. It is the ultimate bearing capacity divided by a factor of safety:

$$q_a = \frac{q_u}{F.S.}$$

(10.2)

where

$F.S.$ = Factor of safety; this is typically between 1.5 and 3, depending on the situation.

10.3.1.1 Estimation of Ultimate Bearing Capacity for Shallow Foundations

The following is a summary of a method commonly used for assessing the bearing capacity of foundations on land (see Hicks, 2016). It is known as Terzaghi ultimate bearing capacity theory, after the researcher who first developed it. It was subsequently updated by Meyerhof, and many of the correction factors are named after him. A similar equation would be used for gravity-based structures (GBSs) offshore, although the details could well be different. This method consists of a suite of equations for estimating q_u. The primary equation, for circular foundations, is Equation (10.3). Note that there are three terms in this equation, one associated with cohesion, the next with foundation depth, and the third with foundation width. Also note the bearing capacity factors, N_c, N_q, and N_γ, which are empirically derived terms that are related to the angle of internal friction of the soil, ϕ. Finally, there are a number of correction factors, which have to do with the shape of the foundation (s_c, s_d, s_γ), the depth of the foundation (d_c, d_d, d_γ), and whether there is inclined loading (i_c, i_d, i_γ); see Section 10.3.1.2.

$$q_u = c\,N_c s_c d_c i_c + \gamma D\,N_q s_q d_q i_q + 0.5\gamma\,B_{ef}\,N_\gamma s_\gamma d_\gamma i_\gamma \tag{10.3}$$

where

D = depth of foundation, m
B_{ef} = effective width of foundation, m

Bearing capacity factors commonly used are from Meyerhof (see Hicks, 2016) and may be found from Equations (10.4–10.6), where ϕ is in degrees. Table 10.5 illustrates these capacity factors.

$$N_q = \frac{1 + \sin\phi}{1 - \sin\phi}\,e^{\pi\tan\phi} \tag{10.4}$$

$$N_c = \frac{N_q - 1}{\tan\phi}\ \ \text{if}\ \ \phi > 0;\ \ N_c = 5.7\ \ \text{if}\ \ \phi = 0 \tag{10.5}$$

$$N_\gamma = \left(N_q - 1\right)\tan(1.4\,\phi) \tag{10.6}$$

Table 10.5 Meyerhof's bearing capacity factors

ϕ	N_c	N_q	N_γ
0	1.0	5.7	0
5	1.6	6.5	0.4
10	2.5	8.3	1.1
15	3.9	11.0	2.4
20	6.4	14.8	4.7
25	10.7	20.7	9.3
30	18.4	30.1	18.7
35	33.3	46.1	39.1
40	64.2	75.3	87.0
45	134.9	133.9	211.8

(*Source:* From Equations 10.4–10.6)

The effective width of a foundation B_{ef}, when it is subject to a significant moment, such as that of the wind turbine, is found from the actual length, L, adjusted by the eccentricity, e. The eccentricity takes into account the moment due to the thrust loading, F_T:

$$e = \frac{F_T H}{W} \tag{10.7}$$

$$B_{ef} = B - 2e \tag{10.8}$$

where H = effective tower height, W = weight of entire structure, including the foundation. The distance B for an operating wind turbine is the dimension in line with the incident wind. The distance L is perpendicular to that. For a square foundation, B and L are the same. For a circular foundation of radius R, $B = L = 2R$

The eccentricity is used in conjunction with an effective area to calculate a nominal loading. For a square foundation, A_{ef} is found from the effective width B_{ef} and effective length L_{ef}:

$$A_{ef} = B_{ef} L_{ef} \tag{10.9}$$

where $L_{ef} = L$.

For a circular foundation of radius R, the effective area is given by:

$$A_{ef} = 2\left[R^2 \cos^{-1}(e/R) - e\sqrt{R^2 - e^2}\right] \tag{10.10}$$

In addition, the effective widths and lengths are as follows:

$$L_{ef} = \sqrt{A_{ef} l_e / B_e} \tag{10.11}$$

$$B_{ef} = L_{ef} b_e / l_e \tag{10.12}$$

where b_e and l_e are elliptic axes and given by:

$$b_e = 2(R - e) \tag{10.13}$$

$$l_e = 2R\sqrt{1 - \left(1 - \frac{b_e}{2R}\right)^2} \tag{10.14}$$

For foundations of either type, the nominal pressure on the soil is the weight divided by the effective area:

$$p_{nom} = W / A_{ef} \tag{10.15}$$

The eccentricity must be within the area of the base, that is $e < L/2$ or $e < R$ as appropriate. For circular foundation, it should also be less than 30% of the radius.

The effective length and width, B_{ef} and L_{ef}, can also be used to adjust the bearing capacity equations, as discussed next.

10.3.1.2 Correction Factors
Equations for the correction factors are given below (see Svensson, 2010).

Shape correction

$$s_c = \begin{cases} 1 + 0.2\dfrac{B_{ef}}{L_{ef}}, & \text{for } \phi = 0 \\[2ex] 1 + \dfrac{N_q}{N_c}\dfrac{B_{ef}}{L_{ef}}, & \text{for } \phi > 0 \end{cases} ; \quad s_q = 1 + (\tan\phi)\frac{B_{ef}}{L_{ef}}; \quad s_\gamma = 1 - 0.4\frac{B_{ef}}{L_{ef}} \tag{10.16}$$

Depth correction

$$d_c = d_q = 1 + 0.35\frac{d}{B_{ef}}, \quad \text{where} \quad d_c, d_q \leq 1.7; \quad d_\gamma = 1 \tag{10.17}$$

Inclination correction

$$i_c = \begin{cases} 1 - \dfrac{mF_T}{B_{ef}L_{ef}cN_c}, & \text{for } \phi = 0 \\[2ex] i_q - \dfrac{1 - i_q}{N_c \tan\phi}, & \text{for } \phi > 0 \end{cases} ; \quad i_q = \left(1 - \frac{F_T}{W + B_{ef}L_{ef}c\cot\phi}\right)^m; \quad i_\gamma = \left(1 - \frac{F_T}{V + B_{ef}L_{ef}c\cot\phi}\right)^{m+1}$$

$$\tag{10.18}$$

where:

$$m = \begin{cases} m_B = \dfrac{2 + B_{ef}/L_{ef}}{1 + B_{ef}/L_{ef}} \\[2ex] m_L = \dfrac{2 + L_{ef}/B_{ef}}{1 + L_{ef}/B_{ef}} \end{cases} ; \text{ the subscripts } B \text{ and } L \text{ indicate the direction of loading} \tag{10.19}$$

There are additional correction factors if the foundation is inclined, but those are omitted here for clarity. Some other adjustments may also be made depending on the details of the situation, but those are omitted here as well.

10.3.1.3 Bearing Capacity Factors for Other Types of Foundations

The same basic equation as Equation (10.3) can be used for other types of foundations, including deep foundations and piles, but different constants must be used. As in Table 10.5, the bearing capacity factors will depend on the friction angle.

10.3.2 Gravity Foundations

A gravity foundation is one in which the weight alone is sufficient to prevent overturning under any plausible situation. If the foundation is buried deep enough, the soil will augment its weight. An additional important consideration is that the base area be such that the bearing capacity of the soil is not exceeded under any conditions. Most land-based wind turbines use gravity foundations in the form of a steel-reinforced concrete pad. Offshore, gravity foundations may be more suitable economically in shallow water; see Figure 10.2. This section provides a summary of the design issues of gravity foundations. More details may be found in Svensson (2010).

A substantial number of offshore wind turbines also use gravity foundations. In that case, they often go under the name of "gravity-based structure"; see Section 10.7.2.

10.3.2.1 Required Weight and Dimensions of Gravity Foundation

The primary considerations in the design of a gravity foundation are the following:

- Foundation weight
- Foundation dimension (width)
- Bearing capacity of soil

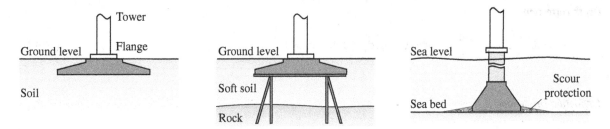

Figure 10.2 Typical gravity foundation. Left: land-based on hard soil (conventional). Middle: land-based on soft soil and hard rock. Right: offshore

Estimation of the soil's bearing capacity was discussed previously. In this section, we consider the foundation weight and width, and how those produce stress on the soil.

In the simplest analysis (static and rigid analysis) of a gravity foundation for a wind turbine, the sum of the moments due to the weight of the RNA, the tower, and the foundation must be equal to or greater than that resulting from the thrust. See Figure 10.3 for a free-body diagram of the situation in which the moment due to the weight is just equal to that due to the thrust.

In this figure, the weights of the RNA, tower, and foundation are W_{RNA}, W_{tower}, and $W_{foundation}$, the thrust due to the rotor is F_T, the tower height is H, and the foundation width is B. Summing the moments about an assumed hinge at O yields Equation (10.20). If everything but the foundation weight is known, then the minimum acceptable value can be easily found.

$$F_T H - \left(W_{RNA} + W_{tower} + W_{foundation}\right)B/2 = 0 \tag{10.20}$$

Note that the reaction forces indicated by the arrows at the hinge are simply equal to the sum of the respective forces in the x and z directions. Equation (10.20) is not realistic, however, since the reaction force appears to be applied only at O and the distribution of forces on the soil is not included.

When an ideal soil is taken into account, then the following simplified approach can be used. In this method, the turbine is supported by a square foundation of width B as shown in Figure 10.4.

The vertical reaction force is now the integral of the pressure from the soil $p(x)$ over the area of the foundation (creating an elementary force $dF = p(x) B\,dx$). Also, to simplify the expressions $W = W_{RNA} + W_{tower} + W_{foundation}$. We also assume that the reaction force is linear. With these assumptions and notations, the moment balance is:

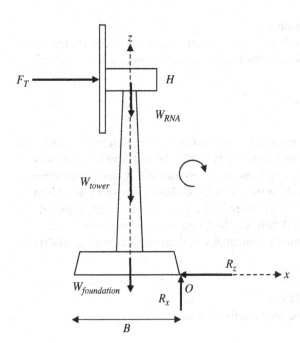

Figure 10.3 Wind turbine on gravity foundation (simplified)

$$F_T H + \int_{-B/2}^{B/2} B\,p(x)x\,dx = 0 \tag{10.21}$$

where $p(x) = (a + bx)$ and x is the distance from point O (here at the center of the tower base).

Solving yields

$$a = \frac{W}{B^2} \qquad (10.22)$$

$$b = \frac{12F_TH}{B^4} \qquad (10.23)$$

From this, the pressure at $x = B/2$ is

$$p(B/2) = \frac{W}{B^2} + \frac{6F_TH}{B^3} \qquad (10.24)$$

Similarly, the pressure at $x = -B/2$ is:

$$p(-B/2) = \frac{W}{B^2} - \frac{6F_TH}{B^3} \qquad (10.25)$$

The loading on the soil should always be compressive so

$$W > \frac{6F_TH}{B} \qquad (10.26)$$

Equation (10.26) is particularly useful in that it illustrates clearly that the required weight of the foundation is increased by the thrust force and the height at which it acts and is decreased by the width of the foundation.

The ultimate bearing capacity q_u should always exceed the greatest value of the pressure thus:

$$q_u > \frac{W}{B^2} + \frac{6F_TH}{B^3} \qquad (10.27)$$

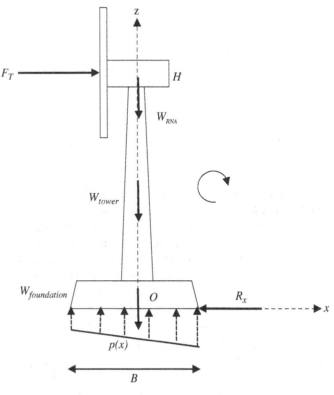

Figure 10.4 Wind turbine on gravity foundation with linearly varying soil reaction

10.3.2.2 Gravity Foundation Above Rock

Sometimes wind turbines are installed on the soil above rock, and the soil is unsuitable to keep the foundation in place. In this case, offshore developers would likely consider suction bucket designs (see below). Onshore, the foundation may include rods grouted into holes drilled deep into the rock. A concrete pad may be used to provide a level surface and to distribute the compressive loads, but any tensile loads are taken ultimately by the rods and then by the rock. If the soil is shallow enough, multiple individual piers may be used instead of a single pad. A comparison of a conventional gravity foundation and one above rock was illustrated in Figure 10.2.

10.3.2.3 Examples of Foundation Design

In this section, we consider some simplified illustrations of foundation design, using the NREL 5-MW reference turbine with an onshore tower, the properties of which were given in Chapter 1. The relevant ones are repeated in Table 10.6. The total height is thus 90 m.

Table 10.6 Relevant dimensions on NREL 5-MW reference turbine with onshore tower

Property	Value	Units
RNA mass	350	t
RNA height	2.4	m
Tower height	87.6	m
Tower mass	347.4	t

Example 10.1 Pier Foundation-Compressive Loading

The NREL 5-MW reference turbine on shallow soil above rock is supported by four small diameter piers, 10 m apart, located on the corners of a square steel transition piece whose mass is 42 500 kg. The total weight (assuming $g = 9.81$ m/s^2) is thus 7258 kN. The wind is blowing perpendicular to one side. The rotor thrust is initially 100 kN. The reaction force in the vertical direction from the two upwind piers can be found by taking moments about a line through the downwind piers using Equation (10.20). That force is 1415 kN/pier. The reaction force from the downwind piers is 2215 kN/pier. The loading on all the piers is compressive (Note that the mass of the piers is irrelevant).

Example 10.2 Pier Foundation-Tension/Compression

The situation is the same as Example 10.1, except that the thrust is now 800 kN, which is close to the nominal thrust at rated wind speed. Applying the same approach, the reaction force from the upwind piers would be −1785 kN/pier. This situation would only work if the piers could apply a tension force, for example, by steel rods passing through the piles and into rock underneath the turbine. The reaction force from the upwind piers would be 5415 kN/pier.

Example 10.3 Gravity Based Foundation

In this case a purely gravity-based foundation is chosen. Assume that the foundation is a 20 m × 20 m square of reinforced concrete with a height of 2 m. The density of the concrete is 2500 kg/m^3, such that its mass is 2 000 000 kg and weight is 19 600 kN. The total weight of the turbine and foundation is 26 461 kN. From Equations 10.24 and 10.25, the pressure at the upwind edge is 12.2 kPa, and at the downwind edge the pressure is 120.2 kPa. The bearing capacity of the soil would need to be at least that value. The eccentricity (Equation 10.7) is 2.72 m.

Example 10.4 Circular Gravity Based Foundation

In this case, a circular reinforced concrete foundation of the same mass as that in Example 10.3 is used. It has the same height and area, so its radius is 11.28 m. From Equation (10.7), the eccentricity is 2.72 m, and from Equation (10.10), the effective area is 278.4. The nominal pressure on the soil from Equation (10.15) is then 95.1 kPa.

Example 10.5 Soil Bearing Capacity

In this example, we evaluate the allowed pressure on the soil supporting the circular foundation. Assume that the foundation is covered with soil to a depth of 2 m and that the soil has the following properties: weight density γ is 15 kN/m^3, friction angle $\phi = 29°$, cohesion $c = 20$ kPa. From Equations 10.3–10.6 and 10.16–10.19, the factors listed in Table 10.7 may be found. The total value of the bearing capacity is 2384 kPa. This value is much higher than the nominal pressure from the turbine and foundation, so soil properties would not be a limiting factor in this situation, even after considering a conservative factor of safety. On the other hand, it should be noted that with different soil conditions, the bearing capacity could be far less, in fact less than the nominal pressure in

Table 10.7 Parameters and factors for soil bearing in Example 10.5

Cohesion		Depth		Width	
c	20	q	30	B_{ef}	14.755
N_c	27.860	N_q	16.443	N_γ	13.237
s_c	1.174	s_q	1.694	s_γ	0.687
d_c	1.047	d_q	1.047	d_γ	1.000
i_c	0.941	i_q	0.943	i_γ	0.908
q_{uc}	645	q_{uq}	825	$q_{u\gamma}$	914

Example 10.4. For example, for an unburied foundation in cohesionless soil with a friction angle of $\phi = 10°$, the predicted ultimate bearing capacity, using the same equations as above, would be only 30 kPa.

10.3.3 Pile Foundations and *p–y* Curves

Piles are relatively long (columnar) objects that are driven or drilled into the soil for structural purposes. Piles may be used to support wind turbines on soft soils both on land and offshore. In that situation, stability is provided by a combination of the properties of the soil. In any case, they may be loaded in either compression or tension.

Pile foundations can take a variety of forms, but basically, they consist of either a single pile (monopile), such as is common offshore, or multiple piles (at each leg of a jacket or truss-type substructure). Pile foundations are less common on land, but an example of a foundation constructed with multiple piles is one at Hull, MA; see Manwell *et al.* (2006).

A pile in soft soil may be subject to loading either axially (typically vertically) or laterally. How much vertical loading is allowed depends on the properties of pile itself (e.g., resistance to buckling; see Section 10.4.4) and of the soil. The bearing capacity of the soil (for vertical loading) comes from two sources: the soil below the lower end of the pile and skin friction along the length of the pile.

Lateral loading is a particularly important consideration with wind turbines; the soil itself provides the majority of resistance to overturning. The so-called "*p–y* method," discussed next, is widely used to assess the ability of piles to resist loads applied in the lateral direction.

More details are also available in texts on pile design, e.g., USACE (2005).

10.3.3.1 *P–y* Method for Piles

A common approach to assessing the resistance of the soil to lateral loading from a pile or other deep foundation is known as the *p–y* method. In this method, "*p*" is the force that the foundation applies to the soil, and "*y*" is the lateral displacement. Both the force and displacement depend on the distance along the foundation. The relation between *p* and *y* is that of a spring, which may be nonlinear, through Equation (10.28) and as illustrated in Figure 10.5.

$$p = ky \tag{10.28}$$

where k is the spring constant in N/m.

As shown on left side of the figure, the soil is modeled as a series of independent springs along the length of the pile. The right side of the figure illustrates that the force increases with deflection, but only up to a point. After that, the soil begins to give way and the pile deflects with no increase in force. It must also be noted that the soil stiffness

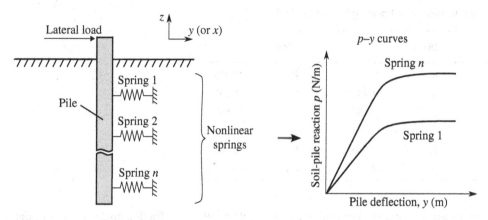

Figure 10.5 *p–y* curve

values depend on the soil type and must be empirically determined. For example, as summarized previously (Table 10.4), clays and sands can have values of elasticity ranging from less than 1 MPa to nearly 200 MPa.

The *p–y* method can be implemented in a straightforward finite difference form. An Excel implementation is provided by *LatPile.xls* (NewtonExcelBach (2009)).

The *p–y* method has been widely used in the offshore oil and gas industry and is now also used with offshore wind turbines. Some initial experience was discussed by Byrne and Houlsby (2003). More recent guidance is provided by Byrne *et al.* (2015).

10.3.3.2 *P–y* Calculation Example

The *p–y* method can be used to estimate the effect of soil stiffness on the offset of the reference 5-MW turbine supported by a monopile foundation. Suppose that the turbine is simplified to consider only the moment resulting from the reference thrust (800 kN) at the center of the RNA, 90 m above the top of a monopile. The monopile is initially assumed to be 30 m long with a diameter of 6 m and a wall thickness of 0.11 m. When the monopile is assumed to be fixed rigidly to the seabed, the deflection can be found without consideration of the soil using the finite difference method discussed in Chapter 4. In this case, the deflection at the top of the pile is approximately 0.02 m and that of the RNA is 0.72 m. Next, suppose that the monopile extends 20 m into the seabed (for a total length of 50 m) and that the soil has a stiffness of 50 MPa. *Latpile.xls* can be used to predict that the top of pile deflects 0.18 m.

If the soil were less stiff, for example with a stiffness of 5 MPa, the deflection would be nearly 1.0 m. The additional pile top deflection will result in more tower top deflection as well. Without considering the soil, the pile top would be inclined only slightly (0.04° from vertical). With soil stiffness of 5 MPa, the angle would be 1.38°, resulting in an additional 2.1 m of deflection at the top of the tower.

10.3.4 Suction Caissons (Buckets)

Suction caissons (or buckets) are a type of foundation that has been used in some offshore wind installations. This foundation is appealing since the installation process does not require pile driving and the total depth of the foundation can be considerably less than for a monopile. The caisson may also be removed relatively easily when the turbine is eventually decommissioned.

A suction caisson resembles an inverted bucket (hence its alternate name), in which the lid corresponds to the bottom of the bucket and the skirt corresponds to the sides. A caisson can be used under each leg of a jacket or as the foundation of non-soil penetrating monopile. Figure 10.6 illustrates suction caissons in place.

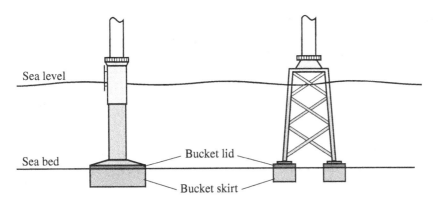

Figure 10.6 Suction buckets with monopile and jacket

To install the caisson, it is lowered to the seabed, where it begins to sink into the soil. A valve on the top of the caisson is opened so that water trapped inside can escape as the caisson continues to sink. The remaining water is pumped out of the inside of the caisson and the resulting pressure difference on the top of caisson forces it further down into the soil until the lid comes to rest on the seabed. External water pressure as well as friction between the skirt and the soil keep the caisson in place when under load.

10.4 Support Structure Requirements

Regardless of the type of support structure being used or considered for a given application, there are some basic, or "global," considerations that will apply in every case. An important category of these are the load effects, which were introduced in Chapter 8, due to axial forces and those from bending moments. In summary, these global requirements include:

- Stresses
- Deflection
- Natural frequency
- Buckling
- Resistance to environmental degradation

Briefly, the stress in every part of the structure must be below certain limits, both in terms of ultimate loading and fatigue. No part of the structure must be allowed to deflect more than a certain amount beyond its unde-flected norm. Vibrations due to whatever source cannot be allowed to cause excessive motions. There must be no chance that the structure will buckle. Finally, the structure must be resistant to environmental degradation throughout its useful life. A corollary of that requirement, however, is that after its useful life has ended, the structure should be capable of being recycled or safely disposed of. The following sections discuss each of these topics in more detail.

10.4.1 Stress

Stresses in the support structure must be assessed to ensure that they do not exceed certain values, either for ulti-mate loading or fatigue.

10.4.1.1 Stresses Due to Combined Axial Forces and Moments

The combined effect of stresses due to axial forces and bending moments is of particular importance in the design of towers and supports structures generally. The relationship of those stresses to material strength is expressed by Equation (10.29).

$$\frac{N_d}{A} + \frac{M_d y}{I} \leq \sigma_d \tag{10.29}$$

where

N_d = design axial load (N)
M_d = design moment (Nm)
A = cross-sectional area (m^2)
I = cross-sectional area moment of inertia (m^4)
y = distance from the center line (m)
σ_d = allowed stress in material (Pa)

For a structure with a circular cross-section, such as monopiles and most towers, it should be remembered that the cross-sectional area and area moment of inertia are given by Equations 10.30 and 10.31, and that $y = R$.

$$A = \pi\left(R^2 - R_{in}^2\right) \tag{10.30}$$

$$I = \pi\left(R^4 - R_{in}^4\right)/4 \tag{10.31}$$

where R is the radius and R_{in} is the inner radius.

Sometimes buckling stresses due to the axial load are more significant than the normal stress. In this case, Equation (10.29) may be modified as described in Section 10.4.4.

10.4.2 Deflection

There are limits to the allowed deflection of various components of the wind turbine, including the support structure. While there is no firm rule for maximum allowed tower-top deflection, the industry consensus at this point is that the maximum deflection of the tower top should not be greater than 1% of the tower height for fixed support structures.

Estimates of deflection of the turbine are made by applying an extension of the discussion of cantilever in Chapter 4. An essential input to the calculations is the stiffness of the entire support structure as a function of height above some reference point. For a land-based turbine, the reference height will be the top of the foundation. For an offshore turbine, particularly one supported by a monopile, the effect of the soil must be considered. The *p–y* method, as discussed previously, is generally used for this. Alternatively, the reference height may be taken to begin somewhat lower than the seabed by an amount known as the fixity length. This is roughly 3–4 times the pile diameter.

10.4.3 Natural Frequency

The vibrational natural frequency of the entire wind turbine structure is of great importance, particularly the first mode. There are two primary considerations in this regard: (1) the natural frequency in relation to sources of excitation and (2) the possibility of damping the vibrations. To the extent possible, the natural frequency of the structure should not coincide with the frequency of strong excitations. In any case, excitation frequencies should generally not be within 5% of any of the support structure's primary natural frequencies during prolonged operation.

The most significant sources of excitation are due to the rotation of the rotor. As discussed in Chapter 4, the frequency associated with one complete rotation is referred to as 1P (i.e., once per revolution); the blade passage frequency for an N-blade turbine is NP, or 3P for a 3-blade turbine. When the natural frequency is less than 1P, the structure is soft. When higher than NP, it is stiff, and when between 1P and NP it is called soft-stiff. Experience has shown that the natural frequency of wind turbine should be between 1P and NP, i.e., soft-stiff. A stiff structure would be heavier (and thus more expensive) than it needs to be. A soft structure is prone to exhibit excessive deflections. It should be noted that variable-speed wind turbines would have corresponding ranges of 1P and NP, and so the band between the two could be relatively small.

Example: Consider the NREL reference 5-MW turbine. It has an operating speed range of 6.9–12.1 rpm, so the 1P range is 0.115–0.202 Hz and the 3P range is 0.345–0.605 Hz. On the reference tower on land, the turbine's natural frequency is 0.32 Hz, which is in the desired soft-stiff range.

Another source of excitations for all turbines is turbulence in the wind. Most of the energy associated with turbulence is at a relatively low frequency (0.001–0.1 Hz), so the excitation itself is not of great concern for the tower. Offshore turbines are also subject to excitation due to the passage of waves, which typically have a frequency of about 0.1 Hz (see Chapter 6). This is high enough to be of concern with large turbines, which may have 1P close to that value.

10.4.3.1 Calculation of Natural Frequencies

The natural frequencies of a land-based turbine with a tapered tubular tower as well as such a turbine on a monopile can be readily calculated with methods discussed in Chapter 4. (It should be noted, however, that the calculation of natural frequencies of structures in the water, such as monopiles, should include the effect of added mass; see Chapter 11.)

The rule of thumb suggested by Sadegh and Worek (2018) may be used when the land-based turbine/support structure can be approximated by a uniform cantilever with a point mass on the top. It is given by:

$$f_n = \frac{1}{2\pi} \sqrt{\frac{3\,E\,I}{(0.23\,m_{SS} + m_{RNA})\,H^3}} \tag{10.32}$$

where f_n is the fundamental natural frequency (Hz), E is the modulus of elasticity, I is the area moment of inertia of the support structure cross-section, m_{SS} is the mass of the support structure, m_{RNA} is the mass of the rotor nacelle assembly, and H is the height of the support structure. The assumed shape function approach presented in Chapter 4 can also be used to find the natural frequency as $f_n = \frac{1}{2\pi} \sqrt{\frac{K}{M + m_{RNA}}}$, where K is the generalized stiffness and M is the generalized mass of the support structure. In the case of monopiles, the generalized added mass should also be added to the denominator, reducing natural the frequency.

Calculation of the natural frequency of jacket structures (see Section 10.7.3) is more complicated and is outside the scope of this text.

10.4.3.2 Damping

When operation is intended in a region where the excitation frequencies are between 30% and 140% of the support structure's natural frequency, damping may be particularly important. In these situations, a dynamic magnification factor, D (Equation 10.33), may be used to multiply the design loads in evaluating the structure. The magnification factor is determined by the damping properties of the structure and the relation between the excitation frequencies. It is equivalent to the nondimensional amplitude which was introduced in Chapter 4 for forced harmonic vibrations of a single degree of freedom system:

$$D = \frac{1}{\sqrt{\left[1 - (f_e/f_n)^2\right]^2 + \left[2\ \zeta(f_e/f_n)\right]^2}}$$ (10.33)

where f_e is the excitation frequency, f_n is the natural frequency, and ζ is the damping ratio.

Except for aerodynamic damping, wind turbines have little intrinsic damping. Other possible sources of damping are in the structure itself (steel or concrete) or in the soil. For offshore turbines, the water itself is a possible source. These are all discussed briefly below.

Aerodynamic damping can be quite significant. As discussed in Chapter 4, it is associated with the operation of the rotor and so is wind speed dependent. Note that aerodynamic damping only occurs in the fore-aft direction and diminishes to often negligible values when the turbine is stopped, as during a storm.

The damping ratio is approximately 0.05 for reinforced concrete and 0.02 for steel.

In the case of an offshore turbine supported by a monopile, the soil can provide some damping via its interaction with the monopile. Soil damping is usually quite difficult to model, however, and is often frequency dependent. That topic is beyond the scope of this text, but some discussion can be found in Carswell (2015).

Hydrodynamic damping originates due to the movement of the structure in the water. Because the structural velocities are quite low, the damping is low as well.

10.4.4 Buckling

Slender columns, such as a tower or monopile, may possibly buckle due to the weight they carry. Buckling is a phenomenon in which a slight deflection in the top of the column results in continuously increasing deflection due to the weight above and eventual collapse, rather than a tendency to return to the vertical. The effect is related to the stiffness of the column rather than the yield stress. Wall thickness is an important factor as well; thin-walled cylinders can buckle even if they are not slender.

Consideration of the potential effect of buckling begins with determining the critical buckling load N_{cr} with Euler's column formula, which is given by Equation (10.34).

$$N_{cr} = 0.25\pi^2 EI/H^2$$ (10.34)

where E = modulus of elasticity and H = length of column.

When the critical buckling load is exceeded, the column is susceptible to buckling. Buckling may be accounted for by including a reduction factor, χ in Equation (10.29), dividing both sides by σ_d and expressing the result as Equation (10.35) (adapted from the Eurocode 3: CEN (2016)).

$$\frac{N_d}{\chi N_{pl,Rd}} + \frac{kM_d}{M_{el,Rd}} \leq 1.0$$ (10.35)

where

$N_{pl,Rd} = A\sigma_d$ is the material-based yield axial load.

$M_{el,Rd} = \dfrac{I\,\sigma_d}{y\gamma_{M0}}$ is the material-based resistance moment.

The variable k is an interaction factor and γ_{M0} is a safety factor, normally set to 1.0.

The reduction factor is found from the "slenderness ratio" $\bar{\lambda}$ of the column as follows:

$$\bar{\lambda} = \sqrt{Af_{yd}/N_{cr}}$$ (10.36)

$$\phi = 0.5\left[1 + \alpha(\bar{\lambda} - 0.2) + \bar{\lambda}^2\right] \qquad (10.37)$$

$$\chi = \frac{1}{\phi + \sqrt{\phi^2 - \bar{\lambda}^2}} \qquad (10.38)$$

where α is an imperfection factor.

It is worth noting that the slenderness ratio $\bar{\lambda}$ is typically in the range of 0–3 (more slender columns have higher values), and ϕ is equal to or greater than 0.5. The imperfection factor α depends on the material, with values in the range of 0.2–0.76 for steel; higher values indicate more imperfections. The reduction factor χ ranges from 1 (no reduction) to 0.1; smaller values result in a greater load effect and thus may result in the requirement for a larger cross-sectional area than originally intended. More details on buckling are given in Eurocode 3 (CEN (2016)).

The method for accounting for buckling strictly speaking only applies to a uniform column with all the mass at the top. For a tapered column of significant mass, such as a tower, a simplified and conservative approach is to evaluate the slenderness ratio at multiple positions on the structure for finding χ and basing N_d on all the mass above that position; see Leite (2015) and the example in Section 10.7.1.1.

10.4.5 Resistance to Environmental Degradation

It goes without saying that the support structure should be able to resist environmental degradation from whatever source may be significant in the environment. This is largely related to choice of materials; see Chapter 7. The potential effects of corrosion, especially offshore, are particularly important.

10.5 Loads on the Support Structure

Wind turbine support structures are exposed to loads from a variety of sources. Whether land-based or offshore, towers experience aerodynamic forces indirectly from the RNA as well as the weight of the RNA. Like any other structure in the environment, the towers also are subject to direct wind-induced forces.

In addition to the forces from the RNA and the tower, offshore wind turbine substructures are subject to loading from the marine environment, most notably waves, but also from currents and sometimes from floating ice. These loadings are summarized in Sections 10.5.1–10.5.8.

10.5.1 Aerodynamic Loading from the Rotor

The primary source of loads on land-based turbines and a major source on offshore wind turbines is the rotor. There are two major types of aerodynamic loads: (1) steady and (2) dynamic. The steady loads are associated with the thrust and torque, which were discussed in Chapters 3 and 4. It will be recalled that thrust is normally largest at, or close to, rated wind speed. It is possible, however, that the force on the stopped rotor could be higher than at rated during an extreme storm.

In addition, aerodynamics is responsible for cyclic excitation once per revolution and at the blade passing frequency. These are among the dynamic effects that can be a significant source of loads, especially insofar as they can excite the natural frequencies of the structure.

10.5.2 Tower Drag

The tower will experience drag forces due to the wind. This force is in general much less than that due to the rotor, although it can be comparable when the rotor is stopped in an extreme storm. Accurate prediction of the tower drag force is difficult, but in general a reasonable approximation is sufficient. It should be recalled that the wind speed and thus the force varies significantly with height about the surface. In addition, the drag coefficient of the tower will depend on the Reynolds number.

10.5.3 Gravity Loads

The wind turbine support structure will be subject to the weight of the RNA as well as its self-weight. With many turbines, the weight of the RNA and the tower are roughly comparable, and for offshore turbines, the substructure is likely to weigh even more. As noted above, buckling is possible, especially of slender monopiles with high D/t ratios, and should be evaluated.

10.5.4 Wave Loads

Substructures of offshore wind turbines are subject to large forces due to waves. These forces, which were discussed in Chapter 6, will vary with the period of the waves and so can be a source of excitation leading to vibration of the entire structure.

10.5.5 Currents

Marine currents may also be a source of loading on the substructure, although usually the effect is small compared to that of waves. On the other hand, currents can be a significant source of erosion of soil around the foundation and can then lead to unfortunate consequences if not properly accounted for. Currents were also discussed in Chapter 6.

10.5.6 Floating Ice Loads

In some locations, floating ice may be a major source of loading on the substructure and must be taken into account. Sometimes the impact can be minimized by proper design of the substructures, such as the provision of "ice cones." Ice loads were summarized in Chapter 6 and more details may be found in IEC (2019b).

10.5.7 Soil Loads

The soil on which the foundation is supported is a source of reaction loads. This topic was discussed in relation to bearing capacity of gravity-base foundations and lateral loads for piles in Section 10.3.

10.5.8 Combined Wind and Wave Load Loading

As has already been pointed out, offshore wind turbines are subject to loading from both the wind and the waves. Sometimes the combined effect has particular significance and must be considered. For example, when the wind and the waves come from the same direction, the total loading may be additive. Conversely, when the wind and waves come from different directions and with different frequencies, the response may be harder to predict. This topic was also discussed in Chapters 6 and 8.

10.6 Towers

Towers are structures that hold the main part of the turbine up into the air. A tower must be at least as high as the blade radius plus some minimum acceptable clearance above the water or the ground for the blade tip as the rotor turns (typically, the tower height is 1.2–1.6 times the rotor radius). For smaller turbines, the tower may be much higher than that. Generally, tower height for larger turbines is not less than 24 m plus the rotor radius because the wind speed is lower and more turbulent closer to the ground or sea surface. For example, the tip clearance of the NREL 5-MW turbine is 27 m.

10.6.1 General Tower Issues

There are three types of towers in common use for horizontal axis wind turbines:

- cantilevered tubular tower;
- free-standing lattice (truss);
- guyed lattice or pole.

Tubular towers are presently the most common. These towers are typically tapered, with the base diameter larger than the tower top. Wall thickness is also likely to vary, with a thicker wall at the base than at the top. Tubular towers are fabricated in sections that are short enough to be readily transported. The sections are combined into a single tower at the site where the turbine is to be installed.

Tubular towers have numerous advantages. Unlike lattice towers, they do not rely on many bolted connections which need to be torqued and checked periodically. They provide a protected area for climbing to access the machine. Aesthetically, they have a shape that is considered by some to be visually more pleasing than an open truss. In addition, they do not provide perches for birds, which can be problematic in some locations.

Free-standing lattice towers were common until the mid-1980s. For example, the Smith–Putnam, US Department of Energy MOD-0, and early US Windpower turbines all used towers of this type. One of the reasons for the move away from lattice towers was that the cross members provided perch locations for birds, which contributed to their mortality in some locations; see Chapter 14. Additionally, monopiles offer more weather protection for access and for ancillary components than do lattice towers. More recently, however, there has been renewed interest in lattice towers, particularly for tall, MW-scale turbines on land at sites that are difficult to access. The reason is that there is a diameter limitation of tubular tower base sections of 4.27 m on most public roads and the tower base of multi-MW turbines should be larger than that. Lattice towers with larger bases can be assembled on site, although when possible individual sections or parts thereof will be prefabricated before being transported to the installation site.

Guyed towers utilize guy wires and anchors to provide stability to the turbine. They are often used for small wind turbines, but not those of medium size (500 kW) or larger.

Examples of the three main tower options are illustrated in Figure 10.7.

(a)   (b)   (c)

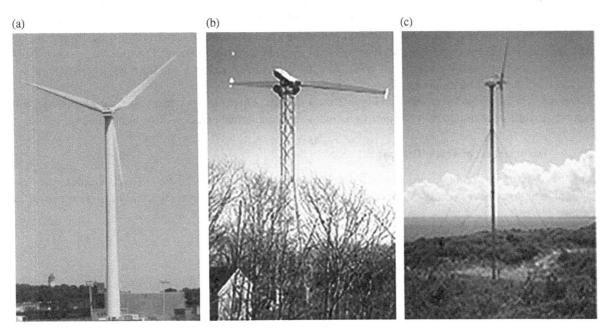

Figure 10.7 Tower options, (a) tubular tower, (b) truss tower, (c) guyed tower (*Source:* (a) and (b) is the author (J. F. Manwell). Option (c) is from Vergnet)

10.6.2 Tower Materials

Wind turbine towers are usually made of steel, although sometimes reinforced concrete is used. When the material is steel, it is normally galvanized or painted to protect it from corrosion. Sometimes Cor-Ten® steel, which is inherently corrosion resistant, is used.

10.6.3 Tower Climbing Safety

Nearly all wind turbines must be climbed occasionally for inspection or maintenance. Provision must be made in the tower design for safe climbing. This typically includes a ladder or climbing pegs and an anti-fall system. Figure 10.8 illustrates tower climbing safety equipment.

10.6.4 Tower Top

The tower top provides the interface for attaching the main frame of the rotor nacelle assembly to the tower. The stationary part of the yaw bearing is attached to the tower top. The shape of the tower top depends on the type of tower. It is usually made from cast steel.

10.6.5 Tower Erection

The intended method of tower erection will have a direct impact on the design of the tower. The towers of larger turbines typically consist of three or four sections. Each section is lifted into place with a crane and connected to the one below, with the base section having been installed first on the foundation (on land) or on a transition piece on top of the substructure (offshore).

Figure 10.8 Tower climbing safety equipment (*Source:* Reproduced by permission of Vestas Wind Systems A/S)

Small- and medium-sized turbines may be self-erecting. The most common method of self-erection is to use a gin pole or "A-frame" at a right angle to the tower. The A-frame is connected to the top of the tower by a cable. A winch is then used, in conjunction with sheaves, to raise the tower. With such a method of erection, the tower base must include hinges as well as a way of securing the tower in place once it is vertical. The RNA itself is connected to the tower before it is raised. Two methods for erecting towers are shown in Figure 10.9.

Regardless of the method of erection, an important consideration in the design of the tower is the load that it will experience during the installation. Chapter 15 includes some additional details on tower erection.

10.6.6 Optimization of Tower Design

A wind turbine tower should be adequate for the application but cost as little as possible. That is simple to state but more difficult to achieve because there are many factors to consider. The most important of these are listed as follows:

- First mode natural frequency
- Maximum tower top deflection
- Blade tip clearance
- Buckling
- Diameter to thickness ratio

(a)   (b)

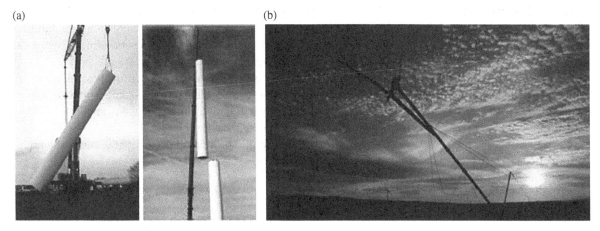

Figure 10.9 Tower erection methods (a) crane erection of tubular tower (*Source:* Reproduced by permission of Vestas Wind Systems A/S), (b) tilt up with gin pole (*Source:* Reproduced by permission of Vergnet SA)

- Ease of fabrication
- Transportation
- Installation
- Total mass
- Total cost
- Ease of access for maintenance
- Minimal perching opportunities for birds

Natural frequency The first mode natural frequency has already been discussed previously. As indicated then, it is presently considered most desirable for the tower to be "soft-stiff," that is for first mode natural frequency to be between 1P and 3P (or 2P for a two-blade turbine). There are arguments to be made for having a softer (i.e., lower frequency) tower, but these would require detailed analysis. More details on natural frequency calculations were given in Chapter 4.

Maximum tower top deflection was discussed in Section 10.4.2.

Blade tip clearance It is critical that the blades do not hit the tower during operation. This can be avoided to some degree by coning the blades upwind, tilting the rotor axis, using pre-bent blades, and increasing blade stiffness, but it becomes more difficult when the upper part of tower has a large diameter. Maximum tower top diameter may be a design constraint.

Buckling was discussed in Section 10.4.4.

Diameter to thickness ratio (D/t) is relevant to natural frequency, mass, buckling, and fabrication. Higher values correspond to thinner walls and so reduced mass and reduced natural frequency. Structures with thinner walls are also more prone to buckling, and there can be a variety of issues during fabrication and transport. All other things being equal, it is recommended that D/t for tubular structures be less than 80. Some wind turbine towers, such as the NREL 5-MW, have D/t values of approximately 160.

Ease of fabrication is a consideration in tower design since only a limited number of facilities may have the necessary manufacturing capabilities such as for rolling thick steel plates.

Transportation may create a limit for the tubular tower base diameter of a land-based turbine, since the maximum clearance under highway bridges in the United States is 4.27 m. The maximum allowed width (with a special permit) is 4.57 m. If the tower base can be transported by rail and the remainder of the route is on private roads without overpasses, the height clearance is greater (7.16 m), but width limits may still be a constraint. The height of the truck bed or railcar must also be considered. The dimensional limit for offshore turbines is less restrictive but is still relevant. It is also possible that large-diameter tubular tower sections could be assembled on-site, but that could also result in additional costs due to fabrication issues. At a certain point, tubular towers become impractical on land, and the effect could be to create a limit on the size of land-based turbines.

Installation may be a limiting factor in tower design or selection due to transportation difficulties, site access, or crane availability. Mass and dimensions of the tower sections as well as the diameter-to-thickness ratio are important considerations in this regard. This limit may result in an overall limit to turbine size in general.

Total mass of the tower is important since the cost is strongly related to the mass.

Total cost is a prime consideration in tower design and selection. The amount of material is one of the prime determinants of the cost, but fabrication can be a significant factor as well.

Ease of access for maintenance is a self-evident benefit.

Minimal perching opportunities for birds reduce bird mortality and facilitate project permitting.

10.6.7 Tower Optimization Example

In the following examples, tower mass and D/t ratios are found as a function of base diameter. The RNA is based on the NREL 5-MW turbine and towers are chosen such that the nominal hub heights are 90 and 120 m. The tower top diameter is 3.87 m and the height of the RNA above the tower top was 2.4 m. The tower natural frequency is held fixed at 0.328 Hz (close to that of the NREL 5-MW on an onshore tower). Calculations were done with the Myklestad method described in Chapter 4. The results are shown in Figure 10.10.

As is apparent, the mass is strongly a function of tower height and base diameter. For example, a tower with a height of 87.6 m with a base diameter of 5.48 m and a D/t ratio of 120 has a mass of 423 tonnes, while one with a base diameter of 5.95 m and a D/t ratio of 160 would have 17% less mass. Correspondingly, a tower with a height of 117.6 m with a base diameter of 7.5 m and a D/t ratio of 120 has a mass of 867 tonnes, while one with a base diameter of 8.1 m and a D/t ratio of 160 would have 16% less mass. Overall, by increasing the hub height by one-third, the tower mass was approximately doubled.

A more detailed tower optimization exercise is provided by Dykes *et al.* (2018).

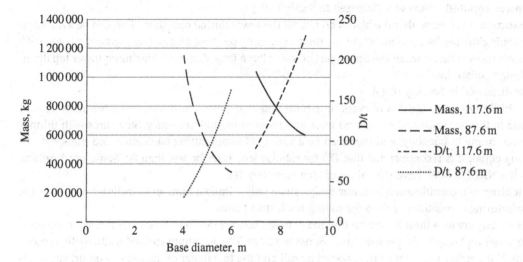

Figure 10.10 Example tower mass and D/t vs. base diameter for two tower heights

10.7 Substructures for Fixed Offshore Wind Turbines

As previously indicated, the substructure of a fixed offshore wind turbine is that portion of the turbine that is below the tower and above the foundation (although in some cases the foundation is actually a section of the substructure). There are basically three types of substructures for fixed offshore wind turbines: (1) monopiles and other tubular structures), (2) gravity-based structures (GBSs), and (3) jackets and other multimember structures. Each of these is discussed in Sections 10.7.1–10.7.3. The primary considerations as to which type of substructure is chosen are water depth and soil conditions. Generally speaking, monopiles are preferred when the conditions allow for it. GBSs are used in shallow water, especially where installing a monopile would be difficult. Jackets are used in deeper water and in softer soils where lateral stiffness is low. One important concern has to do with the structure's natural frequency. Long substructures in deep water will tend to decrease the structure's overall natural frequency. The substructure must be stiff enough to avoid an unsatisfactory decrease in this behavior.

There are many options for offshore wind turbine substructures and foundations and sometimes it is difficult to say where one part begins and ends. In the following, however, the intent is to follow the IEC distinction, namely, that the foundation is in contact with the soil.

Regardless of types, important considerations include water depth (shallow, transitional, and deep), soil type, overall stiffness, and fabrication and installation options. Here, shallow refers to water less than 30 m deep. Transitional depth is between 30 and 60 m. Beyond that, the water is considered to be deep.

10.7.1 Monopiles

Monopiles are steel tubes that are embedded in the seabed and extend above the sea surface to provide a platform on which to affix a wind turbine's tower; see Figure 10.11. The diameter of the pile is normally at least as large as that of the tower base. The wall thickness of the monopile is on the order of 1.5–2% of the diameter. Monopiles are

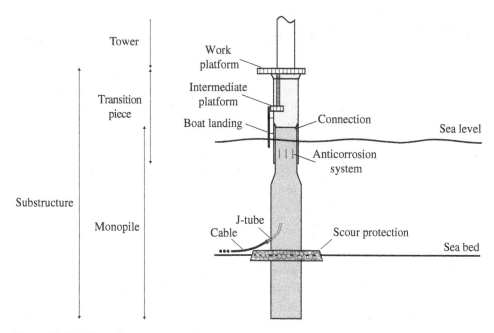

Figure 10.11 Monopile

Table 10.8 Sample monopile for NREL Reference 5-MW turbine

Property	Value	Units
Water depth	20	m
Length above sea level	10	m
Monopile diameter	6.5	m
Monopile wall thickness	0.11	m
Monopile embedment depth	20	m
Monopile mass	938.5	t

generally of a constant diameter, although there may be a taper near the top to facilitate the transition to the tower. At the top of the pile there is, in fact, normally a transition piece; see Section 10.7.1.1.

In accordance with IEC terminology, the part of the monopile above the seabed would be considered the substructure, while the part below the seabed would be the foundation.

Monopiles are most often driven into the seabed with a hydraulic hammer. In harder soils or rock, drilling may be necessary. Little or no seabed preparation is needed, although heavy-duty pile-driving equipment is required. Monopiles may not be suitable in: seabeds with large boulders, soft soils or soils with Glauconite, locations with shallow bedrocks, and sites with important breaking waves which may result in excessively large loads.

Monopiles have been used primarily in shallow to moderate depths (40 m or less), but greater depths are certainly possible. Due to natural frequency considerations, as the depth increases, so too will the diameter and the wall thickness.

In order to provide stability and ensure against leaning, monopiles may extend beneath the seabed a distance greater than the water depth, although the actual distance will depend on soil conditions. Table 10.8 summarizes some dimensions of a monopile that could support the NREL 5-MW reference turbine in 20 m water depth. (As elsewhere, this assumes a nominal steel density of 8500 kg/m^3).

As indicated above, the monopile is driven into the seabed soil to a sufficient depth that it provides a nearly rigid support for the tower. The required depth is a function of the soil characteristics. As discussed previously (Section 10.3.3.1), the resistance of the soil to movement of the pile could be assessed using p–y method.

There continues to be discussion regarding monopile design, turbine size, and water depth, and it is by no means certain how large turbines can be and how deep the water can be before monopiles cannot be used. In any case, the following need to be considered: fabrication, transportation, installation, soil properties, metocean conditions, tower requirements, and turbine design. Drivability is also a major design consideration. Driving large piles not only requires very large pile drivers and associated vessels, but it also creates a significant amount of noise, which can be harmful to marine mammals; see Chapter 14.

Sometimes large diameter steel tubes, referred to as monotubes, are attached at the upper end of the tower and at the lower end to a separate foundation, such as a GBS or a suction caisson. They are in many ways similar to their cousins, the monopiles, but have not been as widely used.

10.7.1.1 Transition Piece

The support structure of an offshore wind turbine based on a monopile normally includes a transition piece. This is a tubular part that provides the connection between the pile and the tower. The installation of the pile typically involves driving it into the seabed with a hydraulic pile driver, a process that can deform the top of the pile. It is also likely that the pile does not end up being perfectly vertical. The purpose of the transition piece is to accommodate the resulting discrepancies. Prior to the erection of the tower, the transition piece is slid over the top of the monopile. Its inner diameter is sufficiently larger than the pile's outer diameter that it can be aligned so that its top surface is horizontal. The transition piece is grouted into place. The base of the tower is then lifted into place and bolted to the now horizontal top surface. See Figure 10.11.

10.7.1.2 Appurtenances

There are several appurtenances associated with the monopile that are worthy of note. These include the J-tube, boat landing, and work platform. The J tube provides a guide and protection for the electrical power cable from the turbine's generator down to the seabed. The boat landing facilitates personnel access from a service vessel to the turbine's work platform. Personnel may undertake activities on the work platform itself or enter the tower through an entrance hatch. Many of the appurtenances are attached to the transition piece during fabrication and so are installed with it.

10.7.1.3 Design Variables

The primary design variables of the monopile and transition piece reflect the considerations discussed previously. These include:

- Height above mean sea level
- Water depth
- Embedment depth
- Diameter
- Wall thickness

10.7.1.4 Monopiles for Deeper Water

As of the present time, the majority of monopiles have been installed in water depths less than 40 m. Those in deeper water may be large enough to be classified as XXL monopiles. That means that the pile is one with a diameter between 7.5 and 11 m and has a mass between 1000 and 2400 tonnes. There is continued interest, however, in using monopiles in deeper water, up to at least 65 m. Such piles would be even larger than XXL. For example, consider a pile 8 m in diameter in 60 m of water, embedded 20 m in the seabed and extending 10 m above the surface. The total length would be 90 m. Assuming a minimum steel density of 7850 kg/m³ and a slender diameter-to-thickness ratio of 120, the pile would have a mass of 4700 tonnes, which is far larger than XXL. Even making the pile more slender by using a diameter-to-thickness ratio of 160, the pile would have a mass of 3530 tonnes. It is clear that special design, fabrication, and installation methods would be needed for such a structure.

10.7.1.5 Example for a Turbine on a Monopile

This example illustrates the load effects, including stability (buckling), on the NREL 5-MW turbine when supported on a monopile. The situation is simplified and considers only the weights of the various main parts of the turbine and a nominal thrust of 800 kN at the rotor center. The relevant parameters of the turbine itself (RNA) were included in Table 10.6. This example uses the offshore tower, whose dimensions were given in Chapter 1 and are repeated here in Table 10.9. Those of the monopile were shown in Table 10.8, but only the portion above the seabed is considered here. The assumed properties of the steel are given in Table 10.10. The interference factor k is assumed to be 1.0. The results of the stress analysis are given in Table 10.11 and the stability analysis in Table 10.12. The calculations here use Equations 10.34–10.38.

The natural frequency for the turbine in this case is 0.336 Hz. Note that this is significantly lower than 0.381 Hz for the turbine with the tower connected directly to a rigid foundation. (Both calculations use the Myklestad method for calculating the frequency.)

Table 10.9 Dimensions of the NREL 5-MW offshore tower

Property	Value	Units
Tower height	77.6	m
Tower top diameter	3.87	m
Tower base diameter	6.5	m
Tower top wall thickness	0.019	m
Tower base wall thickness	0.0247	m

Table 10.10 Properties of steel for monopile example

Property	Symbol	Value	Units
Density	ρ_{steel}	8500	kg/m^3
Modulus of elasticity	E	2.1×10^{11}	Pa
Allowed design stress	σ_d	2.35×10^8	Pa
Imperfection factor	α	0.21	—

Table 10.11 Results of analysis for monopile example

Location (units)	R m	R_{in} m	A m^2	I m^4	N_{Ed} kN	M_{Ed} kN m	N_{Ed}/A kPa	$M_{Ed}R/I$ kPa	$N_{Ed}/A+M_{Ed}R/I$ kPa
Tower top	1.935	1.916	0.230	0.426	3434	1920	14 937	8718	23 655
Tower base	3	2.973	0.507	2.259	5759	64 000	11 367	84 975	96 342
Seabed	3	3.140	2.208	11.274	11 283	88 000	5109	25 368	30 477

Table 10.12 Results of analysis for monopile example (continued)

Location	N_{cr} (kN)	$\bar{\lambda}$	ϕ	χ	$N_{pl,Rd}$	$M_{el,Rd}$	$\dfrac{N_{Ed}}{\chi N_{pl,Rd}} + \dfrac{kM_{Ed}}{M_{el,Rd}}$
Tower top	19 071	1.683	2.065	0.307	54 019	51 753	0.24
Tower base	101 122	1.085	1.177	0.612	119 062	176 993	0.44
Seabed	504 566	1.014	1.096	0.662	518 933	815 208	0.14

10.7.2 Gravity-Based Structures

GBSs have been used with a significant number of offshore wind turbines. Typically, this has been the case in relatively shallow water when the soil is not conducive to installing monopiles. They may form a monolithic component with respect to the substructure, or they may have separate substructures attached to them. An offshore GBS is in principle similar to the type of foundation used most commonly on land, that is a block of sufficient mass to resist the overturning of the wind turbine under any realistic set of conditions. The structure is typically made of

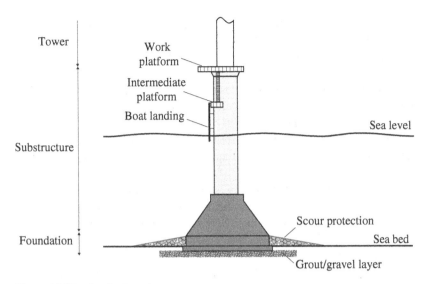

Tower

Work platform

Intermediate platform

Boat landing

Sea level

Substructure

Scour protection

Sea bed

Foundation

Grout/gravel layer

Figure 10.12 Gravity-based structure

reinforced concrete, most often in a dry dock. When ready to be installed, it is loaded onto a barge and towed to the site of installation where it is lowered to the ocean floor with a large sea-going crane. In other cases, if of a suitable design, the GBS can be floated out without a barge. When on-site, it is flooded and lowered to the seabed without the need for a crane. In either case, the seabed must be prepared beforehand to make it relatively smooth and level. Once the GBS is in place additional ballast is often added to bring the total weight up to the value required by the situation.

In addition to the weight of the structure, the other important design consideration is the bearing capacity of the soil: a firm soil will result in the need for a smaller footprint than a weaker soil; see Section 10.3.1. Regardless, it must be recalled that the effective weight of the structure is reduced in proportion to the buoyancy of the water in which it is immersed. The buoyancy force is given by the displaced volume times the weight density of the water. A typical GBS is illustrated in Figure 10.12.

Example: A gravity structure is designed here to be able to support the NREL 5-MW turbine in water depth of 7 m. The foundation will be assumed to be made of reinforced concrete. It is designed so as to float when empty and provide sufficient stability when the inner part is partially filled with ballast when in place. The structure has a base with a diameter of 20 m, a base thickness of 1 m, and conical top of 6 m high whose top diameter is 6.5 m. The wall thickness is 0.3 m. Using equations for a frustum, the volume of concrete is 387.4 m^3. Assuming the concrete has a density of 2500 kg/m^3, the weight in air of the empty GBS is 9501 kN. The structure will float if the displaced volume is such that the weight of the water displaced is the same as that of the empty structure. It may be shown that that will be true if the top of cone protrudes 3.27 m above the sea surface, assuming that the density of the sea water is 1035 kg/m^3. The structure will be sunk on site, and the interior is filled 58% with hematite whose density is 4770 kg/m^3. The net weight of the structure, considering buoyancy, is then 19593 kN. This value is comparable to the circular foundation in Section 10.3.2, Example 10.3.

10.7.3 Jackets and Other Multimember Substructures

A variety of multimember substructures have been used or proposed for offshore wind applications. These include jackets, tripods, and quadrapods. Multimember structures are particularly applicable in deeper water and for larger turbines. The most common is the jacket, which is a welded structure consisting of three or more (normally four) tubular legs with a bracing system between the legs. Tripods, as the name implies, have three legs. The legs are connected to a central tube that projects above the water level. The tower is bolted to the top of the central tube. Quadrapods are similar but have four legs.

Multimember substructures are held in place and supported by piles or suction caissons (see Frieze, 2023). They have the advantage that they require significantly less steel than monopiles in deep water. Another advantage is that foundation piles are much smaller than monopiles and can be driven more easily, with smaller equipment and creating less noise in the process. On the other hand, the cost of fabrication is substantially greater, at least on a per unit weight basis: substantial intricate cutting of steel is required, and the welding is more difficult as well.

Jackets for offshore wind turbines are based on designs adapted from the offshore oil and gas industry, where they have been used for many years. The first use of jackets for supporting offshore wind turbines was at the Beatrice project off the coast of Scotland. The water depth there is approximately 45 m. The jackets were fabricated on shore, brought to the site by heavy lift crane barge, and lowered into place. They were secured to the seabed by piles at each of the legs. Figure 10.13 illustrates a typical jacket.

At the present time, jackets are used primarily in transitional depths, i.e., greater than 30 m. It is easier to make a stiff structure for deep water with a jacket design than a monopile, so the natural frequency issue noted above for monopiles may be less serious. With respect to jackets vs. monopiles, however, a prime consideration will likely be soil conditions rather than depth *per se*. The upper limit of water depth for jackets has not yet been determined, but it will likely be based on the comparative economics with floating wind turbines. The limitation will likely not be technical since fixed platforms in the oil and gas industry are presently used in waters more than 500 m deep. For a comparison of jackets and monopiles, see Damiani *et al.* (2016).

Jackets are essentially trusses and so are designed and analyzed accordingly, but that topic is outside the scope of this text.

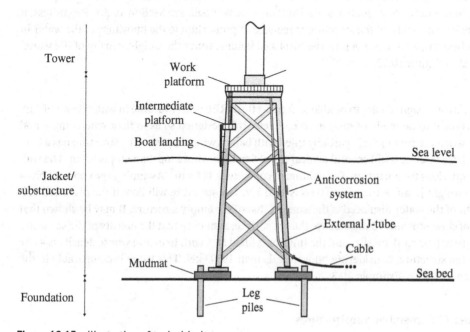

Figure 10.13 Illustration of typical jacket

10.8 Environmental Considerations Regarding Substructures and Foundations

It is important to recognize that there are characteristics of some substructures and foundations that need to be considered with respect to the environment. First of all, as noted above, driving a monopile results in a substantial amount of noise due to the hammering. This noise can be injurious to fish or marine mammals. Special precautions must be taken to mitigate this problem (see Chapter 14). In addition, once a monopile has been driven it cannot be readily removed when the turbine is decommissioned. It must be cut off and left in place.

These issues are less serious for other types of foundations. The piles in foundations for jackets are much smaller than monopiles so driving them produces far less noise. There will also be less material left in the seabed at the end of the project life.

With GBSs and suction caissons, there is no pile driving at all, hence very little noise is associated with installation. Both GBSs and suction caissons can also eventually be removed from the site. One contravening factor to be considered is that GBSs have larger footprints than monopiles. They require site preparation before installation which will disturb whatever is living there. On the other hand, once installation is complete, they may serve as artificial reefs and attract sea life (see Chapter 14 as well.)

References

ASTM (2018) *Standard Practice for Description and Identification of Soils (Visual-Manual Procedures)*. ASTM D2488-17e1, American Society for Testing and Materials, West Conshohocken, PA.

Bowles, J. E. (1996) *Foundation Analysis and Design*, 2nd edition. McGraw Hill, New York.

Byrne, B. W. and Houlsby, G. T. (2003) Foundations for offshore wind turbines. *Philosophical Transactions of the Royal Society of London*, **361**, 2909–2930, available from http://rsta.royalsocietypublishing.org/content/361/1813/2909.full.pdf.

Byrne, B. W., Mcadam, R., Burd, H., Houlsby, G., Martin, C., Zdravkovic, L., Taborda, D. M. G., Potts, D., Jardine, R., Sideri, M., Schroeder, F., Gavin, K., Doherty, P., Igoe, D., Wood, A., Kallehave, D. and Gretlund, J. (2015) New design methods for large diameter piles under lateral loading for offshore wind applications. *Proc. of Third International Symposium on Frontiers in Offshore Geotechnics* (ISFOG 2015), Oslo.

Carswell, W. (2015) *Soil-Structure Modeling and Design Considerations for Offshore Wind Turbine Monopile Foundations*, Doctoral Dissertation. University of Massachusetts Amherst, https://scholarworks.umass.edu/dissertations_2/531. Accessed 12/3/2023.

CEN (2016) *Eurocode 3: Design of Steel Structures – Part 1-1: General Rules and Rules for Buildings*, 2nd edition. European Convention for Structural Steelwork.

Damiani, R., Dykes, K and Scott, G. (2016) A comparison study of offshore wind support structures with monopiles and jackets for U.S. waters. *Journal of Physics: Conference Series*, **753**, 092003.

Dykes, K., Damiani, R. Roberts, O. and Lantz, E. (2018) *Analysis of Ideal Towers for Tall Wind Applications: Preprint*, NREL/CP-5000-70642, National Renewable Energy Laboratory, Golden, CO.

Frieze, P. A. (2023) *Ships and Offshore Structures –Vol. II -Offshore Structure Design and Construction*, https://www.eolss.net/Sample-Chapters/C05/E6-177-OD-01.pdf. Accessed 12/3/2023.

Fugro Marine GeoServices, Inc (2017) *Geophysical and Geotechnical Investigation Methodology Assessment for Siting Renewable Energy Facilities on the Atlantic OCS*, OCS Study BOEM 2017-049, US Department of the Interior, Bureau of Ocean Energy Management, Office of Renewable Energy Programs, available from https://www.boem.gov/sites/default/files/environmental-stewardship/Environmental-Studies/Renewable-Energy/G-and-G-Methodology-Renewable-Energy-Facilities-on-the-Atlantic-OCS.pdf. Accessed 12/3/2023.

Hicks, T. G. (2016) *Handbook of Civil Engineering Calculations*, 3rd edition. McGraw Hill, New York.

IEC (2019a) *Wind Turbines Part 1: Design Requirements*, 61400-1, 4th edition. International Electrotechnical Commission, Geneva.

IEC (2019b) *Wind Turbines Part 3: Design Requirements for Fixed Offshore Wind Turbines, 61400-3-1*, 2nd edition. International Electrotechnical Commission, Geneva.

ISO (2017), *Geotechnical Investigation and Testing – Identification and Classification of Soil – Part 1: Identification and Description*, ISO 14688-1:2017, International Standards Organization, Geneva

Leite, O. (2015) *Review of Design Procedures for Monopile Offshore Wind Structures*, M.Sc. Thesis, Dept. of Civil Engineering, University of Porto.

Manwell, J. F., MacLeod, J., Wright, S. and McGowan, J. G. (2006) Hull Wind II: A Case Study of the Development of a Second Large Wind Turbine Installation in the Town of Hull, MA. *Proc. AWEA 2006 Conference*, available from https://www.researchgate.net/publication/251986656_Hull_Wind_II_A_Case_Study_of_the_Development_of_a_Second_Large_Wind_Turbine_Installation_in_the_Town_of_Hull_MA. Accessed 12/3/2023.

NewtonExcelBach (2009) *LatPile – Analysis of Lateral Loads on Piles*, https://newtonexcelbach.com/2009/09/26/lpile-analysis-of-lateral-loads-on-piles/. Accessed 12/3/2023.

Sadegh. A. M. and Worek, W. M. (Ed.) (2018) *Marks' Standard Handbook for Mechanical Engineers*, 12th edition. McGraw Hill, New York.

State Materials Office (2022) *Soils and Foundations Handbook*, Gainesvilled, FL, available from http://www.dot.state.fl.us/structures/Manuals/SFH.pdf. Accessed 12/3/2023.

Svensson, H. (2010) *Design of Foundations for Wind Turbines*, Master's Dissertation, Lund University, Sweden, https://www.byggvetenskaper.lth.se/fileadmin/byggnadsmekanik/publications/tvsm5000/web5173.pdf. Accessed 12/3/2023.

USACE (1990) *Engineer Manual 1110-1-1904*, available from https://www.publications.usace.army.mil/Portals/76/Publications/EngineerManuals/EM_1110-1-1904.pdf. Accessed 12/3/2023.

USACE (2005) *Design of Pile Foundations*. University Press of the Pacific.

11

Floating Offshore Wind Turbines

A floating offshore wind turbine (FOWT) is a wind turbine whose support structure floats in the body of water in which it is located. The idea of FOWTs is quite attractive and has been so for some time. This is because most of the ocean is too deep for fixed offshore wind turbines, and the area potentially available for energy production will be greatly expanded as FOWTs become feasible options. At the present time, however, there are few commercial-scale FOWTs. Despite the small number of existing FOWTs, there are several possible configurations that are quite plausible and are being further developed. In all cases, the rotor nacelle assemblies and the tower are comparable to those of fixed offshore wind turbines. The main difference is in the support structures. Enough experience has been gained that there are already international recommended practices in place to help guide the design (IEC, 2023). In this chapter, we examine the most promising options.

The chapter begins with historical precedents in Section 11.1, followed by Section 11.2 which provides basic definitions relevant to the topic. Then Section 11.3 gives an overview of FOWT topologies, and Section 11.4 discusses fundamental principles. Section 11.5 is about floating substructure/hull design; Section 11.6 concerns hydrostatics. Section 11.7 discusses motions of FOWTs. Section 11.9 gives worked examples of typical barges, spars, semisubmersibles, and TLPs, and, finally, Section 11.10 introduces the coupled aero/hydro/structural dynamics of FOWTs.

11.1 Historical Precedents

As early as 1893, Fridtjof Nansen of Norway employed a wind turbine on his ship *Fram* in his attempt to reach the North Pole (Nansen, 1897); see Figure 11.1. This turbine did generate electricity, but the power was all used on board.

In the 1930s, Hermann Honnef (Honnef, 1932) proposed an anchored offshore pontoon platform supporting a wind turbine using two rotors, see Figure 11.2.

The vision for large-scale floating offshore wind farms is credited to William Heronemus from the University of Massachusetts (Heronemus, 1972) who in the early 1970s proposed large arrays of floating wind turbines (see Figure 11.3). In his case, the idea was to produce electricity and then utilize that electricity to electrolyze water into hydrogen for use as a fuel. These turbines were to be supported by floats and stabilized by counterweights.

Also of relevance is the experience of the offshore oil and gas industry, especially since the 1940s. Although most oil and gas platforms are fixed bottom jacket structures, floating substructures have become more widely used in deep water.

Wind Energy Explained: On Land and Offshore, Third Edition.
James F. Manwell, Emmanuel Branlard, Jon G. McGowan, and Bonnie Ram.
© 2024 John Wiley & Sons Ltd. Published 2024 by John Wiley & Sons Ltd.

Figure 11.1 Wind turbine on Nansen's ship Fram, 1895 (*Source:* Flickr/Fridtjof Nansen/National Library of Norway)

Figure 11.2 H. Honnef's concept of an offshore wind turbine, 1932 (*Source:* Reproduced by permission of the heirs of H. Honnef through https://virtuellesbrueckenhofmuseum.de)

Figure 11.3 Heronemus' concept for floating offshore wind turbines, 1972 (*Source:* University of Massachusetts Amherst)

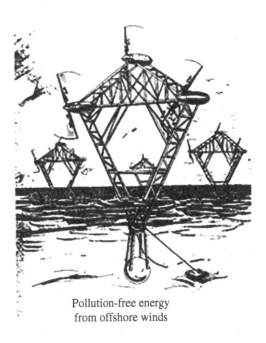

Pollution-free energy
from offshore winds

11.2 Definitions

There are a number of terms that are used with maritime applications, including floating offshore wind turbines (FOWTs), but are less common elsewhere. The most important ones are defined in Sections 11.2.1–11.2.5 below. See IEC (2023) for more complete definitions. All of these terms will be discussed in more detail later in this chapter.

11.2.1 Main Components

A FOWT will include a vessel (also called a platform or floating substructure) and a stationkeeping system (see Figure 11.4):

Vessel. Anything floating, constructed for some utilitarian purpose.
Stationkeeping system. A system for limiting the excursions of a vessel; for instance, a passive system (mooring lines or tendons), or an active system (e.g., using active thrusters); see Section 11.2.3. Passive stationkeeping systems are also known as **mooring systems**.

11.2.2 Generic Description of a Vessel and a Floating Substructure

The following definitions are borrowed from naval architecture to describe a vessel (see Figure 11.4):

Hull. Main body of a vessel, specifically the component that provides the buoyancy.
Deck. The top of the hull.
Keel. The lowest structural member on a vessel.
Ballast. Heavy material added to a low point in a vessel to improve its stability.
Pontoon. A long horizontal member that provides buoyancy.
Draft. The vertical distance from mean sea level to the lowest point on a vessel (always a positive number).
Freeboard. Height of deck above sea level.

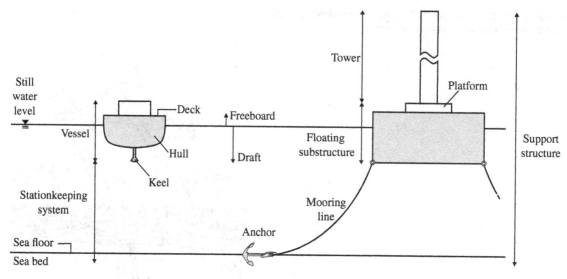

Figure 11.4 Main definitions from naval architecture (left) and for floating offshore wind turbines (right)

For a floating offshore wind turbine, the following definitions are also introduced (see Figure 11.4):

Offset column. Secondary vessel used where buoyancy is required; closed hull.
Platform. Generic term for floating substructure that supports the tower.
Floating substructure. The hull or hulls and associated structural members of a floating offshore wind turbine.
Support structure. The entire system below the rotor-nacelle assembly: tower, platform, floating substructure, and stationkeeping system.

A vessel can move in one or more of six directions: three in translation: **surge, sway, heave** and three in rotation: **roll, pitch**, and **yaw**. These motions were introduced in Chapter 4 and are further described in Section 11.7.

11.2.3 Definitions for Stationkeeping Systems

This chapter only considers passive stationkeeping systems, which consist of lines (mooring lines or tendons) that are attached to an anchor at the seabed and on the vessel. The following definitions are given; they are illustrated together with the different types of floating offshore technologies in Figure 11.5:

Anchor. Device attached to the end of the mooring line or tendon and partially or fully buried in the seabed to limit the movement of the mooring line or tendon and to transfer loads to the seabed.
Cleat. A device used as the attachment point of a line to a vessel or dock.
Mooring. A passive type of stationkeeping system that typically comprises mooring lines (or tendons), anchors, and associated hardware; the mooring line itself is chain, wire rope, or synthetic rope.
Mooring line. A generic term for lines used to keep a vessel in place.
Fairlead. An object or device, such as ring or roller, which changes the direction of a mooring line; last point of contact of a mooring line on a vessel; used here as the reference point of attachment of mooring lines on a vessel.
Point of touchdown. Location where mooring line first contacts the seabed.

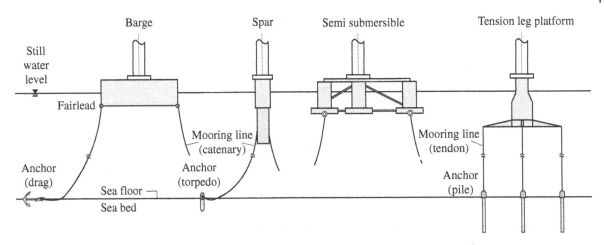

Figure 11.5 Main technologies for floating offshore wind turbines and illustration of different passive stationkeeping system (mooring lines and anchors)

The following types of moorings are defined:

Catenary mooring. Mooring in which the restoring force on the vessel is provided by the distributed weight of the mooring line; often assumed to be attached at an anchor at sufficient distance that the mooring line is parallel to the seabed at that point of touchdown.

Semi-taut mooring. Mooring in which the mooring line rises above horizontal at the anchor, but less so than for a taut mooring.

Slack mooring. Mooring in which the mooring line lies on the sea floor for some distance from the anchor before rising up to meet the vessel.

Taut mooring. A mooring system in which the line is nearly straight from the anchor to the point of attachment.

Tendon. Component of a stationkeeping system, comprised of steel tubes, under tension, which connects a vessel to an anchor. Vessels using tendons are referred to as tension-leg platforms.

11.2.4 Main Types of Technologies for Floating Offshore Wind Turbines

In recent years, a number of topologies for FOWTs have been considered; many of them have been informed by oil and gas industry experience. They may be classified as one of four types: (1) barges, (2) spars, (3) semisubmersibles, and (4) tension leg platforms (TLPs). They are illustrated in Figure 11.5, and they will be described further in Section 11.3.

Barge. A shallow vessel, typically with a flat bottom.

Spar. A type of floating substructure in which the buoyancy is provided by a vertical cylindrical hull, which is deep enough when ballasted to provide floating stability.

Semisubmersible. A type of floating substructure in which the buoyancy is provided by one or more offset columns, spaced so as to provide floating stability.

Tension leg platform (TLP). A vessel maintained in place by tendons.

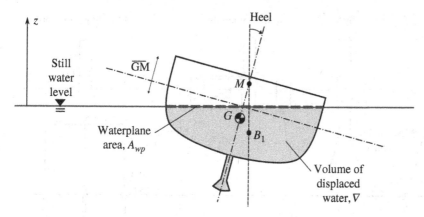

Figure 11.6 Important geometric definitions for the hydrostatics of vessels

11.2.5 Important Geometric Definition for the Study of Vessels

In the study of vessels and in particular their hydrostatics (buoyancy), the following definitions are used (see Figure 11.6):

Center of buoyancy (B). The centroid (geometric center) of the water displaced by a floating vessel.

Displaced center of buoyancy (B_1). Centroid (geometric center) of the water displaced by a floating vessel when the vessel is heeling.

Center of gravity (G). Center of mass of the vessel and whatever it carries.

Heel or heeling angle (θ). Temporary inclination of a vessel from its normal position.

Metacenter (M). The point of intersection of a vertical line passing through the center of buoyancy of a vessel that is tilting, with the tilted line passing through the original center of buoyancy and the center of gravity.

Metacentric height ($\overline{GM}$). Distance on a vertical line from the center of gravity to the metacenter.

Water plane area (A_{wp}). Cross-sectional area of the vessel where it intersects the still water level (see Figure 11.6).

The centers of buoyancy and gravity and the metacenter are often measured with respect to the:

Still water level (SWL). Wave-averaged sea level height; often used as relative reference height, in which case SWL = 0.

11.3 Topology Options for Floating Offshore Wind Turbines

In this section, we study the main aspects of the four main topologies of FOWTs: barge, spar, semisubmersible, and tension leg platform (more succinctly known as TLP). The key factors to consider are buoyancy and floating stability: the floating substructure must be able to support the weight of the RNA, tower, and floating substructure, it must keep the turbine nearly vertical during operation, and it must remain stable even under the most severe conditions.

11.3.1 Overview of Key Features

The key features of the four main types of FOWTs are summarized in Table 11.1. At the present time, any of them are most likely to be used in water deeper than 50 m (i.e., not in shallow water). Spars are best suited for water greater than 150 m deep because of their long hull. The barge is stabilized by buoyancy and ballast, the spar by

Table 11.1 Key features of the main floating offshore wind turbine topologies

	Barge	Spar	Semisubmersible	TLP
Depth	>50 m	>150 m	>50 m	>50 m
Stability	Buoyancy, ballast	Ballast	Buoyancy, ballast, mooring lines	Mooring lines, ballast
Mooring lines	Catenary	Catenary	Catenary	Tendons
Anchors	Drag, gravity, suction caisson	Drag, gravity, suction caisson	Drag, gravity, suction caisson	Heavy gravity, pile
Assembly	In port	Deep port or special design	In port	In port
Transport	Towing	Towing (deep water required for vertical towing)	Towing	Special vessel

ballast alone, the semisubmersible by a combination of buoyancy, ballast, and mooring lines and the TLP primarily by its tendons. All the designs, with the exception of the TLP, could use catenary, or possibly semi-taut moorings. Similarly, all the designs except the TLP could use drag anchors or possibly gravity anchors or suction caissons (see Chapter 10), depending on the soil. The TLP would require heavy gravity anchors, piles, or other anchors designed for large vertical force. All the designs could be assembled in port, but the spar would require a very deep port or a special design. The barge or the semisubmersible could be towed from port by conventional means. The TLP would typically require a special vessel to prevent its overturning before the tendons are connected. As with assembly, the spar would need either a very deep port or a special design to allow it to be towed nearly horizontal and then righted in place.

11.3.2 Barge

One of the first types of support structure options to be seriously considered for a FOWT was the barge. A barge as conceived of for offshore wind turbines is a wide but shallow, flat bottom vessel that is either square or round.

A conceptual rectangular barge that would support the NREL 5-MW turbine was developed by Jonkman (2007). The turbine and barge are illustrated in Figure 11.7. Section 11.9.1 provides sample calculations for a similar, but simplified example.

A conventional barge has the advantage of having a simple design but its behavior in a severe sea state is a significant concern due to the effect of the waves. For this reason, conventional barge-supported offshore wind turbines do not appear to be promising options in the open ocean; they could perhaps be used in more sheltered waters.

11.3.3 Spar

In the case of the spar, the wind turbine is supported by a long steel cylindrical hull floating vertically in the water. The spar needs to be heavily ballasted in order to remain nearly vertical during operation. In one conceptual design (Jonkman, 2010), the reference NREL 5-MW turbine was supported by a spar in which the steel spar had a diameter of 9.4 m and was 120 m long; see Figure 11.8. To maintain stability 6600 metric tonnes of ballast were required to be placed in the bottom of the spar. The displaced volume of the spar was sufficient for the entire structure to float at a reasonable depth. The length and ballasted weight of the spar kept the heel angle of the spar within an acceptable range during operation. In this design, the upper portion of the spar has a tapered section with upper diameter smaller than the rest of the spar. This reduced diameter helps to decrease wave loads. It also facilitates

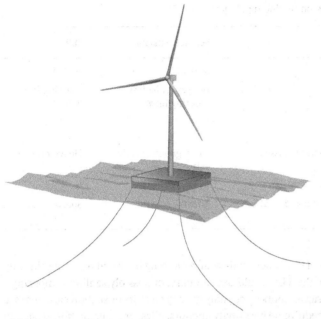

Figure 11.7 NREL 5-MW wind turbine supported by a barge

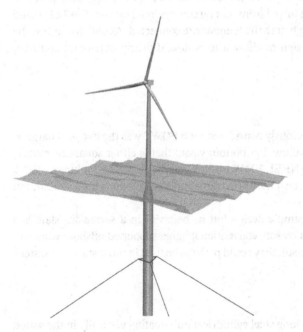

Figure 11.8 NREL 5-MW turbine on spar

connection to the tower, often via a distinct transition piece. The mooring system of a spar typically includes a delta-shaped connection, as can be seen in Figure 11.8, to increase yaw stiffness.

A prototype spar-supported wind turbine was designed and tested by the Norwegian company Statoil (now Equinor); see Skaare *et al.* (2015). The first commercial-scale FOWTs using spars were installed off the coast of Scotland in 2017, also by Equinor.

11.3.4 Semisubmersible Platform

A semisubmersible platform is composed of multiple hulls, or offset columns, spaced sufficiently far apart to provide stability for the overall system. Steel structural members, roughly comparable to those used in jackets, hold the offset columns in place and provide for the attachment of the turbine's tower. An example of a semisubmersible is shown in Figure 11.9. In this version, the buoyancy and water plane area are provided by three offset columns. The semisubmersible substructure facilitates the deployment of the turbine, in that it can be towed from port to the site of installation. In the configuration shown, the offset columns are larger at the base than at the top. The larger base helps to reduce heave (vertical) motion; the smaller tops reduce wave loads. The first commercial semisubmersible wind turbines, known as WindFloat®, were designed, built, and installed by Principle Power, Inc. Analysis of semisubmersibles is discussed by Robertson *et al.* (2014).

11.3.5 Tension Leg Platform

The fourth type of floating substructure is the tension leg platform (TLP). In this configuration, the turbine is mounted on a single hull with spokes for the attachment of tendons. The tendons are in turn secured to anchors on the seabed. When installation is complete, the tendons are under substantial tension. The tension in the tendons keeps the hull in place and serves to counteract any overturning moments, either from the weight of the turbine (gravity), aerodynamic loads (in particular, thrust loading associated with operation) or wave loads. The first prototype TLP-supported wind turbine was constructed by Blue H Engineering in 2007. Detailed analyses of TLPs have also been carried out by Withee (2002), Tracy (2007), and Matha (2009), among others. A schematic of one conceptual TLP is illustrated in Figure 11.10.

Figure 11.9   NREL OC4-DeepCwind semisubmersible

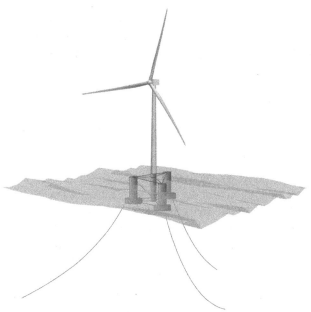

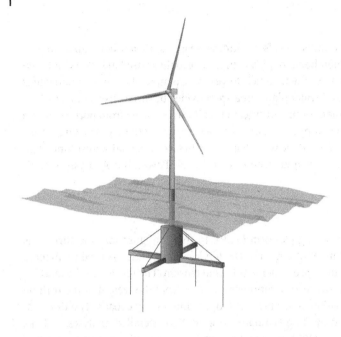

Figure 11.10 Conceptual tension leg platform (TLP)

11.4 Fundamental Principles

There are four fundamental requirements for a FOWT: (1) a floating substructure, consisting of one or more hulls, must provide sufficient buoyancy for the entire structure to float, (2) the dimensions of the hull(s), the distribution of weight, and the mooring lines (or tendons) must ensure that the structure is stable, i.e., not capsize, (3) the tower should remain nearly vertical during operation, and (4) a stationkeeping system must keep the structure in place so that it does not move more than a prescribed amount from its intended location. The principles behind these requirements are discussed in Sections 11.4.1–11.4.4.

11.4.1 Buoyancy

The buoyancy force, F_B, (N) acting on a (partially or fully) submerged body is given by Archimedes' principle:

$$F_B = \rho_w g \nabla \tag{11.1}$$

where ∇ (inverted delta) is the volume of the fluid (here, water) being displaced by the body (m^3), ρ_w is the density of the fluid (kg/m^3), and g is the acceleration of gravity (m/s^2). The mass of displaced fluid is sometimes called "displacement," but this term will not be used in this book to avoid confusion.

The equilibrium condition of buoyancy (floatation) of an isolated body (i.e., without a stationkeeping system) is reached when the weight of the body (mg) equals the buoyancy force, leading to the condition $\rho_w \nabla = m$; that is, a floating body will displace a volume of fluid whose mass is equal to the mass of the body. Consider, for example, a barge that would support the NREL 5-MW turbine. One such barge when fully ballasted and supporting the turbine would have a mass of 5200 tonnes. This mass also corresponds to $\nabla = m/\rho_w = 5.07 \times 10^6$ m^3 of water, assuming a density of seawater of 1025 kg/m^3. It should also be noted that the weight and potential pretension of the stationkeeping system will affect the stability of the body.

It is important to observe that a body that is immersed in water will always experience the buoyancy force but might not be "floating." This is important when considering the effective weight of submerged gravity foundations and mooring lines.

11.4.2 Floating Stability

Floating stability of a floating body refers to the ability of the body to remain upright when subject to a relatively small displacement. This stability can be determined by considering the moments due to the forces acting on the system, the positions of the center of buoyancy, center of gravity, and mooring lines (if any). Any unmoored vessel is stable as long as the center of mass is below the center of buoyancy. To achieve such a low center of gravity, the vessel is likely to be ballasted. A vessel *may* be stable if the center of gravity is above the center of buoyancy, but only under certain conditions. This stability would be due to the shift of the center of gravity of the displaced water. In this case, the vessel is considered to be buoyancy stabilized. The effect of this shift is characterized by the metacentric height, which is described in Section 11.4.2.4. The third way that a vessel may be stable is by virtue of mooring lines which apply forces, and thus moments, which counteract the overturning moments from buoyancy and gravity.

In summary, there are three main ways for a floating body to be stable:

- Ballast stabilization (center of gravity below the center of buoyancy);
- Buoyancy stabilization (shift of center of gravity of the displaced water provides stabilizing moment);
- Mooring line stabilization.

These are discussed in more detail below. Note that in the following, the symbol B indicates the initial buoyancy, which will be in line with the center of gravity G.

11.4.2.1 Ballast Stabilization
A ballast-stabilized body is one whose center of gravity is below its center of buoyancy. This usually entails placing a certain amount of heavy material, known as ballast, at the lowest practical point in the hull. During rolling or pitching motion, the body will remain stable as long as the center of gravity remains below the center of buoyancy. With large motions, however, it may be possible for the center of gravity to rise above the center of buoyancy. In that case, the vessel may capsize. The main requirements for ballast are that it have a high density and be inexpensive. Possibilities include rock and iron ore. In some cases, less dense material, such as seawater itself, may be used as ballast.

11.4.2.2 Buoyancy Stabilization
A buoyancy-stabilized body generally has a large water plane area, which is the cross-sectional area of the body at the still water level, i.e., the transition between the water and the air. When the body is displaced, the center of buoyancy moves sideways to a new position, B_1. If the displacement of the center of buoyancy is large enough, the buoyancy force will produce a moment that is opposite in direction to that produced by the weight of the body. In this case, the body will be stable. This is illustrated in Figure 11.11. An important parameter in evaluating buoyancy stabilization is the metacentric height, as discussed in Section 11.4.2.4.

11.4.2.3 Mooring Line Stabilization
The TLP is the quintessential mooring stabilized structure. In this case, the tension in the mooring tendons is sufficient to create a moment to counter the moments due to the thrust force on the rotor as well as the forces due to waves. In general, most mooring systems will provide some stabilization from the weight of the lines (acting similar to ballasting) and the restoring forces that they provide at the attachment points.

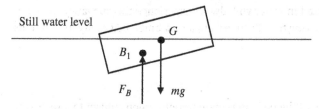

Figure 11.11 Illustration of buoyancy stabilization; G = center of gravity, B_1 = displaced center of buoyancy, m = mass of object, g = gravitational constant, F_B = buoyancy force

11.4.2.4 Metacentric Height

The metacentric height (see Figures 11.12–11.14) is used for assessing the stability of a vessel when the center of gravity G is above the initial center of buoyancy B, and contribution of the mooring lines is minimal. Note that these centers are typically defined along the vertical axis (positive above sea level) and relative to still water level (SWL), so that B is invariably a negative number. The center of gravity may or may not be negative. The process involves determining the metacenter, M, which is the point of intersection of a vertical line passing through the new center of buoyancy with a line that passes through the original center of buoyancy and the center of gravity. The metacentric height is the distance between the metacenter and the center of gravity (see Section 11.2.5); it is designated $\overline{GM}$. The calculation can be complicated for large angles of heel and for vessels with nonuniform cross-sections, but it is simpler for small angles of heel and most floating substructures for offshore wind turbines, where Equation (11.2) can be used. The reader should consult a text on naval architecture, such as Tupper (2013), for information on larger heel angles.

$$\overline{GM} = \frac{I_{yy}}{\nabla} - \overline{BG} \tag{11.2}$$

where:

I_{yy} is the area moment of inertia of the vessel at the water line about the y-axis, m^4,
$\overline{BG}$ is the distance from B to G (positive for B below G), m,
∇ is the displaced volume, m^3,

The important thing to note is that the metacentric height $\overline{GM}$ must be positive for stability. Since the distances to B and G are measured from SWL, $\overline{BG}$ is negative (and stabilizing) when G is below B, as expected.

Stable Situation

When the metacenter is *above* the center of gravity (Figure 11.12), the vessel is stable; the effect of gravitational force (acting at G) and the buoyancy force (acting at B) results in a moment that tends to right the vessel.

Neutral Stability

When the metacenter is *coincident* with the center of gravity (Figure 11.13), the stability is neutral; the moments neither tend to right the vessel nor to capsize it. If the vessel is subject to a moment, however, due for example to the wind turbine's thrust, the vessel would not be stable.

Unstable Situation

The vessel is unstable when the metacenter M is *below* the center of gravity G (see Figure 11.14). The moment due to gravity tends to force the vessel to pitch or roll in an unfavorable direction (i.e., to capsize).

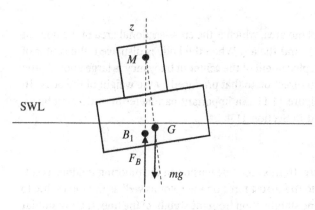

Figure 11.12 Stable situation

11.4.2.5 Maintenance of Near Verticality During Operation

A large positive metacentric height is most relevant for a FOWT when it is operating. It will help counteract the effect of the rotor thrust, which during operation creates a substantial overturning moment on the platform. The platform must remain stable at that time and the point of attachment of the tower to the substructure should remain sufficiently close to horizontal that the tower is nearly vertical. Recall that guidelines for land-based turbines recommend that the tower should not deviate more than 1% from the vertical during operation. Floating turbines can deviate more than that, but that deviation should be minimized to the extent possible. The deviation may be characterized by the pitch angle of the platform. Minimizing the pitch angle of the platform is accomplished by having a substantial amount of ballast, a large water plane area, or a large distance between the fairleads. At present, it may be assumed that the maximum pitch angle during operation should be less than 10°.

11.4.2.6 Other Considerations

In the design for stability of floating offshore turbines, there are some other factors that need to be considered, such as ensuring that the platform continues to float even after an accident. Manufacturability, allowable draft with respect to ports, and methods of installation are also important concerns.

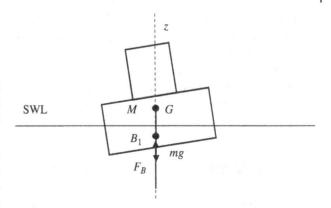

Figure 11.13 Neutral situation

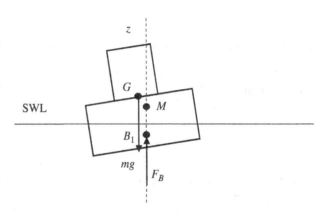

Figure 11.14 Unstable situation

Variation of mooring line tension in severe sea states also needs to be assessed. In particular, slack lines followed by sudden high tension can result in mooring line failure.

11.5 Floating Substructure/Hulls

An essential feature of a FOWT is that it is supported by some type of floating substructure, also referred to as a platform. The principal component(s) of the floating substructure is/are one or more hulls (or offset columns).

11.5.1 Hull Design

A hull of a FOWT is in many ways similar to that of a ship's hull, in that it must keep out the water and must be structurally adequate for the application, but it differs in that the FOWT is intended to remain stationary, so there is little or no directionality in the design.

Most hulls are constructed of steel, although concrete has also been proposed. For a steel hull, a cylindrical or square structure with bottom, deck, and side plating is typically used. Depending on the design, the hull may also have an internal structure to which the plating is attached. The hull may also be divided internally into compartments through the use of bulkheads. The bulkheads provide rigidity to the structure, and the compartments should be watertight so as to prevent loss of floating stability (and sinking in the case of a leak). Insofar as the hull of a FOWT is symmetrical, or nearly so, it does not necessarily have a traditional keel, but the term "keel" is often used to designate the lowest part of the hull.

The wall thickness of a hull must be sufficient to prevent buckling due to external water pressure or damage from a possible collision of moderate severity. Wall thicknesses on the order of 0.015 m are typical.

11.6 Hydrostatics of Floating Offshore Wind Turbines

A primary consideration in the design of a FOWT is to ensure its stability under steady-state conditions. Examining the hydrostatics of the floating turbine is the first step in this process. This topic is the subject of the present section. First, we introduce some general formulae to compute the inertial properties; then we consider the fundamentals of the four platform types.

11.6.1 General Considerations and Assumptions

In the design and analysis of floating substructures, the keel is often used as a reference, since it is the lowest point fixed to the structure. Given the centers of gravity and masses from all the main components of the FOWT (e.g., the tower and main hull), the overall center of gravity with respect to the keel z_{k-cg} may be found from:

$$z_{k-cg} = \sum_{i=1}^{N_s} m_i z_{k-cg,i} / \sum_{i=1}^{N_s} m_i \tag{11.3}$$

where $z_{k-cg,i}$ is the distance from the keel to the center of gravity of the ith component, and N_s is the total number of masses of the structure.

Note that to find the center of gravity of a component with respect to the keel, it is necessary to add the distance from the keel to the base of the component, $z_{k-b,i}$ as shown in Equation (11.4). That distance may be found from the geometry of the structure.

$$z_{k-cg,i} = h_{cg,i} + z_{k-b,i} \tag{11.4}$$

where $h_{cg,i}$ is the height of the center of gravity of a component with respect to its base.

The mass moment of inertia of the entire structure with respect to the overall center of gravity (noted z_{k-cg}) and about the x or y axis is found using Equation (11.5). This is an application of the parallel axis theorem.

$$J_{cg} = \sum_{i=1}^{N_s} J_{cg,i} + m_i \left(z_{k-cg,i} - z_{k-cg} \right)^2 \tag{11.5}$$

In this section, we will assume that the wind is along the x direction, and we are interested in the pitching motion around the y-axis. Therefore, for conciseness, when referring to mass moment of inertia, it will be implied that it is about the y-axis, unless specified otherwise.

For all the platform types, a wind turbine is installed on top, which will be assumed to consist of a rigid, tapered tower, and a rigid rotor-nacelle assembly.

Substructures of FOWTs are typically comprised primarily of components with relatively simple shapes: cuboids, cylinders, and frustums. For reference, the fundamental properties of these components have been collected together in Appendix D.

Example: Consider a hollow frustum of height 10 m, top radius 3 m, base radius 5 m, top wall thickness 0.02 m, base wall thickness 0.03 m, and density of 8000 kg m³. Applying Equations (D.10)–(D.14), (D.16) and (D.17) from Appendix D yields a net material volume of $V_{net} = 6.37$ m³, mass $m = 50\,944$ kg, center of gravity $z_{cg} = 4.26$ m, and mass moment of inertia $J_{cg} = 845\,841$ kg m³.

11.6.2 Barges

In this section, we consider a wind turbine mounted on a barge. The notations are illustrated in Figure 11.15. The analysis here assumes a conventional square barge with width and length W. Some barge configurations incorporate an opening to the water, known as a *moon pool,* in the middle of the hull to enhance stability. Those designs, while interesting, are outside the scope of this text.

11.6.2.1 Displaced Volume
When the turbine is not operating, the barge is horizontal ($\theta = 0$), and the lengths a and b in the figure are the same and equal to the draft d. The displaced volume of water, ∇, is then:

$$\nabla = W^2 d \tag{11.6}$$

11.6.2.2 Mass
The total mass of the turbine and platform is the same as the displaced mass:

$$m = \rho_w \nabla \tag{11.7}$$

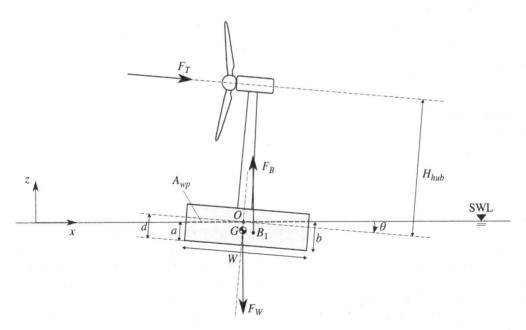

Figure 11.15 Sketch and notations for the study of the barge stability

11.6.2.3 Water Plane Area
The water plane area for the barge is simply:

$$A_{wp} = W^2 \tag{11.8}$$

11.6.2.4 Center of Buoyancy
The barge has vertical sides, so the center of buoyancy without heel is simply:

$$B = -d/2 \tag{11.9}$$

11.6.2.5 Center of Gravity
The center of gravity of the overall structure is obtained by first finding the center of gravity of each of the components with respect to their own base $h_{cg,i}$ and then they are referred to the keel. The center of gravity of the main hull is at its midpoint. The center of gravity of the ballast is also its midpoint, which may be determined from its mass and its density.

$$h_{cg,blst} = \frac{m_{blst}}{2\rho_{blst}W^2} \tag{11.10}$$

The overall center of gravity with respect to the keel may be found from the products of component centers of gravity, as given in Equation (11.3). For the barge, in absence of heel, this becomes Equation (11.11)

$$z_{k-cg} = \frac{m_{RNA}z_{k-cg,RNA} + m_{twr}z_{k-cg,twr} + m_{hull}z_{k-cg,MH} + m_{blst}z_{k-cg,blst}}{m_{RNA} + m_{twr} + m_{hull} + m_{blst}} \tag{11.11}$$

where *twr* and *blst* indicate "tower" and "ballast," respectively.

11.6.2.6 Area Moment of Inertia
The area moment of inertia of a x–y cross-section of the barge with respect to the y axis passing through the center of the cross-section is given by:

$$I_{yy} = W^4/12 \tag{11.12}$$

11.6.2.7 Metacentric Height
The metacentric height is given by Equation (11.2), expressed here as Equation (11.13):

$$\overline{GM} = \frac{I_{yy}}{\nabla} - (B - G) \tag{11.13}$$

11.6.2.8 Platform Pitch Angle
When the wind turbine is operating, the rotor will impose thrust at the top of the tower and thus a moment on the platform. This will cause the platform to pitch in the downwind direction. Finding the steady-state pitch angle θ of the barge involves equating the moment due to the thrust force to the difference in the moments from the weight of the structure acting at the center of gravity and the buoyancy force acting at the displaced center of buoyancy. Note that in order to keep the platform from moving the stationkeeping system must apply a force equal and opposite to the thrust. For the purpose of this example and the following ones, it is assumed that this opposing force is applied at the still water level. It may be shown that, to a good approximation, the platform pitch angle may be found from Equation (11.14):

$$\theta = \tan^{-1}\left[\frac{F_T H_{hub}}{\rho_w g \nabla GM}\right]$$ (11.14)

11.6.2.9 Mass Moment of Inertia

The mass moment of inertia of the barge can be estimated as follows. Assume that the barge is square when viewed from above and that the thickness of the walls, top and bottom are all the same. From Equation (D.4) with $L = W$, the mass moment of inertia about the y axis passing through the center of gravity $J_{cg,barge}$ is given by:

$$J_{cg,barge} = \frac{1}{12}\rho_{steel}\left[W^2 H\left(H^2 + W^2\right) - W_{in}^2 H_{in}\left(H_{in}^2 + W_{in}^2\right)\right]$$ (11.15)

where H is the height of the barge.

The mass moment inertia of the ballast in the barge may be found assuming that it takes the shape of the bottom of the barge and has the same width and length. Its height, H_{blst}, is found from its density, ρ_{blst}, given that the total mass m_{blst} is known. The mass moment of inertia is then:

$$J_{cg,blst} = \frac{1}{12}m_{blst}\left(H_{blst}^2 + W^2\right)$$ (11.16)

where $H_{blst} = m_{blst}/(W^2 \rho_{blst})$.

The mass moment of inertia for the entire structure is then found in Equation (11.5).

11.6.3 Spars

The spar platform typically consists of three distinct components: (1) the main hull, (2) a taper, and (3) a smaller diameter transition piece to which the tower is attached. There will also be ballast at the bottom of the main hull. The notations used in this section are illustrated in Figure 11.16.

For the spar to float at the desired depth, the total mass of the structure must be equal to the mass of the displaced water (Equation (11.17)), so the mass of the ballast can be estimated.

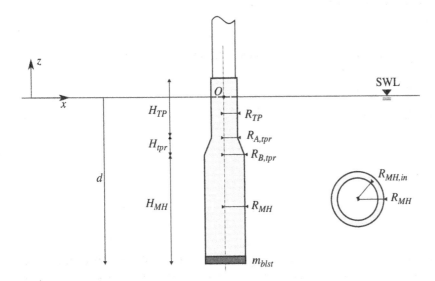

Figure 11.16 Sketch and notations for the study of the spar stability

$$\rho_w \nabla = m_{RNA} + m_{twr} + m_{TP} + m_{tpr} + m_{MH} + m_{blst} \tag{11.17}$$

Where subscripts, *TP*, *tpr*, *MH*, and *blst* refer to the transition piece, taper, main hull, and ballast, respectively.

11.6.3.1 Displaced Volume

The displaced volume of water for the structure is the summation of the volumes of the components that are submerged (main hull and taper) and the portion of the transition piece which is below sea level. The main hull and transition piece can be assumed to be cylindrical, and the taper is a frustum. The total displaced volume is then:

$$\nabla = \pi R_{MH}^2 H_{MH} + \frac{\pi H_{tpr}\left(R_{A,tpr}^2 + R_{A,tpr}R_{B,tpr} + R_{B,tpr}^2\right)}{3} + \pi R_{TP}^2 H_{TP,sub} \tag{11.18}$$

where R indicates radius, H is height; the subscripts A and B are for the top and bottom of the taper. The subscript *sub* indicates that portion of the transition piece length that is submerged.

When the draft of the spar is specified:

$$H_{TP,sub} = H_{MH} + H_{TP} - d \tag{11.19}$$

where d is the draft (a positive number).

11.6.3.2 Mass

The total mass of the turbine and platform is the same as the displaced mass:

$$m = \rho_w \nabla \tag{11.20}$$

The ballast mass is found from:

$$m_{blst} = m_{RNA} + m_{twr} + m_{tpr} + m_{TP} + m_{MH} - \rho_w \nabla \tag{11.21}$$

Most of the mass of the main hull is in the walls, so from Equation (D.7) the spar mass is:

$$m_{MH} = \rho_{steel}\pi H_{MH}\left(R_{MH}^2 - R_{MH,in}^2\right) \tag{11.22}$$

where $R_{MH,in} = R_{MH} - t_{wall,MH}$ is the inner radius of the spar, H is its height, and the mass of the base is ignored.

The mass of the transition piece is found similarly, so:

$$m_{TP} = \rho_{steel}\pi H_{TP}\left(R_{TP}^2 - R_{TP,in}^2\right) \tag{11.23}$$

The mass of the taper is found from Equations (D.10) to (D.12) with the appropriate values inserted:

$$m_{tpr} = \frac{\pi \rho H_{tpr}}{3}\left[\left(R_{A,tpr}^2 + R_{A,tpr}R_{B,tpr} + R_{B,tpr}^2\right) - \left(R_{A,tpr,in}^2 + R_{A,tpr,in}R_{B,tpr,in} + R_{B,tpr,in}^2\right)\right] \tag{11.24}$$

11.6.3.3 Water Plane Area

The water plane area for the spar is simply:

$$A_{wp} = \pi R_{TP}^2 \tag{11.25}$$

11.6.3.4 Center of Buoyancy

Since the main hull, taper and transition piece have different diameters, finding the center of buoyancy requires the displaced volume and the center of gravity of the displaced volume of each of them. In the absence of heeling, the distance from the keel to the center of buoyancy is:

$$z_{k-cb} = \left(V_{MH} z_{k-cg,MH} + V_{tpr} z_{k-cg,tpr} + V_{TP} z_{k-cg,TP} \right) / \left(V_{MH} + V_{tpr} + V_{TP} \right) \tag{11.26}$$

$$B = z_{k-cb} - d \tag{11.27}$$

11.6.3.5 Center of Gravity

The center of gravity of the overall spar-supported wind turbine is obtained analogously to that of a barge. The centers of gravity of the main hull and the transition piece are at their respective midpoints (Equation (D.8)). The center of gravity of the ballast (without heel) is also its midpoint as is given by Equation (11.28).

$$h_{cg,blst} = \frac{m_{blst}}{2\rho_{blst}\pi R_{MH}^2} \tag{11.28}$$

The center of gravity of the taper, $h_{cg,tpr}$, is found from Equation (D.14).

As with the barge, the centers of gravity of all the components are then referred to keel, $z_{k-cg,i}$.

The overall center of gravity with respect to the keel may be found from the products of component centers of gravity, as given in Equation (11.3). It may then be referred to the still water level if desired by subtracting the draft. For the spar, this is:

$$z_{k-cg} = \frac{m_{RNA} z_{k-cg,RNA} + m_{twr} z_{k-cg,twr} + m_{TP} z_{k-cg,TP} + m_{tpr} z_{k-cg,tpr} + m_{MH} z_{k-cg,MH} + m_{blst} z_{k-cg,blst}}{m_{RNA} + m_{twr} + m_{TP} + m_{tpr} + m_{MH} + m_{blst}} \tag{11.29}$$

11.6.3.6 Area Moment of Inertia

The area moment of inertia of the spar's transition piece where it passes through the SWL is small enough that it is often ignored, but when desired it can be found from:

$$I_{yy} = \pi R_{TP}^4 / 4 \tag{11.30}$$

11.6.3.7 Metacentric Height

The metacentric height is found from Equation (11.2), but because the water plane area is so small the value is very close to the difference between the center of buoyancy and the center of gravity:

$$\overline{GM} = \frac{I_{yy}}{\nabla} - (G - B) \cong B - G \tag{11.31}$$

11.6.3.8 Platform Pitch Angle

The pitch angle of the spar is found by equating the righting moments to the overturning moments. For the spar, the primary righting moment is due to the difference between the center of gravity and the center of buoyancy. As with the barge, a reaction force from the stationkeeping system is assumed to be applied at SWL. Accordingly, the angle is given by:

$$\theta = \tan^{-1}\left[\frac{F_T H_{hub}}{\rho_w g \nabla (B - G)} \right] \tag{11.32}$$

where F_T is the thrust force, H_{hub} is the height of the rotor above SWL, and B and G are the centers of buoyancy and gravity, respectively (both negative).

11.6.3.9 Mass Moments of Inertia

The mass moment of inertia of the taper may be found from Equations (D.15) or (D.17) with the appropriate values substituted.

The mass moments of inertia of the transition piece and the main hull about their centers may be found from Equation (D.9) for a cylinder, again with appropriate values substituted.

The mass moment of inertia of the ballast is found in Equation (D.9), except that in this case the hull is $R_{in} = 0$. Accordingly:

$$J_{cg,blst} = \frac{1}{12} m_{blst} \left(H_{blst}^2 + 3R_{MH}^2 \right) \tag{11.33}$$

where $H_{ballast} = m_{blst} / \left(\pi \rho_{ballast} R_{MH}^2 \right)$.

The mass moment of inertia of the entire turbine is found from Equation (11.5), expressed here as Equation (11.58).

$$\begin{aligned} J_{cg} = {} & J_{cg,RNA} + J_{cg,twr} + J_{cg,TP} + J_{cg,tpr} + J_{cg,MH} + J_{cg,blst} \\ & + m_{RNA} \left(z_{k-cg,RNA} - z_{k-cg} \right)^2 + m_{twr} \left(z_{k-cg,twr} - z_{k-cg} \right)^2 + T_{TP} \left(z_{k-cg,TP} - z_{k-cg} \right)^2 \\ & + m_{tpr} \left(z_{k-cg,tpr} - z_{k-cg} \right)^2 + m_{MH} \left(z_{k-cg,MH} - z_{k-cg} \right)^2 + m_{blst} \left(z_{k-cg,blst} - z_{k-cg} \right)^2 \end{aligned} \tag{11.34}$$

11.6.4 Semisubmersibles

Semisubmersible support structures, as previously discussed, consist of multiple hulls (offset columns), typically 3, spaced far enough apart that the water plane area is a major source of stability. In addition, there may be a close central column or hull. In the following discussion, we assume the 3-hull configuration with a central column. The notations used in this section are illustrated in Figure 11.17. The draft of each hull is d, the freeboard is f, and the length is $H_{hull} = d + f$.

In the following discussion, we assume that the wind is blowing parallel to a line connecting one of the hulls and the central column and perpendicular to a line between the other two hulls, which are downwind of the tower. Since the hulls are each on the vertex of an equilateral triangle the distance from any hull to the center of the platform is $R_{c-h} = R_{h-h} \sqrt{3}/3$, where R_{h-h} is the spacing between the centers of the hulls. Accordingly, the perpendicular distance from the center to the line of the two downwind hulls is $R_{h-h} \sqrt{3}/6$.

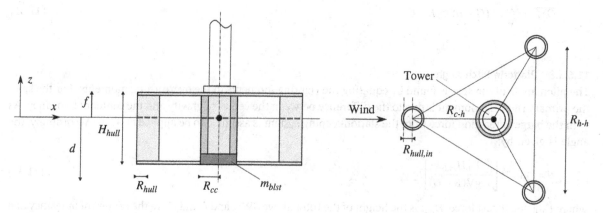

Figure 11.17 Sketch and notations for the study of the semisubmersible stability

11.6.4.1 Displaced Volume

The total displaced volume of the three main hulls and the central column, when the structure is horizontal, is:

$$\nabla = \pi\left(3R_{hull}^2 + R_{cc}^2\right)d \tag{11.35}$$

where R_{hull} and R_{cc} are the outer radii of the main hulls and central column, respectively.

11.6.4.2 Mass

The total mass of the turbine and platform is the same as the displaced mass:

$$m = \rho_w \nabla \tag{11.36}$$

The mass of the three hulls and central column (including top and bottom) is found using Equation (D.7), expressed here as Equation (11.37):

$$m_{all\ hulls} = \rho_{steel}\pi\left[3\left(H_{hull}R_{hull}^2 - H_{hull,in}R_{hull,in}^2\right) + \left(H_{hull}R_{cc}^2 - H_{hull,in}R_{cc,in}^2\right)\right] \tag{11.37}$$

where $R_{hull,in}$ and $R_{cc,in}$ are the inner radii of the main hulls and the central column.

Given the basic parameters of the platform and desired draft, the mass of the ballast may be found from:

$$m_{blst} = \rho_w \nabla - m_{RNA} - m_{tower} - m_{all\ hulls} \tag{11.38}$$

11.6.4.3 Water Plane Area

The water plane area is found by analysis of the geometry of the system, as:

$$A_{wp} = 3\pi R_{hull}^2 + \pi R_{cc}^2 \tag{11.39}$$

11.6.4.4 Center of Buoyancy

The hull and central column sides are all vertical without heel, so the center of buoyancy is simply:

$$B = -d/2 \tag{11.40}$$

11.6.4.5 Center of Gravity

By applying Equation (11.3), the overall center of gravity of the entire semisubmersible turbine, in absence of heel, is obtained as:

$$z_{k-cg} = \frac{m_{RNA}z_{k-cg,RNA} + m_{twr}z_{k-cg,twr} + m_{all\ hulls}z_{k-cg,all\ hulls} + m_{blst}z_{k-cg,blst}}{m_{RNA} + m_{twr} + m_{all\ hulls} + m_{blst}} \tag{11.41}$$

$$G = z_{k-cg} - d \tag{11.42}$$

11.6.4.6 Area Moment of Inertia

The area moment of inertia of the platform at the water line is:

$$I_{yy} = \frac{3\pi R_{hull}^4}{4} + 1.5\pi R_{hull}^2 R_{c-h}^2 + \frac{\pi R_{cc}^4}{4} \tag{11.43}$$

11.6.4.7 Metacentric Height

The metacentric height is given by Equation (11.2), expressed here as Equation (11.44).

$$\overline{GM} = \frac{I_{yy}}{\nabla} - (B - G) \tag{11.44}$$

11.6.4.8 Platform Pitch Angle
The structure will pitch slightly in the downwind direction when the turbine is operating. The upwind hull will rise slightly to have a decreased draft and the downwind hulls will submerge to have a greater draft. The net effect is to shift the center of buoyancy in the downwind direction.

Analogously to the barge, assuming that the stationkeeping system applies a reaction force at SWL, the platform pitch angle may be found from Equation (11.45):

$$\theta \cong \tan^{-1}\left[\frac{F_T H_{hub}}{\nabla \rho_w g \overline{GM}}\right] \tag{11.45}$$

11.6.4.9 Mass Moments of Inertia
The mass moment of inertia of each hull about its center of gravity and the central column are found from Equation (D.9), expressed here as Equations (11.46) and (11.47):

$$J_{cg,per\ hull} = \frac{1}{12} m_{hull}\left(3R_{hull}^2 + H_{hull}^2\right) - \frac{1}{12} m_{hull,in}\left(3R_{hull,in}^2 + H_{hull}^2\right) \tag{11.46}$$

$$J_{cg,cc} = \frac{1}{12} m_{cc}\left(3R_{cc}^2 + H_{hull}^2\right) - \frac{1}{12} m_{cc,in}\left(3R_{cc,in}^2 + H_{hull}^2\right) \tag{11.47}$$

From the parallel axis theorem and geometry, the total mass moment of inertia about the center of gravity of the three main hulls and the central column is:

$$J_{cg,all\ hulls} = 3J_{cg,hull} + 1.5 m_{hull} R_{c-h}^2 + J_{cg,cc} \tag{11.48}$$

Finding the mass moment of inertia of all the ballast about the ballast center of gravity involves a couple of steps. Assuming that the depth of ballast is the same in all the hulls and central column, then the ballast mass will be distributed among the hulls and the central column in proportion to their area as shown in Equations (11.49) and (11.50):

$$m_{blst,hull} = m_{blst} R_{hull}^2 / \left(3R_{hull}^2 + R_{cc}^2\right) \tag{11.49}$$

$$m_{blst,cc} = m_{blst} R_{cc}^2 / \left(3R_{hull}^2 + R_{cc}^2\right) \tag{11.50}$$

Analogously to Equation (11.48), the mass moment of inertia of the ballast is then:

$$J_{cg,blst} = 3J_{blst,hull} + J_{blst,cc} + 1.5 m_{blst,hull} R_{c-h}^2 \tag{11.51}$$

From Equation (11.5), the overall mass moment of inertia of the entire turbine with respect to its center of gravity is obtained as:

$$J_{cg} = J_{cg,twr} + J_{cg,blst} + J_{cg,all\ hulls} + m_{RNA}\left(z_{k-cg,RNA} - z_{k-cg}\right)^2 + m_{twr}\left(z_{k-cg,twr} - z_{k-cg}\right)^2$$
$$+ m_{all\ hulls}\left(z_{k-cg,all\ hulls} - z_{k-cg}\right)^2 + m_{blst}\left(z_{k-cg,blst} - z_{k-cg}\right)^2 \tag{11.52}$$

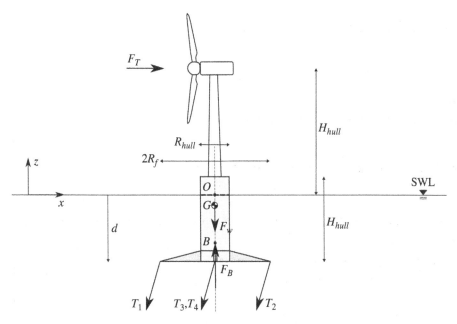

Figure 11.18 Free body diagram and notations for tension leg platform (TLP)

11.6.5 Tension Leg Platforms

The TLP is different from the others in that a primary source of stability is the force from the tendons. There is some contribution from the ballast, but the water plane area has little effect on stability. Typically, the single hull is of relatively small diameter, but the spokes extend well beyond the hull. These serve as attachments for the tendons and increase the effectiveness of the tendon forces in creating the moments to resist overturning. There may be a transition piece that extends above the top of the hull. For the discussion here, we ignore the mass of the spokes and tendons. A distinct transition piece is not considered.

The basic parameters of the TLP are summarized in Figure 11.18.

11.6.5.1 Displaced Volume
The displaced volume of the TLP (ignoring the spokes) is:

$$\nabla = \pi R_{hull}^2 d \tag{11.53}$$

11.6.5.2 Mass
The mass of the hull, including the top and bottom, is:

$$m_{hull} = \rho \pi \left(H_{hull} R_{hull}^2 - H_{hull,in} R_{hull,in}^2 \right) \tag{11.54}$$

where $R_{hull,in} = R_{hull} - t_{wall}$, $H_{hull,in} = H_{hull} - 2t_{wall}$.

The total mass of the platform, RNA, and tower is:

$$m_{TLP} = m_{RNA} + m_{twr} + m_{hull} + m_{blst} \tag{11.55}$$

Note that in the case of the TLP, the total mass is different from the displaced mass, $\rho_w \nabla$.

11.6.5.3 Water Plane Area

The water plane area is found analogously to the spar:

$$A_{wp} = \pi R_{hull}^2 \tag{11.56}$$

11.6.5.4 Center of Buoyancy

The hull sides are vertical, so the center of buoyancy, without heel, is:

$$B = -d/2 \tag{11.57}$$

Note that B will be lower when the tendons are connected than otherwise.

11.6.5.5 Center of Gravity

By applying Equation (11.3), the overall center of gravity of the TLP when the tendons are not connected to the turbine, and without heel, is:

$$z_{k-cg} = \frac{m_{RNA}z_{k-cg,RNA} + m_{twr}z_{k-cg,twr} + m_{hull}z_{k-cg,all\ hulls} + m_{blst}z_{k-cg,blst}}{m_{RNA} + m_{twr} + m_{hull} + m_{blst}} \tag{11.58}$$

$$G = z_{k-cg} - d \tag{11.59}$$

As with the center of buoyancy, G will be lower when the tendons are connected.

11.6.5.6 Area Moment of Inertia

The area moment of inertia of the platform at the water line is:

$$I_{yy} = \frac{\pi R_{hull}^4}{4} \tag{11.60}$$

11.6.5.7 Metacentric Height

The metacentric height is given by Equation (11.2), expressed here as Equation (11.61).

$$\overline{GM} = \frac{I_{yy}}{\nabla} - (B - G) \tag{11.61}$$

The metacentric height of the TLP is primarily relevant when the tendons are not connected.

11.6.5.8 Forces in the Tendons

For this discussion, we assume that all of the parameters have been specified initially, including the mass of the ballast. In the no thrust condition, the total force in all the tensions is set to obtain the desired draft (Equation (11.62)).

$$F_{tendons} = g(\rho_w \nabla - m_{TLP}) \tag{11.62}$$

Depending on the amount of ballast and thus the location of the center of gravity, a TLP may or may not be stable when the turbine is not operating without the tendons in place. That will be apparent by calculating the metacentric height. During operation, however, the tendons must be connected. When the tendons are in tension, they will stretch somewhat. Under these conditions, the tendons will stretch more on the upwind side of the turbine than on the downwind side. The difference in the amount of extension of the tendons will determine the platform pitch angle. In the following, we assume that there are four tendons, and that the wind is parallel to pairs of tendons on each side.

Referring to Figure 11.18, for static stability the sum of the forces in all directions must equal to zero. Therefore, $F_B - [F_W + T_{1,z} + T_{2,z} + 2T_{3,z}] = 0$ and $F_T - [T_{1,x} + T_{2,x} + 2T_{3,x}] = 0$. Here, $F_W = g \, m_{TLP}$ is the weight of the entire structure, $F_B = \nabla \rho_w g$ is the buoyancy force, R_f is the distance from the center of the hull to the fairleads, and T_1, T_2, T_3, and T_4 are the tensions in the tendons. Note that $T_3 = T_4$. Assuming that the platform pitch angle θ is small, we can take the moments about the upwind fairlead, resulting in:

$$F_B R_f - [F_W + 2T_{3,z} + 2T_{2,z}]R_f - F_T H_{hub} = 0 \tag{11.63}$$

where the magnitudes of the forces are used, and the signs indicate the direction of the moments.

The elongation of the tendons is a function of the stiffness as well as the applied force, so in general:

$$\Delta l = T/k \tag{11.64}$$

where $k = EA/l$, N/m, E = modulus of elasticity, N/m²; A = cross-sectional area, m² and l = tendon length, m (Note that the product EA is known as extensional stiffness).

The elongation on the upwind tendons will be:

$$\Delta l_1 = T_1/k \tag{11.65}$$

On the downwind side, the elongation will be:

$$\Delta l_2 = T_2/k \tag{11.66}$$

11.6.5.9 Platform Pitch Angle

The difference in the upwind and downwind extensions will determine the platform pitch angle:

$$\theta = \tan^{-1}\left(\frac{T_1 - T_2}{k2R_f}\right) \tag{11.67}$$

11.6.5.10 Mass Moment of Inertia

The mass moment of inertia of the TLP is found analogously to that of the spar. It is less directly relevant than the mass moments of inertia of the other platforms during operation because of the effect of the tendons.

11.7 Motions of Floating Wind Turbines

A floating wind turbine can move back and forth in one of the six major directions, or a combination of them. Taken individually, these are surge, sway, heave, roll, pitch, and yaw (see also Chapter 4). For ships, these are defined relative to the normal direction of motion. For a wind turbine, they are defined relative to wind direction, with the x (longitudinal) axis parallel to the SWL and in alignment with the wind, y (lateral) axis also parallel to the sea surface but perpendicular to the x axis and the vertical axis z, which is perpendicular to both. Using these assumptions, surge is motion in the x direction, sway is in the y direction (sideways), heave is vertical. Roll is motion around the x axis, pitch is motion around the y axis, and yaw is motion around the vertical (z) axis. See also Figure 11.19. The pitch and yaw motions of the platform should not be confused with the blade pitch and the nacelle yaw.

There are numerical indices that indicate the direction of motion; by convention the indices are those indicated in Table 11.2.

The indices listed in Table 11.2 are convenient to describe the motion of the six degrees of freedom, placed into a six-vector. One can define mass, stiffness, and damping matrices of dimension 6×6 to describe the rigid body

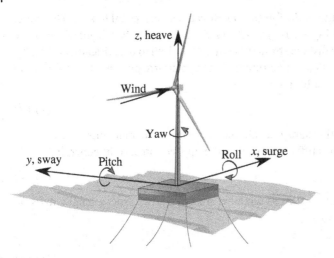

Figure 11.19 Degrees of freedom of floating offshore wind turbine

Table 11.2 Indices of motion of floating object

Index	Motion
1	Surge
2	Sway
3	Heave
4	Roll
5	Pitch
6	Yaw

motion of a FOWT (see also Chapter 4 for the dynamics of multiple degrees of freedom written in matrix form). The matrix elements will have double subscripts (for row and column). The subscript "*ij*" indicates the load in the degree of freedom "*i*," for a motion (displacement, velocity, and acceleration) in the degree of freedom "*j*." For example, the subscripts for heave force related to heave motion will be "33."

11.7.1 Oscillatory Motions of the Floating Platform

The fundamental oscillatory motions of a floating platform can be described by the same equation as for spring-mass vibration introduced in Chapter 4, with some adjustments. The equation is repeated here in general form:

$$f_n = \frac{1}{2\pi}\sqrt{\frac{k}{m}} \tag{11.68}$$

where f_n is the natural frequency (Hz), k is a spring constant (N/m), and m is mass (kg).

For roll, pitch, and yaw, the relevant motion is in rotation, so the appropriate equation is:

$$f_n = \frac{1}{2\pi}\sqrt{\frac{k_\theta}{J}} \tag{11.69}$$

where k_θ is the rotational spring constant, or stiffness, (Nm/rad) and J is the mass moment of inertia with respect to the center of gravity (kg m^2).

For an object in water, the stiffness depends on the effect of the motion. For example, the stiffness in heave motion is essentially the force of gravity acting on the mass which, as we will see, comes into play via the water-plane area. If a stationkeeping system is present, it will also contribute to the heave stiffness. The stiffness in surge of a moored floater is the restoring force of the moorings. The stiffness in pitch or roll is the differential force due to the rotational motion and is essentially the same as that described above for stability of a platform in pitch.

The mass in Equation (11.68) includes the mass of the floating object itself, but it also includes the "added mass," which to some extent, represents the mass of the water that is pushed out of the way as a result of the motion, as discussed in the next section 11.7.1.1. Similarly, there is added mass moment of inertia to consider.

11.7.1.1 Added Mass

When a floating object is accelerated in the water, some of the water adjacent to the object must be moved as well. It requires additional force to move this water. In the analysis of the motion, the mass of the water affected is accounted for by the term "added mass." In more rigorous terms, added mass results from pressure variations around the body which are caused by radiation of outgoing waves away from the body. Added mass is typically dependent on the frequency of the motion. The force required to move the body, F_{water}, is given by Equation (11.70):

$$F_{water} = (\rho_w \nabla + m_{add})a \tag{11.70}$$

where $\rho_w \nabla$ is displaced mass, m_{add} = added mass and a is acceleration.

The added mass can be accounted for by using an added mass coefficient, which is defined by:

$$c_{add} = m_{add}/\rho_w \nabla \tag{11.71}$$

The added mass coefficient depends on the shape of the object, the direction of motion, and the frequency of oscillation. Typical values are $c_{add,11} = 1$ for a cylinder in surge, $c_{add,33} = D/3H$ for a vertical cylinder in heave, where D = diameter and H = height, and $c_{add} = 0.5$ for a sphere. A comprehensive table of added mass coefficients may be found in DNV-RP-C205 (DNV, 2014).

There is a similar effect in rotational motion, which is accounted for by an added mass moment of inertia for pitch and roll. This term is on the order of 0.5 and is referred to the moment of inertia of the platform. Again, the actual value depends on the circumstances.

11.7.2 Natural Frequencies of Floating Offshore Wind Turbines

11.7.2.1 Heave

When an object is pushed down somewhat, the force to do so is given by the weight of the additional water displaced. This force will be proportional to the water plane area A_{wp}. The implied spring constant (stiffness) is then as given in Equation (11.72).

$$K_{33} = \rho_w g A_{wp} \tag{11.72}$$

The natural frequency in heave f_{n3} is given by Equation (11.73):

$$f_{n3} = \frac{1}{2\pi}\sqrt{\frac{K_{33}}{\rho_w \nabla + A_{33}}} = \frac{1}{2\pi}\sqrt{\frac{\rho_w g A_{wp}}{\rho_w \nabla + A_{33}}} \tag{11.73}$$

where $A_{33} = c_{add,33}\rho_w \nabla$.

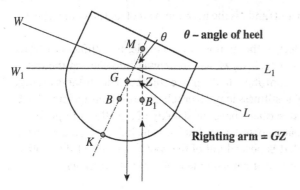

θ – angle of heel

Figure 11.20 Righting arm, metacentric height and centers of gravity and buoyancy

Righting arm = GZ

11.7.2.2 Roll Natural Frequency

The restoring force in roll is given by the buoyancy force $\rho_w g \nabla$ times the righting arm $\overline{GZ}$, as illustrated in Figure 11.20, where $\overline{GZ} = \overline{GM} \sin\theta$.

For small angles $\sin\theta \approx \theta$ and the stiffness K_{44} becomes

$$K_{44} = \rho_w g \nabla \overline{GM} \tag{11.74}$$

The natural frequency of the roll motion f_{n4} is given by Equation (11.75).

$$f_{n4} = \frac{1}{2\pi}\sqrt{\frac{K_{44}}{J_4 + A_{44}}} = \frac{1}{2\pi}\sqrt{\frac{\rho_w g \nabla \overline{GM}}{J_4 + A_{44}}} \tag{11.75}$$

where J_4 is the mass moment of inertia of the entire structure about the roll axis (which can be assumed to pass through the center of mass) and $A_{44} = c_{add,44} J_4$.

11.7.2.3 Pitch

The stiffness for pitch motion is analogous to that for roll. Since FOWTs are normally symmetrical, stiffnesses and natural frequencies for pitch and roll will be the same (subject to the orientation of the RNA). That is, $K_{55} = K_{44}$, $f_{n5} = f_{n4}$ and $J_5 = J_4$.

11.7.2.4 Surge/Sway/Yaw

Surge, sway, and yaw natural frequencies are largely dependent on the characteristics of the mooring system and are beyond the scope of this text.

11.8 Stationkeeping Systems for Floating Offshore Wind Turbines

Regardless of the type of support structure that may be used with a FOWT, excessive lateral movement of the structure must be prevented. This is accomplished by stationkeeping facilities, which include anchors and mooring lines or tendons.

11.8.1 Mooring Lines

A mooring line is a means of connecting a floating platform to an anchor on the seabed. For an offshore wind turbine, the mooring line could be a steel chain, a wire rope, or a synthetic fiber rope. One or more of these could be used in a particular application. Generally, chains are used in shallower water (<100 m), whereas either chains or wire ropes are used at greater depths.

Two chain constructions are stud link chain and studless link chain. The studless link chain is most common for permanent moorings.

Wire rope is comprised of strands of steel wire, wrapped together around a core. In general, wire rope is lighter than the equivalent chain but is more elastic; a disadvantage of wire rope is that wire rope is more vulnerable to corrosion than is chain.

Synthetic fiber rope mooring lines have many promising attributes, but they are generally considered too expensive and unnecessary for shallow water. Whether they will be used for mooring offshore turbines in the foreseeable future is unknown.

11.8.1.1 Mooring Line Connections

A mooring line is connected at two points: at a fairlead on the vessel (or at a cleat just inboard of the fairlead) and the anchor. There are many types of anchors (see Section 11.8.2), but one of the most widely used types is referred to as a "drag anchor." Moorings using drag anchors are often referred to as slack moorings. This is the type that is commonly used on ships. The force applied by the mooring line to the anchor in this case is parallel to the seabed so that it tends to "drag" the anchor along the bottom. The anchor is usually placed some distance from the vessel being anchored so that the length of the line is considerably greater than the water depth. The ratio between the line length and the water depth is known as the scope. In general, scopes in slack moorings are between 3 and 10. The shape of the portion of a slack mooring line of uniform mass density that is not on the seabed is a catenary. In an idealized slack mooring, the tangent of the mooring line at the anchor is parallel to the seabed. The distance between the attachment points (and thus the length of the mooring line) will depend on the tension in the line as well as its weight. Thus, greater tension is associated with longer lines (and correspondingly greater scope). The characteristics of catenary lines are described in more detail in Section 11.8.1.2.

For other types of moorings, there may be a significant vertical component in the mooring line at the point of connection to the anchor. These are known as semi-taut or taut moorings. TLPs have taut moorings (tendons) with only a vertical force component.

11.8.1.2 Catenary Mooring Lines

Mooring lines that have a uniform mass density and are subjected only to tension forces at the ends form a catenary shape (sometimes referred to as a skipping rope shape); see Figure 11.21. Ship-type anchors typically do not support vertical forces; therefore, it is usually required to have a large section of the mooring line resting on the seabed.

The following discussion presents the mooring line equations that are solved to find the shape of the mooring line between the fairlead and the point of touchdown on the sea bed. These equations are nonlinear and algebraic and typically involve solving for the horizontal and vertical tension in the mooring line as a function of the horizontal and vertical distance between the fairlead and point of touchdown. These distances will vary throughout a time domain simulation. The mooring line is tangent to the sea floor at the point of touchdown. (Henceforth, the term catenary mooring will apply to this condition, unless otherwise noted.) In this situation, the height, z, of a point on the line above the horizontal at a distance x from the point of touchdown is given by Equation (11.76):

$$z = a\left(\cosh\left(\frac{x}{a}\right) - 1\right) \tag{11.76}$$

The catenary parameter a (m) is the ratio of the tension in the line at the point of touchdown, T_a (N) to the unit weight density of the mooring line; that is:

$$a = \frac{T_a}{\bar{\rho}'g} \tag{11.77}$$

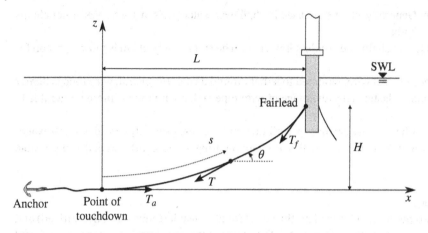

Figure 11.21 Catenary mooring line

where $\tilde{\rho}'$ is the effective mass density per unit length (kg/m) of the mooring line. Note that the expression for a takes into account the buoyancy of the water. The effective density is derived from the mass density in air $\tilde{\rho}$ as given in Equation (11.78):

$$\tilde{\rho}' = \tilde{\rho}(\rho_{steel} - \rho_w)/\rho_{steel} \tag{11.78}$$

where ρ_{steel} is the density of the line and ρ_w is the density of seawater. The distance along the line $s(x)$ is given by Equation (11.79):

$$s(x) = a\sinh\left(\frac{x}{a}\right) \tag{11.79}$$

From Equation (11.76), it is apparent that the horizontal distance L from the fairlead to the point of touchdown is:

$$H = a\left(\cosh\left(\frac{L}{a}\right) - 1\right) \tag{11.80}$$

Similarly, from Equation (11.79), the total length of the suspended portion (i.e., not lying on the seabed) of the mooring line s is:

$$s = a\sinh\left(\frac{L}{a}\right) \tag{11.81}$$

In some situations, the length of the mooring line and its mass density are known, as are the horizontal and vertical distances L and H from the point of touchdown to the fairlead on the platform, but the tension in the mooring line at the point of touchdown, T_a, is unknown. In this case, the tension may be found from the following implicit equation:

$$\sqrt{s^2 - H^2} = 2a\sinh\left(\frac{L}{2a}\right) \tag{11.82}$$

The tension in the mooring line, T_f, and its angle below horizontal, θ, at the fairlead may be found by applying the principles of statics. The force components are given in Equations (11.83) and (11.84); the tension and the angle are found in Equations (11.85) and (11.86).

$$T_{f,x} = -T_a \tag{11.83}$$

$$T_{f,z} = -\tilde{\rho}'gs \tag{11.84}$$

$$T_f = \sqrt{T_{f,z}^2 + T_{f,x}^2} \tag{11.85}$$

$$\theta = \tan^{-1}(T_{f,z}/T_{f,x}) \tag{11.86}$$

The weight (in water), W, of the suspended portion of the line is the magnitude of $T_{f,z}$

$$W = |T_{f,z}| = \tilde{\rho}'gs \tag{11.87}$$

Example: Consider the case of a wind turbine mounted on a spar and held in place with catenary mooring lines in which the seabed is 200 m below the fairleads. The lines consist of 87 mm chain, which has a linear mass density of 176.7 kg/m. The seawater density is 1025 kg/m³ and that of the steel chain is 7861 kg/m³. Suppose that the turbine is not operating and that the horizontal pretension at the point of touchdown is $T_a = 1000$ kN. It may be readily verified using Equations (11.76)–(11.85) that the constant $a = 663.4$ m, the distance to the points of touchdown is 503.0 m, the length of the mooring line is 552.6 m, their effective weight (submerged) is 833.0 kN, and the resultant tension in the mooring lines at the fairleads is 1301.5 kN.

11.8.1.3 *Catenary Mooring Lines under Differential Loading

If a FOWT is held in place by multiple mooring lines of equal length and if the turbine is not operating, the lines will be equally loaded, and the turbine will remain in the middle of all the anchors. If the turbine is operating, however, then there will be a differential load on the mooring lines, and the turbine will be displaced in the downwind direction. The upwind lines will become more taut and will be lifted off the seabed closer to the anchor; the downwind lines will sink to the seabed farther the anchor.

Suppose that the platform is displaced some distance due to a net external force F_T (due, for example, to the wind) along a line between two anchors in the direction of anchor 2 (downwind). The mooring line attached at the upwind anchor (anchor 1) will begin to lift off the seabed, and the tension there will increase. We may assume that the horizontal force in the mooring line is now $T_1 = T_{a,0} + F_T$ where $T_{a,0}$ is the initial horizontal pretension. Assuming that the depth from the fairlead to the seabed does not change, Equation (11.80) can be used for the new shape of mooring line 1, as shown in Equation (11.88):

$$H = a_1\left(\cosh\left(\frac{L_1}{a_1}\right) - 1\right) \tag{11.88}$$

where $a_1 = T_1/\tilde{\rho}'g$ and L_1 is the distance from the new point of touchdown to fairlead 1.

Similarly, the equation for the downwind mooring line is Equation (11.89):

$$H = a_2\left(\cosh\left(\frac{L_2}{a_2}\right) - 1\right) \tag{11.89}$$

where $a_2 = T_2/\tilde{\rho}'g$, $T_2 = T_{a,0} - F_T$, and L_2 is the distance from fairlead 2 to the new point of touchdown of mooring line 2.

The length of each mooring line on the seabed will also change. There will be less of line 1 on the seabed and more of line 2, so the total weight of the lines supported by the floating platform will change as well. In addition, there will be a net moment applied due to the differential forces. In most cases, however, the effect will be minor.

The length, s_1, of the portion of the upwind line that is suspended from the platform may be found from Equation (11.90):

$$s_1 = a_1 \sinh\left(\frac{L_1}{a_1}\right)$$

(11.90)

The suspended portion, s_2, of line 2 may be found analogously.

Example: Consider the same situation as the previous example, but now suppose that the thrust due to the turbine when operating is such that the net increase in force on the spar in the downwind direction is 500 kN. Assume that the total length of the mooring lines is 900 m, which is more than long enough that there is always a portion on the seabed. The solution is implicit but straightforward. The results are summarized in Table 11.3. The shapes of both lines are illustrated in Figure 11.22, together with that of the lines from the symmetrical situation.

11.8.1.4 Semi-taut Mooring Lines

Semi-taut mooring lines are basically similar to catenary lines, but they rise above the seabed at the anchor. The effect is that there is always a vertical component of the force at the anchor, so that needs to be considered in the selection of an anchor. The benefit is that the distance from the moored vessel to the anchors is invariably less than with catenary lines. In addition, vessels are subject to shorter excursions.

Table 11.3 Summary results for catenary mooring lines example

Line	A (m)	L (m)	s (m)	W (kN)	T_f (kN)
1	995.1	620.8	661.8	997.7	1,801
2	331.7	348.0	415.5	626.4	801

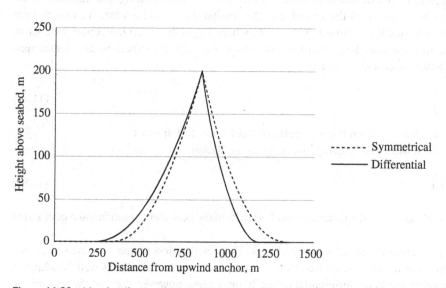

Figure 11.22 Mooring lines under symmetrical and differential loading

11.8.1.5 Tendons

Mooring lines for TLPs are known as tendons. Tendons for TLPs used in the oil and gas industry are typically fabricated from steel tubes, 0.75–1.0 m in diameter and 10 cm thick. On the order of 16 tendons may be used. Tendons are used in taut moorings and allow little or no horizontal displacement of the vessel from the anchors that are holding it in position. The dimensions of tendons for a wind turbine platform would depend on the design; see, for example, Global Security (2023).

11.8.2 Anchors

Anchors are used to attach the mooring line to the seabed. The resisting force provided by an anchor is often referred to as a multiple of the weight of the anchor. Typical ranges are from 1 to 10 times, depending on the type of anchor and the soil.

There are a wide variety of anchors, but there are two basic types: temporary and permanent. The ones that are likely to be used with offshore wind turbines are of the permanent type, whereas installation ships and service vessels may use temporary anchors. A few of the options for permanent anchors are described briefly here and some are illustrated in Figure 11.23.

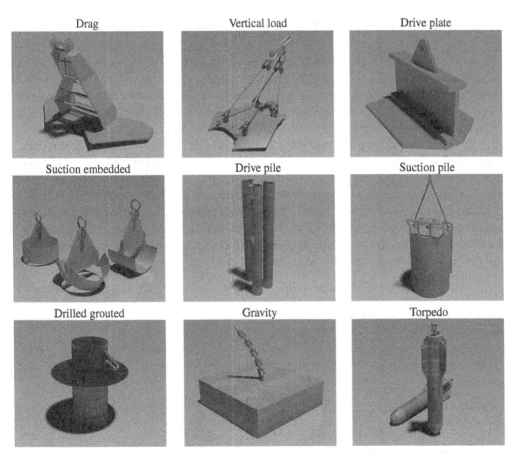

Figure 11.23 Typical anchors considered for offshore wind turbines (*Source:* Joshua Bauer. Reproduced from U.S. Department of Energy)

The **mushroom anchor**, as its name implies, has a mushroom shape. It is particularly suitable in sandy soil. The head becomes buried in the sand, and resistance is provided by the weight of the sand above the anchor.

A **deadweight**, "clump" or gravity, anchor relies on the weight alone to provide resistance to the motion of the ship or floater. The earliest anchors were actually of this type. Modern gravity anchors are usually large blocks of reinforced concrete.

A **screw** anchor is screwed into the seabed to form a permanent mooring.

A **steel plate** anchor is a type of anchor that may be driven into the sea floor to a considerable depth. They provide resistance in any direction including uplift. They are installed by attaching the anchor to a structural steel member, called a follower, and driving the follower with conventional pile driving equipment. The follower is removed after installation. Dimensions can exceed 5 m and the weight can be over 22 tonnes.

Propellant anchors are anchors that are installed by free fall or by explosive. The flukes are set to resist pull-out once installed. These anchors can be installed at considerable depth and may have capacities of up to 300 tonnes (2943 kN).

Suction anchors are essentially the same as suction caissons, but smaller; these were described in Chapter 10.

Piles may be used as anchors. They may be either driven or drilled, depending on the seabed conditions. See Chapter 10 for more information on piles.

11.9 Sample Calculations for Typical Floating Offshore Wind Turbines

The following are examples in which a wind turbine is supported by various types of platforms. In all cases, the turbine used is the offshore version of the NREL 5-MW turbine on a 77.6 m, uniformly tapered steel tower with base diameter of 6.5 m, mounted on the platform at 10 m above the sea surface. The basic parameters of the turbine were given in Chapter 1; the important ones for these examples are repeated in Table 11.4. The table also includes derived values, such as tower mass, center of gravity (with respect to the base), and mass moment of inertia (about the center of gravity). The latter were found from Equations (D.12), (D.14), and (D.17). The mass of the RNA can be assumed to be concentrated at one point 2.4 m above the top of the tower, i.e., 90 m above the SWL. When applicable, the turbine is assumed to be operating with a rotor thrust of 800 kN.

For the following examples, it assumed that the density of the steel in the platform is 8500 kg/m³, that of the ballast is 7700 kg/m³ and that of seawater is 1025 kg/m³. The thickness of the walls, top and bottom of the hulls in all cases are assumed to be 0.015 m. Freeboard f for all examples is 10 m. In any figures or equations, it is assumed that the x axis points in the downwind direction, the z axis is vertical, and the y axis is defined by the right-hand rule. All centers of gravity, cg_i, are referred to the still water level.

Table 11.4 Parameters for reference offshore wind turbine

Parameter	Symbol	Value	Units
RNA mass	m_{RNA}	350	t
RNA height	H_{RNA}	2.4	m
Tower mass	m_{twr}	249.6	t
Tower center of gravity	cg_{twr}	33.35	m
Tower mass moment of inertia	$J_{cg,twr}$	1.202×10^8	kg m²
Nominal thrust	F_T	800 000	N

11.9.1 Barge

The barge in this example is square with width and length equal to 32 m, the freeboard is 10 m, and the draft is 5 m. All the important parameters are summarized in Table 11.5.

Calculations were done using the equations in Sections 11.6.2 and 11.7.2. The added mass coefficient in heave was 2.0 and that for pitch was 0.5. The added mass moment of inertia in this case was based on that for the total structure. The results are summarized in Table 11.6.

11.9.2 Spar

This example considers the reference offshore turbine mounted on a simplified spar, similar to the one prepared for the IEA Offshore Code Collaboration (OC3) project; see Jonkman (2010). The properties of the spar are summarized in Table 11.7.

Calculations were done with equations in Sections 11.6.3 and 11.7.2. The added mass coefficients were $c_{33} = 0.0261$ and $c_{55} = 0.5$, as suggested following Equation (11.71). Results are summarized in Table 11.8. All elevations are given with respect to still water level. It may be noted in this case that the contribution of the water plane area was very small.

Table 11.5 Parameters for barge example

Parameter	Symbol	Value	Units
Draft	d	5	m
Width	W	32	m
Length	L	32	m
Steel plate thickness: wall, top, bottom	t	0.015	m

Table 11.6 Summary of results for barge

Parameter	Symbol	Value	Units
Displacement	∇	5120	m³
Hull mass	m_{hull}	505	t
Ballast mass	m_{blst}	4143	t
Total mass	m	5248	t
Center of gravity overall	G	4.23	m
Center of buoyancy	B	−2.5	m
Area moment of inertia at waterline	I_{yy}	87 381	m⁴
Metacentric height	$\overline{GM}$	10.3	m
Platform pitch	θ	7.8	deg
Mass moment of inertia	J_{cg}	3.56×10^9	kg m²
Heave natural frequency	f_{n3}	0.129	Hz
Pitch natural frequency	f_{n5}	0.05	Hz

Table 11.7 Properties of simplified spar

Parameter	Symbol	Value	Units
Draft (depth to bottom of spar relative to SWL)	d	120	m
Transition piece height	H_{TP}	10	m
Transition piece radius	R_{TP}	3.25	m
Transition piece wall thickness	t_{TP}	0.015	m
Taper height	H_{tpr}	10	m
Taper upper radius	$R_{A,tpr}$	3.25	m
Taper lower radius	$R_{B,tpr}$	4.7	m
Taper wall thickness	t_{tpr}	0.015	m
Main hull height	H_{MH}	110	m
Main hull radius	R_{MH}	4.7	m

Table 11.8 Summary of results

Parameter	Symbol	Value	Units
Displacement	∇	7638	m^3
Hull mass (inc. taper and transition piece)	m_{hull}	474	t
Ballast mass	m_{blst}	6759	t
Total mass	m	7829	t
Center of gravity of hull	cg_{hull}	−65.0	m
Center of gravity of ballast	cg_{blst}	−113.6	m
Center of gravity overall	G	−96.1	m
Center of buoyancy	B	−65.0	m
Area moment of inertia at waterline	I_{yy}	87.62	m^2
Metacentric height	$\overline{GM}$	31.2	m
Steady state platform pitch angle	θ	1.72	deg
Mass moment of inertia of entire turbine	J_{cg}	2.065×10^{10}	kg m^2
Heave spring constant	K_{33}	3.34×10^5	N/m
Pitch spring constant	K_{55}	2.39×10^9	N/m
Heave natural frequency	f_{n33}	0.032	Hz
Pitch natural frequency	f_{n55}	0.044	Hz

11.9.3 Semisubmersible

The semisubmersible wind turbine in this example is a simplified version of the one introduced in Section 11.3.1. It consists of three hulls and a central column on which the turbine is mounted. The hulls all extend 10 m above sea level and have a draft of 20 m. The hulls are all centered on the vertices of an equilateral triangle. The hulls are also assumed to be connected by a steel framework, but that is not included in this example. There is ballast in each hull. The key parameters are summarized in Table 11.9.

Table 11.9 Parameters of semisubmersible platform

Parameter	Symbol	Value	Units
Draft	d	20	m
Hull radius	R_{hull}	7.5	m
Central column radius	R_{cc}	6.5	m
Hull center-center	$R_{h\text{-}h}$	29	m
Wall thickness	t_w	0.015	m

Table 11.10 Summary of results for the semisubmersible

Parameter	Symbol	Value	Units
Mass of all hulls	m_{hull}	865	t
Ballast mass	$m_{ballast}$	12 125	t
Total mass	m	13 589	t
Displacement	∇	13 258	m³
Center of gravity of all hulls	cg_{hull}	−10	m
Center of gravity of ballast	cg_{blst}	−18.8	m
Center of gravity overall	G	−14.0	m
Center of buoyancy	B	−10.0	m
Area moment of inertia	I_{yy}	80 999	m⁴
Metacentric height	$\overline{GM}$	10.1	m
Platform pitch	θ	1.57	deg
Mass moment of inertia	J_{cg}	6.81×10^9	kg m²
Heave natural frequency	$f_{n,33}$	0.091	Hz
Pitch natural frequency	$f_{n,55}$	0.058	Hz

Calculations were done in accordance with equations in Sections 11.6.4 and 11.7.2. The main results are summarized in Table 11.10.

11.9.4 Tension Leg Platform

The TLP in this example is a simplified version of the one in Matha (2009). The single hull provides the buoyancy, and the extended arms help to minimize the required force in the tendons to keep the platform pitch to a suitable value. The parameters of platform are summarized in Table 11.11.

Calculations were done in accordance with Section 11.6.5. Assessment of the natural frequency of the structure was not undertaken due to the effect of the tendons. Results are summarized in Table 11.12.

11.9.5 Observations on the Platforms

The platforms in the examples in Sections 11.9.1–11.9.4 are only illustrations and by no means optimized. Nonetheless, there are some interesting observations that can be made. Most obviously, all of the platforms are initially

Table 11.11 Properties of sample tension leg platform

Parameter	Symbol	Value	Units
Water depth	$depth$	−200	m
Draft	d	30	m
Hull radius	R_{hull}	9	m
Arm radius	R_{spoke}	18	m
Ballast mass	m_{blst}	6000	t
Tendon length	L	170	m
Tendon extensional stiffness	EA	1.5×10^8	N

Table 11.12 Summary of results from TLP example

Parameter	Symbol	Value	Units
Displacement	∇	7634	m^3
Mass of hull	m_{hull}	353	t
Ballast mass	m_{blst}	6000	t
Total mass	m	6953	t
Center of gravity overall	G	−19.0	m
Center of buoyancy	B	−15	m
Area moment of inertia	Iyy	5153	m^4
Metacentric height	$\overline{GM}$	4.66	m
Tension, upwind tendons	$F_{tdn,up}$	2414	kN
Tension, downwind tendons	$F_{tdn,down}$	1863	kN
Platform pitch	θ	1.0	deg
Surge	Δx	2.95	m

stable, and they all have platform pitch angles during operation of less than 10°; all except the barge have a platform pitch angle of less than 2°. Total mass ranged from a low of 5248 tonnes for the barge to 13 589 tonnes for the semisubmersible. The hull mass ranged from a low of 353 tonnes for the TLP to a high of 865 tonnes for the semisubmersible. The ballast in every case accounted for the majority of the mass, ranging from a low of 4143 tonnes for the barge to a high of 13 589 tonnes for the semisub. These results indicate that structural mass of the floating platforms should be roughly comparable to that of the rotor nacelle assembly and the tower, but the ballast accounted for at least 75% of the total mass.

Another observation has to do with draft. Most deep-water ports have channels that are less than 16 m deep; few are deeper than 20 m. The barge has a very shallow draft and could be towed out from nearly any port, even when ballasted. The spar's draft is so deep that there are very few ports in the world from which it could be deployed while vertical. The semisubmersible could be towed out from most ports if only partially ballasted. The TLP could not be deployed from most ports while ballasted, and without ballast, it would likely be unstable and thus require a special vessel for assistance (as noted previously).

Optimization would likely improve the designs significantly in many ways, but the fundamental requirements of buoyancy, stability, and structural integrity need to be accommodated in any case. The examples did not include the stationkeeping system, so those would also need to be considered in any complete design.

11.10 *Coupled Aero/Hydro/Structural Dynamics of Floating Offshore Wind Turbines

The full dynamic behavior of FOWTs can be very complicated, because the support structure can move in any of the six ways summarized in Section 11.7. That motion could also affect the aerodynamics as well as the structural response. For example, if the rotor is moving upwind the wind speed would be higher than it otherwise would be, if moving downwind the wind speed would be lower. These motions need to be accounted for in any structural dynamics model for the results to be meaningful. The controller could have a significant effect here.

One important topic in assessing the response of a FOWT is the response amplitude operator or RAO. The RAO quantifies the dynamic response of the system to varying inputs. It is equivalent to a transfer function as used in the analysis of linear systems; see Appendix B. It is used to elucidate the response as a function of frequency of the excitation. As will be recalled from Chapter 6, the external conditions (primarily wind and waves) have stochastic characteristics that are described by power spectral densities (psd). The response to those conditions may be described by a cross-spectral density. The RAO provides a measure of the response to the input as a function of frequency. It is defined as the ratio of the output cross-spectral density to the input power spectral density for a given type of excitation. For example, the RAO of platform pitch in response to wave motion would be defined as:

$$RAO = \frac{S_{pw}(f)}{S_{ww}(f)} \tag{11.91}$$

where $S_{ww}(f)$ is the power spectral density of the waves, and $S_{pw}(f)$ is the cross-spectral density of the waves and platform pitch.

An example of the magnitude of an RAO for platform pitch, surge, and heave for a spar is shown in Figure 11.24. The responses shown in this case are only in response to the waves and for a rigid turbine. As may be expected, the RAO has peaks at the corresponding natural frequencies of the structure. The situation is more complicated when the turbine is flexible and the wind (and thus rotor thrust) is included in the input. An illustration of the RAO, in that case, is shown in Figure 11.25. These examples are for the NREL 5-MW wind turbine on a spar.

Further discussion of this subject is beyond the scope of this text. The reader could refer to Sebastian and Lackner (2013), Jonkman (2010), and Jonkman and Matha (2010) among other sources for more discussion of this topic.

11.10.1 Example: Two-degrees of Freedom Model of a Floating Wind Turbine

In this section, we model a floating wind turbine using two degrees of freedom (DOF): the platform surge (horizontal displacement), x, and the platform pitch, θ. The vertical coordinate is noted z, assumed to be zero at the mean sea level. The vertical motion of the structure is neglected. The notations adopted are presented in Figure 11.26.

The equations of motion are written about the point O, located at the intersection of the structure with the mean water level. The mass of the structure is noted m. The moment of inertia of the structure about the point O is written as J. The following points are introduced: the lowest point of the spar, L, the center of mass, G, the center of buoyancy, B, the attachment point to the mooring system, A, and the hub, H. For a given point P, the horizontal and vertical coordinates are written as x_P and z_P. All points are assumed to be on the mean line of the structure for simplicity. Using a small angle approximation, the vertical coordinate of a point is always z_P, and the horizontal

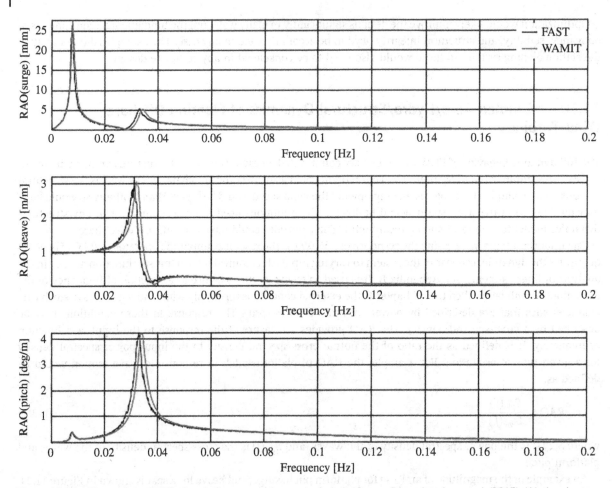

Figure 11.24 Illustration of RAOs for platform surge, heave, and pitch (*Source:* Ramachandran *et al.* (2013) / National Renewable Energy Laboratory)

coordinate is $x_P = x + z_P\theta$. The forces acting on the system are the hydrostatic (buoyancy), hydrodynamic, gravitational, mooring, and aerodynamic forces. A restoring hydrostatic moment is present in response to pitching motion. The equations of motion are:

$$m\ddot{x}_G = F_{aero} + F_{moor} + F_{hydro} \tag{11.92}$$

$$J\ddot{\theta}_G = M_{aero} + M_{moor} + M_{hydro} + M_{buoy} \tag{11.93}$$

where $\ddot{x}_G = \ddot{x} + z_G\ddot{\theta}$.

The mooring lines are assumed to act as a restoring spring with constant k_{moor}. The hydrostatic moment is $M_{buoy} = -g[\rho_w I_{yy} + mz_B]\theta$ where I_{yy} is the area moment of inertia about the y-axis. The horizontal hydrodynamic force on a section of height dz is given by the Morison equation (see Chapter 6), where the structural velocity of the section is $\dot{x}_P = \dot{x} + z\,\dot{\theta}$:

$$dF_{hydro} = \rho_w\Big[\pi R^2\big(-C_a(\ddot{x} + z\ddot{\theta}) + (1 + C_a)\dot{U}\big)$$
$$+ C_D R(U - \dot{x} - z\,\dot{\theta})|U - \dot{x} - z\,\dot{\theta}|\Big]dz \tag{11.94}$$

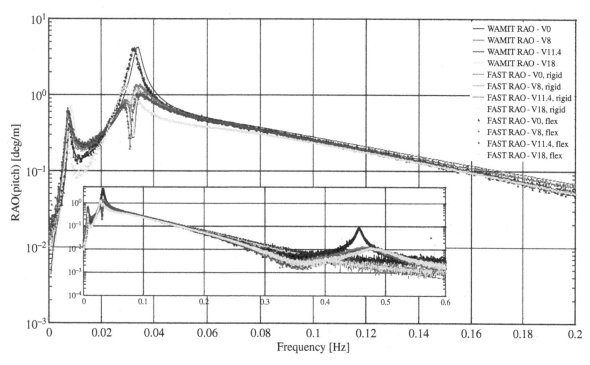

Figure 11.25 Examples of RAO for platform pitch for a flexible turbine (*Source:* Ramachandran *et al.* (2013) / National Renewable Energy Laboratory)

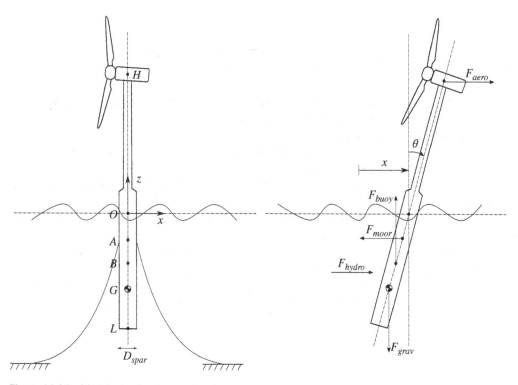

Figure 11.26 Model of a floating wind turbine using 2-DOF: platform surge (*x*) and pitch (*θ*)

where $U(z)$ is the wave velocity; C_a and C_D are the added mass and drag coefficients of the Morison equation; see Chapter 6. This horizontal force is integrated over the submerged length of the spar to obtain the total hydrodynamic force and moment. The resulting equations of motion may be written in vectorial form by introducing the vertical vector $\boldsymbol{x} = [x \ \theta]^T$ as follows:

$$\left(\boldsymbol{M} + \boldsymbol{M}_{hydro}\right)\ddot{\boldsymbol{x}} + \boldsymbol{K}\boldsymbol{x} = \boldsymbol{f} \tag{11.95}$$

where

$$\boldsymbol{M} = \begin{bmatrix} m & mz_G \\ mz_G & J_{yy} \end{bmatrix} \tag{11.96}$$

$$\boldsymbol{M}_{hydro} = \begin{bmatrix} -\rho_w C_a \pi R^2 z_L & -\rho_w C_a \pi R^2 \dfrac{z_L^2}{2} \\ -\rho_w C_a \pi R^2 \dfrac{z_L^2}{2} & -\rho_w C_a \pi R^2 \dfrac{z_L^3}{3} \end{bmatrix} \tag{11.97}$$

$$\boldsymbol{K} = \begin{bmatrix} k_{moor} & k_{moor}z_A \\ k_{moor}z_A & k_{moor}z_A^2 + \rho_w g I_{yy} + mg(z_B - z_G) \end{bmatrix} \tag{11.98}$$

$$\boldsymbol{f} = F_{aero}\begin{bmatrix} 1 \\ z_H \end{bmatrix} + \rho_w \left[\int_{z_L}^0 \pi R^2 (1 + C_a)\dot{U} + C_D R(U - \dot{x} - z\dot{\theta})|U - \dot{x} - z\dot{\theta}|\right]\begin{bmatrix} 1 \\ z \end{bmatrix} dz \tag{11.99}$$

We recall that z_L is negative, therefore $\boldsymbol{M}_{hydro}$ consists of positive terms.

References

DNV (2014) *Environmental Conditions and Environmental Loads*, DNV-RP-C205, Det Norske Veritas AS, Oslo.

Global Security (2023) http://www.globalsecurity.org/military/systems/ship/platform-tension-leg.htm (Accessed 1/23/2023).

Heronemus, W. E. (1972) Pollution-free energy from offshore winds. *Proceedings of 8th Annual Conference and Exposition*, Marine Technology Society: Washington, DC.

Honnef, H. (1932) *Windkraftwerke*. Friedrich Vieweg & Sohn, Braunschweig, Germany.

IEC (2023) *Wind Energy Generation Systems – Part 3-2: Design Requirements for Floating Offshore Wind Turbines*, 2nd edition, International Electrotechnical Commission, Geneva.

Jonkman, J. (2010) *Definition of the Floating System for Phase IV of OC3*, NREL/TP-500-47535 National Renewable Energy Laboratory, Golden, CO.

Jonkman, J. (2007) *Dynamics Modeling and Loads Analysis of an Offshore Floating Wind Turbine*, NREL/TP-500-41958 National Renewable Energy Laboratory, Golden, CO.

Jonkman, J. and Matha, D. (2010) *A Quantitative Comparison of the Responses of Three Floating Platforms*, NREL/CP-500-46726, National Renewable Energy Laboratory, Golden, CO.

Matha, D. (2009) *Model Development and Loads Analysis of an Offshore Wind Turbine on a Tension Leg Platform, with a Comparison to Other Floating Turbine Concepts*, NREL/SR-500-45891, National Renewable Energy Laboratory, Golden, CO.

Nansen, F. (1897) *Farthest North*. MacMillan & Co. Ltd, London.

Ramachandran, G. V. K, Robertson, A., Jonkman, J. M. and Masciola, M. D. (2013) *Investigation of Response Amplitude Operators for Floating Offshore Wind Turbines*, NREL/CP-5000-58098 National Renewable Energy Laboratory, Golden, CO.

Robertson, A., Jonkman, J., Vorpahl, F., Popko, W., Qvist, J., Frøyd, L., Chen, X., Azcona, J., Uzunoglu, E., Guedes Soares, C. and Luan, C. (2014) *Offshore Code Comparison Collaboration Continuation Within IEA Wind Task 30: Phase II Results Regarding a Floating Semi-submersible Wind System*, NREL/CP-5000-61154, National Renewable Energy Laboratory, Golden, CO.

Sebastian, T. and Lackner, M. (2013) Characterization of the unsteady aerodynamics of offshore floating wind turbines. *Wind Energy*, **16**(3), Wiley, 339–352.

Skaare, B., Nielsen, F. G., Hanson, T. D., Yttervik, R., Havmøller, O. and Rekdal, A. (2015) Analysis of measurements and simulations from the Hywind Demo floating wind turbine. *Wind Energy*, **18**(6), 1105–1122.

Tracy, C. (2007) *Parametric Design of Floating Wind Turbines*, MSc Thesis, Massachusetts Institute of Technology, Cambridge, MA.

Tupper, E. C. (2013) *Introduction to Naval Architecture*, 5th edition, Butterworth-Heinemann, Oxford, UK.

Withee, J. E. (2002) *Fully coupled dynamic analysis of a floating wind turbine system*, Ph.D. Dissertation, Massachusetts Institute of Technology, Cambridge, MA.

12

Wind Farms and Wind Power Plants

These days, large wind turbines are almost always installed in groups of several to many. Initially, these groups were referred to as wind farms in analogy to growing crops. This analogy has become less useful as these groups of turbines became larger and began to take on features that are different than the sum of the individuals. The term "wind power plant" is now used to describe those large groups of wind turbines. These large groups of turbines are different in four fundamental ways. First, their cumulative rated power is frequently commensurate with that of other generating plants that supply power to large regional electricity networks and so are best considered in that context. Second, the spatial and temporal partial correlation of the wind results in the combined power being different from that of scaling the power of a single turbine. Third, the wakes created by the turbines can have a significant effect on the other turbines, both decreasing their power output and increasing fatigue damage. Fourth, wind plants as distinct entities can affect the wind resource in the surroundings, primarily downwind, but also upwind.

This chapter, then, considers large groups of wind turbines as wind power plants operating in electricity networks, focusing on two basic topic areas: electrical and aerodynamic. It is divided into fifteen sections. Section 12.1 provides a summary of the relevant aspects of electricity networks in general. Section 12.2 discusses conventional generators. Section 12.3 is about the electrical load. Section 12.4 describes the fundamental principles of transmission systems in general and land-based transmission in particular. Section 12.5 is about offshore transmission. Section 12.6 focuses on how wind turbines fit into electrical networks. Section 12.7 examines how the combined electrical power of multiple turbines is affected by the presence of other turbines. Section 12.8 provides a brief overview of wind plants in power markets. Section 12.9 introduces the aerodynamic aspects of wind turbines in large arrays. Section 12.10 provides background on wind turbine inflow and how that relates to wakes. Section 12.11 is about array losses. Section 12.12 describes a number of wake models. Section 12.13 is about wake effect mitigation. Finally, Section 12.14 discusses blockage effects.

In this chapter, wind power plants are viewed as contributors to the total electricity supply, but not dominant. A useful metric is the term "penetration," which is defined here as the ratio of maximum energy that can be generated by the wind power plants (installed capacity) to the total energy produced by all generators during a given time period. Thus, the focus here is on power systems in which the wind energy penetration is well under 100%.

As wind power continues to play an even larger role in the world's transition away from fossil fuels, wind power plants may become even more significant, with even higher penetration levels. In that case, there are other considerations. These are discussed in Chapter 16.

Because of the widespread use of the term "wind farm," this text will continue to use it interchangeably with "wind power plant" when the focus is on performance and interaction of the wind turbines themselves and use "wind power plant" when the focus is on power production and the electrical network.

Wind Energy Explained: On Land and Offshore, Third Edition.
James F. Manwell, Emmanuel Branlard, Jon G. McGowan, and Bonnie Ram.
© 2024 John Wiley & Sons Ltd. Published 2024 by John Wiley & Sons Ltd.

12.1   Electrical Grids – Overview

Electrical grids can be divided into four main parts: generation, transmission, distribution (including feeders), and loads (i.e., uses of the electricity), as illustrated in Figure 12.1 for a typical situation. The primary distinction has to do with voltage, as is described in more detail below.

The fundamental principles of electricity were reviewed in Chapter 5. Here, the concern is networks, in which the electricity is generated, transmitted, distributed, and used. Conventional generators are discussed next in Section 12.2. The electrical loads are the subject of Section 12.3. The basic characteristics of the transmission and distribution systems are described in Section 12.4.

12.2   Conventional Electricity Generators

The generation function in utilities has historically been provided by conventional, electrically excited synchronous generators (EESGs) driven by a prime mover (such as the main rotor of a steam, gas, or hydro turbine). In the context of this text, prime movers in large systems include those supplied by fossil fuels (coal, natural gas), nuclear reactors, or water power. Coal and nuclear plants use a steam-based (Rankine) power cycle. There is a variety of types of natural gas plants, including open cycle, closed cycle (Brayton), and combined cycle (Brayton plus Rankine). These power plants respond to load variations, keep the system frequency stable, and adjust the voltage and power factor at the generating station as needed. The real and reactive power is supplied by the generators. This is done in conjunction with voltage regulator on the generators. The frequency of the AC power is maintained by governors on the prime movers, which adjust the fuel, steam, or water input as necessary. Conventional large-scale prime movers and generators are discussed in other sources, and additional details will not be given here except where directly relevant.

Generators in large, central power plants produce power at relatively high voltage (up to 25 kV). These generators feed current into an even higher voltage transmission system (110–765 kV) used to distribute the power over large regions. The transmission systems use high voltage to reduce the losses in the transmission lines. The distribution system operates at a lower voltage (10–69 kV), distributing the power to neighborhoods. Locally, the voltage is reduced again, and the power is distributed through feeders to one or more consumers. Industrial loads in the United States typically consume 480 V power while commercial and residential consumers in the

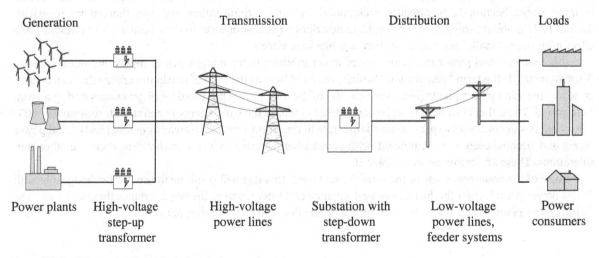

Figure 12.1   Schematic of a typical electrical power system

United States and much of the rest of the world connect to 120 or 240 V systems. In Europe, industrial loads generally connect at 690 V and residential loads use 230 V power.

Despite differences in type and size, nearly all conventional power plants have some similar defining characteristics. These include rated power, fuel vs. power relations, no-load fuel consumption, minimum allowed power level, start-up time, and minimum run time. Definitions are provided in Sections 12.2.1–12.2.6.

12.2.1 Rated Power

A generator's rated power is the maximum power level at which the generator can run for a sustained period. Output power is typically given in kW or MW, although generators are often specified in kVA or MVA, together with the rated power factor (usually 0.8).

12.2.2 Fuel vs. Power Relations

Fuel consumption of power plants is often reported in heat rate (GJ/kWh). An often more useful description is a curve or equation of fuel units as a function of power output. An example is provided in Figure 12.2 for a typical open-cycle natural gas turbine-driven generator rated at 132 MW. These curves are often nearly linear (a purely linear relationship is used in this example). The slope of the line is the heat rate, sometimes called the marginal fuel consumption. The line does not go through zero because of inherent losses in the system. The y-intercept is referred to as the no-fuel load consumption. The figure also presents the efficiency curve. Typically, the efficiency of the power plant decreases when at partial load (that is, when the power produced is less than the rated power). This is also illustrated in Figure 12.2.

12.2.3 No-load Fuel Consumption

The no-load fuel consumption is the fuel consumption at zero power, that is, the fuel needed to keep the power plant running. The ratio of the no-load fuel consumption to that at rated power is an indicator (between 0 and 1) of the part load efficiency of the prime mover/generator combination. For example, the ratio for the gas turbine in Figure 12.2 is just under 0.3 (400/1475≅0.3) while efficiency at half load has dropped to less than 80% of that at full

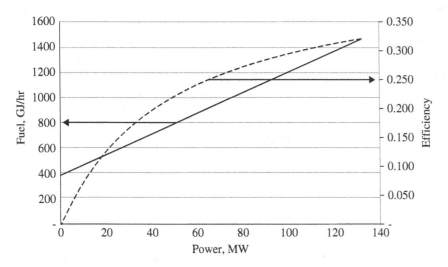

Figure 12.2 Typical gas turbine generator fuel consumption (solid line) vs. power (approximated as a linear relation). Efficiency vs. power (dashed line)

load. Regardless of the actual number, the most effective way to reduce fuel consumption is to shut off the generator altogether. But when a conventional generator is coupled with renewable energy sources, the generator may need to keep running at no (or partial) load to provide safety margins. No-load fuel consumption and the reduction in fuel efficiency at partial load are therefore important considerations in determining the amount of fuel that can be saved when introducing renewable energy solutions.

12.2.4 Minimum Allowed Power Level

In general, there is a minimum allowed power level for any generator. This ensures that the efficiency will be as high as possible and also avoids damage that can occur due to low-load operation.

12.2.5 Start-up Time

Many power plants take a significant time (many hours) to come to operating temperature, so they cannot be brought online immediately. For that reason, forecasting is used to anticipate when the plant will be needed so that it can be prepared well in advance. The longest start-up times are for nuclear plants (12 hours or more) and coal plants (9 hours or more). These plants typically operate to supply the base load (normal minimum load of a power system) and are seldom shut down.

12.2.6 Minimum Operating Time

Many power plants perform best at their design (rated) operating conditions, which include design temperature, and they can suffer thermal fatigue damage when cycled on and off. Therefore, once these plants are brought online, they are kept operating for some significant amount of time. Depending on the type of plant, the minimum operating time will be on the order of hours to days.

12.3 Electrical Loads

The term electrical load is used to describe a sink for power or a specific device that absorbs power. Electrical loads are of two types: resistive or inductive. Resistive loads include incandescent light bulbs and space and water heaters. Devices with electric motors are both resistive and inductive. They are a major cause of the need for a source of reactive power. The total electrical load on the system is the sum of the many fluctuating end loads. This load generally increases in the morning and decreases in the evening with similar patterns each day (associated with patterns of human activity). The daily load pattern may change with changes in seasons (primarily associated with heating and cooling) and on weekends and holidays. The magnitude of the load or its average will also change over these time scales. Figure 12.3 shows an example of what the load pattern of a large grid system might look like. As can be seen, the system load is never zero and fluctuates over the range between the minimum (base load) and the maximum (peak load). It is the goal of the managers of electrical utilities to be able to supply these loads under all circumstances, or as is often stated, "to supply firm power on demand." As discussed subsequently, the generators are dispatched to closely follow the fluctuating load and minimize voltage and frequency fluctuations as well as to minimize costs.

The frequency of occurrences of loads of different magnitudes is commonly presented by load duration curves, which are derived from the cumulative distribution function (CDF) of the load. These are comparable to wind speed duration curves which were introduced in Chapter 2. A typical load duration curve of a regional power system is shown in Figure 12.4. The figure shows the percentage of time that loads greater than a given magnitude occur. It can be seen that there is a load of at least 8 GW at all times; this is the base load. The intermediate load in this case varies from 8 to 20 MW. It corresponds to the range of the typical daily fluctuations seen in Figure 12.3.

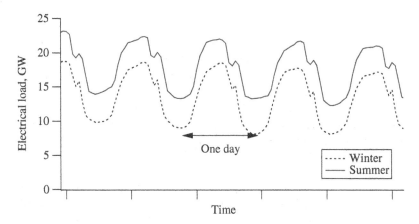

Figure 12.3 Sample utility load curve for a utility with different load levels in winter and summer

Figure 12.4 Sample load duration curve for a utility

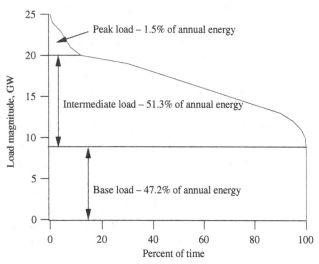

Finally, peak loads are those that occur rarely or infrequently, perhaps on hot days in the summer. In this example, the peak loads, those over 20 GW, occur 15% of the time and account for only 1.5% of the annual energy.

Power plants with different characteristics are used to supply power at these different power levels. Nuclear and coal-fired plants are often used for base load. Intermediate load power stations use technologies that can change generation levels more rapidly. Combined-cycle gas turbines are well suited for this. Peak-power stations need to be able to be ramped up very quickly. Open-cycle gas turbines and hydropower plants can serve this function. Peak-power stations may only operate occasionally and only in some seasons. Wind power plants can contribute to the electricity supply in a variety of ways, primarily by reducing the power that would otherwise be supplied by conventional intermediate-load plants. How much they can contribute will be determined by many factors, including the penetration (see Chapter 16 for high penetration situations). It can be seen that each power system must have a mix of generators that, together, can supply any load that may occur and can control the system voltage and frequency on sufficiently fast time scales.

12.3.1 Grid Control/Power Balancing

The power being consumed in the grid must be balanced by the power that is being generated. In order to maintain that power balance, the grid operator must have some generation capacity that is able to vary the power output to match the load as closely and quickly as possible. There are three major mechanisms by which this is accomplished: regulation or primary control, load following or secondary control, and scheduling or unit commitment summarized as follows; see also Ackerman (2012) and Weedy and Cory (2012).

As the load changes during the day, generators are brought online, but large prime movers may take a while to prepare for generation. Thus, a certain amount of unused generation capacity, called "spinning reserve," is kept online ready to respond to rapid load fluctuations.

Regulation. Power plants used for primary control or regulation are the first to be called on to increase or decrease output when there is a mismatch between the load and the power generated. These units can respond within seconds to minutes. In this category, generators are kept on spinning reserve. These machines are run at minimum or very low output, where they are quite inefficient, ready to ramp up output when conditions call for it. Regulation is usually tied to grid frequency, increasing power output when the frequency begins to fall, and decreasing output when frequency rises. Peak load plants may serve this function.

Load following. Load following, or secondary control, generation capacity also helps to balance the load and the generation but has a slower reaction time. Generators in this category might have response times of tens of minutes to hours. For example, load-following generators are called upon when the average power demand is increasing, in order to free up plants performing regulation to follow rapid fluctuations in load. The capabilities of load-following units will be most challenged during times when the energy demand is changing most rapidly. The most rapid increase that a power system is likely to experience in a given year is called the maximum ramp rate, usually measured in megawatts per minute. The higher this rate, the more control capacity is needed to cover potential load increases. Load following can also be difficult during times of rapid decreases in load. In order to prevent an unacceptable increase in frequency, generation must be decreased to match demand at times when energy use is decreasing. Under these circumstances, control generators must produce enough power at any given time that they can cover a decrease in load by decreasing their power output. It is possible that during a sustained drop in load or a grid fault, there might be an insufficient margin for decrease in generation to maintain power balance. Intermediate load plants fit into this category.

Unit commitment. Scheduling or unit commitment is the process of scheduling base load, intermediate-load, and peak-load generators so they will be available when needed in the upcoming hours or days. Utilities also try to ensure that there is enough reserve power to cover the load after the loss of the largest single generator in the power system. Thus, the outage of any one plant should not require the disconnection of customers.

Dispatch. The term "dispatch" is also commonly used in association with connecting or disconnecting generators to the power system. Dispatch is similar to unit commitment, but it normally includes a reference to power level as well as whether a generator is on or off.

12.3.2 Frequency/Voltage Control

System operators control the grid frequency and voltage to be within a narrow range of the nominal system values. Frequency in large electrical grids in industrialized countries is maintained at less than $\pm 0.1\%$ of the desired value. Depending on the country, voltages at distribution points are allowed to fluctuate from $\pm 5\%$ to up to $\pm 7\%$ of the nominal value, but allowable customer-induced voltage variations are often less (Patel, 1999).

The grid frequency is controlled by the power flows in the system. The torque on the rotor of any given generator consists of the torque of the prime mover, Q_{PM}, the electrical torques due to loads in the system, Q_L, and other generators in the system, Q_O. The equation of motion for the generator, with rotating inertia J (kg m^2) and angular speed Ω (rad/s), follows:

$$J\frac{d\Omega}{dt} = Q_{PM} + Q_L + Q_O \tag{12.1}$$

Each of the generators in the system is synchronized with the other generators in the system. Thus, this same equation of motion can represent the behavior of the system as a whole if each term represents the sum of all of the system inertias or loads.

Recognizing that power is just the product of torque and speed, one can derive the following equation:

$$\frac{d\Omega}{dt} = \frac{1}{J\Omega}(P_{PM} + P_L + P_O) \tag{12.2}$$

where P_{PM} is the prime mover power, P_L is the power from electrical loads (a negative number here), and P_O is the power from other generators.

Here, changes in the rotational speed of the generators (which is proportional to the grid frequency) are expressed as functions of the input power from the prime movers, the system load, and any power flow from other connected equipment. If the system load changes, then the power from the prime movers is adjusted to compensate and keep the system frequency stable.

The system voltage is normally regulated by controllers on the field excitation circuits of the system's synchronous generators (EESGs; see Chapter 5). Changing the field excitation changes both the terminal voltage and the power factor of the power delivered to the load. When the fields are controlled to stabilize system voltage, the power factor is determined by the connected loads. Voltage is controlled, in this manner, at each generating station.

It should also be noted that many modern wind turbines – with advanced power electronics – can contribute to grid frequency and voltage control. They may provide fast response, and by using down-regulation (reducing power), provide a contingency. See Chapter 9 for more details.

The presence of wind turbines on the network can also result in certain electrical issues that need to be accounted for. These include possible variations in voltage levels from the planned values at certain points on the system. The variations could be at various time scales: steady state, slowly varying, flicker, transient, and harmonics. On the other hand, modern wind turbines can contribute to grid stability in a number of ways. Some of these topics were discussed in Chapter 9; some additional details are given in Section 12.6.1.

12.4 Transmission and Distribution Systems

The power generated by any type of generator (wind turbines or otherwise) must be transmitted to the point of consumption by means of an electrical network or grid. This network consists of a transmission system and a distribution system. The transmission system operates at higher voltages and typically over longer distances, while the distribution system operates at lower voltages and shorter distances. These systems are basically of two types, overhead (above ground) and buried. Not surprisingly, the overhead systems are used on land; buried cables may be used on land or offshore (submarine cables). In the offshore case, special measures are taken to ensure that cables function properly in a submerged environment. Wind power plants, consisting of many large turbines, are invariably connected to a transmission system. It is common for one or a few smaller wind turbines to be connected to a distribution system directly. The following sections provide a summary of both overhead transmission systems and buried cables. Most transmission systems are AC based, but some use high-voltage direct current (HVDC).

12.4.1 General Properties of Transmission Systems

Transmission systems normally use three phases (see Chapter 5). In accordance with the three phases, there are three conductors. Sometimes there is a fourth (neutral) conductor that serves as the ground. The conductors are comprised of multiple smaller wires, which may be either copper or aluminum, depending on the application. Transmission lines consist of series resistances R and inductances L and parallel capacitances C, as illustrated in Figure 12.5.

In most wind energy applications, the transmission line associated with the project is less than 70 km long, and a more simplified equivalent circuit may be applied, as illustrated in Figure 12.6. Furthermore, for overhead lines, the capacitance is minimal. Capacitance may be significant for buried cables, however. Note that in Figures 12.5 and 12.6, the actual values of "C" are the total capacitance of the line divided by the number of occurrences.

12.4.1.1 Resistance
The resistance of a conductor depends on its length, l, cross-sectional area, A_c, and the resistivity of the material, ρ_m, as shown in Equation (12.3).

$$R = \rho_m l / A_c \tag{12.3}$$

The resistivity is also a function of temperature. In fact, the resistance considered in this way is actually DC resistance. An AC phenomenon called the skin effect increases the resistance somewhat, depending on the frequency. Actual resistances of conductors per unit length are provided by manufacturers.

12.4.1.2 Inductance
Inductance is a consideration for all transmission lines. For a single, long wire, it may be shown that the inductance per unit length, $\tilde{L}$ (H/m), is:

$$\tilde{L} = \frac{\mu_0}{2\pi}\left(\ln d - \ln\left(re^{-\frac{\mu_r}{4}}\right)\right) = \frac{\mu_0}{2\pi}\ln\frac{d}{r} \tag{12.4}$$

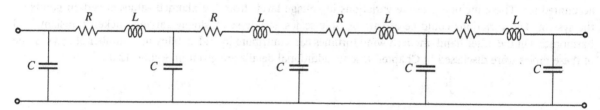

Figure 12.5 Equivalent circuit of generic transmission line, R is resistance, L is inductance, and C is capacitance

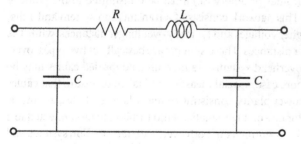

Figure 12.6 Equivalent circuit of medium-length transmission line

where d is the distance from the conductor (m), r is the radius of conductor (m), μ_0 is the permeability of free space, $\mu_0 = 4\pi \times 10^{-7}$ (H/m), and $r' = re^{-\frac{\mu_r}{4}} \approx 0.78r$. Equation (12.4) also applies for equally spaced conductors (e.g., equilateral triangle of side d).

Example: For a three-phase transmission line with a conductor radius of 0.01 m and an equal spacing of 2 m from Equation (12.4), the inductance per phase is 1.1×10^{-5} H/m.

12.4.1.3 Capacitance
Capacitance can also be significant for transmission lines. This is most likely to be true for buried or submarine cables. The capacitance C between conductors of radius r spaced d apart is given by Equation (12.5) (assuming equal radius and spacing).

$$C = \frac{2\pi\varepsilon}{\ln(d/r)} \tag{12.5}$$

where $\varepsilon = \varepsilon_0\varepsilon_r$ is the dielectric constant, $\varepsilon_0 = 8.854 \times 10^{-12}$ F/m is the dielectric constant of free space, and ε_r is the relative permittivity. Typical values of relative permittivity are given in Table 12.1.

Example: For a typical case of three overhead conductors of radius 0.01 m and spacing 2 m, the capacitance is approximately 0.01 μF/km. By comparison, submarine cables have a capacitance in the range of 0.2–0.4 μF/km (ABB, 2023). The difference is due to the closer spacing of the conductors and the higher relative permeability of the insulating material in the cable.

12.4.1.4 Reactance and Impedance
As discussed in Chapter 5, inductance and capacitance are accounted for in AC circuits through reactances and impedances. The relations for inductive and capacitive impedances are repeated here as Equations (12.6) and (12.7).

$$X_L = 2\pi f L \tag{12.6}$$
$$X_C = 1/2\pi f C \tag{12.7}$$

where C is capacitance, L is inductance, and f is AC frequency in Hz.

Table 12.1 Typical values of relative permittivity

Material	Relative permittivity
Air	1
Paper	3.5
Rubber	7
Glass	4–10
Plastics	2–3

The complex impedance of the transmission line is found as in other AC circuits; see Chapter 5. For example, the complex impedance $\hat{Z}$ of the line illustrated in Figure 12.6 is:

$$\hat{Z} = 1/\left[\frac{1}{R + j(X_L - X_C)} - \frac{1}{X_C}\right] \tag{12.8}$$

In many situations, the inductive reactance is much larger than the capacitive reactance, so the complex impedance of the transmission line is given approximately by:

$$\hat{Z} \cong R + jX_L \tag{12.9}$$

12.4.2 Overhead AC Transmission Systems

The most common method of transmitting large amounts of electricity over large distances on land is via overhead AC, three-phase, high-voltage transmission lines. The voltages of such lines are in the range of 138–765 kV.

An overhead transmission line normally consists of three conductors. The conductors are uninsulated themselves but are suspended from poles or transmission towers by insulators. The spacing of the conductors ensures that they do not touch under any circumstances and also serves to minimize capacitance effects. The conductors are made up of aluminum conductor steel-reinforced cable (ACSR), in which the inner steel wires provide strength and the outer aluminum wires carry the current. See Figure 12.7.

There are many considerations in the design of overhead transmission systems. Among these are the sagging of lines when they become hot. These are beyond the scope of this text, however. The reader should consult a text on transmission system design, such as Gönen (2014) for more information.

12.4.3 High Voltage Direct Current (HVDC) Transmission

Historically, DC has not been used for long-distance power transmission due to the difficulty of increasing the voltage to sufficiently high levels that currents could be kept low enough that line losses would not be a problem.

With advances in power electronics, since the 1960s, it has become progressively more feasible to use HVDC instead of alternating current. In particular, transistor-based power converters are key to this development. See Chapter 5 or Masters (2013) for more discussion of power converters.

The main advantages of HVDC are:

- Fewer conductors are required
- There are no reactive power requirements of the lines
- There are lower losses
- Transmission with minimal losses over large distances is possible.

Other advantages of power converter-based HVDC include:

- Independent control of active and reactive power
- Frequency control

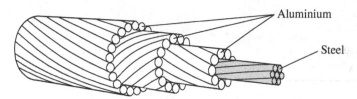

Aluminium

Steel

Figure 12.7 Aluminum conductor steel reinforced cable

- Black-start capability
- Multi-terminal operation
- Bipolar operation possible for increased reliability
- Electrical fault isolation between sending and receiving ends

Disadvantages of HVDC include:

- Power converters are required at the end of each end of the line
- Power converters are expensive
- Special circuit breakers are needed

More details on HVDC transmission may be found in a variety of sources, including Beaty and Fink (2013). A brief discussion of HVDC transmission in offshore power systems is given in Section 12.5.2.

12.5 Offshore Electricity Transmission

A transmission system is required to transmit to shore the electricity generated by offshore wind turbines. An illustration of the transmission system of a large-scale offshore wind plant is illustrated in Figure 12.8.

The primary components are submarine cables, of which there are normally two main types: array cables, which connect the turbines to the substation, and an export cable, which transmits the power to shore. In addition, a transformer is typically installed at each turbine. For large offshore wind plants, an offshore substation is installed, which includes additional transformers and various ancillary equipment. For smaller offshore wind power plants, close to shore, the submarine cables will be of medium voltage, on the order of 34.5 kV (US). In larger wind plants, medium voltage array cables will connect the turbines to the substation. One or more high voltage (115 kV) export cables will then transmit the collected power to shore from the offshore substation. A typical layout is illustrated in Figure 12.9. Note that the location where the export cable is connected to the grid is the POC . An overview of submarine cables is provided below. More information may be found in Wright *et al.* (2002) or Worzyk (2009).

12.5.1 Submarine Cables

The most important consideration in the design of submarine transmission cables is that they be able to transmit the desired amount of electricity with minimal losses. They must also be designed to withstand the effect of the water in which they are immersed as well as to avoid damage from the surroundings (anchors, fishing nets, seabed movement, etc.)

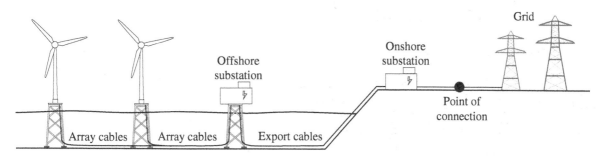

Figure 12.8   Offshore electricity transmission

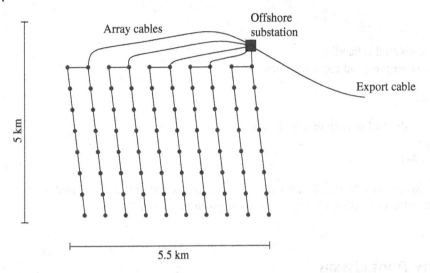

Figure 12.9 Cable system of the Horns Rev offshore wind plant (each dot indicates a wind turbine)

12.5.1.1 Construction of Cables/Cable Components

The primary components of submarine cables are the conductors. Each conductor is insulated. The entire assembly is surrounded by more insulation and protection against damage from the environment. Cross-sections of typical three conductor cables are shown in Figure 12.10.

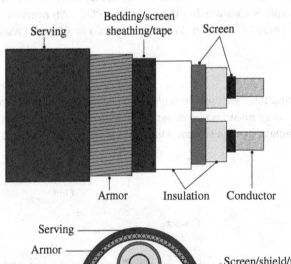

Figure 12.10 Cross sections of typical three conductor submarine cables

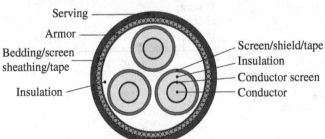

12.5.1.2 Cable Conductors

Conductors in submarine cables are made of copper or aluminum. As with above-ground conductors, individual strands are used to make them flexible. The size of the conductor is determined in accordance with the current that is intended to pass through it. The greater the cross-sectional area, the lower the losses, but the higher the cost. Thus, selecting the size involves trade-offs. Conductors are typically specified according to their cross-sectional area (mm^2), rather than their diameter.

12.5.1.3 Cable Insulation

Three types of insulation are in common use for submarine transmission cables. While insulation construction and thickness vary based on voltage, all three types discussed here are used for both medium and high voltages. Insulation is characterized by the material, implementation details, and whether it is lapped or extruded.

Low-pressure oil-filled (LPOF), or fluid-filled (LPFF) cables, insulated with fluid-impregnated paper, have historically been the most common for submarine AC transmission. The insulation is impregnated with synthetic oil whose pressure is typically maintained by pumping stations on either end. The pressurized fluid prevents voids from forming in the insulation when the conductor expands and contracts as the loading changes. LPFF cables run the risk of fluid leakage, however, which is an environmental hazard. Fluid-filled cables can be made up to about 50 km in length.

Solid, mass-impregnated paper-insulated cables are another option. These are often used for HVDC transmission. The lapped paper insulation is impregnated with a high-viscosity fluid and these cables do not have the LPOF cable's risk of leakage.

Extruded insulation is now replacing lapped installation as the favored option in many applications. Cross-linked polyethylene (XLPE) is less expensive than LPOF of a similar rating and has lower capacitance. XLPE can also be manufactured in longer lengths than LPFF (Gilbertson, 2000.)

Another extruded insulation used in submarine cables is ethylene propylene rubber (EPR), which has similar properties to XLPE at lower voltages, but at 69 kV and above has higher capacitance (Gilbertson, 2000)

Additional layers are typically present around the insulation layer:

- **Screening**. A layer of screening is placed around the conductor to smooth the electric field and avoid concentrations of electrical stress and also to assure a complete bond of the insulation to the conductor. Figure 12.10 shows a three-core cable with screening on both the individual conductors and the three-core bundle.
- **Sheathing**. Outside the screening of all the conductors is a metallic sheathing, which plays several roles. It helps to ground the cable as a whole and carries fault current if the cable is damaged. It also creates a moisture barrier. In AC cables, current will be induced in this sheath, leading to circulating sheath losses; various sheath-grounding schemes have been developed to reduce circulating currents that arise in the sheath. Unlike other cable types, EPR insulation does not require a metal sheath.
- **Armor**. An overall jacket and then armoring complete the construction. Corrosion protection will be applied to the armor; this may include a biocide to inhibit destruction by marine creatures such as marine borers that are present in some locations.

12.5.1.4 Electrical Properties of Submarine Cables

A key issue in the design or selection of submarine cables is their total impedance, the primary factor of which is resistance. It is desired to keep the losses to a minimum, so overall resistance needs to be low. Lowering the resistance indefinitely results in higher cable costs, so trade-offs need to be made.

Electrical properties of cables are similar in many ways to overhead cables. Resistance is found analogously to overhead cables. As in that case, resistance is determined by:

- Resistivity of the material
- Length of the cable

Calculating inductance and capacitance is more complex than it is for overhead cables and is beyond the scope of this text. Typical values for some commercially available cables are given in Section 12.5.1.8.

12.5.1.5 Power Losses in Cables
As is apparent from previous discussions, there are power losses associated with current flow in any real conductor. The most important of these is the well-known I^2R loss in the conductors. In addition, there are other losses, for example, due to eddy currents in metallic shields. Locations of these other losses include:

- Dielectric (insulation material)
- Sheath
- Armor

The total of these other losses can be significant, in some cases comparable to the conductor loss.

12.5.1.6 Additional Considerations
Fiber optic cables for communications and control can be bundled into the cables. Systems to protect the cable under certain conditions may also be integrated with the cables. These may include a temperature monitoring system or a cable overload protection system.

12.5.1.7 Commercially Available Cables
Table 12.2 summarizes the characteristics and limitations of some commercially available AC and HVDC cables.

12.5.1.8 Examples of Electrical and Mechanical Properties
Submarine cables tend to be quite heavy, due primarily to the density of the conductor and the required dimension of the conductors for a given amount of current. The overall dimensions of cables are larger than those of the conductors due to the insulation and sheathing. Table 12.3, based on manufacturer's data sheets (ABB, 2023), provides some examples of the properties of a typical medium voltage (30 kV) XLPE submarine cable. These examples are for cables with copper conductors and copper screening. Cables with lead sheaths are heavier than these. Higher voltage cables are also larger and heavier. For example, an 800 mm^2 three-core, lead-sheathed cable rated at 110 kV from the same manufacturer has an external diameter of 185 mm and a weight of 69.5 kg/m.

Table 12.2 Characteristics and limitations of some high voltage cables

	AC		HVDC		
Insulation	XLPE	LPOF	LPOF	Paper	XLPE
Max. voltage	400 kV	500 kV	600 kV	500 kV	150 kV
Max. power or MVA	1200 MVA	1500 MVA	2400 MW	2000 MW	500 MW
Max. length	100 km	60 km	80 km	Unlimited	Unlimited

(*Source:* Adapted from Wright *et al.*, 2002)

Table 12.3 Properties of typical submarine cables

Cross-section mm^2	Conductor diameter mm	Cable diameter mm	Cable weight kg/m	Cable capacitance μF/km	Cable inductance mH/km	Current rating A
95	11.2	104	19.5	0.18	0.44	300
300	20.4	123.9	28.2	0.26	0.36	530
500	26.2	137.3	36	0.32	0.34	655
800	33.7	154.4	47.2	0.38	0.31	775

12.5.1.9 Environmental Factors in Submarine Cables

There are numerous environmental factors associated with submarine cables. The primary concern is the disturbance to the seabed during the laying and burying of the cable. Another important concern is the maximum allowed seabed temperature. This is closely related to the seabed thermal resistivity, which is also a factor in evaluating ampacity, as discussed in the next section 12.5.1.10.

12.5.1.10 Ampacity of Submarine Cables

Ampacity refers to the maximum allowed current in the cable, subject to a specified maximum cable temperature. In particular, temperatures within the cable must be kept within prescribed limits in order not to damage the insulation. A typical maximum allowed temperature is 90 °C. The actual internal temperature of the cable is a function of:

- Current
- Thermal properties of the cable
- Temperature of the seabed
- Thermal properties of the seabed

More information may be found in Neher and McGrath (1957) or de León (2006).

12.5.1.11 Overload

An electrical cable must be able to withstand a certain amount of overload current. This allowed overload current will vary depending on what the current had been previous to the beginning of the overload. Allowed overload current may initially be two to three times the rated current but decreases with time, and after a few hours must eventually approach the rated current.

12.5.2 HVDC Offshore

All offshore wind plants to date have used AC cables for transmitting the power to shore. However, HVDC has frequently been considered, and it is expected that it will be used in some applications in the near future.

A conceptual schematic for a possible HVDC transmission system for an offshore wind plant is illustrated in Figure 12.11. In this example, turbines would send AC power at medium voltage to an offshore substation where it would be transformed to higher voltage. That power would then be rectified to HVDC and transmitted to shore, where it would be inverted back to AC, first to a moderately high voltage then transformed up to the transmission system voltage.

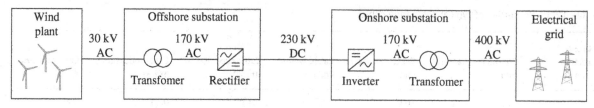

Figure 12.11 Conceptual wind plant with HVDC cable

12.6 Wind Turbines and Wind Power Plants in Power Systems

Large wind power plants are normally connected into the transmission system. Individual wind turbines or small wind farms are typically connected to the distribution system or, in the case of very small wind turbines, into the feeder system. The actual location of the interconnection is known as the POC. Closely related to the POC is the point of common coupling, or PCC, which signifies ownership changes as much as physical features. In some cases, there are multiple connections at the PCC. There are both technical and economic aspects to consider regarding wind turbines and wind plants in power systems. This section provides an overview of the technical aspects. Section 12.8 does the same for the economic ones.

12.6.1 *Electrical/Technical Aspects of Wind Power Plants

This section provides an overview of the electrical/technical aspects of wind power plants in power systems. The focus is mainly on interactions that affect the local system voltages and currents on medium- to short-term time scales. Topics considered are steady-state voltage levels, voltage flicker, harmonics, and grid capacity limits. The discussion builds on the material in Chapter 5.

Large numbers of wind turbines can have a significant effect on the power system to which they are connected. On the one hand, the introduction of wind turbines into an electrical network can, at times, lead to problems, potentially including instability, that limit the magnitude of the wind power that can be connected to the grid. On the other hand, depending on the grid and the turbines, the introduction of turbines may help support and stabilize the grid. Turbine–grid interactions depend on the electrical behavior of (1) the turbines under consideration and (2) the electrical grids to which the turbines are connected. Important aspects of these have been summarized above. Interconnection issues include steady-state voltage levels, flicker, harmonics, and grid capacity limits. As noted earlier, high penetration systems, in which wind turbines provide a very large fraction of the electrical network, are discussed in Chapter 16.

12.6.1.1 Wind Turbines/Steady-State Voltage Levels

Electrical grids, like other electrical circuits, provide an impedance to current flow that causes voltage changes between the generating station and other connected equipment. This can be illustrated by considering a wind turbine generator connected to a grid (see Figure 12.12) with a line-to-neutral voltage, V_S. The voltage at the wind turbine, V_G, will be higher than V_S. The difference in voltage is caused by the electrical collection system impedance, which consists of the collection system resistance, R, that causes voltage changes primarily because of the real power flowing in the system and collection system reactance, X, that causes voltage changes because of the reactive power flowing in the system (see Chapter 5 for a definition of electrical terms). The magnitudes of R and X, which are functions of the collection system, and the magnitudes of the real generated power, P, and reactive power requirements, Q, of the wind turbine or wind plant will determine the system voltage at the wind turbine.

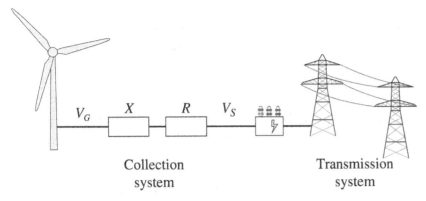

Figure 12.12 Collection system schematic: *R*, resistance; *X*, reactance; V_S and V_G, grid and wind turbine voltage, respectively

The voltage at the generator can be determined from Equation (12.10) (Bossanyi *et al.*, 1998):

$$V_G^4 + V_G^2\left[2(QX - PR) - V_S^2\right] + (QX - PR)^2 + (PX + QR)^2 = 0 \tag{12.10}$$

In lightly loaded circuits, the voltage change can be approximated by Equation (12.11):

$$\Delta V = V_G - V_S = \frac{PR - QX}{V_S} \tag{12.11}$$

It can be seen that the voltage increases due to the real power production (*P*) in the system. However, the voltage decreases when reactive power (*Q*) is consumed by equipment in the system.

These voltage changes can be significant. Transformers equipped for automatic voltage control (AVR) are used to provide reasonably steady voltages to end-users. These transformers have multiple taps on the high-voltage side of the transformer. The current flow is switched automatically from one tap to another as needed. The different taps provide different turn ratios and therefore different voltages (for more information, see Rogers and Welch, 1993).

The same cable resistance that causes voltage variations also dissipates energy. The electrical power losses in the collection system, P_{Loss}, can be expressed as:

$$P_{Loss} = \frac{(P^2 + Q^2)R}{V_S^2} \tag{12.12}$$

Finally, grid "strength" or stiffness is characterized by its fault level, *M*. The fault level at any location in the grid is the product of the system voltage and the current that would flow if there were a short circuit there. Using the example above, if there were a short circuit at a wind turbine or plant, the fault current, I_F, would be:

$$I_F = \frac{V_S}{\left(R^2 + X^2\right)^{1/2}} \tag{12.13}$$

and the fault level (in volt-amperes), *M*, would be:

$$M = I_F V_S \tag{12.14}$$

This fault level is an indication of the strength of the network, with higher fault levels indicating stronger networks and lower fault levels a "weaker" grid.

Voltage Changes Using Phasors

The voltage changes along a power line (collection system) may also be found using phasors, which were discussed in Chapter 5. This method facilitates calculations. For a wind turbine with a conventional induction generator, as shown in Chapter 5, the current $\hat{I}$ is:

$$\hat{I} = \hat{V}_G/\hat{Z}_{IM} \tag{12.15}$$

where $\hat{V}_G$ is the voltage at the terminals and $\hat{Z}_{IM}$ is the impedance of the induction generator (per phase, as given in Chapter 5). The voltage at the generator is also given by:

$$\hat{V}_G = \hat{I}\hat{Z}_{line} + \hat{V}_S \tag{12.16}$$

where $\hat{Z}_{line} = R + jX$ is the impedance of the line (per phase), $\hat{V}_S$ is the voltage of the power system, R is the resistance of the line per phase, and X is the reactance of the line per phase. The current is found from:

$$\hat{I} = \hat{V}_S/(\hat{Z}_{line} + \hat{Z}_{IM}) \tag{12.17}$$

The real loss (per phase) in the lines is given by:

$$P_{Loss} = \mathrm{Re}\left[(\hat{V}_G - \hat{V}_S)\hat{I}^*\right] \tag{12.18}$$

When all the line and generator parameters as well as the slip s (see Chapter 5) are known, finding the voltage drop along the line and losses is straightforward. When the wind turbine rotor power is known rather than the slip, an iterative search is needed.

Example 12.1   Line Loss

Consider the wind turbine with an induction generator which formed the basis of Example 5.3 in Chapter 5. In this case, it is connected to an electrical grid through a power line with resistance and inductive reactance per phase of $R = 0.04\,\Omega$ and $X = 0.04\,\Omega$. The line-neutral voltage of the grid is $\hat{V}_S = 389.25\angle 0°$. The magnitude of the total power from the turbine's rotor is assumed to be the same as in the Chapter 5 example, namely, 323.6 kW. Similarly, the mechanical efficiency is assumed to be 1.0. The slip in the original example was −0.025; in this case, however, s must be found iteratively.

Performing the algebraic calculations, the slip is now −0.0244, $\hat{Z}_{IM} = -1.010 + j0.7035\,\Omega$, and the line impedance is $\hat{Z}_{line} = 0.04 + j0.04\,\Omega$. As a result, the total impedance is $\hat{Z}_{total} = -0.970 + j0.7435\,\Omega$. The current is $\hat{I} = -252.76 - j193.75$ A and the voltage at the generator terminals is $\hat{V}_G = 391.6 + j17.860$ V. The magnitude of the voltage drop between the SCIG and the power system is $\Delta V = |\hat{V}_G| - |\hat{V}_S| = 2.77$ V. The total real power produced $3\,\mathrm{Re}\left(\hat{V}_G\hat{I}^*\right)$ is 307.3 kW, the total line loss ($3P_{Loss}$) is 12.2 kW, the power delivered to the grid is 295.2 kW, the power factor is 0.82, and the overall electrical efficiency is 0.912 (see Chapter 5 for the definition of these terms).

The situation could be improved by adding capacitors in parallel with the generator. For example, it can be shown that 1269 µF of capacitors per phase (corresponding to 72.5 kVAR at the system voltage) would bring the power factor to near unity. The line loss would decrease to 8.2 kW, and the power delivered to the grid would increase to 300.3 kW.

12.6.1.2   Wind Turbines/Unsteady Voltage Levels

Wind turbine operation results in varying real and reactive power levels and may result in voltage and current transients or harmonics, depending on the type of generator.

As discussed in Chapter 5, constant-speed turbines connected to electrical grids generally use squirrel cage induction generators (SCIG). These generators provide real power (P) to the system and absorb reactive power (Q) from the system. The relationship of real to reactive power is a function of the generator design and the power being produced. Both real and reactive power are constantly varying during wind turbine operation. Low-frequency real power variations occur as the average wind speed changes. Reactive power needs are approximately constant or increase slowly over the operating range of the induction generator. Thus, low-frequency reactive power variations are usually smaller than low-frequency real power fluctuations. Higher frequency fluctuations of both real and reactive power occur as a result of turbulence in the wind, tower shadow, and dynamic effects from drive train, tower, and blade vibrations.

Variable-speed turbines have a power electronic converter between the generator and the grid. Details depend on the type of generator used and the type of converter. For example, with SCIGs the generator side of the converter may supply the reactive power to the generator (rather than the grid). With wound rotor induction generators (WRIG), the effect on the grid is a mixture of stator currents (much like those in a conventional SCIG) and the currents from the power electronics associated with the generator's rotor. In the case of variable-speed turbines with synchronous generators, all the currents on the grid side originate in the power converters. In general, the grid side of the converter can also provide current to the grid at any desired power factor. This capability may be used to improve grid operation, if desired.

When the generators are connected or disconnected from a power source, voltage fluctuations and transient currents can occur. As explained in Chapter 5, connecting a SCIG to the grid results in a momentarily high starting current as the magnetic field is energized. Also, if the generator is used to speed up the rotor from speeds far from synchronous speed (high slip operation), significant currents can occur. These high currents can be limited, but not eliminated, with the use of a "soft-start" circuit, which limits the generator current. When SCIGs are disconnected from the grid, voltage spikes can occur as the magnetic field decays.

Changes in the mean power production and reactive power needs of a turbine or a wind plant can cause quasi-steady state voltage changes in the connected grid system. These changes occur over numerous seconds or more as explained previously. The X/R ratio (reactance divided by resistance) of the distribution system and the generator operating characteristics (the amount of real and reactive power at typical operating levels) determine the magnitude of the voltage fluctuations. It has been found that an X/R ratio of about two results in the lowest voltage fluctuations with typical fixed-speed turbines with SCIGs. The X/R ratio is typically in the range of 0.5–10 (Jenkins, 1995).

The weaker the grid is, the greater are the voltage fluctuations. "Weak" grids that can cause problematic turbine–grid interactions are those grid systems in which the wind turbine or wind plant-rated power is a significant fraction of the system fault level. Studies suggest that problems with voltage fluctuations are unlikely with turbine ratings of 4% of the system fault level (Walker and Jenkins, 1997). Germany limits renewable power generator ratings to 2% of the fault level at the POC, and Spain limits them to 5% (Patel, 1999).

Power factor correction capacitors may be installed at the grid connection of SCIGs to reduce system voltage fluctuations and the reactive power needs of the turbine (see Example 12.1). Power factor correction capacitors need to be chosen carefully, however, to avoid self-excitation of the generator. This occurs when the capacitors are capable of supplying all of the reactive power needs of the generator and the generator becomes disconnected from the grid. In this case, the capacitor–inductor circuit, consisting of the power factor correction capacitors and the generator coils, can resonate, providing reactive power to the generator and resulting in possibly very high voltages.

12.6.1.3 Voltage Flicker

Voltage flicker consists of disturbances to the network voltage that occur faster than steady state voltage changes, but which are fast enough and of a large enough magnitude that lights noticeably change brightness. These disturbances can be caused by the connection and disconnection of generators or by torque fluctuations in fixed-speed turbines as a result of turbulence, wind shear, tower shadow, and pitch changes. The human eye is most sensitive to brightness variations around frequencies of 10 Hz. The blade-passing frequency in some wind turbines may also be the source of voltage flicker. The magnitude of the flicker due to wind turbulence depends on the slope of the real vs. reactive power characteristics of the generator, the slope of the power vs. wind speed characteristics of the turbine, and the wind speed and turbulence intensity. Variable-speed system power electronics usually do not impose rapid voltage fluctuations on the network but still may cause flicker when turbines are connected or disconnected. Flicker does not damage equipment connected to the grid, but in weak grids, where the voltage fluctuations are greater, it may become an annoyance to consumers. Many countries have standards for quantifying flicker and limits for allowable flicker and step changes in voltage (see, for example, CENELEC, 1993).

12.6.1.4 Harmonics

Power electronics in variable-speed wind turbines may introduce sinusoidal voltages and currents into the grid at frequencies that are multiples (harmonics) of the grid frequency; see also Chapter 5. Because of the problems associated with harmonics, utilities have strict limits on the harmonics that can be introduced into the system by power producers such as wind turbines.

In both the United States and Europe, many power companies use the IEEE 519 Standard (ANSI/IEEE, 1992) to determine allowable total harmonic distortion (THD) at the PCC. Minimizing problems at this point minimizes problems for other electrical customers. The allowable THDs of the voltage according to IEEE 519 are detailed in Table 12.4. Similar restrictions on current harmonics, which depend on the ratio of maximum demand load current to maximum short circuit current at the PCC, can be found in IEEE 519.

12.6.1.5 Grid Capacity Limitations

As indicated earlier, utilities limit the capacity of wind turbines that may be connected to their network as a function of the nominal voltage of the grid at the POC. In addition, there may be times when the current-carrying capacity of the grid is limited, possibly due to outages of a line or hot weather when the grid capacity needs to be de-rated. Under these conditions, if it is windy, turbines may be required by the local utility to curtail their output. This might be accomplished by operating only some wind turbines or by operating all of the wind turbines at a reduced power level. See also Section 12.6.3.2 for more on curtailment.

12.6.2 Grid Connection Equipment

The turbine–grid connection consists of equipment to connect and disconnect the wind plant from the larger grid, equipment to sense problems on the grid or the turbine side of the connection, and transformers to transform power between different voltage levels. This equipment is in addition to the electrical equipment associated with each wind turbine which was described in Chapter 5.

Table 12.4 Maximum allowable total harmonic distortion (THD) of voltage at the point of common coupling (PCC)

PCC voltage	Individual harmonic, %	THD, %
2.3–69 kV	3.0	5.0
69–138 kV	1.5	2.5
>138 kV	1.0	1.5

- **Switchgear.** Switchgear to connect and disconnect wind power plants from the grid usually consists of large contactors controlled by electromagnets. Switchgear should be designed for fast automatic operation in case of a turbine problem or grid failure.
- **Protection equipment.** Protection equipment at the POC needs to be included to ensure that turbine problems do not adversely affect the grid and vice versa. This equipment must include provision for rapid disconnection in case of a short circuit or over-voltage situation in the wind plant. The wind plant should also be disconnected from the grid in case of a deviation of the grid frequency from the rated frequency due to a grid failure or a partial or full loss of one of the phases in a three-phase grid (see Chapter 5). The protection equipment includes sensors to detect problem conditions. Outputs from these sensors control contactor magnets or additional solid-state switches such as silicon-controlled rectifiers (SCRs). The ratings and operation of protection equipment should be coordinated with that of other local equipment to ensure that no problems occur. For example, in the case of a momentary grid failure, the disconnect at the wind plant should react fast enough to prevent currents from flowing into the grid fault and should remain off long enough to ensure that reconnection only occurs after the other grid faults have been cleared (Rogers and Welch, 1993).
- **Grounding.** Turbines, wind plants, and substations need grounding systems to protect equipment from lightning damage and short circuits to ground. Providing a conductive path for high currents to the earth can be a significant problem in locations with exposed bedrock and other nonconductive soils. Lightning strikes or faults can result in significant differences in ground potential at different locations. These differences can disrupt grid protection equipment and pose a danger to personnel.

12.6.3 Wind Plant Operational Considerations

There are a number of aspects of wind turbine operation that may be significant in power systems. These include:

- Net load
- Power smoothing
- Spatial separation and wind plant variability
- Curtailment
- Ancillary services
- Ramp rates

Each of these are described in Sections 12.6.3.1–12.6.3.5.

12.6.3.1 Net Load

When wind turbines supply a significant fraction of an electrical system's requirement, the difference between the system load, P_{load}, and the wind power, $P_{turbines}$, is known as the net electrical load or just "net load." It is this net load that must be supplied by conventional generators. The net load is defined by Equation (12.19):

$$P_{net} = P_{load} - P_{turbines} \tag{12.19}$$

The reference time period for the net load may be either the utility's time period for demand projections, which is 1 hour, or 10 minutes, which is the same as that used for wind speed.

The net load may fluctuate within that interval due to variations in wind (turbulence) as well as in the load itself. Assuming that the fluctuations of the wind power and the load are normally distributed and uncorrelated the standard deviation of the net load, σ_{net} is given by:

$$\sigma_{net} = \sqrt{\sigma_{load}^2 + \sigma_{P,turbines}^2} \tag{12.20}$$

where σ_{load} is the standard deviation of the system load and $\sigma_{P, turbines}$ is the standard deviation of the wind generation.

Those net load fluctuations are generally not of great concern in large networks except in conjunction with unusually high ramp rates (see Section 12.6.3.4). They may be important in smaller systems, however, as discussed in Chapter 16.

12.6.3.2 Curtailment

Curtailment refers to situations in which some or all wind turbines are purposely shut down or set to operate at lower power levels than the wind speed would allow. For example, in some power systems, the available wind power may sometimes exceed the load or an allowed set-point which is based on the operational requirements of the other generators in the system. In that case, the simplest remedy may be to reduce ("curtail") the production from the wind turbines. If the turbines have conventional induction generators, they may need to be shut down altogether, because there is no other way to reduce their power output. If they have pitch control, the blade pitch can be adjusted to accomplish this. In general, curtailment is seen as a first step to avoid occasional excess wind power as penetration increases. It can also provide means of supplying contingency spinning reserve and help the grid stability (see Section 12.3.1). For higher penetration, other, more proactive measures would likely be taken such as using energy storage; see Chapter 16.

Curtailment may also be employed for other reasons. It can be used to mitigate environmental issues such as noise or shadow flicker; see Chapter 14. Curtailment can minimize fatigue damage by purposely setting some turbines to operate at reduced power, thereby decreasing the wake deficit (see Section 12.13). This can allow downstream turbines to generate more power and suffer less fatigue damage due to the fluctuating flow coming from the upstream wake.

12.6.3.3 Ancillary Services

Most large turbines today include power electronics that have capabilities beyond that which are needed to ensure proper operation of the turbines themselves. These capabilities may be of use to the grid as a whole and in that case are referred to as ancillary services. These could include providing reactive power, frequency regulation, or facilitating the "black start" of other generators (black start refers to restarting generators after a power system blackout or other major system failure). As more and more turbines are added to some networks, it is anticipated that such ancillary services could become even more important, or that additional equipment would be added to the system to help relieve some of the burden on conventional generators, allowing more of them to be shut down. See Rebello *et al.* (2020) for one example.

12.6.3.4 Ramp Rates

As already discussed, variability in the wind can result in changes in power from one minute to the next. From the point of view of the utility system operator, these changes are referred to as ramp rates; they are commonly measured in MW/minute. High ramp rates can be problematic for the operation of the grid. Spatial separation as described previously can reduce ramp rates, however. For example, wind plants of 200 MW have had measured ramp rates of 20 MW/min, whereas 1000 MW of wind generation consisting of 10–20 MW plants were estimated to have a maximum ramp rate of only 6.6 MW/min (Ackermann, 2012).

12.6.3.5 Wind Power Variability and Unit Commitment

Not only does variability of the net load increases due to the variability of wind power, but the uncertainty of the net load in the future also increases. Unit commitment decisions are made based on projections of the net load, taking into account the uncertainties involved. With wind plants in the system, there may be more uncertainty. The increase in the variability and uncertainty of the net load may require the commitment of additional spinning reserves to handle larger increases and decreases in net load from one interval to the next than would otherwise be needed.

The effect of the wind's variability on system operation costs depends on the mix and characteristics of the conventional generators in the system, fuel costs, the nature of the wind resource, the characteristics and geographical dispersion of the wind power plants, and the regulatory environment. Numerous studies of the cost of the inclusion of large amounts of wind power into grid systems have been completed (see DOE, 2008 for a summary). The effects depend on which grid system is being considered. These studies have generally shown that having wind energy capacities up to 20% of the system load increases system operating costs by less than $0.005/kWh. Most of these costs are for committing additional generators. Load following and regulation costs are usually a small proportion of the total cost.

12.7 Power from Wind Plants

As discussed in Chapter 2, the 10-minute average power production from a single wind turbine can be readily estimated from the wind speed and the wind turbine's power curve. The power from multiple wind turbines or even multiple wind plants within a utility's purview has some additional considerations. These have to do with the turbulence in the wind, the effect of wakes, and the spatial separation between turbines and plants.

12.7.1 Moderate Spatial Separation and Power Smoothing

Although not explicitly discussed in Chapter 2, the variability of the wind power within any 10-minute interval is, to a first approximation, proportional to the variability of the wind. In large networks, this short-term variability is seldom an issue. In small/medium power systems, however, the result is that the short-term variability in the net load can be considerably higher than that of the load itself, and this may sometimes be significant (see Chapter 16.) This is most likely to be the case when there are a few wind turbines and if the turbines are relatively close together (less than 1000 m apart). On the other hand, some separation between the turbines does reduce the variability (because spatial correlation of the wind decreases with distance). This is known as short-term power smoothing and is described below.

The total output power of a group of turbines is the sum of the power produced by the individual turbines (subject to the effect of wakes, which is discussed later). Turbulent wind fluctuations result in fluctuating power from each of the wind turbines and, thus, from the group of turbines. The nature of turbulence is such that unless the turbines are close to each other, the fluctuations will be only partially correlated or not at all. The effect is to reduce wind farm power fluctuations compared with the power that would be expected from turbines all experiencing the same wind.

For example, assume that one wind turbine produces an average power P_1 over some time interval with a standard deviation of $\sigma_{P,1}$. Then, if N wind turbines in an ideal wind farm experienced the same wind, the total wind farm production would be $P_N = NP_1$ and the standard deviation of the wind farm electrical power output would be $\sigma_{P,N} = N\sigma_{P,1}$. Conversely, it can be shown that if N wind turbines experience wind with the same mean speed but uncorrelated turbulence, then the mean power output of the N turbines is still $P_N = NP_1$, while the standard deviation of the resulting aggregated power is reduced, as shown in Equation (12.21):

$$\sigma_{P,turbines} = \frac{N\sigma_{P,1}}{\sqrt{N}} \tag{12.21}$$

Thus, the fluctuation of the total power from the wind farm is less than the fluctuation of the power from individual wind turbines and so the power is said to be "smoothed."

In reality, the wind at two different turbine sites in a wind farm is neither perfectly correlated nor perfectly uncorrelated. The degree of correlation depends on the distance between the two locations and on the spatial

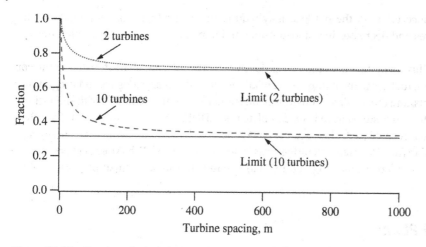

Figure 12.13 Fraction of wind plant power variability as a function of turbine spacing

and temporal character of the wind field. With a spectrum of turbulent wind and an appropriate coherence function, as described in Chapter 6, it is possible to develop an expression for the variance in wind farm output power as a function of the turbulent length scale and the number of and distance between wind turbines. This is known as a "wind farm filter," and it characterizes the effect on smoothing when the distances between turbines are on the order of 1000 m or less.

As an example, Figure 12.13 illustrates the effect of spacing for 2 and 10 wind turbines, assumed to be equally spaced along a line perpendicular to the wind direction. For this example, the turbulent length scale L is 100 m and the wind speed U is 10 m/s. The figure shows the fractional reduction in power variability as a function of cross-wind spacing. See Beyer *et al.* (1989) for more information.

12.7.2 Wind Farm Power Curves

In larger wind plants, wake effects, as discussed in detail in Sections 12.9–12.14, can significantly reduce the power from many of the turbines with respect to the overall prevailing wind speed. To illustrate the effect on power performance, we consider a situation in which the wind speed increases progressively from zero, covering the full range of operation of the turbine. As the wind speed increases, it will reach the cut-in wind speed, when the first row of turbines will start to produce power. That power production will reduce the wind speed behind the first row and no other turbines will operate. As the wind increases, more and more rows of turbines will produce power until all of the turbines are producing power, with the front row producing the most power per turbine. Once the wind reaches rated wind speed, only the first row of turbines will produce rated power. Other turbines will be producing rated power only after the winds are somewhat higher than rated for the turbines in the wind farm. Thus, not only is the total wind farm energy production lower than that of multiple isolated turbines, but the energy production as a function of wind speed has a different shape for the wind farm as a whole than for an individual turbine (see Figure 12.14). The example in the figure assumes that all wind turbines are

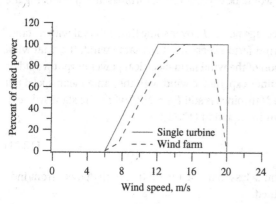

Figure 12.14 Comparison of single turbine and wind farm power curves

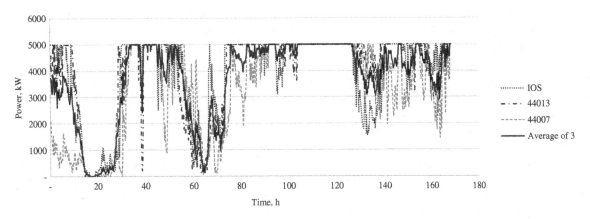

Figure 12.15 Example of effect of spatial separation on wind power

operating correctly. If some turbines are out of service, the effective wind farm power will be shifted down. The reduction in power at below-rated wind speed is referred to as "array losses." See Section 12.11 for more details.

12.7.3 Larger Spatial Separation and Wind Plant Variability

In large electrical networks, there may be multiple wind power plants, located at significant distances from one another. This separation can reduce the medium-term variability (minutes to hours) of the wind power, depending largely on the distances involved and the characteristics of the wind.

Example 12.2 Wind Plant Separation

For an illustration of this effect, wind data for 2021 was used to estimate power from NREL 5-MW turbines at 3 offshore sites. The data came from 3 National Data Buoy Center (NDBC, 2022) sites off the coast of New England: Isle of Shoals (IOS) and buoys 44013 and 44007. These buoys are on a nearly straight line, with a spacing of 71 km between Isle of Shoals and both buoys and 137 km between the buoys which are farthest apart. The data was scaled from the buoy anemometer to hub height, yielding a nominal average wind speed of 8 m/s. For the entire year, the average capacity factor was 0.418. The effect of the separation was to reduce the variability by approximately 14%. A one-week period of the result is illustrated in Figure 12.15. As can be seen, all the turbines have the same power when the winds are high. During periods of lower wind, the variability of the power is reduced.

12.8 Wind Power Plants in the Power Market

The discussion so far in this chapter has focused on the characteristics of conventional generators, the power system and the electrical aspects of the wind turbines which may provide some of the required power. When viewed as true power plants, wind turbines are treated as such to the extent possible. That is, the operators of the wind plants may enter them into the power market such that they are considered to be part of the forecasted required energy supply for the upcoming time period of interest.

Traditionally, electric power systems have been vertically integrated. That is, the utility generated the power, transmitted it, distributed it, and sold it to retail customers. In such utilities, power balancing is performed by the utility. In many large power systems now, however, this function is under the auspices of an independent system operator (ISO) or regional transmission organization (RTO). These entities coordinate the operation of the generators available in the system using a market-based approach. The process involves forecasting the load over various time periods, such as a day ahead or a week ahead. The ISO/RTO solicits offers from generator operators regarding how much power they can supply, when they can supply it and how much it will cost. Generators that can meet the required technical requirements are then chosen based primarily on cost, with the lowest cost generators chosen sequentially until there are enough selected to be able to supply the forecasted load. The offered price is based on the *marginal cost* of operating the generators. This is the cost which varies with the amount of electricity produced and is closely related to the cost of fuel. The final price assigned for all the power will normally be that which is offered by the most expensive generator that is accepted. This is referred to as the *clearing price*. Power plants that participate in the market are referred to as "merchant plants." Throughout all of this, the ISO/RTO ensures that the technical requirements of the grid, such as maintaining voltage and frequency within prescribed limits, are provided for. Details of real power markets can get quite complicated and are outside the scope of this text. For more information on the policies of one ISO, see ISO-NE (2022).

The marginal cost for wind generation will be very low, since turbines do not consume any fuel. Accordingly, the wind turbines are considered "must-run" and will always be chosen to be part of the generation mix. In this process, it is also incumbent on the plant operators to be as certain as possible that the turbines will actually be able to supply the amount of power that they are prepared to offer. In order to be able to do this, they will need forecasts of the wind conditions for the upcoming time period. This is an important topic itself and is discussed in Chapter 15. If the actual production later differs from the forecasts there will likely be financial penalties. In addition, the ISO/RTO will plan to have sufficient spinning reserve available to take care of any differences that may arise.

There are other financial instruments that may be used for valuing the energy produced by the wind power plants, such as power purchase agreements (PPAs). These and some others are discussed in Chapter 13. Regardless of the financial methods used, however, the ISO/RTO must be certain that the total power supply online at any given time will be sufficient to meet the demand.

As the wind energy penetration increases, electricity prices can become negative in periods of strong wind and system operators may require shutting down the wind plants. Conversely, there is also an increasing need for generating more wind energy at low wind speed and moving toward lower specific power turbines (bigger rotor area for the same power rating).

12.9 Wind Farm Aerodynamics: Overview

Numerous issues arise with the close spacing of multiple wind turbines. These challenges have to be tackled when designing the wind farm layout to answer the question of where to locate and how closely to space the turbines. Because substantial distances may be involved and the terrain may change across and upstream of a wind farm, the wind resource may vary as well. In addition, as noted earlier, the extraction of energy by those wind turbines that are upwind of other turbines results in lower wind speeds at the downwind turbines and increased turbulence. These wake effects can decrease energy production and increase fatigue in turbines that are downwind of other machines. Wake effects will also result in power fluctuations in those turbines, and hence the whole wind farm. As described above, the fluctuating power from a wind farm may affect the local electrical grid to which it is connected. The energy extraction from a wind turbine or wind farm also induces a reduction of the upstream wind speed. This effect is referred to as wind farm blockage and the area which is affected is called the induction zone. The blockage effect can have a negative impact on the overall performance of the wind farm. This and the following sections describe the wake and blockage effects and present engineering models to account for them. A review of wind farm flows is given by Porté-Agel *et al.* (2020). Figure 12.16 illustrates what wakes look like in a typical array.

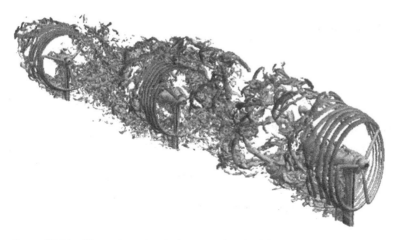

Figure 12.16 Illustration of typical array wake. Vorticity contours are used to visualize the wakes (*Source:* C. Bottaso)

12.10 Characteristics of the Wind Turbine Inflow and Wakes

The primary features of the flow field around a wind turbine can be explained by momentum theory (Chapter 3): the velocity is reduced due to the energy extraction, the rotor induces a rotation in the wake, and the wake expands to satisfy conservation of mass. The actual flow field, however, is more complicated. Some of the details of the flow around and behind a horizontal axis wind turbine are described below and illustrated in Figure 12.17. The consequences of these flow patterns affect downwind turbines and may cause increased fluctuating loads on them compared to the unwaked situation.

12.10.1 Main Flow Features of the Near Wake and Far Wake

Wind turbine wakes can be thought of as consisting of a near wake region, where the velocity deficit is at its strongest, and a far wake region, where the velocity deficit progressively diffuses. The near wake typically extends within one to three rotor diameters downstream of the rotor but there is no clear boundary and the transition between the two regions is progressive.

In the near wake, after the drop of pressure introduced by the rotor, the pressure progressively recovers to reach a value close to the ambient pressure (see middle of Figure 12.17). The change of pressure is accompanied by an expansion of the near wake, most pronounced close to the rotor. According to momentum theory, at the end of the near wake region, the axial induction is approximately twice the axial induction at the rotor. The wake deficit in the near wake region follows the distribution of induction at the rotor, therefore wake deficits are stronger when the turbine operates at higher thrust coefficients. In the near wake, the average wake deficit has a characteristic "double hump" shape (see the right of Figure 12.17). The average wake deficit is lower in the middle ($r \sim 0$) because the airfoils near the root of the blades produce less lift (typically, they are cylindrical sections).

Within the near wake, the vortex structures that are continuously shed from the blade (see Chapter 3) tend to "roll up" into a tip and root vortex. The turbulence kinetic energy in the wake is strong near the tip and root due to these large, concentrated vortices. In the near wake close to the rotor, the distinct wake behind each blade is clearly visible in cross-sections (see bottom of Figure 12.17). Further downstream, the tip vortices of each blade progressively merge and form a cylindrical sheet of rotating turbulence as they mix and diffuse through the flow; see Sørensen and Shen (1999). Therefore, the wake becomes more axisymmetric. In the absence of atmospheric turbulence external to the wake, the shape and stability of the near wake is a function of the operating condition of

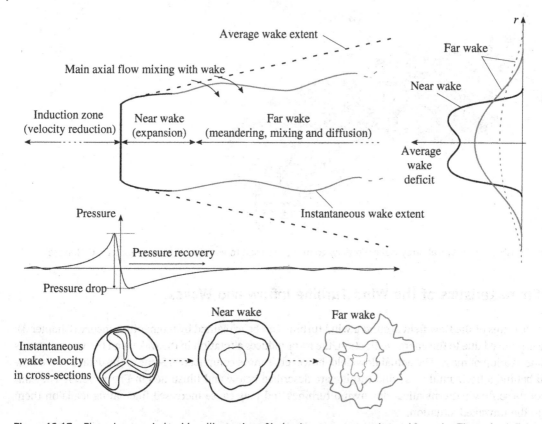

Figure 12.17 Flow about a wind turbine. Illustration of induction zone, near wake, and far wake. The wake deficit progressively diffuses and meanders downstream with the effect of the surrounding turbulence

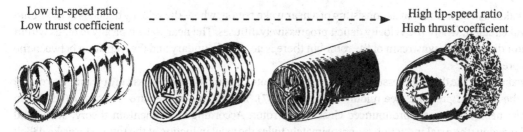

Figure 12.18 Near wake of a wind turbine as function of the operating conditions (*Source*: Adapted from Troldborg *et al.*, 2010)

the rotor (see Figure 12.18). At low thrust coefficients ($C_T \sim 0.4$), the wake can retain a helical shape for long downstream distances. As the thrust coefficient increases, the helical vortices get closer together and merge into a vortex sheet. At very high-thrust coefficients ($C_T > 1$), the wake becomes quickly unstable and breaks down into smaller turbulence structures.

In the far wake region, the free stream flow progressively enters the low-velocity wake region and re-energizes the flow, further accelerating the mixing. Because of this mixing, much of the quasi-periodic nature of the flow (due to similar flow behind each blade) is lost. It becomes impossible to distinguish the vortices from the individual blades and the turbulence and velocity are more evenly distributed in space; see Ebert and Wood (1994). The large scales of turbulence also cause the wake to meander, making the instantaneous definition of the wake boundary and wake center difficult. The time-averaged lateral extent of the wake increases linearly due to its meandering and through mixing with the main flow (see the "average wake extent" in Figure 12.17). The spatial distribution of the time-averaged wake deficit (taken along the vertical or lateral directions) approximates the shape of a Gaussian distribution (see the right of Figure 12.17 and Appendix C). Mixing and diffusion continue in the far wake until the flow field exhibits velocity and turbulence levels representative of the ambient atmospheric conditions without wind turbines. The process of "wake recovery" can extend tens to hundreds of diameters downstream. The rate of wake recovery varies based on the state of the atmospheric boundary layer and has been found to be correlated with turbulence intensity and atmospheric stability. Lower turbulence intensity and stable conditions will lead to slow wake recovery, whereas unstable conditions and high turbulence intensities will lead to faster recovery due to increased mixing.

12.10.2 Impact of Wakes on Downstream Turbines

The most obvious effects of wakes are the decreased wind speed and increased turbulence in the flow at turbines that are downwind of other turbines in a wind farm. The consequences of vorticity and turbulence in turbine wakes are usually increased loads and fatigue (see Hassan *et al.*, 1988) on downstream wind turbines.

12.11 Array Losses

As noted earlier, lower wind speeds within a wind farm result in less energy capture by the downstream turbines. Thus, a wind farm cannot produce 100% of the energy that a similar number of isolated turbines would produce for the same wind speed. The energy loss is termed "array loss" or wake loss. The term "array" was historically used because wind turbines were commonly arranged onto a grid layout, but a modern wind farm can have a more irregular layout, optimized to specific local conditions. Common terms for referring to wind turbine array spacing are illustrated in Figure 12.19. The net effect of the reduced power of many of the turbines in a wind plant is captured by the wind farm power curve, which was introduced above in Section 12.7.1.

Array losses are mainly functions of:

- wind turbine spacing (downwind, crosswind, and in any other directions);
- wind turbine operating characteristics (e.g., the thrust coefficient and tip-speed ratio);
- the number of turbines and size of the wind farm;

Figure 12.19 Wind farm array schematic

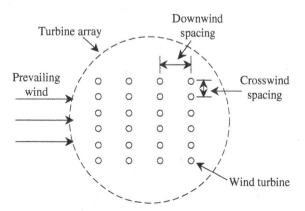

- atmospheric conditions, quantifiable, e.g., by turbulence intensity, atmospheric stability, atmospheric coherence, and turbulence length scale;
- frequency distribution of the wind direction (the wind rose).

Array losses can be reduced – but not eliminated– by optimizing the geometry of the wind farm. Different distributions of turbine sizes, the overall shape and size of the spatial distribution of wind turbines in the farm, and turbine spacing within the wind farm all affect the degree to which wake effects affect energy capture.

The momentum exchange between the turbine wake and the incoming wind is accelerated when there is higher turbulence in the wind field. The higher mixing reduces the velocity deficits downstream, and therefore also reduces array losses. Typical turbulence intensities are between 10% and 15%, but they may be as low as 5% over water or as high as 50% in rough terrain (see Chapter 6). Turbulence intensity also increases through the wind farm due to the interaction of the wind with the turning rotors and the strong vortex structures generated in the turbine wake.

Finally, total array losses are also a function of the annual wind direction frequency distribution. The distances between wind turbines will vary depending on the geometry of the wind turbine locations and the direction of the wind. Thus, array losses need to be calculated based on representative annual wind direction data in addition to wind speed and turbulence data. Additionally, interannual variability and longer-term changes to the wind climate need to be considered.

The geometry of turbine placement and ambient turbulence intensity have been shown to be the most important parameters affecting array losses. Studies have shown that for turbines that are spaced 8–10 rotor diameters, D, along the prevailing wind direction and five rotor diameters along the crosswind direction, array losses may be less than 10%, depending on the wind rose; see Lissaman *et al.* (1982). Figure 12.20 illustrates array losses for a hypothetical 6 × 6 array of turbines with a downwind spacing of 10 rotor diameters. The graph presents array losses as a function of crosswind spacing and turbulence intensity. Results are shown for conditions when the wind is only parallel to turbine rows (turbines directly in the wake of other turbines) and when wind is evenly distributed from all directions.

Array losses may also be expressed as array efficiencies where:

$$\text{Array efficiency} = \frac{\text{Annual energy of whole array}}{(\text{Annual energy of one isolated turbine})(\text{total number of turbines})} \tag{12.22}$$

The array efficiency is 100% minus the array losses in percent.

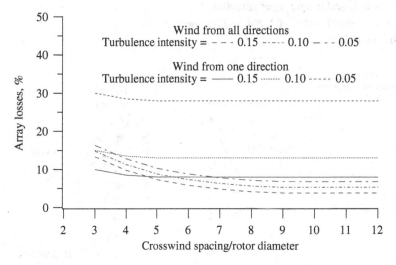

Figure 12.20 Wind farm array losses (after Lissaman *et al.*, 1982) (*Source:* Reproduced by permission of BHR Group Limited)

The design of a wind farm requires careful consideration of array effects in order to maximize energy capture. For example, closer spacing of wind turbines may allow more wind turbines on the site but will reduce the average energy captured from each turbine in the wind farm.

12.12 Wake Models

Wake modeling is a topic of active research due to the negative impact of wakes on power production and wind turbine fatigue. Wake models are used to compute loads, power production, and array losses in a wind farm, based on the turbine locations, wind turbine characteristics, and the wind regime. A number of turbine wake models have been proposed. These include steady-state and time-averaged models, dynamic wake (also called wake tracking) models, and computational fluid dynamics (CFD) models.

12.12.1 Steady-State and Time-Averaged Wake Models

Steady-state wake models provide an account of the mean wake deficit profiles, their diffusion, recovery, and expansion with downwind distance, and they require special methods to merge different wakes together when wakes of different wind turbines interact in a wind farm. The most popular model is the one developed by N.O. Jensen and Katić; see Jensen (1983), Katić *et al.* (1986), and we will present this model in detail in Section 12.12.2. In the present section, the different models are classified as surface roughness models, semi-empirical models, and eddy viscosity models. All models provide descriptions of the energy loss for individual turbines operating in the wake of others.

The surface roughness models are based on data from wind tunnel tests. The first models to attempt to characterize array losses were of this type. A review by Bossanyi *et al.* (1980) describes a number of these models and compares their results. These models assume a logarithmic wind velocity profile upstream of the wind farm. They characterize the effect of the wind farm as a change in surface roughness that results in a modified velocity profile within the wind farm. This modified velocity profile, when used to calculate turbine output, results in a lower power output for the total wind farm. These models were usually derived for arrays of turbines in flat terrain.

Examples of semiempirical models include the ones by Lissaman and Bates (1977), Vermeulen (1980), Jensen and Katić (Jensen, 1983; Katić *et al.*, 1986), and Frandsen *et al.* (2006). These models are based on simplified assumptions about turbine wakes which in turn are based on observations and conservation of momentum. They may include empirical constants derived either from wind tunnel data or wind turbine field tests.

Time-averaged, eddy viscosity models are based on solutions to simplified Navier–Stokes equations. These equations mathematically describe the conservation of momentum and mass of a fluid with constant viscosity and density. The use of the Navier–Stokes equations to describe time-averaged turbulent flow results in terms that characterize the turbulent shear stresses. These stresses can be related to flow conditions using the concept of eddy viscosity. Eddy viscosity models use simplifying assumptions such as axial symmetry and analytical models to determine the appropriate eddy viscosity. These models provide fairly accurate descriptions of the velocity profiles in turbine wakes without a significant computational effort and are also used in array loss calculations. Examples include the model of Ainslie (1985, 1986), Smith and Taylor (1991), and the model *FUGA* (Ott *et al.*, 2011). The curled wake model is another example of an approach that uses simplified Navier–Stokes equations to obtain the evolution of the wake profile in yawed configurations (Martinez-Tossas *et al.*, 2019).

Different wake calculation frameworks have been implemented to compute wind farm performances using various wake models. For instance, the numerical tool *FLORIS* (Annoni *et al.*, 2018) is a controls-oriented modeling tool that determines the steady-state wake characteristics in a wind farm. Several wake models are implemented in *FLORIS*, including the models called: N. O. Jensen model, Gaussian model, and curl wake model. The framework also includes different methods to merge wakes together (e.g., linear superposition or energy superposition).

Figures 12.21–12.23 illustrate wind speed measurements in the far wake of a wind turbine operating at a tip-speed ratio of $\lambda = 4$. The graphs also include the results of the eddy viscosity wake model by Smith and Taylor (1991). Figure 12.21 shows non-dimensionalized vertical velocity profiles at various distances (measured in rotor diameters) behind a wind turbine. The velocity deficit and its dissipation downwind of the turbine are illustrated.

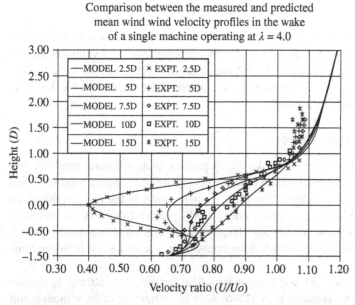

Figure 12.21 Vertical velocity profiles downwind of a wind turbine (Smith and Taylor, 1991); λ, tip speed ratio, U_0, free stream wind speed, D, height with respect to turbine rotor centerline (*Source:* Reproduced by permission of Professional Engineering Publishing)

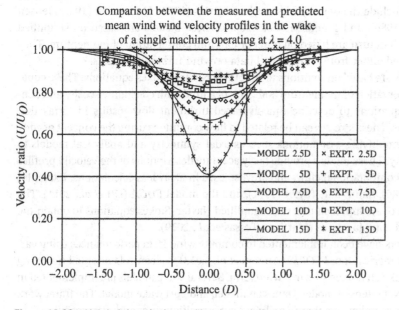

Figure 12.22 Hub height velocity profiles downwind of a wind turbine (Smith and Taylor, 1991); λ, tip speed ratio (*Source:* Reproduced by permission of Professional Engineering Publishing)

Figure 12.23 Turbulence intensity downwind of a wind turbine (Smith and Taylor, 1991); λ, tip speed ratio (*Source:* Reproduced by permission of Professional Engineering Publishing)

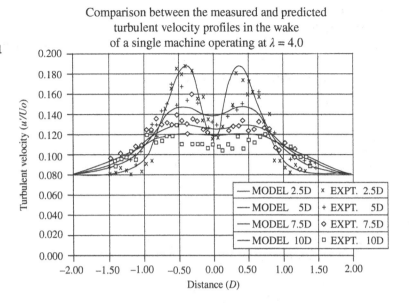

Figure 12.22 shows the hub height velocity profiles as a function of distance from the rotor axis for the same conditions. The Gaussian shape of the hub height velocity deficit in the far wake can be clearly seen. Figure 12.23 shows measured and predicted turbulence intensities at various downstream distances for a wind field with an ambient turbulent intensity of 0.08 and a tip-speed ratio $\lambda = 4$. The double-hump shape is due to the presence of the strong tip-vortices surrounding the overall turbulent core of the wake.

12.12.2 Jensen–Katić Model

The Jensen–Katić semi-empirical model (Katić *et al.*, 1986) is a time-averaged model that is often used for siting and wind farm output predictions. For historical reasons, this model is also referred to as the "N.O. Jensen" model. The model attempts to characterize the energy content in the flow field and ignores the details of the exact nature of the flow field. As seen in Figure 12.24, the flow field is assumed to consist of a linearly expanding wake with a uniform velocity deficit that decreases with distance downstream. The initial free stream velocity is U_0 and the turbine diameter is D. The velocity in the wake at a distance X downstream of the rotor is U_X with a diameter

Figure 12.24 Schematic view of wake description (after Katić *et al.*, 1986); U_0, initial free stream velocity; D, turbine diameter; U_X, velocity at a distance X; D_X, wake diameter at a distance X; k, wake decay constant (*Source:* Katić *et al.*, 1986/with permission of Elsevier)

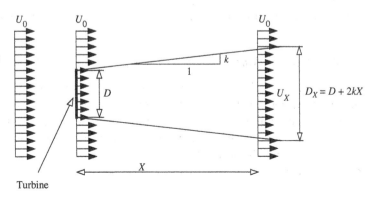

of $D_X = D + 2kX$. The wake decay constant, k, determines the rate at which the wake diameter increases in the downstream direction.

In this and many other semiempirical models, the initial nondimensional velocity deficit (the axial induction factor), a, is assumed to be a function of the turbine thrust coefficient (see Chapter 3). The equation for the axial induction factor is then given by Equation (12.23).

$$a = \frac{1}{2}\left(1 - \sqrt{1 - C_T}\right) \tag{12.23}$$

where C_T is the turbine thrust coefficient. Assuming conservation of momentum, one can derive the following expression for the velocity deficit at a distance X downstream:

$$1 - \frac{U_X}{U_0} = \frac{\left(1 - \sqrt{1 - C_T}\right)}{\left(1 + 2k\frac{X}{D}\right)^2} \tag{12.24}$$

The model assumes that the kinetic energy deficit of interacting wakes is equal to the sum of the energy deficits of the individual wakes (indicated by subscripts 1 and 2). Thus, the velocity deficit at the intersection of two wakes is:

$$\left(1 - \frac{U_X}{U_0}\right)^2 = \left(1 - \frac{U_{X,1}}{U_0}\right)^2 + \left(1 - \frac{U_{X,2}}{U_0}\right)^2 \tag{12.25}$$

The only empirical constant in the model is the wake decay constant, k, which is a function of numerous factors, including the ambient turbulence intensity, turbine-induced turbulence, and atmospheric stability. Katić notes that in a case in which one turbine was upstream of another, $k = 0.075$ adequately modeled the upstream turbine, but $k = 0.11$ was needed for the downstream turbine, which was experiencing more turbulence. He also notes that the results for a complete wind farm with wind coming from multiple directions are relatively insensitive to minor changes in the value of k. A small constant gives a large power reduction in a narrow zone, while a large value gives a smaller reduction in a wider zone. The net effect of varying this parameter, when analyzing wind farm performance at many wind speeds from a variety of directions, is small.

The following steps are used to determine the output of a wind farm using the model:

1) The wind turbine radius, hub heights, power, and thrust characteristics are determined.
2) The wind turbine locations are determined such that the coordinate system can be rotated for analysis of different wind directions.
3) The site wind data are binned by wind direction with, for example, 30-degree wide bins. Weibull parameters are determined for each bin together with the frequency of the wind occurring from each sector.
4) The annual average wind power is calculated by stepping through all wind speeds and directions. Thrust coefficients are determined from the operating conditions at each turbine.

A generalization of the model can be found in the work of Frandsen *et al.* (2006).

12.12.3 Dynamic Wake Models

Dynamic wake (or wake tracking) models are an engineering approach to capture the time evolution of a wake at a lower computational cost than solving the full Navier–Stokes equations. The models track "slices" of wake deficits that diffuse with time and are convected downstream by the local wake velocity and sideways ("meandering") by the large scales of turbulence. The dynamic wake meandering model developed by Larsen *et al.* (2008) is the first model of this kind. The open-source tool *FAST.Farm* provides an alternative formulation with several improvements (see Jonkman *et al.*, 2017).

12.12.4 CFD Models

A variety of CFD approaches exist to solve the complete set of Navier–Stokes equations. These models require a significant computational effort and may use additional models to describe the transport and dissipation of turbulent kinetic energy (the k–ε model) to converge to a solution. In the past, these models were mostly used for research purposes to obtain detailed descriptions of wake behavior and to guide the development of simpler models. Examples include models by Crespo *et al.* (1985), Voutsinas *et al.* (1993), Sørensen and Shen (1999), and Porté-Agel *et al.* (2020). With the advance in computer technologies, CFD models are becoming more and more accessible and adaptable beyond research.

12.12.5 Challenges for Wake Modeling

Several factors affect the accuracy of engineering wake models (i.e., non-CFD models). For these models, decisions must be made about how to handle the superposition of multiple wakes and effects specific to the surrounding environment (various terrain types, offshore, or coastal conditions) on wake decay and ambient wind speed. Typically, multiple wakes are combined by summing the energy in the wakes, although some models assume linear superposition of velocities. The effects of complex terrain (mountainous/forrested) may be significant (see Smith and Taylor, 1991) but are more difficult to address in engineering models and are often ignored. Other effects that are difficult to consider and include in engineering models are ground effects (interaction between the wake and the ground) and deep-array effects (interaction between the atmosphere and multiple wakes at larger scales).

The challenges of CFD wake modeling are typically the same as the ones listed for CFD simulations in Chapter 3. Accurate modeling of the turbulence and the atmospheric boundary layer is of particular importance because wake dynamics are a strong function of the stability of the atmosphere. The stability is influenced by temperature gradients (see Chapter 6) and the surface boundary conditions. Therefore, modeling of the surface roughness, the surface orography, and the heat flux at the domain boundaries are important for the predicting capability of the models. Simulations in complex terrain are still highly challenging for CFD practitioners.

Validation of CFD models is now eased with the advancements of remote sensing technologies such as lidars, which offer the opportunity to measure the wind speed at a variety of points in space. Current research attempts to better understand wake interactions and the wake dynamics under different stability conditions and under complex terrain. The reader is referred to reports from the Wakebench project (Doubrawa *et al.*, 2020), and the review paper by Porté-Agel *et al.* (2020) for more details on the status of CFD wake modeling.

12.13 Wake Effect Mitigation

As a consequence of wake turbulence and its negative effect on the fatigue life of wind turbines, different mitigation strategies may be adopted by wind farm operators and wind turbine manufacturers. Optimization of the wind farm layout in the design phase is a passive way to minimize the negative impacts of wakes. Another way to address these effects passively is to use a different wind turbine technology altogether: downwind rotor, have negative tilt angles and thereby steer the wake toward the ground and reduce wake effects. Active strategies, on the other hand, consist of performing specific maneuvers during operation. Common active strategies consist in reducing the wake deficit (typically via curtailment, also called "down-regulation"), introducing cyclic pitch to destabilize the wake faster, or steering the wakes away from downstream turbines (referred to as "wake steering") by yawing the turbine. Wake steering can be achieved actively by intentionally operating the upstream turbine at a given yaw offset, making use of the fact that the wake of a yawed turbine deflects away from the mean wind direction (see the section on skewing of the wake in Chapter 3). Different numerical and experimental studies have shown that the overall wind farm power production in wind plants can be increased by implementing a yaw angle control strategy throughout the wind farm (e.g., Campagnolo *et al.*, 2016; Fleming *et al.*, 2018). Implementing such strategies is a compromise between the power production and the load effects from wakes but also from the strategy itself (e.g., wear of actuators, increased loads when operating in yawed conditions, etc.).

12.14 Wind Farm Wakes and Blockage Effects

Multiple wind turbines spaced relatively close together in a wind farm will result in downwind and upwind effects. In the downwind direction, there is a wind farm wake, which is basically similar to the wakes of individual turbines, but larger in scale. The wind farm wake is a zone of reduced wind speed and increased turbulence that can extend tens to hundreds of diameters downstream. Examples and analyses of these wind farm wakes are provided by Bodini *et al.* (2021).

The upwind effect of a wind farm is also a reduction in wind speed; it is referred to as wind farm blockage, with the area of influence referred to as the induction zone (see Figure 12.17). The analytical vortex cylinder and vortex doublet models developed by Branlard and Forsting (2020) can be used to provide the velocity field in front of a single turbine, or a wind farm (by superposition), to assess the blockage effect. Examples of velocity fields obtained with the vortex cylinder model for a single turbine are given in Figure 12.25 for an aligned and yawed actuator disk. The model predicts the expected velocity reduction in front of the turbine and some speed up on the sides of the rotor. The expression of the axial velocity along the rotor axis obtained using this model was given in Chapter 3.

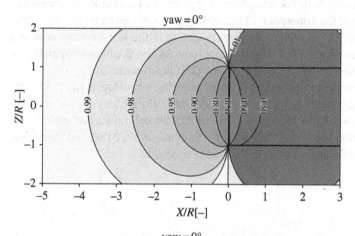

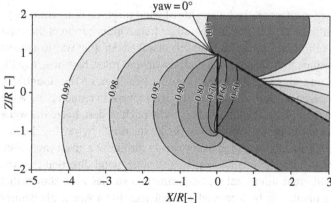

Figure 12.25 Velocity field in the induction zone for aligned (left) and yawed (flows), as predicted using the vortex cylinder model for a thrust coefficient of $C_T = 0.95$. Contours indicate the axial velocity normalized by the free stream.

Current research investigates the effect of atmosphere stability on blockage effects and implements dynamic blockage models.

References

ABB (2023) *XLPE Submarine Cable Systems- Attachment to XLPE Land Cable Systems - User's Guide*, https://new.abb.com/docs/default-source/ewea-doc/xlpe-submarine-cable-systems-2gm5007.pdf Accessed 6/6/2023.

Ackerman (2012) *Wind Power in Power Systems*. John Wiley & Sons, Ltd, Chichester.

Ainslie, J. F. (1985) Development of an eddy-viscosity model for wind turbine wakes. *Proc. 7th British Wind Energy Association Conference*. Multi-Science Publishing Co, London.

Ainslie, J. F. (1986) Wake modelling and the prediction of turbulence properties. *Proc. 8th British Wind Energy Conference*, Cambridge, pp. 115–120.

Annoni, J., Fleming, P., Scholbrock, A., Dana, S., Adcock, C., Porte-Agel, F., Raach, S., Haizmann, F. and Schilpf, D. (2018) Analysis of control oriented wake modeling tools using lidar field results. *Wind Energy Science*, **3**, 819–831.

ANSI/IEEE (1992) *IEEE Recommended Practices and Requirements for Harmonic Control in Electrical Power Systems*. ANSI/IEEE Std 519-1992.

Beaty, H. W. and Fink, D. G. (2013) *Standard Handbook for Electrical Engineers*. McGraw-Hill Education.

Beyer, H. G., Luther, J. and Steinberger-Willms, R. (1989) Power fluctuations from geographically diverse, grid coupled wind energy conversion systems. *Proc. of the 1989 European Wind Energy Conference*, Glasgow, pp. 306–310.

Bodini, N., Lundquist, J. K. and Moriarty, P. (2021) Wind plants can impact long-term local atmospheric conditions. *Scientific Reports*, **11**(1), 1–12.

Bossanyi, E. A.,Maclean, C., Whittle, G. E., Dunn, P. D., Lipman, N. H. and Musgrove, P. J. (1980) The efficiency of wind turbine clusters. *Proc. Third International Symposium on Wind Energy Systems*, Lyngby, DK.

Bossanyi, E., Saad-Saoud, Z. and Jenkins, N. (1998) Prediction of flicker produced by wind turbines. *Wind Energy*, **1**(1), 35–55. John Wiley & Sons, Chichester, UK.

Branlard, E. and Meyer Forsting, A. (2020) Assessing the blockage effect of wind turbines and wind farms using an analytical vortex model. *Wind Energy*, **23**(11), 2068–2086.

Campagnolo, F., Petrovic, V., Schreiber, J., Nanos, E. M., Croce, A. and Bottasso, C. L. (2016) Wind tunnel testing of a closed-loop wake deflection controller for wind farm power maximization. *Journal of Physics Conference Series*, **753**(3), 1–9. 032,006.

CENELEC (1993) *Flickermeter – Functional and Design Specifications*, European Norm EN 60868: 1993 E: IEC686:1986 +A1: 1990. Brussels.

Crespo, A., Manuel, F., Moreno, D., Fraga, E. and Herandez, J. (1985) Numerical analysis of wind turbine wakes. *Proc. of the Workshop on Wind Energy Applications,* Delphi.

de León, F. (2006) Calculation of Underground Cable Ampacity. *Proceedings North American Transmission and Distribution Conference*, Montreal, Canada, June 13–15.

DOE (2008), *20% Wind Energy by 2030*, Report DEO/GO-102008-2567, US Department of Energy, Washington, DC.

Doubrawa, P., Quon, E. W., Martinez-Tossas, L. A., Shaler, K., Debnath, M., Hamilton, N., Herges, T. G., Maniaci, D., ChristopherKelley, C. L., Hsieh, A. S., Blaylock, M. L., van der Laan, P., Andersen, S. J., Krueger, S., Cathelain, M., Schlez, W., Jonkman, J., Branlard, E., Steinfeld, G., Schmidt, S., Blondel, F., Lukassen, L. J., and Moriarty, P. (2020) Multimodel validation of single wakes in neutral and stratified atmospheric conditions. *Wind Energy* **23**(11) 2027–2055.

Ebert, P. R. and Wood, D. H. (1994) Three dimensional measurements in the wake of a wind turbine, *Proc. of the 1994 European Wind Energy Conference, Thessalonika*, pp. 460–464.

Fleming, P., Annoni, J., Churchfield, M., Martinez-Tossas, L. A., Gruchalla, K., Lawson, M. and Moriarty, P. (2018) A simulation study demonstrating the importance of large-scale trailing vortices in wake steering. *Wind Energy Science* **3**(1):243–255.

Frandsen, S., Barthelmie, R., Pryor, S., Rathmann, O., Larsen, S., Højstrup, J. and Thøgersen, M. (2006) Analytical modelling of wind speed deficit in large offshore wind farms. *Wind Energy* **9**(1–2), 39–53.

Gilbertson, O. I. (2000) *Electrical Cables for Power and Signal Transmission.* Wiley-Interscience, Hoboken, NJ.

Gönen, T. (2014) *Electrical Power Transmission System Engineering: Analysis and Design.* CRC Press.

Hassan, U., Taylor, G. J. and Garrad, A. D. (1988) The impact of wind turbine wakes on machine loads and fatigue. *Proc. of the 1998 European Wind Energy Conference,* Herning, DK, pp. 560–565.

ISO-NE (2022) *Day-Ahead and Real-Time Energy Markets,* https://www.iso-ne.com/markets-operations/markets/da-rt-energy-markets/. Accessed 6/9/2023.

Jenkins, N. (1995) Some aspects of the electrical integrations of wind turbines. *Proc. of the 17th BWEA Conference,* Warwick, UK.

Jensen, N. O., (1983) A note on wind generator interaction. *Risø National Laboratory –Risø-M-2411,* Roskilde, DK.

Jonkman, J., Annoni, J., Jonkman, B., Hayman, G. and Purkayastha, A. (2017) *Development of FAST.Farm: A New Multiphysics Engineering Tool for Wind Farm Design and Analysis.* NREL https://www.nrel.gov/docs/fy17osti/67528.pdf. Accessed 6/9/2023.

Katić, I., Højstrup, J. and Jensen, N. O. (1986) A simple model for cluster efficiency *Proc. of the 1986 European Wind Energy Conference,* Rome.

Larsen, G. C., Madsen, H. A., Thomsen, K. and Larsen, T. J. (2008) Wake meandering. *Wind Energy* **11**(4), 377–95.

Lissaman, P. B. S. and Bates, E. R. (1977) *Energy Effectiveness of Arrays of Wind Energy Conversion Systems.* AeroVironment Report AV FR 7050. Pasadena, CA.

Lissaman, P. B. S., Zaday, A. and Gyatt, G. W. (1982) Critical issues in the design and assessment of wind turbine arrays. *Proc. 4th International Symposium on Wind Energy Systems,* Stockholm.

Martinez-Tossas, L., Annoni, J., Fleming, P. A., Chruchfield, M. J. (2019) The aerodynamics of the curled wake: a simplified model in view of flow control. *Wind Energy Science,* **4**, 127–138.

Masters, G. M. (2013) *Renewable And Efficient Electric Power Systems,* 2nd edition. John Wiley & Sons.

NDBC (2022) *National Data Buoy Center,* https://www.ndbc.noaa.gov/ Accessed 3/1/2023.

Neher, J. H. and McGrath, M. H. (1957) The calculation of the temperature rise and load capability of cable systems. *AIEE Transactions Part III – Power Apparatus and Systems,* **76**, 752–772.

Ott, S., Berg, J. and Nielsen, M. (2011) Linearised CFD models for wakes. *Technical report Risø–R–1772(EN).* Roskilde, Denmark: Risø National Laboratory.

Patel, M. R. (1999) Electrical system considerations for large grid-connected wind farms. *Proc. of the 1999 American Wind Energy Assoc. Conference.* Burlington, VT.

Porté-Agel, F., Bastankhah, M. and Shamsoddin, S. (2020) Wind-Turbine and wind-farm flows: a review. *Boundary-Layer Meteorology* **174**, 1–59 https://doi.org/10.1007/s10546-019-00473-0 Accessed 6/12/2023.

Rebello, E., Watson, D. and Rodgers, M. (2020) Ancillary services from wind turbines: automatic generation control (AGC) from a single Type 4 turbine, *Wind Energy Science,* **5**, 225–236.

Rogers, W. J. S. and Welch, J. (1993) Experience with wind generators in public electricity networks. *Proc. of BWEA/RAL Workshop on Wind Energy Penetration into Weak Electricity Networks,* Rutherford Appleton Laboratory, Abington, UK.

Smith, D. and Taylor, G. J. (1991) Further analysis of turbine wake development and interaction data. *Proc. of the 13th British Wind Energy Association Conference,* Swansea.

Sørensen, J. N. and Shen, W. Z. (1999) Computation of wind turbine wakes using combined Navier–Stokes actuator-line methodology. *Proc. 1999 European Wind Energy Conference, Nice,* pp. 156–159.

Troldborg, N., Sørensen J. N. and Mikkelsen, R. (2010) Numerical simulations of wake characteristics of a wind turbine in uniform inflow. *Wind Energy,* **13**, 86–99. https://doi.org/10.1002/we.345.

Vermeulen, P. E. J. (1980) An experimental analysis of wind turbine wakes. *Proc. Third International Symposium on Wind Energy Systems,* Lyngby, DK, pp. 431–450.

Voutsinas, S. G., Rados, K. G. and Zervos, A. (1993) Wake effects in wind parks. A new modelling approach. *Proc. of the 1993 European Wind Energy Conference,* Lübeck, pp. 444–447.

Walker, J. F. and Jenkins, N. (1997) *Wind Energy Technology*. John Wiley & Sons, Chichester, UK.

Weedy, B. M. and Cory, B. J. (2012) *Electric Power System*, 5th edition. John Wiley & Sons, Chichester, UK.

Worzyk, T. (2009) *Submarine Power Cables: Design, Installation, Repair, Environmental Aspects*. Springer-Verlag, Berlin.

Wright, S. D., Rogers, A. L., Manwell, J. F. and Ellis, A. (2002) Transmission Options for Offshore Wind Farms in the United States, *Proceedings of the AWEA Annual Conference*, Portland, OR.

13

Wind Energy System Economics

13.1 Introduction

In the previous chapters, the main emphasis has been on the technical and performance aspects of wind turbines and their associated components. As discussed there, in order for a wind turbine to be a viable contender for producing energy, it must: (1) produce energy, (2) survive, and (3) be cost effective.

Assuming that one has designed a wind energy system (turbine or wind farm) that can reliably produce energy, one should be able to predict its annual energy production. With this result and the determination of the manufacturing, installation, operation and maintenance (O&M), and financing costs, the cost effectiveness can be addressed. As shown in Figure 13.1, in discussing the economic aspects of wind energy, it is also important to treat the costs of generating wind energy and the market value of the energy produced (its monetary worth) as separate subjects. The economic viability of wind energy depends on the match between these two variables. That is, the market value must exceed the cost before the purchase of a wind energy system can be economically justified.

The economic aspects of wind energy vary depending on the application. Grid-connected wind turbines will undoubtedly make a larger contribution to the world's energy supply than turbines on isolated networks. Thus, the primary emphasis in this chapter will be on wind systems supplying electricity to consumers on the main grid. Much of the material here, however, can be used for small to medium-sized wind energy installations that are isolated from a large electrical grid.

This chapter concentrates on the economics of larger wind energy systems, first introducing the subjects shown in Figure 13.1. Following an overview of the economic assessment of wind energy systems (Section 13.2), details of the capital, operation, and maintenance costs of wind energy systems are summarized in Sections 13.3 and 13.4. Then, Sections 13.5 and 13.6 discuss the value of wind energy and the variety of economic analysis methods that can be applied to determine the economic viability of wind energy systems. These methods range from simplified procedures to detailed life cycle costing (LCC) models. The chapter concludes with Section 13.7, reviewing wind energy system market considerations.

13.2 Overview of Economic Assessment of Wind Energy Systems

This section discusses the overall economics of wind energy systems, covering the topics shown in Figure 13.1.

Wind Energy Explained: On Land and Offshore, Third Edition.
James F. Manwell, Emmanuel Branlard, Jon G. McGowan, and Bonnie Ram.

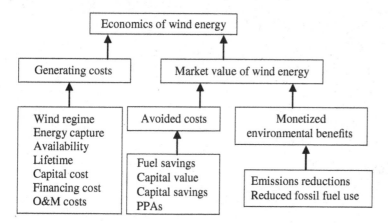

Figure 13.1 Components of wind system economics

13.2.1 Generating Costs of Grid-connected Wind Turbines: Overview

The total generating costs/MWh for an electricity-producing wind turbine system are determined by the following factors:

- wind regime;
- energy capture efficiency of the wind turbine(s);
- availability of the system;
- lifetime of the system;
- capital costs;
- financing costs;
- O&M costs.

The first two factors have been addressed in detail in previous chapters. The remaining factors are summarized below. Chapter 14 discusses regulatory compliance and stakeholder engagement that also add to costs. As pointed out in Chapter 2, variation in the long-term wind resource can cause uncertainty in production and resulting generation costs (Raftery *et al.*, 1999). Thus, these topics should be considered in more detailed economic analyses.

13.2.1.1 Availability

The availability is the fraction of the time that a wind turbine is able to generate electricity. The times when a wind turbine is not available include downtime for periodic maintenance or unscheduled repairs. As noted by DNV GL (2017), wind turbine availability terms differ widely with intended use; thus, when an availability term is used, it should be carefully defined. For example, Pfaffel *et al.* (2017) define time-based, technical, and energetic availability terms. They observed that the time-based availability of onshore wind turbines is, with a few exceptions, close to 95%. They also found that, due to harsh environmental conditions and more difficult access, offshore wind turbines tend to have lower availability when compared to onshore wind turbines.

13.2.1.2 Lifetime of the System

It is common practice to equate the design lifetime with the economic lifetime of a wind energy system. Typically, a period of 20 years is assumed for the economic assessment of wind energy systems (WEC, 1993), although a 2017 NREL report (Mai *et al.*, 2017) used a 24-year lifetime for wind turbines. Also, following improvements in wind turbine design, an operating life of 30 years has been used for detailed US economic studies. This assumption requires that adequate annual maintenance be performed on the wind turbines and that 10-year major mainte-nance overhauls be performed to replace key parts.

13.2.1.3 Capital Costs

Details on the determination of the capital cost of wind energy systems are given in Section 13.3. These costs are primarily the cost of the wind turbine(s) and the cost of the remaining installation (often referred to as "balance of system"). Wind turbine costs can vary significantly. For example, costs vary significantly for each rated generator size. This may be due to differing tower heights, rotor diameters, and different wind turbine designs.

In generalized economic studies, wind turbine installed costs are often normalized to cost per rated kW. An example of normalized onshore wind turbine costs is shown in Figure 13.2.

The normalization of wind turbine costs to installed cost per unit of rated power (e.g., $/kW) follows conven-tional large-scale electrical generation practice and is often used. For commercial turbines used in wind farms, there has been a steady decrease in cost per unit of power since the 1980s. The cost of wind turbines, however, can vary, primarily due to higher materials and energy input prices (e.g., steel, aluminum, copper, and rare earth minerals). This point was illustrated in Figure 13.2, in which installed US wind turbine costs ($/kW) are given as a function of time (Wiser and Bolinger, 2018).

As discussed previously, the costs of a wind farm installation include costs beyond those of the wind turbines themselves. For example, in general, the wind turbine represents only approximately 65–75% of the total invest-ment costs. Wind farm costs also include costs for infrastructure and installation, as well as electrical grid connec-tion costs. Costs per kW (unit cost) of single wind turbine installations are generally much higher, which is the reason for wind farm development in the first place.

As will be discussed later in this chapter, capital costs for offshore wind turbines are generally higher than those for land-based turbines. As stated by the International Renewable Energy Agency, the costs for offshore wind in the early 2000s increased as projects shifted to deeper water, primarily raising the support structure, installation, and grid connection costs (International Renewable Energy Association (IRENA), 2012). They also note that, since then, the totaled costs have decreased and are expected to further decrease to a range of $1400 to $2800/kW in 2050.

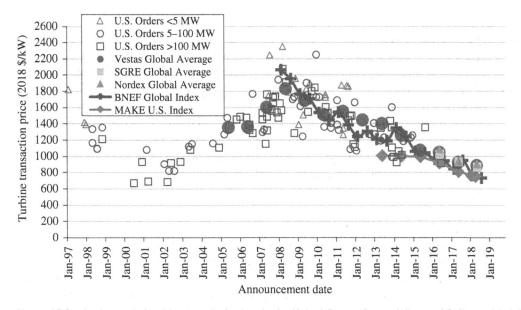

Figure 13.2 Onshore wind turbine installed prices in the United States (*Source:* Wiser and Bolinger, 2018/Public Domain)

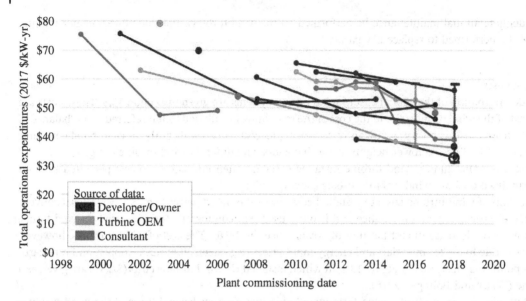

Figure 13.3 Operational expenditures, based on plant commissioning data (*Source:* Wiser and Bolinger, 2019/with permission of Elsevier)

13.2.1.4 Financing Costs

Wind energy projects are capital-intensive, and the majority of the costs must be borne at the beginning. For that reason, the purchase and installation costs are largely financed. The purchaser or developer will pay a limited down payment (perhaps 10–20%) and finance (borrow) the rest. The source of capital may be a bank or investors. In either case, the lenders will expect a return on the loan. The return in the case of a bank is referred to as the interest. Over the life of the project, the cumulative interest can add up to a significant amount of the total costs. Other methods of financing are also used, but they are beyond the scope of this text.

13.2.1.5 Operation and Maintenance Costs

Many authors prefer to use a set cost per kWh of output for their O&M cost estimates. Figure 13.3 gives results from an onshore US study on the operating costs from the late 1980s to 2018 (Wiser and Bolinger, 2019). This work shows that the levelized O&M costs have averaged about $80/kW-yr for projects built in the late 1990s to below $40/kW-yr for recent projects.

For offshore wind turbines, the operating and maintenance costs are generally higher than for onshore systems. These costs can vary considerably between projects for a number of reasons, and the two largest cost variables are the distance from the maintenance facility to the project site and the meteorological climate at the site (Stehly *et al.*, 2018). Section 13.4 gives more details on the range of values for O&M costs.

13.2.2 Value of Grid-Connected Wind Energy: Overview

The value of wind energy depends on the application and on the costs of alternatives to produce the same output. To determine value, one may try to figure "what the market will bear." For example, a manufacturer may ask for a price and adjust it until he or she finds a buyer (negotiated price). Alternatively, a buyer, such as an electric utility, may ask for competitive bids to supply power and select those with the lowest prices that look qualified.

For a utility, the value of wind energy is primarily a function of the cost of the fuel that would not be needed or the amount of new system-generating capacity that could be deferred. For society as a whole, however, the environmental benefits can be quite significant (see Section 14.4). When these benefits are monetized, they can contribute significantly to the market value of electricity. More details on the determination of wind energy's value are given in Section 13.5.

13.2.3 Economic Analysis Methods: Overview

A wind energy system can be considered to be an investment that produces revenue. The purpose of an economic analysis is to evaluate the profitability of a wind energy project and to compare it with alternative investments. The alternative investments might include other renewable electricity-producing power systems (photovoltaic power, for example) or conventional fossil fuel systems. Estimates of the capital costs and O&M costs of a wind energy system, together with other parameters that are discussed below, are used in such analyses.

In Section 13.6, the following three different types of overall economic analysis methods are described:

- Simplified models;
- Detailed life cycle cost models;
- Electric utility economic models.

Each of these economic analysis techniques requires its own definition of key economic parameters, and each has its particular advantages and disadvantages.

13.2.4 Market Considerations: Overview

At the present time, the market for utility-scale wind energy systems is rapidly expanding. The world market demand has grown from about 200 MW/year in 1990 to more than 45 000 MW/year through 2018 (IRENA, 2019a). In Section 13.7, the market considerations for wind systems are discussed. They include the following issues:

- Potential market for wind systems;
- Barriers to expansion of markets;
- Incentives for market development.

13.3 Capital Cost Estimation of Wind Turbines

13.3.1 General Considerations

The determination of the capital costs of a wind turbine is often a challenging one. The problem is complicated because wind turbine manufacturers are usually not anxious to share their own cost figures with the rest of the world, or, in particular, with their competitors. Cost comparisons of wind turbine system research and development projects are particularly difficult; that is, the development costs cannot be compared consistently. Also, costs other than the turbine may be site-specific. This problem can further complicate capital cost estimates. This is especially true for offshore projects, in which support structure costs may be large relative to those of the rotor nacelle assembly and the rest of the system.

In determining the cost of the wind turbine itself, one must distinguish between the following types of capital cost estimates:

- **Cost of wind turbine(s) today.** For this type of cost estimate, a developer or engineer can contact the manufacturer of the machine of interest and obtain a formal price quote.

- **Cost of wind turbine(s) in the future.** For this type of cost estimate (assuming the present wind turbine state of the art), one has a number of tools that can be applied, namely: (1) historical trends, (2) learning curves (see next section), and (3) detailed examination of present designs (including the total machine and components) to determine where costs may be lowered.
- **Cost of a new (not previously built) wind turbine design.** This type of capital cost estimate is much more complex since it must first include a preliminary design of the new turbine. Price estimates and quotes for the various components must be obtained. The total capital cost estimate must also include other costs such as design, fabrication, testing, etc.
- **Future cost of a large number of wind turbines of a new design.** This type of cost estimate will involve a mixture of the second and third types of cost estimates.

It is assumed that one can readily obtain the first type of capital cost estimate. Accordingly, the following discussion concentrates on information that can be used for the last three types of cost estimates.

Recent worldwide experience with wind turbine fabrication, installation, and operation has yielded some data and analytical tools that can be used for capital cost estimates of wind turbines and/or of the supporting components that are required for a wind farm installation. For example, there have been numerous capital cost estimates for wind turbines based on various simplified scaling techniques. These usually feature a combination of actual cost data for a given machine and empirical equations for the cost of the key components based on a characteristic dimension (such as rotor diameter) of the wind turbine. The same type of generalized scaling analysis has been used to predict wind farm costs.

13.3.2 Use of Experience (or Learning) Curves to Predict Capital Costs

One unknown in capital cost estimates is the potential reduction in costs of a component or system when it is produced in large quantities. One can use the concept of experience (or learning) curves to predict the cost of components when they are produced in large quantities. The experience curve concept is based on over 40 years of studies of manufacturing cost reductions in major industries (Johnson, 1985; Cody and Tiedje, 1996).

The experience curve gives an empirical relationship between the cost of an object $C(V)$ as a function of the cumulative volume, V, of the object produced. Functionally, this is expressed as:

$$\frac{C(V)}{C(V_0)} = \left(\frac{V}{V_0}\right)^b \tag{13.1}$$

where the exponent b, the learning parameter, is negative, and $C(V_0)$ and V_0 correspond to the cost and cumulative volume at an arbitrary initial time. From Equation 13.1, an increase in the cumulative production by a factor of 2 leads to a reduction in the object's cost by a progress ratio, s, where $s = 2^b$. The progress ratio, expressed as a percentage, is a measure of the technological progress that drives cost reduction.

A graphical example of this relationship is presented in Figure 13.4, which gives a plot of normalized cost $C(V_0) = 1$ for progress ratios ranging from 70% to 95%. As shown, given the initial cost and an estimate of the progress ratio, one can estimate the cost of the tenth, hundredth, or whatever unit of production for the object.

Experience has shown that s can range from 70% to 95%. For example, Johnson (1985) gives values of s of 95% for electrical power generation, 80% for aircraft assembly, and 74% for hand-held calculators. Various parts of a wind turbine would be expected to have different progress ratios; that is, components such as the blades and hub that are essentially unique to wind turbines would have smaller values of s. Towers, however, could represent a more mature technology with higher progress ratio values. For wind turbines, Neij (1999) gives a summary of progress ratios for various utility-scale machines. Global experience curves for wind farms are given by Junginger *et al.* (2005), and Rubin *et al.* (2015) present a detailed review of learning rates for onshore wind farms.

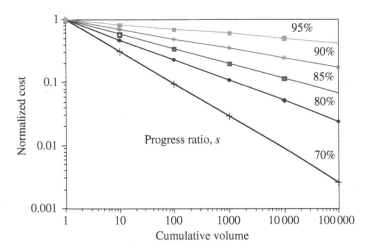

Figure 13.4 Normalized cost vs. cumulative volume for varying progress ratios

13.3.3 Detailed Models for Estimation of Wind Turbine Costs

13.3.3.1 Historical Overview

One way to estimate the capital costs of a new wind turbine is to use cost data for smaller existing machines normalized to a machine size parameter. Here, the usual parameters that are used are unit cost per kW of rated power or unit cost per area of rotor diameter. Although this methodology can be used for energy generation systems or planning studies, it yields little information as to the details of a particular design and the potential cost reduction for new designs of wind turbines.

A more fundamental way to determine the capital cost of a wind turbine is to divide the machine into its various components and to determine the cost of each one. This method represents a major engineering task, however, and there are relatively few such studies documented in the open literature. Historically interesting examples here include US work on a conceptual 200 kW machine (NASA, 1979) and work on the MOD2 (Boeing, 1979) and MOD 5A (General Electric Company, 1984) machines. Other work, supported under the WindPACT (Wind Partnerships for Advanced Component Technology) program sponsored by NREL in the United States, produced a number of reports containing methods for the estimation of wind turbine components and total wind turbine costs (e.g., Fingersh *et al.*, 2006). More recent NREL work has used an integrated system design and engineering model (WISDEM) to produce capital cost estimates for wind turbines (see Dykes *et al.*, 2017).

In an effort to quantify the capital costs for horizontal axis wind turbines, a detailed cost model for horizontal axis wind turbines was developed at the University of Sunderland (Harrison and Jenkins, 1993; Hau *et al.*, 1996; Harrison *et al.*, 2000). This approach has formed the basis for NREL cost models in the United States (see Fingersh *et al.*, 2006). Starting with an outline specification of a proposed wind turbine and using some basic design options, a computer code can provide a capital cost estimate of a specific wind turbine design. An overview of the key features of the (weight-based) model's operation is given in Figure 13.5.

Based on the input data, the model uses first principles to develop estimates of the most important influences on the cost of the proposed machine. As shown in Figure 13.5, these are called "design drivers" and include such variables as the blade loadings, the horizontal thrust on the machine, gearbox requirements, and generator specifications. Next, the model estimates the loads on the major subsystems of the wind turbine allowing the approximate sizes of the major components to be determined. For example, analytical expressions are used to estimate the

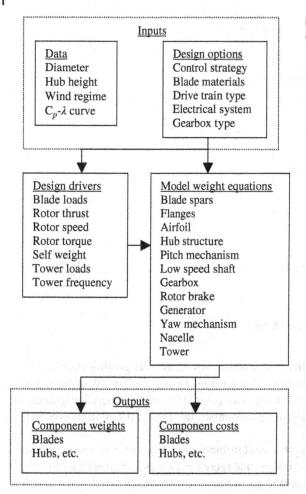

Figure 13.5 Flowchart of Sunderland cost model; C_p, power coefficient; λ, tip speed ratio

low-speed shaft diameter such that the shaft will be able to carry both the torsional and axial loading exerted on it by the rotor. For more complex components, where analytical expressions are not sufficient, look-up tables relating component size to loadings are used. In certain cases, components are assigned a complexity rating in order to reflect how much work is involved in their manufacture.

Using the calculated size of each component, the weights of the components (assuming a known material density) are then determined. It should be noted that some of the weights are used to calculate loadings on other components, but their primary purpose is to allow the estimation of the wind turbine cost. The cost of each subsystem is then determined via the following expression:

$$\text{Cost} = \left(\begin{array}{c}\text{Calibration}\\\text{coefficient}\end{array}\right) \times (\text{Weight}) \times \left(\begin{array}{c}\text{Cost per}\\\text{unit weight}\end{array}\right) \times \left(\begin{array}{c}\text{Complexity}\\\text{factor}\end{array}\right) \qquad (13.2)$$

The calibration coefficient is a constant for each subsystem. It is determined by a statistical analysis of existing wind turbine cost and weight data. The complexity factor is the value assigned during the component sizing phase to reflect the amount of work required for the subsystem's construction.

The total capital cost of the wind turbine is calculated from the sum of the cost estimates for each subsystem. To validate the model's cost prediction, the code has been used to predict the weight and capital cost of a wide range of actual wind turbines (Harrison *et al.*, 2000). Figure 13.6 gives a comparison between the model predictions of the relationship between blade weight and rotor diameter for three types of blade material (solid or dotted lines) and data from a number of wind turbines. Figure 13.7 presents a comparison of total capital cost predictions and actual factory costs of early small machines and current large machines. Both graphs show good agreement between the model and actual weight and cost data. (WEGA is a German acronym for large wind turbines; GRP is an acronym for glass reinforced polyester and GRE for glass reinforced epoxy.)

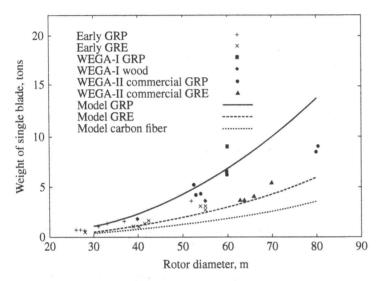

Figure 13.6 Validation of Sunderland cost model weight predictions (*Source:* Harrison *et al.*, 2000/John Wiley & Sons)

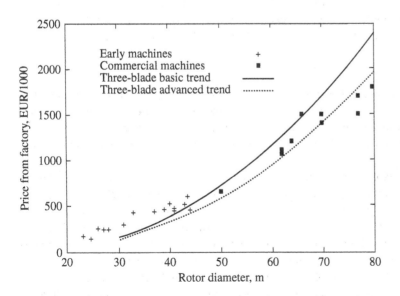

Figure 13.7 Validation of Sunderland wind turbine cost predictions (*Source:* Harrison *et al.*, 2000/John Wiley & Sons)

13.3.4 Wind Farm Costs

13.3.4.1 Onshore Wind Farms

The determination of total wind energy project capital cost, or installed capital cost, involves more than just the cost of the turbines themselves. For example, the installed capital cost of an onshore wind farm could include the following:

- Wind resource assessment and analysis;
- Permitting and public involvement, surveying, and financing;
- Construction of service roads;
- Construction of foundations for wind turbines, pad-mount transformers, and substations;
- Wind turbine and tower delivery to the site and installation;
- Construction and installation of wind speed and direction sensors;
- Construction of a power collection system, including the power wiring from each wind turbine to the pad mount transformer and from the pad mount transformer to the substation;
- Construction of operations and maintenance facilities;
- Construction and installation of a wind farm communication system supporting control commands and data flow from each wind turbine to a central operations facility;
- Provision of power measurement and wind turbine computer control, display, and data archiving facilities;
- Integration and checkout of all systems for correct operation;
- Commissioning;
- Final turnover to owner or operating agency.

A detailed discussion of the determination of each of these cost components is beyond the scope of this text, but an example of a capital cost breakdown ($/kW) for a 200 MW wind farm using 2.4 MW wind turbines is given below in Table 13.1.

An estimate of the future capital cost of onshore wind farms is shown in Figure 13.8 (IRENA, 2019a). This work noted that the total construction costs of onshore wind fell by an average of 22% between 2010 and 2018 and further declined by 6% in 2018 as compared to 2017. This more rapid decline rate was due to deployment in China and India. Furthermore, they predicted that the installed cost is expected to drop further in the next three decades, reaching an average of $650–$1000/kW in 2050 (compared to a 2018 average level of about $1500/kW).

13.3.5 Offshore Wind Farms

The main differences in costs between onshore and offshore wind farms are due to support structure and electrical connection issues (EWEA, 2009). For example, Table 13.2 gives a breakdown of the average investment costs for typical Danish offshore wind farms (Horns Rev and Nysted). As shown, the support structure and electrical connection cost components are each approximately 21% of the total and are thus considerably more expensive than onshore sites.

The Opti-OWECS project (Kühn *et al.*, 1998) investigated the cost difference between offshore and onshore wind farms based on the European experience with such projects. They found that the per-kW cost difference between offshore and onshore wind farms was generally in the range of 25% more to three times as much, depending on such factors as distance from shore, water depth, environmental conditions, soil type, maturity of the technology, and project size. Although their cost estimates for some of the components are somewhat dated, the overall methodology for this study served as a guideline for future offshore wind farm capital cost models.

As with onshore wind farms, a detailed discussion of the determination of the individual cost components of offshore wind farms is beyond the scope of this text. There are a number of recent studies that center on or include the determination of the capital cost of offshore wind farms. They include the work of the International Renewable Energy Association (IRENA, 2012, 2019b) and the US National Renewable Energy Laboratory (Stehly *et al.*, 2018). For example, based on the work of Stehly *et al.* the average cost breakdown percentages for fixed and floating offshore wind farms are shown in Figures 13.9 and 13.10.

Table 13.1 Example of wind farm capital cost breakdown

Subsystem	Component	$/kW	$/kW
Rotor	Blades		188
	Pitch assembly		61
	Hub assembly		45
Rotor total		293	
Nacelle	Nacelle structural assembly		100
	Drivetrain assembly		195
	Nacelle electrical assembly		170
	Yaw assembly		33
Nacelle total		498	
Tower		219	
Turbine total capital cost		1011	
Balance of system	Development cost		16
	Engineering and management		19
	Foundation		60
	Site access and staging		45
	Assembly and installation		45
	Electrical infrastructure		148
Balance of system total		332	
Financial costs	Construction financing cost		39
	Contingency fund		88
Financial costs total		127	
Total capital expenditures		1470	

(*Source:* Adapted from Stehly and Beiter, 2019)

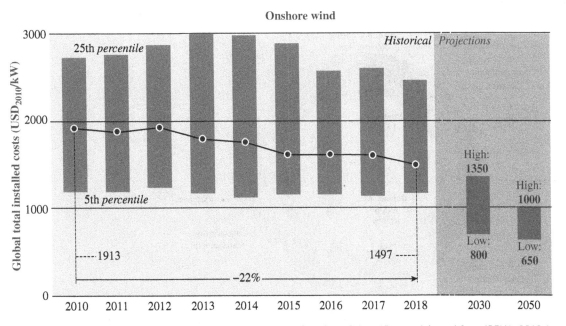

Figure 13.8 Total installed cost of onshore wind farms as a function of time (*Source:* Adapted from IRENA, 2019a)

Table 13.2 Average investment costs related to offshore wind farms in Horns Rev and Nysted

Component	Cost (1000€/MW)	Percentage of total
Turbines (including transport and erection)	815	49.0
Transformer station and main cable to coast	270	16.0
Internal grid between turbines	85	5.0
Support structure	350	21.0
Design and project management	100	6.0
Environmental analysis	50	3.0
Miscellaneous	10	<1

(*Source:* Adapted from EWEA, 2009)

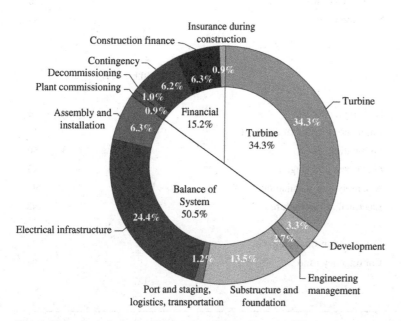

Figure 13.9 Capital cost breakdown for a fixed reference offshore wind farm (*Source:* Stehly *et al.*, 2018/National Renewable Energy Laboratory)

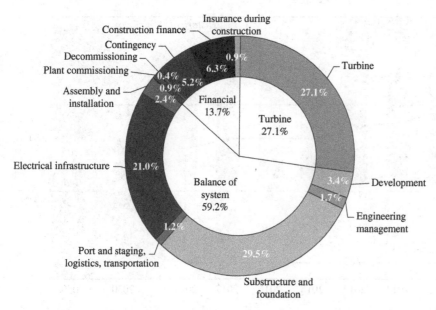

Figure 13.10 Capital cost breakdown for a floating reference offshore wind farm (*Source:* Stehly *et al.*, 2018/National Renewable Energy Laboratory)

13.4 Operation and Maintenance Costs

13.4.1 Land-Based Wind Operation and Maintenance Costs

Next to the purchase and installation of the wind turbines, O&M is the most significant source of costs. It is especially important to project O&M costs if the owner of a wind farm seeks to refinance or sell the project. The operation costs can include insurance on the wind turbines, taxes, and land rental, while the maintenance costs can include the following typical components:

- Routine checks;
- Periodic maintenance;
- Periodic testing;
- Blade cleaning;
- Electrical equipment maintenance;
- Unscheduled maintenance costs.

Experience indicates that there is a wide range of O&M costs for wind turbines (see IRENA, 2012 and Stehly *et al.*, 2018). Note also that O&M costs are not evenly distributed over time. That is, they tend to increase as the age of the project increases (Vachon, 1996; IRENA, 2012).

O&M costs can be divided into two categories: fixed and variable. Fixed O&M costs are yearly charges unrelated to the level of plant operation. These must be paid regardless of how much energy is generated (generally expressed in terms of $/kW installed or percentage of turbine capital cost). Variable O&M costs are yearly costs directly related to the amount of plant operation (generally expressed in $/MWh). Probably the best estimates of O&M costs are a combination of these two categories. For example, Stehly *et al.* (2018) stated that the average fixed costs were $43.6/kW per year and the variable costs were $12/MWh.

13.4.2 Offshore Wind Operation and Maintenance Costs

O&M costs for offshore wind farms are generally higher than those for onshore wind. This is primarily due to the higher costs of access to the site and to the need to perform maintenance on the support structures and the connecting cables. Also, the marine environment is more complex than dry land to operate within, further increasing the costs. O&M costs for offshore wind farms can vary considerably, especially between the fixed and floating designs. Recent estimates range between $0.02 and $0.05/kWh (IRENA, 2019b). As reviewed by Costa *et al.* (2021), O&M costs typically account for 20–25% of the total levelized cost of electricity (LCOE) and are a top priority in wind turbine maintenance strategies. These authors also note that in the 2007–2018 time frame, the O&M costs of onshore projects have declined by 52% and the O&M costs of offshore projects have declined by 45%.

There are numerous offshore wind operations and maintenance models. A detailed discussion of them is beyond the scope of this text, however. For example, European researchers have studied the subject of offshore O&M costs in detail for some time, starting with the Opti-OWECS project (see Kühn *et al.*, 1998, Rademakers *et al.*, 2003, 2008, 2009, and van de Pieterman *et al.*, 2010). The structure of the ECN O&M Cost Estimator (OMCE) is shown in Figure 13.11 (van de Pieterman *et al.*, 2010). This tool requires information on three types of maintenance for the estimation of O&M costs:

1) **Unplanned corrective maintenance**. Due to unexpected failures leading to immediate shut-downs, unforeseen downtimes, and unplanned maintenance actions.
2) **Condition-based maintenance**. Not foreseen initially, but when it has to be carried out during the lifetime, it generally will be planned with a minimum turbine shutdown.
3) **Calendar-based maintenance**. Preventative maintenance.

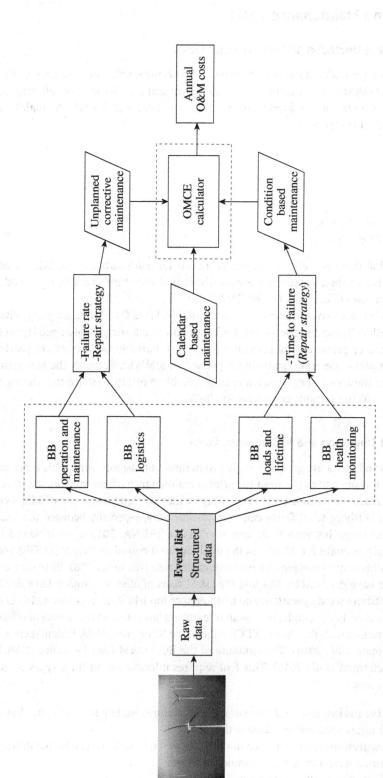

Figure 13.11 OMCE model showing data flow from "raw data" to estimated O&M costs (*Source:* van de Pieterman *et al*, 2010/Energy Research Centre of the Netherlands)

13.5 Value of Wind Energy

13.5.1 Power Purchase Agreements (PPA)

At the present time, the economic value (and viability) of large wind farm projects is generally based on power purchase agreements (PPAs). A PPA is a long-term contract between the seller of wind energy and a buyer. Economic terms are the most important elements of a PPA, but typical PPAs also include sections addressing issues such as agreement length, the commissioning process, the purchase and sale of energy, curtailment agreements, transmission issues, credit, insurance, and environmental benefits or credits. Furthermore, Miller *et al.* (2017) point out that an accurate estimate of the levelized cost of energy (LCOE) is needed to establish the terms of a PPA.

A variety of ways to structure PPA are possible. The three main options are: (1) traditional PPA, (2) virtual power purchase agreement (vPPA), and (3) proxy generation PPA, as summarized below. More information on PPAs may be found in Tang and Zand (2019) and Fan (2021).

13.5.1.1 Traditional Power Purchase Agreement

In a traditional PPA, the owner/developer of a wind energy project has a contract to sell electricity to a utility or an intermediary ("buyer") and physically deliver that electricity over wires to that utility. In the case of a buyer, that entity would make money by buying and reselling the electricity. In a traditional PPA, the project is on the same electrical grid as the utility. The buyer in this situation is likely to be an entity whose primary purpose is to facilitate the transactions and make a profit in the process. The price is set with respect to anticipated future production and sale of the electricity, i.e., a "futures market." The price takes into account the anticipated avoided costs, as discussed in Section 13.5.2.

13.5.1.2 Virtual Power Purchase Agreement (vPPA)

Virtual power purchase agreements (vPPAs) are contracts for differences relative to some reference price. The actual energy is delivered to the grid at the point of interconnection (POC) by the wind project owner/developer, who sells the energy and is paid for the energy by the utility at the spot market (immediate) price. The developer is also given the renewable energy certificates ("RECs"; see Section 13.5.4.3). The contract involves the difference between the value of the energy plus the RECs and the reference price. At the same time, the buyer purchases electricity from the spot market, possibly at a different location and from a different utility. If the buyer pays more than the reference price, the developer pays the buyer the difference (that is OK with the seller because s/he has made money by selling into the spot market at a higher than anticipated price). If the buyer pays less than the reference price, the buyer pays the developer the difference. The buyer in the vPPA case could also be an entity, such as a company or university, that wants to meet sustainability goals by getting their electricity from a renewable source. There may also be buyers who just want the RECs for a similar purpose.

13.5.1.3 Proxy Generation Power Purchase Agreement (Proxy Gen PPA)

A proxy generation PPA is based on a reference to the energy that the wind energy project should have produced, considering what the wind was measured or estimated to have been during every time interval of interest (for example, each hour over a certain period). Details are outside the scope of this text see Fan (2021).

13.5.2 Avoided Cost Valuation

The traditional way to assess the value of wind energy was to equate it to the direct savings that would result from the use of the wind rather than the most likely alternative. These savings are often referred to as "avoided costs." Avoided costs result primarily from the reduction of fuel that would be consumed by a conventional generating plant. They may also result from a decrease in total conventional generating capacity that a utility requires. That is, they have "capacity value." These topics are discussed immediately below. Environmental benefits can also result in avoided costs for society as a whole. These benefits are the subject of Section 13.5.3. (See also Chapter 14).

13.5.2.1 Fuel Savings

The inclusion of wind turbines in an electrical network can reduce the demand for other generating plants that require a fossil fuel input. It might appear that the calculation of fuel savings would be quite straightforward, but this is not generally the case. The type and amount of fuel savings depends on such factors as the mix of different fossil fuel and nuclear plants in the electrical generating system, requirements for "spinning reserves," and operating characteristics of the fossil fuel components (such as efficiency or heat rate as a function of component load). These topics were also discussed in Chapter 12.

The electrical system as a whole must be modeled in order to accurately estimate the avoided fuel consumption. For example, Figure 13.12 summarizes the results of a European Wind Energy Association study on avoided fuel costs from wind energy (Arapogiannia and Moccia, 2014). As shown, wind energy avoided about 10 billion euros in 2012 and was expected to avoid over 50 billion euros in 2030.

13.5.2.2 Capacity Value

The capacity value of a wind system is defined as: "the amount of conventional capacity which must be installed to maintain the ability of the power system to meet the consumers' demand if the wind power installation is deleted" (Tande and Hansen, 1991). Despite the apparent simplicity of the concept, some authors (see Fox *et al.*, 2007) note that few topics in wind energy generate more controversy than the determination of capacity value and its potential economic benefits.

At one extreme, if the utility were absolutely certain that a wind power plant would produce its full rated power during peak demand hours, its capacity value would equal its rated power. At the other extreme, if the utility were certain that the wind energy plant would never operate during peak hours, then its capacity value would be zero. For a utility, the higher the capacity value of the connected wind power plants, the less other generation capacity is

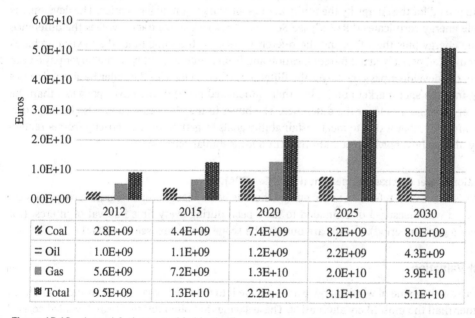

	2012	2015	2020	2025	2030
▨ Coal	2.8E+09	4.4E+09	7.4E+09	8.2E+09	8.0E+09
= Oil	1.0E+09	1.1E+09	1.2E+09	2.2E+09	4.3E+09
▧ Gas	5.6E+09	7.2E+09	1.3E+10	2.0E+10	3.9E+10
▥ Total	9.5E+09	1.3E+10	2.2E+10	3.1E+10	5.1E+10

Figure 13.12 Annual fuel cost avoided in the European Union (*Source:* Adapted from Arapogiannia and Moccia, 2014)

required. In practice, the capacity value of a wind power plant is somewhere between zero and its average power generation during some time interval of high electrical demand in the system. Accordingly, the actual value will depend on the seasonal match of the wind resource to the utility's demand. There are two main methods for the calculation of capacity value (Walker and Jenkins, 1997):

1) The contribution of the wind turbines during peak demand on a utility is assessed over a period of years, and the average power at these times is defined as the capacity value.
2) The loss of load probability (LOLP) or loss of load expectation (LOLE) is calculated, initially, with no wind turbines in the system (Billinton and Allan, 1984). It is then recalculated with wind turbines in the system, and then conventional plant capacity is subtracted until the initial level of LOLP is obtained. The subtracted conventional plant capacity is the capacity value of the wind turbines.

Both methods give similar results. Also, when the wind provides a relatively small amount of the total generation, the capacity value is generally close to the average output of the wind turbines.

An example of the first method, using an analytical equation, is given in the work of Voorspools and D'haeseleer (2006). Most often, the capacity value is calculated by the second method (generally using statistical techniques). For example, as shown in Figure 13.13, the capacity value of wind energy for selected worldwide locations is given (Milligan *et al.*, 2019).

Some studies have produced capacity values ranging from a few percent to values approaching 30–40%. In the middle are other researchers, who propose that renewables with somewhat predictable schedules deserve partial capacity payments (Perez *et al.*, 1997). Also, in the United States, as reviewed by Milligan *et al.* (2019), the determination of the capacity value of wind is a regional function and can vary considerably depending on the regional definition of capacity reserve requirements and capacity resources. More information on the capacity value of wind turbines may also be found in Frew (2016).

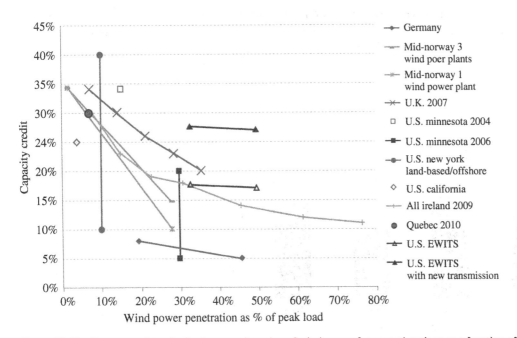

Figure 13.13 Summary of results for the capacity value of wind power for several regions as a function of the installed wind power share (*Source:* Milligan *et al.*, 2019/John Wiley & Sons)

13.5.3 Environmental Value of Wind Energy

Given the cost reductions in wind turbines themselves over the last several decades, there are already many locations where wind may be the most economic option for a new generation. Basing wind energy's value exclusively on the avoided costs, however, would result in many potential applications being uneconomic.

Equating wind energy's value exclusively to fuel savings or capacity value misses the substantial environmental benefit that results from its use. The environmental benefits are those that arise because wind generation does not result in any significant amount of air emissions during operations, particularly carbon dioxide and oxides of nitrogen and sulfur. Reduction in emissions translates into a variety of health benefits. It also decreases the concentration of atmospheric chemicals that cause acid rain and global warming.

Converting wind energy's environmental benefits into monetary form (known as "monetizing") can be difficult. Nonetheless, the results of doing so can be quite significant, because when that is done, many more potential projects can be economically viable.

The incorporation of environmental benefits into the market for wind energy is done through two steps: (1) quantifying the benefits and (2) monetizing those benefits. Quantifying the benefits involves identifying the net positive effects on society as a result of the wind's use. Monetizing involves assigning a financial value to the benefits. It allows a financial return to be captured by the prospective owner or developer of the project. Monetization is usually accomplished by governmental legislation or regulation. The process frequently considers the cost of alternative measures to reduce emissions (such as scrubbers on coal plants) to assign a monetary value to the avoided emissions. Some methods used are discussed in Section 13.5.4.3.

The effect of monetizing environmental benefits is to create an additional category of potential revenue for a wind project. Figure 13.14 illustrates what the relative magnitude of this revenue source might be, together with those from avoided fuel and capacity. In the cases shown, the effect of including the monetized environmental benefits ("avoided emissions") is to increase the total value by approximately 50% above that based on avoided fuel and capacity alone.

Because of the very real environmental benefits of wind energy and because of the enormous impact that including these benefits in economic assessments could have, many countries have developed laws and regulations to facilitate the process. These are discussed in Section 13.7.5.

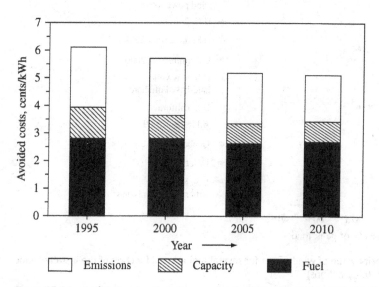

Figure 13.14 Avoided costs of wind energy for the Netherlands for 1995–2010 (WEC, 1993) (*Source:* Reproduced by permission of the World Energy Council, London)

How revenues based on avoided costs and monetized environmental benefits may factor into an economic assessment of a particular project depends on what may be referred to as the "market application" and the relation of the project or developer to that application; see Section 13.5.4.

13.5.3.1 Sources of Environmental Benefits

The primary environmental benefit of electricity generated from wind energy projects is that the wind offsets emissions that would be generated by conventional fossil fuel-based power plants. These emissions include sulfur dioxide (SO_2), oxides of nitrogen (NO_x), carbon dioxide (CO_2), particulates, slag, and ash. The amount of emissions saved via the use of wind energy depends on the types of power plants that are replaced by the wind system and the particular emissions control systems currently installed on the various fossil-fired plants.

Numerous studies have been carried out to determine the emissions savings value of wind power systems (e.g., EWEA, 2004). In general, depending on the country and/or region, there are significant differences in the amount of emissions that can be saved per unit of electricity generated. This depends mainly on the mix of conventional generating units and the emissions control status of the individual units. For example, in the United States, each MWh of electricity generated by conventional means results in the emission of about 608 kg of carbon dioxide, 3.4 kg of sulfur dioxide, and 1.6 kg of oxides of nitrogen (Reeves, 2003). More recent work by the International Renewable Energy Agency (IRENA, 2019a) has concluded that wind power could contribute 6.3 Gt of CO_2 emissions reduction in 2050 (which represents 27% of the overall emissions reduction needed to meet the Paris climate goals). A detailed review of the external costs (costs that are not incorporated into the cost streams of a project) for utility-scale energy systems is given in a European Commission report by Bickel and Friedrich (2005).

There are also benefits associated with wind energy production other than the direct environmental ones. These include such indirect benefits as improved public health and political benefits that would accrue by reducing fossil fuel imports. All the benefits (and the corresponding costs) from all sources of energy production are often considered under the general topic of "social costs of energy production." The earliest studies in this area were from Hohmeyer (1990) and reviewed by WEC (1993) and Gipe (1995). In general, the results show higher social or "external" costs for conventional electrical power generation plants. As with the value of direct emissions reduction, there is as yet no agreement as to how to quantify the exact monetary value. In general, however, estimates of the external costs are considered in developing incentives for wind energy, such as renewable portfolio standards (RPS), tax credits, or guaranteed prices. These are discussed in Section 13.7.5.

13.5.4 Market Value of Wind Energy

The market value of wind energy is the total amount of revenue one will receive by selling wind-generated electricity or will avoid paying through its generation and use. The value that can be "captured" depends strongly on three considerations: (1) the "market application," (2) the project owner or developer, and (3) the types of revenues available.

The market application affects the type of project that might be built and the magnitude of the avoided costs. The relationship of the owner to the market application will determine whether the avoided costs will be accounted for directly or indirectly in the sale of the electricity. The types of revenues are basically in two categories. The first category, which includes the sale of electricity and reduced purchases, relates to the market value of the energy itself. The second category includes revenues that are based on monetized environmental benefits and are derived from governmental policies. All of these topics are discussed briefly below.

13.5.4.1 Market Applications

The market application refers to the system in which the wind project is located. The most common market applications include: (1) traditional utilities, (2) restructured utilities, (3) customer-owned generation, (4) remote power systems, and (5) grid-independent applications. The relation between the market application and the person or entity that may own or develop the project has the important effect of determining whether the revenues appear in the form of savings in the costs of production or as sales of energy. A more recent discussion of the market value of wind energy was carried out by the International Renewable Energy Agency in 2019 (IRENA, 2019b).

Traditional Utilities

Traditional electric utilities are entities that generate, transmit, distribute, and sell electricity. They may be privately or publicly owned. When privately owned, they are considered a "natural monopoly" and are normally regulated by a state or national government. Although most traditional utilities generate at least some of the electricity they sell, many also purchase electricity from other utilities. Two types of publicly owned utilities are (1) municipal electric companies and (2) rural electric cooperatives. These utilities are owned and operated by cities or towns or by their customers (in the case of the co-ops). Frequently, they buy most or all of their power from large, publicly owned generating entities, such as the Tennessee Valley Authority in the southern United States.

Restructured Utilities

In recent years, many electric utilities have been "restructured" or "deregulated." They have typically been broken into three parts: (1) a generating company, (2) a "wires" company, and (3) a marketing company. The wires company only owns and maintains the transmission and distribution wires in a given service area. It does not have its own generation (at least for that service area). Generation is owned separately. The marketer of energy, not necessarily the utility itself, will purchase energy from a generator, "rent" the wires for moving the electricity, and then sell the electricity to its customers. Customers are, in principle, free to choose their supplier (marketer). The marketer sometimes offers electricity "products" with particular attributes that may be of interest to the customer. These might have to do with reliability, interruptability, or fuel source. The proper functioning of the system is maintained by an "independent system operator" (ISO), which oversees the process of energy purchase and sale. The wire companies are normally still regulated. See also Chapter 12.

Customer-Owned Generation

Customers of utilities, whether traditional or restructured, may wish to generate some of their own electricity. In this case the customer may purchase a wind turbine and connect it to the local distribution lines, normally on his or her side of the meter. For this reason, the customer-owned generation is also known as the "behind-the-meter" generation. To what extent this is possible or economically worthwhile depends greatly on the specific laws and regulations that may be in place. In projects of this type, most of the energy is used to reduce purchases of electricity, but in some cases, any excess generation may be sold into the electrical grid.

Remote Power Systems

Remote power systems are similar in many ways to larger utilities. The main differences are: (1) the type of generators that they use (normally diesel) and (2) their inability to exchange power with other utilities. When wind turbines are added to such systems, they result in a hybrid power system.

Grid-Independent Systems

Grid-independent wind energy systems typically consist of a wind turbine, battery storage, a power converter, and electrical loads. They are normally relatively small and self-contained. They may be installed as an alternative to interconnecting to a utility or in lieu of a diesel or gasoline generator.

13.5.4.2 Wind Project Owners or Developers

The nature of owners or developers, and their relation to the market application, will have a significant impact on the type of revenues available. Types of owners or developers may include: (1) traditional utilities themselves, (2) merchant power plants (3) independent power producers (IPPs), (4) electricity consumers wishing to reduce their purchases, (5) entities responsible for isolated power systems, or (6) individuals needing energy in a grid-independent application. Some more details on the first four types are given below. An isolated power system is basically comparable to a conventional utility, just smaller, but the options are similar. Owners of grid-independent applications are comparable to residential utility customers; the difference is that none of their generation would go into an external electrical network.

Traditional Utilities

A traditional utility could develop and own its own wind generation capacity. The wind generation would save some fuel or reduce purchases of electricity from other utilities. The wind project might also provide some capacity value, as described in Section 13.5.2.2.

Merchant Wind Power Plants

A merchant wind power plant is one that is intended to operate in a competitive (restructured) electricity market. Participation may take a variety of forms, including supplying base load, peaking, capacity, or ancillary services, either through PPAs or by selling at the spot market price. The wind power plant market structure described in Chapter 12 was essentially of the merchant plant type. Insofar as they produce electricity from a renewable source, they are also able to take advantage of incentives not offered to generators that use fossil fuels.

Independent Power Producers

IPPs are similar to merchant plants in that they own and operate their own generating facilities and sell the electricity to a utility. They may operate as either traditional or restructured utilities. Wind project IPPs are similar to merchant plants, although they may not be as large or need not necessarily play an active role. IPPs normally require a multi-year contract ("power purchase agreement" or PPA) to secure financing. The details of the PPA are often very important in determining whether or not a project is economically viable.

Customer Generators

The customer is normally (but not always) the owner and developer of a behind-the-meter wind energy project. Typical examples include residential customers, farmers, and small businesses.

13.5.4.3 Types of Revenue

As described above, there are basically two types of revenue that may be available: those based on avoided costs and those based on monetized environmental benefits. The first category includes: (1) reduction in purchases of electricity, fuel, capital equipment, or other expenses (avoided costs themselves) and (2) sale of electricity (the price of which will reflect the avoided costs). The second category includes: (1) sale of RECs, (2) tax benefits, (3) guaranteed, above-market rates, and (4) net metering.

Reduced Purchases

When the owner of the turbine(s) is the same as the traditional generator, the revenue from the wind project is actually a saving. In the case of a traditional utility or the manager of a remote power system, the revenue would appear as reduced fuel or electricity purchases, reduced O&M costs on the other generators, and possibly reduced costs for replacements or upgrades to the other generators. The value of the savings (the "avoided cost") depends on the fuel source and type of conventional generators used. For the owner of a "behind-the-meter" project, the value of the reduced purchases would be the retail price of electricity. Generally, this is much higher than the wholesale price.

Electricity Sale

A merchant wind power plant or IPPs would normally derive revenue by selling electricity into the grid. As indicated above, what is actually sold could take a variety of forms, but often the rate would be set by a long-term contract. This rate is close to the expected "avoided cost" of energy, as discussed previously, for the utility into which the electricity will be sold. The contract price is generally the highest for generators that are dispatchable.

A wind project operator may also gain a better price if production can be forecast to some extent. Electricity can also be sold on a day-to-day basis (the "spot market"), as previously indicated.

Renewable Energy Certificates

Over the last several decades, new incentives have been developed to facilitate the introduction of more renewable energy into restructured utilities. One of these incentives is the renewable energy or "green" certificate, often referred to by the acronym REC. In the system on which this incentive is based, the desired attribute of "greenness" is assumed to be separated from the energy itself and is used to assign value to a certificate. The certificates then have value, typically in the range of a few US cents per kWh. Such certificates may be used to ensure compliance, for example, with mandates known as "renewable portfolio standards" (RPS). The wind project will receive a quantity of certificates in proportion to the energy generated. The certificates can then be sold, and the proceeds will augment the revenue from the sale of the electricity itself. More details can be found in Section 13.7.5.

Tax Benefits

Over the years, a number of types of tax credits have been used to foster the development of wind energy. At one time, the investment tax credit (ITC) was widely used. This provides a source of revenue to the project developer based on the cost of the project rather than its production. A more common incentive nowadays is the production tax credit (PTC). It is based on the actual energy produced, with typically amounts (in the Unites States) in the range of $0.03/kWh (the rate may be updated annually).

Guaranteed, Above-market Rates

Some countries, such as Germany and Spain, have provided guaranteed, above-market rates for generators of wind energy. This scheme is known as a "feed-in tariff" (see Ringel, 2006 and Section 13.7.5). This arrangement functions similarly to the energy sale method described above. The difference is that the price is set by the government at a value higher than one could expect to receive from conventional sales into the electricity market. The rate for land-based wind energy in these situations has typically been close to the retail price.

Net Metering

The situation faced by a small behind-the-meter generator is actually a little more complicated than it first appears. Due to the variability of the wind and the load, it may frequently be the case that the instantaneous generation from a turbine is more than the electrical load, even if the average generation is well below the average load. Depending on the utility rules and type of metering, it is possible for an operator to receive little or no revenue from the energy that is produced in excess of the load. This could seriously affect the effective average value for the generation.

In an effort to address this situation, and to provide additional incentives to small generators, a number of states in the Unites States have developed "net metering" rules. Under these rules, the key consideration is the net energy that is produced, in comparison to the consumption, over some extended period of time (typically a month or more). As long as the net generation is less than the consumption, all the generation will be valued at retail. Any generation in excess of the consumption is subject to the same rules as would apply to an IPPs.

13.5.4.4 Examples

It should be apparent that there is a wide range of revenue or value that might apply to a unit of wind-generated electricity, depending on the details of the application. The following are some hypothetical examples of this:

1) A private utility using coal-fired generators has a fuel cost of $0.03/kWh and needs no new capacity. There are no monetized environmental benefits in the state in which it operates. The value of wind energy would thus be the avoided fuel costs, or $0.03/kWh.

2) A wind energy project developer in a state with restructured utilities can sell his energy into the grid for $0.03/kWh. The state provides RECs, which he can sell for $0.035/kWh. He can also take advantage of a federal production tax credit valued at $0.03/kWh. The total value of his wind-generated electricity would be $0.095/kWh.

3) A municipal electric utility buys all its electricity at a fixed rate of $0.07/kWh. The value of wind energy would thus be the avoided energy purchase costs, or $0.07/kWh.

4) A farmer is considering whether to install a 50 kW wind turbine to supply part of the farm's electricity. She presently pays $0.15/kWh. She can take advantage of net metering and a production tax credit of $0.03/kWh. The value of her wind-generated electricity would be $0.18/kWh.

13.6 Economic Analysis Methods

As stated previously, a wind energy system can be considered to be an investment that produces revenue. An economic analysis is used to evaluate the profitability of a wind energy project and to compare it with alternative investments. Economic analysis methods can be applied for wind energy systems, assuming that one has a reliable estimate for the capital costs and O&M costs. The general purpose of such methods is not only to estimate the economic performance of a given design of wind energy system but also to compare it with conventional and other renewable energy-based systems. Three different types of economic analysis methods will be described in this section. They are:

- Simplified models;
- Life cycle cost models;
- Electric utility economic models.

Each of these economic analysis techniques has its own definition of key economic parameters, and each has its own particular advantages and disadvantages.

It is important to clarify who the owner or developer is and what market value can be expected for the energy, as discussed in Section 13.5.4. Depending on the application, one or more of the economic evaluation methods discussed below may be appropriate.

13.6.1 Simplified Economic Analysis Methods

For a preliminary estimate of a wind energy system's feasibility, it is desirable to have a method for a quick determination of its relative economic benefits. Such a method should be easy to understand, free of detailed economic variables, and easy to calculate. Two methods that are often used are: (1) simple payback and (2) cost of energy (*COE*).

13.6.1.1 Simple Payback Period Analysis

A payback calculation compares revenue with costs and determines the length of time required to recoup an initial investment. The payback period (in years) is equal to the total capital cost of the wind system divided by the net average annual return from the energy produced. In its simplest form (simple payback period), it is expressed in equation form as:

$$SP = C_c/AAR \tag{13.3}$$

where SP is the simple payback period, C_c is the installed capital cost, AAR is the net average annual return, and $C_{O\&M}$ is the annual O&M cost. The latter can be expressed by:

$$AAR = E_a P_e - C_{O\&M} \tag{13.4}$$

where E_a is the annual energy production (kWh/year) and P_e is the price obtained for electricity (\$/kWh). Thus, the simple payback period is given by:

$$SP = \frac{C_c}{E_a P_e - C_{O\&M}} \tag{13.5}$$

Consider the following example: C_c = \$40 000, E_a = 100 000 kWh/yr, P_e = \$0.10/kWh, $C_{O\&M}$ = \$2000. Then SP = 40 000/(100 000 × 0.10–2000) = 5 years.

Note that the calculation of the simple payback period omits many factors that may have a significant effect on the system's economic cost effectiveness. These include changing fuel and loan costs, depreciation on capital costs, variations in the value of delivered electricity, and lifetime of the system.

13.6.1.2 Cost of Energy Analysis

The cost of energy, *COE*, is defined as the unit cost to produce energy (in \$/kWh) from the wind energy system. That is:

$$\text{Cost of energy} = (\text{Total costs})/(\text{Energy produced}) \tag{13.6}$$

Based on the previous nomenclature, the simplest calculation of *COE* is given by:

$$COE = [(C_c \times FCR) + C_{O\&M}]/E_a \tag{13.7}$$

where *FCR* is the fixed charge rate. The fixed charge rate is a term that reflects the interest one pays or the value of interest received if money were displaced from savings. For utilities, *FCR* is an average annual charge used to account for debt, equity costs, taxes, etc. (see Section 13.6.3).

Using the values from the previous numerical example and assuming that *FCR* = 10% and $C_{O\&M}$ = \$2000/yr:

$$COE = [(40\,000 \times 0.10) + 2000]/100\,000 = \$0.06/\text{kWh}$$

This simplified calculation is based on a number of key assumptions, most of which neglect the time value of money. They will be addressed in the next section.

13.6.2 Life Cycle Costing Methods

LCC is a commonly used and generally preferred method for the economic evaluation of energy-producing systems based on the principles of the "time value" of money. The LCC method summarizes expenditure and revenue occurring over time into a single parameter (or number) so that an economically based choice can be made. A common result of this type of analysis is the calculation of the (levelized cost) of energy (*COE* or COE_L).

In analyzing future cash flows, one needs to consider the time value of money. An amount of money can increase in quantity by earning interest from some investment. Also, money can have a reduced value over time as inflation forces prices upward, making each unit of currency have lower purchasing power. As long as the rate of inflation is equal to the return on investment (*ROI*) for a fixed sum of money, purchasing power is not diminished. As is usually the case, however, if these two values are not equal, then the sum of money can increase in value (if investment return is greater than inflation) or decrease in value (if the inflation rate is greater than investment return).

The concept of LCC analysis is based on accounting principles used by organizations to analyze investment opportunities. The organization seeks to maximize its *ROI* by making an informed judgment on the costs and benefits to be gained by the use of its capital resources. One way to accomplish this is via an LCC-based calculation of the *ROI* of the various investment opportunities available to the organization.

To determine the value of an investment in a wind power system, the principles of LCC costing can be applied to its costs and benefits; that is to say, its expected cash flows. The costs include the expenses associated with the purchase, installation, and operation of the wind system (see Sections 13.3 and 13.4). The economic benefits of a wind system include the use or sale of the generated electricity (Section 13.5) as well as tax savings or other financial incentives (see Section 13.7.5). Both costs and benefits may also vary over time. The principles of life cycle costing can consider time-varying cash flows and refer them to a common point in time. The result will be that the wind system can be compared to other energy-producing systems in an internally consistent manner.

LCC methodology, as described in this section, takes the parameters of inflation and interest applied to money and uses a model based on the "time value of money" to estimate a "present value" for an investment at any time in the future. The important variables and definitions used in LCC analysis follow.

13.6.2.1 Overview and Definition of Life Cycle Costing Concepts and Parameters
In LCC analysis, some key concepts and parameters include:

- Time value of money and present worth factor;
- Levelizing;
- Capital recovery factor;
- Net present value.

A summary description of each follows.

Time Value of Money and Present Worth Factor
A unit of currency that is to be paid (or spent) in the future will not have the same value as one available today. This is true even if there is no inflation, since a unit of currency can be invested and bear interest. Thus, its value is increased by the interest. For example, suppose an amount with a present value (*PV*) (sometimes called present worth) is invested at an interest (or discount) rate *r* (expressed as a fraction) with annual compounding of interest. (Note that in economic analysis, the discount rate is defined as the opportunity cost of money. This is the next best rate of return that one could expect to obtain.). At the end of the first year, the value has increased to $PV(1 + r)$, after the second year to $PV(1 + r)^2$, etc. Thus, the future value, *FV*, after *N* years is:

$$FV = PV(1 + r)^N \tag{13.8}$$

The ratio *PV/FV* is defined as the present worth factor *PWF*, and it is given by:

$$PWF = PV/FV = (1 + r)^{-N} \tag{13.9}$$

For illustration purposes, the numerical values of *PWF* are given in Table 13.3.

Levelizing
Levelizing is a method for expressing costs or revenues that occur once or at irregular intervals as equivalent, equal payments at regular intervals. A good way to illustrate this variable is by using the following example (Rabl, 1985). Suppose one wants loan payments to be arranged as a series of equal monthly or yearly instalments. That is, a loan of *PV* is to be repaid in equal annual payments *A* over *N* years. To determine the equation for *A*, first consider a loan of value PV_N that is to be repaid with a single payment F_N at the end of *N* years. With *N* years of interest *r* on the amount PV_N, the payment is: $F_N = PV_N(1 + r)^N$. In other words, the loan amount PV_N equals the *PV* of the future payment F_N; $PV_N = F_N(1 + r)^{-N}$

A loan that is to be repaid in *N* equal installments can be considered as the sum of *N* loans, one for each year, with the jth loan being repaid in a single installment *A* at the end of the jth year. Thus, the value, *PV*, of the loan equals the sum of the *PV*s of all loan payments:

$$PV = \frac{A}{1+r} + \frac{A}{(1+r)^2} + \dots + \frac{A}{(1+r)^N} = A\sum_{j=1}^{N}\frac{1}{(1+r)^j} \tag{13.10}$$

Table 13.3 Present worth factor for discount rate r and number of years N

Discount rate, r	Number of years, N					
	5	10	15	20	25	30
0.01	0.9515	0.9053	0.8613	0.8195	0.7798	0.7419
0.02	0.9057	0.8203	0.7430	0.6730	0.6095	0.5521
0.03	0.8626	0.7441	0.6419	0.5537	0.4776	0.4120
0.04	0.8219	0.6756	0.5553	0.4564	0.3751	0.3083
0.05	0.7835	0.6139	0.4810	0.3769	0.2953	0.2314
0.06	0.7473	0.5584	0.4173	0.3118	0.2330	0.1741
0.07	0.7130	0.5083	0.3624	0.2584	0.1842	0.1314
0.08	0.6806	0.4632	0.3152	0.2145	0.1460	0.0994
0.09	0.6499	0.4224	0.2745	0.1784	0.1160	0.0754
0.10	0.6209	0.3855	0.2394	0.1486	0.0923	0.0573
0.11	0.5935	0.3522	0.2090	0.1240	0.0736	0.0437
0.12	0.5674	0.3220	0.1827	0.1037	0.0588	0.0334
0.13	0.5428	0.2946	0.1599	0.0868	0.0471	0.0256
0.14	0.5194	0.2697	0.1401	0.0728	0.0378	0.0196
0.15	0.4972	0.2472	0.1229	0.0611	0.0304	0.0151
0.16	0.4761	0.2267	0.1079	0.0514	0.0245	0.0116
0.17	0.4561	0.2080	0.0949	0.0433	0.0197	0.0090
0.18	0.4371	0.1911	0.0835	0.0365	0.0160	0.0070
0.19	0.4190	0.1756	0.0736	0.0308	0.0129	0.0054
0.20	0.4019	0.1615	0.0649	0.0261	0.0105	0.0042

Or, using an equation for a geometric series:

$$PV = A \left[1 - (1 + r)^{-N} \right] / r \tag{13.11}$$

It should be noted that this equation is perfectly general and relates any single present value, PV, to a series of equal annual payments A, given the interest or discount rate, r, and the number of payments (or years), N. Also note that when $r = 0$, $PV = A \times N$.

Capital Recovery Factor

The capital recovery factor, CRF, is used to determine the amount of each future payment required to accumulate a given PV when the discount or interest rate and the number of payments are known. The CRF is defined as the ratio of A to PV and, using Equation (13.11), is given by:

$$CRF = \begin{cases} r / \left[1 - (1 + r)^{-N} \right], \text{if } r \neq 0 \\ 1/N, \text{if } r = 0 \end{cases} \tag{13.12}$$

The inverse of the CRF is sometimes defined as the series present worth factor, SPW.

Net Present Value

The net present value (*NPV*) is defined as the sum of all relevant *PV*s. From Equation (13.8), the present, or levelized, value of a future cost, *C*, evaluated at year *j* is:

$$PV = C/(1+r)^j \tag{13.13}$$

Thus, the *NPV* of a cost *C* to be paid each year for *N* years is:

$$NPV = \sum_{j=1}^{N} PV_i = \sum_{j=1}^{N} \frac{C}{(1+r)^j} \tag{13.14}$$

If the cost *C* is inflated at an annual rate *i*, the cost C_j in year *j* becomes:

$$C_j = C(1+i)^j \tag{13.15}$$

Thus, the net present value, *NPV*, becomes:

$$NPV = \sum_{j=1}^{N} \left(\frac{1+i}{1+r}\right)^j C \tag{13.16}$$

As will be discussed in the next section, *NPV* can be used as a measure of economic value when comparing investment options.

13.6.2.2 Life Cycle Costing Analysis Evaluation Criteria

Earlier in this chapter, two economic parameters, simple payback and *COE*, were introduced. These can be used for a preliminary economic analysis of a wind energy system. With LCC analysis, one can apply a number of other economic figures of merit or parameters to evaluate the feasibility of a wind energy system. Among others, these include *NPV* of cost or savings and the LCOE.

Net Present Value of Cost, Savings, or Profit

The *NPV* of a particular parameter is generally used as a measure of economic value when comparing different investment options in a life cycle cost analysis. Note that it is important to define the *NPV* clearly, since various authors have used the term "net present value" to define a variety of life cycle cost analysis parameters. First, one can define the savings version of net present value, *NPVs*, as follows:

$$NPV_s = \sum_{j=1}^{N} \left(\frac{1+i}{1+r}\right)^j (S_i - C_i) \tag{13.17}$$

where S_i and *C* represent the yearly gross savings and costs during a project's lifetime (S_i–C_i is the yearly net saving).

The *NPV* of savings applies to smaller projects in which the wind turbine(s) are used to reduce energy purchase and thus "save" money. Larger projects, such as wind farms, are commercial enterprises in which the financial purpose is to generate a profit by selling electricity and possibly RECs. In this case, the term of relevance is the net present value of the profit, NPV_P, as given in Equation (13.18)

$$NPV_P = \sum_{j=1}^{N} \left(\frac{1+i}{1+r}\right)^j (R_i - C_i) \tag{13.18}$$

Where R_i is the annual revenue

In evaluating various systems using this criterion, one would look for the systems with the largest value of *NPVs*. In practice, a spreadsheet format can be used to evaluate the yearly gross savings (S) and cost (C) and to calculate the sum of the annual levelized values. As will be shown next, however, it is also possible to use closed form analytical expressions for the direct calculation of the *NPV*. These expressions are valuable tools for use in generalized parametric studies of wind energy systems.

If only cost factors are considered, then a cost version of net present value, NPV_C, may be used. It is the sum of the levelized costs of the energy system. For this version of this parameter, when comparing a number of different systems, the design with the lowest NPV_C is desired.

As an example of costs for wind energy systems, the *NPV* of costs can be found by calculating the total costs of a system for each year of its lifetime. The annual costs should then be levelized to the initial year (commonly, year "zero") and then summed up. If one assumes that both the general and energy inflation rates are constant over the system's life and that the system loan is repaid in equal instalments, NPV_C may be found from the following equation:

$$NPV_C = P_d + P_a Y\left(\frac{1}{1+r}, N\right) + C_c f_{OM} Y\left(\frac{1+i}{1+r}, L\right) \tag{13.19}$$

where:

P_d = down payment on system costs (in year zero)
P_a = annual payment on system costs = $(C_c - P_d)CRF$
CRF = capital recovery factor, based on the loan interest rate, b, rather than r
b = loan interest rate
r = discount rate
i = general inflation rate
N = period of loan
L = lifetime of system
C_c = capital cost of system
f_{OM} = annual operation and maintenance cost fraction (of system capital cost)

The variable $Y(k, \ell)$ is a function used to obtain the *PV* of a series of payments. It is determined by:

$$Y(k,\ell) = \sum_{j=1}^{\ell} k^j = \begin{cases} \dfrac{k - k^{\ell+1}}{1-k}, \text{if } k \neq 1 \\ \ell, \text{if } k = 1 \end{cases} \tag{13.20}$$

Using Equations (13.18–13.20), the *NPV* of the profit from a commercial project can be expressed as Equation (13.21).

$$NPV_P = RY\left(\frac{1}{1+r}, L\right) - P_d - P_a Y\left(\frac{1}{1+r}, N\right) + C_d f_{OM} Y\left(\frac{1+i}{1+r}, L\right) \tag{13.21}$$

Where R is the annual revenue, which is based on the annual energy and the contracted value of that energy. For this equation, R is assumed to be constant over the lifetime of the project. If not, Equation (13.21) could be adjusted or, more likely, a cash flow method could be used.

Cost of Energy/Levelized Cost of Energy

In its most basic form, the cost of energy, *COE*, is given by the sum of the annual levelized costs for a wind energy system divided by the annual energy production. Thus:

$$COE = \frac{\sum(\text{Levelized annual costs})}{\text{Annual energy production}} \tag{13.22}$$

This type of definition is generally used in a utility-based calculation for *COE*. Sometimes, the LCOE is defined as the value of energy (units of $/kWh) that, if held constant over the lifetime of the system, would result in a cost-based *NPV*, such as the value calculated via Equation (13.19). Using this basis, the COE_L is given by:

$$COE_L = \frac{(NPV_C)\,(CRF)}{\text{Annual energy production}} \tag{13.23}$$

Note that the capital recovery factor, *CRF*, is here based on the lifetime of the system, *L*, and the discount rate, *r*. On this basis, the LCOE multiplied by the annual power production would equal the annual loan payment needed to amortize the *NPV* of the cost of the energy system.

Other Life Cycle Analysis Economic Parameters

There are a number of other economic performance factors that can be used for evaluating the life cycle-based economic performance of an energy system. Two of the more common parameters are the benefit–cost ratio (*B/C*) and the internal rate of return (*IRR*). The benefit-cost ratio is given by Equation (13.24).

$$B/C = \frac{\text{Present value of all benefits (income)}}{\text{Present value of all costs}} \tag{13.24}$$

The *IRR* is defined as the discount rate (*r*) for which NPV_S or NPV_P is equal to zero. In the case of NPV_P, the IRR comes from Equation (13.21) and becomes Equation (13.25).

$$RY\left(\frac{1}{1+r},L\right) - P_d - P_a Y\left(\frac{1}{1+r},N\right) + C_c f_{OM} Y\left(\frac{1+i}{1+r},L\right) = 0 \tag{13.25}$$

The *IRR* is often used by utilities or businesses in assessing investments and is a measure of profitability. As noted by Masters (2013), the IRR is perhaps the most persuasive measure of the value of an energy generation project and the trickiest to compute. Figure 13.15 illustrates that it can be calculated as a function of the simple payback period and the project lifetime (see Masters, 2013). This example uses the simple payback period based on the value of the energy alone and does not consider financing, inflation, or O&M. In this case, Equation (13.25) reduces to Equation (13.26).

$$RY\left(\frac{1}{1+r},L\right) - C = 0 \tag{13.26}$$

Where *C* is the project cost. The principle should be clear, however, that the higher the *IRR*, the better the economic performance of the wind energy system in question.

Generally, systems with a benefit–cost ratio greater than one are acceptable, and higher values of the B/C value are desired.

13.6.3 Electric Utility-Based Economic Analysis for Wind Farms

From a power-generating business or utility perspective, the previous definition of *COE* can be used for a first estimate of utility generation costs for a wind farm system. In this application, *COE* can be calculated from:

$$COE = \frac{[(C_c)\,(FCR)\ +\ C_{O\&M}]}{\text{Annual energy production}} \tag{13.27}$$

where C_c is the capital cost of the system, *FCR* is the fixed charge rate (a present value factor that includes utility debt and equity costs, state and federal taxes, insurance, and property taxes), and $C_{O\&M}$ is the annual operation and maintenance cost.

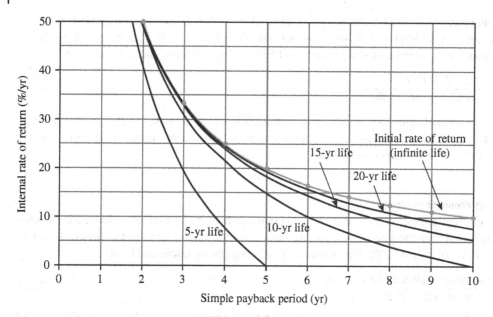

Figure 13.15 Internal rate of return (IRR) as a function of simple payback period and project lifetime (*Source:* Masters, 2013/ John Wiley & Sons)

In the United States, electric utilities and the wind industry commonly use either of two methods to estimate the *COE* from a utility-sized wind energy system (Karas, 1992): (1) the Electric Power Research Institute, Technical Analysis Group (EPRI TAG) method (EPRI, 1979, 1989) or (2) a cash flow method. Summary details of each follow.

13.6.3.1 EPRI TAG Method

This method produces a levelized cost of energy (*COE*), and in its simplest form for wind energy systems, *COE* ($/kWh) is calculated from:

$$COE = FCR \left(\frac{\overline{C}_c}{8760 \times CF} \right) + \overline{C}_{O\&M} \tag{13.28}$$

where:

$\overline{C}_c$ = total cost of constructing the facility normalized by rated power ($/kW)
CF = capacity factor
$\overline{C}_{O\&M}$ = cost of operation and maintenance normalized per unit of energy ($/kWh)

Since this method produces a levelized energy cost, it can be applied to a number of technologies, including conventional power plants (with the addition of fuel costs), for a useful comparison index. Some limitations of the EPRI TAG method include:

- it assumes a loan period equal to the life of the power plant;
- it does not readily allow for variable equity return, variable debt repayment, or variable costs.

The second limitation generally excludes the use of this method for merchant power plants or IPPs, under which many wind farm projects fall.

13.6.3.2 Cash Flow Method

The cash flow method is based on the use of an accounting-type spreadsheet that requires an annual input of estimated income and expenses over the lifetime of the project. The COE is calculated via the following operations:

- each cost component of the plant and operation is identified (e.g., plant construction, O&M, insurance, taxes, land, power transmission, administrative costs, etc.);
- a cost projection of each of the above components for each year of the plant's service life;
- an annual estimate of depreciation, debt services, equity returns, and taxes;
- discounting of the resulting cash flows to PVs using the utility's cost of capital (as established by appropriate regulatory agency);
- levelizing the cost by determining the equal payment stream (annuity) that has a *PV* equal to the results of the previous calculations.

The cash flow method allows for the real variations that can be expected in cost, operational, and economic data, such as price increases, inflation, and changing interest rates. A detailed description of several cash flow models used to assess the effects of ownership structure and financing on utility-scale wind systems is given by Wiser and Kahn (1996).

Karas (1992) notes that the *COE* estimates can be determined in a wide variety of ways. His examples show that wind energy system *COE*s are extremely sensitive to actual plant cost and operating assumptions, especially capacity factor and plant installation costs.

Regardless of the method used to calculate a levelized or other type of *COE*, it should be emphasized that one plant can have a number of equally valid *COE*s depending on the assumptions used. Thus, it is important that in the determination of any *COE* for a wind system, the terminology, methods of parameter determination, and assumptions be made explicit.

13.6.3.3 Economic Performance Sensitivity Analysis

The previous sections have described a number of methods for the determination of economic performance parameters (such as *COE*) that can be used to evaluate various wind systems or to compare their performance with other types of power systems. In practice, however, one soon realizes that a variety of system performance and economic input variables exist that can influence the desired parameter's value. In addition, the potential variation or range of uncertainty of many input parameters may be large.

A complete economic study of a wind energy system should include a series of sensitivity studies for the calculation of the output variable of interest. Specifically, each parameter of interest is varied around some central or "best estimate." The desired figure of merit (such as *COE*) is then calculated. The calculations are repeated for a range of assumptions. For example, the following economic and performance variables could be varied:

- Plant capacity factor (or average wind speed at the site);
- Turbine availability;
- O&M costs;
- Capital or installed cost;
- Lifetime of system and components;
- Length of loan repayment;
- Interest or discount rate.

With calculations of this type, the results are often summarized graphically. This can be accomplished via the use of a "spider" or "star" diagram, illustrated in Figure 13.16. It is intended to show how a percentage change in a given parameter changes the *COE*, expressed as a percentage. This type of diagram readily shows the more important variables by lines with steeper slopes. Note, however, that it may not be reasonable to vary all parameters by the same percentage amount since some may be known quite accurately (Walker, 1993).

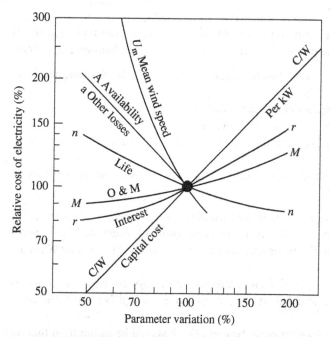

Figure 13.16 Sensitivity of cost of electricity to various parameters (Ferris, 1990). (*Source:* Reproduced by permission of Pearson Education Limited)

13.7 Wind Energy Market Considerations

13.7.1 Overview

There exists a large potential worldwide market for wind energy systems. In this section, a review of market considerations for wind energy systems is presented. There are three major issues to consider:

1) Potential market for wind systems.
2) Barriers to expansion of markets (see also Chapter 14).
3) Incentives for market development.

There are numerous technical publications on these subjects in the open literature. This is especially true for utility-scale or wind farm applications (e.g., OECD/IEA, 1997; EWEA, 2004; IRENA, 2019a). The potential market for small or dispersed wind hybrid systems has also been addressed (for example, see WEC, 1993). A summary of some of this work on marketing issues (for utility-scale systems) follows.

13.7.2 The Market for Onshore Utility-scale Wind Systems

An overview of onshore utility-scale market projections from an international and US perspective is presented in this section.

13.7.2.1 International Onshore Wind Energy Market

The prediction of long-term trends in worldwide installed wind power capacity is a complex task. For example, work by the European Wind Energy Association (EWEA, 2004) noted that predictions for onshore wind power development in Europe were significantly underestimated, even by wind power advocates. In recent times, a number of organizations have made detailed long-term predictions for the growth of wind power installations in the

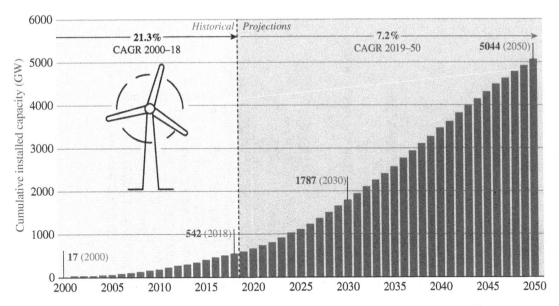

Figure 13.17 Predicted cumulative worldwide onshore wind capacity (*Source:* IRENA, 2019a/International Renewable Energy Agency)

world at large and in Europe. For example, a summary of predictions for the worldwide growth of wind power capacity is shown in Figure 13.17 (IRENA, 2019a). As shown in this International Renewable Energy Agency Report, at the end of 2018, the global cumulative installed capacity of onshore wind reached over 540 GW. The authors also predicted that the total capacity of onshore wind would grow to about 1800 GW in 2030 and then nearly ten-fold to over 5000 GW in 2050.

13.7.2.2 United States Onshore Wind Energy Market

Recent work (Wiser and Bolinger, 2019) has produced a forecast for US onshore wind energy capacity shown in Figure 13.18. The expected capacity additions were predicted to increase from 9–12 GW in 2019 to 11–15 GW in 2020. The actual amounts (9 GW in 2019 and 13.2 GW in 2020) were close to the projections. Forecasts from 2021 to 2028, however, were expected to decrease in part due to a decrease in production tax credits (PTC).

13.7.3 International Offshore Wind Energy Market

In the past, there have been numerous predictions for the growth of offshore wind energy, and many have faced similar problems as onshore wind with their predictions. In recent times the IRENA has noted that the expansion of offshore wind energy has emerged as a significant renewable energy resource, with several emerging markets setting targets for offshore wind system deployment (IRENA, 2019a). Their predictions for the cumulative installed capacity of offshore wind are given in Figure 13.19. As shown, offshore wind market would grow significantly from 2020 to 2050, rising about 10-fold from about 23 GW in 2018 to about 1000 GW in 2050.

13.7.4 Barriers to Wind Energy Deployment

In spite of the potential for wind energy development, there exists a number of barriers that will slow the planning process. The barriers to the deployment of utility-scale wind energy can be divided into the following areas:

- **Direct costs**. A general disadvantage of wind turbines is that their total cost is highly concentrated at the time of initial construction as a result of high capital costs.

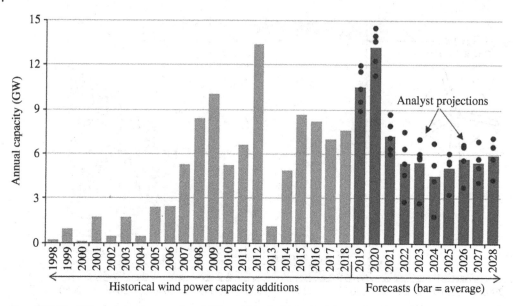

Figure 13.18 US wind power capacity additions: historical installations and projected growth (*Source:* Wiser and Bolinger, 2019/U.S. Department of Energy)

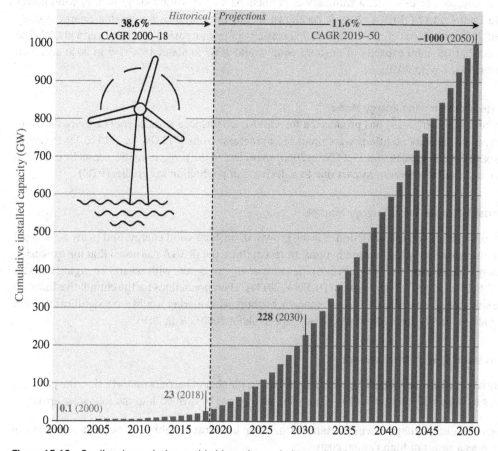

Figure 13.19 Predicted cumulative worldwide onshore wind capacity (*Source:* IRENA, 2019a/International Renewable Energy Agency).

- **System integration**. For low-penetration systems, the integration of wind power into electrical power systems is readily accomplished with existing engineering tools, but there may be major costs involved. High penetration system integration is more challenging; see Chapter 16.
- **Regulatory and planning processes**. The responsibilities for approving the siting and other aspects of land based wind project implementation are usually assigned to existing bodies (i.e., local zoning boards) that may have had little experience with wind power – with corresponding regulatory and institutional bottlenecks. Regulatory and stakeholder issues are particularly challenging for offshore projects; see Chapter 14.
- **Environmental impacts**. All wind energy projects have effects on the environment (as will be discussed in Chapter 14) and potential barriers that must be avoided or mitigated.
- **Electricity system planning**. One barrier to the introduction of large-scale wind generation into some utility systems is its novelty. Those utilities in particular may lack appropriate planning experience. Other utilities may already have wind generation in their service area but have not yet addressed the issue of how much more wind generation they can absorb before they need to consider additional measures, such as forecasting or energy storage.

13.7.5 Incentives to Wind Energy Development

The largest amount of wind energy generation has been installed in countries that have offered various incentives for its market deployment. These have either been in the form of financial incentives (e.g., capital tax credits, production tax credits, or guaranteed energy prices) or regulatory options (e.g., RPS and net metering). These are mechanisms that monetize the environmental benefits of wind energy, as discussed in Section 13.5.4.3. Countries where such incentives have made the largest impact have included some in Europe (Denmark, Germany, Spain, the Netherlands, and the United Kingdom) and the United States. A detailed discussion of the various incentives for the first three countries is included in an OECD/IEA (1997) publication.

A number of national support programs have been implemented in Europe. Many different incentives have been used, and the level of economic support has differed considerably from country to country and with time. Some of the most common incentives for market stimulation follow (EWEA, 2004):

- Public funds for R&D programs;
- Public funds for demonstration projects;
- Direct support of investment cost (% of total cost or based on kW installed);
- Support through the use of guaranteed price of electricity from wind systems (i.e., feed-in tariffs);
- Financial incentives – special loans, favorable interest rates, etc.;
- Tax incentives (e.g., production tax credit, favorable depreciation);
- Renewable energy or "green" certificates.

Regulatory and institutional measures carried out in Germany are of particular note. The German system has provided a framework that has resulted in a rapid increase in the amount of installed wind generation. Among other things, the German system mandates a guaranteed, above-market rate for wind-generated electricity (feed-in tariffs). See Gutermuth (2000) for more details on these measures. One interesting feature of the German feed-in tariff incentive is that two rates are provided over the life term of the project. The duration of the initial, higher, rate is a function of the energy production of the wind turbine. The rates are set such that a turbine will be economical in most locations in Germany. The effect of the two-rate tariff is that extensive site monitoring is not needed before the decision to proceed with the wind project is made. Also, in the European Union, both feed-in tariffs and green certificates have been used as incentives for renewable energy development (see Ringel, 2006).

In the US, significant wind energy development occurred in California wind farms in response to federal and state legislation that provided a market and favorable tax incentives that attracted private capital. Incentives for wind farms were accomplished via three separate, yet coincidental, federal government actions and one act passed by the state of California. They were the Public Utilities Regulatory Policies Act (PURPA) of 1978, the Crude Oil Windfall Profits Act of 1980, the Economic Recovery Tax Act of 1981, and California state tax credits. These incentives were either in the form of investment or production tax credits.

At the present time, the major incentives for wind energy development in the United States are the PTC, ITC, RECs, and RPS. As noted previously, an RPS is a policy mechanism to ensure a minimum level of renewable energy generation in the portfolios of power suppliers in an implementing jurisdiction (e.g., a state or nation). Rules for the implementation of RPS have been adopted by a number of states in the United States (as well as several European countries and Australia; see Berry and Jaccard, 2001). Some states are using RECs to track and account for renewable electrical generation. These certificates are issued to generators in an amount proportional to production and are tradable on a commodities exchange. One state that has adopted a certificate-based RPS is Massachusetts. Details can be found at Mass.gov (2023).

References

Arapogiannia, A. and Moccia, J. (2014) *Avoiding Fossil Fuel Costs with Wind Energy*. European Wind Energy Association Report, March.

Berry, T. and Jaccard, M. (2001) The renewable portfolio standard: design considerations and an implementation survey. *Energy Policy*, **29**, 263–277.

Bickel, P. and Friedrich, R. (2005) *ExternE Externalities of Energy – Methodology 2005 Update*, European Commission Report EUR 21951.

Billinton, R. and Allen, R. (1984) *Reliability Evaluation of Power Systems*. Plenum Press, New York.

Boeing (1979) *MOD-2 Wind Turbine System Concept and Preliminary Design Report*, Report No. DOE/NASA 0002-80/2, Boeing Engineering and Construction Co.

Cody, G. and Tiedje, T. (1996) A learning curve approach to projecting cost and performance in thin film photovoltaics. *Proc. 25th Photovoltaic Specialists Conference.*, Washington, DC, IEEE, pp. 1521–1524.

Costa, A. M., Orosa, J. A., Vergara, D., and Fernandez-Arias, P. (2021) New tendencies in wind energy operation and maintenance. *Applied Sciences*, **11**, 1386.

DNV GL (2017) *Definitions of Availability Terms for the Wind Industry*, Report No. EAA-WP-15.

Dykes, K., Damiani, R., Ning, A., Graf, P., Scott, G., King, R., Guo, Y., Quick, J., Sethuraman, L., and Veers, P. (2017) *Wind Turbine Optimization with WISDEM*, NREL/PR-5000-70652. National Renewable Energy Laboratory, Golden, CO.

EPRI (Electric Power Research Institute) (1979) *Technical Assessment Guide*, Vols 1–3. EPRI Report: EPRI-PS-1201-SR, EPRI.

EPRI (Electric Power Research Institute) (1989) *Technical Assessment Guide*, Vol 1, Rev. 6. EPRI Report: EPRI P-6587-L, EPRI.

EWEA (2004) *Wind Energy, The Facts*, Vols **1–5**. European Wind Energy Association, Brussels.

EWEA (2009) *The Economics of Wind Energy*. European Wind Energy Association, Brussels.

Fan, S. (2021) *Wind Turbine Power Production Estimation for Better Financial Agreements*, MSc Thesis, University of Massachusetts Amherst, https://scholarworks.umass.edu/masters_theses_2/1147, Accessed 8/15/2023

Ferris, L. L. (1990) *Wind Energy Conversion Systems*. Prentice Hall, New York.

Fingersh, L., Hand, M. and Laxon, A. (2006) *Wind Turbine Design Cost and Scaling Model*, NREL/TP-500-40566. National Renewable Energy Laboratory, Golden, CO.

Fox, B., Flynn, D., Bryans, L., Jenkins, N., Milborrow, D., O'Malley, M., Watson, R., Anaya-Lara, O. (2007) *Wind Power Integration: Connection and System Operational Aspects*. The Institution of Engineering and Technology, London.

Frew, B. (2016) *Assessing Capacity Value of Wind Power*, NREL/PR-6A20-67501, National Renewable Energy Laboratory, Boulder, CO.

General Electric Company (1984) *MOD-5A Wind Turbine Generator Program Design Report – Vol II, Conceptual and Preliminary Design.* Report No. DOE/NASA/0153-2, NTIS.

Gipe, P. (1995) *Wind Energy Comes of Age.* John Wiley & Sons, Inc., New York.

Gutermuth, P-G. (2000) Regulatory and institutional measures by the state to enhance the deployment of renewable energies: German experiences. *Solar Energy,* **69**(3), 205–213.

Harrison, R. and Jenkins, G. (1993) *Cost Modelling of Horizontal Axis Wind Turbines.* UK DTI Report ETSU W/34/00170/REP.

Harrison, R., Hau, E. and Snel, H. (2000) *Large Wind Turbines: Design and Economics.* John Wiley & Sons, Ltd, Chichester.

Hau, E., Harrison, R., Snel, H. and Cockerill, T. T. (1996) Conceptual design and costs of large wind turbines. *Proc. 1996 European Wind Energy Conference,* Göteborg, pp. 128–131.

Hohmeyer, O. H. (1990) Latest results of the international discussion on the social costs of energy – how does wind compare today? *Proc. 1990 European Wind Energy Conference,* Madrid, pp. 718–724.

IRENA (2012) *Renewable Energy Technology: Cost Analysis Series, Volume 1: Power Sector Issue 5/5- Wind Power.* International Renewable Energy Agency, Abu Dhabi.

IRENA (2019a) *Future of wind Deployment, Investment, Technology, Grid Integration and Socio-economic Aspects (A Global Energy Transformation Paper).* International Renewable Energy Agency, Abu Dhabi.

IRENA (2019b) *Renewable Power Generation Costs in 2018,* International Renewable Energy Agency, Abu Dhabi.

Johnson, G. L. (1985) *Wind Energy Systems.* Prentice-Hall, Englewood Cliffs, NJ.

Junginger, M., Faaij, A. and Turkenburg, W. C. (2005) Global experience curves for wind farms. *Energy Policy,* **33**, 133–150.

Karas, K. C. (1992) Wind energy: what does it really cost? *Proc. Windpower '92,* American Wind Energy Association, Washington, DC, pp. 157–166.

Kühn, M., Bierbooms, W. A. A. M., van Bussel, G. J. W., Ferguson, M. C., Göransson, B., Cockerill, T. T., Harrison, R., Harland, L. A., Vugts, J. H. and Wiecherink, R. (1998) *Opti-OWECS Final Report,* Vol. 0–5. Institute for Wind Energy, Delft University of Technology, Report No. IW-98139R.

Mai, T., Lanz, E., Mowers, M. and Wiser, R. (2017) *The value of Wind Technology Innovation: Implications for the U.S. Power System, Wind Industry, Electricity Consumers, and Environment,* NREL/TP-6A20-70032. National Renewable Energy Laboratory, Golden, CO.

Mass.gov (2023) *Renewable Energy Portfolio Standard,* https://www.mass.gov/renewable-energy-portfolio-standard, Accessed 8/13/2023.

Masters, G. M. (2013) *Renewable and Efficient Electric Power Systems.* Wiley, N.J.

Miller, L., Carriveau, R., Harper, S., and Singh, S. (2017) Evaluating the link between LCOE and PPA elements and structure for wind energy. *Energy Strategy Reviews,* **16**, 33–42.

Milligan, M., Frew, B., Ibanez, E., Kiviluoma, J., Holttinen, H. and Soder, L. (2019) *Capacity Value Assessment of Wind Power,* Chapter 22 in *Advances in Energy Systems: The Large-scale Renewable Energy Integration Challenge.* John Wiley & Sons, Hoboken, N.J.

NASA (1979) *200 kW Wind Turbine Generator Conceptual Design Study.* Report No. DOE/NASA/1028-79/1, NASA Lewis Research Center, Cleveland, OH.

Neij, L. (1999) Cost dynamics of wind power. *Energy,* **24**, 375–389.

OECD/IEA (1997) *Enhancing the Market Deployment of Energy Technology: A Survey of Eight Technologies.* Organisation for Economic Co-Operation and Development, Paris.

Perez, R., Seals, R., Wenger, H., Hoff, T. and Herig, C. (1997) Photovoltaics as a long-term solution to power outages *Proc. ASES Annual Conference,* Washington, DC.

Pfaffel, S., Faulstich, S., and Rohrig, K. (2017) Performance and reliability of wind turbines: a review. *Energies,* **10**,1904.

Rabl, A. (1985) *Active Solar Collectors and Their Applications.* Oxford University Press, Oxford.

Rademakers, L. W., Braam, H., Zaaijer, M. B. and Van Bussel, G. J. (2003) Assessment and optimisation of operation and maintenance of offshore wind turbines. *Proc. 2003 European Wind Energy Conference,* Madrid.

Rademakers, L. W., Braam, H. and Obdam, T. S. (2008) Estimating costs of operation and maintenance for offshore wind farms. *Proc. EWEC 2008,* European Wind Energy Association, Brussels, Belgium.

Rademakers, L. W., Braam, H., Obdam, T. S., and Van de Pieterman, R. P. (2009) Operation and maintenance cost estimator (OMCE) to estimate the future O&M costs of offshore wind farms. *Proc. EWEC 09 Offshore Conference,* Stockholm.

Raftery, P., Tindal, A., Wallenstein, M., Johns, J., Warren, B. and Vaz, F. D. (1999) Understanding the risks of financing wind farms. *Proc. 1999 European Wind Energy Conference,* Nice, pp. 496–499.

Reeves, A. (2003) *Wind Energy for Electric Power: A REPP Issue Brief.* Renewable Energy Policy Project, Washington, DC.

Ringel, M. (2006) Fostering the use of renewable energies in the European Union: the race between feed-in tariffs and green certificates. *Renewable Energy,* **31**, 1–17.

Rubin, E. S., Azevdo, I. M. L, Jaramillo, P. and Yeh, S. (2015) A review of learning rates for electricity supply technologies. *Energy Policy,* **86**, 198–218.

Stehly, T. and Beiter, P. (2019) *2018 Cost of Wind Energy Review,* NREL/TP-5000-74598. National Renewable Energy Laboratory, Golden, CO.

Stehly, T, Beiter, P., Heimiller, D. and Scott, G. (2018) 2017 *Cost of Wind Energy Review,* NREL/TP-6A20-72167, National Renewable Energy Laboratory, Golden, CO.

Tande, J. O. G. and Hansen, J. C. (1991) Determination of the wind power capacity value. *Proc. 1991 European Wind Energy Conference,* Amsterdam, pp. 643–648.

Tang, C. and Zhang, F. (2019) Classification, Principle and Pricing Manner of Renewable Power Purchase Agreement, *IOP Conf. Series: Earth and Environmental Science,* p. 295.

Vachon, W. A. (1996) Modeling the Reliability and Maintenance Costs of Wind Turbines Using Weibull Analysis. *Proc. Wind Power 96,* American Wind Energy Association, Washington, DC, pp. 173–182.

van de Pieterman, R. P., Braam, H., Rademakers, L. W. M. M. and Obda, T. S. (2010) *Operation and Maintenance Cost Estimator (OMCE)- Estimate future O&M cost for offshore wind farms,* ECN Report ECN-M-10-089.

Voorspools, K. R. and D'haeseleer, W. D. (2006) An analytical formula for the capacity credit of wind power. *Renewable Energy,* **31**, 45–54.

Walker, S. (1993) Cost and Resource Estimating. *Open University Renewable Energy Course Notes – T521 Renewable Energy.*

Walker, J. F. and Jenkins, N. (1997) *Wind Energy Technology.* John Wiley & Sons, Ltd, Chichester.

WEC (1993) *Renewable Energy Resources: Opportunities and Constraints 1990–2020.* World Energy Council, London.

Wiser, R. and Bolinger, M. (2018) *2018 W*ind Technologies *Market Report,* Lawrence Berkeley National Laboratory Report.

Wiser, R. and Bolinger, M. (2019) *2018 Wind Technologies Market Report,* U.S. Department of Energy, Lawrence Berkeley National Laboratory Report.

Wiser, R., Bolinger, M., and Lantz, E. (2019) Assessing wind power operating costs in the United States: Results from a survey of wind industry experts, *Renewable Energy Focus,* **30**, 46–57.

Wiser, R., and Kahn E. (1996) *Alternative Windpower Ownership Structures: Financing Terms and Project Costs.* Lawrence Berkeley National Laboratory Report: LBNL-38921.

14

Project Development, Permitting, Environmental Considerations, and Public Engagement

14.1 Overview of the Chapter

In order for the wind turbines to fulfill their purpose, they must be installed somewhere. Someone or some group has to make that happen. Rights to the land or offshore location must be secured. Installation and operation of a wind plant will invariably have some impacts, both social and environmental, and those impacts will need to be addressed. There are often competing uses for the land or the sea, and potential conflicts will need to be resolved. Permit and public engagement processes will normally be part of this. The steps that are required to move an envisioned project from the conceptual stage to where construction can begin are called project development, and that is the subject of this chapter.

Section 14.2 provides an overview of the development process for land-based projects. Section 14.3 does the same for offshore. There are a number of topic areas that are commonly encountered in wind turbine siting. These are discussed in the rest of the chapter. The first group deals with topics that are important, at least on land and often offshore as well. The remainder are primarily offshore. Section 14.4 gives a broad overview of the environmental topics that need to be considered. The first group after that (Sections 14.5–14.9) deals with topics that are important, at least on land and often offshore as well. The remainder (Sections 14.10–14.12) are primarily offshore. Section 14.5 focuses on visual impact. Section 14.6 is a primer on noise, which can be a significant concern both on land and offshore. Section 14.7 reviews potential issues with birds and bats. Section 14.8 is about aviation safety. Section 14.9 has to do with a phenomenon called "shadow flicker." Section 14.10 discusses marine mammals. Section 14.11 has to do with the relationship between offshore wind energy development and commercial fisheries. Section 14.12 provides a synopsis of issues regarding electromagnetic fields (EMFs) and interference.

It should be noted that for wind energy to develop at the scale that it has over the last several decades, policies were needed to make that possible. That topic is discussed immediately below.

14.1.1 National Policies Fostering Wind Energy

Until relatively recently, electricity production was under the purview of national or regional utilities. For both historical and technical reasons, they were not structured to readily incorporate electricity from external generators into their networks. National policies needed to be designed and implemented to facilitate and then encourage that to take place. For example, in the United States, Congress created the Public Utility Regulatory Policy Act (PURPA, 1978) that required regulated utilities to accept all sources of electricity at a fair rate. One result of the law was to open the utility markets for renewable energies, including wind power (APPA, 2023). All nations with mature wind power industries have enacted similar legislation to restructure power markets and allow wind

Wind Energy Explained: On Land and Offshore, Third Edition.
James F. Manwell, Emmanuel Branlard, Jon G. McGowan, and Bonnie Ram.
© 2024 John Wiley & Sons Ltd. Published 2024 by John Wiley & Sons Ltd.

energy and other low-carbon sources to diversify their electricity supplies. These new national policies have focused on the following elements:

- First, a portfolio of policy incentives was developed. For example, subsidies were introduced by governments when the costs of wind were higher than conventional electricity supplies. There are various subsidies that were designed to increase the market share of wind in electricity markets, including production tax credits and renewable energy certificates, feed-in tariffs, and later, in the offshore market, offshore renewable energy credits and contracts for difference. Those incentives originated in Europe, the United Kingdom, and the United States but have been used elsewhere as well.
- Second, regulatory frameworks have been designed and implemented to govern the life cycle of the wind plant from siting and construction through decommissioning. The legal system also governs siting, environmental permitting, and public involvement. In addition, national governments assess available wind resources (see Chapter 2), transmission line availability, and geographic project areas such as renewable energy zones on land and ocean leases [see Section 14.3.1 on Marine Spatial Planning (MSP)]. Both public and private entities are guided by regulatory frameworks affecting how power is purchased and distributed by utilities.
- And finally, national governments set emission reductions and installation goals, including climate change metrics, to signal to the market that policy incentives and capital investments will support the deployment of these low-carbon technologies over a particular period (see Chapter 13).

National policies play a pivotal role in shaping the trajectory of wind project development, their feasibility, and successful integration, as will be seen in the next section.

14.2 Project Development

This section provides an overview of the key aspects of the development of a land-based wind energy project.

14.2.1 Project Development Process and the Development Team

The development process can be discussed from the perspective of the developer, the regulator, the community, the utility, or the financial sector. Each of these perspectives will vary based upon the interests and legal requirements of the institutions, their stakeholders, and the public. Figure 14.1 provides an overview of the wind project development process, which can take several years to complete. The timeframes may vary depending on various factors, including the size of the project, the type of land use (private, public, or conservation lands), and interaction with the community around the site.

The project development process for land-based wind involves the following steps:

- **Prospecting and market assessment**. Land parcels are identified and secured for the entire site footprint of the project with a lease and/or rental agreement. Also, the developer may compensate landowners and/or farmers.
- **Resource review, siting, and land lease.** The developer conducts a desktop study and obtains or purchases data about the site, including wind resource maps and topological data. Also, the wind resource estimates are used to identify whether the project is "bankable" (i.e., investment-worthy). See Chapter 2.
- **Stakeholder engagement and communication strategy.** Beyond the landowners, engagement with residents, politicians, and other interested business owners near the site should be implemented. This may include indigenous communities in some areas.

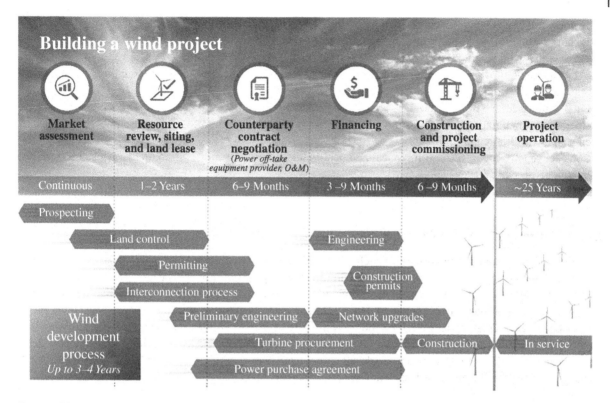

Figure 14.1 Example of development process for a land-based wind power plant (*Source:* U.S. Department of Energy / https://www.nrel.gov/docs/fy21osti/78591.pdf / last accessed 09/01/2023)

- **Counterparty contract negotiation.** Along with preliminary engineering and design calculations, simple calculation procedures are used to evaluate the potential wind farm yield, the wind turbine class, and potential construction challenges.
- **Analyze the interconnection process.** The developer and/or owner secures a customer for the expected power [e.g., a power purchase agreement (PPA)] and applies to the relevant utility for permission to connect (referred to as joining an interconnection queue). Securing access to utility markets with off-take agreements and PPAs varies widely by country and region (see Chapters 12 and 13).
- **Permitting and consultations.** Throughout the development process, a suite of permits and consents are prepared, including zoning and local ordinances and consultations with relevant government authorities. (see Sections 14.2 and 14.3.)
- **Environmental assessment.** An environmental assessment (EA) and/or an environmental impact statement (EIS) will need to be prepared, depending on the state and national laws and the size of the project (See Section 14.3.1 below). These permits often trigger biological studies (e.g., the presence of endangered bird or plant species).
- **Secure funding and investment capital**. The financial feasibility considers potential subsidies and/or tax credit mechanisms (see Chapter 13).

When the steps above are completed, the development process, as considered in this chapter, is considered complete. After that comes the actual installation and operation. That process is summarized below and discussed in more detail in Chapter 15.

- Perform full engineering design; in particular, design load cases are run in this phase.
- Procure turbines, transformers, and other components, and secure the supply chain.
- Finalize the wind turbine foundation design.
- Site construction involves building roads, installing the turbines and cables, and constructing a substation (if required by the utility).
- Final commissioning of the project.
- Operate and maintain the wind farm, including post-construction monitoring for environmental effects and engaging with host communities.
- Consider repowering or retrofitting (once the operational lifetime of a project is complete, typically 25 years).
- Decommission and recycle.

14.2.2 Siting of Land-Based Wind Projects: Challenges and Opportunities

Planning and regulatory compliance vary widely across the globe due to different national policies, political structures, land use management, and a diversity of environmental and cultural values. The local planning process and siting authorities of each country depend upon their administrative and regulatory structures and, in some cases, the maximum size of the project. There could be combinations of local, state, and county authorities involved in project approvals. Larger projects may be the responsibility of public utilities, national grid operators, and/or county and state siting authorities. Understanding the institutional context and legal responsibilities for how decisions are made is critical up front when the development team is in preliminary phases (UK Government, 2017; DOE, 2022). Also, this dictates how local landowners and stakeholders may get involved in or oppose the decision process.

The early stages of securing land rights can be a complex process in densely populated areas or those sites that require securing of private land rights by negotiating property lease agreements and compensation payments. Often, developers must weave together rights across dozens of properties to secure enough land for the project. There may be partnership opportunities for agricultural landowners and farmers who can negotiate a lease agreement with annual revenues and continue to use the land around the turbine location for crops and livestock (NREL, 2021). In Denmark, developers have offered ownership shares of land-based turbines. This has been a model of how to involve communities and make potential benefits more explicit beyond job creation. As turbine sizes have grown along with the capital costs, however, these co-ownership schemes have become more difficult to realize. Also, landowners that are not included in a lease agreement or will not benefit financially may pose challenges in the project approval processes. Some of the regulatory aspects of project approvals that might need to be granted and may generate delays or controversies include:

- Local zoning regulations, such as setbacks from residents, the tower height, distance between turbines, and shadow flicker (see Section 14.9).
- Noise restrictions, including both low frequency and amplitude (see Sections 14.4.2 and 14.6).
- Limitation of construction schedules during local recreational events or tourist seasons.
- Environmental restrictions, including effects on endangered species and sensitive habitats (see Section 14.4).
- Transmission interconnection queues (see below) and construction of substations.

A major bottleneck to meeting the development timeline related to the amount of new power generation and energy storage is the transmission interconnection queues. For example, in the United States, as of 2022 over 2000 GW of total generation and storage capacity were seeking connection to the grid (Rand *et al.*, 2023). See Figure 14.2 for more details about this process.

- A project developer initiates a new *interconnection request (IR)* and thereby enters the *queue*
- A series of *interconnection studies* establish what new transmission equipment or upgrades may be needed and assigns the costs of that equipment
- The studies culminate in an *interconnection agreement (IA):* a contract between the ISO or utility and the generation owner that stipulates operational terms and cost responsibilities
- Most proposed projects are *withdrawn*, which may occur at any point in the process
- After executing an IA, some projects are built and reach *commercial operation*

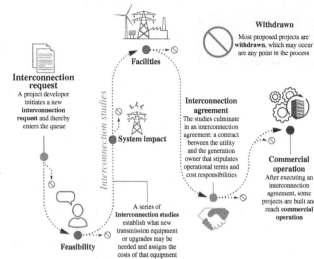

Source: Derived from image courtesy of Lawrence Berkeley National Laboratory and used with permission. | GAO-23-105583

Notes: These steps are in accordance with Federal Energy Regulatory Commission (FERC) approved open-access transmission tariffs and generator interconnection procedures. Image from Government Accountability Office report GAO-23-105583: Utility-Scale Energy Storage, used with permission.

Figure 14.2 Typical interconnection study process and timeline in the United States (*Source:* Rand *et al.*, (2023) / U.S. Department of Energy / Public Domain)

14.3 Offshore Project Development

The burden of permitting procedures is greater for offshore wind projects because they tend to have a larger footprint than land-based wind farms, which straddle different jurisdictions and stakeholder users. The ocean is a very busy public resource and triggers a panoply of national regulatory requirements governing its uses, ecology, and coastal areas (see Section 14.3.1). The development of a land-based project could take 12 months to several years of planning to deploy. Offshore, this planning horizon may be 7–10 years (although Asian developers are deploying more rapidly than their counterparts elsewhere).

The offshore development team is usually much larger than for land-based projects. The regulatory issues are more challenging at sea, and the national government is the lead agency. This is in contrast to land-based wind projects where local jurisdictions are in charge (unless the project is on public lands). Table 14.1 provides an overview of the development process classification criteria and the start and end stages of an offshore project.

Some important functions of the offshore project development team include:

- Legal and permitting experts who understand the ownership and lease fees (or royalties) for using the ocean spaces as well as the rules governing MSP.
- Stakeholder engagement coordinators for informal dialogues, outreach, and required public comment management for EIS or EIAs.
- Wind resource specialists who understand wind and wave interactions.
- Financial advisors who can provide the capital investment advice needed to bid on the ocean spaces and manage the project costs, including the PPA.
- Geotechnical engineers to analyze the soil conditions prior to finalizing the design of the substructure and its foundation.

Table 14.1 Offshore wind project development process classification criteria

Step	Phase name	Start criteria	End criteria
1	Planning	Starts when a developer or regulatory agency initiates the formal site control process (e.g., designation of a wind energy area [WEA])	Ends when a developer obtains control of a site (e.g., through competitive auction or a determination of no competitive interest in an unsolicited lease area (United States only)
2	Site Control	Starts when a developer obtains site control (e.g., a lease or other contract)	Ends when the developer files major permit applications (e.g., a Construction and Operations Plan [COP] for projects in the United States)
3	Permitting = Site Control + Offtake Pathway	Starts when the developer files major permit applications (e.g., a COP or an offtake agreement for electricity sales)	Ends when regulatory entities authorize the project to proceed with construction and certify its offtake agreement
4	Approved	Starts when a project receives regulatory approval for construction activities and offtake agreement certification	Ends when the sponsor announces a "financial investment decision" and has signed contracts for construction work packages
5	Financial Close	Starts when the sponsor announces a financial investment decision and has signed contracts for major construction work packages	Ends when the project begins major construction work
6	Under Construction	Starts when construction is initiated	Ends when all wind turbines have been installed and the project is connected and generating power to an electrical grid
7	Operating	Starts when all wind turbines are installed and transmitting power to the grid; commercial operation date marks the official transition from construction to operation	Ends when the project has begun a formal process to decommission and stops feeding power to the grid
8	Decommissioned	Starts when the project has begun the formal process to decommission and stops transmitting power to the grid	Ends when the site has been fully restored and lease payments are no longer being made

(*Source:* Musial *et al.*, 2023/U.S. Department of Energy/Public Domain)

- Certified staff from captains to project engineers to manage the preliminary survey work and the construction at the site.
- Marine biologists and other environmental specialists who can design and oversee any necessary pre- and post-construction biological surveys.

Successful site planning and the tools associated with these plans are related to MSP. The next section summarizes and highlights why it is important.

14.3.1 Marine (Maritime) Spatial Planning: Overview

Marine (or maritime) spatial planning (MSP) is a process by which the relevant authorities analyze and organize human activities in marine areas to achieve ecological, economic, and social objectives. In practice, it involves both formal and informal public initiatives on how to use the ocean space in line with societally agreed goals, values, and targets. Ocean rights, restrictions, and responsibilities in the marine environment are contingent on designated national jurisdictions and administrative boundaries.

The ocean spaces where wind energy projects may be sited are governed by complex and overlapping authorities that regulate many industrial activities and recreational uses. Many of the complexities do not involve engineering challenges *per se* but rather spatial, legal, social, and cultural considerations that may pose difficult stakeholder conflicts and tradeoffs about the use of these areas. The co-existence of these ocean uses, along with offshore wind energy activities, is a central challenge of MSP. In addition to wind energy projects, some of the significant competing uses for coastal ecologies and communities include:

- Navigation, including shipping, fishing, recreation, and other vessels
- Mineral extraction (e.g., sand and gravel)
- Oil and gas exploration/extraction
- Fishing and aquaculture
- Tourism and recreational activities
- Marine archeological resources and cultural heritage
- Nature protection and management of endangered species (see Section 14.4)
- Military exercises and security apparatus
- Submarine cables and pipelines

In the United States, the federal government administers the waters to the boundary of the Outer Continental Shelf (OCS), but various government bodies have rights over the resources. All lease payments are directed to the US Treasury. In the United Kingdom, the Crown Estate owns the seabed and manages it on behalf of Parliament. The lead agency, along with the developer team, is responsible for consulting with many other environmental institutions and nongovernmental organizations (NGOs), representing a diverse set of stakeholders using the ocean spaces and living along the coastlines. For an understanding of the legal milieu for individual countries, refer to the regulations, authorities, and planning processes for each country (see, for example, European MSP Platform, 2023). A couple of important examples of MSP processes are described briefly to provide an overview of these planning regimes.

14.3.2 Navigational Risk Characterization

Maritime transport can come into conflict during the entire lifecycle of offshore wind plants, including expanding existing ones. Most of the conflicts are triggered by concerns about possible accidents and route diversion, particularly where shipping activity is intense. Some larger vessels cannot pass through an offshore wind project and need to chart a course a safe distance away from the turbines. Smaller vessels, such as fishing boats, may be able to transit through a wind farm, but there is a risk of collision with the offshore structures. European conflicts between offshore wind farming and shipping occur mainly in the North Sea, Irish Sea, and Baltic Sea, where many projects already exist and are planned. In Belgium and Germany, wind farms are considered maritime exclusion zones, a policy designed to prevent accidents. In the United Kingdom and Denmark, wind farms are open for transit for both commercial and recreational use. No special requirements regarding vessel equipment or limits on the vessel size are imposed (European MSP Platform, 2023). The risk characterization may include the following:

- Risk of accidents is increased by greater traffic density and reduced sea space, which might lead to the creation of choke points. This can become an issue in case there are problems with a ship's on-board navigation equipment (i.e., radar) and/or weather events. Accidents may lead to large financial losses, human casualties, and serious environmental damage.
- Diversion can lead to the following problems for the shipping sector: (1) increased time and fuel spent, more greenhouse gas emissions, and higher wages for the crew; (2) financial penalties (3) higher insurance costs due to riskier routes; (4) conflicts between national and international law.
- Most nations in the European Union (EU) have stringent processes in place requiring offshore wind project developers to demonstrate that they have thoroughly assessed the maritime risks and implemented adequate

risk management measures with a Navigational Risk Assessment (NRA). Traditionally, an NRA is conducted in the licensing and approval stages of a project, but the process may be more effective in the MSP stage and/or when designating the lease area. Examples of mitigation strategies include: Co-design shipping routes and offshore wind plant layouts in a collaborative process.

- Utilize an NRA during the part of the MSP process that considers documented experiences and guidelines.
- Use existing design guidelines for the layout and placement of offshore wind plants.
- Consider the seasonality of the shipping sector when planning offshore construction.
- Use technical means to increase safety within wind farms, including safe crossings for specialized vessels (Salerno *et al.,* 2019; European MSP Platform, 2023).

14.3.3 Marine Archaeological Resources and Cultural Heritage

Submerged archaeological marine resources include shipwrecks, sunken aircraft, war graves, coastal historical and heritage sites, and religious and ceremonial areas. Ancient, submerged landforms, such as paleochannels, are marine cultural resources of importance to Native American tribes off the Atlantic Coast of the United States. Early identification of important heritage sites through MSP is recommended to minimize harm and local conflict. It is possible, however, that important sites and finds may arise during the permitting phase and from stakeholder engagement. Protection of underwater archaeology and historical settings usually requires a permitting process and verification by national historic preservation agencies. Sunken military craft also remain the sovereign property of the national government. In the United States, for example, the Bureau of Ocean Energy Management (BOEM) requires a marine archaeological resource assessment as part of their environmental review process.

14.3.4 Regulatory Structure for Offshore Wind – The US Case

This section illustrates the project development process of offshore wind, using the United States as a case study. In many respects, the US case mirrors the European planning approaches that launched the offshore wind market.

The US regulatory structure operates at the federal and state levels (when cables or construction come ashore). BOEM is the federal regulatory entity charged with leasing and permitting wind farms in federal waters. The federal government retains the power to regulate commerce, navigation, power generation, national defense, and international affairs throughout state waters. For most states, jurisdiction extends to 3 nautical miles (5.56 km) from the shoreline, and federal jurisdiction extends out to at least 200 nautical miles (370 km) on the OCS. The jurisdictions of the Great Lakes, on the other hand, fall under state agencies until they meet the international borders with Canadian provinces. State and local authorities govern the approvals of the cable connections to shore.

The US legal framework, passed in 2005, establishes a "cradle to grave" approach to regulate offshore renewable energy activity from lease selection and auctions, site assessment, construction and operations, and decommissioning (Energy Policy Act of 2005). Key elements support safety, environmental protection, state and local coordination with affected agencies, financial assurance, fair return for use of submerged public lands, and revenue sharing with the states (BOEM, 2023a). The BOEM planning process for leasing offshore wind in US federal waters is a multi-year process with four phases, as shown in Figure 14.3.

Planning and Analysis. In the first phase, ocean wind energy lease areas are identified and then auctioned to private bidders. This phase involves engaging with a panoply of stakeholders, including state regulators and state and federal tribal governments, business interests, and environmental NGOs to consider which locations would have the least number of environmental conflicts and enhance co-benefits. After the identification of the lease area, BOEM begins the competitive process by issuing a "Call for Information and Nominations" to announce interest in leasing an area and to solicit relevant financial assurances.

Leasing. BOEM evaluates the nominations and information received during the call, determines the leasing area, issues a proposed sale notice to inform the public, and requests comments on lease terms and conditions.

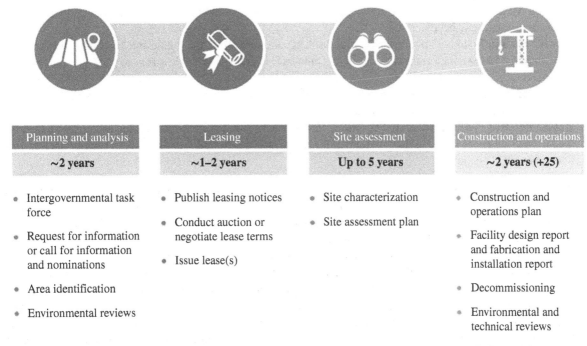

Figure 14.3 BOEM regulatory roadmap for offshore wind commercial lease (*Source:* Adapted from BOEM, 2023a)

Ultimately, BOEM issues a final sale notice announcing the details of the auction at which bids for leases will be accepted.

Site Assessment. After the auction and sale of the lease area, the developer prepares a site assessment plan (SAP) for geotechnical and other environmental surveys and assesses the wind resources. The developer has five years to complete the SAP activities. Typically, before the developer begins the SAP and the permitting process, a state agency or public utility will issue a competitive solicitation to buy power from the lease areas. BOEM is not involved in this planning step. The developers bid to obtain a PPA before they invest further in the planning process (see Chapters 12 and 13).

Construction and Operations Plan (COP). This document details the project proposal, including a range of alternatives. Once the COP is deemed "sufficient and complete," BOEM triggers the preparation of a project-specific EIS (see Section 14.3.5) with a notice of intent. The European process does not require a COP.

14.3.5 Environmental Impact Statement/Assessment

For both land-based and offshore projects, governments require the project developers to address potential environmental impacts. In the United States, the National Environmental Policy Act (NEPA), enacted in 1970, requires the preparation of an EIS. EU nations have enacted similar forms of legislation, but the document is called an environmental impact assessment (EIA).

Both laws require and promote the concept of public involvement and "environmental democracy" (European Commission, 2023). In brief, the steps in this process involve the following actions:

- **Scope** to assess and quantify the potential socioeconomic, cultural, physical, and biological impacts and benefits that could result from the siting, construction, O&M, and decommissioning.
- **Consult** to coordinate and consult with dozens of other national and regional government agencies with jurisdiction over natural resources in offshore leases and coastal and land-based areas.

- **Develop Alternatives Analysis** to consider project alternatives, including the option of no construction across all resources that may be affected.
- **Involve the Public** to hold public hearings and solicit comments from the public and key stakeholders on the draft and final EIS. An important objective of this process is to encourage local citizens to participate in the decision-making process.
- **Avoid and Mitigate** to identify and design mitigation strategies to avoid and/or reduce potential significant environmental and socio-economic impacts.
- **Define Benefits** to identify potential benefits from the low-carbon electricity source, including health benefits (i.e., lives saved because of a reduction in greenhouse gases and other pollutants) and environmental and economic benefits (e.g., jobs).

It is beyond the scope of this text to describe how risks are characterized for each resource topic, receptor (e.g., marine species, humans), and stressor (e.g., noise) in a NEPA document. As Figure 14.4 shows, there is a wide range of biological and physical sciences and social, economic, and cultural considerations. Sections 14.4–14.12 will provide an overview of selected environmental considerations, however.

Both the United States and EU nations characterize the potential adverse or beneficial impacts with a four-level classification scheme as either negligible, minor, moderate, or major. Both quantitative studies (e.g., fishery stocks, bird abundance) and qualitative analyses (e.g., stakeholder comments, value of cultural heritage) come into play. Determining the magnitude of each effect on the resources requires defining the spatial extent, duration, frequency, and severity of all the resources addressed. Many of these EIAs and EISs are several thousand pages long.

The following mitigation hierarchy is widely used to minimize or avoid potential impacts from the proposed project. In its simplest form, risks are managed in a structured set of steps that includes three stages:

1) Avoid creating impacts from the outset,
2) Minimize the impacts that cannot be avoided, and
3) Compensate for or offset the impacts that cannot be minimized.

Compensating and offsetting are usually more costly and have a higher risk of failure (Arlidge *et al.*, 2018). Proper application of the hierarchy should decrease the impacts of the project over time, such that most of the impact is alleviated through avoidance, leaving a modest amount remaining to minimize and only a residual

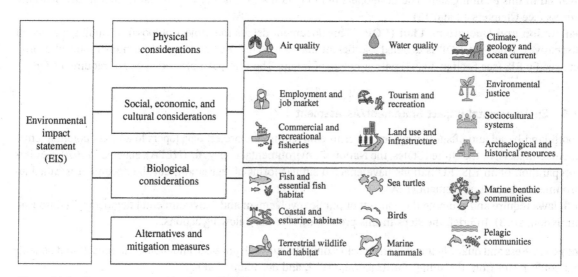

Figure 14.4 Resources to consider for an EIS/EIA

to compensate. In practice, avoidance should be prioritized and can be a cost-effective means of mitigation. Many organizations and countries already have guidance for the mitigation hierarchy, but there is no standard terminology within or among stages, which may cause confusion (Dempsey *et al.*, 2023). Traditionally, the stages of the hierarchy corresponded to the phase of the project (i.e., avoidance during project planning and restoration during decommissioning). However, a recent focus on adaptive management has expanded the options on how the mitigation hierarchy can be applied iteratively through every stage of the development cycle (see Table 14.2 below).

The mitigation hierarchy is used for identifying and prioritizing mitigation actions within the EIA/EIS process. The most extensive work has taken place in the United States and Europe, where wind deployments are subject to extensive regulations and government oversight. Therefore, the examples of mitigation strategies in the following section are mostly from these nations. To design mitigation strategies, in-depth knowledge of the baseline and the extent of the impacts on various resources is needed. For example, to mitigate potential impacts on bird or bat species, data are gathered from pre- and post-construction fatality surveys; see Section 14.7.5.

The last stages of the EIS process involve the review and approval of the draft and final EISs, and then BOEM issues a record of decision (ROD). The final steps in the US process, as shown in Figure 14.3 above, involve the preparation and approvals of a facility design report and fabrication and installations report before project construction can begin.

14.3.6 Public Engagement and Communities

Although the EIS process requires citizen engagement through public hearings and comment processes, this is not sufficient for social acceptance of projects in many cases. Social acceptance is a multi-dimensional concept with inter-related socio-political, market, and community aspects (Wüstenhagen *et al.*, 2007). The distributed nature of wind power plants requires the development of many sites across a variety of ecosystems on land and at sea. Siting, constructing, and operating a wind power plant will affect the host communities and a range of stakeholders. The project may change the landscape and/or the seascape and impact wildlife and habitats. To help assess these complex and interrelated factors, an energy systems perspective is needed. Often, project siting is compartmentalized, where engineers work separately from the siting experts and from the citizens that host the project. Also, each country and locality may have a slightly different position on climate change policies and the need for wind energy. These views and values influence how stakeholders will perceive the risks and benefits of wind energy. With the urgency for a clean electricity transition, decision-makers are advocating for accelerated permitting timeframes and construction schedules. However, this acceleration needs to be balanced with conservation objectives and public involvement activities. This section highlights some aspects of public engagement and community involvement in decision-making.

Table 14.2 Applying the mitigation hierarchy through different phases of project development

Project stages	Mitigation hierarchy application			
	Avoidance	Reduction/Minimization	Restoration	Compensation/Offset
Planning	✓		✓	
Project Design	✓	✓		
Construction	✓	✓	✓	✓
Operation		✓		✓
Decommissioning	✓	✓	✓	✓

(*Source:* Dempsey *et al.*, 2023/U.S. Department of Energy/Public Domain)

Public opinion surveys across the world usually indicate a high degree of support for renewable energy technologies (Leiserowitz, 2007). As deployments on land and sea have increased, the number of individuals and groups that may be skeptical or oppose specific projects has also increased. A variety of reasons may trigger opposition due to potential conflicts over habitats, endangered species, conservation areas, viewsheds, etc. (Gross, 2020). It should also be noted, however, that some of the opposition originates from other sources, for example, from those who believe that their own political or economic interests may be threatened by the success of renewable energy projects; see, for example, Oreskes and Conway (2011). On the other hand, many members of the public do have legitimate concerns and risk perceptions that need to be addressed in a balanced manner. As noted in the IPCC (2011), "...environmental concerns, public acceptance, and/or the infrastructure required to manage system integration are likely to limit the deployment of individual renewable energy technologies before absolute technical resource potential limits are reached" (Bruckner *et al.*, 2014). For more information on how to effectively approach public engagement with wind energy, see IEA Wind (2023). Some useful principles are summarized below and shown below in Figure 14.5:

- Two-way communication is essential for effective public engagement, where citizen views and concerns are addressed.
- Vocal minorities in public hearings may not reflect community-wide views.
- Risks, uncertainties, and benefits should be communicated with transparency.
- Engagement does not always guarantee that the project is acceptable, but the absence of engagement often leads to failure.
- Spokespeople communicating project information, other than the developer, need to be trusted by the community and perceived as independent from the developer.
- Citizen views are important in formulating risk information.
- Understanding the history of the project area, including indigenous communities and cultural heritage, provides needed insights on demographics and social justice.

Figure 14.5 Aspects of meaningful public involvement (*Source:* DOT, 2022/The United States Department of Transportation/Public Domain)

Communities are interested in how they will share in the potential benefits of these projects. Accordingly, issues of social justice and equity will arise as citizens want to know how project benefits will be shared with the host community, as well as with citizens that have been historically disadvantaged. Equity involves distributing benefits and ensuring that the planning process is transparent and fair (Toke and Breukers, 2015). Project developers may collaborate with fishermen, for example, by using local ships and work boats, and this may yield significant economic benefits to these communities. Research has shown that if the area has had a wind project sited there and the community is satisfied with the process, there is more social acceptance for the next deployment (Hoen *et al.,* 2019). And the opposite could be true as well.

14.4   Environmental Considerations: Overview

Many of the stakeholder and regulatory concerns that were discussed above relate to environmental and siting considerations, and they are often the most challenging barriers to wind turbine deployments – whether on land or at sea. As Chapter 1 noted, the reason societies want to accelerate the deployment of wind power is to address the consequences of climate change and reduce the pollution associated with fossil fuel electricity production (IPCC, 2021). Yet it is inevitable that concepts of human–wildlife co-existence and conflicts (IUCN, 2022) will become more central to wind deployments as the energy transition accelerates.

Environmental considerations related to wind power need to be understood within the context of tradeoffs and benefits from all electricity supplies and other ecological risks to species and habitats (e.g., other anthropogenic threats to birds). One must be able to compare risks to understand the significance of these potential threats from wind projects. Some of these environmental considerations have regulatory requirements (e.g., permits) designed and enforced by national and sub-national government institutions. These laws may require a variety of pre- and post-construction studies and activities that must be conducted by the developer for the lifecycle of the project – from siting to repowering or decommissioning. Environmental risks require experts across various scientific fields, for example, biologists and forest managers for land-based wind and marine scientists and fisheries managers for offshore wind. These experts work with the development team to ensure that the project complies with the regulations and that schedules can be met. Early consultations with regulatory agencies, affected stakeholders, and communities help to ensure an efficient pathway to construction and operation.

Engineers may play an important role in communicating environmental impacts to various stakeholders. Table 14.3 highlights an example of how some environmental considerations may affect different aspects of project development and engineering decisions.

In the sections that follow, some of the key environmental topics are discussed. These include: (1) visual impact, (2) noise, (3) effects on birds and bats, (4) aviation safety, and (5) shadow flicker. Those topics are of definite relevance on land but are often important offshore as well. For offshore in particular, the following are discussed: (1) marine mammals, (2) commercial fisheries, and (3) electromagnetic fields (EMFs). These overviews will include a brief characterization of the issues, mitigation strategies that minimize the potential threats, and some references for further investigation.

14.5   Visual Impact of Wind Turbines

Aesthetic issues are often one of the most contentious issues with stakeholders when siting wind turbines on land as well as offshore. These concerns have increased as larger turbines have become available. An important challenge for developers and regulators is the assessment of potential visual impacts that may affect important scenic, historic, and recreational resources as well as indigenous lands and seascapes. Turbine siting may trigger strong

Table 14.3 Environmental aspects of project development

Selected environmental topics	Environmental aspects of project development						
	Permit req.	Stakeholders/Host communities' views and values	Siting and layout	Construction schedules and costs	Turbine controls and operations	Cumulative impacts	Co-benefits/ Co-existence
Avian – Landbased	X	X	X	X	X	X	No
Avian Offshore	X	X	X	X	X	X	Uncertain
Bats – Land-based	X	X	X	X	X	X	Uncertain
Bats – offshore	X	X	X		X	X	Uncertain
Habitat changes and displacement on land	X	X	X	X	X	X	X
Changes and displacement -offshore	X	X	X	X	X	X	X

public reactions even when it is relatively far from population centers or when it alters the character of a residential neighborhood or recreational areas. In addition to turbines, associated infrastructure, such as transmission lines and substations, may create visual impacts.

It is important to understand the role of visual effects in shaping public attitudes toward wind power and community acceptability of the project. In addition, visual effects may cause aspects of project development to be modified, including turbine selection, the number and layout of turbines, and the cost of energy. Developers employ sophisticated spatial mapping and photomontages to assess the potential impacts. Visual appearances must be considered very early in the design process and involve local stakeholders wherever possible.

Early in the history of utility-scale wind deployments, there were few regulatory processes and scientific methods that adequately addressed aesthetic issues (Pasqualetti *et al.*, 2002; Gipe, 2013). At present, visual impact methodologies are both more sophisticated and required by regulatory agencies in the United States and the EU. Although the evaluation of impact level ultimately involves judgment, the method by which the evaluation is made is now systematic, based on accepted criteria, and part of the formal permitting processes. Qualitative methods for viewshed assessments in the United States are conducted by qualified professionals in visual resource management, landscape architecture, or related fields. Methods and criteria are similar in the EU and United Kingdom for both land-based and offshore projects. The next section summarizes some aspects of the guidance for offshore projects.

14.5.1 Visual Impacts: Risk Characterization of Offshore Sites

One approach to characterizing the visual impact of an offshore project is with a Seascape, Landscape, and Visual Impact Assessment (SLVIA). The methodology evaluates the impacts on people of adding the proposed development to views from selected viewpoints and assesses how the people who are likely to be at that viewpoint may be affected by the change to the view. The enjoyment of a particular view depends on the viewers and the impact on them (See Sullivan *et al.*, 2013). The viewers include local residents, travelers, tourists and vacationers, recreational users, and fishing communities (recreational, boats for hire, and commercial).

The SLVIA process in the United States includes six major phases and is included in the EIS or EIA for offshore projects (adapted from Sullivan, 2021), as summarized below:

1) **Describe the proposed project in detail**, any alternatives under consideration, and the technology proposed, including turbine and substation sizes (at sea and on land) and cable connections to shore. Key stakeholders are involved early in this process.

2) **Identify the geographic scope of the project area and describe the impact receptors** (humans). Review applicable laws, ordinances, and regulations; for example, visually sensitive public resources are identified (e.g., historic sites, culturally important views, parks, public beaches, conservation areas, etc.).

3) **Determine the locations from which the wind turbines will be visible and the people or entities (i.e., receptors) that may be affected** by the visual impacts. Visit key observation points at different times of the day. The magnitude, duration, frequency, and spatial extent of the visual impacts of the wind turbine project are assessed.

4) **Employ visual simulation tools, such as field photography**. Tools include digital photography, GIS software, 3D modeling tools, video animation, and photomontages.

5) **Determine the levels of the potential impacts.** This requires combining expert and stakeholder judgments about the sensitivity of the seascape/landscape and visual impact receptors and the magnitude of the impacts. Then the sensitivity and magnitude judgments are combined to determine the overall level of the impacts. Assessment of impacts from reasonably foreseeable future actions (also referred to as cumulative impacts) for both seascape/landscape and visual resources are conducted.

6) **Identify potential measures to minimize or mitigate the visual impacts** and consider best management practices (BMPs), for example, a radar-based obstacle avoidance system to reduce safety lighting effects on coastal communities (see Section 14.5.5).

7) **Communicate the results.** The developer presents the findings of the visual impact assessment to stakeholders, including the public and regulatory agencies.

14.5.2   Visual Impacts: Land-Based Risk Characterization

Visual impacts can be affected by the characteristics of the viewer, viewshed limiting factors, lighting, atmospheric conditions, distance, viewing geometry, object visual characteristics and backdrop, as shown in Figure 14.6 below.

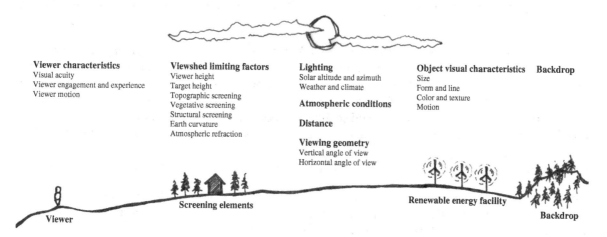

Figure 14.6   Schematic diagram of visibility factors that affect visual perception of landscape elements (*Source:* Lindsey Utter. ANL. Sullivan and Meyer, 2014/U.S. Department of the Interior/Public Domain)

14.5.3 Delineation of Distance Zones or Zone of Theoretical Visibility

Government agencies devise mapping guidelines that describe the area where a development can theoretically be seen. For US land-based projects, the Bureau of Land Management (BLM) uses a Zone of Theoretical Visibility (ZTM) when considering land use decisions for managing visually sensitive resources (BLM, 2023). The following ranges are defined: foreground to middle ground views extend from the viewing location out up to 8 km, background views range from 8 to 24 km, and views beyond 24 km are classified as the "Seldom Seen" zone (Sullivan *et al.*, 2012; DOI, 2013). The Landscape and Visual Impact Assessment (LVIA) is an example of a European method. The recommended ZTV distance for defining the study area is based on turbine height. For example, a turbine height of 131–150 m would require a 40 km distance from the nearest turbine or outer circle of the wind farm. Cumulative impact assessments should also anticipate future actions that could affect the landscape and account for future generations of viewers. This guidance includes impacts on landscape character, wild lands, and special landscape interests (Scottish Natural Heritage, 2017).

14.5.4 Lighting Impacts

Lighting and marking of turbine structures and substations on land and at sea will impact the viewshed of communities. These lighting regulations are needed for navigation, as mentioned above (Section 14.3.2) and aviation safety (see Section 14.8). Observations of existing offshore facilities suggest that night visibility of aviation hazard signals is visible at distances greater than 39 km (Sullivan *et al.*, 2013) and onshore wind turbine aviation lighting is seen at distances greater than 58 km (Sullivan *et al.*, 2012). Light pollution is a very sensitive issue for coastal residents and tourists. Moreover, the concept of sustaining the "night sky" is an historic problem due to light from a variety of sources; wind energy facilities add to these challenges.

14.5.5 Visual Impact Mitigation Strategies – Offshore and Land-Based

Mitigation strategies are developed as part of the environmental planning requirements, whether offshore or on land. For land-based wind, local zoning laws typically have siting regulations for wind energy systems on private land that are intended to mitigate visual impacts. If the wind energy project is in or near public lands, scenic preserves, a national park, or a historic monument (e.g., a lighthouse, a wildlife refuge), then more rigorous evaluations would be conducted, and a national governing agency may be involved. The potential for visual impacts from the project on historic properties triggers a more complex environmental process and mitigation requirements.

The general principles of the mitigation hierarchy are applicable to visual impacts, as illustrated by the following examples:

- **Avoidance.** Moving a project (or its components) to take advantage of screening topography or vegetation.
- **Minimization.** Painting structures to match their backgrounds to minimize visual contrast with the existing landscape. For offshore, layouts of the turbine rows that are perpendicular to the horizon or curved in an arc (as with the Middelgrunden, DK offshore wind project) may minimize some impacts from the shoreside.
- **Rectification.** Revegetation of an area disturbed during project construction (on land).
- **Reducing or eliminating over time.** An audio–visual warning system (AVWS) may reduce lighting impact (see Section 14.5.5.1)
- **Compensation.** Community benefits to host communities may compensate for adverse viewshed impacts.

Mitigation measures can be used to reduce the visual impacts of wind power projects. Table 14.4 provides some general best management practices (BMPs) that need to be considered throughout the project lifecycle, i.e., siting and design, construction, operations, repowering, and decommissioning.

A specific offshore example involves two near-shore projects in Denmark – Vesterhav Syd (20 turbines/168 MW) and Vesterhav Nord (21–8.4 MW turbines, 176 MW) – that are between 9 and 8 km from shore. There was

Table 14.4 Wind energy facility BMPs for mitigating visual effects across project lifecycle

General best management practices (BMPS)	Siting and design	Construction	Operations	Repowering/ Decommission-ing
Consider topography (i.e., flat or rolling terrain)	√			√
Consider landscaping for camouflage and possible native plantings	√	√	√	√
Cluster or group turbines to break up overly long lines of turbines	√	√		√
Create visual order and unity among turbine clusters and building and project infrastructure in and around the site	√	√		√
Consider weather and atmospheric conditions	√	√	√	√
Avoid impacting sensitive sites (i.e., historic, cultural, indigenous)	√	√		√
Site wind turbines to minimize shadow flicker (see section 14.9)	√	√		√
Relocate turbines to avoid visual impacts	√	√		√
Consider lighting and AVWS to reduce night sky impacts			√	√
Create visual uniformity in shape, color, and size (both towers and blades)	√	√		√
Use fewer, larger turbines* and increase distance between them	√			√
Use nonreflective coatings to the extent possible	√			√
Keep wind turbines in good repair			√	
Clean nacelles and towers			√	
Communicate visual options to local stakeholders	√	√	√	√
Consider cumulative impacts	√	√	√	√

* *Note:* Larger turbines may also have a more significant impact on viewsheds.
(*Source:* Adapted from DOI, 2013. Page 43).

considerable public opposition to the near-shore developments because of the significant visual impacts expected and the presence of thousands of summer homes along the coast (Ram *et al.*, 2018). With public and government consultation, the developer redesigned the layout (using single rows) and the distance to shore before the approvals were final. In addition, a radar system that allows the lights to be deactivated at night to reduce visual impacts for coastal residents is under consideration.

14.5.5.1 Audio-Visual Warning System (AVWS)

An AVWS is a radar-based system that is intended to minimize night-sky impacts from hazard navigation lighting required by national agencies. The purpose is to only illuminate obstruction lighting on the turbines and other structures when aircraft are approaching the project area, whether on land or at sea. This technology would reduce the impacts on communities that can see the turbines while not compromising the safety of aircraft in the area. To date, innovative lighting systems that reduce the visual impact at night are the most effective mitigation technology commercially available in terms of visual impacts to shoreside and land-based stakeholders.

14.5.6 Insights from the Social Sciences

The attitudes of the host community toward viewsheds are related to qualitative factors in the social sciences, such as risk perceptions and "place attachment" (or sense of place). These theories help to understand better individual and community views and values related to visual effects (Devine-Wright and Howes, 2010; Ram and Webler, 2022). The relationship between wind energy siting and "sense of place" is complex as it relates to a range of social science disciplines, including risk management and decision science (Gregory *et al.*, 2012) and social psychology (Devine-Wright, 2005). Place attachment is concerned with the cultural, heritage, and recreational values of a landscape to an individual or a community, such as a fisherman. In some cases, these views may influence acceptance of a new wind project (Jobert and Laborgne, 2012).

As mentioned in Section 14.3.6, local citizens may have differing opinions of the project and the developer, including the visual effects, depending on whether they were involved in the decision-making process (Hoen *et al.*, 2019). Some people may view wind turbines as a symbol of progress toward the clean energy transition and switching away from fossil fuels as a positive effect on their community (Devine-Wright, 2005). Wind turbines have a clean sculptural form that is considered attractive by some viewers (Pasqualetti *et al.*, 2002).

Studies in Denmark (Ladenburg and Dubgaard, 2009; Ladenburg, 2015), where the population has considerable experience with wind power and a deep commitment to climate change actions, revealed a generally positive attitude toward offshore wind farms, even among respondents who can see offshore installations from their residence or summer house. The studies also indicated that offshore wind farms are preferred to land-based developments, and only 5% of the respondents indicated a negative attitude toward the development of more offshore wind farms. However, various US and European studies indicate that viewer attitude toward offshore projects vary widely according to age, gender, income, education, length of residence, frequency of exposure, and experience with operating wind projects (South Fork Wind Farm, 2020). More discussion of the social science perspectives may be found on the IEA Wind Task 28 platform (IEA Wind, 2023).

14.6 Wind Turbine Noise

As already been noted and is illustrated in Figure 14.7, noise is an important part of the environmental impact of wind turbines. In this section, we begin by introducing the engineering aspects of noise before introducing the environmental concerns related to noise and their mitigations. Other sources of noise (e.g., pile driving noise) are discussed in subsequent sections.

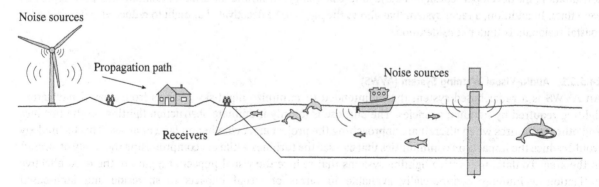

Figure 14.7 Wind energy noise sources and the propagation of sound to receivers.

14.6.1 Noise and Sound Fundamentals

14.6.1.1 Characteristics of Sound and Noise

Sound is generated by numerous mechanisms and is associated with rapid, small-scale pressure fluctuations (which the human ear perceives). Sound waves are characterized in terms of their wavelength, λ, frequency, f, and velocity of sound, c, where c is related to λ and f by Equation (14.1):

$$c = f\lambda \tag{14.1}$$

The velocity of sound is a function of the medium through which it travels, and it generally travels faster in denser media. The velocity of sound is about 340 m/s in air; in water, its speed is 1500 m/s.

Sound frequency determines the "note" or pitch that one hears, which, in many cases, corresponds to notes on the musical scale (Middle C is 262 Hz). An octave denotes the frequency range between sound with one frequency and one with twice that frequency. The human hearing frequency range is quite wide, generally ranging from about 20 Hz to 20 kHz (about ten octaves). Sounds experienced in daily life are formed from a mixture of numerous frequencies from numerous sources.

Sound turns into noise when it is unwanted. Whether sound is perceived as noise depends on the response to subjective factors such as the level and duration of the sound. There are numerous physical quantities that have been defined that enable sounds to be compared and classified, and which also give indications for the human perception of sound. They are discussed in numerous texts on the subject such as Beranek and Ver (2005).

14.6.1.2 Sound Power and Pressure Measurement Scales

It is important to distinguish between *sound power level* and *sound pressure level (SPL)*. Sound power level is a property of the source of the sound. It gives the total acoustic power emitted by the source. SPL is a property of sound at a given observer location and can be measured there by a microphone. Because of the wide range in sound level, the magnitude of acoustical quantities is given in logarithmic form, expressed as a level in decibels (dB) with respect to a reference level.

The sound power level of a source, L_W, in units of decibels (dB), is given by Equation (14.2):

$$L_W = 10 \, \log_{10}(W/W_0) \tag{14.2}$$

where W is the source sound power and W_0 is a reference sound power (usually 10^{-12} W).

SPL, L_p, in units of decibels (dB), is given by Equation (14.3):

$$L_p = 20 \, \log_{10}(p/p_0) \tag{14.3}$$

where p is the instantaneous sound pressure and p_0 is a reference sound pressure. The reference sound pressure for air is 20×10^{-6} Pa $= 20$ μPA and 1×10^{-6} Pa $= 1$ μPA for sound in water. (Note that the 20 in Equation (14.3) arises because sound power is proportional to the square of the pressure.) It may also be shown that a sound power intensity level of 0 dB will correspond to a 0 dB SPL at a distance of 1 m.

As a result of the difference in reference pressures, decibels in water have different significance than in air. Accordingly, measurements in water are usually stated as "dB re 1 μPa," where 1 μPa $= 1 \times 10^{-6}$ Pa.

Because the decibel scale is logarithmic, it has the following characteristics:

- Except under laboratory conditions, a change in sound level of 1 dB cannot be perceived.
- Outside of the laboratory, a 3 dB change in sound level is considered a barely discernible difference.
- A change in sound level of 5 dB will typically result in a noticeable community response.
- A 10 dB increase is subjectively heard as an approximate doubling in loudness and almost always causes an adverse community response.

Low-frequency noise is also of some interest: low-frequency pressure vibrations are typically categorized as *low-frequency sound* when they can be heard near the bottom of human perception (20–100 Hz) and *infrasound* (below 20 Hz) when they are at the limit of human perception. This is also the type of noise that presents the most difficulties in its measurement and assessment (BWEA, 2005).

Infrasound is always present in the environment and stems from many sources, including ambient air turbulence, sea waves, or traffic. Infrasound propagates further (i.e., with lower levels of dissipation) than higher frequencies. The primary human response to perceived infrasound is annoyance, with resulting secondary effects. Some characteristics of the human perception of infrasound and low-frequency sound are:

- Lower frequencies must be of a higher magnitude (dB) to be perceived. (e.g., the threshold of hearing at 10 Hz is around 100 dB, whereas at 50 Hz it is 40 dB).
- Infrasound may not appear to be coming from a specific location because of its long wavelengths.

Figure 14.8 gives some examples for various sound pressure levels on the decibel scale. The threshold of pain for the human ear is about 200 kPa, which corresponds to a sound pressure level of 140 dB.

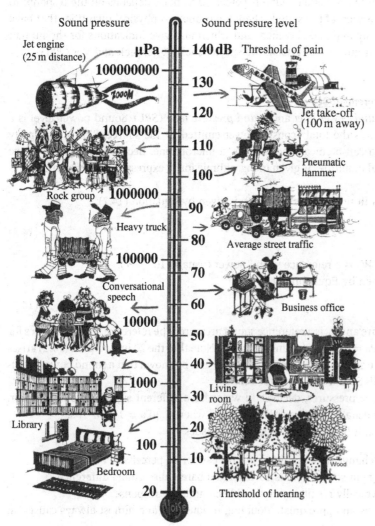

Figure 14.8 Sound pressure level (SPL) examples (*Source:* Reproduced by permission of Bruel and Kjaer Instruments)

14.6.1.3 Measurement of Sound or Noise

SPLs are measured via the use of sound level meters. These devices make use of a microphone, which converts pressure variations to a voltage signal, which is then recorded on a meter (calibrated in decibels).

A sound level measurement that combines all frequencies into a single weighted reading is defined as a broadband sound level. Humans do not perceive all frequencies the same way. Therefore, for the determination of the human ear's response to changes in noise, sound level meters are generally equipped with filters that give less weight to the lower frequencies. Figure 14.9 illustrates three of these filters (referenced as A, B, and C). There is also a G-weighting scale specifically designed for infrasound. The most common scale used for noise in air is the A scale; see Equation (14.4). Measurements made using this filter are expressed in units of dB(A).

$$A(f) = 2.0 + 20 \log_{10}(R_A(f)) \tag{14.4}$$

Where $R(f)$ is given by Equation (14.5)

$$R_A(f) \frac{(12\,200)^2 f^4}{(f^2 + 20.6^2)\sqrt{(f^2 + 107.7^2)(f^2 + 737.9^2)}(f^2 + (12\,200)^2)} \tag{14.5}$$

Once the A-weighted sound pressure is measured over a period of time, it is possible to determine a number of statistical descriptions of time-varying sound and to account for the greater sensitivity to night-time noise levels. Common descriptors, which are often referred to in regulations, include the following:

- L_{10}, L_{50}, and L_{90}. These represent the A-weighted noise levels that are exceeded 10%, 50%, and 90% of the time, respectively. During the measurement period, L_{90} is generally taken as the background noise level.
- L_{eq} (**equivalent noise level**). The average A-weighted SPL gives the same total energy as the varying sound level during the measurement period of time.
- L_{dn} (**day–night noise level**). The average A-weighted noise level during a 24-hour day, obtained after adding of 10 dB to levels measured at night between 10 p.m. and 7 a.m.

Figure 14.9 Definition of A, B, and C frequency weighting scales (*Source:* Beranek and Ver, 2005/John Wiley & Sons)

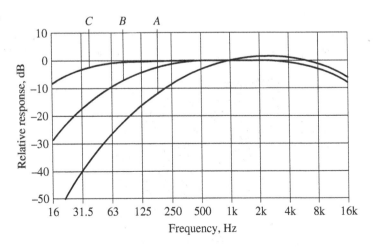

14.6.2 Noise Mechanisms of Wind Turbines

There are four types of noise that can be generated by wind turbine operation: tonal, broadband, low-frequency, and impulsive. These are described below:

- **Tonal.** Tonal noise is defined as noise at discrete frequencies. It is caused by wind turbine components such as meshing gears, nonlinear boundary layer instabilities interacting with a rotor blade surface, vortex shedding from a blunt trailing edge, or unstable flows over holes or slits.
- **Broadband.** This is noise characterized by a continuous distribution of sound pressure with frequencies greater than 100 Hz. It is often caused by the interaction of wind turbine blades with atmospheric turbulence and is also described as a characteristic "swishing" or "whooshing" sound.
- **Low-frequency.** This describes noise with frequencies in the range of 20–100 Hz, mostly observed with down-wind turbines, as the blades interact with the wake of the tower (see Madsen *et al.*, 2008).
- **Impulsive.** Short acoustic impulses or thumping sounds that vary in amplitude with time characterize this noise. They may be caused by the interaction of wind turbine blades with disturbed air flow around the tower or the sudden deployment of tip breaks or actuators.

The causes of noise emitted from operating wind turbines can be divided into two categories: (1) aerodynamic and (2) mechanical. Aerodynamic noise is produced by the flow of air over the blades. The primary sources of mechanical noise are the gearbox and the generator. Mechanical noise is transmitted along the structure of the turbine and is radiated from its surfaces. A summary of each of these noise mechanisms follows. More details are given in the texts of Wagner *et al.* (1996) and Bowdler and Leventhal (2012).

14.6.2.1 Aerodynamic Noise

Aerodynamic noise originates from the flow of air around the blades. As shown in Figure 14.10, a large number of complex flow phenomena generate this type of noise. Such noise generally increases with tip-speed or tip-speed ratio. It is broadband in character and is typically the largest source of wind turbine noise.

The various aerodynamic noise mechanisms are shown in Table 14.5 (Wagner *et al.*, 1996). They are divided into three groups: (1) low-frequency noise, (2) inflow turbulence noise, and (3) airfoil self-noise. A detailed discussion of the aerodynamic noise generation characteristics of a wind turbine is beyond the scope of this text.

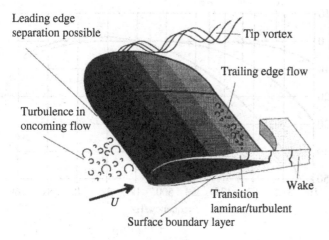

Figure 14.10 Schematic of flow around a rotor blade (Wagner *et al.*, 1996); *U*, wind speed (*Source:* Reproduced by permission of Springer-Verlag GmbH and Co.)

Table 14.5 Wind turbine aerodynamic noise mechanisms (Wagner *et al.*, 1996)

Type or indication	Mechanism	Main characteristics and importance
Low-frequency noise		
Steady thickness noise; steady loading noise	Rotation of blades or rotation of lifting surfaces	Frequency is related to blade passing frequency and is not important at current rotational speeds
Unsteady loading noise	Passage of blades through tower velocity deficits or wakes	Frequency is related to blade passing frequency, small in cases of upwind turbines/possibly contributing in case of wind farms
Inflow turbulence noise	Interaction of blades with atmospheric turbulence	Contributing to broadband noise; not yet fully quantified
Airfoil self-noise		
Trailing-edge noise	Interaction of boundary layer turbulence with blade trailing edge	Broadband, main source of high-frequency noise (770 Hz < f < 2 kHz)
Tip noise	Interaction of tip turbulence with blade tip surface	Broadband; not fully understood
Stall, separation noise	Interaction of turbulence with blade surface	Broadband
Laminar boundary layer noise	Nonlinear boundary layer instabilities interacting with the blade surface	Tonal, can be avoided
Blunt trailing edge noise	Vortex shedding at blunt trailing edge	Tonal, can be avoided
Noise from flow over holes, slits, and intrusions	Unstable shear flows over holes and slits, vortex shedding from intrusions	Tonal, can be avoided

Source: Reproduced by permission of Springer-Verlag GmbH and Co.

14.6.2.2 Amplitude Modulation

One feature of wind turbine sound is known as amplitude modulation. This refers to an increase and then decrease in the amplitude of the sound from the blades as they rotate. The frequency associated with modulation corresponds to the blade passage frequency, which in turn depends on the rotational speed of the turbine and the number of blades. A more detailed description may be found in Bowdler and Leventhal (2012).

14.6.2.3 Mechanical Noise

Mechanical noise originates from the relative motion of mechanical components and the dynamic response that results. The main sources of such noise include gearbox, generator, yaw drives, cooling fans, and auxiliary equipment (e.g., hydraulics).

Since the emitted noise is associated with the rotation of mechanical and electrical equipment, it tends to be tonal (of a common frequency) in character, although it may have a broadband component.

In addition, the hub, rotor, and tower may act as amplifiers, transmitting the mechanical noise and radiating it. The transmission path of the noise can be air-borne (a/b) or structure-borne (s/b) first and then air-borne. Figure 14.11 shows the type of transmission path and examples of sound power levels for the individual components determined at a downwind position (115 m) for a 2 MW wind turbine (Wagner *et al.*, 1996). Note that for this turbine, the main source of mechanical noise was the gearbox, which radiated noise from the nacelle surfaces and the machinery enclosure.

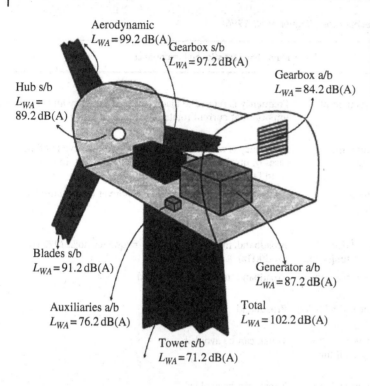

Aerodynamic
$L_{WA} = 99.2\,\text{dB(A)}$

Gearbox s/b
$L_{WA} = 97.2\,\text{dB(A)}$

Gearbox a/b
$L_{WA} = 84.2\,\text{dB(A)}$

Hub s/b
$L_{WA} = 89.2\,\text{dB(A)}$

Blades s/b
$L_{WA} = 91.2\,\text{dB(A)}$

Generator a/b
$L_{WA} = 87.2\,\text{dB(A)}$

Auxiliaries a/b
$L_{WA} = 76.2\,\text{dB(A)}$

Total
$L_{WA} = 102.2\,\text{dB(A)}$

Tower s/b
$L_{WA} = 71.2\,\text{dB(A)}$

Figure 14.11 Components and total sound power level for wind turbines (Wagner *et al.*, 1996); L_{WA}, predicted A-weighted sound power level; a/b, air-borne; s/b, structure-borne (*Source:* Reproduced by permission of Springer-Verlag GmbH and Co.)

14.6.3 Noise Prediction from Wind Turbines

14.6.3.1 Single Wind Turbines

The prediction of noise from a single wind turbine under expected operating conditions is an important part of an environmental noise assessment. The calculations are quite involved and are still the topic of active research. Despite these problems, researchers have developed analytical models and computational codes for noise prediction. Wagner *et al.* (1996) describe different models based on their levels of fidelity. A detailed discussion of higher-fidelity models is beyond the scope of this text, but the reader is referred to Bortolotti *et al.* (2020) for the models implemented in *OpenFAST*.

A simple model for the prediction of sound power level is given by Equation (14.6).

$$L_{WA} = 50\left(\log_{10} V_{Tip}\right) + 10(\log_{10} D) - 4 \tag{14.6}$$

where L_{WA} is the overall A-weighted sound power level, V_{Tip} is the tip speed at the rotor blade (m/s) and D is the rotor diameter (m).

This equation embodies the observation that aerodynamic noise varies approximately as the fifth power of the tip speed.

14.6.3.2 Multiple Wind Turbines

Doubling the number of wind turbines at a given location does not double the sound level. Since the decibel scale is logarithmic, the relation to use for the addition of two SPLs (L_1 and L_2) is given as Equation (14.7):

$$L_{total} = 10\log_{10}\left(10^{L_1/10} + 10^{L_2/10}\right) \tag{14.7}$$

This equation has two important implications:

- Adding SPLs of equal value increases the noise level by 3 dB.
- If the absolute value of $L_1 - L_2$ is greater than 15 dB, the addition of the lower level has negligible effects.

This relation can be generalized for N noise sources in Equation (14.8):

$$L_{total} = 10 \log_{10} \sum_{i=1}^{N} 10^{L_i/10} \tag{14.8}$$

14.6.4 Noise Propagation from Wind Turbines

In order to predict the SPL at a distance from a source with a known power level, one must consider how the sound waves propagate. Details of sound propagation in general are discussed in Beranek and Ver (2005). For the case of a stand-alone wind turbine, one might calculate the SPL by assuming spherical spreading, which means that the SPL is reduced by 6 dB per doubling of distance. If the source is on a perfectly flat and reflecting surface, however, then hemispherical spreading has to be assumed, which leads to only a 3 dB reduction per doubling of distance.

The effects of atmospheric absorption and the ground effect, both dependent on frequency and the distance between the source and observer, have to be considered. The ground effect is a function of the reflection coefficient of the ground and the height of the emission point.

Wind turbine noise also exhibits some special features (Wagner *et al.*, 1996). First, the height of the source is generally higher than conventional noise sources by an order of magnitude, which makes noise screening less significant. In addition, the wind speed has a strong influence on the generated noise. The prevailing wind directions can also cause considerable differences in SPLs between upwind and downwind positions.

The development of an accurate noise propagation model generally must include the following factors:

- source characteristics (e.g., directivity, height, etc.);
- distance of the source to the observer;
- air absorption;
- ground effect (i.e., reflection of sound on the ground, dependent on terrain cover, ground properties, etc.);
- propagation in complex terrain;
- weather effects (i.e., change of wind speed or temperature with height).

A discussion of complex propagation models that include all these factors is beyond the scope of this work. A discussion of work in this area is given by Wagner *et al.* (1996) and Bowdler and Leventhall (2012). For estimation purposes, a simple model based on hemispherical noise propagation over a reflective surface, including air absorption, is given as Equation (14.9):

$$L_p = L_W - 10 \log_{10}(2\pi R^2) - \alpha R \tag{14.9}$$

where L_p is the SPL (dB) at a distance R from a noise source radiating at a power level L_W (dB) and α is the frequency-dependent sound absorption coefficient.

This equation can be used with either broadband sound power levels and a broadband estimate of the sound absorption coefficient [$\alpha = 0.005$ dB(A) m^{-1}] or more preferably in octave bands using octave band power and sound absorption data.

14.6.5 Noise Reduction Methods for Wind Turbines

Turbines can be designed or retrofitted to minimize mechanical noise. This can include special finishing of gear teeth, using low-speed cooling fans, mounting components in the nacelle instead of at ground level, adding baffles and acoustic insulation to the nacelle, using vibration isolators and soft mounts for major components, and designing the turbine to prevent noise from being transmitted into the overall structure.

Aeroacoustic emissions from the blades are the main source of noise for modern wind turbines. It has been previously noted that the following three mechanisms of aerodynamic noise generation are important for wind turbines (assuming that tonal contributions due to slits, holes, trailing edge bluntness, control surfaces, etc. can be avoided by proper blade design):

- trailing edge noise;
- tip noise;
- inflow turbulence noise.

A review of work in these three areas is beyond the scope of this text, and here the reader is again referred to the text of Wagner *et al.* (1996). It should be noted that noise has been reduced in modern turbine designs via the use of lower tip speed ratios, lower blade angles of attack, upwind designs, and, most recently, by using specially modified blade trailing edges.

14.6.6 Noise Assessment, Standards, and Regulations

An appropriate noise assessment study should contain the following four major types of information:

- An estimation or survey of the existing ambient background noise levels.
- Prediction (or measurement) of noise levels from the turbine(s) at and near the site.
- Identification of a model for sound propagation (sound modeling software will include a propagation model).
- An assessment of the acceptability of the turbine(s) noise level via comparison of calculated SPLs from the wind turbine(s) with background SPLs at the location of concern.

There are standards for measuring sound power levels from utility-scale wind turbines as well as national and local standards for acceptable noise power levels. The internationally accepted standard used to measure utility-scale wind turbine sound power levels is the IEC Standard: 61400-11; see IEC (2012).

At the present time, there are no common international noise standards or regulations for SPLs. In most countries, however, noise regulations define upper bounds for the noise to which people may be exposed. These limits depend on the country and are different for daytime and nighttime. The limits are between 30 and 60 dB(A), depending on the country, time of day, and type of habitat (commercial, mixed, residential, or rural).

It should also be pointed out that imposing a fixed noise level standard might not prevent noise complaints. This is due to the relative level of broadband background turbine noise compared to changes in background noise levels. That is, if tonal noises are present, higher levels of broadband background noise are needed to effectively mask the tone(s). Accordingly, it is common for community noise standards to incorporate a penalty for pure tones, typically 5 dB(A). Therefore, if a wind turbine meets a sound power level limit of 45 dB(A), but produces a strong whistling, 5 dB(A) are subtracted from the limit. This forces the wind turbine to meet a real limit of 40 dB(A).

A discussion of noise measurement techniques that are specific to wind turbines or regulations is beyond the scope of this text. The reader is referred to Bowdler and Leventhal (2012).

14.6.7 Wind Turbine Sound – Example

The following is an example of wind turbine sound data. The data was taken at the Rhede wind farm in Germany in 2002 and sampled at 44 100 Hz. An extract of two seconds of data is shown in Figure 14.12.

The SPL as a function of frequency, obtained from 10 minutes of data, is shown in Figure 14.13.

Figure 14.12 Sound pressure segment

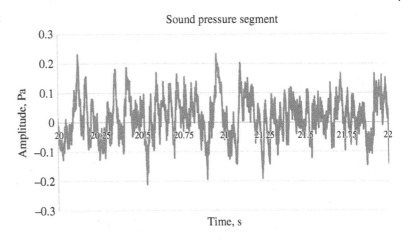

Figure 14.13 Sound pressure level, dB, and dB(A)

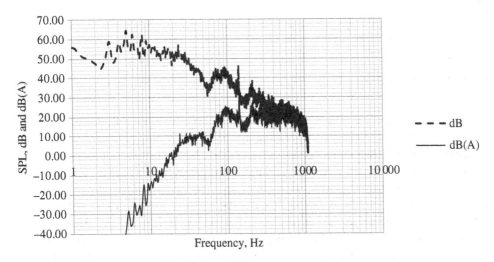

14.7 Wind Turbines and Effects on Birds and Bats

14.7.1 Avian Interactions with Land-Based Wind Turbines: Overview

Birds and wind turbines are often discussed in the context of their potential negative impacts. Understanding the most significant risks to birds, however, begins with other anthropogenic impacts, including climate change. Many peer-reviewed articles have stated that climate change poses unprecedented adverse risks to our avian habitats, including shifting habitats and extinction of dozens of species (Audubon, 2019). North America's birds are disappearing from the skies at a rate that is shocking even to ornithologists. Since the 1970s, the continent has lost three billion birds, nearly 30% of the total, and even common birds such as sparrows and blackbirds are in decline (Pennisi, 2019). Also, it is important to frame wind turbine impacts with how they compare to other electricity sources as well as other causes of fatalities. Comparative frameworks that examine these deleterious effects on habitats and birds from other anthropogenic activities need to be part of the scientific and stakeholder discussions.

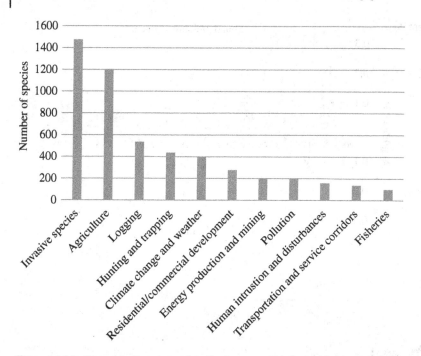

Figure 14.14 The main threats to globally threatened birds worldwide (*Source:* Adapted from BirdLife International, 2023)

Figure 14.14 illustrates the threats to birds globally. From Figure 14.14, it is apparent that the potential impacts on birds from wind turbines (included as part of the threat from all energy production and mining), although important to understand and to mitigate, pale in comparison to adverse effects from other anthropogenic activities. The overall impact of wind energy on bird populations is yet very difficult to assess (USFWS, 2016). Even with the most accurate numbers of collision risk estimates from wind power sites, decision-makers still face significant uncertainties about what these numbers mean for risk management and mitigation strategies (Ram and Stephens, 2017). An important fact to be aware of is that inappropriate siting decisions can pose significant risks to certain species, have long-term reputational impacts on the developer, and impact the cost of energy. The next section provides a brief overview of how the wind energy community currently characterizes the potential risks to avian populations and habitat fragmentation.

14.7.2 Avian Risk Characterization – Land Based

After decades of wind energy research about avian species (i.e., birds) and their habitats, there is a robust knowledge base that has defined some of these challenges and potential mitigation measures to reduce the adverse impacts (Tethys, 2023). However, there is not complete agreement on how to characterize the long-term risk of bird population viability, including songbirds, grouses, and raptors, and how to design and implement mitigation strategies.

It has been learned, however, that certain types of sites pose higher risks. Among these are mountain passes that are preferred routes for migratory birds. In addition, some species, such as raptors, are long-lived with low reproductive rates and low natural mortality rates. A significant increase in those risks could lead to a regional population decline. Moreover, pre-construction studies, mandated by some endangered species regulations, are designed to identify high-risk species and locations that can be avoided by wind developers. The estimates of collisions of birds vary widely because project sites and national data across countries are often not comparable, and

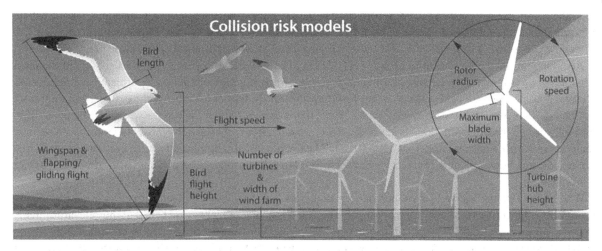

Figure 14.15 Parameters used in collision risk models (*Source:* SEER, 2022a. Illustration by Alfred Hicks, NREL/National Renewable Energy Laboratory/Public Domain)

estimating mortality by individual site does not necessarily indicate regional or populations impacts (Loss *et al.*, 2019). In the United States, most adverse impacts affect smaller songbirds, and the second-most impacted are raptors – including eagles (Smallwood and Bell, 2020).

The impact on birds is generally divided into the following two categories:

- **Direct impacts**, including collision risks (see Figure 14.15).
- **Indirect impacts**, including total or partial displacement of the birds from their habitats, particularly when they are breeding, staging, or foraging. These impacts may involve deterioration or destruction of the habitats, habitat disturbances, and obstructions (i.e., turbines or meteorological towers) that may affect ground nesting species, such as grouse. These potential effects on avian populations are especially fraught when endangered and/or threatened species are involved.

14.7.3 Bat Risk Characterization – Land-Based

It is now well known that bats may collide with wind turbines at some locations and under some circumstances. Most species of bats are nocturnal and use echolocation (emit ultrasound and interpret the returning echoes) to navigate and find prey (mostly insects) in the dark. Like birds, bats face many other stressors, including climate change, habitat alteration, and disease. Many species are already in decline, making it difficult for decision-makers to weigh the projected population impacts of various bat species compared to other ongoing risks in comparison to wind turbines. An example of these difficult decisions relates to White-nose Syndrome (WNS) in North America. First documented in 2006, WNS is a deadly fungal disease that has killed millions of cave-hibernating bats in North America (Cheng *et al.*, 2021) and has spread to Europe and Asia. This environmental disaster has resulted in the listing of the northern long-eared bat as endangered, and two other species, the tri-colored bat and little brown bat may soon be listed. Although these species make up a small percentage of fatalities at wind energy facilities, their listing entails strict regulatory requirements and will impact siting and operational decisions for the wind energy industry.

The species that are most vulnerable to wind turbine collisions are migratory tree-roosting bats, including the hoary bat, eastern red bat, and silver-haired bat. These species appear to be attracted to wind turbines and may perceive these structures as a potential resource. A recent study (Guest *et al.*, 2022) divided the attraction hypothesis into five hypotheses, including attraction based on sound, roost sites, foraging and water, mating behavior, and lights, and one new hypothesis regarding olfaction. Biologists believe that identifying the causes of bat interactions with wind turbines is critical for developing effective impact minimization strategies.

According to Bat Conservation International (BCI), collisions with wind energy turbines are one of the leading causes of bat mortality in North America and are a growing concern as more projects are developed. Observed fatality rates of bats at wind energy facilities have the potential to cause declines in certain migratory bat populations (e.g., hoary bats), which some experts believe may reduce the estimated size of the population and increase the risk of extinction (Frick *et al.*, 2017). To date, hoary bats are the most affected at wind energy facilities in North America (Arnett *et al.*, 2016a; O'Shea *et al.*, 2016). High fatality rates are not unique to the United States, however, and there is evidence from Canada, Europe, Africa, and Latin America (Arnett *et al.*, 2016b).

Whether these fatalities could drive populations to dangerously low levels or even extinction is a critical question. As noted by Frick *et al.* (2017), addressing this question is challenging because bats that migrate latitudinally over long distances have the highest fatalities at wind energy facilities and are among the least studied (Kunz *et al.*, 2007). Moreover, scientists do know that reproductive rates for bats are low, which can impact their ability to respond to mortality threats (Barclay and Harder, 2003). To better characterize the threats and the significance of the collision risks, basic demographic parameters and even rough empirical estimates of population size are needed but do not exist (Lentini *et al.*, 2015). Lack of empirical demographic and population data for migratory bats, especially for non-colonial species, limits the ability to quantitatively assess the potential impact of wind energy on these species (Hein *et al.*, 2021; Hale *et al.*, 2022).

The challenges associated with empirical estimation will likely remain insurmountable into the foreseeable future, given the ecology of these species. In the United States of America, the Bats and Wind Energy Cooperative (BWEC) was formed to address the research needs of this new potential risk with land-based turbines in collaboration with industry, government, and scientific stakeholders (BWEC, 2004). Currently, BWEC is focusing on the following areas of risk characterization research (BWEC, 2022): bat behavior, population data, minimization strategies, and post-construction fatality surveys (see Section 14.7.6 regarding the GenEst estimator for conducting fatality surveys).

14.7.4   Bird and Bat Monitoring Strategies

To assess the potential impacts of wind energy development on birds and bats, various monitoring methods and technologies are applied to or are being tested for improved assessments (SEER, 2022a), including:

- **Acoustic detectors** record the vocalizations of bats within approximately 40 m and birds within 100 m. The range of detection will vary based on the frequency of vocalization (i.e., lower-frequency sounds travel farther), intensity, and orientation of the animal to the microphone. Data can be used to identify species or species groups, characterize seasonal and temporal activity patterns at a local scale, and relate such patterns to weather and wind turbine operational conditions.
- **Boat surveys** record observations of individual animals along transects. Data can be used to determine presence, density, behavior, and flight height.
- **High-resolution digital aerial surveys** collect imagery from fixed-wing aircraft surveys over large distances (e.g., a wind energy facility and surrounding area). Surveys are conducted during daylight hours and require high-visibility conditions. Analysis can provide species abundance and distribution.
- **Visual and thermal cameras** are used to record data for animals active during the day or night, respectively. Cameras can be positioned on the tower, pointing up toward the blades or out toward the water, and on the rotor, pointing along the length of the blades. Data can be used to assess behavioral interactions (e.g., attraction or avoidance) between bats or birds and wind turbines, but this technology cannot easily identify types of species.

- **Radar** tracks moving objects from a few to several hundred km away from a wind turbine, depending on the type of radar technology (e.g., antenna type, power of the radar signal, and radar wavelength). Data on flight height, flight direction, flight speed, and passage rates are recorded. Note, however, that radar technologies cannot distinguish between bats and birds.
- **Strike indicators** record collision events using sensors installed along the length of a blade. Sensors can include microphones and accelerometers. Data can be used to determine the timing of collisions, the size of the object striking the blade, and where along the blade the collision occurred. Indicators cannot identify the type of animal (bird or bat) or species colliding with the blade (see Figure 14.15 for collision risk parameters).
- **Telemetry** provides information on movement patterns of tagged animals over time. Automated radio telemetry requires specialized receiving stations to record the presence of tagged individuals within a few to several km. Radio tags are small enough to be used on most bats and birds. Satellite telemetry does not require receiving stations, but the weight of the tags limits their use to larger species and/or shorter tracking duration. Data can be used to identify flight paths, movement corridors, and high-use areas (SEER, 2022a).

Since avian collisions and habitat fragmentation can pose significant risks, the following factors need careful consideration and study throughout the lifecycle of the wind energy project:

1) **Screening prospective sites and predicting risks to wildlife before operations**. Pre-construction studies can inform project design, are key to avoiding or minimizing habitat-based impacts, and can save developer costs (Milieu *et al.*, 2016). In some cases, project design, including exact placement of individual turbines, can also help to minimize other risks for some species, such as displacement. This is an area where biologists and developers can collaborate in an effective manner to reduce potential adverse impacts.

2) **Before and after construction impact (BACI) studies** are the most rigorous approaches to address potential direct and indirect impacts. Regulatory agencies almost always require a site-specific standardized mortality study to periodically estimate the mortality of avian species before the construction and then from the operation of the wind site. This helps determine if the collision fatality predictions pre-construction match post-construction realities. Thereafter, appropriate mitigation measures may be identified and employed (see Section 14.7.6 on post-construction studies and fatality estimators).

3) **Endangered and threatened species** are particularly challenging risk factors, as they imply that an animal (or plant) species is in danger of extinction. Given that these species are also rare, it may be difficult to confirm their presence in pre-construction or post-construction studies.

4) **Seasonal and migration patterns and behaviors** can increase the risk of avian collisions with wind turbines. For example, some species of birds migrate during the night, which can increase the risk of collisions with wind turbines that are not visible in low-light conditions. The risk also depends on factors such as the bird's flight height, flight speed, migratory routes, habitat preferences, and breeding seasons.

5) **Habitat loss, fragmentation, and displacement**. Wind power projects are generally spread over large land areas. Even though the turbines themselves have a relatively small footprint, there could be habitat loss or fragmentation. Construction noises and disturbances caused by wind developments can result in the displacement of bird species from their habitats during breeding and nesting periods. Associated infrastructure, such as roads and transmission lines, can create barriers to movement, which may impact species' ability to access food, water, and suitable habitat.

6) **Location and placement of turbines and meteorological towers**. The location of wind turbines can also affect the risk of avian collisions. For example, placing turbines in areas that are known to be important bird habitats or migration corridors can increase the risk of collisions. Un-guyed meteorological and electrical distribution poles are preferred over guyed structures that can create barriers to avian species.

7) **Turbine choice and design**. The design of wind turbines can also play a role in the risk of avian collisions. For example, turbines with smaller and/or slower rotors may reduce the risk of collisions. Taller wind turbines may change the risk factors as well.

Transmission lines and substation upgrades are critical aspects of site infrastructure and can affect avian species. Wherever feasible, electrical utility lines should be underground to avoid collision risks. Techniques that reduce bird electrocutions need to be implemented (APLIC, 2012). Some of the key recommendations of the new guidelines include:

- Conducting pre-construction surveys to identify and evaluate potential risks to birds.
- Installing bird-safe power poles and conductors that minimize the risk of electrocution.
- Ensuring that power lines are well-insulated and properly grounded.
- Installing bird flight diverters on power lines will make them more visible to birds and prevent collisions.
- Conducting regular maintenance and monitoring of power lines to identify and address potential hazards.

The guidelines also emphasize the importance of collaboration between utilities, wildlife agencies, and other stakeholders to develop and implement effective strategies for reducing bird mortality on power lines. By implementing these guidelines, utilities can help protect bird populations and ensure the safe and reliable operation of their electrical infrastructure (APLIC, 2012). Developers also can reap benefits by following these guidelines and monitoring methods. They can better understand the feasibility of a wind power project, minimize potential environmental impacts, and ensure that the project is designed and constructed to maximize its energy production potential while minimizing risks to avian species and habitats.

The next section will discuss the potential mitigation strategies and post-construction surveys that can minimize and avoid the potential risks of collision and habitat fragmentation.

14.7.5 Avian Mitigation Concepts

Deterrent strategies already in use and under development, including audible or visual signals that can deter some species and thereby help minimize bird collision fatalities with operating turbines. To make the turbine more visible and elicit avoidance behaviors in avian species, scientists have experimented with passive visual cues, such as painting one or all blades black (Hodos, 2003) or painting black and white stripes on the blades. It is believed that these visual cues may reduce collision susceptibility to tall structures while foraging and breeding. Recently, a Norwegian BACI study repeated this experiment and reduced the annual fatality rate by 70% relative to the control turbine, with the largest reduction in raptor mortalities (May *et al.*, 2020).

Prey base management could be an option to avoid the congregation of certain species around wind turbines. A humane program, such as live trapping, has been tried to remove unwanted prey from existing wind farms and thereby reduce the risks to raptors (Loss *et al.*, 2019).

Other mitigation approaches may consider the following:

- Avoid known migration corridors, micro-habitats, and flyways.
- Larger turbines may reduce the total number of structures at the site. On the other hand, taller towers may present new risks to different species and have adverse effects on the viewshed (see Section 14.5.1).
- Only use unguyed meteorological towers and electric distribution poles.
- Bury electrical lines wherever possible.
- Remove raptor nests pre-construction wherever possible.

Compensatory mitigation offsets may require specific actions, such as making payments or other contributions to conserve at-risk species or restore a habitat when it is impossible to avoid or minimize adverse impacts (REWI, 2023). Section 14.7.7 briefly discusses selected risks and mitigations for bat species that differ from those discussed for birds.

Curtailment strategies are control techniques to reduce collisions whereby the rotor speed is slowed automatically when wildlife is considered at risk (McClure *et al.*, 2022). High-risk periods include migration seasons, weather conditions in which birds are most active, or fog that degrades visibility. The strategies have evolved from blanket curtailment (i.e., based on time of day, season, and wind speed) to smart curtailment (i.e., blanket curtailment plus additional variables such as temperature or bird activity); see Hein and Straw (2021).

The most effective fatality-reduction measure has been curtailment, which has been documented for bats but not for many bird species. In one BACI study, it was found that because the migration season is relatively brief, seasonal curtailment would greatly reduce bat fatalities for a slight loss in annual energy generation, but it might not benefit many bird species (Smallwood and Bell, 2020). Curtailment may become a standard industry mitigation option in high-risk habitats with protected and regulated species such as bald and golden eagles. The US Migratory Bird Treaty Act and the Bald and Golden Eagle Protection Act (1940) make it illegal to kill or disturb these species or their nests without a permit. This is important to understand, as micro-siting strategies and automated control systems are employed to manage these solutions.

More innovative mitigation strategies are always sought out to reduce potential risks to avian species and bats. The wider accessibility of artificial intelligence (AI) is one of the latest applications. One such possibility is the "Eagle Eye," which uses computer vision and machine learning algorithms to analyze video footage, identify eagles, and send real-time alerts to a controller, which can then shut down individual turbines or take other steps to minimize the collision risk (McClure *et al.*, 2022). The experiment showed that the number of eagle fatalities declined by 82% at the treatment site relative to the control site.

The cost of curtailment and tradeoffs with power production are also being assessed by project development teams to effectively balance the costs associated with mitigating wildlife mortality while also maximizing energy production. Some studies conducted simulations with a range of curtailment scenarios across the contiguous United States of America to examine sensitivities of annual energy production (AEP) loss and potential impacts on economic metrics for future wind energy deployment. The studies found that high levels of curtailment may substantially reduce the future of financially viable wind energy sites (Allison *et al.*, 2019; Maclaurin *et al.*, 2022).

14.7.6 Site-Specific Post-construction Monitoring Studies for Birds and Bats

Post-construction studies can be used to estimate fatalities, compare fatalities across project sites, and determine patterns of fatality in relation to weather and habitat variables (e.g., habitat heterogeneity, ridge lines, topographical effects). Estimating fatalities begins by conducting carcass searches around a subset of wind turbines. According to the USFWS (2016), 30% of the wind farm or 30 turbines – whichever is less – need to be searched. The field methodologies for collecting data, statistical theory, and calculations for estimating indirect risks, such as habitat fragmentation, are beyond the scope of this text. A conceptual model for a post-construction study that estimates fatalities is explained below, however.

Carcass searches can be performed by humans or trained dogs. The raw carcass count, obtained from these searches, is then adjusted by the probability of detection, which includes biases associated with carcasses that were missed by searchers or removed by scavengers prior to a search, to calculate total mortality ($\hat{m}$). The generalized mortality estimator (GenEst) was developed in 2018 to improve estimates of mortality and provide clear guidance on data requirements and use of the software (GenEst – A Generalized Estimator of Mortality (2018); Dalthorp *et al.*, 2018; Rabie *et al.*, 2021).

A conceptual model for estimating mortality, known as the Horvitz–Thompson estimator, is given by Equation (14.10), which essentially states that mortality can be estimated by the number of observed carcasses, x, divided by the probability of their being detected.

$$\hat{m} = \frac{x}{\hat{r}\,\hat{d}_{wp}\hat{p}} \tag{14.10}$$

Where:

$\hat{m}$ = estimate of the total number of fatalities
$\hat{r}$ = estimate of fraction of carcasses persisting until the next search (carcass persistence)
$\hat{d}_{wp}$ = estimate of fraction of carcasses landing in the search area
$\hat{p}$ = estimate of fraction of carcasses observed (searcher efficiency)

14.7.7 Bat Fatality Mitigation Concepts

Over the last several decades, researchers have been developing measures to reduce bat fatalities that involve deterrents and/or curtailment strategies, as discussed above for bird species. A recent workshop (Hein and Straw, 2021) documented some studies related to these mitigation strategies. Existing technology for deterrents includes:

- Ultrasonic deterrents, ultraviolet light, and texture coating.
- Integrating deterrent technologies with wind turbines (either by retrofit or during manufacture or installation).

Ultrasonic deterrents are intended to generate a disorienting airspace around wind turbines by disrupting bat echolocation capabilities. To date, these deterrents have varied in their effectiveness among species (Romano *et al.*, 2019; Weaver *et al.*, 2020), and ongoing work is focused on the stimuli that work best for as many species as possible. Ultraviolet (UV) lights have also been investigated (Cryan *et al.*, 2021).

As discussed above for birds, blanket curtailment is applicable to bats. Two independent studies showed that raising the turbine cut-in speed to 5.0 m/s significantly reduced bat mortality by an average of nearly 50% (Adams *et al.*, 2021; Whitby *et al.*, 2021). Validation studies for smart curtailment are ongoing (Hein and Straw, 2021; BWEC, 2022).

In addition to regulatory compliance (i.e., whether the species is under legal protection), project developers must assess the tradeoffs with how these curtailment strategies may impact energy generation and costs. One recent study (Maclaurin *et al.*, 2022) examined the cost impacts from various curtailment strategies. The result was that curtailment was beneficial but was accompanied by a significant reduction in energy production and a corresponding increase in the cost of energy.

14.7.8 Characterizing Offshore Wind Risks

One of the major challenges in risk characterization of birds and bats offshore is how to understand the additive risks of offshore wind turbines in the ocean because marine ecosystems are expected to change dramatically over the next few decades from other anthropogenic impacts, including climate change (Lenton *et al.*, 2023; Stenseth *et al.*, 2020). Since there is a 30-year operating history of offshore wind, there are hundreds of environmental effects studies in Europe and the US, including post-construction monitoring (see, e.g., Degraer *et al.*, 2019; DEA 2017). This section only highlights several areas of concern that have substantial evidence.

Early MSP and mapping of sensitive species and protected habitats are critical to ensure responsible planning and management across multiple projects. BACI studies are central to understanding project-level and cumulative impacts across regional ocean ecosystems, such as the North and Baltic Seas. Also, standardizing study methods and data collection is important so that biological and geographic information can be shared across sites and locations (Nordic Energy Research, 2022).

Cumulative impacts are particularly difficult to address. Cumulative impacts are defined as the impact on the environment that results from the incremental impact of an action or project when added to other past, present, and reasonably foreseeable future actions. Figure 14.16 shows the iterative scientific process for analyzing cumulative impacts on bats, where the outcomes of initial studies will inform future steps and should serve to refocus research efforts as needed.

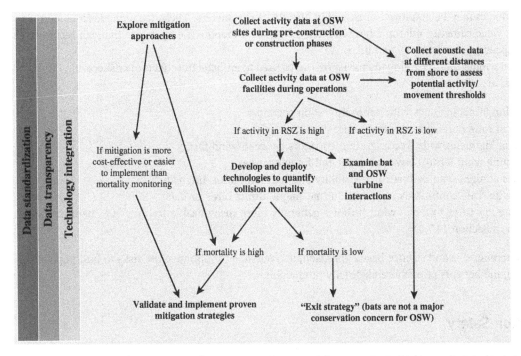

Figure 14.16 Proposed order of operations to understand the cumulative biological impacts of offshore wind developments to bats (*Source:* Hein *et al.*, 2021/Biodiversity Research Institute) RSZ = Rotor-swept zone

A summary of selected adverse effects from offshore wind projects follows:

- **Noise impacts.** The construction and O&M of offshore wind turbines can generate significant amounts of noise that may impact seabirds.
- **Increased vessel traffic** during the lifecycle of the project may also amplify noise levels from shoreside to the site, particularly traffic during construction and O&M. This can disrupt the migration patterns of bird and bat species.
- **Collision risks** are difficult to assess, given the difficult challenge of tracking carcasses offshore. Fortunately, researchers have conducted over a decade of studies and devised collision risk models, with a range of parameters, as noted in Figure 14.15. There is a focus on risks to seabirds, migrating passerines, waterbirds, and bats, particularly during migration periods. (Hüppop *et al.*, 2006; Thaxter *et al.*, 2017). Some biologists are exploring how high-resolution tracking data can be used to improve our knowledge of avoidance/attraction behavior of offshore species around established wind farms and so inform assessments of collision impacts (Johnston *et al.*, 2021). Collision detection systems are also being developed and tested. These technologies combine sensors installed on the blades that detect collision events with acoustic detectors and camera systems to help identify the object (e.g., bird or bat).
- **Habitat fragmentation and ecosystem modifications** The construction and operation of offshore wind plants can cause disturbance to the seabed and marine habitats, which may impact marine biodiversity. Structures across large areas of the ocean can create a barrier effect for migratory birds. The species, however, may adapt to these structures in some habitats, as several studies have shown. For example, at the Nysted offshore wind farm in Denmark, traveling birds (e.g., sea ducks) displayed profound avoidance behavior, with the number of birds entering the area declining dramatically following the construction of the wind farm (Desholm and Kahlert, 2005; Petersen *et al.*, 2019).

Various technologies may be employed in assessing bird flight parameters. These technologies include radar, thermal imaging, video cameras, microphones, GPS telemetry, ornithodolite (i.e., a device that can record large samples of bird speed measurements), and lidar.

As in land-based wind farms, the following measures can be used to mitigate the effects of offshore wind turbines on birds (Köller *et al.*, 2006):

- Avoid siting wind farms in zones with dense migration corridors.
- Align turbines in rows parallel to the main migration direction.
- Establish several kilometer-wide free migration corridors between wind farms.
- Avoid constructing wind farms between resting and foraging areas.
- Curtail turbines at night with bad weather/visibility and high migration intensity.
- Refrain from large-scale continuous illumination that might attract birds or bats.
- Continue testing measures to make wind turbines generally more recognizable to birds, e.g., painting blades black (as noted in Section 14.7.5).

By using a comprehensive and science-based approach, it is possible to minimize the risks to bird populations while maximizing the benefits of offshore electricity production.

14.8 Aviation Safety

Aviation safety considerations involving the siting and operation of wind turbines include physical size, lighting, and potential radar interference. A summary description of each of these effects follows:

- **Physical size.** As an example for the United States, the Federal Aviation Administration (FAA) regulates structures 200 feet (61 m) above ground level, such as utility-scale wind turbines, to ensure that they do not influence aviation safety or other uses of airspace, such as radar. With modern land-based wind turbines reaching heights of up to and over 250 m above ground level, as physical obstacles, they may create hazards for low-flying aircraft. Often, the greatest problems are near small aviation airports. In addition, larger airports, low-flying zones and corridors, as well as helicopter pads, need to be considered. Another problem that needs to be considered is that the rotating blades of wind turbines create turbulence downwind. This potential problem is especially important for light aircraft and helicopters.
- **Lighting.** Obstacle lighting is of high importance for aviation safety. Using the United States as an example, the FAA has developed standards for marking and lighting obstructions that are a hazard to navigable airspace (US Department of Transportation, 2015). It should be noted that in some locations, the lights on wind turbines are often considered annoying to residents (see Section 14.5).
- **Potential radar interference.** Wind turbine blade rotation can cause interference when located within the line-of-sight of a radar system. An example of siting and mitigation methods that reduce or eliminate the adverse effect of wind turbines on radar systems is presented by Karlson (2021).

14.9 Shadow Flicker

Shadow flicker is a phenomenon in which the moving blades of a wind turbine rotor cast moving shadows that cause a flickering effect. This flicker can disturb people living close to the turbine during the hours of the year (on the order of tens of hours) where their home is in the shadow. Similarly, it is possible for sunlight to be reflected from gloss-surfaced turbine blades and cause a "flashing" effect. Shadow flicker will occur during a limited amount of time in a year, depending on the altitude of the sun, α_s, the tower height, H, the rotor radius, R, and the height, direction, and distance to the viewing point. At any given time, the maximum distance from a turbine that a flickering shadow will extend is given by:

$$x_{shadow,\max} = (H + R - h_{view})/\tan(\alpha_s) \tag{14.11}$$

Where h_{view} is the height of the viewing point.

The solar altitude depends on the latitude, the day of the year, and the time, as given in the following equations (Duffie and Beckman, 2006):

$$\alpha_s = 90° - \cos^{-1}[\cos(\delta)\cos(\varphi)\cos(\omega) + \sin(\delta)\sin(\varphi)] \tag{14.12}$$

Where

δ = declination of the earth's axis
φ = latitude
ω = the hour angle

The declination is found from the following equation:

$$\delta = 23.45\sin(360(284 + n)/365) \tag{14.13}$$

Where n = day of the year.

The hour angle is found from the hours from noon (solar time, negative before noon, positive after noon), divided by 15 to convert to degrees.

Another relevant angle is the solar azimuth. This indicates the angle of the sun with respect to a certain reference direction (usually north) at a particular time. For example, the sun is always in the south at solar noon, so its azimuth is 180° at that time. The solar azimuth is important since it determines the angle of the wind turbine's shadow with respect to the tower. See Duffie and Beckman (2006) for details on calculating the solar azimuth.

For illustration, consider a location that has a latitude of 43°. Assume that the day is March 1 (day 60) and the time is 3:00 in the afternoon. Also assume that the turbine has a tower height of 80 m and a radius of 30 m, and that the viewing height is 2 m. From the equations above, the declination is −8.3°, the solar altitude is 24.4°, and the solar azimuth is 50.2° W of S. The maximum extent of the shadow is 238 m from the turbine. The angle of the shadow is 50.2° E of N.

Sites are typically characterized by charts such as the one illustrated in Figure 14.17 for a location in Denmark (EWEA, 2004). The chart gives the number of hours per year of flicker shadow as a function of direction and distance (measured in units of hub height). In the example shown, two viewing points are considered. One of them (A) is directly to the north of the turbine at a distance of six times the hub height. The other (B) is located to the southeast at a distance of seven times the hub height. The figure shows that the first viewing point will experience shadow flicker from the turbine for five hours per year. The second point will experience flicker for about 12 hours per year.

Note that the equations and calculations above assume a clear sky and the absence of rain, clouds, etc., which will further reduce the number of hours with flicker.

14.9.1 Mitigation Possibilities

Most modern wind turbines allow for real-time control of turbine operation by computer in order to shut down during high shadow flicker times, if necessary. In addition, computer software such as *windPRO* (emd-international.com, 2023) can be used ahead of time when planning a wind turbine project to assess sites to know what the shadow flicker impact is likely to be. A commonly used guideline is that siting or control needs to be implemented such that shadow flicker at a location of concern occurs no more than 30 h/yr; see World Bank (2015).

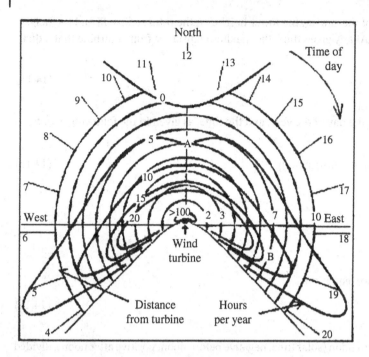

Figure 14.17 Diagram of shadow flicker calculation (*Source:* EWEA, 2004/European Wind Energy Association); A and B are viewing points.

In terms of safe distances to reduce shadow flicker, these are often project-specific because it depends on whether there are residences or roadways present and what the geographic layout is. This could be particularly important in areas with more forestry and existing shadow, which could reduce nuisance from turbine-produced shadow flicker or in an otherwise open land area such as farmland that would be more susceptible to the annoyance of shadow flicker. A general rule of thumb for assessing shadow flicker risk is to evaluate a circular zone around the turbine with a diameter 10 times that of the rotor diameter.

14.10 Marine Mammals

Offshore wind development may pose risks to certain marine mammals, in particular some cetaceans (whales, dolphins, and porpoises) and pinnipeds (seals, sea lions, and walruses). The primary concern is noise, but there may be other considerations as well. Depending on the location of the project, different species may be present at various times of the year. This topic is an important risk due to the endangered species involved, the value of certain iconic species, such as North Atlantic right whales (NARWs), and the difficulty in assessing the impacts the species may experience. The determination of how, when, and to what degree marine mammals are exposed to underwater noise that could result in a physiological and/or behavioral impact is very complex. To assess the impacts, scientists consider sound propagation, specific marine animal densities and movements around the project area, and animal behavior when exposed to project-related noise.

14.10.1 Marine Mammals – Risk Characterization

Due to the endangered species status of some marine mammals and the difficulty in assessing the impacts, this is considered a significant risk factor throughout the lifecycle of the wind plant. The presence of marine mammals in European waters is limited to harbor porpoise and seals in the North and Baltic Seas. The critically endangered

Baltic Proper harbor porpoise population has been severely reduced by other ecological pressures and therefore has been extensively studied since the start of offshore wind developments. As a result, various studies have been designed to understand the population's presence and behaviors over time with acoustic monitoring technologies (Carlén, 2017).

Ambient noise from natural and anthropogenic sources can result in auditory masking for marine mammals. Masking refers to the reduction of an individual's ability to effectively communicate and detect important predators, prey, and other environmental features associated with spatial orientation (Clark *et al.*, 2009). Anthropogenic sources known to contribute to ambient sound levels include boats and ships, sonar (military and commercial), geophysical surveys, acoustic deterrent devices, construction noise, and scientific research sensors. In coastal waters, noise from boats and ships, particularly commercial vessels, is the predominant source of anthropogenic noise. Cumulative effects of sounds in the ocean space are also important.

14.10.2 Sources of Noise Affecting Marine Mammals

Underwater sounds are generated by biological, physical, and anthropogenic sources. Sound propagates farther and faster in water than in air, which may result in unintended consequences for the marine environment. The primary noise sources of concern are described below. Refer to Section 14.6 for technical details on noise in general.

14.10.2.1 Pile-Driving Noise
One noise source of particular significance in offshore wind energy is pile driving during installation. This is greatest for monopiles, but it is also a consideration for the smaller piles used with jackets. Pile driving sound levels vary with the technology (monopiles vs. pin piles for jackets), pile size (diameter and wall thickness), subsurface and geotechnical characteristics, hammer energy, and type of pile driver. Pile-driving sounds propagate both above and below the sea surface, although sound transmission is different in water than in air, making it difficult to compare airborne and underwater sound levels. Impact pile-drivers typically utilize a weight (referred to as a piston or hammer) to impact the top of a pile and force it into the seafloor, as described in Chapter 10. The primary sources of noise associated with impact driving are the impact of the hammer on the pile/drive cap and the noise radiated from the pile. (Koschinski and Lüdemann, 2013; Norro, 2020).

In a typical monopile installation, the hammer is dropped 15–60 times per minute, depending on the soil. The resulting underwater sound has a peak frequency in the range of 100–400 Hz and a sound exposure level (SEL) of approximately 160 dB re 1 μPA at 400–500 m from the hammering, as illustrated in Figures 14.18 and 14.19. See Matuschek and Betke (2009) for details.

The noise from pile driving is in the hearing range of marine mammals and is known to have physiological (e.g., auditory), behavioral, and potentially injurious effects on harbor porpoises, dolphins, and grey seals in the North and Baltic Seas. Similar effects are anticipated in the Atlantic Ocean off North America. Thresholds to prevent behavioral effects have not yet been set for all species, although research is underway to better understand the issues and develop effective mitigation measures (SEER, 2022a; Tougaard, 2021). Fortunately, some studies have found that marine mammals (e.g., porpoises and seals) are highly mobile and likely to avoid the construction area temporarily without any long-term effects at the population level, particularly when construction periods are relatively short (MMO, 2014). Other research studies concluded that the impacts are temporary, during construction, and then the species return to some of their feeding and habitat areas (Holm *et al.*, 2019; Tougaard *et al.*, 2019).

14.10.2.2 In-Air Noise from Pile Driving
Pile driving will generate in-air impulse sounds as the hammer strikes the pile, but the effect is expected to be minimal. In-air noise is expected to decrease to 60 dB(A) at 732 m from the source. No-pile driving noise is expected to reach the shore.

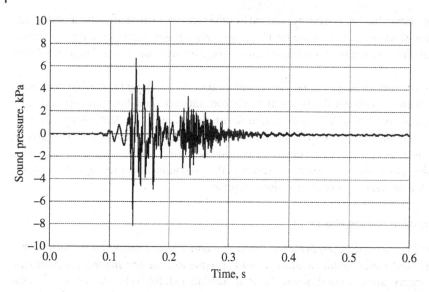

Figure 14.18 Illustration of an underwater impulse due to an impact pile driver (*Source:* Matuschek and Betke (2009)/U.S. Department of Energy/Public Domain)

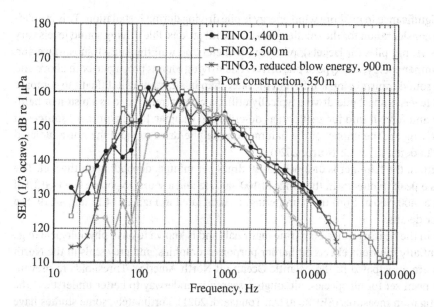

Figure 14.19 Spectra from pile driving (*Source:* Reproduced by permission of Matuschek and Betke (2009)/U.S. Department of Energy/Public Domain

14.10.2.3 Underwater Noise from Vessels

Vessels of various sorts are used during all phases of offshore projects from initial siting through decommissioning. These all produce some sound, which contributes to the ambient level in the ocean. One source of particular significance are vessels collecting soil data with acoustic sub-bottom profilers. These are sonar devices, similar in principle to sodar (Chapter 2). They emit sound pulses aimed at the seabed. Some of the pulses are reflected back, received, and analyzed on board, providing information about the underlying substrate.

14.10.3 The North Atlantic Right Whale (NARW)

US offshore ocean planning challenges are unique because of the presence of one of the world's most endangered whale populations, the NARW. Only about 350 remain in the Atlantic Ocean, including fewer than 100 breeding females. NARWs primarily occur in Atlantic coastal waters on the continental shelf, in and around shipping lanes and fishing areas, although they are also known to travel far offshore, over deep water. Some lease areas identified for offshore wind along the Atlantic Ocean are vulnerable areas because they include the NARWs' foraging ground (e.g., New England), where they feed during the fall months before heading south to calve (S.E. Coast). Other whale species, including fin and humpback, are present in these regions as well.

This environmental risk has a long, sordid history, as commercial whaling in the 1800s decimated a range of whale species. The whale also has a low birth rate, so regeneration of the population is quite slow and perhaps tied to food scarcity. In addition, climate change impacts on the ocean may affect species birth rates and recovery as habitats change and their distribution across migration corridors shift (NOAA, 2023). Significantly, historical research from necropsies of carcasses determined cause-specific mortality rates in NARW are linked to seriously injured whales from fishing gear entanglement and ship strikes (Pace *et al.,* 2021; Knowlton *et al.,* 2022; NOAA, 2023). This is important for characterizing the cumulative risk framework related to the additional potential risks posed by the offshore wind industry.

As construction ramps up along the Atlantic coast, pile driving and increased vessel traffic are major risk factors. Although offshore wind may not cause any additional mortality cases for this endangered species, the national regulations and policies are restricting developer operations at sea both before and after construction. A draft strategy released by the federal government outlines how to protect the endangered NARW in relation to offshore wind development and to engage with the public and ocean stakeholders (BOEM and NOAA, 2022; CBI, 2023). The main goal of the policy is to recover the population and mitigate any potential threats.

14.10.4 Mitigation Strategies

There are extensive mitigation strategies required by regulatory agencies to reduce the potential impacts of offshore wind developments. These strategies must be designed, implemented, and managed across the developer project team. Examples of these extensive requirements in the United States of America follow (adapted from the South Fork Wind Farm, 2021; US Wind, 2023):

- **Bubble curtains** can be employed around the pile driving to reduce the sound that is emanated. These are now commonly required.
- **"Soft start"** of pile driving will likely be required to give marine mammals the opportunity to leave the area where the process begins in earnest.
- **Exclusion and monitoring zones** for marine mammals will be established for pile driving and high-resolution geophysical (HRG) survey activities. Timing of construction may have to be modified or re-scheduled during migration seasons.
- **Pile driving and HRG mitigation measures** for survey activities will be required. These measures will include soft-start measures, shut-down procedures, protected species monitoring protocols, use of qualified and NOAA-approved protected species observers (PSOs), and noise attenuation systems such as bubble curtains, as appropriate.
- **Restricted construction schedules.** Impact pile driving activities should not occur from January 1 to April 30 to minimize potential impacts on the NARW, which will also have a protective effect on other marine mammal species.
- **Vessel operation regulations and marine mammal protection** will need to follow national (e.g., NOAA) guidelines for marine mammal strike avoidance measures, including vessel speed restrictions. Vessels should maintain a specified separation distance from any sighted cetacean (whale, dolphin, or porpoise).

- **PSOs will be required to be on these vessels.** All personnel working offshore should receive training on marine mammal awareness and marine debris awareness.
- **To prevent and control accidental spills, discharges,** and releases of oils or other hazardous materials, all construction and operations vessels will need to comply with the relevant regulatory requirements, including the oil spill response requirements.

Other countries have similar regulations and procedures. For example, in order to protect the harbor porpoise, the German government requires that sound may not exceed 160 dB re 1 μPa at a distance of 750 m from the construction site (German Environment Agency, 2023). In general, mitigation strategies for harbor porpoise focus on reducing the high-frequency part of the noise spectrum and using pingers to disperse any species before construction activities would begin (Kastelein *et al.*, 2022).

The second most abundant group of marine mammals in the North Sea and Baltic Sea are seals (harbor and grey seals). These animals have been extensively studied, and there is now a consensus throughout Europe that pile-driving noise is the most potentially harmful impact that may cause lethality, auditory injury, or behavioral disturbance and displacement.

A Marine Mammal Mitigation Plan is required for most EU sites with a mitigation monitoring zone of 500 m radius. In Belgium, Denmark, and Germany, impact monitoring has been required as a condition to proceed with construction (Verfuss *et al.*, 2015; Degraer *et al.*, 2019). In all four countries, impact studies focused on the harbor porpoise. Most studies were conducted in a BACI design, in which data were collected during a sufficient period before and after the potential impact, within the wind farm area as well as in the designated control area (MMO, 2014).

Marine mammals in Asia are quite prevalent and involve the protection of endangered species. For example, there are the Chinese White Dolphins in the Taiwan Straits, where a significant offshore wind market is operating and under construction. The current underwater noise threshold is the same as the German stipulations. Some citizens and private organizations have expressed concern about whether the current threshold is enough to protect Chinese White Dolphins from harmful effects. Hence, their environmental agency is examining industry best practices for stricter underwater noise thresholds. The Taiwan EPA is enforcing these provisions (EPA-Taiwan, 2023).

The most prevalent and most effective mitigation strategy for noise attenuation is the use of a bubble curtain, developed in the oil and gas sector. They have been designed and implemented successfully in Germany and Denmark during monopile installations, although they add significant costs (Dähne *et al.*, 2017).

14.11 Commercial Fisheries: Risk Characterization

The potential effects of offshore turbine structures on groundfish (i.e., species living at or near the seabed such as halibut, sole, and turbot) and crustaceans (e.g., crab, lobster, and shrimp) have been addressed through multiple studies in Europe (EU Parliament, 2019). In general, these studies have shown that fish stocks may increase because of the artificial reef effect and restrictive fishing zones that allow species to replenish. However, fisheries remain a sensitive stakeholder concern, given the multiple anthropogenic threats to family and commercial businesses, including ocean acidification and other climate change effects.

Some of the following possible impacts have been suggested (Gill *et al.*, 2020):

- Certain fishing practices and vessel traffic may be restricted for safety reasons during survey work, construction, and future decommissioning.
- Habitat modification from the presence of turbines and foundations may change the spatial distribution and abundance of commercially fished marine species.

- Safe operations may impose a change in fishing activity or method, such as from active (trawls, dredges) to passive (e.g., traps).
- Noise and substrate vibrations may injure or kill some fish species and chase away others for a time until construction is completed (see Section 14.11).
- EMF fields from subsea cables create uncertainties related to migration barriers for species, such as skates, sharks, and rays (see Section 14.12).
- Changes to the hydrodynamic regime (see below) may influence the level of scour protection.

Some stakeholders have suggested that changes in certain oceanographic processes could affect fish presence and fishery stocks. These processes generally fall into three categories:

- Wind energy extraction reducing surface wind stress and altering water column turbulence, potentially influencing the nutrient cycle.
- Wind farm wake-driven divergence and convergence are driving upwelling and downwelling, possibly affecting bedload sediment transport.
- Turbulence generated in currents passing by turbine foundations.

Potential direct or indirect impacts to fisheries from these potential processes are largely unknown, including cumulative effects (MMO, 2014; NOAA, 2023).

European post-construction studies have found higher abundances of biomass, including pelagic fish and shellfish, close to offshore turbine support structures than further away (see, e.g., Hutchison *et al.*, 2020).

14.11.1 Mitigation Hierarchy Strategies

The overall objective of the mitigation strategies for the fishing industries is for developers to pose as little disturbance as possible to their activities and to keep fishermen informed throughout the project lifecycle. Recent innovations to address mitigation strategies include earlier and more meaningful inclusion of fisheries representatives in planning and decision-making, designating and involving fisheries liaisons in the process, conducting more cumulative studies, and taking collaborative approaches to assessing the potential impacts on fishing (Haggett *et al.*, 2020). Since there is often no single umbrella organization that represents these groups, regulators have often required a fishing liaison officer or representative integrated into the project team. This individual helps to ensure effective communication between developers and the fishing industry (FLOWW, 2014). An example is the issuance of mariner warnings regarding the presence of survey or construction vessels in and around fishing zones.

To reduce impacts on the fisheries, the placement of turbines may have to modified or the total number of turbines reduced. Location-specific benthic and habitat characterization, in collaboration with regulatory agencies, would assist in the placement. US developers have reckoned with wider navigation corridors between shore and out to sea for large and small fishing vessels – a major stakeholder concern in the busy ocean spaces. A group of developers agreed to 1.6 km between turbines and modifications to the layouts of the project (Vineyard Wind, 2019), allowing for safer navigation routes for fishermen. (Revolution Wind Farm, 2023).

Apart from prohibitions and restrictions on fishing activities, fishing vessels tend to avoid navigating near a wind farm even if access is permitted because of the risk of accidental damage, snagging, and loss of fishing gear. Consequently, the fear of potential loss is a source of concern that may hinder co-existence. Therefore, the International Hydrographic Organization (IHO) recommends avoiding fishing activities at a minimum distance of 0.25 nautical miles (0.46 km) on either side of a submarine cable to minimize risks (EU Parliament, 2019). In addition, power cables need to be trenched adequately to avoid exposure from scouring, sand migration, or trawling activities.

When and if displacement of fishing grounds takes place, fishermen may not have the capacity to move to fishing grounds further afield or to change fishing methods so developers and fishing associations are calling for appropriate compensation as a last resort. Some possible issues associated with compensation schemes involve:

- **Loss and/or damage to gear**. Affected parties should understand the various types of gear, the cost of replacement, and the process for claiming compensation before an incident occurs.
- **Restriction zones**. A plan should be in place for whether some fishing activities or vessels in and around the lease areas may have to be restricted for a time for safety reasons and whether there are economic consequences to these restrictions.
- **Compensation schemes**. At this time, there is not a standardized compensation scheme across fishing groups, types of gear, or incidences that may occur and compensation options. A uniform regional approach would be helpful to all parties.

A life-cycle approach from construction through operation and decommissioning is needed to avoid the potential negative long-term impacts caused by offshore wind turbines on certain habitats and ecosystems, fish stocks and biodiversity, and consequently on lifestyles and incomes of fishermen. This approach requires stakeholder engagement, information sharing, and negotiation on potential risks and benefits with regulatory agencies and conservation groups.

14.12   Electromagnetic Fields and Electromagnetic Interference

EMF radiation is a phenomenon that is sometimes brought up in relation to wind turbines. EMFs are present in the environment across a spectrum of frequencies, including radio waves, microwaves, visible and ultraviolet light, and X-rays. Low-frequency EMF radiation is generated from such sources as the earth's geomagnetic field, thunderstorms, power cables, and electronics. Wireless communication relies on electromagnetic waves of various levels of energy and frequencies. This section discusses the impact of EMFs on the subsea environment and the potential effect of electromagnetic interference from wind turbines on land.

14.12.1   EMF Impacts on the Subsea Environment and Communities

Naturally occurring EMFs are present throughout the oceans, and some marine animals, such as sharks, whales, and sea turtles, can detect them; see Figure 14.20. Although the biological mechanisms of species are not completely understood, it is known that these fields aid various species in searching for prey and predators, other organisms, and navigation and orientation to breeding sites (e.g., loggerhead sea turtles and sockeye salmon).

Electric power cables interact with and add to the naturally occurring fields on land, underwater, and in the atmosphere. Submarine inter array cables in offshore wind plants, those from the plant to shore, and the substations will add some low-level EMFs to these undersea habitats.

According to BOEM (2023b), three major factors determine the exposure of marine organisms to magnetic and induced electric fields from undersea power cables: (1) the amount of electrical current being carried by the cable, (2) the design of the cable, and (3) the distance of marine organisms from the cable.

The possible effect of low-level EMFs is not well characterized, although some anecdotal reports have suggested that EMFs may create an artificial barrier within the habitat and on the seabed. Several European offshore wind studies have, however, found no conclusive evidence of risks from undersea cable EMFs. A study of 18 EU offshore wind plants, for example, concluded that EMFs post-consent monitoring should not ordinarily be required. (MMO, 2014; WOZEP, 2016). Whether or not cumulative effects from EMFs could be significant is also not apparent (SEER, 2022b).

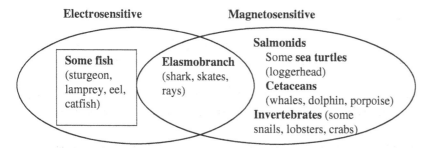

Figure 14.20 Examples of marine wildlife with the ability to sense EMFs. This diagram does not include all electrosensitive and magneto-sensitive marine life (*Source:* SEER, 2022b/U.S. Department of Energy/Public Domain)

Emissions from the cables are reduced by an outer metal sheath. In addition, cables are normally buried 1–2 m below the seafloor or are covered by concrete mattresses for protection. This provides physical separation between the highest levels of EMFs adjacent to the cable and organisms near the bottom. Cable burial, however, does not necessarily eliminate all the potential EMF impacts, as some of the field passes through the seabed. More information may be found at SEER (2022b).

In addition to some concerns with the marine environment impacts, there is a risk perception of EMFs affecting the health of coastal residents from the submarine cable under their beaches, parks, or ports to the point of connection to the grid, although there is no evidence of this health concern. Yet, ongoing scientific studies find that cell phones, household appliances, and overhead power lines expose people to much stronger electrical and magnetic fields than buried cables. According to the World Health Organization (WHO), none of those other sources have been shown to have clear health effects (WHO, 2016).

14.12.2 Electromagnetic Interference

Electromagnetic interference (EMI) is any type of interference that can potentially disrupt, degrade, or interfere with the effective performance of an electronic device. Modern society is dependent on the use of devices that utilize electromagnetic energy, such as power and communication networks, electrified railways, and computer networks. During the generation, transmission, and utilization of electromagnetic energy, the devices generate electromagnetic disturbance that can interfere with the normal operation of other systems.

Wind turbines can potentially disrupt electromagnetic signals used in telecommunications, navigation, and radar services. The degree and nature of the interference will depend on:

- The location of the wind turbine between receiver and transmitter.
- Characteristics of the rotor blades.
- Characteristics of receiver.
- Signal frequency.
- The radio wave propagation in the local atmosphere.

Interference can be produced by four elements of a wind turbine: tower, rotating blades, generator, and power cables (as discussed previously). Towers and blades may obstruct, reflect, or refract the electromagnetic waves. Modern blades are typically made of synthetic materials, which have a minimal impact on the transmission of electromagnetic radiation. The electrical system is not usually a potential problem in telecommunications because interference can be eliminated with proper nacelle insulation and good maintenance. The reader is referred to the second edition of this book, where more detail on the topic is provided.

References

Adams E.M., Gulka J, Williams, K.A. (2021) A review of the effectiveness of operational curtailment for reducing bat fatalities at terrestrial wind farms in North America. *PLoS ONE*, **16**(11), e0256382. https://doi.org/10.1371/journal. pone.0256382 Accessed 12/9/2023

Allison, T. D., Diffendorfer, J. E., Baerwald, E. F., Beston, J. A., Drake, D., Hale, A. M., Hein, C. D., Huso, M. M., Loss, S. R., Lovich, J. E., Strickland, M. D., Williams, K. A. and Winder, V. L. (2019) Impacts to wildlife of wind energy siting and operation in the United States. *Issues in Ecology*, **21**, 2–22.

APLIC (2012) *Suggested practices for avian protection on power lines: the state of the art. Reducing Avian Collisions with Power Lines: State of the Art in 2012*, https://www.aplic.org/mission Accessed 5/2023.

APPA (2023) *American Public Power Association. The Public Utility Regulatory Policies Act of 1978 Issue Brief,* https:// www.energy.gov/oe/articles/reference-manual-and-procedures-implementation-purpa-standards-epact-2005-march-2006 Accessed 12/9//2023.

Arlidge, W. N. S., Bull, J. W., Addison, P. F. E., Burgass, M. J, Gianuca, D., Gorham, T. M., Jacob, C., Shumway, N., Sinclair, S. P. and Watson, J. E. M. (2018) A Global Mitigation Hierarchy for Nature Conservation, *BioScience*, **68**(5), https://doi.org/10.1093/biosci/biy029.

Arnett, E. B., Huso, M. M. P., Schirmacher, M. R. and Hayes, J. P. (2016a) Effectiveness of selective cut-in of wind turbines to reduce bat fatalities. *The Journal of Wildlife Management*, **80**(5), 875–890.

Arnett, E. B., Baerwald, E. F., Mathews, F., Rodrigues, L., Rodríguez-Durán, A., Rydell, J., Villegas-Patraca, R. and Voigt, C. C. (2016b). Impacts of Wind Energy Development on Bats: A Global Perspective, in: C. Voigt and T. Kingston (Eds.), *Bats in the Anthropocene: Conservation of Bats in a Changing World*. Springer, Cham, https://doi.org/10.1007/978-3-319-25220-9_11.

Audubon (2019) Survival by Degrees. 389 Bird Species on the Brink. https://www.audubon.org/climate/ survivalbydegrees Accessed 12/9/2023.

Barclay, R. and L. Harder (2003). Life Histories of Bats: Life in the Slow Lane, in: M.B. Fenton and T. Kunz (Eds.), *Bat Ecology*, University of Chicago Press, USA, 799pp.

Beranek, L. L. and Ver, I. L. (2005) *Noise and Vibration Control Engineering: Principles and Applications*, 2nd edtion. John Wiley and Sons, Ltd, Chichester, UK.

BirdLife International (2023) A range of threats drives declines in bird populations, http://datazone.birdlife.org/sowb/ casestudy/a-range-of-threats-drives-declines-in-bird-populations Accessed 8/18/2023.

BLM (2023) *Bureau of Land Management. Visual Resources Clearinghouse*, https://blmwyomingvisual.anl.gov Accessed 12/9/2023.

BOEM and NOAA (2022) *BOEM and NOAA Fisheries North Atlantic Right Whale and Offshore Wind Strategy. Docket BOEM-2022-0066. Created by the Bureau of Ocean Energy Management*, https://www.regulations.gov/docket/BOEM-2022-0066 Accessed 12/9/23.

BOEM (2023a) *Regulatory Roadmap and Guidelines*, https://www.boem.gov/renewable-energy/regulatory-framework-and-guidelines Accessed 12/9//2023.

BOEM (2023b) *Environmental Studies. Electromagnetic Fields (EMF) and Marine Life. Factsheet,* https://www.boem.gov/ sites/default/files/documents/renewable-energy/state-activities/Electromagnetic-Fields-Marine-Life-web.pdf Accessed 8/2/2023.

Bortolotti, P., Branlard, E., Platt, P., Moriarty, P. J., Sucameli, C. and Bottasso, C. (2020) *Aeroacoustics Noise Model of OpenFAST*, NREL/TP-5000-75731. National Renewable Energy Laboratory, Golden, CO.

Bowdler, D. and Leventhal, H. (2012) *Wind Turbine Noise*, Multi-Science Pub, New York.

Bruckner, T., Bashmakov, I. A., Mulugetta, Y., Chum, H., de la Vega Navarro, A., Edmonds, J., Faaij, A., Fungtammasan, B., Garg, A., Hertwich, E., Honnery, D., Infield, D., Kainuma, M., Khennas, S., Kim, S., Nimir, H. B., Riahi, K., Strachan, N., Wiser, R. and Zhang, X. (2014) Energy Systems. *Climate Change 2014: Mitigation of Climate Change. Contribution of Working Group III to the Fifth Assessment Report of the Intergovernmental Panel on Climate Change*. Cambridge University Press, Cambridge, United Kingdom and New York, NY, USA.

BWEA (2005) *Low Frequency Noise and Wind Turbines Technical Annex*. British Wind Energy Association, London.

BWEC (2004) *Proceedings Bats and Wind Power Generation Technical Workshop*. February 19-20, 2004, https://www. batsandwind.org/docs/batsandwindenergycooperativelibraries/assets/finalbatpro2004.pdf?sfvrsn=cc8eb1fa_1 Accessed 8/2/2023.

BWEC (2022) *Bats and Wind Energy Cooperative*. 6th Science and All Committees Meeting. BWEC 2022 Workshop Proceedings. 7, 8, 25 February; 4 March; and 7 September 2022, https://www.batsandwind.org/docs/ batsandwindenergycooperativelibraries/assets/bwec-2022-research-priorities-508.pdf?sfvrsn=335b0f8f_1 Accessed 12/9/2023.

Carlén, I. (2017) *SAMBAH – Static Acoustic Monitoring of the Baltic Sea Harbour Porpoise: Bathymetric derivatives*. https:// doi.org/10.5879/9q6e-vr77 Accessed 12/9/23.

CBI (2023). *Consensus Building Institute. Stakeholder Perspectives on the Development of a North Atlantic Right Whale Vessel Risk Reduction Strategy*. Report prepared by Consensus Building Institute for the U.S. Department of the Interior, Bureau of Ocean Energy Management, Sterling, VA. OCS Study BOEM 2023-004.

Cheng, T. L., Reichard, J. D., Coleman, J. T. H., Weller, T., Thogmartin, W. E., Reichert, B., Bennett, A., Broders, H. G., Campbell, J., Etchison, K., Feller, D. J., Geboy, R., Hemberger, T., Herzog, C., Hicks, A. C., Houghton, S., Humber, J., Kath, J. A., King, A. L., Loeb, S. C., Masse, A., Morris, K. M., Niederriter, H., Nordquist, G. E., Perry, R. W., Reynolds, R., Sasse, D. B., Scafini, M. R., Stark, R. C., Stihler, C. W., Thomas, S. C., Turner, G. G., Webb, S., Westrich, B. and Frick, W. F. (2021) The scope and severity of white-nose syndrome on hibernating bats in North America. *Conservation Biology*, **35**(5), 1586–1597. https://www.ncbi.nlm.nih.gov/pmc/articles/PMC8518069/ https://doi.org/10.1111/ cobi.13739.

Clark, C. W., Ellison, W. T., Southall, B. L., Hatch, L., van Parijs, S. M., Frankel, A. and Ponirakis, D. (2009) Acoustic masking in marine ecosystems: Intuitions, analysis, and implication. *Marine Ecology Progress Series*, **395**, 22.

Cryan, P., Gorresen, P., Straw, B., Thao, S., DeGeorge, E. (2021) Influencing activity of bats by dimly lighting wind turbine surfaces with ultraviolet light. *Animals*, **12**(1), 23. https://doi.org/10.3390/ani12010009.

Dähne, M., Tougaard, J., Carstensen, J., Rose, A., Nabe-Nielsen, J. (2017) Bubble curtains attenuate noise from offshore wind farm construction and reduce temporary habitat loss for harbour porpoises. *Marine Ecology Progress Series*, **580**, 221–237. https://doi.org/10.3354/meps12257.

Dalthorp, D., Madsen, L., Huso, M., Rabie, P., Wolpert, R., Studyvin, J., Simonis, J., and Mintz, J. 2018 GenEst statistical models—A generalized estimator of mortality: U.S. *Geological Survey Techniques and Methods*, book 7, chap. A2, 13, https://doi.org/10.3133/tm7A2. https://code.usgs.gov/ecosystems/GenEst.

DEA (2017) *Danish Energy Agency*. Danish Experiences from Offshore Wind Development. Copenhagen.

Degraer, S., Brabant, R., Rumes, B., Vigin, L., Derous, S. and Van Lancker, V. (2019) Environmental Impacts of Offshore Wind Farms in the Belgian Part of the North Sea: Assessing and Managing Effect Spheres of Influence, in: S. Degraer, R. Brabant, B. Rumes, L. Vigin (Eds.) *Offshore Wind Energy Technology*, Springer, 295–318.

Dempsey, L., Hein, C. and Münter, L. (2023) *The Mitigation Hierarchy. Working Together to Resolve Environmental Effects of Wind Energy (WREN) Short Science Summary*. PNNL and NREL.

Desholm, M. and Kahlert, J. (2005) Avian collision risk at an offshore wind farm. *Biology Letters*, **1**, 296–298, https://doi. org/10.1098/rsbl.2005.0336.

Devine-Wright, P. (2005) Beyond NIMBYism: towards an integrated framework for understanding public perceptions of wind energy. *Wind Energy*, **8**, 125–139, https://doi.org/10.1002/we.124.

Devine-Wright, P. and Howes, Y. (2010). Disruption to place attachment and the protection of restorative environments: A wind energy case study. *Journal of Environmental Psychology*, **30**(3), 271–280.

DOE (2022) *Wind Energy Technology Office. Land Based Wind Market Report: 2022 Edition*. August 16, 2022.

DOI (2013) *Department of the Interior. Best Management Practices for Reducing Visual Impacts of Renewable Energy Facilities on BLM-Administered Lands*. Bureau of Land Management. Cheyenne, Wyoming. April 2013.

DOT (2022) *Promising Practices for Meaningful Public Involvement in Transportation Decision-Making*. Department of Transportation, Washington, DC.

EPA-Taiwan (2023) *EPA Reviews and Supervises Offshore Wind Power Projects*. Environmental Protection Administration of Taiwan https://www.epa.gov.tw/eng/F7AB26007B8FE8DF/1d49e06a-db00-495e-8daa-1cfb174372a0 Accessed 12/9/23.

Duffie, J. A. and Beckman, W. A. (2006) *Solar Engineering of Thermal Processes*, 3rd edition. John Wiley & Sons, Chichester, UK.

Emd-international.com (2023) *windPRO*, https://www.emd-international.com/windpro/ Accessed 8/17/2023.

European Commission (2023). *Environmental Assessments in the EU's environmental policy*. Developed by the Academy of European Law. https://www.era-comm.eu/EU_Legislation_on_Environmental_Assessments/part_1/index.html Accessed 12/9/23.

European MSP Platform (2023) https://maritime-spatial-planning.ec.europa.eu/msp-eu/introduction-msp Accessed 12/9/23.

EU Parliament (2019). Report on the impact on the fishing sector of offshore wind farms and other renewable energy systems. A9-0184/2021. https://www.europarl.europa.eu/doceo/document/A-9-2021-0184_EN.html Accessed 12/9/2023.

EWEA (2004) *Wind Energy, The Facts*, **1–5**. European Wind Energy Association, Brussels.

FLOWW (2014) FLOWW Best Practice Guidance for Offshore Renewables Developments: Recommendations for Fisheries Liaison. Fishing Liaison with Offshore Wind and Wet Renewables Group (FLOWW), Edinburgh, Scotland, FLOWW, http://dx.doi.org/10.25607/OBP-956.

Frick, W. F., Baerwald, E. F., Pollock, J. F., Barclay, R. M. R., Szymanski, J. A., Weller, T. J., Russell, A. L., Loeb, S. C., Medellin, R. A. and McGuire, L. P. (2017). Fatalities at wind turbines may threaten population viability of a migratory bat. *Biological Conservation*, **209** (2017), 172–177.

GenEst – A Generalized Estimator of Mortality (2018) USGS Digital Object Identifier Catalog. https://www.usgs.gov/software/genest-a-generalized-estimator-mortality. Accessed 2/13/2024.

German Environment Agency (2023). *(Umweltbundesamt – UBA). Underwater Noise.* https://www.umweltbundesamt.de/en/underwater-noise#underwater-noise-a-man-made-problem Accessed 12/9/23.

Gill, A. B., Degraer, S., Lipsky, A., Mavraki, N., Methratta, E. and Brabant, R. (2020) Setting the context for offshore wind development effects on fish and fisheries. *Oceanography*, **33**(4), 118–127, https://doi.org/10.5670/oceanog.2020.411.

Gipe, P. (2013) *Wind Power in View: Energy Landscapes in a Crowded World*. Chelsea Green Publishing.

Gregory, R., Failing, L., Harstone, M., Long, G., McDaniels, T. and Ohlson, D. (2012) *Structured Decision Making: A Practical Guide to Environmental Management*. Wiley-Blackwell, West Sussex, UK.

Gross (2020) *Renewables, Land Use, and Local Opposition in The United States*. Brookings Institution. https://www.brookings.edu/wp-content/uploads/2020/01/FP_20200113_renewables_land_use_local_opposition_gross.pdf. Accessed 12/9/23.

Guest, E. E., Stamps, B. F., Durish, N. D., Hale, A. M., Hein, C. D., Morton, B. P., Weaver, S. P. and Fritts, S. R. (2022) An updated review of hypotheses regarding bat attraction to wind turbines. *Animals (Basel)*, **12**(3), 343, https://doi.org/10.3390/ani12030343.

Haggett, C., Brink, T., Russell, A., Roach, M., Firestone, J., Dalton, T. and McCay, B. J. (2020) Offshore wind projects and fisheries: conflict and engagement in the United Kingdom and the United States. *Oceanography*, **33**(4), 38–47, https://doi.org/10.5670/oceanog.2020.404.

Hale, A. M., Hein, C. D. and Straw, B. R. (2022). Acoustic and genetic data can reduce uncertainty regarding populations of migratory tree-roosting bats impacted by wind energy. *Animals*, **2022**(12), 81. https://doi.org/10.3390/ani12010081.

Hein C., Williams, K. A. and Jenkins, E. (2021) Bats Workgroup Report for the State of the Science Workshop on Wildlife and Offshore Wind Energy 2020: Cumulative Impacts. Report to the New York State Energy Research and Development Authority (NYSERDA). Albany, NY, https://www.nyetwg.com/2020-workgroups Accessed 8/2/2023.

Hein, C. and Straw, B. (2021) *Proceedings from the State of the Science and Technology for Minimizing Impacts to Bats from Wind Energy*. NREL/TP-5000-78557. National Renewable Energy Laboratory. Golden, CO, https://www.nrel.gov/docs/fy21osti/78557.pdf Accessed 12/9/23.

Hodos, B. (2003) *Minimization of Motion Smear: Reducing Avian Collisions with Wind Turbines.* NREL/SR-500-33249. National Renewable Energy Laboratory. Golden, CO. https://www.nrel.gov/docs/fy03osti/33249.pdf Accessed 12/9/23.

Hoen, B., Firestone, J., Rand, J., Elliot, D., Hübner, G., Pohl, J., Wiser, R., Lantz, E., Haac, T. R. and Kaliski, K. (2019) Attitudes of U.S. Wind Turbine Neighbors: analysis of a nationwide survey. *Energy Policy*, **134**. https://www.sciencedirect.com/science/article/pii/S0301421519305683 Accessed 12/9/2023.

Holm, P., Rasmussen, M. H., Noerregaard, J. and Kildegaard, A. (2019) Environmental impacts of offshore wind farms in the Danish part of the North Sea. *Renewable and Sustainable Energy Reviews*, **103**, 100–114.

Hüppop, O., Dierschke, J., Exo, K.-M., Fredrich, E. and Hill, R. (2006) Bird migration studies and potential collision risk with offshore wind turbines. *Ibis*, **148**, 90–109, https://doi.org/10.1111/j.1474-919X.2006.00536.x.

Hutchison, Z. L., LaFrance Bartley, M., Degraer, S., English, P., Khan, A., Livermore, J., Rumes, B. and King, J. W. (2020) Offshore wind energy and benthic habitat changes: Lessons from Block Island Wind Farm. *Oceanography*, 33(4), 58–69, https://doi.org/10.5670/oceanog.2020.406.

IEA Wind (2023) *Wind Technology Collaboration Program. Social Acceptance of Wind Energy Projects. Task 28 Publications.* International Energy Agency, https://iea-wind.org/task28/t28-publications/ Accessed 12/9/2023.

IEC (2012) *Wind Turbines – Part 11: Acoustic noise measurement techniques, IEC 61400-11*, 3rd edition. International Electrotechnical Commission, Geneva.

IPCC (2011) *Renewable Energy Sources and Climate Change Mitigation. IPCC, 2011.* O. Edenhofer, R. Pichs-Madruga, Y. Sokona, K. Seyboth, P. Matschoss, S. Kadner, T. Zwickel, P. Eickemeier, G. Hansen, S. Schloemer and C. von Stechow (Eds.), Cambridge University Press, Cambridge, UK.

IPCC (2021) *Climate Change 2021: The Physical Science Basis.* Contribution of Working Group I to the Sixth Assessment Report of the Intergovernmental Panel on Climate Change [V. Masson-Delmotte, P. Zhai, A. Pirani, S. L. Connors, C. Péan, S. Berger, N. Caud, Y. Chen, L. Goldfarb, M. I. Gomis, M. Huang, K. Leitzell, E. Lonnoy, J. B. R. Matthews, T. K. Maycock, T. Waterfield, O. Yelekçi, R. Yu and B. Zhou (Eds.)]. Cambridge University Press, Cambridge, UK, https://doi.org/10.1017/9781009157896.

IUCN (2022) The IUCN Red List of Threatened Species. Version 2022-2. https://www.iucnredlist.org Accessed 12/9/2023.

Jobert, A. and Laborgne, P. (2012) Local acceptance of wind energy: factors of success identified in French and German case studies. *Energy Policy*, **42**, 215–226.

Johnston, D., Thaxter, C., Boersch-Supan, P., Humphreys, E. M., Bouten, W., Clewley, G. Scragg, E. S., Masden, E., Barber, L., Conway, G., Clark, N. A., Burton, N. H. K. and Cook, A. S. C. P. (2021) Investigating avoidance and attraction responses in lesser black-backed gulls *Larus fuscus* to offshore wind farms. *Marine Ecology Progress Series*, **686**, https://doi.org/10.3354/meps13964.

Karlson, B. (2021) *Wind Turbine Radar Interference Mitigation*, Sandia National Laboratory Report SAND2021-9336PE, https://www.osti.gov/servlets/purl/1882288, Accessed 8/2/2023.

Kastelein, R., de Jong, C. T., Tougaard, J., Hoek, L. and Defillet, L. (2022) Behavioral responses of a harbor porpoise (*Phocoena phocoena*) depend on the frequency content of pile-driving sounds. *Aquatic Mammals*, **48**, 97–109, https://doi.org/10.1578/AM.48.2.2022.97.

Knowlton, A. R., Clark, J. S., Hamilton, P. K., Kraus, S. D., Pettis, H. M., Rolland, R. M. and Schick R. S. (2022) Fishing gear entanglement threatens recovery of critically endangered North Atlantic right whales. *Conservation Science and Practice*, **4**, e12736.

Köller, J., Köppel, J. and Peters, W. (2006) *Offshore Wind Energy: Research on Environmental Impacts.* Springer, Berlin, Heidelberg.

Koschinski, S. and Lüdemann, K. (2013) *Development of Noise Mitigation Measures in Offshore Wind Farm Construction.* Federal Agency for Nature Conservation (Bundesamt für Naturschutz, BfN).

Kunz, T., Arnett, E., Cooper, B., Erickson, W., Larkin, R., Mabee, T., Morrison, M., Strickland, M., Szewczak, J. (2007) Assessing impacts of wind-energy development on nocturnally active birds and bats: a guidance document. *Journal of Wildlife Management*, **71**, 2449–2486, https://doi.org/10.2193/2007-270.

Ladenburg, J. *and* Dubgaard, A. (2009). Preferences of Coastal Zone User Groups Regarding the Siting of Offshore Wind Farms. *Ocean & Coastal Management*, **52**(5), 233–242. https://doi.org/10.1016/j.ocecoaman.2009.02.002 Accessed 12/9/23

Ladenburg, J. (2015) *Energy Research & Social Science*, 10, 26–30. https://doi.org/10.1016/j.erss.2015.06.005 Accessed 12/9/23

Leiserowitz, L. (2007) American opinions on global warming: A Yale University / Gallup / Clear Vision Institute Poll. https://climatecommunication.yale.edu/visualizations-data/ycom-us/ Accessed 12/9/23.

Lentini P.E., Bird Tomas J., Griffiths Stephen R., Godinho Lisa N. and Wintle Brendan A. (2015). A global synthesis of survival estimates for microbats. *Biological Letters*, 11 (8). https://doi.org/10.1098/rsbl.2015.0371 Accessed 12/9/2023.

Lenton, T. M., Armstrong McKay, D. I., Loriani, S., Abrams, J. F., Lade, S. J., Donges, J. F., Milkoreit, M., Powell, T., Smith, S. R., Zimm, C., Buxton, J. E., Bailey, E., Laybourn, L., Ghadiali, A., Dyke, J. G. (eds), 2023, *The Global Tipping Points Report 2023*. University of Exeter, Exeter, UK.

Loss, S. R., Will, T., and Marra, P. P. (2019) Estimates of bird collision mortality at wind facilities in the contiguous United States. *Biological Conservation*, **237**, 230–239.

Maclaurin, G., Hein, C., Williams, T., Owen, R., Lantz, E., Buster, G. and Lopez, A. (2022) National-scale impacts on wind energy production under curtailment scenarios to reduce bat fatalities. *Wind Energy*, **25**(9), 1514–1529, https://doi.org/10.1002/we.2741.

Madsen, H. A. (2008) *Low frequency noise from MW wind turbines – Mechanisms of generation and its modeling*. Technical University of Denmark, Denmark. Forskningscenter Risoe. Risoe-R No. 1637(EN).

Matuschek, R. and Betke, K. (2009) Measurements of Construction Noise During Pile Driving of Offshore Research Platforms and Wind Farms, *Proc. of the International Conference on Acoustics* (NAG/DAGA 2009), https://tethys.pnnl.gov/publications/measurements-construction-noise-during-pile-driving-offshore-research-platforms-wind Accessed 8/17/2023.

May, R, Nygård, T, Falkdalen, U, Åström, J, Hamre, Ø. and Stokke, BG. (2020). Paint it black: Efficacy of increased wind-turbine rotor blade visibility to reduce avian fatalities. *Ecology and Evolution*, **10**, 8927–8935. https://doi.org/10.1002/ece3.6592 Accessed 12/9/2023.

McClure, C. J. W., Rolek, B. W., Dunn, L., McCabe, J. D., Martinson, L. and Katzner, T. (2022) *Confirmation that eagle fatalities can be reduced by automated curtailment of wind turbines*. British Ecological Society. First published: 26 August 2022. Volume 3, Issue 3. July–September 2022, https://illumination.duke-energy.com/articles/how-duke-energy-is-using-technology-to-save-eagles-at-a-wyoming-wind-site and https://www.identiflight.com/eagle-eye/ Accessed 12/5/2023.

Milieu, IEEP and ICF (2016) *Evaluation Study to Support Fitness Check of Birds and Habitats Directives*. March 2016, https://op.europa.eu/en/publication-detail/-/publication/f020d8df-07cb-11e8-b8f5-01aa75ed71a1/language-en/format-PDF/source-281369228, Accessed 8/1/2023.

MMO (2014) *Review of post-consent offshore wind farm monitoring data associated with license conditions*. Marine Management Organization MMO Project No: 1031. ISBN: 978-1-909452-24-4. https://assets.publishing.service.gov.uk/government/uploads/system/uploads/attachment_data/file/317787/1031.pdf Accessed 12/9/2023.

Musial, W., Spitsen, P., Beiter, P., Duffy, P., Mulas Hernando, D. M., Hammond, R., Shields, M., Marquis, M. and Sriharan, S. (2023) *Offshore Wind Technologies Market Report: 2023 Edition*. U.S. Department of Energy Office of Energy Efficiency & Renewable Energy, Washington, D.C.

NOAA (2023) *National Oceanic and Atmospheric Administration*. North Atlantic Right Whale, https://www.fisheries.noaa.gov/species/north-atlantic-right-whale, Accessed 12/9/2023.

Nordic Energy Research (2022) *Accommodating Biodiversity in Nordic Offshore Wind Projects. A collaborative project between Nordic Energy Research and DNV*. https://pub.norden.org/nordicenergyresearch2022-01/#88955 Accessed 12/9/23.

Norro, A. (2020) An Evaluation of the Noise Mitigation Achieved by Using Double Big Bubble Curtains in Offshore Pile Driving in the Southern North Sea. https://www.vliz.be/imisdocs/publications/357629.pdf Accessed 8/2/2023.

NREL (2021). *Land Based Wind Energy Siting: A Foundational and Technical Resource.* National Renewable Energy Laboratory, Golden, CO. https://www.nrel.gov/docs/fy21osti/78591.pdf Accessed 8/2/2023.

Oreskes, N. and Conway, E. M. (2011) *Merchants of Doubt: How a Handful of Scientists Obscured the Truth on Issues from Tobacco Smoke to Global Warming.* Bloomsbury Publishing, London.

O'Shea, T. J., Cryan, P. M., Hayman, D. T. S., Plowright, R. K. and Streicker, D. G. (2016) Multiple mortality events in bats. *Mammal Review*, **46**, 175–190. https://doi.org/10.1111/mam.12064. .

Pace, R. M., Williams, R., Kraus, S. D., Knowlton, A. R., Pettis, H. M. (2021) Cryptic mortality of North Atlantic right whales. *Conservation Science and Practice*, **3**, e346.

Pasqualetti, M., Gipe, P. and Righter, R. (2002) *Wind Energy in View: Landscapes of Power in a Crowded World.* Academic Press. ISBN: 0125463340.

Pennisi, E. (2019) Billions of North American birds have vanished. *Science*, **365** (6459), 1228–1229. https://doi.org/10.1126/science.365.6459.1228.

Petersen, I. K., Fox, A. D. and Delany, S. N. (2019). Collision risk of birds with offshore wind turbines in the North Sea. *Biological Conservation*, **234**, 139–149.

PURPA (1978) Public Utility Regulatory Policies Act of 1978. Public Law 95-617 (92 Stat. 3117) enacted November 9, 1978. 16 U.S. Code Chapter 46.

Rabie, P. A., Riser-Espinoza, D., Studyvin, J., Dalthorp, D. and Huso, M. (2021) *Performance of the GenEst Mortality Estimator Compared to the Huso and Shoenfeld Estimators.* AWWI Technical Report. Washington, DC. https://pubs.usgs.gov/publication/70217783 Accessed 8/2/2023.

Ram, B. and Stephens, J. (2017) Risk Conundrums of the Renewable Energy Transition: Can We Balance Opportunities, Optimism, and Challenges. R. E. Kasperson (Ed.) in *Chapter in Risk Conundrums: Solving Unsolvable Problems*, Earthscan Routledge.

Ram, B., Anker, H. T., Clausen, N.-E., and Nielsen, T. R. L. (2018). Public Engagement in Danish Near Shore Wind Projects: In Law and Practice. *Wind 2050 Task 3,* Copenhagen University, and University of Delaware EPSCoR RII, Wind2050.org/publications Accessed 12/9/23.

Ram, B. and T. Webler (2022). Social Amplification of Risks and the Clean Energy Transformation: Elaborating on the Four Attributes of Information, *Risk Analysis.* https://onlinelibrary.wiley.com/doi/10.1111/risa.13902 Accessed 8/2/2023.

Rand, J., Strauss, R., Gorman, W., Seel, J., Kemp, J. M., Jeong, S., Robson, D. and Wiser, R. (2023) *Queued Up: Characteristics of Power Plants Seeking Transmission Interconnection as of the End of 2022.* Lawrence Berkeley National Laboratory, https://emp.lbl.gov/sites/default/files/queued_up_2022_04-06-2023.pdf Accessed 12/9/2023.

Revolution Wind Farm (2023) Revolution Wind Export Cable Project Final Environmental Impact Statement.

REWI (2023) Compensatory Mitigation. Updated December 2022. https://rewi.org/guide/guide-references/ Accessed 12/6/2023.

Romano, W.B., Skalski, J.R., Townsend, R.L., Kinzie, K.W., Coppinger, K.D. and Miller, M.F. (2019), Evaluation of an acoustic deterrent to reduce bat mortalities at an Illinois wind farm. *Wildlife Society Bulletin*, **43**, 608–618. https://doi.org/10.1002/wsb.1025 Accessed 12/9/2023.

Salerno, J., Krieger, A., Smead, M. and Veas, L. (2019) *Supporting National Environmental Policy Act.* OCS Study BOEM 2019- 011. U.S. Department of the Interior, Bureau of Ocean Energy Management. Washington.

Scottish Natural Heritage (2017) *Visual Representation of Wind Farms. Guidance.* Version 2.2. February 2017. https://www.nature.scot/doc/visual-representation-wind-farms-guidance Accessed 12/8/2023.

SEER (2022a) *Environmental Effects of U.S. Offshore Wind Energy Development: Compilation of Educational Research Briefs* [Booklet]. U.S. Offshore Wind Synthesis of Environmental Effects Research. 2022. Report by National Renewable Energy Laboratory and Pacific Northwest National Laboratory for the U.S. Department of Energy, Wind Energy Technologies Office. https://tethys.pnnl.gov/seer Accessed 12/9/23.

SEER (2022b) *Electromagnetic Field Effects on Marine Life.* U.S. Offshore Wind Synthesis of Environmental Effects Research. 2022 Report by National Renewable Energy Laboratory and Pacific Northwest National Laboratory for the U.S. Department of Energy, Wind Energy Technologies Office. https://tethys.pnnl.gov/seer Accessed 12/9/23.

Smallwood, K. S. and Bell, D. A. (2020), Effects of wind turbine curtailment on bird and bat fatalities. *Journal of Wildlife Management*, **84**, 685–696. https://doi.org/10.1002/jwmg.21844.

South Fork Wind Farm (2020) *Construction and Operation Plan (COP). Appendix V. Visual Impact Assessment*. Prepared by Environmental Design and Research (EDR). https://www.boem.gov/sites/default/files/documents/renewable-energy/state-activities/App-V-SFWF-VIA_0.pdf Accessed 12/5/2023.

South Fork Wind Farm (2021) *Wind Farm Construction and Operations Plan. APPENDIX P1. Assessment of Impacts to Marine Mammals, Sea Turtles, and Sturgeon*, https://www.boem.gov/sites/default/files/documents/oil-gas-energy/MarineMammalSeaTurtleSturgeon_Report.pdf Accessed 12/5/2023.

Stenseth, N. C., Payne, M. R., Bonsdorff, E., Dankel, D. J., Durant, J. M., Anderson, L. G., Armstrong, C. W., Blenckner, T., Brakstad, A., Dupont, S., Eikeset, A. M., Goksøyr, A., Jónsson, S., Kuparinen, A. and Våge, K. (2020) Attuning to a changing ocean. *Proceedings of the National Academy of Sciences of the United States of America*, **117**(34), 20363–20371.

Sullivan, R. G., Kirchler, L., Lahti, T., Roché, S., Beckman, K., Cantwell, B. and Richmond, P. (2012) Wind Turbine Visibility and Visual Impact Threshold Distances in Western Landscapes. *Proc. Of the National Association of Environmental Professionals 37th Annual Conference*, Portland, OR.

Sullivan, R., Kirchler, L., Cothren, J. and Winters, S. (2013) Offshore wind turbine visibility and visual impact threshold distances. *Environmental Practice*, **15**, 33–49. https://doi.org/10.1017/S1466046612000464.

Sullivan, R. and Meyer, M. (2014) Guide to evaluating visual impact assessments for renewable energy projects 10.13140/2.1.3216.5767.

Sullivan, R. (2021). *Assessment of Seascape, Landscape, and Visual Impacts of Offshore Wind Energy Developments on the Outer Continental Shelf of the United States*. OCS Study. BOEM 2021-032. Bureau of Ocean Energy Management. Washington.

Tethys Knowledge Base (2023). https://tethys.pnnl.gov/knowledge-base-wind-energy. Accessed 12/9/2023.

Thaxter, C. B., Buchanan, G. M., Carr, J., Butchart, S. H. M., Newbold, T., Green, R. E., Tobias, J. A., Foden, W. B., O'Brien, S. and Pearce-Higgins, J. W. (2017) Bird and bat species' global vulnerability to collision mortality at wind farms revealed through a trait-based assessment. *Proceedings of the Royal Society B: Biological Sciences*, **287**(1925), 20201848. https://doi.org/10.1098/rspb.2017.0829.

Toke, D. and Breukers, S. (2015) The politics of procedural and distributive justice in community renewable energy. *Energy Policy*, **86**, 586–594.

Tougaard, J., Jensen, F. H., Carstensen, J., Bech, N. I., and Nabe-Nielsen, J. (2019). Acoustic impact assessment of offshore wind farms on harbour porpoises: acoustic monitoring of harbour porpoises in the vicinity of Horns Rev 3 offshore wind farm. *Environmental Monitoring and Assessment*, **191**(3), 146.

Tougaard, J. (2021). *Thresholds for Behavioral Responses to Noise in Marine Mammals: Background note to revision of guidelines from the Danish Energy Agency*. Aarhus University. Danish Centre for Environment and Energy. Technical Report from Danish Centre for Environment and Energy. Aarhus University, Department of Ecoscience.

UK Government (2017) *Guidance. Environmental Impact Assessment. Town and Country Planning* https://www.gov.uk/guidance/environmental-impact-assessment Accessed 12/9/23.

US Wind (2023). *Whales and Offshore Wind Development. US Wind's plans to protect marine mammals before, during, and after construction*, https://uswindinc.com/wp-content/uploads/2022/06/us-wind-whale-fact-sheet.pdf Accessed 12/4/2023.

U.S. Department of Transportation (2015) *Advisory Circular No.70/7460-1L: Obstruction Marking and Lighting*, https://www.faa.gov/documentLibrary/media/Advisory_Circular/AC_70_7460-1L_-_Obstuction_Marking_and_Lighting_-_Change_2.pdf Accessed 8/2/2023.

USFWS (2016) *Land-Based Wind Energy Guidelines: Minimizing Impacts to Wildlife During Site Selection, Construction, and Operation*. U.S. Fish and Wildlife Service. Washington, DC. https://www.fws.gov/sites/default/files/documents/land-based-wind-energy-guidelines.pdf Accessed 12/9/2023

Verfuss, U. K., Sparling, C. E., Arnot, C., Judd, A. and Coyle, M. (2015). Review of offshore wind farm impact monitoring and mitigation with regard to marine mammals. *Advances in Experimental Medicine and Biology*, **875**, 1175–1182. https://doi.org/10.1007/978-1-4939-2981-8_147.

Vineyard Wind (2019) *Project Update: Vineyard Wind Finalizes Turbine Array to Boost Mitigation for Fishermen and Historic Preservation on Nantucket and Vineyard*, https://www.vineyardwind.com/press-releases/2019/6/24/project-update-vineyard-wind-finalizes-turbine-array-to-boost-mitigation-for-fishermen-and-historic-preservation-on-nantucket-and-vineyard Accessed 8/2/2023.

Wagner, S., Bareib, R. and Guidati, G. (1996) *Wind Turbine Noise*. Springer, Berlin.

Weaver, S.P., Hein, C. D., Simpson, T.R., Evans, J.W. and Castro-Arellano, I. (2020). Ultrasonic acoustic deterrents significantly reduce bat fatalities at wind turbines, *Global Ecology and Conservation*, **24**. ISSN 2351-9894. https://doi.org/10.1016/j.gecco.2020.e01099 Accessed 12/9/2023.

Whitby, M. D., Schirmacher, M. R. and Frick, W. F.(2021) The State of the Science on Operational Minimization to Reduce Bat Fatality at Wind Energy Facilities. A report submitted to the National Renewable Energy Laboratory. Bat Conservation International. Austin, Texas. https://tethys.pnnl.gov/publications/state-science-operational-minimization-reduce-bat-fatality-wind-energy-facilities Accessed 12/9/2023

WHO (2016) World Health Organization. *Radiation: Electromagnetic fields*, https://www.who.int/peh-emf/about/WhatisEMF/en/index1.html Accessed 8/17/2023.

World Bank (2015) *Environmental, Health, and Safety Guidelines: Wind Energy*, https://documents1.worldbank.org/curated/en/498831479463882556/pdf/110346-WP-FINAL-Aug-2015-Wind-Energy-EHS-Guideline-PUBLIC.pdf Accessed 8/17/2023.

WOZEP (2016) *Dutch Offshore wind energy ecological programme (Wozep). Monitoring and research programme 2017–2021*. Published by Rijkswaterstaat, https://a6481a0e-2fbd-460f-b1df-f8ca1504074a.filesusr.com/ugd/78f0c4_52be3e158cfa467bb5e73bc2625f81dc.pdf Accessed 12/9/2023.

Wüstenhagen, R., Wolsink, M. and Bürer, M. J. (2007). Social acceptance of renewable energy innovation: An introduction to the concept. *Energy policy*, **35**(5), 2683–2691.

15

Installation, Operation, and Maintenance of Wind Turbines

This chapter provides an overview of the installation, operation, and maintenance of wind turbines, both on-land and offshore. The turbines may be either individual or in wind power plants. The chapter concerns those aspects of wind energy projects that take place after the development phase, in which land or ocean rights have been secured, permits have been obtained, the sale of the electricity has been contracted for, and arrangements have been made for interconnection with the utility. Those topics were the subject of Chapter 14. Throughout this chapter, when possible, activities that are common to both land-based and offshore turbines are discussed in a single section. Otherwise, separate sections are used for land-based and offshore turbines. This chapter includes the following sections: installation of land-based wind turbines (15.1), installation of offshore wind turbines (15.2), installation of offshore electrical system (15.3), vessels for offshore wind (15.4), operation and maintenance: all turbines (15.5), repair and preventive maintenance (15.6), additional considerations for offshore (15.7), operation in severe climates (15.8), decommissioning and recycling (15.9).

15.1 Installation of Land-Based Wind Turbines

There are a number of steps involved in the installation of wind turbines on land. These include:

- Site preparation
- Foundation installation
- Turbine transportation
- Turbine assembly and erection
- Grid connection
- Commissioning

Before the actual installation can begin, all the major components, such as the tower, nacelle, hub and blades, need to be fabricated. Below is a discussion of each of these topics.

15.1.1 Fabrication

A wind turbine is composed of many components; according to DOE (2023) there are approximately 8,000. Fabrication of the nacelle, hub, and blades (rotor nacelle assembly or RNA) takes place in accordance with the overall design process, which was discussed in Chapter 8. The components within the nacelle, such as the gearbox (when applicable), generator, and associated items, are normally fabricated by specialized industries. The hub and blades may be made by the original equipment manufacturer (OEM) of the turbine, or they also may be fabricated by another company. A supply chain is created which brings together all the nacelle components to the OEM, which then assembles them into a single unit.

Wind Energy Explained: On Land and Offshore, Third Edition.
James F. Manwell, Emmanuel Branlard, Jon G. McGowan, and Bonnie Ram.
© 2024 John Wiley & Sons Ltd. Published 2024 by John Wiley & Sons Ltd.

Most importantly, the generator, gearbox, main bearings, shafts, brakes etc., are all affixed to the main frame and then aligned and tightened into place. As many other components are added as possible, and the nacelle cover is put into place. The nacelle is then prepared for shipment. Depending on the design details, there may be some differences depending on the turbine manufacturer, but the principles are the same.

The components of the support structure are fabricated in accordance with the design, as explained in Chapter 10. Towers for any turbine are produced by a manufacturer specialized in their fabrication; this could be the original turbine OEM, but it is more likely to be a separate supplier.

15.1.2 Site Preparation

Prior to installation of the turbine itself, the site needs to be prepared. Roads may need to be built; the site needs to be cleared for delivery, assembly, and erection of turbines; power lines need to be installed and foundations need to be built. The extent and difficulty of the site preparation will depend on the site location, proximity to power lines, the turbine design and site terrain. Turbine foundation design (see Chapter 10) needs to be site and turbine specific. Expected turbine loading, tower design, and soil properties (sand, bedrock, etc.) will affect the type and size of foundation needed.

15.1.3 Roads in Land-Based Wind Plants

Access roads between wind turbines and maintenance and connecting roads to main highways may represent a significant cost, especially in environmentally sensitive areas with rough terrain. Road design will also depend on the size and weight of the loads to be transported, the terrain, local weather conditions, soil properties, and any environmental restrictions. Obviously, rugged terrain can make all aspects of site preparation difficult and costly. Roads need to be constructed in a manner that disturbs the landscape as little as possible, and that does not result in erosion. Grades and curves should be gentle enough that heavy equipment can reach the turbine sites without difficulty. The lengths of the blades or tower sections are important considerations in this regard.

15.1.4 Foundation Installation

The main steps in the installation of the most common type of wind turbine (gravity base) foundation are the following:

- Excavation
- Installing the form work
- Installing reinforcing rod
- Installing the tower anchor bolts
- Pouring concrete

15.1.4.1 Excavation
Excavation involves digging a hole larger than the foundation itself to allow access and assembly of the form work. The depth of the hole is normally such that when the foundation is completed, its surface will be sufficiently below ground level that when it is back-filled, the soil can be revegetated.

15.1.4.2 Form Work
Form work is used to hold the concrete in place while it is cured. The form work is normally made of wood. It is braced as necessary to maintain its position for the amount of time required for curing.

15.1.4.3 Reinforcing Rod

Any foundation of reinforced concrete will need a substantial amount of reinforcing steel rod to carry the tensile loads. The exact amount and location of the rods will be determined during the foundation design process. The rods must all be installed and fastened in place prior to pouring the concrete.

15.1.4.4 Anchor Bolts

High-strength anchor bolts are used to connect the base of the tower to the foundation. The upper parts of the bolts are threaded to accept nuts of the appropriate type and size. The lower parts are integrated with the reinforcing rods and placed in such a way that the threaded parts will extend above the concrete for the requisite distance. The anchor bolts are typically medium-carbon steel or medium-carbon alloy steel with a carbon content of 0.25–0.55%; see, for example, Anyang (2023).

15.1.4.5 Mixing and Pouring the Concrete

The next step is mixing the concrete and pouring it into the form. This is normally done using multiple conventional cement trucks; concrete is poured continuously until the forms are filled to the required depth. After the concrete is sufficiently cured, the formwork is removed, and the foundation is backfilled with earth. The wait time for removing the forms is at least 24 hours, and normally considerably longer. It takes the concrete around 28 days after it is poured to attain its maximum strength. The tower base can then be installed on the foundation.

15.1.5 Turbine Transportation

The primary components of the wind turbine are the nacelle, blades, and tower sections. Transporting these to the installation site can be a significant undertaking, more so with larger turbines and sites with difficult access, such as those on ridge tops. Smaller turbines can often be packed in containers for easy transport over roads. Larger turbines must be transported in subsections and assembled at the site. In remote locations, difficult access may limit the feasible turbine size or design or may require expensive transportation methods such as helicopters or special-purpose vehicles. It is often the case that the turbine does not originate from a single location: the nacelle and drive train may come from one factory, the blades from another and the tower sections from yet one more.

Size and weight limits for road or rail transportation are a particular concern. In the United States, for example, an oversized load is a vehicle and/or load that is wider than 2.59 m. Each state has different requirements regarding height and length; in most states, the height limit is 4.11 m. The length limit for a typical tractor-trailer combination is 19.8 m. There are also limits regarding weight. For loads that are longer or heavier than normal, particular "wide load" permits are required. There may be limitations on the times that a wide load or extra-long load can be on the road; nighttime travel, for example, may be required. Even those unusually wide or heavy loads allowed by special permits, there still may be ultimate physical limitations due to the clearance under bridges along certain routes. When oversized loads are transported on public roads, permits often require that the load be specially marked and accompanied by escort vehicles. In some cases, overhead electrical lines must be temporarily removed until the load passes. The regulations in other countries are similar.

15.1.5.1 Blades

On-land transport of wind turbine blades can be complicated, due to both the chord and the overall length. The chord may be such that passing under bridges must be avoided. The length of the blades may sometimes be accommodated by using extra-long or extendable trailers with a separate set of steerable wheels at the back. Despite such arrangements, some curves may be too constrained for a long blade to negotiate. Often, alternative routes must be found.

15.1.5.2 Towers

Tower length may not pose a serious transportation issue since towers are typically made in sections of 20–30 m each. The diameter of a tower, however, can be problematic in terms of height as well as width if it is on the order of 4 m or more. It may be possible to use a smaller diameter tower with thicker walls, but as discussed in Chapter 10, that may result in a heavier tower and thus more expensive than it would be otherwise.

15.1.5.3 Nacelle Components

The nacelle, when fully assembled, may be too wide, tall, or heavy for transporting it to a particular site. It may be possible, however, to ship various components individually and then mount them to the main frame near the base of the tower. That is less than ideal for many reasons but may be unavoidable in some cases.

15.1.6 Turbine Assembly and Erection

Once at the site, the turbine must be assembled and erected. Issues related to assembly and erection need to be considered during the design phase to minimize installation costs. Ease of erection depends on the turbine size and weight, the availability of an appropriately sized crane, the turbine design, and site access. Small to medium-sized turbines can often be assembled on-site with a truck-mounted telescopic boom crane. Some smaller turbines can even be assembled on the ground, and the whole tower and turbine are placed on the foundation with such a crane. Where site access is difficult, or cranes are not available, such as in some developing countries, it may be advantageous to use smaller turbines with a tilt-up tower. The complete turbine and tower are assembled on the ground, and hydraulic pistons or winches are used to raise the tower about a hinge. This erection method can make maintenance easy, as the turbine can be lowered to the ground for easy access.

The erection of large wind turbines can present more of a challenge. The tower sections, blades, and the nacelle are all very heavy. For example, the tower, nacelle, and blade weights of a typical 3 MW turbine are 156, 88, and 6.6 metric tonnes, respectively. A lattice boom crane, either wheeled or track-mounted, is often needed.

Once the foundation is ready, the next step is to attach the tower base section. That piece is lowered onto the center of the foundation in such a way that the tower bolts pass through an external flange welded to the bottom. The tower nuts are then screwed onto the bolts and torqued down. Then, the next section is added on top of the base section. In this case, there are internal flanges in both sections. The holes are aligned, bolts are passed through the holes and nuts are added to the ends opposite the heads and torqued in place. The process is repeated until the tower is at the desired height. Typically, three or four sections would be used to form the complete tower. See Figure 15.1.

After the tower is assembled, the nacelle assembly is lifted as a single unit and placed on top. In many cases, the rotor hub is also attached to the main shaft before the nacelle is lifted; in other cases, it is attached later. As with the tower sections, the main frame of the nacelle has a ring that mates with an internal ring on the top of tower. The holes are aligned, and the rings are bolted together.

The next major step is installing the blades and the rotor hub, too, if it was not included with the nacelle. Particularly for small to medium size turbines, the blades may be attached to the hub on the ground. Then, the entire rotor is lifted and attached to the main shaft. If the hub is already in place, the blades may be lifted individually and connected to their respective positions on the hub. The blades typically have bolts embedded in their roots. Those bolts are aligned with holes in flanges on the hub, and the nuts are added. Guide pins may be used to help with the alignment.

In all the above steps, careful work with the crane is required. Jigs and fixtures may be used to facilitate lifting and aligning the various components. Normally, only a single crane is required, but tag lines may be used to adjust the position of the component as necessary. Installation of the hub and blades of a medium size turbine is illustrated in Figure 15.2.

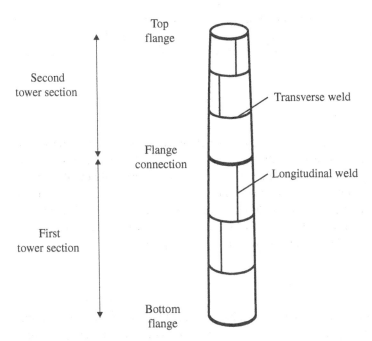

Figure 15.1   Tubular steel tower in two sections and ring flange (*Source:* Based on an example from Öztürk, 2016)

Figure 15.2   Installation of hub and rotor (*Source:* L. Fingersh/U.S. Department of Energy/Public Domain)

Once all the main components of the turbine have been lifted into place and bolted together, any remaining ancillary parts are added or connected. These will include at least the main power cables. Internal access structures, such as ladders and elevators, have already been included to the extent possible, but final details are taken care of at this juncture.

15.1.7 Grid Connection

Once the wind turbine has been fully assembled, its generator contactor (switch) must be readied for connecting to the electric utility grid. The turbine–grid connection consists of electrical conductors, transformers, and switchgear to enable connection and disconnection, as discussed in Chapter 5. All this equipment must be thermally rated to handle the expected current, and the electrical conductors must be sized large enough to minimize voltage drops between the turbine and the point of connection (POC) to the electrical grid (see Figure 15.3 and Chapter 12). Within a wind farm, electrical power is transmitted to the POC via cables from each turbine. These cables are normally buried. Once the turbine is installed and connected to the grid, it is ready for commissioning.

15.1.8 Commissioning

Wind turbines need to be "commissioned" before the turbine owner takes control of the turbine operation. Commissioning consists of (1) appropriate tests to ensure correct turbine operation and (2) maintenance and operation training for the turbine owner or operator. The extent of the commissioning process depends on the technical complexity of the turbine and the degree to which the design has been proven in previous installations. For mature turbine designs, commissioning includes tests of the lubrication, electrical, and braking systems, operator training, confirmation of the power curve, and tests of turbine operation and control in a variety of wind speeds. Commissioning of (one-of-a-kind) research or prototype turbines will involve tests of the various subsystems while the turbine is standing still (lubrication and electrical systems, pitch mechanisms, yaw drives, brakes, etc.) before any tests with the operating turbine are performed.

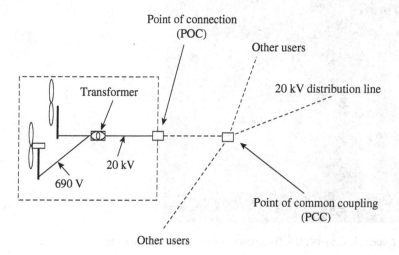

Figure 15.3 Schematic of typical grid connection (with voltages used in Europe)

15.2   Installation of Offshore Wind Turbines

Installation of offshore wind turbines is more complex than with land-based turbines, with great attention given to the supply chain, ports, vessels, reliability, and access. In addition, offshore turbines are nearly always installed as part of a wind power plant with multiple turbines since overall project costs would be prohibitively high for individual offshore turbines. Thus, the location of each turbine is determined in accordance with other turbines in the plant. A particular challenge may have to perform multiple repeated operations in the offshore environment. Installation is significant contributor to the capital cost, and the installation process may drive the selection of substructure technology (EWEA, 2009)

The offshore wind power plant will also include submarine cables from each turbine; these normally terminate at an offshore substation, where the power is consolidated and then transformed to a higher voltage before being transmitted to shore using a smaller number of cables; see Section 15.3.

The overarching factor is that since the turbines are offshore, a maritime infrastructure is required to support their assembly, transportation, installation, and servicing. The process requires adequate harbors, ports, and vessels of a variety of types. The requirements will, in turn, depend on the type and size of the turbines being used and their substructures. A supply chain that extends far beyond the port is also a key part of the process.

Harbors are relatively protected indentations in a coastal area; ports are facilities located within harbors constructed for some commercial purpose. It is at the ports where much of the infrastructure for developing offshore wind plants will be based. For offshore wind plants, important considerations regarding the port are the size of the facilities and water depth and whether the egress is height-limited in any way (for example, by bridges). These can all affect the size of the turbine and the type of substructure that can be deployed from the port. The channel depth of most commercial ports worldwide is less than 17 m, and only a few are more than 20 m. This can impose a limitation on the type of substructure that can be deployed, particularly for floating turbines.

A suitable port is needed as a staging area to pre-assemble many of the individual components into larger units, such as the nacelle assembly, and then undertake the operations to prepare those units for shipment to the installation site. A variety of options may be considered, depending on the port. Individual units could be loaded on to a transport vessel, or in some cases, the rotor/nacelle assembly and tower could be attached to the substructure before transport. For floating offshore wind turbines, as discussed in Chapter 11, certain types of platforms would allow the entire turbine to be assembled and then towed out to the installation site. The ports would then need suitable quays and cranes to allow these operations to be carried out. Lay down yards, staging areas, and transportation hubs (rail, truck, and ship) in close proximity would also be needed. New Bedford, Massachusetts, provides an example of a port where a facility specifically for the offshore wind industry has been constructed. The marine commerce terminal there consists of 8.5 hectares with the ability to support uniform loads of 20 tonnes/m^2 and concentrated loads of 100 tonnes/m^2. This loading capacity allows for large mobile cranes and self-propelled modular transporters (SPMTs) to operate throughout the site. Such cranes could unload large components from ocean-going ships for temporary storage or load them onto barges for transit to the wind plant.

A range of vessels may be used in the transportation and construction phases, all depending on the type and size of turbine, substructure, distance from shore, water depth and what is available. The primary considerations regarding vessels are their carrying capacity, stability, speed, and the specific requirements for which they are intended. See Section 15.4 for more details.

The order of events for the installation is normally the following:

- Substructure installation
- Offshore substation installation
- Turbine (rotor/nacelle assembly and tower) installation
- Cable installation
- Shore-side electrical connection

15.2.1 Preparing for Offshore Installation

The challenges involved in preparing for the installation of an offshore wind plant are considerable. They include organizing or planning for at least the following:

- Overall logistics
- Component supply chain
- Contingency
- Access to necessary port facilities
- Installation vessels
- Substructure fabrication and installation
- Phased delivery of the tower sections, nacelle, hub, and blades
- Offshore electrical system
- Workforce for the various aspects of the project

15.2.1.1 Overall Logistics

Logistics can present significant challenges. The various steps of the entire process must be anticipated and planned for to avoid unnecessary delays and costs. Sufficient space for stockpiling and then assembly of components needs to be identified. There must be flexibility built into the process to accommodate the vagaries of the weather. Some vessels and tooling, such as heavy lift offshore cranes or hammers for driving piles, have long lead times, and in some cases, very few exist anywhere (such as the *Rambiz;* see Section 15.4).

15.2.1.2 Supply Chain Constraints

Even more than with land-based turbines, the supply chain for offshore projects may pose challenges. There may be significant lead times for major equipment and materials. In addition, commodity costs tend to fluctuate with the state of the economy and can result in unexpected changes in the cost of some of the most expensive components (e.g., towers, substructures, etc.)

15.2.1.3 Contingency Planning

When undertaking any large project, especially offshore, contingency planning is essential. In particular, weather can have a major impact on the installation timing. Installation vessels have limited sea states (wind, waves, current) in which they can operate safely. Sometimes, they cannot operate for days or even weeks at a time.

15.2.1.4 Port Facilities

As already mentioned, securing the use of suitable port facilities is essential. These include marshaling yards, quays, cranes, and channels of adequate depth.

15.2.1.5 Installation Vessels

Arranging for installation vessels well in advance is a critical part of the process. Some of the vessels that may be required are discussed in Section 15.4.

15.2.1.6 Substructure Fabrication and Installation

Depending on the capabilities of the port selected, it may or may not be possible to fabricate the substructure there. Usually, it will not be. In that case, arrangements must be made to have the substructures fabricated elsewhere and have them delivered to the project. Most likely, they would be brought first to the port, then to the site of installation. In other cases, they could be brought directly to the site and transferred to an installation vessel there. Since a large wind plant may have varying water depths, it must be ensured that the dimensions of the substructures (for fixed turbines) match the intended locations.

15.2.1.7 Phased Delivery of the Tower Sections, Nacelle, Hub, and Blades
In most cases, the tower sections, nacelle, hub, and blades will be brought to the port before being installed. To the extent possible, delivery should be phased to minimize time in temporary storage and congestion during the loading of the installation vessels. The delivery planning would be adapted to the installation process, for which there are a number of options.

15.2.1.8 Offshore Electrical System
The offshore electrical system consists of an offshore substation, submarine cables from the turbines to that platform, other cables from the substation to the shore and the shoreside cable connection. As with the wind turbine substructures, the offshore substation will likely be constructed at a facility different from the port. Arrangements would be required to have the substation ready at the same time as the turbine substructures so as to take advantage of the same installation vessels. Submarine cables would be laid after the turbine substructures and the offshore substation are installed. By doing this then, it would be easier to connect to the substation's switchgear and transformer as well as the wiring of the turbines. Cable laying includes trenching and backfilling the trenches, so preparations for those steps would need to be made well in advance. Shoreside cable connection is another step with multiple parts that need to be planned will in advance. The greatest challenge is often making the transition from the sea to the land. This typically involves trenching and directional drilling under a beach or inhabited areas. Once the cables are on land, connection to the substation is similar to that of other power plants, but still requires planning. See Section 15.3 for more details on these topics.

15.2.1.9 Workforce for the Various Aspects of the Project
In all aspects of any offshore wind project, a trained workforce will be needed. A range of skills will be involved, including site engineers and specialized tradespeople, such as crane operators, welders, electricians, maritime crew, and wind turbine technicians. All these people must be ready to go well in advance of the start of installation.

15.2.2 Fixed Offshore Turbine Substructure Installation

The type of substructure to be used – jacket, gravity base, suction bucket, and monopile – will determine the majority of the installation procedures for fixed turbine substructures. Below is a description of them.

15.2.2.1 Monopile
As discussed in Chapter 10, monopiles are essentially very large steel cylinders. They are fabricated in a facility that can roll thick sheets of steel into a section of the required diameter and then weld together the resulting joints. Multiple sections are welded together at their ends to create a single structure of the desired length. The process of rolling and welding a cylinder of this type is illustrated in Figure 15.4.

Monopiles may be transported on a barge, or they may be sealed at both ends and towed to the site without the use of a separate vessel. At the installation site, the pile is raised to the vertical position, lowered to the seabed and then hammered until it reaches its required depth in the seabed. This depth will depend on the pile length, which in turn will have been determined in accordance with the soil conditions, water depth, and size of the wind turbine. As discussed in Chapter 10, criteria include preventing excessive deflection of the pile and ensuring the proper natural frequency of the entire structure.

Generally speaking, very little soil preparation is needed before the monopile is driven into place. At some sites, however, hammering may not be possible because the soil is too hard. Drilling is one option to still allow a pile to be installed.

Figure 15.4 Rolling and welding of tubular steel member (*Source:* J. F. Manwell (Book author))

One important consideration in the installation of monopiles is the noise that the hammering can produce, which can be harmful to various aquatic species in the vicinity. There are measures that can be taken to reduce the noise, such as surrounding the site with a bubble curtain during the hammering; see Chapter 14.

The pile may not be completely vertical after the hammering, and the top of the pile may have suffered some distortion. To remedy those possible effects a transition piece (see Chapter 10) is lowered onto the monopile, leveled and then grouted into place. An anode to slow corrosion is often installed on the monopile at this time; see Figure 15.5 for an illustration. The tower will then be bolted to the transition piece in a manner analogous to that of land-based turbines.

15.2.2.2 Gravity Base Structure (GBS)

A gravity base structure (GBS) is typically fabricated out of reinforced concrete in a dry dock; see Figure 15.6. The process is similar to that of a conventional gravity foundation on land, except that there is no excavation. Depending on the design, the GBS is either transported to the site on a barge or floated out to the site.

The seabed needs to be prepared beforehand. Boulders and other large objects on the seabed must be removed and the soil leveled. Once on site, the GBS is lowered into position (see Figure Figure 15.7) and then ballasted, normally with rock. The tower is subsequently bolted directly to the top of the GBS, which would be above the surface of the water, also analogously to that of a land-based foundation. In some cases (deeper water), an extension is attached to or integrated with the rest of the concrete structure. This would be done before the GBS is installed.

15.2.2.3 Jacket

A jacket substructure would be built in a special fabrication facility, which could not only roll and weld the steel tubes but could also cut and weld the angled joints on the cross members. The jacket would be transported to the

Figure 15.5 Cleaning monopile and installing anode (*Source:* Reproduced by permission of Osbit Limited)

Figure 15.6 Fabrication of gravity base structure (GBS) (*Source:* Reproduced by permission of Per Vølund)

Figure 15.7 Installation of GBS (*Source:* Reproduced by permission of Per Vølund)

site on a barge and lowered to the site with a jack-up or floating crane, depending on the water depth. Some site preparation is necessary beforehand in order that the legs all be aligned on the same plane. The pile would be used to secure the jacket to the seabed. These may be hammered into place before lowering the jacket. The process would use a template to ensure proper spacing of the piles. Alternatively, the piles could be driven through the legs after lowering. Hammering of the piles will result in some noise, but considerably less than with monopiles, since these piles will have much smaller diameters. Figure 15.8 shows a schematic of one version of the process; see also Jiang (2021). Figures 15.9 and 15.10 illustrate an actual installation in more detail.

15.2.2.4 Suction Bucket

As discussed in Chapter 10, the suction bucket is essentially an inverted can that is lowered to the sea floor and then evacuated so that water pressure will force the walls into the seabed until the lower side of the top (lid) rests on the soil. One of the advantages of the suction bucket is the reduced noise associated with the installation in comparison to the monopile or even the jacket. Another is that the bucket can be removed at the end of the project life.

15.2.3 Installation of Floating Wind Turbines

The installation of floating wind turbines is significantly different than that of fixed turbines, but the specific method would depend on the type of floater and the parameters of the port and harbor. In many situations, the entire turbine, including the rotor/nacelle assembly and tower, could be floated out as well. The following section discusses the three primary options: semisubmersible, tension leg platform, and spar.

15.2.3.1 Spar

If the water depth at the port and the entire route to the installation site is deep enough, then the entire turbine could be towed vertically. That is not normally possible, however. In most situations, an arrangement would be needed whereby the turbine could be towed in the nearly horizontal position for some distance and then raised to vertical once the water becomes deep enough. One option is shown in Figure 15.11. This figure shows the process

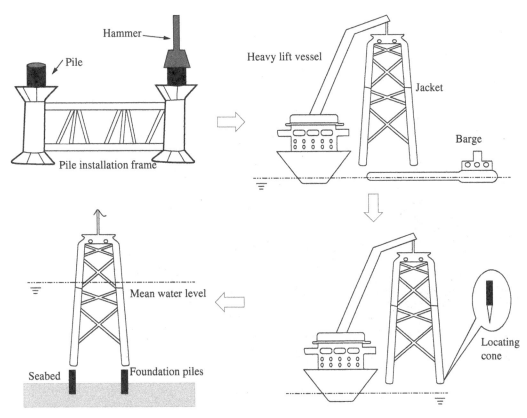

Figure 15.8 Installation of jacket (*Source:* Reproduced by permission of Zhiyu Jiang, University of Agder)

Figure 15.9 Transport of jackets prior to installation (*Source:* Copenhagen Infrastructure Partners)

Figure 15.10 Installation of jacket (*Source:* Copenhagen Infrastructure Partners)

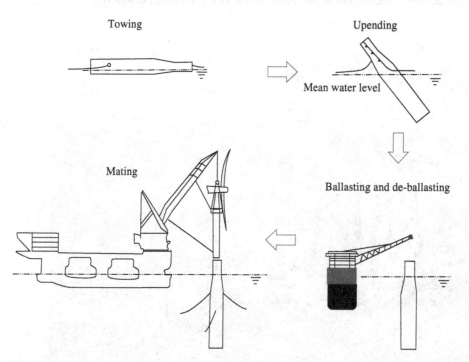

Figure 15.11 Stages in the installation of the Hywind Scotland spar-type platforms (*Source:* Reproduced by permission of Zhiyu Jiang, University of Agder)

up to and including the placement of the turbine on a spar. Photographs of the completed turbine being towed to the installation site may be seen at Offshorewindbiz.com (2017). This method has the drawback that a heavy lift floating crane was used, and that was very expensive. Another option is a special-purpose vessel. It must be presumed, however, that a special-purpose vessel would need to be used for many installations to make it worth the cost.

15.2.3.2 Semisubmersible

An offshore wind turbine on a semi-submersible platform has a number of advantages in terms of installation. In particular, it could be assembled entirely in port, as illustrated in Figure 15.12. In addition, it could be towed directly from the port to the site of installation; see Figure 15.13. Note, however, that the platform might have to be unballasted to transit some harbor channels because the draft would otherwise be too great. The wide spacing between the hulls should ensure that the metacentric height would still be positive, i.e., stable, during transport since the turbine would not be operating.

15.2.3.3 Tension Leg Platform (TLP)

Transportation and installation of tension leg platforms with the turbine already installed would be challenging because, depending on the specifics of the design, the structure is not intrinsically stable. In fact, the center of mass could be well above the center of buoyancy. This would even be true for a deep draft, normally ballasted hull, since it would be necessary to necessary to omit most or all the ballast for the draft to be acceptable for leaving port. Therefore, a special-purpose vessel or temporary pontoons would be needed to support the TLP while under tow and before the tendons are attached. One option is shown in Figure 15.14. If the turbine and RNA were not already in place, however, it may be possible to float out the substructure alone, without a special vessel, install it and then add the tower and RNA afterward.

Figure 15.12 Installation of blades on semisubmersible in port (*Source:* Photo of the WindFloat Atlantic project, courtesy of Principle Power/Ocean Winds)

Figure 15.13 Semisubmersible under tow (*Source:* Photo of the WindFloat Atlantic project, courtesy of Principle Power/ Ocean Winds)

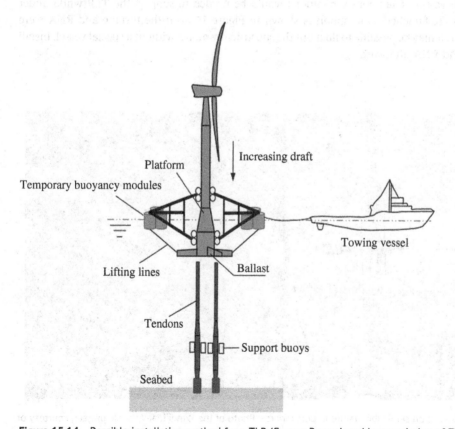

Figure 15.14 Possible installation method for a TLP (*Source:* Reproduced by permission of Zhiyu Jiang, University of Agder)

15.2.3.4 Anchors and Mooring Lines

For most floating wind turbines, the anchors would likely be installed before bringing the floating platform into position, whether or not the turbine and tower were already in place. A special vessel may be used for this; see Section 15.4.7.4. The mooring lines or tendons would also be connected to the anchors. The other ends of the lines could be held near the water surface with pontoons. They would then be ready to be attached to the connection points on the platform with a minimum of delay. For the TLP, the platform would be ballasted with extra water so it would sink below its normal operating draft, and then the tendons would be attached. The extra water would be pumped out to increase the buoyancy, thereby putting the tendons under tension.

15.2.4 Installation of Offshore Turbine Tower and Rotor/Nacelle Assembly

Regardless of whether an offshore wind turbine is to be fixed or floating, it is preferable to do as much of the assembly on shore as possible. How much of that is possible is related to the facilities of the port, the carrying capacity of the vessels and the lifting capabilities of jack-up or floating cranes. The most common method is piecewise installation offshore, in which as many of the components as possible are preassembled onshore into subunits and then all connected at the offshore site. The other option is to preassemble the entire tower and rotor nacelle onshore and transport it as a single unit to the offshore site.

15.2.4.1 Piecewise Installation Offshore

In this approach, the offshore turbine above the substructure is divided into subunits, or pieces, most often consisting of tower sections, the hub and nacelle, and the blades. Once the substructure is ready, the tower is assembled offshore, section by section, with the base section attached first to the substructure, normally via a transition piece. The hub and nacelle are lifted to the top of the tower, then lowered and bolted into place. Then, the blades are attached one by one. Figure 15.15 illustrates the blade assembly stage of this approach for a fixed turbine.

Figure 15.15 Installation of blades offshore (*Source:* Copenhagen Infrastructure Partners)

Figure 15.16 Installation of entire turbine on jacket (*Source:* Reproduced by permission of Scaldis Salvage and Marine Contractors)

Variants of the piecewise installation method include assembling the entire tower on shore and preassembling some or all of the rotors. Preassembly of the entire rotor is problematic for very large rotors due to transportation and maneuvering difficulties, but the so-called "bunny-ears" approach has been used in some situations. In this method, two blades are attached to the hub and the assembly is loaded for transport vertically, hence the name. As turbines get larger, this method, too, becomes difficult.

15.2.4.2 Entire Turbine Assembled Onshore

In this approach, the entire turbine, including the tower and RNA, but without a substructure, is pre-assembled onshore, transported as a unit and lowered on to the support structure. This method avoids much of the assembly work offshore and also facilitates commissioning of the turbine. The downside is that the entire turbine will be so heavy as to require a large crane vessel to transport and position it. Figure 15.16 illustrates the near-final step of the first such large turbine to be installed that way. Coincidentally, the turbine in the figure is the REpower 5M. This is one after which the NREL 5 MW reference turbine, which is discussed throughout this text, was patterned. Note that the installation vessel is the *Rambiz*, one of the largest of its type in the world.

15.3   Installation of Offshore Electrical Systems

The offshore electrical system was introduced in Chapter 5 and summarized in Section 15.2. It consists of transformers at each turbine, cables within the plant, an offshore substation/transformer platform, one or more cables to shore and a shore side utility connection. A variety of special accessories may be used as well. This section focuses on the installation of the electrical system.

15.3.1 Installation of Offshore Substation Transformer Platform

The offshore substation, first of all, requires a support structure. This is typically either a monopile or jacket. These are installed in a manner similar to that of the corresponding substructures for offshore wind turbines themselves. The jacket is often preferable because the footprint of the offshore substation is relatively large, especially in comparison to a turbine's tower base.

Once the support structure is installed, the electrical components, including the transformer, must be lowered on to it. Ideally, these components have all be preassembled into a single unit. Figure 15.17 shows the installation of a typical substation.

15.3.2 Submarine Cable Installation

Installation of submarine cables typically involves burying the cables in the seabed. This is done to minimize the chance of damage either during installation or later during operation. The cables are usually buried in trenches at least one meter deep.

The process involves bringing the cables to the site on a special-purpose cable-laying ship; see Figure 15.18 and BOEM (2022). The ship pays off the cable along the selected route, feeding it to a jet plow which travels over the seabed. An example of such a plow is shown in Figure 15.19. The plow uses high-pressure water to create a

Figure 15.17   Installation of offshore substation (*Source:* BOEM (2019))

Figure 15.18 Typical cable laying ship (*Source:* Reproduced courtesy of LS Cable Systems America and Kokosing Industrial)

Figure 15.19 Jet plow for laying cable (*Source:* Courtesy of Osbit Limited)

trench in the soil, laying the cable into the trench as it goes. Considerations in the preparation for cable laying include:

- Identification of potential hazards to cables
- Site investigation to identify seabed properties
- Scour protection
- Cable route selection
- Cable transport
- Vessel and equipment selection

15.3.3   Shore-Side Connection of the Main Transmission Cable

The shore side connection of the main transmission cable(s) also requires considerable planning and, quite possibly, specialized equipment. The concern is often the transition from below the seabed to the land since the cable typically must pass underneath an environmentally sensitive area, which, in addition, may be subject to erosion by the sea. Most often, directional drilling is used and is done from inland to the seabed well beyond the shore.

A detailed description of the cable laying process for the Block Island (RI) offshore wind farm is given by Elliott *et al.* (2017)

15.4   Vessels for Offshore Wind

A major difference between offshore and onshore wind is the need for a variety of vessels, some of which may be of large size and special purpose design. This is particularly true for installation, but it will also be true for other phases. This section provides an overview of vessel characteristics. In the following, we use the common terminology that ships are large, self-propelled vessels used for carrying cargo (or people) over long distances, such as across oceans. Boats are smaller, self-propelled, and used for a variety of purposes, such as towing, servicing, etc. Barges are flat-bottomed vessels used for transport or installation; typically, they are not self-propelled, although sometimes dynamic positioning is used to keep them in place. There is much in common with the required properties of vessels for offshore wind installation and service with the floating substructures, which were discussed in Chapter 11, so reference will be made to that chapter where appropriate.

15.4.1   Important Terms

Some important terms applied to ships include the following. These are in addition to or elaborated on definitions of vessels given in Chapter 11.

Beam. The beam of a vessel is width at its widest point.
Draft. This is the maximum distance below the water line and the bottom of the hull or keel. It will vary depending on whether the vessel is loaded or not.
Length. The length of a vessel refers to the fore-aft direction; due to the shape of the bow and stern, the actual values used may depend on the position with respect to the water line, although the overall length refers to the longest dimension. Typically, there is a distinction made between the overall length and the water line length. See Tupper (2013) or similar text for more details.
Water line. The highest location where the still water meets the side of the vessel.

15.4.2 Carrying Capacity

One of the primary considerations in the selection or design of a vessel of any sort is its carrying capacity. The term "deadweight tonnage" (DWT) is used to characterize that property. The carrying capacity is distinguished from the displacement, which is a measure of the total weight of the vessel. The displacement weight when the ship is loaded is greater than when it is empty "light." The difference is the carrying capacity. These terms are, of course, related to the buoyant force resulting from the displaced volume. As will be recalled from Chapter 11, the buoyant force F_B (N) is given by Equation (15.1).

$$F_B = g\rho_w \nabla \tag{15.1}$$

where ∇ is the displaced volume (m^3), g is the gravitational constant (m/s^2) and ρ_w is the water density (kg/m^3)

The total weight of the vessel, including the cargo, must equal the buoyant force. The vessel will sink lower in the water when it is loaded due to the additional weight. For example, consider a barge that is $15 \times 15 \times 5$ m deep with a mass of 100 metric tonnes in seawater of density 1025 kg/m^3. When the barge is empty, it will have a draft of 0.43 m. Suppose it was determined that the barge could carry 700 tonnes. If that amount were loaded onto the barge, its draft would increase to 3.5 m. In this situation, the loaded barge would have a displacement weight of 800 tonnes, while the carrying capacity (DWT) would be 700 tonnes.

15.4.3 Stability

As explained in Chapter 11, a vessel is definitively stable is its center of gravity is below the center of buoyancy. It may also be stable, even if that is not the case, depending on its metacentric height. The floating platforms discussed in Chapter 11 were relatively simple in shape so that the metacentric height could be found relatively easily. Some vessels have a complicated hull shape, so that determination may be a bit more difficult, but still straightforward. Ships are also nonsymmetric in so far as they are intended to go in one direction. They are more likely to heel over sideways if overloaded than in the fore-aft directions. In the case of floating cranes, an important consideration is stability when the crane is placing a load that it has lifted.

15.4.4 Hulls

The general characteristics of hulls for floating offshore wind turbines were discussed in Chapter 11. Those were all symmetrical. In the case of nonstationary vessels, such as ships and boats, the shape is more complex in order to reduce drag. An important consideration in the design of a vessel whose length is relatively long compared to its beam is strength in the waves. One possible effect due to the hull just being supported, even if momentarily, by a wave underneath the middle is "hogging." A hull pushed up by waves under both ends but less so under the middle is subject to "sagging."

The hulls of larger vessels, including barges, are typically manufactured from steel. Those smaller ones may be constructed from fiberglass.

15.4.5 Vessel Motion

As discussed in Chapter 11, vessels may experience a variety of motions. In this case, the orientation of the axes is with respect to the long axis of the vessel rather than the wind direction. For example, roll is rotation about longitudinal axis of the vessel; pitch is rotation about the transverse axis. Otherwise, the discussion in Chapter 11 applies here as well.

15.4.6 Forward Motion: Resistance and Propulsion

When discussing floating offshore wind turbines, resistance and propulsion were of no real concern. For ocean-going ships and installation and service vessels, however, they are quite significant, whether the vessel is self-propelled or towed.

Propulsion is most commonly accomplished by means of engine-driven propellers. The primary considerations in estimating the amount of power required are the resistance of the vessel, which is, in turn, a function of its displacement, its shape, and its speed. In general, slower vessels have a single hull with a relatively simple shape. For greater speed with less power consumption, a more streamlined shape at the bow may be used. For even higher speeds, especially for crew transport, two hulls (catamarans), three hulls (trimarans), or even hydrofoils may be used. The power required to move the hull through the water P_{hull} is given by Equation (15.2):

$$P_{hull} = V_S R_{total} \tag{15.2}$$

where:

V_S = the vessel's velocity, m/s
R_{total} = total resistance to forward motion, N

15.4.6.1 Resistance to Forward Motion

The resistance of a vessel to forward motion has three major components: (1) viscous drag of a smooth hull, (2) waves produced by the vessel itself as it moves through the water, (3) additional resistance due to hull roughness and drag due to motion of the upper part of the vessel in the air.

These components are summarized as shown in Equation (15.3). (Note: depending on the vessel, there may be other contributors as well, but they are, in any case, less significant.)

$$R_{total} = R_f(1 + k_1) + R_w + R_A \tag{15.3}$$

where

R_f = viscous (drag) resistance of the hull, N
R_w = resistance due to wave drag
R_A = resistance due to other sources
k_1 is a form factor related to the shape of the vessel, typically in the range of 0.1–0.2.

Viscous Resistance

The most significant source of resistance at low to moderate vessel speeds is due to the viscous drag of the hull. As in other situations where drag is relevant, the force varies with the square of the velocity and the area affected. In the case of a monohull vessel, the following equation is commonly used:

$$R_f = C_F \frac{1}{2} \rho_w S_S V_S^2 \tag{15.4}$$

where

S_S is the wetted area of the hull, m^2
ρ_w is the density of the water, kg/m^3

C_F is the coefficient of friction, commonly given by $C_F = \dfrac{0.075}{(\log_{10} \mathrm{Re}_{L_s} - 2)^2}$

Here Re_{L_S} is the Reynolds number based on the vessel's length L_S:

$$\mathrm{Re}_{L_S} = \frac{V_S L_S \mu_w}{\rho_{sw}}$$

where μ_w = the viscosity of water (Pa s); a typical value at 20 °C is 1.09×10^{-3}Pa s

The wetted area is the actual area in contact with the water. Due to the complex shape of some hulls, this value can be difficult to determine accurately, but there are approximations that can be used. One of them is given in Equation (15.5).

$$S_S \cong 1.025 L_{PP}(C_B B + 1.7T) \cong 1.025\left(\frac{\nabla}{T} + 1.7 L_{PP} T\right) \tag{15.5}$$

where

L_{pp} = Length between perpendiculars, m (this is somewhat less than the overall length of the vessel)
B = The waterline breadth (beam) of the hull, m
T = The draft amidships, m
∇ = The displacement (displaced volume of water) of the hull, m^3
C_B = The block coefficient of a ship is the ratio of the underwater volume of the ship to the volume of a rectangular block having the same overall length, breadth and depth (typical value ~0.6).

The actual friction resistance is likely to be somewhat higher than R_f due to the shape of the hull. This is accounted for by the form factor k_1.

Wave Resistance

A major factor limiting the speed of a vessel is the resistance due to the waves which are produced by its motion. These waves can interfere with each other and increase the drag in the process. In fact, the hull speed is the speed of a vessel at which the bow and stern waves begin to interfere and result in a significant increase in drag. For example, the hull speed of a conventional monohull vessel, V_{hull}, is approximately proportional to the square root of the water line length, L_{WL}, as shown in Equation (15.6). Vessels are not actually limited by the hull speed, but they do require progressively more power. (Note that catamarans and trimarans are inherently faster than monohull vessels, and so are often used for crew transport)

$$V_{hull} = C\sqrt{L_{WL}} \tag{15.6}$$

where C is a constant in the range of 1.25–1.41 m/s m$^{1/2}$.

A method to estimate wave resistance has been developed by Holtrop and Mennen (1982) in a classic paper on the subject; see Equation (15.7).

$$R_W = c_1 c_2 c_5 \nabla \rho g \exp\left[m_1 F_n^{-0.9} + m_2 \cos\left(\lambda F_n^{-2}\right)\right] \tag{15.7}$$

where

F_n is the dimensionless Froude number given by $F_n = \dfrac{V_S}{\sqrt{gL_S}}$

$c_1, c_2\, c_5$ are constants in the vicinity of 1
m_1, m_2, λ are constants which depend on the shape of the hull.

Holtrop and Mennen (1982) also provide a detailed example of estimating the various resistances for a ship 205 m long with a displacement of 37 500 m^3 at a speed of 25 knots (12.9 m/s). Power may be found from those resistances may be found by multiplying the resistances by the vessel's speed. Figure 15.20 illustrates an extension of that example for speeds ranging from 2 to 16 m/s. The example shows the power required resulting from viscous, wave

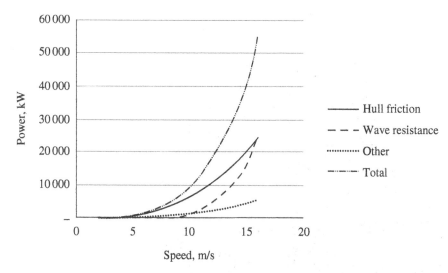

Figure 15.20 Sample ship power vs. speed

and total resistance. The resistance from other sources (drag due to air, extra hull roughness, etc.) amounts to less than 15% of the total. (In this example, $k_1 = 0.157$, $c_1 = 1.4012$, $c_2 = 0.7595$, $c_5 = 0.9592$, $m_1 = -2.1248$, $m_2 = -0.2379$, $\lambda = 0.6551$.) The hull speed for this ship would be approximately 18–20 m/s, depending on the value of C Equation (15.6). As can be seen from the figure, the wave resistance will have already increased drastically, exceeding hull friction, before the hull speed is reached. Results for other vessels would be different according to their size and shape, but the general trends would be similar.

15.4.6.2 Propulsion

The prime movers that drive ship propellors these days are either diesel engines or gas turbines. When it comes to fuel consumption, it should be noted that propellers and the rest of the drive train have combined efficiencies in the range of 70%, and diesel engine efficiency is on the order of 50%. Further discussion is beyond the scope of this text. More details may be found in Tupper (2013) or similar references.

15.4.7 Commonly Used Vessels for Offshore Wind

A wide variety of vessels may be used in offshore wind energy situations. Some of these are discussed briefly. Photographs of such vessels can be readily found on-line. A few of these are also included.

15.4.7.1 Barges

Barges are flat-bottom vessels, usually not self-propelled but towed by tugboats. A number of the illustrations in this chapter include components carried on barges. Special types of barges include jack-up barges, crane barges, and launch barges, as described next.

Jack-up barges. Jack-up barges have three or four legs which are raised during transport. The legs are pressed down on the seabed to raise the barge above the water and provide stability during use. Examples of jack-up barges may be seen in Figures 15.15, 15.21, and 15.23.

Crane barges. Crane barges are barges with cranes mounted on them. They are often used for installing substructures or other large components such as the tower or RNA. Crane barges may either be floating or jack-up. If floating, they will employ thrusters and dynamic positioning systems to keep them in place during use. An

Figure 15.21 Jack-up leg backhoe dredger (*Source:* Ana 2016/Wikimedia Commons/CC BY-SA 4.0.)

example of such a vessel is the *Rambiz*, which can lift to 3300 tonnes 78 m above the surface; see Scaldis-smc.com (2023). That vessel was also shown lifting a turbine with a tower whose combined mass was approximately 600 tonnes in Figure 15.16. Another example can be seen in Figure 15.17.

Jack-up crane barges are often used for installing offshore wind turbines in relatively shallow water.

Launch barge. A launch barge may be used to carry an offshore turbine's substructure to the installation site and then lower it into the water. It may include a ramp at one end to facilitate the launching of the substructure. Winches and cables are also used.

15.4.7.2 Dredges
A dredge may be used in site preparation, particularly for gravity base structures. See Figure 15.21 for an example.

15.4.7.3 Supply Boats
Supply boats carry a wide variety of supplies to wherever they are needed. These boats are relatively small, and they may be monohulls or catamarans.

15.4.7.4 Anchor Handling Boats
Anchor handling boats carry anchors to the point of use and facilitate the placement of the anchors. See Figure 15.22.

15.4.7.5 Tugboats
Tugboats are used to move barges, floating offshore wind turbines, or other vessels that are not self-propelled. An example can be seen in Figure 15.13.

15.4.7.6 Drilling Vessels
Drilling vessels may be used to take soil samples in preparation for pile driving or as part of the site assessment process. See Figure 15.23 for an example.

Figure 15.22 Anchor handling boat (*Source:* BoH/Wikimedia Commons/CC BY-SA 4.0.)

Figure 15.23 Jack up vessel taking soil samples (*Source:* University of Massachusetts Amherst)

Figure 15.24 Survey vessel (*Source:* University of Massachusetts Amherst Wind Energy Center)

15.4.7.7 Survey Vessels
Survey vessels may take a variety of forms, depending on the situation. They are used primarily during the site assessment process; see Figure 15.24.

15.4.7.8 Maintenance Crew Boats
Maintenance crew boats carry crew to the site where they will work. The primary concerns are speed, safety and relative comfort of the crew. Catamarans such as the one depicted in Figure 15.25 are commonly used for this purpose.

15.5 Operation: All Turbines

It is one thing to install wind turbines. It is another thing to operate and maintain them so that they continue to operate satisfactorily over their design lifetimes. This section summarizes some of the issues in this regard.

15.5.1 General Operational Issues

The operation of wind turbines in the sense of turning them on or off is taken care of automatically by the control system. There is more to the operation than that, however. Human operators are also required to ensure not only that the turbines are operating when the wind conditions are right, but also that the energy production and revenue are optimized, that the plant availability remains high and that the conditions of the permits and the contracts are met. Accordingly, the operation of a turbine or wind plant requires (1) information systems to monitor turbine performance, (2) an understanding of factors that reduce turbine performance, (3) measures to maximize turbine

Figure 15.25 Catamaran for maintenance crew (*Source:* J. F. Manwell (Book author))

productivity, and (4) consideration of environmental issues (such as bird or bat migration, shadow flicker, etc.; see Chapter 14).

As noted in Chapter 13, the availability of wind turbines with mature designs is typically on the order of 95%, although it may fall below that under adverse conditions. Reduced availability is caused by scheduled and unscheduled maintenance and repair periods, power system outages, and control system faults. For example, the inability of control systems to properly follow rapid changes in wind conditions, imbalance due to blade icing, or momentary high component temperatures can cause the controller to stop the turbine. The controller usually clears these fault conditions, and the operation is resumed. Repeated tripping usually causes the controller to take the turbine offline until a technician can determine the cause of anomalous sensor readings. This results in decreased turbine availability.

As discussed elsewhere in this text, power curves are used to represent turbine power output as a function of wind speed. Various factors may reduce the energy capture of a turbine or wind farm from that expected, based on the published power curve and the wind resource at the site. These include reduced availability, poor aerodynamic performance due to soiled blades and blade ice, lower power due to yaw error, control actions in response to wind conditions, and interactions between turbines in wind farms (see Chapter 12). Soiled blades have been observed to degrade aerodynamic performance by as much as 10–15%. Airfoils that are sensitive to dirt accumulation require either frequent cleaning or replacement with airfoils whose performance is less susceptible to degradation by the accumulation of dirt and insects. Blade ice accumulation can, similarly, degrade aerodynamic performance. Energy capture is also reduced when the wind direction changes. Controllers on some turbine designs might wait until the magnitude of the average yaw error is above a predetermined value before adjusting the turbine orientation, resulting in periods of operation at high yaw errors. This results in lower energy capture. Turbulent winds can also cause the generator to trip offline. For example, in turbulent winds, sudden high yaw errors might cause the system to shut down and restart, also reducing energy capture. In high winds, gusts can cause the turbine to shut down for protection when the mean wind speed is still well within the turbine's operating range. These problems may reduce energy capture significantly below projected values. Operators should not only be prepared to minimize these problems but should also anticipate them in their financing and planning evaluations.

15.5.2 Safety Issues During Operation

The installed wind turbine needs to provide a safe work environment for operating and maintenance personnel. The turbine also needs to be designed and operated in a manner that is not a hazard to neighbors. Safety issues include such things as protection against contact with high voltage electricity, protection against lightning damage to personnel or the turbine, protection from the effects of ice build-up on the turbine or the shedding of ice (see Section 15.5.5), the provision of safe tower-climbing equipment, and lights to warn local night-time air traffic of the existence of the wind turbine. Maintenance and repairs may be performed by on-site personnel or turbine maintenance contractors.

15.5.3 Forecasting of Wind Power Production for Operation

Electrical generation from wind turbines is dependent on the wind speed at any given time. Nonetheless, from the point of view of economic operation, it is often useful to be able to predict the possible generation in the near term. Wind energy forecasts are increasingly being integrated into the control systems of grids with large amounts of installed wind power. These predictions are important on several time scales. Short-term forecasting of loads can give regulation control equipment a head-start on covering increases or decreases in load. Day-ahead or two-day-ahead forecasts of grid conditions allow system operators to plan unit commitment (i.e., which generators, conventional or otherwise, can be planned on for supplying the load).

Utility system operators routinely forecast the load on their system based on detailed statistics of past power demand. Successful prediction of load includes knowledge of seasonal variations such as heating and lighting demands and how these change with weather conditions, daily variations in residential and industrial power use, and small but statistically significant short-term variations such as television use during popular sporting events.

When wind turbines are a significant factor in a utility system, it is desirable to be able to forecast wind power as well. Wind power forecasts for 6–36 hours in the future generally use mesoscale numerical weather prediction (NWP) models, such as those described in Chapter 2. The method uses the NWP model to predict wind speeds and then applies that information on the installed wind turbines in the region together with historical and recent wind power generation to predict overall wind power generation over the balancing area. These predictions are complicated by a number of factors:

- NWP models may not have the spatial and time resolution to accurately predict the timing or magnitude of wind speed changes of fast-moving weather systems or systems with large changes in wind conditions over short distances.
- NWP models may have difficulty predicting wind speed and direction variations over complex terrain and at specific geographic locations.
- NWP model results become more uncertain as the time horizon increases, and different NWP models may provide different predictions.
- Actual generation depends on wake losses, which may be sensitive to wind direction.
- Low wind turbine availability, power curtailments due to transmission constraints, and energy losses due to icing, high-speed cut-out hysteresis, direction sector management, or other control actions may limit power.
- Power generation changes with air density.

- Power levels are very sensitive to wind speeds below rated.
- It may be difficult to obtain operational information about all the wind turbines or wind farms in a region, and predictions may be based on a sample of the total population of wind turbines.
- Future turbine availability may be unknown.

For these reasons, forecasting systems often employ methods for incorporating current operating data, historical operating data, relationships between historical NWP predictions and measured wind speeds and directions or measured generation at each wind farm and density corrections to power curves, or relationships between wind speed and power.

Of most value to grid operators are predictions of the exact magnitude and timing of sudden changes in generation levels. These are difficult to provide, given the spatial and temporal resolution of NWP models. A technique known as ensemble forecasting is sometimes used to provide better estimates of generation and the uncertainty of generation levels or the timing of fronts moving through the region. Ensemble forecasting consists of integrating two or more forecast methods to mitigate the shortcomings of individual methods. Especially when the different forecast methods are quite different, and their errors are uncorrelated, this method can yield significantly better results than the individual methods that are its constituents. When combining the predictions, the individual predictions can also be weighted based on their performance over time. These weights might change depending on the season, the characteristics of the input data, or the time horizon for the prediction.

15.5.4 Control, Monitoring, and Data Collection Systems

Modern wind plants include systems for controlling individual turbines and displaying and reporting information on wind farm operations. These systems, introduced in Chapter 9, are called supervisory control and data acquisition (SCADA) systems. SCADA systems display operating information on computer screens. Information about the whole wind plant, sets of turbines, or one individual turbine can be displayed. The information typically includes turbine operating states, power level, total energy production, wind speed and direction, and maintenance and repair notes. SCADA systems also display power curves or graphs of other information and allow system operators to shut down and reset turbines. Newer SCADA systems connected to modern turbines may also display oil temperatures, rotor speed, pitch angle, etc. SCADA systems also provide reports on turbine and wind farm operation to system operators, including information on operation and revenue from each turbine based on turbine energy production and utility rate schedules.

15.5.5 Ice Throw/Falling Objects

Safety problems can occur when low temperatures and precipitation cause a build-up of ice on turbine blades. As the blades warm, the ice melts and either falls to the ground or can be thrown from the rotating blade. Falling ice from nacelles or towers can also be dangerous to people directly under the wind turbine. This problem was reviewed by NWCC (2005), and a detailed technical discussion was given by Bossanyi and Morgan (1996).

Another safety risk is that a blade or blade fragment or other object would be thrown from a rotating machine. In actuality, these events are rare and usually occur in extreme wind conditions when other structures are also susceptible. The distance a blade, or turbine part, may be thrown depends on many variables (e.g., turbine size, height, size of broken part, wind conditions, topography, etc.).

In any case, the likely extent of distance that an object could travel can be estimated in accordance with basic physics, as summarized next. The equations of motion of an object capture the essential facts. Equations (15.8) and (15.9) describe the motion of the object of mass m in the vertical (z) and horizontal (x) directions, respectively. (Note that x here is the direction of travel when $\Psi = 0$.)

$$m\ddot{z} = -mg - \text{sgn}(V_z)C_D\frac{1}{2}\rho A V_z^2 \tag{15.8}$$

$$m\ddot{x} = -C_D \frac{1}{2}\rho A V_x^2 \tag{15.9}$$

where V_x and V_z are the velocities of the object in the x and z directions, ρ is the density of the air, g is the gravitational constant, A is the projected area of the object, C_D is the drag coefficient, and sgn() is the signum function, which is used to ensure that the drag force is always opposite to the velocity.

The drag coefficient depends on the Reynolds number and the shape of the object. For a smooth sphere under these conditions, $C_D \cong 0.5$, while for an irregular object, C_D can be as much as 1.5 (see McCleskey, 2008).

The coordinates of the object when it is released from the moving blade at time $t = 0$, distance r from the axis of rotation and blade azimuth angle ψ are:

$$z_0 = H + r\cos\psi; \ x_0 = r\sin\psi \tag{15.10}$$

The velocity components at that time are:

$$V_{z,0} = -\Omega r\cos\psi; \ V_{x,0} = \Omega r\sin\psi \tag{15.11}$$

In the absence of drag and assuming a flat surface, Equations (15.8) and (15.9) can be solved analytically to yield Equation (15.12), which gives the maximum distance from the tower, x_{max}, as a function of the initial velocity and the position of release. When drag is considered, Equations (15.8) and (15.9) must be solved numerically.

$$x_{max} = x_0 + V_{x,0}\left[\frac{V_{z,0} + \sqrt{V_{z,0}^2 + 2z_0 g}}{g}\right] \tag{15.12}$$

Example: A piece of ice the size of a baseball is on the tip of the blade of a wind turbine with hub height of 100 m, radius of 50 m, and rotational velocity of 10 rpm. The ice is assumed to be spherical with radius of 0.365 m, a density of 920 kg/m^3, and a drag coefficient of 1.0 or 0.5. It is released while rising at an angle of 345°. Figure 15.26 illustrates the trajectories (found numerically from Equations (15.8) and (15.9)) of the ice before it hits the surface 127.5 m ($C_d = 1.0$) or 179.2 m ($C_d = 0.5$) from the tower. The figure also includes the trajectory when drag is ignored.

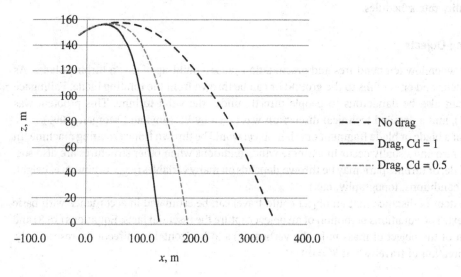

Figure 15.26 Illustration of ice throw

The maximum distance from the tower without drag, 343.7 m, may also be found in Equation (15.12). Note that in this example, the object falls on the opposite side of the tower from which it was released.

Additional discussion on this topic is given in Larwood (2005) and Madsen *et al.* (2005).

15.6 Maintenance and Repair

15.6.1 Overview of Maintenance and Repair

In any machine or construction, there are likely to be some repairs or maintenance needed over its operational lifetime. A common situation is that there are some adjustments or repairs needed early on, then a relatively constant rate over a substantial period, followed by increasing failure rates as time goes on, parts wear, and fatigue begins to take its toll. Some failures may occur relatively often but are quick and easy to fix; others may occur more infrequently but result in more downtime. Three approaches to the maintenance and repair of wind turbines are common: corrective, preventive, and predictive. In the corrective approach, the turbine is run until there is a failure or a problem has been identified. The problem would likely have been found by the SCADA system (Section 15.5.4) or direct inspection (Section 15.6.2). When the component fails, it is repaired or replaced. In the preventive method, a component is replaced or repaired according to a schedule. The schedule will have been developed based on experience. In the predictive approach, various techniques are applied to assist in anticipating when a failure could occur, and repair or replacement timing is based on that prediction. The corrective approach is conceptually simplest, but it has the disadvantage that the failure may occur at an inopportune time, or it may result in additional collateral damage. Preventive maintenance avoids many of the disadvantages of corrective repair, but it could also be overly conservative and result in repairs that are not actually needed or, conversely, not being conservative enough to avoid unexpected failures. Predictive maintenance (Section 15.6.3) can result in a more optimized approach and is increasingly being used with wind turbines. Predictive maintenance is frequently associated with condition monitoring, which is described in Section 15.6.4. Vibration signature analysis (see Section 15.6.5) may be part of this approach.

15.6.2 Direct Inspection

Direct inspection can provide useful information on the condition of many important components. This is particularly true of the blades. For example, a small crack might be a minor problem itself and could be repaired easily if found in time. If it is not repaired, it could grow and eventually lead to serious structural problems. Routine inspections could find such cracks and lead to their repair. Inspections of blades can be done in a variety of ways. The conventional method is visual. In this case, a technician with climbing experience could rappel down the blade of the turbine (while the rotor is stopped!) and examine the blade during the descent or photograph it for detailed examination later. Another method is via unmanned aerial vehicle (UAV, a.k.a. drone). The UAV would fly close to the blade, directing a camera at the surface. The UAV operator would search for defects using the image on a screen. This approach is illustrated in Figure 15.27. A third option is a robot that would climb along the blade, searching for defects in a manner

Figure 15.27 UAV inspecting a wind turbine (*Source:* Photo: Werner Slocum/The National Renewable Energy Laboratory [NREL])

similar to that of the UAV. At least one example of this method is under development at the time of writing; see Bladebug (2023).

15.6.3 Predictive Maintenance

Predictive maintenance involves three major processes: (1) collection of data from the operating turbine, (2) analyzing that data to search for changes, and (3) applying information processing techniques to anticipate when a failure might occur and issue recommendations for timing the repair. Predictive maintenance can be thought of as an extension of condition monitoring. Examples of algorithms that may be applied in such situations are given by Udo and Muhammad (2021).

15.6.4 Condition Monitoring

Condition monitoring refers to real-time collection and observation of operating data to determine if there is a component that needs to be repaired. The intent is to correct a problem before it becomes more serious and perhaps catastrophic. Condition monitoring systems are already a well-developed technology for many other industries but have been applied relatively recently for wind turbines. An example could be gearbox oil temperature measurement, in which the temperature could be an indicator of a problem. More information on this topic can be found in comprehensive reviews on the subject by Tchakoua *et al.* (2014) and Badihi *et al.* (2022). See also Sparkcognition. com (2023).

15.6.5 Vibration Signature Analysis

Vibration signature analysis is an example of a technique that could be applied in either real-time condition monitoring or predictive maintenance. The approach involves converting time series data into the frequency domain, using techniques such as those described for turbulent wind or ocean waves in Chapter 6. The power spectrum is then examined for abnormalities. For example, the power spectrum of vibration data from a properly functioning gearbox would show a baseline energy density at distinct frequencies corresponding to the rotational speed of the gears. If a problem in the gears arose, such as pitting, then it would be expected that the amplitude of the corresponding frequency would increase. In conventional condition monitoring, if the value of the spectrum were to rise above a threshold, that would indicate that a repair should be made. In predictive maintenance, an artificial intelligence algorithm could be used to track the changes in the spectrum over time and provide a recommendation for when the repair should be done.

15.6.6 Support Personnel

Once a certain number of turbines are placed in a wind farm, it becomes economical to provide a dedicated operating and maintenance staff. The staff needs to be appropriately trained and provided with suitable facilities. For servicing smaller projects, technicians will be dispatched from an off-site location.

A key consideration in the maintenance of turbines is safety since most of the issues that are likely to be encountered are associated with the blades or in the nacelle. Accordingly, the technicians must be trained to work safely and effectively at height.

15.6.7 Wind Turbine Component Failure Rates and Effects

An important consideration in wind turbine maintenance is the likelihood of component failures and the resulting downtime. Studies have been done to evaluate such failures. The results of one of them (Hahn *et al.*, 2007) are illustrated in Figure 15.28. As can be seen, most failures are in the electrical or control systems. Much smaller fractions are with major components such as the blades, generator, or gearbox.

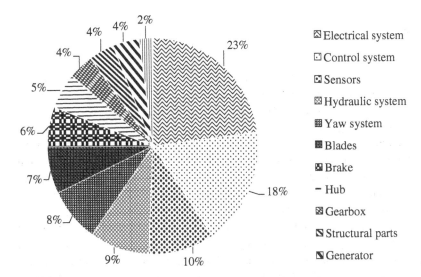

Figure 15.28 Share of wind turbine component failures (*Source:* Adapted from Hahn *et al.*, 2007)

Conversely, a study of downtime resulting from component failure found that the major sources of downtime were due to less frequent failures in the gearbox or the blades (see Hahn *et al.*, 2007). What these studies indicate is that while failures can occur in many parts of a wind turbine, some failures, even when uncommon, can still result in the most significant downtime. For example, failures in sensors, electronic control and the electrical system were the most common failures, but they accounted for little downtime. On the other hand, failures in the gearbox and generator were infrequent, but resulted in the most downtime. It is necessary to take these observations into account when working to minimize these adverse effects and ensure high turbine availability. It is in this regard that condition monitoring and predictive maintenance are most helpful.

15.7 Additional Considerations for Offshore Wind Turbines

Operating and maintaining offshore wind turbines is considerably more challenging than onshore. Downtime can result in significantly reduced revenue and threaten the economic viability of a project. Fortunately, much of the actual operation can be done remotely, but maintenance and repair often require that technicians and their equipment visit the site. This section provides a summary of some of the key issues.

15.7.1 SCADA and Condition Monitoring Systems Offshore

Even more than for land-based turbines, a fundamental requirement for any offshore project is a comprehensive SCADA system. Condition monitoring systems and predictive maintenance are also likely to be used. An effective SCADA system helps to minimize the number of site visits. See Chapter 9 for more information.

15.7.2 Access for Operation and Maintenance

Access to offshore wind turbines for operation and maintenance is a significant issue. The primary concern needs to be safety, but it is also important that trips to the turbine not be excessively expensive and that the vessels involved be able to carry the necessary equipment and spare parts. See Feuchtwang and Infield (2009) for more discussion of this topic.

15.7.2.1 Requirements for Access

There are three major areas of concern regarding access: (1) environmental conditions, (2) technical requirements, and (3) economic cost factors.

Environmental conditions. The primary consideration is the sea state. If the waves are too high, it will be impossible to board the turbine from a service vessel. In very high seas, it would be possible to access the turbine only by helicopter. Helicopters are expensive, and approaching the turbine can be difficult in high winds. It is also difficult and, in some cases, impossible to bring along heavy tools or spare parts by helicopter.

Technical requirements. A vessel should be able to approach the turbine without difficulty and without damaging either the vessel or the turbine, even in moderate seas. Personnel transfer should be straightforward and safe in all situations. It should also be possible to transfer the required equipment without difficulty. As part of the initial design process, maintenance and repair should be easy and lend themselves to being done quickly.

Economic cost factors. A key consideration is that with a robust design, the need for repairs will be minimized, and that immediately helps to keep the availability high. When repairs are needed, good design and ease of repair will allow those repairs to be done quickly, also contributing to high availability. When access is needed, one must be aware of the relation between vessel size and capability and the type of repair that may be needed: larger and more specialized vessels cost more, whether to purchase or lease. They are also likely to be more expensive to operate. The speed of the vessel can also be important, since the technicians are unable to carry out their work until they reach the turbine.

15.7.2.2 Access Vessels

A variety of vessels may be appropriate for accessing and servicing an offshore turbine. For issues that do not require the removal and replacement of large components, small to moderate size vessels may suffice. Examples include Zodiak inflatables, mid-size monohull vessels, catamarans (including the small-waterplane area twin hull, a.k.a. SWATH, ship), and hovercraft. Each of these has some advantages and disadvantages, and whether it would be suitable would depend on the situation.

For removal and replacement of large components, vessels comparable to those that installed the turbine in the first place would likely be needed.

15.7.2.3 Vessel Docking System

Upon arrival at the turbine, the vessel must be able to transfer crew and equipment safely. Typically, this would be done by docking the vessel in some way. This could involve pushing the vessel against the structure with propeller thrust, tying the vessel to the turbine, or keeping the vessel at a short distance from the structure but using an intermediate device to facilitate the transfer, such as a flexible gangway.

15.7.2.4 Personnel Transfer

In terms of personnel transfer, some of the options include the following: vertical ladder, horizontal direct access, walkway access, hoisted vessel, hoisted personnel, helicopter, and motion-compensated offshore access system. Below is an overview of them.

Vertical ladder. A vertical ladder attached to the turbine is the most common means of access to the turbine. Personnel step forward from the deck of the vessel onto a ladder and they then climb about 5 m to a platform at the turbine entrance. On return, personnel climb down and cross to the vessel deck from a backward position.

Horizontal direct access. This is used with vertical-sided, GBSs that stand clear of the water. Personnel cross from one horizontal plane to another with only one stride.

Walkway access. Here, there is an inflated or firm walkway attached either to the vessel or to the substructure. The walkway creates a "bridge" between the vessel and the substructure.

Figure 15.29 Ampelmann offshore access system (*Source:* Reproduced by permission of Ampelmann Operations B.V.)

Hoisted vessel. A hoisted vessel enables personnel to exit directly onto the wind turbine

Hoisted personnel. Personnel are hoisted from the vessel to the turbine by a crane located on the substructure.

Helicopter. A helicopter hovers, with personnel on a hoisting harness, descending to or being lifted from the nacelle roof (the turbine rotor is locked stationary).

Motion-compensated offshore access system. This uses motion compensation such that vessel motions are continuously measured and used to actuate hydraulic cylinders. The effect is to counteract the motions of the deck on which the gangway system is installed, keeping the gangway stable. An example of this is the Ampelmann (https://www.ampelmann.nl/); see Figure 15.29.

15.7.3 Inspection for Offshore Turbines

Advanced inspection techniques are even more important offshore than they are on land, since access is invariably more difficult, and the working conditions are likely to be worse. In addition to the methods described for all turbines, remotely operated offshore vehicles (ROVs) would find many applications. One of those would be to inspect underwater welds, which would otherwise be quite difficult and expensive.

15.7.4 Digital Twins

A digital twin is a virtual (digital) model of a physical system that can be used to help understand the real system and make recommendations as to how it should be operated or maintained. It will include a simulator but differs from a simulation in that it can use data from the real system to inform the calculations. In the case of a wind turbine, data could come from meteorological/oceanographic sensors, strain gauges, vibration or temperature sensors, electrical measurements, etc. More information on digital twins in general can be found in IBM (2023). An example of the use of digital twins for assessing wind turbine operation is given by Sivalingam *et al.* (2018).

15.8 Operation in Severe Climates

The use of wind turbines in severe climates imposes special considerations for both design and operation. Severe climates may include those with unusually high extreme winds, high moisture and humidity, very high or low temperatures, and lightning.

High temperatures and moisture in warm climates can cause a number of problems. High temperatures can thin lubricants, degrade the operation of electronics, and may affect motion in mechanical systems that expand with the heat. Moisture and humidity can corrode metal and degrade the operation of electronics. Moisture problems may require the use of desiccants, dehumidifiers, and improved sealing systems. All of these problems can be solved with site-specific design details, but these should be anticipated before the system is installed in the field.

Operation in cold temperatures also raises unique design considerations. A number of wind turbines have been installed in cold weather regions of the globe, including Finland, northern Quebec, Alaska, and other cold regions of Europe, North America, and Antarctica. Experience has shown that cold weather locations can impose significant design and operating requirements on wind turbines because of sensor and turbine icing, material properties at low temperatures, permafrost, and snow.

Turbine icing can be a significant problem in cold climates, depending on the humidity and other weather conditions. Ice comes in two main forms: glaze ice and rime ice. Glaze ice is the result of rain freezing on cold surfaces and occurs close to 0 °C. Glaze ice is usually transparent and forms sheets of ice over large surfaces. Rime ice results when supercooled moisture droplets in the air come into contact with a cold surface. Rime ice accumulation occurs in temperatures colder than 0 °C. Ice accumulation on aerodynamic surfaces degrades turbine performance and, on anemometers and wind vanes, results in either no information from these sensors or misleading information. Ice can also result in rotor imbalance, malfunctioning aerodynamic brakes, downed power lines, and a danger to personnel from falling ice. Attempts to deal with some of these problems have included special blade coatings (Teflon®, black paint) to reduce ice build-up, heating systems, and electrical or pneumatic devices to dislodge accumulated ice.

Cold weather also affects material properties. Cold weather reduces the flexibility of rubber seals, causing leaks reducing clearances, reducing fracture strength, and increasing lubricating oil viscosity. Each of these can cause mechanical malfunctions or problems in everything from solenoids to gearboxes. Most turbines designed for cold weather operation include heaters on a number of critical parts to ensure correct operation. Materials also become more brittle in cold climates. Component strengths may need to be de-rated for cold climate operation, or special materials may be required for the correct operation of components in cold weather or to ensure adequate fatigue life.

Wind turbine installation and operation may be affected by cold weather climate conditions. Wind turbine access may be severely limited by deep snow. This may result in longer downtimes for turbine problems or delayed and expensive maintenance. In installation in permafrost, the turbine installation season may be limited to the winter when the permafrost is fully frozen and transportation is easier.

More information on wind turbine operation in cold weather may be found in the proceedings of the BOREAS conferences. These are conferences that have a particular focus on wind energy in cold climates. Many of the papers from these conferences are available on the Internet. See also Laakso *et al.* (2003).

Finally, some regions have frequent occurrences of lightning. Lightning can damage blades and mechanical and electrical components as it travels to the ground. Designing for lightning protection includes providing very low-impedance electrical paths to the ground that bypass important turbine components, protecting circuitry with voltage surge protectors, and designing a low-impedance grounding system (IEC, 2019).

15.9 Decommissioning and Recycling

At the end of their useful life, wind turbines need to be decommissioned. The process is essentially the reverse of installation: blades are removed, then the nacelle and finally, the tower is taken down. Decommissioning of off-shore turbines will use the same type of vessels and associated equipment as for the installation. Jackets and suction buckets may be lifted from the seabed. When monopiles are used, they may be cut-off at the sea floor and the upper part removed. The embedded part will likely be left in place. GBSs could be either removed or left in place. Floating offshore wind turbines would be towed back to port, and the substructures would be broken up and recycled.

15.9.1 Recycling

For wind energy to be considered renewable, it is necessary to recycle the materials used in the original construction to the extent possible. Figure 15.30 shows the percentage of the five principal types of materials in a typical land-based 2 MW wind turbine (rotor nacelle assembly and 80 m tower). These materials account for approximately 98% of the total turbine mass. As is apparent from the figure, most (87%) of the materials in a wind turbine are metals: steel, iron, copper, and aluminum. Those are straightforward to recycle. The substructures of offshore turbines, whether fixed or floating, are primarily of steel, so those, too, will be recycled. Turbines with permanent magnet generators would also have some rare earths; these, too, would be recycled.

The foundations of land-based turbines and GBSs offshore are of reinforced concrete. Recycling is possible: the concrete can be broken up, and the reinforcing steel can be removed and recycled with the other steel. The concrete itself can be used as aggregate for other structures. Whether or not recycling is actually done will likely depend on the economics of the situation.

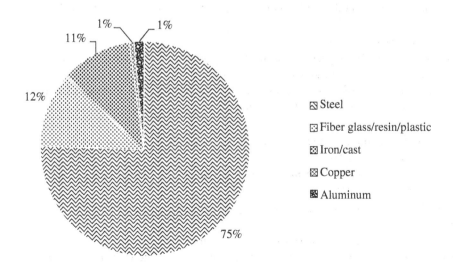

Figure 15.30 Material percentage in typical wind turbine (*Source:* Adapted from Mone *et al.*, 2015)

15.9.2 Blade Recycling

Dealing with wind turbine blades at the end of their design life is a challenge, but not necessarily an unsolvable one. There are a number of approaches that can be taken. First of all, the frequency of blade replacement can be reduced by designing more long-lasting blades and repairing them as necessary. Regardless of how long a blade lasts, however, eventually, it will need to be disposed of. At that point, there are four possibilities: primary recycling or reuse (recycling into wind-related products), secondary recycling (recycling into non-wind-related products), tertiary recycling (recycling to recover plastics and materials), and quaternary recycling (incinerating plastics and materials and recovering energy through heat generation); see Mishnaevsky (2021). Primary recycling essentially means that the blade is reused in some less stressful situation, such as on a turbine in a lower IEC class site. Secondary recycling refers to using a part or all of a wind turbine blade for some other application. Examples given by Mishnaevsky include bus shelters or playgrounds. Tertiary recycling of wind turbine blades or other composites may be carried out by thermal (pyrolysis) or chemical decomposition (solvolysis). In the pyrolysis process, the blade is broken up and heated in the absence of oxygen. The polymer is vaporized and then condensed into one or more usable products. The fibers are left behind and can later be recovered for some other purpose or disposed of. Chemical decomposition involves dissolving the polymer with a chemical reagent, leaving behind the fibers, which may be reusable in some other application. Some of the dissolved polymers may also be recovered. Quaternary recycling is, as given in the definition, self-explanatory: the blade is ground up and incinerated. The waste heat is recovered, but the fibers remain and are presumably disposed of in a landfill.

There are other options for blades as well. One of them is to use biologically based material in the blades (as has been done historically); see Koh (2017). The wind turbine manufacturer Siemens Gamesa also reports that they have developed a recyclable wind turbine blade. See also Chen *et al.* (2019) for an overview of the recycling and reuse of wind turbine blades.

References

Anyang (2023) *Wind Turbine Anchor Bolts*, Anyang General International Co., Ltd., https://www.high-strength-steel.com/wind-turbine-anchor-cage-system/anchor-bolt, Accessed 5/19/2023.

Badihi, H., Zhang, Y., Jiang, B., Pillay, P. and Rakheja, S. (2022) A comprehensive review on signal-based and model-based condition monitoring of wind turbines: fault diagnosis and lifetime prognosis. *Proceedings of the IEEE*, **110**, 754–806.

Bladebug (2023) Bladebug, https://bladebug.co.uk/ Accessed 6/5/2023.

BOEM (2022) *Cable Laying Process*, https://www.boem.gov/sites/default/files/documents/renewable-energy/state-activities/RWF-Scoping-Poster-Cable-Laying.pdf Accessed 6/4/2023.

BOEM (2019) *Installation of Offshore Substation*, https://www.boem.gov/sites/default/files/uploadedImages/BOEM/Renewable_Energy_Program/State_Activities/Nordsee-Ost-offshore-wind-substation-installation-credit-RWE%20(2).jpg Accessed 6/4/2023.

Bossanyi, E. A. and Morgan, C. A. (1996) Wind turbine icing – its implications for public safety. *Proc. 1996 European Union Wind Energy Conference*, Göteborg, pp. 160–164.

Chen, J., Wang, J. and Ni, A. (2019) Recycling and reuse of composite materials for wind turbine blades: an overview. *Journal of Reinforced Plastics and Composites*, **38**(12), 567–577.

DOE (2023) Wind Manufacturing and Supply Chain, https://www.energy.gov/eere/wind/wind-manufacturing-and-supply-chain Accessed 1/19/2023.

Elliott, J., Smith, K., Gallien, D. R. and Khan, A. (2017) *Observing Cable Laying and Particle Settlement During the Construction of the Block Island Wind Farm*. Final Report to the U.S. Department of the Interior, Bureau of Ocean Energy Management, Office of Renewable Energy Programs. OCS Study BOEM 2017-027.

EWEA (2009) *Wind Energy- The Facts*. European Wind Energy Association, Earthscan, London.

Feuchtwang, J. B. and Infield, D. G. (2009) The offshore access problem and turbine availability- probabilistic modelling of expected delays to repairs. *Proc. EWEC 09 Offshore Conference*, Stockholm.

Hahn, B., Durstewitz, M. and Rohrig, K. (2007) Reliability of Wind Turbines, in: J. Peinke, P. Schaumann, S. Barth (Eds.), *Wind Energy*, Springer, Berlin, Heidelberg, 329–332. https://doi.org/10.1007/978-3-540-33866-6_62 Accessed 6/19/2023.

Holtrop, J. and Mennen, G. G. J. (1982) An approximate power prediction method. *International Shipbuilding Progress*, **29**, 166–170.

IBM (2023) *What is a digital twin*, https://www.ibm.com/topics/what-is-a-digital-twin Accessed 6/5/2023.

IEC (2019) *Wind Energy generation systems: Part 24 Lightning Protection*, IEC 61400-24, International Electrotechnical Commission, Geneva.

Jiang, Z. (2021) Installation of offshore wind turbines: a technical review. *Renewable and Sustainable Energy Reviews*, **139**, 110576.

Koh, R. (2017) *Bio-based Wind Turbine Blades: Renewable Energy Meets Sustainable Materials for Clean, Green Power*, Ph. D. Thesis, University of Massachusetts, Amherst, MA.

Laakso, T., Holttinen, H., Ronsten, G., Tallhaug, L., Horbaty, R., Baring-Gould, I., Lacroix, A., Peltola, E., Tammelin, B. (2003) *State-of-The-art of Wind Energy in Cold Climates*. VTT Technical Research Centre of Finland, Vuorimiehen.

Larwood, S. (2005) *Permitting Setbacks for Wind Turbines in California and the Blade Throw Hazard*, California Wind Energy Collaborative Report: CWEC-2005-01. Sacramento.

Madsen, K. H., Frederiksen, S. O., Gjellerup, C. and Skjoldan, P. F. (2005) Trajectories for detached wind turbine blades. *Wind Engineering*, **29**(2), 143–154, John Wiley & Sons, Chichester.

McCleskey, F. (2008) *Drag Coefficients for Irregular Fragments*. Naval Surface Warfare Center, Dahlgren, VA.

Mishnaevsky, L., Jr. (2021) Sustainable end-of-life management of wind turbine blades: overview of current and coming solutions. *Materials*, **14**, 1124. https://doi.org/10.3390/ma14051124.

Mone, C., Hand, M., Bolinger, M., Rand, J., Heimiller, D and Ho, J. (2015) *Cost of Wind Energy Review*, NREL/TP-6A20-66861.

NWCC (2005) *Proceedings: Technical Considerations in Siting Wind Developments*, RESOLVE, National Wind Coordinating Committee, Washington.

Offshorewindbiz.com (2017) *First Hywind Turbine Arrives in Scottish Waters*, https://www.offshorewind.biz/2017/07/24/first-hywind-turbine-arrives-in-scottish-waters/ Accessed 8/13/2023.

Öztürk, F. (2016) *Finite Element Modelling of Tubular Bolted Connection of a Lattice Wind Tower for Fatigue Assessment*, MSc Thesis, University of Coimbra, Portugal.

Scaldis-smc.com (2023) *Rambiz*, https://www.scaldis-smc.com/en/fleet/rambiz/ Accessed 8/13/2023.

Sivalingam, K., Marco Sepulveda, M.; Spring, M.; Davies, P. (2018) A review and methodology development for remaining useful life prediction of offshore fixed and floating wind turbine power converter with digital twin technology perspective. *Proc. 2nd International Conference on Green Energy and Applications* (ICGEA), pp. 197–204, https://doi.org/10.1109/ICGEA.2018.8356292 Accessed 8/13/2023.

Sparkcognition.com (2023) *The Asset Management and Predictive Analytics Solution for Clean Energy*, https://www.sparkcognition.com/wp-content/uploads/2021/06/Asset-Management-and-Predictive-Analytics-Solution-for-Clean-Energy-White-Paper-ENG-EML-WP-051221-v1.0.pdf Accessed 8/14/2023.

Tchakoua, P., Wamkeue, R., Ouhrouche, M., Slaoui-Hasnaoui, F., Tameghe, T. A. and Ekemb, G. (2014) Wind turbine condition monitoring: state-of-the-art review, new trends, and future challenges. *Energies*, **7**(4), 2595–2630.

Tupper, E. (2013) *Introduction to Naval Architecture*, 5th edition. Butterworth-Heineman, Oxford, UK.

Udo, W and Muhammad, Y. (2021) Data-Driven Predictive Maintenance of Wind Turbine Based on SCADA Data, in: *IEEE Access*, vol. **9**, Institute of Electrical and Electronics Engineers, 162370–162388.

16

Wind-Generated Energy: Present Use and Future Potential

16.1 Overview

So far, the focus of this book has been on wind turbines as generators of electricity, with little discussion of how that electricity fits into the larger context of its role in the energy supply. In fact, much of the text related more to the science of producing torque from the wind (consistent with the eponymous title of a biannual conference of the European Academy of Wind Energy). In the context of the need to decarbonize the world's energy supply, however, there is much more to think about. The most obvious point is that since wind energy generation results in very little CO_2 production, it can play a major role in the transition to a low-carbon future. On the other hand, since wind energy is intrinsically variable and non-dispatchable (in the traditional sense), there are some complications involved in using it to supply a large fraction of the energy needed. Fortunately, there are multiple ways to address this issue, and it is the purpose of this chapter to examine some of those methods in more detail.

This chapter, then, extends the discussion of wind energy beyond that of the electricity emanating from the generator to how that energy intersects with the requirements of society. The fundamental concern here is wind energy in high-penetration applications. These are situations in which the rated power of the wind turbines is commensurate with or greater than the immediate electrical load. These high penetration systems can be of many types and in a range of sizes. The turbines could be in a small, isolated network or a large regional power grid. In fact, the turbines need not necessarily be part of any electrical network: they could operate autonomously and serve a load directly or produce fuel.

There has actually been considerable experience with high-penetration wind energy systems over the past several decades. These are isolated electrical networks, typically supplied by diesel generators, into which wind turbines have been introduced originally to reduce fuel consumption and then to extend the capability of the energy supply. The experience with these wind/diesel systems is directly relevant to thinking about how to address similar issues in much larger networks.

Despite the difference in scale between diesel systems and large regional networks, there is enough similarity in issues when large amounts of wind energy are added that they all can be considered hybrid power systems. As used in this text, hybrid power systems are those systems that include conventional generators, wind turbines (and possibly solar photovoltaics) of sufficient capacity that they can have a significant effect on the system. By significant effect, we mean that some aspects of the operation, either technical or economic, are affected by the presence of the variable sources. These hybrid power systems can be considered low penetration, high penetration, or very high penetration. Low penetration systems are those in which the effect on the grid is relatively small in comparison to high penetration, where the effect can be quite significant. Very high penetration systems are those in which the wind generation is frequently in excess of the load.

Wind Energy Explained: On Land and Offshore, Third Edition.
James F. Manwell, Emmanuel Branlard, Jon G. McGowan, and Bonnie Ram.
© 2024 John Wiley & Sons Ltd. Published 2024 by John Wiley & Sons Ltd.

A primary issue in these systems is how to deal with the variability of the wind power and its relation to the variability of the load. As will be discussed, there are a range of options. These include wind turbine control, conventional generator control, load management (smart grids), energy storage, interconnection to other networks ("supergrids"), and "power-to-x," among others.

16.2 Types of Hybrid Power Systems

There are a variety of possibilities for hybrid power systems, but there are two main categories of interest: large regional networks, which include a mix of conventional generators and smaller, isolated wind/diesel systems. The following subsections provide a brief overview.

16.2.1 Regional/National Hybrid Power Systems

Regional or national hybrid power systems are large but otherwise conventional electrical networks into which substantial amounts of wind turbines or solar photovoltaic (PV) panels have been added. The national electrical systems of Denmark and Ireland are prime examples. In the context of this chapter, the electrical grids of Chapter 12 could be considered hybrid power systems if the total capacity of the wind power plants is commensurate with the peak load of the grid.

16.2.2 Isolated Small/Medium Hybrid Power Systems

A substantial number of wind turbines are connected not to large electrical grids, but to small, independent, diesel-powered grids, in which wind generators may be a large fraction of the total generating capacity. These are known as wind/diesel power systems (Hunter and Elliot, 1994) and are the original examples. Sometimes, other renewable generators are added to complement the power from the wind.

Isolated power systems in this category may range in size from relatively large island grids of many megawatts down to systems with a capacity of a few kilowatts. Some isolated grids powered by diesel generators provide power for only a part of the day to conserve fuel. Some have large voltage swings due to the effect of one or two significant loads on the system, such as a sawmill or a fish-processing plant. Large, isolated grids provide power at more stable voltages and frequency. In general, however, isolated grids can be considered weak grids (see Chapter 12) in which voltage and frequency are susceptible to disruption by interconnected loads and generation.

16.3 Hybrid Power System Components

Hybrid power systems typically include the same components as would the corresponding conventional system of equivalent size, namely prime movers, generators, normal electrical loads, transmission and distribution and associated electrical equipment, as discussed in Chapters 5 and 12. In addition, they would likely include some specialized controls, energy storage, and load management to deal with the wind/load mismatch. In summary, hybrid power systems would include one or more of the following:

- Electrical loads
- Conventional generators
- Wind turbines
- Other variable generation, e.g., solar PVs
- Power converters
- Ancillary electrical equipment
- Components to facilitate accommodating the variability of the wind

Most of these components are discussed in the remainder of this section. Components to facilitate accommodating the variability of the wind include energy storage, load management systems with deferrable and/or optional loads, and supervisory control systems. These are discussed subsequently.

16.3.1 Electrical Loads

AC loads were discussed in Chapter 12. Some small hybrid power systems use only direct current (DC) sources, which can only supply resistive loads. DC loads may have an inductive component, but this only causes transient voltage and current fluctuations during changes in system operation.

16.3.1.1 Net Electrical Load

The net electrical load was introduced in Chapter 12 as the actual load minus the wind power. The short-term variability of the net load, σ_{net}/P_{net}, is particularly important in smaller hybrid systems in situations where the average wind power exceeds the average load such that the load could conceivably be supplied by the wind alone over some significant time period (hours or more). The variability in the net load, however, could be great enough that such an operation would not actually be possible unless some short-term energy storage is included in the control system. Flywheels may be attractive options in such situations; see Section 16.9.

16.3.2 Conventional Generators in Hybrid Power Systems

Hybrid power systems, by definition, include one or more conventional synchronous generators driven by some prime mover. Generators for larger systems were discussed in Chapter 12. Diesel generators have some distinct features, which are discussed in Section 16.6.

A particularly important consideration in high-penetration hybrid power systems of any size is the relationship between power production and the fuel consumption of the conventional generators. Ideally, fuel consumption would be proportional to the amount of power that they produce. In reality, the situation is not that simple. This is a result of the reduced efficiency of conventional generators at part load, minimum allowed power levels, and minimum run times. An overview of these characteristics was provided in Chapter 12. An implication is that in order to achieve large fuel savings in high-penetration power systems, it will be necessary to improve the operational characteristics of the conventional generators or be able to shut off some or all of them for much of the time. Additional discussion is given in Manwell and McGowan (2023).

16.3.3 Wind Turbines in Hybrid Systems

Wind turbines used in large hybrid power systems are the commercially available types that have been discussed previously in this text. Land-based wind turbines are typically in the size range of 1–3 MW; offshore turbines are now 10 MW or larger. Most large wind turbines now use pitch control and power electronics. See Section 16.6.2 for wind turbines used in wind/diesel systems.

16.3.3.1 Wind Turbine Operational Considerations

There are a number of aspects of wind turbine operation that may be particularly significant in hybrid energy systems. These were introduced in Chapter 12 and include power smoothing, spatial separation and wind plant variability, curtailment, ancillary services, and ramp rates. See also Section 16.5.3 for more on curtailment.

16.3.4 Solar Photovoltaic Panels in Hybrid Power Systems

Photovoltaic (PV) panels may provide a useful complement to wind turbines in many hybrid power systems. PV panels produce electrical power directly from incident solar irradiance. PV panels are inherently a DC power source. As such, they often operate in conjunction with battery storage, which is also DC. The stored power is

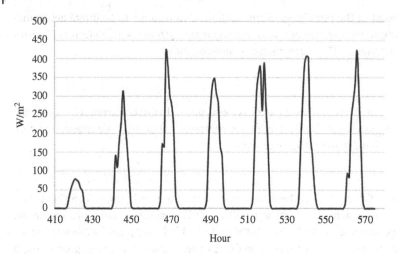

Figure 16.1 Example of solar irradiance on a horizontal surface for a seven-day period, Boston (latitude 42.36°)

subsequently converted into AC by an inverter. In larger systems, panels are typically coupled with a dedicated inverter and thus act as a *de facto* AC power source.

PV panels provide variable power, depending primarily on their orientation (azimuth and slope), the latitude, the day of the year and the time of day. Variations also occur due to the weather, ambient temperature, and the passage of clouds. To a first approximation, the power production from a PV panel is proportional to the solar irradiance, an example of which is shown in Figure 16.1. Further discussion of PV panels is outside the scope of this text.

16.3.5 Power Converters

Power converters are now widely used in hybrid power systems. As discussed elsewhere, most modern wind turbines employ some form of power converter to allow the turbine to operate at variable speeds or to facilitate connection to the grid. In addition, power converters are used with most energy storage technologies. For example, rectifiers are used to charge batteries from an AC source. Inverters are used to supply AC loads from a DC source such as batteries.

As mentioned in Chapter 5, most electronic inverters are of one of two types: line-commutated or self-commutated. Line-commutated inverters require the presence of an external AC line. Thus, they cannot set the grid frequency if, for example, all the conventional generators in the power system are turned off. Self-commutated inverters control the frequency of the output power based on internal electronics. They do not normally operate together with another device that also sets the grid frequency. Some inverters can operate either in a line-commutated or a self-commutated mode. These are the most versatile, but presently, they are also the most expensive.

16.3.6 Ancillary Electrical Equipment

Various ancillary equipment may be included in the power system, including transformers, circuit breakers, and switchgear. Of particular relevance for hybrid power systems are synchronous condensers. These are synchronous electrical machines that are connected to the network and are allowed to spin at a speed determined by the grid frequency. Operating in conjunction with a voltage regulator, they supply additional reactive power to the network. These may be most useful in systems with large numbers of induction machines or power electronic converters, but relatively few conventional synchronous generators, or in situations where the synchronous generators are disconnected.

16.4 Wind Power Variability: Hybrid System Design and Operation

Wind is inherently variable over many time scales, and there is frequently a mismatch between when wind energy is available and what the load is. Several factors, which are discussed below, make the design and functioning of a hybrid system more difficult.

16.4.1 Power System Balancing and Stability

As discussed elsewhere in this text, wind power is inherently variable. Accordingly, the net load on the generators will be variable as well. In order to ensure proper operation of the electrical system, power flows need to be balanced over multiple time frames. Over short time periods, the frequency regulation of an electrical system is a function of the rotating inertia in the system, the fluctuating load, and the responsiveness of the prime mover and its control system. The faster the prime mover can respond to changing power flows, the better the frequency regulation. In high-penetration systems, the prime mover may not have the capability to respond to changing power flows. In that case, additional controllable sources or sinks for power need to be used to control the system frequency. These may be loads that can be switched on to balance power flows, subsystems for the short-term storage and production of power (with a few minutes to an hour of storage), or additional generators that can be brought online. In high-penetration systems, the conventional generators should ideally be turned off when the wind power can supply the entire load. In that case, the system has fairly low inertia, and without another rapidly controllable power source to control frequency, the system frequency could deviate significantly from its set point.

Power or energy flows also need to be balanced over longer time frames. If the load peaks in the daytime and the wind blows only at night, then the wind can be used neither to supply energy to the daytime load nor to save much fuel. In such a case, the addition of longer-term energy storage (corresponding to a few hours or days of storage) would allow the wind energy to be stored and used later. Long periods of lack of wind power would deplete any energy storage and require the use of conventional generation, which would need to be capable of supplying the whole load.

Based on these considerations, it is apparent that a hybrid system might benefit from the addition of energy storage and/or controllable loads. Energy storage could provide power for periods when the wind power is less than the load. When the wind power is greater than the load, energy storage and controllable loads could provide sinks for excess power. With multiple sources and sinks for power, a hybrid power system would also need a system supervisory controller (SSC) to manage power flows to and from system components.

16.4.2 Short-Term Fluctuations in Wind Power (Seconds/Minutes)

The wind speed varies over the short term in a manner that is uncorrelated with the load. The effect also results in short-term mismatch and causes an increase in the short-term variability of the net load (Section 16.3.1.1). The mismatch, which is on the order of minutes, is essentially a technical issue that relates to the stability of the power

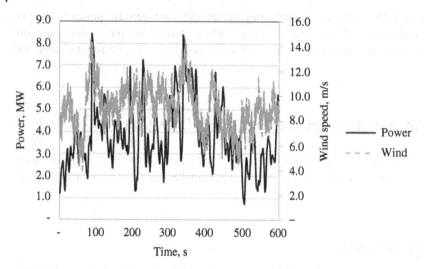

Figure 16.2 Power from a hypothetical 8 MW wind turbine and turbulent wind speed

system. It is likely to be a minimal problem in large grids where the lack of correlation between wind turbines and wind power plants smooths out short-term power fluctuations. It may be more of an issue in smaller networks (wind/diesel systems) where there are relatively few turbines and so less power smoothing.

Turbulence in the wind also results in very short-term fluctuations in the wind power. In fact, the shorter-term variability of wind power may be greater than the short-term variability of the wind (turbulence intensity). This is due to the structural-dynamic response of the turbine. Figure 16.2 illustrates one 10-minute period of wind power and wind speed. In this case, the short variability of the wind power is approximately twice that of the turbulence. This illustration is based on an *OpenFAST* (Chapter 8) simulation of a hypothetical two-blade, downwind 8 MW turbine operating in the wind with a hub height mean of 9 m/s and a turbulence intensity of 0.19. The wind speed was synthesized with *TurbSim* (Chapter 6).

16.4.3 Mismatch Between Load and Wind over the Medium Term (Hours/Days/Weeks)

Daily average energy loads tend to vary relatively gradually over the course of a season, but the wind resource tends to vary with changes in the weather, which may take place over a period of several days. Thus, there may be a number of consecutive days with high winds followed by a few days of lower winds. The result is a medium-term mismatch of load and wind energy availability. Load management or medium-term energy storage would be useful here. Figure 16.3 illustrates the mismatch between the load and the power from a hypothetical wind turbine in an isolated Alaskan community. The case study of Section 16.7.1 provides an illustration of the mismatch as well as the potential benefits of storage for a much larger network.

16.4.4 Mismatch Between Load and Wind Power over the Long Term (Seasonal)

In most locations, the electrical load varies seasonally, and the wind resource varies seasonally as well. These seasonal variations do not necessarily align, and the mismatch needs to be considered for any plausible system design. Some form of long-term energy storage or synthetic fuel would be needed. The long-term mismatch can be seen for the small network in Figure 16.3 as well as in the case study of Section 16.7.1.

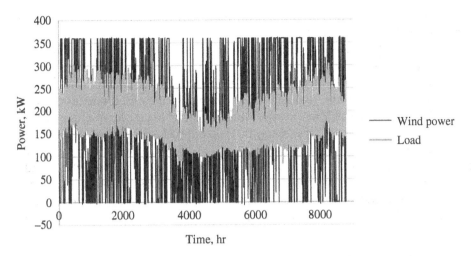

Figure 16.3 Electric load and hypothetical wind power in an isolated network

16.5 Methods to Successfully Implement High Penetration

As discussed in the previous section, the variability of the wind is a key consideration in the design of a hybrid power system, particularly one with high penetration of wind power. Fortunately, there are a variety of methods that can be used to compensate for this variability. These include the following:

- Switching to electricity
- Load management
- Curtailment/dump loads
- Supervisory control
- Forecasting
- Power-to-x (includes fuel)
- Energy storage: long-term, medium-term, short-term
- Super grids
- Energy islands
- Technologies for low inertia electrical grids

16.5.1 Switching to Electricity

Switching loads that do not use electricity to those that do is an obvious way to facilitate the use of energy from the wind. As the total capacity of the grid gets larger, this change will also reduce the penetration associated with a given amount of wind power capacity. Prime opportunities for switching to electricity include heating, cooking, and using electricity for transportation. Electric vehicles have the added advantage that they include energy storage in the form of batteries. These batteries could, in principle, have significant benefits for the grid as a whole. For more information on electric vehicles and the electric grid, see Richardson (2013).

16.5.2 Load Management

One of the most straightforward methods to reduce the effect of the variability of wind power is via load management. Load management consists of adjusting the time at which loads are supplied. Such adjustment can be used most readily to adjust for daily variations in the wind energy supply. In this approach, methods are provided to help customers change the time at which they use certain electrical appliances. These methods typically employ sensors, controllers and financial incentives. For example, the water in a water heater could be heated at a lower cost when there is a lot of wind energy available.

In the context of wind/diesel systems, the term "deferrable loads" has been used to refer to those loads which must be supplied at some point within a longer period (typically a day), but for which the exact time is flexible. An extension of the deferrable load is the optional load. These loads are those that provide a use for surplus power that would otherwise go to waste. An example would be excess energy that was directed toward space heating, which would reduce the need for other fuels.

Grids with load management systems that include two-way communication and control are commonly referred to as "smart grids." Smart grids are electrical systems that utilize deferrable or optional loads and flexible rates to help the load match the available supply. Smart grids are seen by some as an alternative to super grids (see Section 16.5.8). Others think that the approaches are compatible. See Blarke and Jenkins (2013) for a discussion of this topic.

16.5.3 Curtailment/Dump Loads

When the available wind power exceeds the load or is otherwise above some specified limit, the excess power must eliminated, either by curtailment or "dumping" it. Power can be reduced by changing the blade pitch on turbines with adjustable pitch. Fixed-pitch turbines can be stopped altogether. In some cases, such as in many wind/diesel systems, an actual physical device is used. These are based on power electronics or switchable resistors. These loads can react very fast and can be used to control the grid frequency.

16.5.4 Supervisory Control

In high-penetration hybrid power systems, there will need to be significant coordination of the various components (generators, energy storage, load management) in order to keep the system stable. In larger grids, this role will, in most cases, be taken on by the utility or the ISO/RTO (see Chapter 12). Financial incentives will likely be used to encourage consumers to take advantage of load management options. It is also quite possible that third parties may be used to provide energy storage or provide load management opportunities. Coordination with these entities would be required.

16.5.5 Forecasting

Forecasting wind power production can play an important role in hybrid power system operation because it can facilitate dispatching generators, allocating storage and using load management. This topic was discussed in Chapter 2 for the wind and Chapter 12 for the load.

16.5.6 Power-to-X

Power-to-x refers to making fuel or products with renewable energy. The process often uses electrolysis of water to hydrogen as a first step. Such transformation would be most useful to address long-term variability. This topic is discussed in Section 16.10.

16.5.7 Energy Storage

Energy storage may be useful in many situations, including long-term, medium-term and short-term. As may be expected, certain technologies are better suited to some time frames than to others. This topic is addressed in Section 16.9.

16.5.8 Super Grids

Many of the world's nations have their own electrical networks, and sometimes they have more than one. When much of the energy within a network's control area comes from variable sources, such as the wind, it is often desirable to be able to exchange energy with a larger network. Denmark and Norway, for example, often exchange wind and hydro-generated electricity. The concept of super grids is an extension of this idea, and a number of them have been proposed for both Europe and Asia. The term "supergrid" refers to the connecting of large grids together to facilitate the transfer of electricity from one location to another as needed. The purpose is to take advantage of the reduced correlation in wind energy production when there is significant spacing between the wind power plants. The effect is to "smooth" the wind energy production over the time scale of hours or days. The increase in the total load in the expanded network also helps to minimize the amount of production that must be curtailed from a given amount of wind-generating capacity. The use of super grids can thus also reduce the need for energy storage. The fundamental enabling technology for super grids is high-voltage direct current (HVDC), which was introduced in Chapter 5. HVDC cables can carry substantial amounts of power long distances with relatively low losses. For more information, see Blarke and Jenkins (2013).

16.5.9 Energy Islands

Energy islands are a concept first proposed for Denmark, but which could be applicable elsewhere. The idea is to use an island as an intermediary hub for offshore wind energy where the electricity would be consolidated from multiple plants before being transmitted to shore, quite possibly by HVDC. The islands could be either natural or artificial. They might also be used as locations for power-to-x conversion plants. See Tosatto *et al.* (2022) for more information.

16.6 Wind/Diesel Systems

As discussed previously, wind/diesel systems can be considered prototypes for larger, regional hybrid power systems. As such, many of the same issues occur, and similar solutions apply. On the other hand, there are some distinct differences. For one thing, wind/diesel systems, by definition, have only one type of conventional generator, namely diesel. That means that they do not experience the same type of base load generator restriction that larger grids do, and so they can be more flexible. Diesel networks also tend to be relatively small and remote. The following provides an overview of some of the key differences.

16.6.1 Diesel Generators

By definition, wind/diesel systems employ diesel engine-driven synchronous generators. As with the generators in larger grids, important considerations for diesel generators include fuel vs. power curves, no-load fuel consumption, minimum allowed power level, start-up time, and minimum operating time. Figure 16.4 shows a fuel vs. power curve for a small diesel generator. Note the similarity in shape of the curve to that of the much larger generator illustrated in Chapter 12.

The no-load fuel consumption of the diesel in Figure 16.4 is about 50% of the full-load fuel consumption. Fortunately, the relative no-load fuel consumption of most larger diesel generators is well under that level, but it is still significant enough that in order to minimize fuel consumption, diesels should be shut off completely whenever possible.

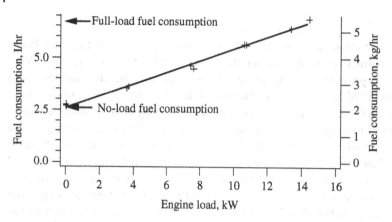

Figure 16.4 Sample diesel engine fuel consumption with linear fit

16.6.2 Wind Turbines for Wind/Diesel Systems

Wind turbines used in wind/diesel systems typically have capacities of 10–500 kW, depending on the load. Historically, most of the wind turbines in this size range have been fixed-speed turbines that use squirrel cage induction generators (SCIG) and so require an external source of reactive power. Thus, they could operate only when at least one diesel generator was operating or in systems that had separate sources for reactive power. The starting current of turbines with SCIGs also needs to be supplied by the system.

It is now more common for wind turbines in larger wind/diesel systems to use power electronic converters similar to those of larger turbines. Depending on the design details, these turbines may also be able to provide reactive power, set the frequency, and supply the grid without a diesel generator being online.

At the present time, many wind turbine manufacturers who formerly made small or medium-sized turbines, such as those that would be used in wind/diesel systems, now only make larger turbines. It can, therefore, be challenging to find suitable turbines for some potential systems.

16.6.3 Other Distinctive Features of Wind/Diesel Systems

- **Electrical load variability**. Many remote power systems have large variations in the electrical load over the year. These may be associated with high latitudes, which are much colder in the winter than in the summer, or due to an island's use as a tourist destination in the summer.
- **Fuel availability**. In some remote locations, such as in the Arctic, fuel can only be delivered in warmer seasons and must be considered in the system design.
- **Maintenance infrastructure**. The availability of trained operating and maintenance personnel affects the long-term operability of the system, operating costs, and installation costs. In many remote locations, access to heavy equipment and cranes may be difficult or impossible. This can be a problem and may affect the type of turbines that can be used.
- **Supervisory control**. A supervisory controller will, in most cases, be a new capability, and it will be under the auspices of the system operator.
- **Frequency control**. When all the diesels are off the grid, frequency may maintained by a power electronic-based dump load; such control in larger grids is possible but unlikely.
- **Reactive power**. In a wind/diesel system in which the wind turbines have conventional induction generators, when all the diesel is off, reactive power may need to be supplied by synchronous condensers, as described

previously. Conversely, in larger networks, the wind turbines would likely be able to provide reactive power themselves.

- **Fuel production**. Whether or not a small wind/hybrid system would be likely to produce its own fuel (hydrogen) is debatable at this time. A research project in Norway demonstrated that it would be possible, although not necessarily practical; see Ulleberg *et al.* (2010). In any case, conventional diesel generators as backups to the wind turbines could not operate on hydrogen directly. They would need to be modified or replaced by fuel cells.
- **Economic development**. In some locations, the impetus for wind/diesel may be related closely to economic development. In sites with an existing generation and distribution system, the system design objective is usually the minimization of the cost of energy by reducing fuel consumption and increasing overall system capacity to enable continued local economic development. New systems are often implemented as an alternative to other options such as grid extension.

16.7 Hybrid System Modeling

As should be apparent, there are many factors affecting the design of a high-penetration hybrid power system and many choices of components that need to be considered to optimize system cost and efficiency. Design choices include the type, size, number of wind turbine solar panels, the instantaneous and long-term energy storage capacity, the size of dump loads, possibilities for other load management strategies, and the control logic needed to decide when and how to use each of the system components. Thus, the problem becomes one of designing an integrated power system with multiple controllable power sources and sinks. The possibilities for controllable loads will depend on the fit of any given load management approach with the daily needs of the consumers. Evaluating all of the parameters is most easily done with a computer model intended for hybrid system design.

Many simulation models have been developed for smaller hybrid power system designs. See, for example, articles by Infield *et al.* (1990) and Manwell *et al.* (1997). Some examples of hybrid power system software for smaller systems in use today include HOMER (https://www.homerenergy.com/) and RETScreen (National Resources Canada). A more recent simulation model, *Hybrid2Grid.xlsm*, which is applicable to larger grids, is described in Manwell *et al.* (2018).

16.7.1 Case Study: Simulation of a High Penetration Regional Electric Grid

The following is an example of a conceptual analysis of introducing a large amount of wind power into a regional electric grid. It is based on and an extension of one given by Manwell *et al.* (2018). The analysis, which used *Hybrid2Grid.xlsm*, illustrates both the opportunities and challenges associated with providing very large fractions of large electrical networks from wind power. The analysis used hourly time series of electric load from the New England electric grid and wind speeds from offshore of that region. For the wind energy contribution, 7000 NREL 5 MW wind turbines were assumed. The wind speeds were adjusted to have a nominal mean at hub height of 9 m/s. The average wind generation over a year was 17 969 MW, slightly more than the average load of 17 489 MW. A time series of wind power and electrical load is shown in Figure 16.5. As can be seen in this situation, the electric load varies between 10 000 and 20 000 MW; the wind power varies from no generation for a few hours to 35 000 MW. As is also apparent from the figure, the hourly wind power and the load are not well correlated.

A number of scenarios were considered in simulations. The load could be supplied directly by wind, indirectly from the wind (via storage) or by conventional generators, either ideal or with some limitations. Four simulations are illustrative: (1) a single ideal generator and no storage, (2) a single ideal generator and 250 GWh of ideal energy storage, (3) multiple non-ideal generators and no storage, (4) multiple nonideal generators with 250 GWh of energy storage. The ideal generator could be turned on and off without limitation and had no minimum power level. The

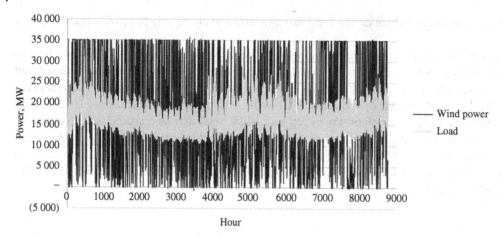

Figure 16.5 Sample hourly electrical load and wind power

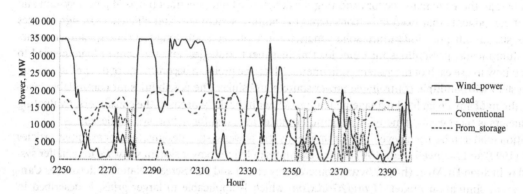

Figure 16.6 Sample utility power flows in high penetration scenario

non-ideal generators included 10 000 MW of base load, which was always running; the minimum allowed power level was 5000 MW. The amount of storage in the scenarios corresponds to approximately 0.6 days at the average load. An example of a 150-hour period for one case is shown in Figure 16.6. In this figure, there is a single ideal generator and 250 GWh of energy storage. The fraction of conventional power that was required ($F_{L,conv}$) and the fraction of available wind power that was dumped ($F_{P,dump}$) are summarized in Table 16.1.

As shown by the summary results, at very high penetration levels, energy storage can have a significant benefit. Although not considered in these examples, deferrable loads could likely have a similar benefit to energy storage for compensating for short-term mismatch between the wind resource and the load. Also, the simulations in this example used a single annual time series for the wind speed. As illustrated in Chapter 12, however, spatial separation between the turbines of 70 km or more could smooth out the medium-term power fluctuations and thus have a beneficial effect comparable to some amount of energy storage. On the other hand, further simulations indicate that it would take a great deal more energy storage to be able to supply the entire electrical load with wind energy alone. This is due to the seasonal mismatch.

In addition, the operational requirements of conventional generators will also have major implications for how much of the available wind energy can be effectively utilized. Requiring a certain base load generator to be on at all times is one of these. Accordingly, improvements to combustion-based generators, such as increasing efficiencies

Table 16.1 Effect of conventional generator capability and energy storage

Scenario	Base load On	Storage, GWh	$F_{L,conv}$, %	$F_{P,dump}$, %
1	N	0	34.1	35.8
2	N	250	17.2	19.5
3	Y	0	47.6	49.0
4	Y	250	29.1	31.1

at part load and decreasing minimum run times, would facilitate the use of wind energy while minimizing the energy that needed to be "dumped" via curtailment.

As noted previously, one way to make more effective use of a given amount of wind energy capacity is through the conversion of other users of energy to electricity, a prime example of which is transportation. It may be shown that in the case of New England if all the land-based motor vehicles were converted to electric, the total electricity consumption could increase by approximately 50%. Extending the example by scaling the load upwards by 50% results in less than 3% of the wind energy being dumped (when there is no base load constraint) or 8% if the base load constraint is kept. Further discussion of the use of electric vehicles, as well as a more in-depth assessment of the benefits of spacing the wind power plants, is given in Manwell and McGowan (2023).

Another observation is that to deal with the seasonal mismatch, a low or no-carbon alternative to fossil fuels, which could be used in conventional generators, could be quite beneficial. See Sections 16.15.4 and 16.15.5.

16.8 Additional Hybrid Power System Topics

16.8.1 Grid Connected Semi-Autonomous Hybrid Power Systems

There are now numerous hybrid power systems that are not isolated, but rather, they are connected to a larger utility network. Such power systems can include a variety of generator types and energy storage, such as wind turbines, PV panels and batteries. These have some similarities to wind/diesel systems, but do not typically include diesel generators because of the presence of the grid. The control of a grid-connected hybrid power system is also simpler than that of isolated systems since the grid provides the frequency set point. The rationale behind systems of this type is that there is an economic benefit to be gained. In particular, when energy storage is available, energy that is produced at one time of day can be sold at a higher rate at a different time. See Wiser *et al.* (2021) for more information.

16.8.2 Distributed Generation

Distributed generation refers to situations in which generators such as wind turbines are connected to a utility's low-voltage distribution system. The voltage in many distribution systems voltages is on the order of 15 kV or less. These systems are designed to accept power flow in one direction. When large amounts of wind generation are added, the remainder of the supply from the local utility can become more variable, and in some cases, generation may be sufficient to send power in the other direction. Adjustments may need to be made to the electrical system to allow this to take place smoothly and also to protect the system in case of an electrical fault. This topic is discussed in more detail by Masters (2013).

Sometimes, distributed generators are augmented by energy storage. In such cases, the overall system would be considered a grid-connected hybrid power system, as discussed previously.

16.8.3 Technical Considerations for Low Inertia Electrical Grids

Conventional electrical networks rely on rotating machinery to provide the inertia, which is used to stabilize the grid frequency and provide other useful services. Power systems in which much of the energy is produced by wind turbines using power electronics do not have as much inertia. That is a significant issue that needs to be considered but is outside the scope of this text. For more information on this topic, see Denholm *et al.* (2020).

16.9 Energy Storage

Energy storage can serve a very useful role in wind energy systems, particularly those in which the wind is to provide a large fraction of the total energy requirement. Energy storage can help to overcome mismatches between the availability of wind energy and the requirement for energy. There are many storage options to consider. These include batteries, compressed air, flywheels, and pumped hydroelectric, among others. These are all discussed in this section. Hydrogen can also be thought of as storage, but since it also has the properties of a fuel, it will be discussed in a later section. There are two characteristics of particular interest: (1) the quantity of energy that can be stored per unit cost and (2) the rate at which energy can be absorbed or delivered (i.e., the power). The time scale of the storage is also important – is it useful for dealing with fluctuations in the order of seconds, minutes, hours, or days? For example, pumped hydroelectric is typically used to cover daily variations in electrical load. Batteries are most effective for dealing with variations on the order of minutes to hours, and flywheels are used for smoothing power fluctuations on the order of seconds to minutes. At the present time, the most common storage medium for wind energy applications is batteries, and there has been renewed interest in compressed air energy storage and flywheels. Pumped hydroelectric also remains an attractive option for some situations. For that reason, the remainder of this section focuses on these four options. Figure 16.7 illustrates in an approximate manner the range and power capacity of a variety of storage options. These are all discussed in the following sections (except for electric double-layer capacitors – EDLC – which are used exclusively for short-term, high-power applications. More information on energy storage may be found in Baxter (2005), Ibrahim *et al.* (2008) and Arellano-Prieto (2022).

16.9.1 Rechargeable Battery Energy Storage

Battery energy storage is very common with smaller hybrid power systems and is also used in larger utility-scale electrical networks (see Rahman *et al.*, 2021). Batteries have proven to be a popular energy storage medium, based primarily on their convenience and cost. Battery storage systems are modular, and multiple batteries can store large amounts of energy. For smaller systems as well as many utility applications, lead acid batteries are most prevalent, although nickel–cadmium batteries are occasionally used. For transportation applications where weight and volumetric energy density are important considerations, other types of batteries are used, particularly lithium-ion. Battery technology is a subject of intensive development at present. Other types include sodium–sulfur and nickel metal hydride. Batteries are inherently DC devices. Thus, battery energy storage in AC systems requires a power converter.

The following discussion focuses on lead-acid batteries that historically have been used for smaller hybrid systems but have also been used in utility-scale systems (see May *et al.*, 2018). Similar work for other competing types of battery storage systems has been presented by Rahman *et al.*, 2021).

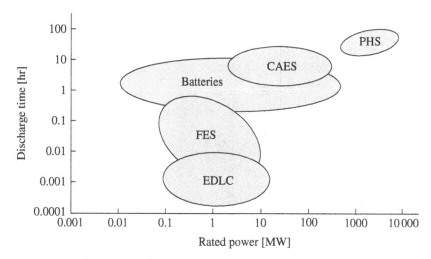

Figure 16.7 Approximate range of powers and discharge time for a few types of energy storage

A number of aspects of lead-acid battery behavior affect their use in power systems (see Manwell and McGowan, 1994). These are summarized below. More details on some of these topics are provided subsequently.

- **Terminal voltage**. Terminal voltage is a function of the state of charge and current level. This affects the operation of the power transfer circuit between the battery storage and the rest of the system.
- **Capacity**. Effective battery capacity is a function of current. Thus, the amount of available storage is a function of the rate at which the storage is used.
- **Efficiency**. Batteries are not 100% efficient. Battery losses can be minimized by intelligent controller operation, but most of the losses are due to differences in voltage during discharging and charging and are inherent to battery operation.
- **Battery life**. Batteries have a limited useful life. Battery life is a function of the number and depth of charge–discharge cycles and is also related to the battery design.
- **Temperature effects**. Battery capacity and life are functions of temperature. Usable battery capacity decreases as the temperature decreases. Typically, battery capacity at 0 °C is only half that at room temperature. Above room temperature, battery capacity increases slightly, but battery life decreases dramatically.

16.9.1.1 Terminal Voltage
An important aspect of all batteries is their terminal voltage, which varies according to current and state of charge. The typical voltage of a lead-acid during a discharge–charge cycle is illustrated in Figure 16.8. It can be seen that the terminal voltage drops as the battery is discharged. When charging is initiated, the terminal voltage jumps to a value above the nominal cell voltage. As the cell becomes fully charged, the terminal voltage increases even more before gassing occurs (the production of hydrogen gas in the cells) and the terminal voltage levels off.

16.9.1.2 Capacity
Total battery capacity is expressed in Amp-hours (Ah), a unit of charge, or kWh. Rated battery capacity is considered to be the Ah discharged at the rated current until the voltage has dropped to 1.75 V per cell (10.5 V in a 12 V battery). Usable battery capacity depends on the charge or discharge rate. High rates of discharge result in early depletion of the battery. The voltage soon drops, and no more energy is available. At low discharge rates, the battery can provide much more total energy before the voltage drops. High charge rates result in rapidly increasing

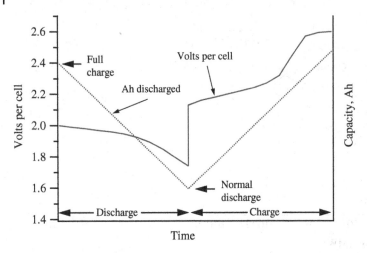

Figure 16.8 Typical lead-acid battery voltage and capacity curve (Fink and Beaty, 1978) (*Source:* Reproduced by permission of McGraw-Hill Companies)

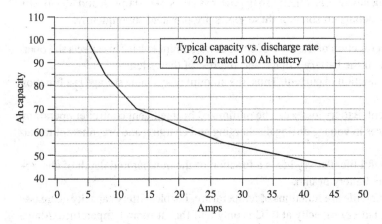

Figure 16.9 Typical battery capacity vs. discharge rate curve

terminal voltage after only a short while. Slower charging rates result in much more charge being returned to the battery. Figure 16.9 illustrates this relationship between usable battery capacity and discharge rate. In this example, the battery energy storage capacity changes by a factor of 2 over the range of currents illustrated. For practical applications, the capacity is often converted to energy (Wh) by multiplying the Amp-hours at a specified discharge current times the nominal voltage.

Battery Capacity Modeling

Battery capacity is known to decline with rising charge and discharge rates, as was previously discussed. This behavior can be modeled by assuming that some of the charge is "available" (that is, immediately accessible) and some of it is "bound." The bound charge can be released at a rate proportional to a rate constant, k. The Kinetic Battery Model embodies these assumptions and can be used to model battery capacity and voltage (Manwell and McGowan, 1993, 1994). In this model, the total charge, q, in the battery at any time is the sum of the available charge and bound charge. This model was developed originally for lead acid batteries, but it can be applied to other types of batteries as well. According to this model, the maximum apparent capacity, $q_{max}(I)$, at current I is given by:

$$q_{max}(I) = \frac{q_{max}kct}{1 - e^{-kt} + c(kt - 1 + e^{-kt})} \tag{16.1}$$

where:

t = charge or discharge time, defined by $q_{max}(I)/I$, hrs
q_{max} = maximum capacity (at infinitesimal current), Ah
k = rate constant, hrs^{-1}
c = ratio of available charge capacity to total capacity.

The three constants, q_{max}, k, and c can be derived from the manufacturer's or test data, as described by Manwell et al. (1997).

Battery Voltage Modeling

Battery terminal voltage, V, as a function of the state of charge and current, can be modeled using the equivalent circuit shown in Figure 16.10 and Equations (16.2) and (16.3). Equation (16.2) describes the terminal voltage in terms of the internal voltage E_0 and the internal resistance R_0. Equation (16.3) relates the internal voltage of the battery to its state of charge.

$$V = E - IR_0 \tag{16.2}$$

$$E = E_0 + AX + CX/(D - X) \tag{16.3}$$

where:

E_0 = fully charged/discharged internal battery voltage (after any initial transient)
A = parameter reflecting the initial linear variation of internal battery voltage with state of charge. "A" will typically be a negative number in discharging and a positive in charging.
C = parameter reflecting the decrease/increase of battery voltage when the battery is progressively discharged/charged. C will always be negative in discharging and positive in charging.
D = parameter reflecting the decrease/increase of battery voltage when the battery is progressively discharged/charged. D is positive and is normally approximately equal to 1.0.
X = normalized capacity at the given current.

The normalized capacity, X, during charging, is defined in terms of the charge in the battery by:

$$X = q/q_{max}(I) \tag{16.4}$$

Figure 16.10 Battery equivalent circuit

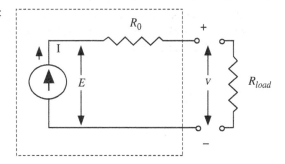

During constant current discharging, X is defined in terms of the charge removed by:

$$X = (q_{max}(I) - q)/q_{max}(I) \qquad (16.5)$$

16.9.1.3 Battery Efficiency

There are two measures of battery efficiency. Coulombic efficiency is the ratio of the charge delivered by the battery during discharging to the charge put into the battery during charging in one complete charge–discharge cycle. Typical coulombic efficiencies range from 90% to 100%. Coulombic efficiency is higher with lower charging currents (and reduced gassing). The second measure of efficiency is energetic efficiency. Energetic efficiency is the ratio of the energy transferred from the battery to the energy provided to the battery in one complete charge–discharge cycle. Energetic efficiency reflects the lower voltages on discharge and the higher voltages required for charging. Energetic efficiencies are usually between 60% and 90%, depending on operating conditions.

16.9.1.4 Battery Life

Unlike other storage media, battery capacity decreases with use. Batteries are typically deemed to be exhausted when their capacity has dropped to 60% of the rated capacity. Battery life is often expressed as the number of charge–discharge cycles to a certain depth of discharge that one can get from the battery. Generally, for a given battery, the deeper the cycle depth of discharge, the shorter the life of the battery. Cycle life also depends on battery construction. Long cycle life lead-acid batteries last 1500–2000 deep discharge cycles, whereas automotive batteries, for example, can only be deep discharged about 20 times. Battery life is sometimes modeled with techniques patterned after those developed for material fatigue. See Wenzl *et al.* (2005) for more information on battery life.

16.9.2 Compressed Air Energy Storage

Compressed air energy storage (CAES) uses an external electricity source, such as wind turbines, to power air compressors. The compressed air is then stored, for example, in an underground storage chamber and used subsequently in a turbine/generator to provide electricity back into the network. In principle, compressed air could be used in an air turbine, but in all of the actually existing CAES systems, the compressed air is used to supply some or all of the air required by a natural gas combustion turbine.

The central issues of compressed air energy storage can be illustrated by reference to the ideal gas law, which relates pressure, volume, and mass. For air, the equation is:

$$pV = mR_a T \qquad (16.6)$$

where p = pressure (kPa), V = volume (m^3), m = mass (kg), R_a = specific gas constant for air (0.287 kPa m^3/kg °K), and T = temperature (°K).

Ideal compressors and turbines are adiabatic. That means that there is no heat transfer to or from them during the compression or expansion. (Strictly speaking, the ideal devices are actually isentropic, so the processes are reversible as well as adiabatic.) Equation (16.7) shows the relationship between temperatures and pressures in isentropic compression or expansion, and Equation (16.8) gives the isentropic workout during the process (by convention, work in is negative).

$$T_{2s} = T_1 \left(\frac{p_2}{p_1}\right)^{(k-1)/k} \qquad (16.7)$$

$$w_s = \frac{kR_a T_1}{k-1}\left[1 - \left(\frac{p_2}{p_1}\right)^{(k-1)/k}\right] \qquad (16.8)$$

where k is the ratio of constant pressure specific heat to constant volume specific heat (k for air is equal to 1.4 at room temperature and varies somewhat from that at other temperatures), w_s is the specific work out of the device (kJ/kg), and the subscripts 1 and 2 refer to the inlet and outlet of the device respectively. The subscript "s" signifies isentropic.

The thing to note is that the temperature of the air increases during compression and decreases during expansion. Once the air is compressed and the temperature has risen, however, the air begins to lose heat to the surroundings. If the heat transfer persists, the temperature will become equal to that of the surroundings. The resultant energy loss will be a significant fraction of the original energy imparted to the air during compression. This "lost" energy will have to be replaced prior to expansion in the turbine, for example, by burning natural gas to heat the air. The overall effect is one of significantly decreasing the round-trip efficiency of the energy storage. There are methods to decrease the effect of heat loss, such as using isothermal compressors and turbines (or approximations thereto, using multiple stages), thermal storage, and regenerators. In this concept, heat is removed from the compressor, stored temporarily, and then transferred to the air before the latter enters the turbine. This still entails losses, and it also increases the complexity of the system. In very large air storage chambers, the surface area (where the heat loss occurs) is relatively small compared to the volume, so when the holding time is short, the effect of the heat loss is reduced.

In actually existing CAES systems, air is compressed and stored at night or other times when the value of the electricity is low. The compressed air is released and supplied to a gas turbine during the day or other times of higher demand for electricity. A typical gas turbine may use approximately 50% of the power it produces for running its own compressor. By being able to use the stored compressed air, the gas turbine will consume substantially less fuel for the same output.

Figure 16.11 is a schematic of a CAES system using an open-cycle gas turbine. As can be seen, air enters the compressor (driven by an electric motor) at "1." It is compressed up to a higher pressure and temperature to state "2" and is stored in the air chamber. The air then enters the combustion chamber, together with the fuel (natural gas). The fuel is combusted, increasing the temperature at the inlet to the turbine, "3." The turbine extracts much of the energy from the combustion gases in the form of shaft work. The pressure and temperature both drop during this part of the process, and the exhaust gases leave at "4." Under conventional conditions (i.e., no wind turbine), much of the shaft work is used to drive the compressor. The remainder is used to turn the electrical generator.

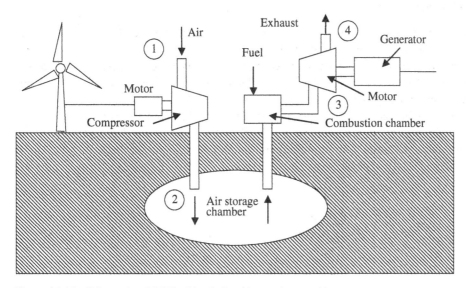

Figure 16.11 Schematic of CAES with wind turbine and gas turbine

For the ideal conventional (no CAES) gas turbine, the efficiency is a function of the ratio of the pressures at state points "3" and "4," as given in Equation (16.9).

$$\eta_{th,Brayton} = \frac{w_{net}}{q_{in}} = 1 - \left(\frac{P_3}{P_4}\right)^{\frac{k}{k-1}} \tag{16.9}$$

where w_{net} = net specific work out of the turbine; q_{in} = heat input from the combustion; P_3 and P_4 are the respective pressures at points 3 and 4.

Equation (16.9) is based on the assumption that the process is that of an ideal Brayton cycle. For practical pressure ratio ranges, the highest efficiency one might get from an ideal Brayton cycle is about 60%. Real gas turbines have lower efficiencies, typically in the range of 35–40%. For more details on this cycle, see a thermodynamics text such as Cengel and Boles (2019).

The net specific work may also be expressed as shown in Equation (16.10), which more clearly illustrates turbine work, w_{turb}, and compressor work, w_{comp}, (which is negative). Turbine and compressor efficiency (η_{turb} and η_{comp}) are also included. For the CAES system, the net-specific work will increase so that the apparent efficiency will increase as well.

$$w_{net} = w_{comp} + w_{turb} = \frac{kR_aT_1}{\eta_{comp}(k-1)}\left[1 - \left(\frac{p_2}{p_1}\right)^{\frac{k-1}{k}}\right] + \eta_{turb}\frac{kR_aT_3}{k-1}\left[1 - \left(\frac{p_4}{p_3}\right)^{\frac{k-4}{k}}\right] \tag{16.10}$$

The first commercial-scale compressed air energy storage facility was built in Huntorf, Germany, in 1978. It has a capacity of 290 MW. The second one, a facility of 110 MW capacity, was constructed in McIntosh, Alabama, in 1991. In this facility, the storage chamber is an abandoned salt cavern with a capacity of 183 000 m³, located 457 m underground. When full, the air pressure is 7.48 MPa, and when "empty" the pressure is 4.42 MPa.

With present designs, the compression stage of the CAES process is actually far from isentropic, largely due to heat loss from the compressed air in the storage chamber (as discussed above). This lost heat needs to be compensated for by burning additional fuel. One approach to improving the efficiency of compressed air energy storage, by making the compression part of the process closer to isentropic, is discussed in Bullough *et al.* (2004).

16.9.3 Flywheel Energy Storage

Energy can be stored in a rotating flywheel by accelerating it and can be recovered by decelerating it. A typical flywheel energy storage (FES) system is shown in Figure 16.12.

As shown in the figure, two power converters, which can direct power either way, are connected to an electrical machine, which is, in turn, connected to the flywheel. The electrical machine can run as either a motor or a generator and so can either accelerate or decelerate the flywheel.

The total energy (J) stored in a flywheel is proportional to the mass moment of inertia, J_{Rot} (kg m²) and the square of the rotational speed (rad/s). The useful energy capacity, E, is limited by the range of allowable rotational speeds, Ω_{min} and Ω_{max}, as shown in Equation (16.11). The round-trip efficiency of FES is approximately 80–90%.

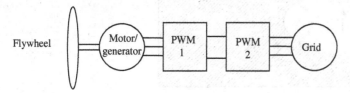

Figure 16.12 Flywheel energy storage system; PWM = power converter using pulse width modulation

$$E = 0.5 J_{Rot}\left(\Omega_{max}^2 - \Omega_{min}^2\right) \tag{16.11}$$

Flywheels are typically made of composite materials. In order to minimize the amount of material in the flywheel itself, FES systems are designed so that the flywheel can run at speeds that are as high as practical. Rotational speeds may be on the order of 20 000–50 000 rpm (2094–5236 rad/s). Flywheels are normally run in a near vacuum to minimize losses. One of the limiting factors is the high internal stress that arises during high rotational speeds.

Flywheel energy storage devices have been successfully applied in some wind/diesel systems, such as the one on Fair Isle, Scotland (Sinclair and Somerville, 1989). Subsequently, research on FES systems was undertaken in Spain (Iglesias *et al.*, 2000).

16.9.4 Pumped Hydroelectric Energy Storage

Energy can be stored by pumping water from a lower reservoir to a higher one. The energy is recovered by allowing the water to return to the lower level, passing through a hydroelectric turbine in the process. The pump and turbine may actually be the same device, but able to function with water flowing in either direction. Round trip energy efficiency is in the range of 65–80%, depending on the details of the system. Figure 16.13 shows a schematic of a typical pumped hydroelectric energy storage system (PHS).

The energy, E (J), that may be stored in a reservoir of volume V (m³) at an elevation z (m) above a reference elevation is:

$$E = V \rho_w g z \tag{16.12}$$

where ρ_w is the density of water (kg/m³), and g is the gravitational constant (m/s²).

Pumped hydroelectric energy storage has been employed in many locations for large-scale energy storage. For example, the pumped storage facility in Northfield, Massachusetts, operates between the Connecticut River and a mountaintop reservoir, 244 m higher, which is capable of holding 21.2 million m³ of water. The effective energy storage capacity is approximately 14 GWh, and the facility can deliver up to 1080 MW of power. At least one pumped hydroelectric facility has been constructed to operate with a wind/diesel system. This was on the island of Foula, off the coast of Scotland (Somerville, 1989).

16.10 Power-to-X

In order to fully decarbonize the energy sector, it is not possible or sufficient to just replace the present methods of electricity generation with wind turbines. It is also necessary to replace liquid, gaseous and solid fossil fuels. Furthermore, energy use in sectors other than the electric grid *per se* must be addressed. These include heating and cooling, commercial, industrial, and transportation. In the process, wind-generated electricity may well be used,

Figure 16.13 Pumped hydroelectric energy storage system

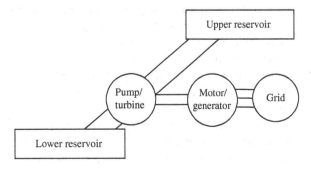

but as an intermediary and not as an end product. The wide range of possible applications has inspired the term "power-to-x" (or P2X) for this approach to decarbonization. Here, the "x" can refer to energy storage, fuel production, electric vehicles, as well as water pumping, desalination, heating, and ice making. In fact, this idea is not new, but it has been re-discovered: wind energy has historically been used for a wide range of applications, ranging from grinding grain to sawing wood, and including many others as well.

It is also important to note, when considering the various power-to-x processes, that the potentially variable input power may be a significant concern for the design and use of the process. In general, this issue is minor for some processes, such as resistive heating or battery charging, but more significant for processes where there are constraints on pressures and flow rate, such as desalination, or on the requirements for chemical reactions, such as the production of ammonia.

In the following sections, we present an overview of the most important instances of the x: "products" (pumped water, desalinated water, heating, cooling); energy storage in gases (hydrogen, synthetic methane) or liquid fuels (ammonia, renewable fuel oil, etc.) and for use in transportation. See also Sterner and Specht (2021) for more on this topic.

16.10.1 Power-to-X when X Includes Carbon

In some of the power-to-x processes, the product contains carbon, and the inputs include carbon dioxide. For these processes to be acceptable in a low-carbon future, the carbon dioxide would need to come from the air, and the carbon dioxide recovered again during or subsequent to the eventual combustion phase.

One of the promising options for obtaining carbon dioxide is known as direct air capture (DAC). An overview of DAC is given in (IEA, 2021a). In one approach, air is blown through a filter containing a chemical adsorbent that absorbs CO_2 from the atmosphere. The adsorbent is then heated to release the CO_2, which is then captured, compressed and stored for subsequent use. The energy consumption in this process is 2.0 kWh/kg, of which 0.4 kWh/kg is electrical and 1.6 kWh/kg is low-temperature thermal; see Beuttler *et al.* (2019). For comparison, the heat of combustion of methane is 15.3 kWh/kg of methane or 5.56 kWh/kg of CO_2 released. This indicates that the energy required to capture the required CO_2 is significant, but since most of that energy is low-temperature thermal, that is less important than it might seem: the low-temperature energy could come from solar thermal collectors, heat pumps, or industrial waste heat.

16.11   Power to Pumped Water

There are millions of people throughout the world who do not have access to water for all of their needs. In many of these situations, water is available in wells or aquifers, but it must be pumped from those sources for it to be usable. Wind has been used for water pumping for many hundreds of years, and it still has a role to play for that purpose today. Historically, wind pumps were purely mechanical devices. Nowadays, mechanical pumps are still a viable option, but there are other possible arrangements as well. These include wind/electric water pumps and conventional water pumps in a hybrid power system.

There are four main types of applications for wind water pumping. These are (1) domestic use, (2) irrigation, (3) livestock watering, and (4) drainage. Another possibility is high-pressure pumping for use in seawater desalination. That topic, however, is discussed in Section 16.12.

A basic introduction to wind water pumping may be found in Manwell (1988). A comprehensive discussion of the topic may be found in Kentfield (1996). More details on pumps in general may be found in Fraenkel (1997). In the 1980s, there was a considerable effort in the Netherlands devoted to developing and investigating "fast-running" mechanical water-pumping windmills. The group that carried out this work was the Consultancy for

Wind Energy in Developing Countries (CWD). Some of the results of that work are still readily available (see Lysen, 1983). The following provides a summary of the key issues.

16.11.1 Mechanical Water Pumping

The most commonly used type of water pumping is completely mechanical. The wind water pump includes at least the following components: (1) the wind rotor, (2) the tower, (3) a mechanical pump, (4) mechanical linkage, (5) a well or other water source, and (6) a water delivery system (piping). Frequently, some form of water storage also accompanies the wind water pump. Tanks, ponds, or reservoirs may be used for storage, depending on the application. A schematic of a typical mechanical water pump is shown in Figure 16.14.

The key considerations in mechanical wind water pump design include (1) selecting or designing the rotor, (2) selecting the pump, (3) matching the rotor to the pump, and (4) safe operation and then shutting down or furling in high winds. These tasks are undertaken with reference to the likely nature of the wind resource.

Figure 16.14 Mechanical water-pumping windmill (*Source:* Manwell, 1988/United States Agency for International Development/Public Domain)

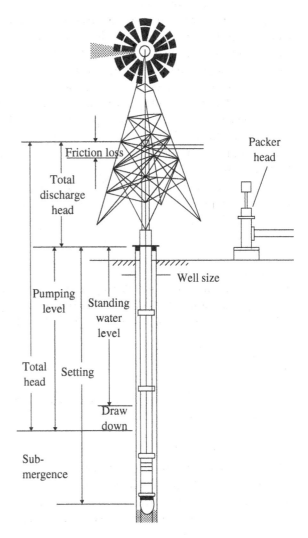

Windmill pumps have historically been of the piston type. These are relatively simple devices that serve the purpose adequately and are straightforward to service. Some early pump designs have used Archimedes screws, and more recently, progressive cavity pumps have been used in some wind pump applications. Centrifugal pumps have historically not been used with mechanical water pumps due to their intrinsically high operating speeds.

Wind water pumping, by its nature, typically requires fairly high torques and low pump operating speeds. Accordingly, wind turbine rotors are generally designed to have their maximum power coefficients at relatively low tip speed ratios. That means that the rotor will have a fairly high solidity. Due to the low speeds, drag will not pose a major limitation on the performance of the rotor, so the airfoil can be of a simple shape.

16.11.1.1 Water Requirements

Water may be needed for domestic use or agriculture. Pumping may also be needed for drainage.

Domestic water use will depend a great deal on the amenities available. A typical villager in a rural community in much of the world may use from 15–30 liters per day. When indoor plumbing is available, water consumption may be substantially higher. For example, a flush toilet consumes 5–15 liters (depending on the design) with each use, and a shower may take 60 liters or more.

Agricultural water use includes livestock and irrigation. Basic livestock requirements range from about 0.2 liters a day for chickens or rabbits to 135 liters per day for a milking cow. Estimating irrigation requirements is more complex and depends on a variety of meteorological factors as well as the types of crops involved. The amount of irrigation water needed is approximately equal to the difference between that needed by the plants and that provided by rainfall. Various techniques may be used to estimate evaporation rates due, for example, to wind and sun. These may then be related to plant requirements at different stages during their growing cycle. By way of example, in one semi-arid region, irrigation requirements varied from 35 000 liters per day per hectare for fruits and vegetables to 100 000 liters per day per hectare for cotton (Manwell, 1988).

Drainage requirements are very site-dependent. Typical daily values might range from 10 000 to 50 000 liters per hectare.

In order to make the estimate for the water demand, each user's consumption is identified and summed to find the total. It is desirable to do this on a monthly basis so that the demand can be conveniently related to the wind resource.

16.11.1.2 Water-pumping Power

The power required to pump water is proportional to its density (1000 kg/m^3), the acceleration of gravity, the total pumping head (m), and the volume flow rate of water (m^3/s). Power is also inversely proportional to the pump efficiency. Note that 1 cubic meter equals 1000 liters. Expressed as a formula, pumping power, P_p, (W) is given by:

$$P_p = \frac{\rho_w g z \dot{V}}{\eta_{pump}}$$

(16.13)

where: ρ_w = density of water, g = acceleration of gravity, z = pumping head, $\dot{V}$ = water volumetric flow rate, and η_{pump} = pump mechanical efficiency.

16.11.1.3 Mechanical Coupling and Control

For any given wind speed, the power from the rotor must match the pumping power. Control of these windmills is also purely mechanical. The analysis is interesting, but it is outside the scope of this chapter. For more information, see the texts referred to above or Le Gouriérès (1982).

16.11.2 Wind Electric Water Pumps

In a number of situations where water is needed, mechanical water pumping may not be suitable. A wind electric pump might be more appropriate. This may be the case, for example, when the wind resource at the site of the well is low, but there may be a site with good exposure to the wind not too far away.

In the wind electric water pump, the wind turbine could be an essentially conventional wind electric wind turbine used for remote applications. Rather than have its generated power used to charge batteries or be integrated into an AC grid, the power can be used directly by an electrically driven water pump.

The wind turbines used in wind-electric water-pumping applications typically have permanent magnet AC generators. Water pumps here are of the centrifugal type, driven by a conventional induction motor. The wind turbine's generator is generally directly connected to the pump motor. Batteries or power converters such as inverters are not used in this arrangement. Pressure and/or float switches and possibly a microcontroller are used for control purposes. With suitable controls, the turbine and pump can operate over a range of speeds. Some wind-electric pumping systems have even been configured to provide both water pumping and electrification. More details on one approach to wind electric pumping are given in Bergey (1998).

16.11.3 Hybrid Grid Water Pump

Water may also be pumped with wind energy in a hybrid power system. In this case, the technical aspects of the pumping are essentially the same as they are in a conventional utility network. The one distinction is that the water pumping may be done in conjunction with water storage so as to assist in load management and thereby make more effective use of the available wind resources.

16.12 Power to Desalinated Water

Much of the world does not have access to potable water. In many locations, however, there is brackish or salt water, which could be converted into drinking water by desalination. Desalination is an energy-intensive process, and often, there is no attractive conventional source of energy to power the desalination process. In many of these situations, the wind is a plausible source of that energy.

Desalination is most often accomplished through either a thermal process or a membrane process. The most common methods are reverse osmosis (RO) or vapor compression systems. Advances in membrane technology over the last several decades have been making reverse osmosis progressively more attractive for desalination, regardless of the energy source. The following section discusses RO-based desalination.

16.12.1 Principles of Reverse Osmosis Desalination

Reverse osmosis is a filtration process in which a semipermeable membrane is used to allow water to pass through while rejecting salts. Osmosis is a naturally occurring phenomenon that can be explained by statistical thermodynamics. If a partially permeable membrane separates a volume of pure water and a salt solution, the result is a gradual flow of freshwater to the salt solution. When mechanical pressure is applied to the salt solution, however, water can be forced the other way through the membrane. The result is that pure water (the "permeate") can be collected on the other side. This is the reverse of the normal osmosis process, hence the origin of the name reverse osmosis.

The pressure required to force a solution through a semi-permeable membrane is called osmotic pressure. For low-concentration solutions, the osmotic pressure, p_π, (Pa), can be calculated by the following equation (also known as the Morse equation):

$$p_\pi = iS\rho_w R_u T/M \tag{16.14}$$

where i = the van't Hoff factor (here, approximately 1.8), S is the salinity (g/kg), ρ_w is the density of water (kg/m³), R_u is the universal gas constant (8.31447 m³ Pa/mol °K), T is the absolute temperature (°K), and M is the molecular weight of salt (58.5 g/mol).

The salinity of seawater varies depending on the location. A typical value is 35 g salt per kg of water. By way of example, this salinity value would correspond to an osmotic pressure of approximately 2.7 MPa, assuming seawater density of 1025 kg/m³ and temperature of 20 °C. This means that the feed water pressure must be more than 26 times atmospheric before any water can permeate the membrane.

The power required to desalinate water is primarily a function of the pressure required to force water through the membrane and the desired flow rate of the permeate (salt-free water). Approximately 7% of the water entering the system passes through the membrane as permeate. The remainder continues through the system and leaves as concentrated brine. Power is required to pump the brine as well as the permeate, so the pressure across the membrane must be higher than the osmotic pressure to ensure a reasonable flow rate. Typically, this is at least twice the osmotic pressure. Common values are in the range of 5–7 MPa (Thompson, 2004). The result is a greater total power expended per unit of permeate than might be expected based on just the permeate flow rate and the osmotic pressure. The total specific work, $w_{desal,tot}$ (in kWh/m³), can be determined from the following equation:

$$w_{desal,tot} = \frac{\left(\Delta p_{L,1} + \Delta p_{L,2} + \Delta p_H\right) + \Delta p_{RO} f_p + \Delta p_B\left(1 - f_p\right)}{\left(f_p\right)(3600)} \tag{16.15}$$

where: $\Delta p_{L,1}, \Delta p_{L,2}, \Delta p_H$ = pressure rise in first and second low-pressure feed pumps and high-pressure feed pump; Δp_{RO} = pressure drop across membrane (approximately twice the osmotic pressure); Δp_B = pressure rise across brine pump, f_p = permeate fraction. (All pressures are in kPa. The 3600 converts kJ/m³ to kWh/m³.) See also Figure 16.15.

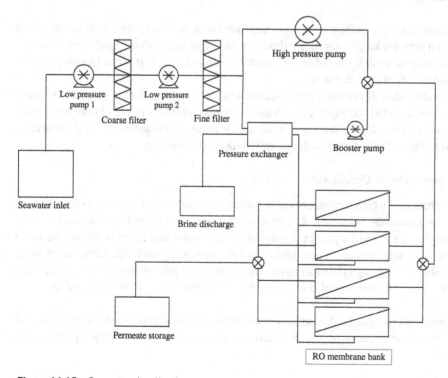

Figure 16.15 Seawater desalination

The net practical effect is that a typical reverse osmosis system nowadays consumes between 3 and 5 kWh per cubic meter of pure water produced (assuming input salinity of 35 g salt/kg water), depending on the design of the system. This is at least four times as much energy as would be expected, considering only the osmotic pressure.

16.12.2 Applications for Wind/Desalination Systems

Wind-powered desalination may be applied in one of four types of applications: (1) within a large central electrical grid, (2) on the distribution system of a central electrical grid, (3) in a hybrid power system network, or (4) in a direct-connected system.

Systems in the first category are physically essentially the same as conventional, grid-connected desalination systems. Insofar as the wind turbines are assumed to be at transmission level voltage, they will, in general, be located some distance from the desalination plant. If the desalination plant were located close to the generators, then they could have some characteristics of distribution-level desalination systems. Otherwise, there are few, if any, issues of substance distinguishing these systems from conventional desalination systems.

16.12.2.1 Distribution Level Desalination Systems

Systems in this category are those in which the wind turbines and the desalination plant are all connected to the distribution system and at a distribution voltage (typically 15 kV or less). In these systems, there could be technical issues associated with the amount of generation that could be connected, and so there could be, in effect, an upper limit imposed on the amount of generation that could be installed or on the amount of power that could be exported from the system during times of high generation and low loads. In practice, limitations could be regulatory as well as technical. Economic factors could also be significant in these systems. In principle, at least, a community in this category could, at times: (1) purchase or sell electricity, (2) produce water by desalination, (3) purchase or sell water, (4) distribute water immediately, and (5) store or release water. At any given time, decisions would need to be made regarding which of these should be done.

16.12.2.2 Wind Hybrid Desalination Systems

Systems in this category include those on islands and those in isolated communities. There could be significant technical issues in such systems, affecting both the electrical stability of the network and the satisfactory operation of the desalination system. Maintenance of electrical stability may be facilitated by the use of equipment common to wind/diesel or hybrid power systems. These could include a supervisory control system, dump loads, power-limiting of the generators (e.g., pitch control on wind turbines), and electrical energy storage.

Load control, particularly of the desalination plant, is also an option. This would presumably be done in RO systems with multiple modules, such that one or more modules would be turned on or off, as called for by the situation. It should be noted, however, that cycling would be done relatively slowly so as not to adversely affect the module membranes. It is also expected that, with the exception of varying the number of operating modules, an RO unit would be run as it normally is. That is, the pressures would remain essentially constant, as would the flow rates within each module. The effect of short-term fluctuations of the wind turbine would, in general, be compensated by a conventional generator (typically diesel in these situations) or through some power-smoothing storage.

A number of analytical investigations or demonstration projects have involved wind-driven reverse osmosis systems for autonomous or non-grid connected situations. The economics of such systems can be strongly affected by the coincidence of the water requirement and the availability of energy. An example of a study of such a system is one undertaken at Star Island in New Hampshire (Henderson *et al.*, 2009).

16.12.2.3 Direct Connected Wind/Desalination Systems

The fourth category of system is the free-standing, direct-coupled wind/desalination system. These are typically smaller systems, and in some ways, the most problematic, in that short-term power fluctuations and their possible impact on the desalination system must be dealt with directly. These are technically very interesting, but they are beyond the scope of this text. Discussion of various aspects of such systems may be found, however, in articles or reports by Miranda and Infield (2002), Thompson and Infield (2002), Fabre (2003), and Carta *et al.* (2003). Adapting RO to work directly with wind turbines was also investigated in the 1980s by Warfel *et al.* (1988).

16.12.3 Desalination Plant Description

A schematic of a typical desalination plant is shown in Figure 16.15. As can be seen, it includes a number of pumps, filters, heat exchangers, and a bank of reverse osmosis modules. Each module contains a wound semipermeable membrane. Pure water passes through the membrane. The remainder continues to the brine discharge.

The majority of seawater RO system designs are single stage with from five to eight membrane elements per pressure vessel (Wilf and Bartels, 2005). There is an obvious cost advantage in the increased number of elements per vessel. For a typical configuration, an RO system with six elements per pressure vessel requires 33% more vessels than one with eight elements per vessel. The actual number of membrane elements that can be "stacked" in a pressure vessel is a unique characteristic of each membrane.

RO membranes are particularly susceptible to failure and excessive fouling when the inlet water is not properly filtered. Accordingly, a pretreatment system with cartridge-type filters is usually installed upstream of the membranes. Low-pressure pumps are used to push water through the filters. The purpose of these filters is to remove any substance that could otherwise damage the membranes. Cartridge filters are not effective in removing pollutants such as oil, however.

In order to reduce the energy requirements of a desalination system, an energy recovery device may be incorporated. Some RO systems include a Pelton water turbine that is connected to an electrical generator, which, in turn, delivers power to a high-pressure pump. Another approach uses a pressure exchanger. This device allows the high-pressure brine discharge to be directly used to pressurize the inlet feedwater (Andrews *et al.*, 2006).

16.13 Power to Heat

Wind energy can be used for space heating, domestic hot water, or other similar purposes. One version of the concept, which originated in the 1970s, was known as the Wind Furnace (Manwell and McGowan, 1981). In this arrangement, a dedicated, specially designed wind turbine was used to provide heat to a residence. The wind turbine's electricity was dissipated in resistance heaters, which were inside a water storage tank. Hot water was circulated from the storage tank to the rest of the house as needed. The purpose was to provide the heat directly, rather than to interconnect to the utility power lines. The advantage was that the turbine could run at a constant tip speed ratio (and thus at maximum power coefficient) for its entire operating range between cut-in and rated wind speed, all without the need for intermediate power converters. A similar concept, known as the Windflower, was investigated in Denmark in the 1980s, and at least one wind heating system was installed in Bulgaria in the 1990s. There has also been some work on the concept in Finland. In spite of these endeavors, wind heating based on dedicated wind turbines has not been widely used.

A variant of the wind heating concept has been more widely used, however, particularly in hybrid power systems. In these cases, a conventional wind turbine is used to supply regular electrical loads and is also used to supply space or domestic hot water heating. In these situations, the heating is normally undertaken with some type of thermal storage, so the heating system serves a load management role. When there is excess energy from the wind turbine, some of it is used to heat up the thermal storage. When heat is needed, it is either taken directly from the wind turbine if there is sufficient available, or heat is taken from the storage. An example of a hybrid power system that utilizes wind energy in its supply of heat and electricity is that of Tanadgusix (TDX Corporation) on the island of St Paul, Alaska (Lyons and Goodman, 2004).

In wind heating systems with storage, the heat is typically either stored in water or in ceramics. One concept of a wind/diesel system with heating and ceramic thermal storage is described by Johnson *et al.* (2002). In this concept, distributed controllers were to be used to facilitate the match between the availability of wind-generated electricity and the requirement to recharge the storage.

One final point to bear in mind is that the conversion efficiency of electricity from the wind turbine to heat is essentially 100%. When heat is the desired product, wind-generated electricity is thus a plausible way to produce it. On the other hand, electricity has a very high availability (in the thermodynamic sense). Accordingly, electricity can be used to produce even more heat than it can by simple dissipation through the use of a heat pump. Just how much more heat can be produced will be a function of the temperature of the heat pump's heat source reservoir. The heat reservoir may be the air, the ground, or water from a well. Depending on the nature of the source, its temperature, and the design of the heat pump, the coefficient of performance (ratio of delivered heat to input electricity) will be in the range of 2.0–5.0. Including a heat pump in the system will thus clearly result in a more effective use of a wind turbine's electricity than just using resistors. Whether it makes economic sense to do so will depend on the cost of the heat pump and the total additional useful energy it will provide over the course of a year.

16.14   Power to Cold

Wind energy can be used for ice making. The premise behind this concept is somewhat similar to others in which wind energy is used directly to supply a particular requirement. Here, the requirement is for ice or the cooling capacity that results from it.

The principle behind the wind-powered ice-making concept is to use the electricity from a wind turbine to drive a conventional Rankine cycle-based freezer. The ice may be made in either a batch or continuous process. If a batch process is used, however, the process must be interrupted when the ice is removed. Ideally, both the wind turbine's rotor and the freezer's compressor should be able to turn at variable speed. This helps to maximize the efficiency of the conversion of wind energy to mechanical form and could obviate the need for maintaining a constant frequency of the electrical power.

As far as is known to the authors, the only wind-driven ice-making systems that have been built were experimental. At least one such system was built in the United States (Holz *et al.*, 1998), and another one was built in the Netherlands in the 1980s. Those experimental systems were small, but there is no reason that much bigger systems cannot be built. It is important to note, however, that if ice is going to be used as a source of cold, then there must be provision for insulated storage, a modern version of the "ice house." This should not be a serious impediment – ice has been used for long-term cooling for generations, in fact even predating ancient Rome.

16.15   Power to Hydrogen

The electricity generated by a wind turbine can be used to produce hydrogen by the electrolysis of water. The fundamental (endothermic) relation is given in Equation (16.16).

$$H_2O \rightarrow H_2 + \frac{1}{2}O_2; \Delta H = 286 \, kJ/mol \qquad (16.16)$$

where

ΔH = change in enthalpy
mol = gram molecular weight

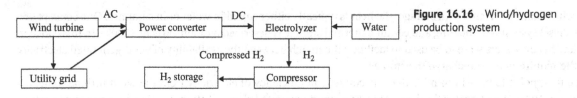

Figure 16.16 Wind/hydrogen production system

The hydrogen that has been produced can be stored in tanks, typically as a compressed gas, but sometimes as a liquid. The hydrogen can be used as a fuel, or it may be converted back to electricity at some later time. The energy content of hydrogen is theoretically the same as that required to electrolyze the water, i.e., 286 kJ/mol. Taking into account the molecular weight of hydrogen (2.016), this converts to 39.4 kWh/kg. On a weight basis, hydrogen has a very high energy density, but on a volume basis, the energy density is very low. Because the volume of a gas varies so much with changes in pressure or temperature, properties are typically referenced to standard conditions, which are 0 °C and 1 atm (100 kPa). Under these conditions, the specific volume of hydrogen (i.e., the reciprocal of the density) is 11.11 Nm^3/kg, where Nm^3 means normal cubic meter. The energy content of the hydrogen at normal conditions is, therefore, 3.55 kWh/Nm^3. It is because the density of hydrogen is so low that hydrogen is compressed before being stored.

A schematic of a hydrogen production, compression, and storage process using a grid-connected wind turbine is illustrated in Figure 16.16. In this configuration, the wind turbine can supply electricity to the hydrogen system or the utility grid. In addition, the grid can be used for the electricity supply as well. See also Bossel (2003) for a discussion of hydrogen as an energy carrier.

16.15.1 Electrolysis of Water

In the electrolysis process, a DC voltage is applied to two electrodes immersed in a solution of water and an electrolyte. Typically, potassium hydroxide is used as the electrolyte, although under some circumstances, sodium hydroxide, sulfuric acid, or common salt (sodium chloride) could be used. The electrolyte facilitates the flow of current through the solution. The electrodes are normally constructed from nickel, although other materials may be used. The voltage difference between the electrodes is on the order of a few volts. For example, in one case, the ideal voltage difference between the two electrodes is 1.23 V. In practice, it is actually higher, typically around 1.76 V. This increased voltage difference is one of the major reasons that the conversion efficiency of electricity to chemical energy is less than 100%. Presently, efficiencies of real electrolyzers are in the range of 65–75%.

During the electrolysis, the negative electrode (cathode) supplies electrons to positively charged hydrogen ions, resulting in the formation of H_2. Similarly, the positive electrode (anode) accepts electrons from the negatively charged hydroxyl ions, resulting in the production of O_2. The complete process involves a number of other steps as well, including maintaining the separation of the hydrogen from the oxygen, separating the gases from the potassium hydroxide, and deoxidizing and drying the hydrogen. A typical electrolysis process, including some of the ancillary components, is illustrated in Figure 16.17.

It is worth noting that electrolysis uses pure water, so if only salt water is available, the water would first be desalinated.

16.15.2 Hydrogen Storage

As indicated, the density of hydrogen is very low. Therefore, it is compressed for storage or transport. Storage tank pressures are in the range of 20–40 MPa. A significant amount of energy is required to compress the hydrogen. Just

Figure 16.17 Hydrogen electrolysis process

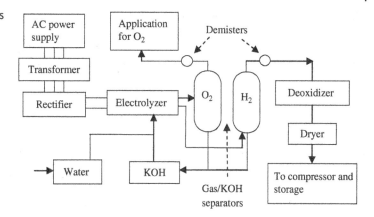

how much this is will depend on the type of process used and the efficiency of the compressor. The process will be between adiabatic and isothermal. The specific work involved in the adiabatic compression (kJ/kg) is:

$$w_{comp,ad} = \frac{p_i v_i k/(k-1)}{\eta_{comp}} \left[\left(\frac{p_f}{p_i} \right)^{\frac{k-1}{k}} - 1 \right] \tag{16.17}$$

where:

p_i = initial pressure, kPa
p_f = final pressure, kPa
k = ratio of constant pressure to constant volume specific heats ($k = 1.407$ for hydrogen)
v_i = initial specific volume of hydrogen, m³/kg.
η_{comp} = compressor efficiency.

The specific work involved in isothermal compression, $w_{comp,iso}$, (kJ/kg) is:

$$w_{comp,iso} = \frac{p_i v_i \ln \left(\frac{p_f}{p_i} \right)}{\eta_{comp}} \tag{16.18}$$

By way of illustration, consider compression from 100 kPa to 40 MPa at 25 °C. The adiabatic specific work would be approximately 5.42 kWh/kg, corresponding to about 14% of the energy in the hydrogen. Isothermal compression under the same conditions would consume approximately 2.03 kWh/kg or 5% of the energy.

Hydrogen could also be liquefied for storage, but this is not commonly done in a stationary application. This is because the energy required for liquefaction is quite large, typically at least 30% of the energy in hydrogen.

Hydrogen could also be stored in metal hydrides, but at the present time, this is not considered practical for stationary applications.

16.15.3 Uses of Hydrogen

Once it is produced, hydrogen may be used as an energy source. There are basically four main ways by which this may be done. The first is to use it as a fuel for the generation of electricity. The second is to use hydrogen as a transportation fuel. The third is that it could be used to create a different fuel, such as ammonia, methane, methanol, or a biomass-based hydrocarbon. The fourth way is to use hydrogen to produce chemicals. These are covered in the section that follows.

Table 16.2 Summary properties of hydrogen and selected derivatives

Compound	Formula	Phase	T (°C)	p (MPa)	g/mol	kJ/mol	kWh/kg	kg/m³
Hydrogen	H_2	Gas	20	0.1	2.016	285.8	39.4	0.083
		Gas	20	35				23.2
		Liquid	−252.9	0.1				71
Ammonia	NH_3	Gas	20	0.1	17	382.8	6.25	0.76
		Liquid	−30	0.1				678
Methane	CH_4	Gas	20	0.1	16.04	891	15.4	0.714
		Gas	20	5				36.1
		Liquid	−164	0.1				425.6
Methanol	CH_3OH	Liquid	20	0.1	32.04	726	6.31	792
Dimethyl ether	CH_3OHCH_3	Liquid	20	0.1	46.07	1460	8.81	729
Octadecane	$CH_{18}H_{38}$	Liquid	20	0.1	254.5	12 008	13.1	777

For comparative purposes, some of the key properties of hydrogen and some of its derivatives are summarized in Table 16.2. These include the formula, the molecular weight (g/mol), the energy density (kJ/mol and kWh/kg), and mass density (kg/m³) at various temperatures (T, °C), pressures (p, MPa) and phase (gas or liquid).

16.15.4 Hydrogen for Stationary Electricity Generation

If hydrogen is used for stationary electricity generation, then the overall effect of the entire process is to use hydrogen as a reversible energy storage medium, similar to what might otherwise be done with batteries, pumped hydroelectric, or compressed air energy storage. As with the other energy storage options, the key questions have to do with overall round-trip efficiency and the economics of the process. As will be seen below, the round-trip efficiency of hydrogen storage used in this way is rather low, presently well under 30% (unless the hydrogen is combusted in a combined cycle power plant). The economics are also still not attractive at the present time.

There are essentially two methods by which hydrogen could be used to produce electricity: (1) gas turbine generators and (2) fuel cells.

16.15.4.1 Gas Turbine Generators

Hydrogen could be used in a conventional gas turbine (perhaps with some modifications) in a similar way to the use of natural gas. The hydrogen could either be used straight, or it could be mixed with natural gas. Gas turbines operate according to the Brayton cycle, which was discussed previously in Section 16.9.2. As also noted there, typical efficiencies of real gas turbines are on the order of 35–40%.

The round-trip efficiency of the hydrogen energy storage when using a gas turbine generator, taking into account the electrolyzer (65–75%), the compression of the hydrogen (85–95%), the gas turbine (35–40%), and the electrical generator (95%), would be between 18% and 27%. Higher efficiencies could be obtained if the hydrogen were combusted in a combined cycle power plant. Such a plant could have an efficiency as high as 60%, resulting in an overall efficiency of 31–41%. The round-trip efficiency is thus, even in the best of cases, quite low. It is, therefore, apparent that both the capital costs of the complete system and the value of electricity during the time of hydrogen production would also have to be low for this to be an economically attractive method of storing energy. It may be noted that, due to the high temperatures involved, hydrogen combustion in gas turbines can also result in the emission of oxides of nitrogen (NO_X), which can result in smog. Oxides of nitrogen are also considered to be greenhouse gases.

16.15.4.2 Fuel Cells

Fuel cells are devices that can be used to directly convert hydrogen into electricity. Conceptually, they may be thought of as the reverse of electrolyzers. In reality, fuel cells come in a variety of different designs. The types of hydrogen fuel cells available today are the following: (1) proton exchange membrane (PEM), (2) phosphoric acid, (3) molten carbonate, (4) alkaline, and (5) solid oxide. The efficiency of conversion of hydrogen energy into electricity may be on the order of 50–60%, depending on the type of fuel cell and the details of its operation. The use of a fuel cell instead of a gas turbine generator in the previous example would result in round trip efficiencies of 28–43%, which would be a distinct improvement. An in-depth discussion of fuel cells is outside the scope of this text, so the reader should consult other sources, such as Larminie and Dicks (2003), for more information on this subject.

16.15.5 Hydrogen Derived Fuels: Production

Hydrogen may be used to create a wide range of fuels. The most important are discussed here.

16.15.5.1 Hydrogen to Ammonia

Ammonia is now widely considered as a possible fuel that could be produced using the wind as the energy source. The process would involve the conversion of hydrogen, produced as described above, into ammonia by the following exothermic reaction:

$$N_2 + 3H_2 \rightarrow 2NH_3; \Delta H = -91.8 \, \text{kJ/mol} \tag{16.19}$$

The method used to implement Equation (16.19) is known as the Haber–Bosch process. It was developed in Germany early in the 20th century. Basically, the process involves passing hydrogen gas and nitrogen gas over an iron catalyst at a temperature of about 450 °C.

Ammonia has some properties that make it more convenient than hydrogen. For example, although it is gas under standard conditions, it can be liquefied at relatively moderate pressures and temperatures, such as 1 MPa and ambient temperature or 100 kPa (ambient pressure) at −28 °C. As a liquid, it has approximately 50% greater volumetric energy density than liquid hydrogen and twice the volumetric energy density of hydrogen at 70 MPa. In many ways, ammonia is comparable to propane.

Ammonia can be burned directly in internal combustion engines or can be used directly in alkaline fuel cells. It can also be used as a source of hydrogen when that is desired, such as for supply to PEM fuel cells.

The process to produce the ammonia is exothermic; thus, the excess energy must be disposed of and will also result in a decrease in overall efficiency unless the heat can be recovered and used in another application. The reaction releases 30.6 kJ of energy for every mol of hydrogen (H_2) converted. This amounts to approximately 10% of the energy in the hydrogen. In addition, the nitrogen gas must have been previously separated from the air. This part of the process is not as energy-intensive as others, but it does need to be considered. Overall, the energy of combustion of the ammonia is 6.25 kWh/kg.

Ammonia also has some disadvantages. For example, it is toxic when released into the air. It is also readily absorbed into water, and this property must be considered in the design of equipment to handle and store the ammonia. If the ammonia is not completely combusted, it can produce nitric oxide (NO) or nitrogen dioxide (NO_2), which are air pollutants instrumental in the formation of smog and acid rain.

More details on ammonia as a possible fuel source may be found in Valera-Medina *et al.* (2021).

16.15.5.2 Hydrogen to Methane

Creating synthetic natural gas using wind energy is an attractive option because the infrastructure for the distribution and use of natural gas is already developed in most countries. It is possible to synthesize methane, which is the primary ingredient of natural gas, by using wind-generated hydrogen in the Sabatier reaction, which is shown in Equation (16.20), where the water produced is assumed to be gaseous.

$$CO_2 + 4H_2 \rightarrow CH_4 + 2H_2O; \Delta H = -165.0 \, kJ/mol \tag{16.20}$$

The combustion reaction of the methane is given in Equation (16.21). In this case, the water produced is assumed to be liquid, so the heat of combustion corresponds to what is known as higher heating value (HHV).

$$CH_4 + 2O_2 \rightarrow CO_2 + 2H_2O; \Delta H = -891 \, kJ/mol \tag{16.21}$$

Note that since the molecular weight of methane is 16.04, the heat of combustion converts to 15.4 kWh/kg (higher heating value).

A discussion of one approach to methane synthesis via the co-electrolysis of H_2O and CO_2 using a solid-oxide electrolyzer may be found in Wang *et al.* (2019).

16.15.5.3 Hydrogen to Methanol

Another possible fuel that could be produced from wind-derived hydrogen is methanol. Methanol is particularly convenient because it is a liquid at room temperature and pressure.

The overall exothermic reaction in this case is shown in Equation (16.22). As with ammonia and methane, the energy released in the reaction represents a loss unless it can be recovered.

$$3H_2 + CO_2 \rightarrow CH_3OH + H_2O \quad \Delta H = -49.2 \, kJ/mol \tag{16.22}$$

The combustion reaction of methanol is similar to that of methane and is shown in Equation (16.23).

$$CH_3OH + \frac{3}{2}O_2 \rightarrow CO_2 + 2H_2O; \Delta H = -726 \, kJ/mol \tag{16.23}$$

Taking into account the molecular weight of methanol (32.04), the heat of combustion (HHV) of methanol is 6.31 kWh/kg.

16.15.5.4 Hydrogen to Dimethyl Ether

Dimethyl ether, CH_3OCH_3, is currently being investigated as an alternative to diesel fuel. It can be produced from methanol by the following reaction:

$$2CH_3OH \rightarrow CH_3OCH_3 + 2H_2O; \Delta H = -23 \, kJ/mol \tag{16.24}$$

Among other advantages, it is liquid at low temperatures. The heat of combustion of dimethyl ether is −1460 kJ/mol or 8.81 kWh/kg.

16.15.5.5 Hydrogen to Biofuels

Another possible wind energy derived hydrocarbon fuel is octadecane, $C_{18}H_{38}$, which is a typical constituent of Number 2 fuel oil. For example, Equation (16.25) illustrates the hydrogenation of the triglyceride triolein, $C_{57}H_{104}O_6$, which can be derived from biomass. A secondary, but important, product of the reaction is propane, C_3H_8, which can also be used as a fuel, accounting for approximately 6% of the useful energy. It is also worth noting that the main effect of this process is to convert the biomass constituent into a more usable form rather than to introduce a great deal of energy content from the hydrogen. In this case, the heat of combustion of the original triolein is −35 100 kJ/mol, and that of the hydrogen is −285.8 kJ/mol. Considering that 15 moles of hydrogen are used for one mole of triolein, approximately 11% of the energy content is from the hydrogen. The rest is from the triolein.

$$C_{57}H_{104}O_6 + 15H_2 \rightarrow 3C_{18}H_{38} + 6H_2O + C_3H_8 \tag{16.25}$$

Note that the heat of combustion of octadecane is $-12\ 008$ kJ/mol (13.1 kWh/kg), and that of propane is -2219 kJ/mol (14.0 kWh/kg).

Although these are all interesting options, further discussion is beyond the scope of this text. See Van Gerpen and He (2014) for some more information.

16.16 Power to Transportation

Wind energy can be used to power transportation in three basic ways: (1) as electricity, (2) as hydrogen, and (3) as hydrogen-derived fuels. Previous sections have discussed electricity, hydrogen, and hydrogen-derived fuels. Here, we consider how these could be applied to transportation on land, sea and air.

16.16.1 Electricity for Land-based Transportation

Electricity has been used by many railways for more than a hundred years, and that use continues to this day. In electrified railways (not considering diesel-electric locomotives), the electricity is produced by an external power plant and then conducted to the electric locomotive via the rails and overhead wires. When the wind-generated electricity goes into the grid, it just becomes part of the normal supply to the locomotives.

Wind-generated electricity can also be used to power electric vehicles that do not run on rails, such as buses and passenger cars. In this application, the vehicles carry batteries that are charged from the grid, which in turn is presumed to derive much of its supply from the wind. In operation, the vehicles are propelled by means of electric motors, which in turn get their power from the batteries. Regenerative braking is usually included as well so that when the vehicle is decelerated, some of the vehicle's kinetic energy is returned to the batteries, thereby increasing the overall efficiency. Charging of the batteries is normally done when the vehicle is not being operated, such as at night or during the workday if the vehicle is used for commuting. Vehicles of this sort are already commercially available, and there were, in fact, one million of them on the road in the US as of 2018 (Cooper and Schefter, 2018). There is still work to be done to reduce the cost, extend the range of electric vehicles, and develop widespread charging infrastructure, but the Cooper and Schefter report projects that, in any case, there will be more than 18 million electric vehicles in the US by 2030. Worldwide, of course, there will be many more.

The presence of batteries on potentially a great many vehicles offers the possibility for using the batteries actively in support of the overall energy supply. The batteries in an electric vehicle typically store between 40 and 100 kWh (E.ON, 2023). Users can be encouraged to charge their batteries during times of high wind and discouraged when less wind energy is available. The electric vehicles, in this case, become a form of deferrable load, such as discussed previously. This concept has been investigated by Kempton and Tomic (2005a, b).

16.16.2 Hydrogen Gas for Land-Based Transportation

At the present time, most of the activity regarding hydrogen as a transportation fuel has focused on public transport, particularly buses. This is because of hydrogen's low density, and the need to carry it in pressurized tanks makes it difficult to use in passenger cars. For example, The Clean Urban Transport for Europe (CUTE) was a European Union project intended to demonstrate the use of hydrogen in buses in a number of European cities. Different hydrogen production and refueling infrastructures were established in each of the cities. More information may be found at (IEA, 2021b)

The hydrogen can be used in either a conventional internal combustion (IC) engine, or it may be supplied to a fuel cell. If an IC engine is used, then the bus is otherwise similar to conventional buses (except for the hydrogen storage tank). If a fuel cell is used, then the electricity from the fuel cell is used to power electric motors, which in turn drive the wheels.

16.16.3 Hydrogen-derived Ammonia for Land-Based Transportation

Ammonia could be used for powering terrestrial vehicles. Ammonia-powered trucks are presently being researched in Canada, among other places; see (Brown, 2019).

16.16.4 Hydrogen-derived Fuels for Marine Transportation

Ammonia has been proposed as a fuel for low-carbon marine transportation. One example is the shipping company Edesvik's vessel, *Viking Energy*. As of 2023, this ship was being retrofitted with an ammonia-based fuel cell to create electricity to power the ship. The ammonia is to be first split into nitrogen and hydrogen. The nitrogen can be released into the air, and the hydrogen will be used by the fuel cell. See Eidesvik (2023).

Ammonia could likely also be used directly in an internal combustion engine for marine applications. A project undertaken to develop such an engine by the manufacturer Wärtsilä is an example of this; see Wartsila (2022).

Methanol is also a possible marine fuel. The Stena ferry line already has such a ship. Maersk has announced plans to order 8 ships; see Carbon Commentary (2021).

16.16.5 Hydrogen or Hydrogen-derived Fuel for Aircraft

Aircraft could be powered directly by hydrogen or a hydrogen-based synthetic fuel. Hydrogen-powered aircraft have already been prototyped, and at least one aircraft manufacturer (Airbus) has announced its intentions to have a hydrogen-powered commercial airplane by 2035. See Airbus (2023).

16.17 Power to Hydrogen-Based Chemical Products

Power to products could take a number of forms, the most obvious being in the direct production of something that is readily storable. Power-to-chemicals is one category.

16.17.1 Synthetic Hydrocarbons

Hydrocarbons have a variety of uses as chemicals in addition to being fuels. Methane and some other compounds, which were already discussed as a possible fuel, are among them. Methane could also be used as a precursor to other synthetic hydrocarbons.

One other route to hydrocarbons is the Fischer-Tropsch process, which is a collection of reactions that yield hydrocarbons from hydrogen and carbon monoxide.

$$(2n + 1)H_2 + nCO \rightarrow C_nH_{2n+2} + nH_2O \tag{16.26}$$

An additional required step would be to obtain the carbon monoxide from the air. There are a variety of options for this, starting with carbon dioxide, presumably extracted from the air, as discussed previously. One possible method is described by Wang *et al.* (2021)

16.17.2 Hydrogen-derived Ammonia-Based Fertilizer

Ammonia-based fertilizer is another obvious product. According to (IPCC, 2007), the fertilizer industry uses 1.2% of the world's energy, and 90% of that is ammonia-based. One type of fertilizer is ammonium nitrate, NH_4NO_3 which is made from the reaction of ammonia with nitric acid, HNO_3 as given in Equation (16.27):

$$HNO_3 + NH_3 \rightarrow NH_4NO_3 \tag{16.27}$$

Nitric acid itself is also made from ammonia, using a three-step process. The overall result is summarized in Equation (16.28).

$$20NH_3 + 21O_2 \rightarrow 8NH_4NO_3 + 4NO + 14H_2O \tag{16.28}$$

Note that nitric oxide is one product of the reaction. It can be recycled back into the process, making additional nitric acid, as shown in Equation (16.29).

$$NO + \frac{3}{4}O_2 + \frac{1}{2}H_2O \rightarrow HNO_3 \tag{16.29}$$

Urea, $CO(NH_2)_2$, is another chemical that can be used as a fertilizer. It is derived from ammonia and carbon dioxide using a two-step process. The overall exothermic reaction is given in Equation (16.30):

$$2NH_3 + CO_2 \rightarrow CO(NH_2)_2 + H_2O \ \Delta H = -101.5\,kJ/mol \tag{16.30}$$

As with synthetic hydrocarbons, the carbon dioxide would ideally come from a carbon capture process.

16.18 The Electricity Grids of the Future

If wind energy is to supply a large fraction of the world's energy in the future and do that using electricity, the electricity grids will need to be significantly upgraded to better incorporate variable generation. Of particular relevance is that the grids will become increasingly converter dominated. There will be fewer rotating generators supplying mechanical inertia to help stabilize the electricity supply than is currently the case. There are numerous technical issues to be addressed to ensure that the grids operate properly: overall stability and reliability; protection against lightning, short circuits and surges; operation under perturbation by transients, resonance, and voltage instabilities; energy demand matching within minutes to hours; and keeping frequency, voltage, and phase within acceptable limits. More details on these and other related issues are provided by Holttinen *et al.* (2016).

16.19 Wind Turbines for the Energy Transition

In order for wind energy to make a major contribution to the transition to a low-carbon energy future, there is still a great deal to be done to facilitate the production of many wind turbines that are as efficient as possible, reliable, use minimal resources themselves, are recyclable and are adaptable to large scale wind power plants. For isolated systems, appropriate wind turbine designs are also needed. Important topics of the future include industrialization of production and circular manufacturing. In this regard, there is still much research to be done. See, for example, Veers *et al.* (2019)

References

Airbus (2023) *ZEROe: Towards the world's first hydrogen-powered commercial aircraft*, https://www.airbus.com/en/innovation/zero-emission/hydrogen/zeroe Accessed 5/19/2023.

Andrews, W. T., Shumway, S. A. and Russell, B. (2006) *Design of a 10,000 cu-m/d Seawater Reverse Osmosis Plant on New Providence Island, The Bahamas.* DesalCo Limited.

Arellano-Prieto, Y., Chavez-Panduro, E., Rossi, P. S. and Finotti, F. (2022) Energy storage solutions for offshore applications. *Energies* **15**, 6153.

Baxter, R. (2005) *Energy Storage: A Non-Technical Guide.* Pennwell Books, Tulsa.

Bergey, M. (1998) Wind electric pumping systems for communities. *Proc. First International Symposium on Safe Drinking Water in Small Systems,* Washington (also available at https://www.bergey.com/wind-school/wind-electric-pumping-systems-for-communities/ Accessed 2/20/2024.)

Beuttler, C., Charles, L. and Wurzbacher, J. (2019) The role of direct air capture in mitigation of anthropogenic greenhouse gas emissions. *Frontiers in Climate,* **1**, 10.

Blarke, M. B. and Jenkins, B. M. (2013) Super grid or smart grid: Competing strategies for large-scale integration of intermittent renewables? *Energy Policy,* **58**, 381–390.

Bossel, U. (2003) The physics of the hydrogen economy. *European Fuel Cell News,* **10**(2), 1–16.

Brown, T. (2019) *Heavy-duty diesel trucks to be converted to use ammonia fuel in Canada,* https://www.ammoniaenergy.org/articles/heavy-duty-diesel-trucks-to-be-converted-to-use-ammonia-fuel-in-canada/ Accessed 12/3/2023.

Bullough, C., Gatzem, C., Jakiel, C., Koller, A., Nowi, A. and Zunft, S. (2004) Advanced adiabatic compressed air energy storage for the integration of wind energy. *Proc. of the 2004 European Wind Energy Conference,* London (also available at https://www.nrc.gov/docs/ML1202/ML12026A783.pdf Accessed 2/20/2024.)

Carbon Commentary (2021) *Maersk's methanol fuelled container ships,* https://www.carboncommentary.com/blog/2021/8/26/maersks-methanol-fuelled-container-ships Accessed 5/19/2023.

Carta, J. A., Gonzáalez, J. and Subiela, V. (2003) Operational analysis of an innovative wind powered reverse osmosis system installed in the Canary Islands. *Solar Energy,* **75**(2), 153–168.

Cengel, Y. and Boles, M. (2019) *Thermodynamics: An Engineering Approach,* 9th edition. McGraw Hill, New York.

Cooper, A. and Schefter, K. (2018) *Electric Vehicle Sales Forecast and the Charging Infrastructure Required Through 2030.* Institute for Electric Innovation, Washington, DC.

Denholm, P., Mai, T. Kenyon, R. W., Kroposki, B. and O'Malley, M. (2020) *Inertia and the Power Grid: A Guide Without the Spin,* NREL/TP-6A20-73856.

Eidesvik (2023) *Viking Energy with ammonia-driven fuel cell,* https://eidesvik.no/viking-energy-with-ammonia-driven-fuel-cell/ Accessed 5/19/2023.

E.ON (2023) Electric car battery capacity and lifespan, https://www.eonenergy.com/electric-vehicle-charging/running-costs-and-benefits/battery-capacity-and-lifespan.html Accessed 12/3/2023.

Fabre, A. (2003) Wind turbines designed for the specific environment of islands case study: experience of an autonomous wind powered water desalination system on the island of Therasia, Greece. *Proc. International Conference RES for Islands, Tourism, and Desalination,* Crete, Greece.

Fink, D. and Beaty, H. (1978) *Standard Handbook for Electrical Engineers.* McGraw-Hill.

Fraenkel, P. (1997) *Water-Pumping Devices: A Handbook for Users and Choosers.* Intermediate Technology Press, Reading.

Henderson, C. R., Manwell, J. F. and McGowan, J. G. (2009) A wind/diesel hybrid system with desalination for Star Island, NH: feasibility study results. *Desalination,* **237**, 318–329.

Holttinen, H., Kiviluoma, J., Forcione, A., Milligan, M., Smith, C. J., Dillon, J., O'Malley, M., Dobschinski, J., van Roon, S., Cutululis, N., Orths, A., Eriksen, P. B., Enrico, Carlini, E. M., Estanqueiro, A., Bessa, R., Söder, L., Farahmand, H., Torres, J. R., Jianhua, B., Kondoh, J., Pineda, I. and Strbac, G. (2016) *Design and Operation of Power Systems with Large Amounts of Wind Power.* International Energy Agency, Geneva.

Holz, R., Drouihlet, S. and Gevorgian, V. (1998) Wind-electric ice making investigation, NREL/CP-500-24622. *Proc. 1998 American Wind Energy Association Conference,* Bakersfield, CA.

Hunter, R. and Elliot, G. (Eds.) (1994) *Wind–Diesel Systems*. Cambridge University Press, Cambridge, UK.

Ibrahim, H., Ilincaa, A. and Perron, J. (2008) Energy storage systems – characteristics and comparisons. *Renewable and Sustainable Energy Reviews*, **12**, 1221–1250.

IEA (2021a) *Direct Air Capture. International Energy Agency*, Geneva, https://www.iea.org/reports/direct-air-capture Accessed 8/3/2023.

IEA (2021b) *Clean Urban Transport for Europe (CUTE) – Hydrogen and Fuel Cell Buses*. International Energy Agency, Geneva https://www.iea.org/policies/3735-clean-urban-transport-for-europe-cute-hydrogen-and-fuel-cell-buses Accessed 8/3/2023.

Iglesias, I. J., Garcia-Tabares, L., Agudo, A., Cruz, I. and Arribas, L. (2000) Design and simulation of a stand-alone wind-diesel generator with a flywheel energy storage system to supply the required active and reactive power *Proc. IEEE Power Electronics Specialists Conference*. IEEE Power Electronics Society, Washington (also available from: https://pdfs.semanticscholar.org/5736/67bc7c300ff96c8eb28299468a4ba93fe489.pdf Accessed 2/20/2024.

Infield, D. G., Lundsager, P., Pierik, J. T. G, van Dijk, V. A. P., Falchetta, M., Skarstein, O. and Lund, P. D. (1990) Wind diesel system modelling and design. *Proc. 1990 European Community Wind Energy Conference*, Madrid, pp. 569–574.

IPCC (2007) *IPCC Fourth Assessment Report: Climate Change 2007*: Fertilizer Manufacture, https://archive.ipcc.ch/publications_and_data/ar4/wg3/en/ch7s7-4-3-2.html Accessed 12/3/2023

Johnson, C., Abdulwahid, U., Manwell, J. F. and Rogers, A. L. (2002) Design and modeling of dispatchable heat storage in remote wind/diesel systems. *Proc. of the 2002 World Wind Energy Conference,* Berlin. WIP-Munich.

Kempton, W. and Tomić, J. (2005a) Vehicle to grid fundamentals: calculating capacity and net revenue. *Journal of Power Sources*, **144**(1), 268–279. https://doi.org/10.1016/j.jpowsour.2004.12.025 Accessed 12/3/2023.

Kempton, W. and Tomić, J. (2005b) Vehicle to Grid Implementation: From stabilizing the grid to supporting large-scale renewable energy. *Journal of Power Sources*, **144**(1), 280–294. https://doi.org/10.1016/j.jpowsour.2004.12.022 Accessed 12/3/2023.

Kentfield, J. (1996) *The Fundamentals of Wind-Driven Water Pumpers*. Gordon and Breach Science Publishers, Amsterdam.

Larminie, J. and Dicks, A. (2003) *Fuel Cell Systems Explained*, 2nd edition. John Wiley & Sons, Ltd, Chichester.

Le Gouriérès (1982) *Wind Power Plants: Theory and Design*. Pergamon Press, London (now Elsevier, Amsterdam).

Lyons, J. and Goodman, N. (2004) TDX Power: St Paul Alaska operational experience. *Proc. 2004 Wind-Diesel Workshop,* Anchorage.

Lysen, E. H. (1983) *Introduction to Wind Energy*. Consultancy Wind Energy in Developing Countries, Amersfoort, Netherlands, https://www.arrakis.nl/documents/Introduction%20to%20Wind%20Energy%20E.H.%20Lysen%20CWD%2082-1%20may%201983%20OCR.pdf, Accessed 8/15/2023.

Manwell, J. F. (1988) *Understanding Wind Energy for Water Pumping*. Volunteers in Technical Assistance, Arlington, VA, (also available at https://pdf.usaid.gov/pdf_docs/PNABC983.pdf Accessed 12/3/2023).

Manwell, J. F. and McGowan, J. G. (1981) A design procedure for wind-powered heating systems. *Solar Energy*, **26**(5), 437–445.

Manwell, J. F. and McGowan, J. G. (1993) Lead acid battery storage model for hybrid energy systems. *Solar Energy*, **50**(5), 399–405.

Manwell, J. F. and McGowan, J. G. (1994) Extension of the Kinetic Battery Model for Wind/Hybrid Power Systems. *Proc. 1994 European Wind Energy Conference,* Thessaloniki.

Manwell, J. F., Rogers, A., Hayman, G., Avelar, C. T. and McGowan, J. G. (1997) *Hybrid2–A Hybrid System Simulation Model, Theory Summary*. NREL Subcontract No. XL-1-11126-1-1. Dept. of Mechanical and Industrial Engineering, University of Massachusetts, Amherst.

Manwell, J. F., McGowan, J. G. and Breger, D. (2018) A design and analysis tool for utility scale power systems incorporating large scale wind, solar photovoltaics and energy storage. *Journal of Energy Storage*, **19**, 103–112.

Manwell, J. F. and McGowan, J. G. (2023) Modelling Energy Storage, Electric Vehicles and Power-to-X in a Large Scale Hybrid Offshore Wind Power System, *Proceedings of the Offshore Energy and Storage Symposium*, Malta.

Masters, G. (2013) *Renewable and Efficient Electric Power Systems*. Wiley-IEEE Press, Chichester.

May, G. J., Davidson, A. and Monahof, B. (2018) Lead batteries for utility energy storage: a review. *Journal of Energy Storage*, **15**, 145–157.

Miranda, M. S. and Infield, D. (2002) A wind powered seawater reverse-osmosis system without batteries. *Desalination*, **153**, 9–16.

Rahman, M. M., Oni, A. O., Gemechu, E. and Kumar, A. (2021) The development of techno-economic models for the assessment of utility-scale electro-chemical battery storage systems. *Applied Energy*, **283**, 116343.

Richardson, D. B., (2013) Electric vehicles and the electric grid: a review of modeling approaches, Impacts, and renewable energy integration. *Renewable and Sustainable Energy Reviews*, **19**, 247–254.

Sinclair, B. A. and Somerville, W. M. (1989) Experience with the wind turbine flywheel combination on Fair Isle. *Proc. 1989 European Wind Energy Conference*, Glasgow.

Somerville, W. M. (1989) Wind turbine and pumped storage hydro generation in Foula. *Proc. 1989 European Wind Energy Conference*, Glasgow.

Sterner, M.; Specht, M. (2021) Power-to-Gas and power-to-X – the history and results of developing a new storage concept. *Energies*, **14**, 6594. https://doi.org/10.3390/en14206594 Accessed 7/3/2023.

Thompson, A. M. (2004) *Reverse-Osmosis Desalination of Seawater Powered by Photovoltaics Without Batteries*, PhD Dissertation, Loughborough University.

Thompson, M. and Infield, D. (2002) A photovoltaic-powered seawater reverse-osmosis system without batteries. *Desalination*, **153**, 1–8.

Tosatto, A., Beseler, X. M., Østergaard, J., Pinson, P. and Chatzivasileiadis, S. (2022) North Sea Energy Islands: Impact on national markets and grids. *Energy Policy*, **167**, 112907.

Ulleberg, Ø., Nakken, T. and Ete, A. (2010) The wind/hydrogen demonstration system at Utsira in Norway: evaluation of system performance using operational data and updated hydrogen energy system modeling tools. *International Journal of Hydrogen Energy*, **35**(5), 1841–1852.

Valera-Medina, A., Amer-Hatem, F., Azad, A. K., Dedoussi, I., De Joannon, M, Fernandes, R. X., Glarborg, P., Hashemi, H., He, X., Mashurk, S., McGowan, J., Mounaim-Rouselle, C., Ortiz-Prado, A., Ortiz-Valera, J. A., Rossetti, I., Shu, B., Yehia, M., Xiao,H., Costa, M. (2021) Review on ammonia as a potential fuel: from synthesis to economics. *Energy & Fuels*, **35**(9), 6964–7029.

Van Gerpen, J. H. and He, B. B., (2014) Biodiesel and renewable diesel production methods, in *Advances in Biorefineries, Biomass and Waste Supply Chain Exploitation*, pp 441–475 Woodhead Publishing

Veers, P., Dykes, K., Lantz, E., Barth, S., Bottasso, C.L., Carlson, O., Clifton, A., Green, J., Green, P., Holttinen, H., Laird, D., Lehtomaki, V., Lundquist, J., Manwell, J., Marquis, M., Meneveau, C., Moriarty, P., Munduate, M., Muskulus, M., Naughton, J., Pao, L., Paquette, J., Peinke, J., Robertson, A., Rodrigo, J.S., Sempreviva, A.M., Smith, J.C., Tuohy, A., Wiser, R. (2019) Grand challenges in the science of wind energy, *Science*, **366**, 6464.

Wang, C., Yang, W. C. D., Raciti, D. *et al.* (2021) Endothermic reaction at room temperature enabled by deep-ultraviolet plasmons. *Nature Materials* **20**, 346–352. https://doi.org/10.1038/s41563-020-00851-x Accessed 12/3/2023.

Wang, L., Rao, M., Diethelm, S., Lind, T-E, Zhang, H., Hagen, A., Maréchal, F., Van herle, J. (2019) Power-to-methane via co-electrolysis of H_2O and CO_2: The effects of pressurized operation and internal methanation, *Applied Energy*, **250**, 1432–1445.

Warfel, C. G., Manwell, J. F. and McGowan, J. G. (1988) Techno-economic study of autonomous wind driven reverse osmosis desalination systems. *Solar and Wind Technology*, **5**(5), 549–561.

Wärtsilä (2022) Wärtsilä coordinates EU funded project to accelerate ammonia engine development, https://www.wartsila.com/media/news/05-04-2022-wartsila-coordinates-eu-funded-project-to-accelerate-ammonia-engine-development-3079950 Accessed 5/19/2023

Wenzl, H., Baring-Gould, I., Kaiser, R., Liaw, B. Y., Lundsager, P., Manwell, J. F., Ruddell, A. and Svoboda, V. (2005) Life prediction of batteries for selecting the technically most suitable and cost effective battery. *Journal of Power Sources*, **144**(2), 373–384.

Wilf, M. and Bartels, C. (2005) Optimization of seawater RO systems design. *Desalination*, **173**, 1–12.

Wiser, R., Bolinger, M. Gorman, W. Rand, J. Jeong, S., Seel, J., Warner, C. Paulos, B. (2021) *Hybrid Power Plants: Status of Installed and Proposed Projects.* Lawrence Berkley National Laboratory, Berkeley, CA.

Appendix A

Nomenclature

A.1 Note on Nomenclature and Units

This text includes material from many different engineering disciplines (e.g., aerodynamics, dynamics, controls, electromagnetism, and acoustics). Within each of these disciplines, commonly accepted variables are often used for important concepts. Thus, for acoustics engineers, α denotes the sound absorption coefficient; in aerodynamics, it denotes the angle of attack; and in one common model for wind shear, it is the power law exponent. In this text, an effort has been made to ensure that concepts that are found in multiple chapters all have the same designation, while maintaining common designations for generally accepted concepts. Finally, in an effort to avoid confusion over multiple definitions of one symbol, the nomenclature used in the text is listed by chapter.

Throughout this text units are in general not stated explicitly. It is assumed, however, that SI units are used, specifically in MKS form. Accordingly, length is in meters (m), mass is in kilograms (kg), and time is in second (s). Other important units are density in kg/m^3, weight and force in newtons (N), stress in pascals (Pa or N/m^2), torque in Nm, and power in watts (W). When other forms are in common usage, such as power in kW or MW and mass in metric tonnes (t), those units are stated. Energy is in general also expressed in kWh or MWh, rather than in kJ or MJ. Angles are assumed to be in radians except where explicitly stated otherwise. There are two possible sources of confusion: rotational speed and frequency. In this text, when referring to rotational speed, n is always in revolutions per minute (rpm) and Ω is in radians/second. With regards to frequency f is always in cycles per second (Hz) while ω is always in radians/second. In other cases where the units are specific to a variable, that is so noted. Finally, advanced topics in this text are indicated with an an asterisk (*) at the beginning of the corresponding sections. Those sections can be bypassed if so desired.

A.2 Chapter 2

A.2.1 English Variables

a	Slope in MCP equation
b	Offset in MCP equation
A	Area
c	Weibull scale factor
c_P	Specific heat at constant pressure
C_f	Capacity factor

(Continued)

Wind Energy Explained: On Land and Offshore, Third Edition.
James F. Manwell, Emmanuel Branlard, Jon G. McGowan, and Bonnie Ram.
© 2024 John Wiley & Sons Ltd. Published 2024 by John Wiley & Sons Ltd.

C_P	Power coefficient
D	Rotor diameter
f	(i) Frequency (Hz)
f	(ii) Coriolis parameter
f_i	Number of occurrences in each bin
F_p	Pressure force
F_c	Coriolis force
$F()$	Cumulative distribution function
$f()$	Probability density function
g	Gravitational acceleration
H_0	Surface heat flux (W/m^2)
h	(i) Enthalpy
h	(ii) Elevation difference
i	Index or sample number
k	Weibull shape factor
ℓ	Mixing length
m	(i) Mass
m	(ii) Direction normal to lines of constant pressure
m_i	Midpoint of bins
N	Number of long-term data points
N_s	Number of samples in short-term averaging time
N_B	Number of bins
N_i	Number of occurrences
p	Pressure
P	Power
P_R	Rated power
$\overline{P}/A$	Average wind power density
$P_w(U)$	Wind turbine power
$\overline{P}_w$	Average wind turbine power
q	Heat transferred
R	(i) Radius of curvature
R	(ii) Radius of wind turbine rotor
R_i	Richardson number
t	Time
T	Temperature
u	Internal energy
$u*$	Friction velocity
U	Mean wind speed (mean of short-term data, typically over 10 minute period)
$\overline{U}$	Long-term mean wind speed (mean of short-term averages)
U_c	Characteristic wind velocity
U_i	Wind speed average over period i (typically 10 minutes)

U_g	Geostrophic wind speed
U_{gr}	Gradient wind speed
v	Specific volume
w_i	Width of bins
x	Dimensionless wind speed
z	Elevation
z_o	Surface roughness
z_r	Reference height

A.2.2 Greek Variables

α	Power law exponent
Γ	Lapse rate
$\Gamma()$	Gamma function
Δt	Duration of short-term averaging time
δt	Sampling period
η	Drive train efficiency
κ	von Karman constant
ρ	Air density
σ	Standard deviation
τ	Shear stress
τ_0	Surface value of shear stress
τ_{xz}	Shear stress in the direction of x whose normal coincides with z
ϕ	Latitude
ω	Angular rotational speed of earth

A.2.3 Acronyms

CDF	Cumulative distribution function
CFD	Computational fluid dynamics
MCP	Measure-correlate-predict

A.3 Chapter 3

A.3.1 English Variables

A	Surface area, rotor swept area
a	Axial induction factor
a_1	(i) Axial induction factor corresponding to $\lambda_r = 0$ (2D momentum theory)

(Continued)

a_1	(ii) Axial induction factor corresponding to λ (2D momentum theory)
a'	Tangential induction factor
b	(i) Axial induction factor in the far-wake
b'	(ii) Tangential induction factor in the far-wake
B	Number of blades
c	Chord, airfoil chord length
c_n	Two-dimensional airfoil coefficient normal to the rotor plane
c_t	Two-dimensional airfoil coefficient tangential to the rotor plane
C_d	Two-dimensional drag coefficient
$C_{d,0}$	Constant drag term
$C_{d,a1}$	Linear drag term
$C_{d,a2}$	Quadratic drag term
C_l	Two-dimensional lift coefficient
$C_{l,a}$	Slope of lift coefficient curve
$C_{l,0}$	Lift coefficient at zero angle of attack
C_m	Two-dimensional pitching moment coefficient
C_P	Power coefficient
C_p	Pressure coefficient
C_T	Thrust coefficient
$C_{T,}$	Local thrust coefficien
d_1	Variable for determining angle of attack with simplified method
d_2	Variable for determining angle of attack with simplified method
dA	Elementary area, area of an annulus
dF_D	Incremental drag force
dF_L	Incremental lift force
dF_n	Incremental force normal to plane of rotation (thrust)
dF_t	Incremental force tangential to circle swept by blade section
dS	Elementary surface
dT	Elementary thrust on an annular control volume or blade cross-section
dT_p	Elementary thrust due to the pressure force
dP	Elementary power on an annular control volume or blade cross-section
dQ	Elementary torque on an annular control volume or blade cross-section
dr	Elementary distance along the radial direction
$\tilde{D}$	Drag force per unit length
f_s	Separation function in dynamic stall model
F	Tip loss correction factor
F_D	Drag force
F_L	Lift force
F_n	Normal force
F_t	Tangential force
k	Factor for the calculation of the axial induction factor with the steady BEM equations

k'	Factor for the calculation of the tangential induction factor with the steady BEM equations
k_ω	Reduced frequency
$\widetilde{L}$	Lift force per unit length
$\widetilde{M}$	Pitching moment per unit length
$\dot{m}$	Flow rate
$\hat{\boldsymbol{n}}$	Unit vector normal to a surface
P	Power; frequency of rotation
p	Pressure
Q	Torque
q_1	Variable for determining angle of attack with simplified method
q_2	Variable for determining angle of attack with simplified method
q_3	Variable for determining angle of attack with simplified method
R	Outer blade radius
r	Radius
r_h	Rotor radius at the hub
T	Thrust
t	Time
$\boldsymbol{u,U}$	Three-dimensional velocity vector
U	Flow velocity in the axial direction; characteristic velocity used for dimensionless factors
U_r	Radial flow velocity
x	Longitudinal distance perpendicular to the rotor plane; Dummy variable

A.3.2 Greek Variables

α	Angle of attack
γ	Vorticity surface intensity
Γ	Circulation
Δp	Pressure jump
η	Efficiency
θ_c	Blade pitch angle as prescribed by the controller.
θ_e	Section torsion angle from the elastic motion of the blade
θ_T	Blade twist angle
θ_p	Section pitch angle (controller pitch, twist, and torsion)
λ	Tip speed ratio
λ_h	Local speed ratio at the hub
λ_r	Local speed ratio
μ	Coefficient of viscosity (dynamic viscosity)
ν	Kinematic viscosity

(Continued)

ξ	Transformed coordinate in conformal mapping
ρ	Air density
τ	Dimensionless time constant
σ	Local blade solidity
σ_{rot}	Rotor solidity
φ, ϕ	Inflow angle
χ	Wake skew angle
ψ	Azimuthal angle
$\boldsymbol{\psi}$	Vector potential
ω	Angular velocity of the wind (swirl)
$\boldsymbol{\omega}$	Vorticity vector
Ω	Angular velocity of the wind turbine rotor

A.3.3 Subscripts

$\perp$	Vector norm in the plane orthogonal to the airfoil section
1	Upstream plane
2	Rotor plane, upstream of actuator disc
3	Rotor plane, downstream of actuator disc
4	Far-wake plane
a	Airfoil coordinate system
atm	Atmospheric (wind), without disturbances and inductions
CV	Control volume
max	Maximum/optimal
n	Normal to rotor plane
$dist$	Disturbed (wind), due to e.g. tower shadow
$elast$	Elastic/structural motion
$elec$	Electrical
ind	Induced (induced velocity)
$mech$	Mechanical
qs	Quasi-steady
$root$	Root of the blade
rel	Relative wind with respect to airfoil section
red	Reduced variable in dynamic inflow model
$skew$	Quantity related to the rotor skew (tilt and yaw)
t	Tangential to rotor plane
tip	Tip of the blade

A.3.4 Acronyms

BEM	Blade element momentum
BET	Blade element theory
CFD	Computational fluid dynamics
HAWT	Horizontal axis wind turbine
VAWT	Vertical axis wind turbine

A.4 Chapter 4

A.4.1 English Variables

A	(i) Area
A	(ii) Constant in Euler beam equation
A	(iii) First axisymmetric flow term in LHSBM (flapping), $= (\Lambda/3) - (\theta_p/4)$
$\boldsymbol{A}$	System state matrix
A_{1b}	First moment of area of one blade
A_3	Axisymmetric flow term in LHSBM (flapping), $= (\Lambda/2) - (2\theta_P/3)$
A_2	Aerodynamic term in LHSBM (lead-lag)
A_4	Aerodynamic term in LHSBM (lead-lag)
a	(i) Offset of center of mass of gyroscope
a	(ii) Axial induction factor
B	(i) Number of blades
B	(ii) Dimensionless gravity term in LHSBM, $= G/2\ \Omega^2$
C	(i) Amplitude
$\mathbf{C}$	(ii) Damping matrix
$C_{l\alpha}$	Slope of angle of attack line
C_P	Rotor power coefficient
C_Q	Rotor torque coefficient
C_T	Rotor thrust coefficient
C_{T_r}	Local thrust coefficient
C_D	Hydrodynamic drag coefficient
C_m	Hydrodynamic added mass coefficient
c	(i) Chord
c	(ii) Damping coefficient
c_b	Maximum distance from neutral axis for a beam
c_c	Critical damping coefficient
c_n	Aerodynamic coefficient normal to rotor plane
c_t	(i) Aerodynamic coefficient tangential to rotor plane
c_t	(ii) Generalized damping constant for tower

(Continued)

D	(i) Diameter
D	(ii) Determinant
$\bar{d}$	Normalized yaw moment arm $= d_{yaw}/R$
d_{yaw}	Yaw moment arm
dF	Elementary force
E	(i) Modulus of elasticity
E	(ii) Energy
e	Nondimensional hinge offset
$\boldsymbol{e}$	Equations of motion
F	External force
$\tilde{\boldsymbol{f}}$	Force vector per unit length
f	Force in generalized equation of motion
G	Gravity term in LHSBM, $= g\ M_B\ r_g/I_b$
g	Gravitational constant
$\boldsymbol{H}$	Angular momentum vector
H_T	Tower height
I	Area moment of inertia
J	Mass moment of inertia of a body with respect to an axis and point
$\mathbf{J}$	Inertia tensor of a body with respect to a point
J_β	Mass moment of inertia of the blade (flapping)
J_ζ	Mass moment of inertia of the blade (lead-lag)
J_r	Mass moment of inertia of the rotor at rotor center with respect to shaft axis
J_g	Mass moment of inertia of the generator on the LSS
$\mathbf{K}$	Stiffness matrix
K	(i) Generalized/effective stiffness
K	(ii) Flapping inertial natural frequency in LHSBM, $K = 1 + \varepsilon + K_\beta/I_b\Omega^2$
K_β	Spring constant in flapping direction
K_ζ	Spring constant in lead-lag direction
K_2	Dimensionless spring constant for lead-lag motion in LHSBM
K_{vs}	Vertical wind shear constant in LHSBM
k	(i) Spring constant
k_t	Generalized spring constant for tower
L	Length
L	Lift force per unit length
M	(i) Moment
M	(ii) Generalized/Effective mass
$\mathbf{M}$	(i) Mass matrix
$\boldsymbol{M}$	(ii) Moment, vector
M_t	Top mass above tower
$M_{X'}$	Yawing moments on tower in LHSBM
$M_{Y'}$	Backwards pitching moments on tower in LHSBM
$M_{Z'}$	Rolling moment on nacelle in LHSBM

M_β	Flapping moment at the root of the blade
M_ζ	Lead–lag moment at the root of the blade
m	Mass
m_B	Mass of blade
m_i	Mass of ith section of beam (Myklestad method)
$\widetilde{m}(z)$	Mass per unit length along a beam
N	(i) Number of teeth on a gear
$N(z)$	(ii) Axial load along a beam
n_g	Gear ratio of the gear box
P	Power
p	Loading per unit length (along a beam)
Q	(i) Torque
Q	(ii) Generalized force
Q_a	Aerodynamic torque
Q_g	Generator torque
$\boldsymbol{q}$	Vector of geenralized degrees of freedom
q	Yaw rate in LHSBM (rad/s)
$\bar{q}$	Normalized yaw rate in LHSBM, $= q/\Omega$
R	(i) Radius of rotor
$\boldsymbol{R_{gb}}$	Rotation/transformation matrix from body b to body g
r	(i) Radial distance from axis of rotation
r	(ii) Taper ratio for a tapered cylinder
r_g	Radial distance from hinge to center of mass in LHSBM
r_{gyr}	Radius of gyration
S	Shear force
S_β	Flapwise shear force
S_ζ	Edgewise shear force
T	(i) Thrust
T	(ii) Kinetic energy for Lagrange's equation
T_a	Thrust
t	Time
U	(i) Free stream wind velocity
U	(ii) Potential energy for Lagrange's equation
$\bar{U}$	Normalized wind velocity, $= U/\Omega\ R = 1/\lambda$
U_{rel}	Relative wind velocity
U_n	Wind velocity normal to rotor plane
U_t	Wind velocity tangential to rotor plane
$\boldsymbol{u}$	Vector of inputs in generalized equations of motion
u	Deflection of a beam (usually in the x direction)
$\mathbf{v}$	Generic vector. In body "b", its coordinates are $\mathbf{v}_b = (v_{x,b},\ v_{y,b},\ v_{z,b})$
W	Total load or weight

(Continued)

x	Motion in surge/fore-aft/downstream direction
$\boldsymbol{x}$	Vector of degrees of freedom
y	Motion in sway/side-side direction
z	(i) Motion in heave/vertical direction.
z	(ii) Distance along a beam, blade or tower

A.4.2 Greek Variables

α	(i) Angular acceleration of rotor, $= \dot{\Omega}$
α	(ii) Mass proportional coefficient in Rayleigh damping
β	(i) Term used in vibrating beam solution
β	(ii) Flapping angle (radians)
β	(iii) Stiffness proportional coefficient in Rayleigh damping
$\dot{\beta}$	Flapping velocity (radians/s)
$\ddot{\beta}$	Second time derivative of flapping angle (radians/s^2)
β_0	Collective flapping coefficient (radians)
β_{1c}	Cosine flapping coefficient (radians)
β_{1s}	Sine flapping coefficient (radians)
β'	Azimuthal derivative of flap angle, $= \ddot{\beta}/\Omega^2$ (radians^{-1})
β''	Azimuthal second derivative of flap angle, $= \dot{\beta}/\Omega$
γ	Lock number, $= \rho\ c\ C_{L\alpha}R^4/I_b$
ε	Offset term in LHSBM, $= 3e/[2(1-e)]$
ζ	(i) Damping ratio
ζ	(ii) Lead-lag/edgewise angle (radians)
θ	(i) Arbitrary angle (radians)
θ	(ii) Slope of deflected beam (radians)
θ_p	Pitch angle (positive toward feathering) (radians)
κ	Curvature of a deflected beam
λ	Tip speed ratio
Λ	Nondimensional inflow, $= U(1-a)/\Omega\ R$
ν	Shaft torsion degree of freedom
ρ	Air density
ρ_m	Material density
ρ_w	Water density
σ	Stress
ϕ	Phase angle (vibrations) (radians)
Φ	Assumed shape function
$\boldsymbol{\Phi}_i$	Assumed shape function (3D vector field)
$\boldsymbol{\Psi}_i$	Assumed rotational field (3D vector field)
ψ	Azimuth angle, $0 =$ up (radians or degrees)
ω	(i) Rate of precession of gyroscope (radians/s)

ω	(ii) Frequency of oscillation (e.g., forced oscillations) (radians/s)
ω_i	(ii) Frequency of oscillation (radians/s) of ith mode
$\boldsymbol{\omega}$	Rate of precession of gyroscope (vector)
ω_d	Damped frequency (radians/s)
ω_n	Natural frequency (radians/s)
ω_{NR}	Natural flapping frequency of nonrotating blade (radians/s)
ω_R	Natural flapping frequency of rotating blade (radians/s)
ω_β	Flapping frequency (radians/s)
ω_ς	Edgewise frequency (radians/s)
Ω	Angular velocity $(= \dot{\psi})$ (radians/s)
$\boldsymbol{\Omega}$	Angular velocity vector

A.4.3 Subscripts and Acronyms

adp	Aerodynamic damping
c	critical
DOF	Degree(s) of freedom
HSS	High-speed shaft
in	Inner, for inner radius of a cylinder
vs	Vertical wind shear
LHSBM	Linear hinge spring blade model
LSS	Low-speed shaft
max	Maximum
min	Minimum
NR	Nonrotating
out	Outer, for outer radius of a cylinder
root	Blade root
R	Rotating
TB	Tower bottom
TT	Tower top

A.4.4 Coordinate Systems

b	Blade
h	Hub
i	Inertial
n	Nacelle
t	Tower

A.5 Chapter 5

A.5.1 English Variables

a	(i) Real component in complex number notation
a	(ii) Turns ratio of transformer
a_n	Coefficient of cosine terms in Fourier series
A	Area
A_c	Cross-sectional area of coil
A_m	Magnitude of arbitrary phasor
A_g	Cross-sectional area of gap
$\hat{A}$	Arbitrary phasor
b	Imaginary component in complex number notation
b_n	Coefficient of sine terms in Fourier series
B	Magnitude of magnetic flux density (scalar)
$\boldsymbol{B}$	Magnetic flux density (vector)
C	(i) Capacitance (F)
C	(ii) Constant
C_m	Magnitude of arbitrary phasor
$\hat{C}$	Arbitrary phasor
e_m	Stored magnetic field energy per unit volume
E	(i) Energy (J)
E	(ii) Induced electromotive force (EMF)
E	(iii) Synchronous generator field induced voltage
$\hat{E}$	Electromotive force phasor
E_m	Stored energy in magnetic fields
E_1	Primary voltage in transformer
E_2	Secondary voltage in transformer
f	Frequency of AC electrical supply (Hz)
$f_{scale}(n)$	Scale factor of nth harmonic
$\boldsymbol{F}$	Force (vector)
g	Air gap width
$\boldsymbol{H}$	Magnetic field intensity (vector) (A-t/m)
H_c	Field intensity inside the core (A-t/m)
HD_n	Harmonic distortion at n^{th} harmonic
i	Instantaneous current
I	Mean current
$\hat{I}$	Current, phasor
I_a	Synchronous machine's armature current
I_f	Field current (a.k.a. excitation)
I_L	Line current in three-phase system

I_{max}	Maximum value of current (AC)
I_M	Magnetizing current
I_P	Phase current in three-phase system
I_{rms}	Root mean square (rms) current
I_R	Rotor current
I_S	Induction machine stator phasor current
j	$\sqrt{-1}$
J	Inertia of generator rotor
k_1	Constant of proportionality in synchronous machine (Wb/A)
k_2	Constant of proportionality (V/Wb-rpm)
ℓ	Path length (vector)
ℓ_c	Length of the core at its midpoint
L	(i) Inductance (H)
L	(ii) Length of coil
n	(i) Actual rotational speed (rpm)
n	(ii) Harmonic number
n_s	Synchronous speed (rpm)
N	Number of turns in a coil
NI	Magnetomotive force (MMF) (A-t)
$N\Phi$	Flux linkages (Wb)
N_1	Number of coils on primary winding
N_2	Number of coils on secondary winding
P	(i) Real power
P	(ii) Number of poles
P_g	Air gap power in induction machine
P_{in}	Mechanical input power input to induction generator
P_{inst}	Instantaneous power
P_{loss}	Power lost in induction machine's stator
$P_{loss,mech}$	Mechanical losses
P_m	Mechanical power converted in induction machine
P_{out}	Electrical power delivered from induction generator
P_1	Power in one phase of three-phase system
PF	Power factor
Q	Reactive power (VA)
Q_e	Electrical torque
Q_m	Mechanical torque
Q_r	Applied torque to generator rotor
r	(i) Radial distance from center of toroid
r	(ii) Radius

(Continued)

r_i	Inner radius of coil
r_o	Outer radius of coil
R	(i) Equivalent induction machine resistance
R	(ii) Reluctance of magnetic circuit (A-t/Wb)
R	(iii) Resistance
Re{ }	Real part of complex number
R_M	Resistance in parallel with mutual inductance
R_s	Synchronous generator resistance
R_S	Stator resistance
R_R'	Rotor resistance (referred to stator)
$sq(t_i)$	Sample approximation of square wave
S	Apparent power (VA)
t	Time
T	Period of fundamental frequency, s
THD	Total harmonic distortion
v	Instantaneous voltage
$v(t)$	Instantaneous voltage at time t
V	(i) Voltage in general
V	(ii) Electrical machine terminal voltage
$\hat{V}$	Voltage, phasor
$\overline{V}_{DC}$	DC rectifier voltage (average)
V_{LL}	Line-to-line voltage in three-phase system
V_{LN}	Line-to-neutral voltage in three-phase system
V_{rms}	Root mean square (rms) voltage
V_{Max}	Maximum voltage
V_{RT}	Rotor terminal voltage (WRIG)
V_1	(i) AC voltage filter input voltage
V_1	(ii) Terminal voltage in primary winding
V_2	(i) AC voltage filter output voltage
V_2	(ii) Terminal voltage in secondary winding
X_C	Capacitive reactance
X_L	Inductive reactance
X_M	Magnetizing reactance
X_R	Rotor leakage inductive reactance (referred to stator)
X_S	(i) Stator leakage inductive reactance (inductance machine)
X_s	(ii) Synchronous generator synchronous reactance
Y	"Wye" connected three-phase system
Y''	Admittance in resonant filter
$\hat{Z}$	Impedance in general
$\hat{Z}_{IM}$	Induction machine equivalent impedance
$\hat{Z}_C$	Capacitive impedance

$\hat{Z}_i$	Impedance with index i
$\hat{Z}_L$	Inductive reactance
$\hat{Z}_M$	Magnetizing (mutual) impedance in induction machine
$\hat{Z}_P$	Parallel impedance
$\hat{Z}_R$	Resistive impedance
$\hat{Z}_s$	(i) Series impedance
$\hat{Z}_s$	(ii) Synchronous impedance
$\hat{Z}_Y$	Impedance of Y connected circuit
Z_1	Series impedance in resonant filter
Z_2	Parallel impedance in resonant filter
$\hat{Z}_\Delta$	Impedance of Δ connected circuit

A.5.2 Greek Variables

δ	Synchronous machine power angle (rad)
Δ	Delta connected three-phase system
η_{gen}	Overall efficiency (in the generator mode)
λ	Flux linkage, $= N\Phi$
μ	Permeability, $\mu = \mu_r\mu_0$ (Wb/A-m)
μ_0	Permeability of free space, $4\pi \times 10^{-7}$ (Wb/A-m)
μ_r	Relative permeability
ϕ	(i) Phase angle (rad)
ϕ	(ii) Power factor angle (rad)
ϕ_a	Phase angle of arbitrary phasor **A** (rad)
ϕ_b	Phase angle of arbitrary phasor **B** (rad)
Φ	Magnetic flux (Wb)
ω_r	Speed of generator rotor (rad/s)

A.5.3 Symbols

$\angle$	Angle between phasor and real axis
μF	MicroFarad (10^{-6} Farad)
Ω	Ohms

A.5.4 Acronyms

AC	Alternating current
DC	Direct current

(Continued)

DD	Direct drive
DFIG	Doubly fed induction generator
EESG	Electrically excited synchronous generator
HS	High speed
MS	Medium speed
PMSG	Permanent magnet induction generator
SCIG	Squirrel cage induction generator
WRIG	Wound rotor induction generator

A.6 Chapter 6

A.6.1 English Variables

$\Pr(U > U_e)$	Probability that $U > U_e$
a	Coherence decay constant
A	Amplitude
A_c	(i) Charnock constant
A_c	(ii) Cross-sectional area of the member
c	(i) Weibull scale factor
c	(ii) Wave celerity
$c_{H_{s,j}}$	Weibull scale factor of significant wave height in j^{th} wind speed bin
C_a	Added mass coefficient
C_d	Drag coefficient
$C_{d,10}$	Drag coefficient at 10 m above surface
C_m	Inertia coefficient
$C(\gamma)$	Normalizing factor in JONSWAP spectrum
D	Diameter (pile)
d	Water depth
f	Frequency (Hz)
f_i	Number of occurrences in each bin
f_p	Peak frequency (waves)
$f()$	Probability density function
F_D	drag force (on pile)
F_I	Inertia force due to waves (on pile)
$\tilde{F}$	Force per unit length
$\tilde{F}_c$	force per unit length due to current
$F()$	Cumulative distribution function
$F(x_2|x_1)$	Conditional cumulative distribution function
g	Gravitational acceleration
H	Wave height
H_{av}	Average wave height

H_B	Breaking wave height
$H_{\max}$	Maximum wave height
H_s	Significant wave height
h	Ice thickness
i, j	index or sample number
I_{ref}	Reference turbulence intensity at 15 m/s
I_{yr}	Intervals/yr
K_0	Regional constant affecting ice thickness
$K_{\max}$	Total hours of mean temperature below freezing point of water
k	(i) Turbulence subscript: u longitudinal, v lateral and w upward
k	(ii) Wave number
k	(iii) Weibull shape factor
$k_{H_{s,l}}$	Weibull shape factor of significant wave heigh in j^{th} wind speed bin
L	(i) Integral length scale
L	(ii) Wavelength
L_p	Peak wavelength
ℓ	Mixing length
M_D	Drag moment at seabed (on pile)
M_I	Inertia moment at seabed (on pile)
N	Number of long-term data points
N_B	Number of bins
N_s	Number of samples in short-term averaging time
N_w	Number of waves in sea state
R_{xy}	Cross-correlation
r	Lag number
S_w	Salinity of water
$S(f), S_{xx}(f)$	Power spectral density function
$S_{JS}(f)$	JONSWAP spectrum
$S_{PM}(f)$	Pierson-Moskowitz spectrum
$S_{xy}(f)$	Cross spectral density function
s	Slope of seabed
T	(i) Period (waves)
T	(ii) Return period (recurrence)
t	Time
T_f	Freezing point of water
TI	Turbulence intensity
T_L	Design life
T_{mean}	Mean temperature
T_p	Peak period (waves)

(Continued)

U	Mean wind speed (mean of short term data)
U_b	Velocity of floating structure
$\dot{U}_b$	Acceleration of floating structure
U_{bw}	Velocity of breaking wave induced surface current
U_{hub}	Wind velocity at hub height
U_i	Wind speed average over period i
U_{gr}	Gradient wind speed
$U_{SS}(z)$	Vertical velocity profile of sub surface current
U_w	Velocity of water
$U_{wind}(z)$	Velocity profile of near surface wind-generated current
$\dot{U}_w$	Acceleration of water
U_{10}	Reference wind speed at 10 m above surface
u_i	Sampled wind speed
u_1, u_2	Transformed variables in IFORM
$\tilde{u}$	Fluctuating wind velocity about short-term mean
u_*	Friction velocity
$u(z, t)$	Instantaneous longitudinal wind speed as function of z and t
$v(z, t)$	Instantaneous lateral wind speed as function of z and t
$\tilde{u}(t_j)$	Fluctuations in turbulent wind with zero mean (sampled)
w_i	Width of bins
$w(z, t)$	Instantaneous vertical wind speed as function of z and t
x	(i) Dimensionless wind speed
x	(ii) Distance
x	(iii) Fetch
$\tilde{x}_i, \tilde{y}_i$	Fluctuating turbulent component with zero mean (sampled)
z	Elevation above ground or still water level
z_r	Reference height
z_o	Surface roughness

A.6.2 Greek Variables

α	Constant in JONSWAP spectrum
α	Power law exponent
β	Reliability index
$\Gamma(x)$	Gamma function
γ	Peak enhancement factor
γ_{xy}	Coherence
Δ_{xy}	Distance between two locations (with reference to coherence)
Δt	Duration of short-term averaging time
δt	Sampling period
$\eta(x, t)$	Sea surface elevation as function of x and t

κ	von Kármán's constant
Λ_1	Scale parameter
$\mu_{\ln x_2}$	Mean of transformed log normal x_2 variable; x_1 analogous
μ_{x_2}	Mean of x_2 variable (log normal); x_1 analogous
ρ_w	Density of water
ρ	Air density
σ	Constant in JONSWAP spectrum
$\sigma_{\ln x_2}$	Standard deviation of transformed log normal x_2 variable; x_1 analogous
σ_{x_2}	Standard deviation of x_2 variable (log normal); x_1 analogous
σ_u	Standard deviation of short-term wind data (turbulence)
σ_U	Standard deviation of long-term wind data
σ_w	Standard deviation of sea surface elevation
σ_x, σ_y	Standard deviation of x and y variables
$\Phi(\ )$	Cumulative distribution function of the standard normal distribution
$\Phi(x, z, t)$	Velocity potential as function of x, z and t
$\Phi^{-1}(\ )$	Inverse cumulative distribution function of the standard normal distribution
ϕ_i	Phase angle (sampled)

A.6.3 Acronyms

ECD	Extreme coherent gust with change in direction
ECG	Extreme coherent gust
EDC	Extreme direction change
EOG	Extreme operating gust
EWM	Extreme wind speed
EWS	Extreme wind shear
HAT	Highest astronomical tide
HWSL	Highest still water level
LAT	Lowest astronomical tide
LWSL	Lowest still water level
MSL	Mean sea level

A.7 Chapter 7

A.7.1 English Variables

A	Fatigue strength coefficient
B	Constant in composites S–N model
b	Width of gear tooth face

(Continued)

C	Fatigue strength coefficient (in two segment S–N curve)
c	(i) Exponent in Goodman's Rule
c	(ii) Tower base radius in bending stress equation, $\sigma_B = Mc/I$
c_j	Weibull cycle range scale factor of interval
D	(i) Dynamic magnification factor
D	(ii) Cumulative damage
D_{Ring}	Diameter of ring gear
D_{Sun}	Diameter of sun gear
D_T	Cumulative damage over lifetime
$D_{T,x}$	Cumulative damage over lifetime in operating condition x
D_y	Damage over a year
d	(i) Pitch diameter
d	(ii) Damage per cycle
d_{del}	Contribution to damage equivalent load
d_j	Damage over j^{th} interval
E	Modulus of elasticity
F_b	Bending load applied to gear tooth
F_t	Tangential force on gear tooth
F_U	Fraction of time wind is in certain range
F_{U_j}	Fraction of time in j^{th} wind speed bin
f_e	Excitation frequency
f_n	Natural frequency
H_{op}	Operating hours per year
h	Height of gear tooth
I	Area moment of inertia
i	Subscript for cycle range bins
j	Subscript for wind speed bins
k	Number of cyclic events per revolution
k_g	Effective spring constant of two meshing gear teeth
k_j	Weibull cycle range shape factor of interval
L	(i) Distance to the weakest point on gear tooth
L	(ii) Height of tower
M	Tower base bending moment
m	Fatigue (Wöhler) exponent
m_{Tower}	Mass of tower
$m_{Turbine}$	Mass of turbine
N	(i) Number of gear teeth
N	(ii) Number of cycles to failure
$N_{B,r}$	Total number of stress range bins in interval
$N_{B,U}$	Total number of wind speed bins in interval
$N_{c,j}$	Number of cycles in j^{th} wind speed bin

N_i	Number of cycles to failure in i^{th} cycle range bin
$N_{ref,tot}$	Reference number of cycles in interval
n	(i) Rotational speed (rpm)
n	(ii) Number of cycles applied
n_{HSS}	Rotational speed of high-speed shaft (rpm)
n_i	Number of cycles in i^{th} stress range bin
n_j	Number of cycles in j^{th} wind speed bin
$n_{j,j}$	Number of cycles in i^{th} stress range bin and j^{th} wind speed bin
n_L	Number of cycles over lifetime
n_{LSS}	Rotational speed of low-speed shaft (rpm)
n_{rotor}	Rotor rotational speed (rpm)
P	Power
P_{rotor}	Rotor power
p	Circular pitch of gear
Q	Torque
R	Reversing stress ratio
r	Fatigue (Wöhler) exponent (in two segment S-N curve)
U	Wind speed
V_{pitch}	Gear pitch circle velocity
y	Form factor (or Lewis factor)
Y	Years of operation
z	Incomplete gamma function transition point

A.7.2 Greek Variables

δ	Logarithmic damping decrement
ξ	Damping ratio
σ	Cyclic stress range
σ_a	Stress amplitude
σ_{al}	Allowable stress amplitude
σ_b	Allowed bending stress in gear tooth
σ_e	Zero mean alternating stress for desired life
σ_{el}	Endurance limit
σ_m	Mean stress
σ_{max}	Maximum stress
σ_{min}	Minimum stress
σ_u	Ultimate strength.
σ_Q	Two segment S-N curve transition point
σ_{del}	Damage equivalent load

(Continued)

$\sigma_{c,j}$	Standard deviation of cycle ranges over j^{th} interval
μ_j	Mean of cycle ranges over j^{th} interval

A.8   Chapter 8

A.8.1   English Variables

A_c	Cross-sectional area
a_i	Constants in linear limit state equation
c	Chord length
C_P	Rotor power coefficient
CDF	Cumulative distribution function
E	Modulus of elasticity
erf()	Error function
F_c	Centrifugal force
$f_x(x)$	Joint probability density function of x (a vector)
k_q	Constant for characteristic value of load effect
k_{1-p}	Constant for characteristic value of resistance
g	(i) Gravitational constant
g	(ii) Gearbox ratio
$g()$	Limit state function
I	Area moment of inertia
M	(i) Moment
M	(ii) Safety margin
M_A	Aerodynamic moment
M_g	Moment due to gravity
n	Rotational speed (rpm)
n_{rated}	Rotational speed of generator at rated power (rpm)
n_{rotor}	Rotor rotational speed (rpm)
n_{sync}	Synchronous rotational speed of generator (rpm)
P	Power
p_f	Probability of failure
$P_{generator}$	Generator power
P_{rated}	Rated generator power
P_{rotor}	Rotor power
Q	Torque
R	(i) Radius
R	(ii) Reversing stress ratio
r_c	Characteristic resistance
r_{cg}	Distance to center of gravity
r_d	Design resistance

S	Load effect
s_c	Characteristic load effect
s_d	Design load effect
t	Thickness
T	Thrust
U	Wind speed
U_{e1}	1-year extreme wind speed
U_{e50}	50-year extreme wind speed
U_{gust50}	50-year return period gust
U_{hub}	Hub height wind speed
U_{ref}	Reference wind speed
U, u	Reduced variables converted to standard normal form
W	Blade weight
$\mathbf{x}$	Random vector representing factors contributing to failure

A.8.2 Greek Variables

β	Reliability index
β_i	Constants in vibrating beam equation
γ	Safety factor
γ_l	Partial safety factor
γ_c	Central safety factor
γ_m	Partial safety factor for materials
γ_n	Consequence of failure safety factor
γ_s	Partial safety factor for loads
η	Overall efficiency of drivetrain
μ_M	Mean of safety margin
μ_R	Mean of resistance
μ_S	Mean of load effect
ρ	Density of air
ρ_b	Mass density of blade
$\tilde{\rho}$	Mass density per unit length
σ_A	Stress due to aerodynamic loading
σ_b	Maximum blade root bending stress
σ_c	Tensile stress due to centrifugal force
σ_g	Stress due to gravity
σ_M	Standard deviation of safety margin
σ_R	Standard deviation of resistance
σ_S	Standard deviation of load effect

(Continued)

$\Phi(\ )$	Cumulative distribution function of standard normal distribution
Ω	Rotor rotational speed
ω_n	Natural angular frequency
ω	Angular frequency

A.9 Chapter 9

A.9.1 English Variables

A	Area
C	Damping coefficient (generic), e.g., of the pitch actuator
C_φ	Damping term in region 3 PI control
C_P	Power coefficient
C_T	Thrust coefficient
c	Slope of the torque-pitch rate curve in pitch actuator model
e	Error between reference signal and signal
f	Frequency
f_{Ny}	Nyquist frequency
GK	Proportional gain-scheduling term
J	Polar mass moment of inertia of the drivetrain with respect to shaft axis, LSS side
J_r	Polar mass moment of inertia of rotor at rotor center with respect to shaft axis
J_g	Polar mass moment of inertia of generator with respect to shaft axis, LSS side
K	Generalized/effective stiffness, spring constant (generic or pitch actuator)
K_a	Proportional constant for aerodynamic torque in region 2
K_D	Derivative gain in PID controller
K_g	Proportional constant for generator torque in region 2
K_I	Integral gain in PID controller
K_P	Proportional gain in PID controller
K_φ	Stiffness term in region 3 PID control
k	Slope of the torque-voltage curve in pitch actuator model
$\mathcal{L}$	Laplace tranform
M	Mass term (generic)
M_φ	Mass term in region 3 PI control
n_g	Gear ratio of the gear box
P	Power (generic)
P_a	Aerodynamic power
Q	Torque (generic)
Q_a	Aerodynamic torque
Q_g	Generator torque
Q_p	Torque about pitch axis
R	Radius of rotor
s	Complex variable

s_n	Signal s, sampled at time index n
T	Thrust
t	Time
U	Free stream wind velocity
v	Voltage signal

A.9.2 Greek Variables

Δ	Small perturbation, with convention: "signal minus reference"
ζ	Damping ratio (generic)
ζ_φ	Damping ratio of Region 3 PI control
η	Efficiency of the drivetrain (gearbox and generator)
θ_k	Parameter in gain scheduling
Θ_p	Laplace transform of pitch signal
θ_p	Pitch angle (positive toward feathering)
λ	Tip speed ratio
ρ	Air density
φ	Time integral of $\Delta\Omega$
ω	Cyclic frequency (generic)
ω_φ	Natural frequency of Region 3 PI control
Ω	Angular velocity

A.9.3 Subscripts and Acronyms

O	Operating point value
HSS	High-speed shaft
LSS	Low-speed shaft
max	Maximum
opt	Optimum
PID	Proportional-Integral-Derivative
ref	Reference value

A.9.4 Symbols

$(\dot{\,})$	Time derivative
$(\,)\vert_0$	Value expressed at operating point

A.10   Chapter 10

A.10.1   English Variables

A	Cross sectional area
A_{ef}	Effective area of foundation
B	Width of foundation (along axis of wind)
B_{ef}	Effective width of foundation
b_e	Elliptic axis of circular foundation
c	Soil cohesion
d_c	Depth correction factor
D	(i) Depth of foundation
D	(ii) Dynamic magnification factor
D	(iii) Base diameter
e	Eccentricity
E	Modulus of elasticity
E_s	Soil modulus of elasticity
$F.S.$	Factor of safety (soil)
f_d	Allowed stress
f_e	Excitation frequency (Hz)
f_n	Natural frequency (Hz)
F_T	Thrust
H	Tower height
i_c	Inclination correction factor
I	Area moment of inertia
k	Interaction factor and
k	Spring constant (in p–y method)
K	Generalized stiffness
l_e	Elliptic axis of circular foundation
L	Length of foundation (perpendicular to B)
L_{ef}	Effective length of foundation
m_B	Term in correction factors
m_L	Term in correction factors
m_{RNA}	Mass of RNA
m_{SS}	Mass of support structure
M	Generalized mass
M_d	Design moment
$M_{el,Rd}$	Material-based resistance moment
$N_{pl,Rd}$	Material-based yield axial load
N_c	Meyerhof bearing capacity factor
N_{cr}	Critical buckling load
N_d	Design axial load

N_q	Meyerhof bearing capacity factor
N_γ	Meyerhof bearing capacity factor
p	Force (in p–y method)
p_{nom}	Pressure on soil (nominal)
q_a	Allowed bearing capacity of soil
$q_{ucs,}$	Unconfined compressive strength of soil
R	(i) Radius of tower
R	(ii) Radius of circular foundation
R_{in}	Inner radius of tower
R_x	Reaction force in x direction
R_z	Reaction force in z direction
s_c	Shape correction factor
s_u	Undrained shear strength of soil
t	Tower base wall thickness
W	Weight of structure
$W_{foundation}$	Weight of foundation
W_{RNA}	Weight of RNA
W_{tower}	Weight of tower
y	Displacement

A.10.2 Greek Variables

α	Imperfection factor
γ	Soil specific weight
γ_{M0}	Safety factor
ζ	Damping ratio
$\bar{\lambda}$	Column slenderness ratio
ϕ	Term in reduction factor calculation
ϕ	Soil friction angle
χ	Reduction factor

A.11 Chapter 11

A.11.1 English Variables

a	(i) Height of the hull that is submerged on the upwind side (for barge calculation)
a	(ii) Catenary parameter
a_1, a_2	Catenary parameters of mooring line 1 or 2 (differential loading)
A	Area
A_{wp}	Water plane area, cross sectional area of the vessel where it intersects the SWL

(Continued)

b	Height of the hull that is submerged on the downwind side (for barge calculation)
c_{add}	Added mass coefficient
C_a	Added mass coefficient in Morison equation
C_D	Drag coefficient in Morison equation
d	Draft, distance from SWL to keel, typically positive
D	Diameter
E	Young's modulus of elasticity
f	(i) Frequency
f	(ii) Freeboard
f_n	Natural frequency
$\boldsymbol{f}$	Force vector
F	Force
F_B	Buoyancy force
F_T	Thrust force
F_W	Weight force
g	Acceleration of gravity
h	Distance above base of a component
$h_{cg,i}$	Height of the center of gravity of a given body i above its base
H	(i) Height of a body (typically in the z direction)
H	(ii) Height above seabed of mooring line attachment
H_{hub}	Height of the hub with respect to SWL
I_{yy}	Area moment of inertia of the vessel at the water line about the y axis
J_{cg}	Mass moment of inertia of the vessel (about the y axis), at its center of gravity
k	Translational or rotational stiffness (e.g. of a mooring line)
$\boldsymbol{K}$	Stiffness matrix
l	Tendon length
L	(i) Length of a body (typically in the y direction)
L	(ii) Horizontal distance from the point of touchdown to the fairlead
L_1, L_2	Horizontal distance from the point of touchdown to the fairleads 1 or 2 (differential loading)
m	Total mass of a body, or water displaced
m_{add}	Added mass
$\boldsymbol{M}$	Mass matrix
N_d	Number of discs (when dividing a cylindrical structure)
N_s	Number of masses/bodies in the structure
R	Radius, typically of a cylindrical body
R_{c-h}	Distance between two bodies (central column and hull)
R_{h-h}	Distance between two hulls
$s(x)$	Curvilinear distance along a mooring line
s_s	Total length of suspended portion of mooring line
s_1, s_2	Total length of suspended portion of mooring line 1 or 2 (differential loading)
$S(f)$	Power spectral density

t	Wall thickness of a given body
T	Tension in a mooring line
T_a	Tension in mooring line at anchor/point of touchdown
V	Volume of a given body
W	Width of a body (typically in the x direction)
x	Longitudinal coordinate, surge
y	Transverse coordinate, sway
z	Vertical coordinate, positive upward, 0 at SWL
z_k	Vertical height with respect to the keel (positive above the keel)

A.11.2 Greek Variables

Δl	Elongation (e.g. for tendon)
θ	(i) Heel angle, temporary inclination of a vessel, same as pitch
θ	(ii) Slope of a mooring line with respect to horizontal
ρ	Material density (of a given body)
ρ	Mass density per unit length of the mooring line (kg/m)
$\widetilde{\rho}'$	Effective mass density per unit length of the mooring line (kg/m)
ρ_w	Water density
ρ_{steel}	Density of steel

A.11.3 Symbols

$\dot{x}$	Time derivative
∇	Volume of displaced water
$\overline{AB}$	Distance from point A to point B
$\overline{GM}$	Metacentric height, distance on a vertical line from the center of gravity G to the metacenter M

A.11.4 Points, Subscripts and Acronyms

A.11.4.1 Points (Also Used as Subscripts)

A	(i) Top of frustrum
A	(ii) Top point of a taper
A	(iii) Point of attachment of mooring lines
B	(i) Center of buoyancy, geometric center of the water displaced
B	(ii) Base of frustrum
B	(iii) Bottom point of a taper

(Continued)

G	Center of buoyancy
M	Metacenter
O	Origin of the body, assumed to be at SWL when undisplaced
P	Arbitrary point

A.11.4.2 Subscripts

1, 2, 3, *etc.*	Indices for the 6 degrees of freedom surge, sway, heave, roll, pitch, yaw
1, 2, 3, 4	Indices for the 4 tendons of a TLP
1, 2	Indices for the two placements of an anchor (point of touchdown) before and after displacements
33	Term at row 3 and column 3 of a matrix (of dimension 6×6)
44, etc.	Term at row 4 and column 4 of a matrix (of dimension 6×6)
a	Anchor/point of touchdown
add	Added mass
aero	Aerodynamic
all hulls	All hulls
barge	Barge
blst	Ballast
buoy	Buoyancy
cc	Central column
$c - h$	Central column to hull (distance)
f	Fairlead
grav	Gravity
$h - h$	Hull to hull (distance)
hull	Hull
hydro	Hydrodynamic
in	Inner (for inner radius of a cylinder)
k	Keel
$k - cb$	Keel to center of buoyancy of a given body
$k - cg$	Keel to center of gravity of a given body
$k - b$	Keel to base of a given body
moor	Mooring
out	Outer (for outer radius of a cylinder)
sb	Seabed
spar	Spar
steel	Steel
sub	Submerged (e.g. submerged part of the TP)
tendons	Tendons
twr	Tower
tpr	Taper
W	Weight

wall	Wall
water	Water

A.11.4.3 Acronyms

MH	Main hull
MSL	Mean sea level
RNA	Rotor nacelle assembly
RAO	Response amplitude operator
SWL	Still water level
TP	Transition piece
TLP	Tension leg platform

A.12 Chapter 12

A.12.1 English Variables

a	Axial induction factor
A_c	Cross sectional area
C	Capacitance
C_T	Thrust coefficient
D	(i) Distance from conductor
D	(ii) Diameter of rotor
D	(iii) Distance
F	Fuel
f	Frequency
I_F	Fault current
$\hat{I}$	Phasor current
J	Generator rotational moment of inertia
k	Decay constant
l	Length (conductor)
L	Inductance
$\tilde{L}$	Inductance per unit length
M	Fault level
N	Number of turbines
P	Power (real)
Q	Reactive power
r	Radius (conductor)
R	Resistance

(Continued)

S	Apparent power
U	Wind speed
U_0	Free stream wind speed
V_G	Generator voltage
V_S	System voltage
$\hat{V}$	Phasor voltage
X_C	Capacitive reactance
X_L	Inductive reactance
Z	Impedance
$\hat{Z}$	Phasor impedance

A.12.2 Greek Variables

A.12.2.1 Subscripts

ε	Dielectric constant
ε_0	Dielectric constant of free space
ε_r	Relative permittivity
μ_0	Permeability of free space
ρ_m	Density of material
σ	Standard deviation
ω	Angular speed

A.12.3 Symbols

A.12.3.1 Subscripts

G	Generator
IG	Induction generator
IM	Induction machine
L	(i) Load, (ii) inductance
$Load$	Load
N	Number
O	Other (generators)
PM	Prime mover
S	System
T	Thrust
X	Distance downwind
$line$	Referring to the line
net	Net (as in net load)
$turbine(s)$	Turbine

A.12.4 Acronyms

AC	Alternating current
CFD	Computational fluid dynamics
DC	Direct current
HVDC	High voltage direct current
ISO	Independent system operator
LPFF	Low-pressure fluid filled
LPOF	Low-pressure oil-filled
PCC	Point of common coupling
POC	Point of connection
RTO	Regional transmission organization
THD	Total harmonic distortion
XLPE	Cross linked polyethylene

A.13 Chapter 13

A.13.1 English Variables

A	Installment
AAR	Average annual return
B/C	Benefit–cost ratio
b	(i) Learning parameter
b	(ii) Interest rate
C	Cost
C_c	Capital cost of system
$\overline{C}_c$	Capital cost of system normalized by rated power
$C_{O\&M}$	Average annual operation and maintenance (O&M) Costs
$\overline{C}_{OM}$	Direct cost of operation and maintenance per unit of energy
$C(V)$	Cost of an object as a function of volume
$C(V_0)$	Cost of an object as a function of initial volume
COE	Cost of energy (per kWh or per MWh)
COE_L	Levelized cost of energy (per kWh or per MWh)
CRF	Capital recovery factor
E_a	Annual energy production (kWh or MWh)
F_N	Payment at end of N years
FC	Capacity factor
FCR	Fixed charge rate
FV	Future value
f_{OM}	Annual operation and maintenance (O&M) cost fraction
IRR	Internal rate of return
i	General inflation rate

(Continued)

j	Index for year
L	Lifetime of system (years)
N	Number of years or Installments
NPV	Net present value
NPV_C	Net present value of costs
NPV_P	Net present value of profit
NPV_S	Net present value of savings
P_a	Annual payment
P_d	Down payment
P_e	Price obtained for electricity
PV	Present value
PV_N	Present value of future payment in year N
PWF	Present worth factor
R	Annual revenue
ROI	Return on investment
r	Discount rate
S	Savings
SP	Simple payback (years)
SPW	Series present worth factor
s	Progress ratio
V	Volume of object produced
V_0	Initial cumulative volume
$Y(k, \ell)$	Function to obtain present value of a series of payments; k, ℓ: arguments

A.13.2 Acronyms

CFM	Cash flow method
GRE	Glass reinforced plastic
GRP	Glass reinforced epoxy
IPP	Independent power producer
ISO	Independent system operator
ITC	Investment tax credit
LOLE	Loss of load probability
LOLP	Loss of load expectation
O&M	operation and maintenance
PPA	Power purchase agreement
PTC	Production tax credit
PURPA	Public Utility Regulatory Policies Act
REC	Renewable energy certificate
RPS	Renewable portfolio standard
vPPA	Virtual power purchase agreement
WEGA	Large wind turbine

A.14 Chapter 14

A.14.1 English Variables

$A(f)$	A weighting scale
c	Speed of sound
$\hat{d}_{wp}$	Estimate of fraction of carcasses landing in the search area
D	Rotor diameter
f	Frequency
L_{10}	Sound pressure level exceeded 10% of the time
L_{50}	Sound pressure level exceeded 50% of the time
L_{90}	Sound pressure level exceeded 90% of the time
L_P	Sound pressure level
L_{total}	Overall sound pressure level (multiple turbines)
L_W	Sound power level
L_{WA}	Overall sound pressure level
n	Day of the year
$\hat{m}$	Estimate of the total number of bird fatalities
$\hat{p}$	Estimate of fraction of observed carcasses
$\hat{r}$	Estimate of fraction of bird carcass persistence
R	Distance from noise source (hemisphere)
R_A	Term in equation for A weighting scale
V_{Tip}	Speed of blade tip
$x_{shadow,max}$	Maximum extent of shadow flicker

A.14.2 Greek Variables

α	Absorption coefficient
α_s	Solar altitude
δ	Solar declination
λ	Wavelength
$\emptyset$	Latitude
ω	Hour angle

A.14.3 Acronyms

BACI	Before and after construction impact
BCI	Bat Conservation International
BMP	Best management practice

(Continued)

BOEM	Bureau of Ocean Energy Management
BWEC	Bats and Wind Energy Cooperative
CEQ	Council on Environmental Quality
COP	Construction and Operations Plan
EIA	Environmental impact assessment
EIS	Environmental impact statement
EU	European Union
GIS	Geographical information system
IHA	Incidental Harassment Authorization
IHO	International Hydrographic Organization
LVIA	Landscape and Visual Impact Assessment
MMPA	Marine Mammal Protection Act
MSP	Marine spatial planning
NEPA	National Environmental Policy Act
NGO	Nongovernmental organization
NOAA	National Oceanographic and Atmospheric Administration
NRA	Navigational Risk Assessment
OCS	Outer Continental Shelf
PPA	Power purchase agreement
PSO	Protected species observer
ROD	Record of decision
SAP	Site assessment plan
SEL	Sound exposure level
SLVIA	Seascape, Landscape, and Visual Impact Assessment
WNS	White-nose Syndrome
ZTM	Zone of Theoretical Visibility

A.15 Chapter 15

A.15.1 English Variables

A	Cross-sectional area of falling object
B	Beam (waterline breadth) of hull
c_1, c_2, c_5	Constants in wave resistance equation
C	Coefficient in the range of 1.25–1.41 m/s m$^{1/2}$
C_B	Block coefficient of a ship
C_D	Drag coefficient of falling object
C_F	Coefficient of friction
F_B	Buoyant force
F_n	Froude number
g	Gravitational constant

k_1	Form factor related to the shape of vessel
L_{pp}	Length of vessel between perpendiculars
L_S	Length of vessel
m	Mass of falling object
m_1, m_2	Hull shape constants in wave resistance equation
P_{hull}	Power required to move hull
r	Radial position on blade
R_A	Resistance of hull due to other sources
Re_{L_S}	Reynolds number based on the vessel's length
R_f	Viscous (drag) resistance of hull
R_{total}	Total resistance of hull to forward motion
R_w	Resistance of hull due to wave drag
S_S	Wetted area of hull
T	Draft amidships
V_{hull}	Hull speed
V_S	Velocity of vessel
V_x	Horizontal velocity of falling object
$V_{x,0}$	Initial horizontal velocity of falling object
V_z	Vertical velocity of falling object
V_{z0}	Initial vertical velocity of falling object
x	Horizontal position of falling object
x_0	Initial horizontal position of falling object
x_{max}	Maximum distance of falling object
z	Vertical position of falling object
z_0	Initial vertical position of falling object

A.15.2 Greek Variables

λ	Hull shape constant in wave resistance equation
μ_w	Viscosity of sea water
ρ	density of air
ρ_w	Density of water
ψ	Azimuthal position
Ω	Rotor rotational speed

A.15.3 Symbols

∇	Displacement; displaced volume

A.16 Chapter 16

A.16.1 English Variables

A	Initial battery voltage parameter
c	Ratio of available charge capacity to total capacity
C	Second battery voltage parameter
C_P	Power coefficient
D	Third battery voltage parameter
E	(i) Useful energy (kJ) stored in a flywheel
E	(ii) Internal battery voltage
E_0	Fully charged/discharged internal battery voltage
$F_{L,conv}$	fraction of conventional power that was required
$F_{P,dump}$	fraction of available wind power that was dumped
f_p	Permeate fraction
g	Gravitational constant
i	Van't Hoff factor
I	Current
J_{Rot}	Mass moment of inertia
k	(i) Battery rate constant
k	(ii) Ratio of constant pressure to constant volume specific heats
m	Mass
mol	Gram molecular weight
M	Molecular weight of salt
p	Pressure
p_f	Final pressure
p_i	Initial pressure
p_π	Osmotic pressure
P_p	Pumping power
q_{in}	Heat input from the combustion
q_{max}	Maximum battery capacity (at infinitesimal current)
$q_{max}(I)$	Maximum battery capacity at current I
R_a	Gas constant for air
R_0	Internal resistance
R_u	Universal gas constant
S	Salinity
t	Charge or discharge time, hrs
T	Absolute temperature
V	(i) Battery terminal voltage
V	(ii) Volume
$\dot{V}$	Volumetric flow rate
v_i	Initial specific volume of hydrogen.

w	Specific work
$w_{comp,ad}$	Specific work involved in adiabatic compression
$w_{comp,iso}$	Specific work involved in isothermal compression
$w_{desal,tot}$	Specific work for desalination
w_{net}	Net specific work out of turbine
X	Normalized capacity at the given current
z	Elevation

A.16.2 Greek Variables

ΔH	Change in enthalpy
Δp	Pressure rise or drop
η_{comp}	Compressor efficiency
η_{pump}	Mechanical efficiency of pump
$\eta_{th,Brayton}$	Brayton cycle efficiency
η_{vol}	Volumetric efficiency of pump
ρ	Density of air
ρ_w	Density of water
Ω_{max}	Maximum rotational speeds
Ω_{min}	Minimum rotational speeds

A.16.3 Acronyms

FES	Flywheel energy storage
P2X	Power-to-X
PEM	Proton exchange membrane
PHS	Pumped hydroelectric energy storage
CAES	Compressed air energy storage
HVDC	High voltage direct current
ISO	Independent system operator
RTO	Regional transmission organization
PV	Photovoltaic
RO	Reverse osmosis

Appendix B

Data Analysis and Data Synthesis

B.1   Overview

This appendix concerns topics in the analysis or synthesis of time series data. Analysis is used to make sense of data from tests and other sources. Conversely, data is often synthesized for input to simulation models. The next two sections (Sections B.2 and B.3) summarize some of the analysis and synthesis methods that are most commonly used in wind energy systems engineering.

B.2   Data Analysis

Much of the data encountered in wind energy engineering has random aspects and/or sinusoidal aspects. This is often the case with wind speed turbulence and its effects, structural vibrations, ocean waves, harmonics in electrical power systems, and noise. This section summarizes some of the methods used in analyzing that data.

Ideally, any data would be continuous and would be denoted by $x(t)$. In fact, real data files as used in wind energy applications are sampled time series rather than being continuous. A single time series is referred to as a data record. The expression for a data record corresponding to $x(t)$ is:

$$x_n = x(n\Delta t); \;\; n = 1, 2, 3, ..., N \tag{B.1}$$

where

N = number of points in data record.
$\Delta t = T/N$ = sample interval.
T = time length of period.
n = index (point number).

Much of the discussion here is, strictly speaking, only applicable to data that is stationary and ergodic. This means that the data record is statistically similar over time and that any data record is statistically similar to any other from the same process. To the extent that the data does not meet those criteria, the analytic results will be somewhat approximate. Noted that in general data records should have their means subtracted out and any trends removed before the analysis described below is undertaken. A trend here refers to a relatively steady

Wind Energy Explained: On Land and Offshore, Third Edition.
James F. Manwell, Emmanuel Branlard, Jon G. McGowan, and Bonnie Ram.
© 2024 John Wiley & Sons Ltd. Published 2024 by John Wiley & Sons Ltd.

increase or decrease in the mean over the time interval of interest. Finally, all real data records have a finite length. This can result in artifacts appearing in the results. These artifacts can be largely eliminated through the use of "windowing" which is also described in Section B.2.3.1.

B.2.1 Autocorrelation Function

The autocorrelation function for sampled data is found by multiplying each value in the record by values in the same record, offset by a time "lag," and then summing the products to find a single value for each lag. The resulting sums are then normalized by the variance to give values equal to or less than one. The normalized autocorrelation function for time series x_n with N points and variance σ_x^2 is given by:

$$R(r\Delta t) = \frac{1}{\sigma_x^2(N-r)} \sum_{n=1}^{N-r} x_n\, x_{n+r} \tag{B.2}$$

where r = lag number and Δt = time interval between lags.

Note that the form of the autocorrelation function used here is sometimes referred to as the normalized autocorrelation function or autocorrelation coefficient. If it is not divided by the variance, then it is said to be unnormalized. In addition, the variance here should be population based rather than sample based. An example of an autocorrelation function was given in Chapter 6.

B.2.2 Cross-correlation Function

The cross-correlation function provides a measure of how similar two records, x_n and y_n, are as they are shifted in time. The equation describing the cross-correlation (Equation B.3) is similar to the autocorrelation, except that two data records are used instead of one.

$$R_{xy}(r\Delta t) = \frac{1}{\sigma_x \sigma_y(N-r)} \sum_{n=1}^{N-r} x_n\, y_{n+r} \tag{B.3}$$

B.2.3 Power Spectral Density

Sinusoidally varying data is often analyzed in the frequency domain. The frequency domain is most often considered via the power spectral density (psd) of a data file. The psd retains information regarding the magnitude of various frequency components but eliminates the phase relations, which are usually irrelevant. In particular, psds are used to identify frequencies of interest and to see which frequencies are most prominent (i.e., have the most energy). With the help of a psd, it is often relatively easy to see where there is an excitation or a natural frequency.

A psd may be found from time series data by applying a Fourier transform. Most commonly an algorithm known as the Fast Fourier Transform (FFT) is used. This method is described in detail in a variety of sources, such as Bendat and Piersol (2010). The following provides a summary.

The basis of this method is the observation that a time series may be expressed as a sum of sines and cosines, as shown in Equation (B.4).

$$x_n = A_0 + \sum_{q=1}^{N/2} A_q \cos\left(\frac{2\pi q n}{N}\right) + \sum_{q=1}^{(N/2)-1} B_q \sin\left(\frac{2\pi q n}{N}\right) \tag{B.4}$$

The coefficients (known as Fourier coefficients) are found as follows:

$$A_0 = \sum_{n=1}^{N} x_n = \bar{x} = 0$$

$$A_q = \frac{2}{N} \sum_{n=1}^{N} x_n \cos\left(\frac{2\pi qn}{N}\right) \quad q = 1, 2, ..., \frac{N}{2} - 1$$

$$A_{N/2} = \sum_{n=1}^{N} x_n \cos(n\pi)$$

$$B_q = \sum_{n=1}^{N} x_n \sin\left(\frac{2\pi qn}{N}\right) \quad q = 1, 2, ..., \frac{N}{2} - 1$$

(B.5)

Equation (B.4) is actually the discrete inverse Fourier transform of $X(f)$, which is shown in Equation (B.7). Equation (B.4) can also be written more succinctly using complex notation as shown in Equation (B.6).

$$x_n = \frac{1}{N} \sum_{k=1}^{N} X_k \exp\left[j\frac{2\pi kn}{N}\right]$$

(B.6)

where X_k are Fourier coefficients corresponding to the A's and B's above. They are explained in more detail below.

The general equation for the Fourier transform, $X(f)$, of a continuous real or complex-valued function, $x(t)$, over an infinite range for some frequency, f, is:

$$X(f) = \int_{-\infty}^{\infty} x(t) e^{-j2\pi f\, t} dt$$

(B.7)

The Fourier transform over some time period, T, is:

$$X(f, T) = \int_{0}^{T} x(t) e^{-j2\pi f\, t} dt$$

(B.8)

The discrete form of Equation (B.8) is a Fourier series:

$$X(f, T) = \Delta t \sum_{n=1}^{N} x_n e^{-j2\pi f\, n\Delta t}$$

(B.9)

Selecting discrete frequencies, $f_k = k/T = k/(N\,\Delta t)$, where $k = 1, 2, ..., N-1$, results in the following expression for the Fourier coefficients:

$$X_k = \frac{X(f_k)}{\Delta t} = \sum_{n=1}^{N} x_n e^{-j\frac{2\pi kn}{N}}$$

(B.10)

The classical method to calculate the Fourier coefficients would be to apply Equation (B.5). This process would involve approximately N^2 multiplications and additions. To speed up the process, the FFT has been developed to calculate the coefficients.

The FFT is far faster than the standard Fourier transform, but it does have some limitations. The most significant of these is that FFTs can only be carried out on data files which are powers of two in length. This could consist of files that have 256, 512, 1024, etc. points. Accordingly, any file that is not that long must be truncated (or sometimes augmented) to have an acceptable length.

The equations for the FFT will not be presented here. It is sufficient to note that they may be found in a number of sources (Bendat and Piersol (2010), for example), and in particular the equations have been incorporated in a number of software packages, such as Microsoft Excel and MATLAB®, among others.

With the Fourier transform at hand, the psd may be readily calculated. The psd of real-valued data, denoted $S_{xx}(f)$, can be obtained from the FFT by multiplying the transform by its complex conjugate, $X^*(f,T)$, and normalizing it as follows:

$$S_{xx}(f) = \frac{1}{T}|X(f,T)|^2 = \frac{1}{T}X(f,T)X^*(f,T) \tag{B.11}$$

Often the data is divided into multiple segments in order to help eliminate spurious results. When the data has been divided into multiple segments, an FFT is performed on each segment. The psd is found by averaging the values obtained from each segment:

$$S_{xx}(f) = \frac{1}{n_d T}\sum_{i=1}^{n_d}|X_i(f,T)|^2 \tag{B.12}$$

where n_d = number of segments into which the data has been divided.

When the psd is calculated with the FFT, half of the points are actually redundant. In particular, $S(f) = S(N-f)$. Therefore, the second half of the points are eliminated from further consideration.

It is also important to note that some software packages do not apply standard normalization to the FFT, which should result in the integral of the psd being equal to the variance. In those situations, the FFT must be normalized separately. For example, the FFT obtained from Excel must be divided by the number of points.

It is also worth noting that the power spectrum is equal to the Fourier transform of the unnormalized autocorrelation function.

B.2.3.1 Side Lobe Leakage and Windowing

All real data time series have a finite length. A result of this is that the Fourier transform will introduce additional energy into certain frequencies. These spurious contributions can be largely eliminated through the use of windowing.

The origin of the problem can be seen as follows. A finite length time series over time T can be considered to be an infinitely long time series, $v(t)$, multiplied by a rectangular window $u(t)$ which is equal to one or zero as shown in Equation (B.13):

$$u(t) = \begin{cases} 1 & 0 \le t < T \\ 0 & \text{otherwise} \end{cases} \tag{B.13}$$

According to this formulation, the time series of interest, $x(t)$, is given by:

$$x(t) = u(t)\,v(t) \tag{B.14}$$

It can be shown that the Fourier transform of $x(t)$, $X(f)$, is given by the convolution of the Fourier transforms of $u(t)$ and $v(t)$, $U(f)$ and $V(f)$, as shown in Equation (B.15) (where α is an integrating variable).

$$X(f) = \int_{-\infty}^{\infty} U(\alpha)V(f-\alpha)d\alpha \tag{B.15}$$

The Fourier transform of a rectangle can readily be shown to be as given in Equation (B.16). The absolute value of Equation (B.16) is illustrated in Figure B.1.

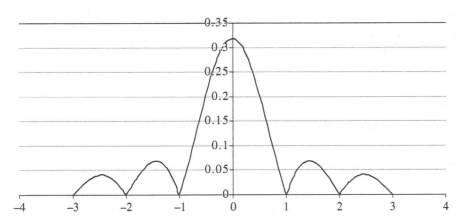

Figure B.1 Magnitude of Fourier transform of rectangle

$$U(f) = T\left(\frac{\sin(\pi fT)}{\pi ft}\right)e^{-j\pi fT} \tag{B.16}$$

The "lumps" on either side of the −1 and 1 are known as "side lobes." It is these side lobes that introduce spurious effects (artifacts) in the psd. Specifically, they cause greater energy to appear at certain frequencies than should actually be the case. These effects can be largely eliminated with specially shaped windows. The windows can take a variety of forms. The basic approach is to taper the time series data near the beginning and end so as to decrease the magnitude of the side lobes. A common window is the cosine squared, or Hanning, window described in Equation (B.17).

$$u_h(t) = \begin{cases} 1 - \cos^2\left(\dfrac{\pi t}{T}\right) & 0 \le t \le T \\ 0 & \text{otherwise} \end{cases} \tag{B.17}$$

B.2.3.2 Graph Formats for Power Spectra

Power spectra may be plotted simply as the spectrum vs. frequency with conventional axes. In that case, the units on the y axis are (original units)^2/Hz, and the x axis units are Hz. In addition, the area under the graph between any two frequencies corresponds to the amount of variance associated with that frequency range. When using that method in some situations, such as with wind data, however, most of the variance is concentrated in a small portion of the graph. In such cases, the spectrum is often plotted with a log x axis and sometimes with a log y axis as well. Another common method of depicting the spectrum is to plot the frequency times the psd on the y axis vs. the natural log of the frequency on the x axis. This method is useful because the areas on the graph between any two frequencies are again equal to the variance associated with that range of frequencies.

B.2.4 Cross-spectral Density

The cross-spectral density provides a measure of the energy at various frequencies of two different records. It is given by:

$$S_{xy}(f_k) = \frac{1}{N\Delta t}[X^*(f_k)Y(f_k)\ k = 1, 2, 3, ..., N] \tag{B.18}$$

where $X(f)$ and $Y(f)$ are the Fourier transforms of $x(t)$ and $y(t)$, and f_k are the discrete frequencies.

B.2.4.1 Coherence

The coherence function is the square of the cross-spectral density normalized by the product of the psds of the two data records:

$$\gamma_{xy}^2(f_k) = \frac{\left|S_{xy}(f_k)\right|^2}{S_{xx}(f_k)S_{yy}(f_k)} \tag{B.19}$$

An exponential decay function for coherence is often assumed, as was the case for wind speed in Chapter 6.

B.2.5 Frequency Response Function/Transfer Function

Frequency response functions, also known as transfer functions, provide a measure of the response of a linear system to an excitation. This measure, $H(f_k)$, which is a function of frequency, is given by the ratio of the cross-spectral density of the response to the power spectral density of the excitation:

$$H(f_k) = \frac{S_{xy}(f_k)}{S_{xx}(f_k)} \tag{B.20}$$

B.2.6 Aliasing

Aliasing is the result of digitally sampling data at frequencies slower than the frequency of the information in the data. Aliasing can result in the representation of higher-frequency data as lower-frequency data. For example, Figure B.2 shows a 2 Hz sinusoid that represents information to be measured and the resulting signals if it is sampled at less than 2 Hz. It can be seen that sampling the 2 Hz sinusoid at 1 Hz provides data that appears to be constant. Sampling the same 2 Hz sinusoid at frequencies near 1 Hz, but not exactly 1 Hz, provides data that appear to vary very slowly.

The effects of aliasing show up in frequency analysis of measured data also. Figure B.3 shows two spectra of an analog signal that includes two frequencies, one at 1 Hz and one at 1000 Hz. The original data were sampled at two different rates (312 and 10 000 Hz) before the spectra were created. The spectrum of the data that has been sampled

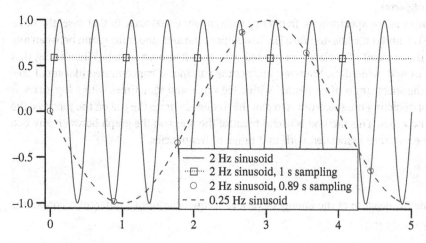

Figure B.2 Simple aliasing example

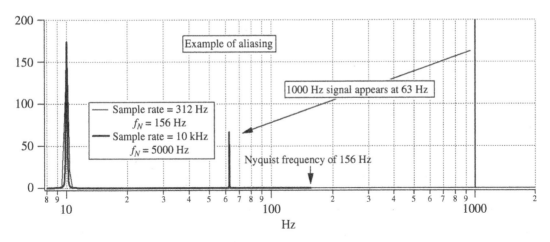

Figure B.3 Example of aliasing

at 10 000 Hz clearly shows that there are variations at 1 and 1000 Hz. In the spectrum of the data that was sampled at 312 Hz, the content of the signal at 100 Hz has been shifted to 63 Hz. In the latter case, the signal has information at frequencies greater than half of the sampling frequency (156 Hz). That frequency, half of the sampling frequency, is called the Nyquist frequency and, in fact, information at frequencies above the Nyquist frequency, f_N, cannot be correctly resolved. Furthermore, any information at frequencies higher than the Nyquist frequency appears as information at lower frequencies. Thus, all information at frequencies above the Nyquist frequency must be removed (or "filtered," see next section) from the data signal before digitization. Once the data have been collected, there is no way to correct the data.

B.2.7 Filtering

Filtering of physical phenomena is frequently observed in the wind energy field. For example, spatially separating multiple wind turbines has the effect of filtering out the large power fluctuations that would occur if a single, large turbine were used instead of the many smaller ones. Sometimes filtering is done intentionally, as when the output of a power electronic converter is filtered to remove frequencies above 50 or 60 Hz, so that the output more closely resembles utility-grade electrical power or where it is necessary to prevent aliasing, as noted in the previous section.

One of the conceptually simplest filters is the ideal low-pass filter. This filter will remove all frequencies above a certain frequency (known as the cut-off frequency) and keep those below that frequency. This filter looks like a rectangle in the frequency domain, with a value of 0 at frequencies above the cut-off frequency and 1 at lower (positive) frequencies.

Real low-pass filters are more gradual than the ideal filter. Nonetheless, they do have the effect of reducing frequencies above the cut-off frequency and only marginally affecting lower frequencies. In electrical circuits, filtering is carried out by means of resistors, capacitors, and inductors. In mechanical systems, filtering is carried out by means of springs, masses, and dampers, and the allowance of some relative motion between various components.

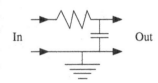

Figure B.4 Simple RC filter

A simple but real low pass filter is the RC filter as illustrated in Figure B.4.

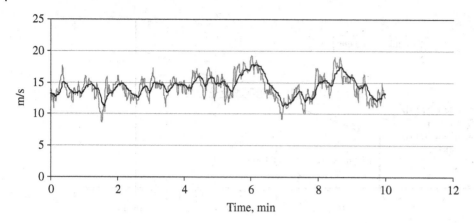

Figure B.5 Sample original and filtered time series data

The RC filter has a cut-off frequency of $f_c = 1/(2\pi RC)$ Hz. Filters are also characterized by their time constant, τ, which for this case is given by $\tau = RC$. Filters are often realized in digital form. The digital expression for the RC filter, for example, is:

$$y_n = ax_n + (1-\alpha)y_{n-1} \tag{B.21}$$

where y_n = the output of the filter, $\alpha = \Delta t/(\tau + \Delta t)$, and Δt = the time step of the data.

For example, Figure B.5 illustrates 10 minutes (600 seconds) of 1 Hz wind speed data. Superimposed on that is a smoother curve of that data filtered, with a filter whose cut-off frequency is 0.02 Hz (such that $RC = 7.9577$ s and $\alpha = 0.11164$).

The data above may also be examined in the frequency domain. Figure B.6 illustrates the psd of the original data and the psd of the filtered data superimposed on it for frequencies less than 0.1 Hz. For comparison, the psd of data filtered with an ideal low pass filter is also shown. The psds shown were obtained by truncating the original time series to 512 points and dividing them into two segments of 256 points each before performing the FFTs. Note that the psd of the ideal filter follows that of the data exactly until the cut-off frequency is reached. The simple low pass

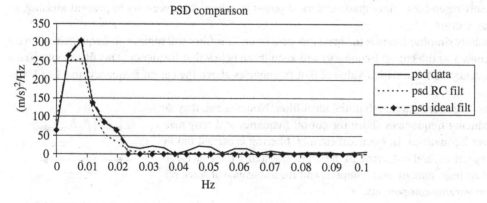

Figure B.6 Power spectral densities of original and filtered data

filter, however, begins to attenuate frequencies well below the cut-off frequency. At higher frequencies, the attenuation increases substantially until there is very little contribution at frequencies higher than twice the cut-off frequency. This is as expected.

B.2.8 Rainflow Cycle Counting

The rainflow cycle counting technique was introduced in Chapter 6 and was also used in Chapter 7. A number of algorithms have been developed that can be incorporated into a computer code to actually do that counting. The essential features of one such algorithm are given below. This is a snippet of code, written in Visual Basic, based on the method suggested by Ariduru (2004). In this code, the vector *peakValleys*(.) contains the peaks and valleys of the data and *numberOfCycles*(.) is an output vector with the number of occurrences in ranges *binWidth* wide. The lowest range value bound is 0; the highest is large enough to incorporate all the data. The role of the other variables should be apparent from their names and context in the code. In this bit of code, any complete cycles are accounted for and then eliminated from the *peakValley*(.) data. The data is checked again for more complete cycles. Any half cycles are also identified and accounted for. This routine as shown only identifies cycle ranges, not their means, but it could be amended to do the latter as well. (Note that the function of the colon is to separate statements that could otherwise be on the next line.)

```
Sub RainFlow(maxRange, minRange, binWidth, nPeakValleyPoints%, peakValleys(),
numberOfCycles())
  ⋮
nRangeBins% = Int(1 + (maxRange - minRange) / binWidth) 'Determine # range bins
remainingPoints% = nPeakValleyPoints%        'Initialize remainingPoints%
startPoint = peakValleys(1): i_point% = 1  'Initialize startPoint, i_point%

Step1:
'Check next peak or valley. If out of data, go to Step 6.
    If i_point% + 1 = remainingPoints% Then GoTo Step6
Step2:
'If there are fewer than three points, go to Step 6. Otherwise, form ranges X 'and
Y using the three most recent not yet discarded peaks and valleys
    If remainingPoints% < 3 Then
        GoTo Step6
    Else
        X_range = Abs(peakValleys(i_point% + 2) - peakValleys(i_point% + 1))
        Y_range = Abs(peakValleys(i_point% + 1) - peakValleys(i_point%))
    End If
Step3:
    If Abs(X_range) < Abs(Y_range) Then
'No complete cycle found yet; advance to next point
        i_point% = i_point% + 1: GoTo Step1
    End If
Step4:
'If range Y_range contains the starting point, startPoint, go to step 5
'otherwise, count Y_range as one cycle;
'discard the peak and valley of Y; and go to Step 2.
```

```
    If i_point% = 1 Then
       GoTo Step5
    Else
'Get rid of two points as cycle
       j% = Int(Y_range / binWidth)
       numberOfCycles(j%) = numberOfCycles(j%) + 1
       remainingPoints% = remainingPoints% - 2
       For i% = i_point% To remainingPoints%
          peakValleys(i%) = peakValleys(i% + 2)
       Next i%
'Go back to beginning of file
       startPoint = peakValleys(1): i_point% = 1: GoTo Step2
    End If
Step5:
'Count range Y_range as one-half cycle; discard first point (peak or valley)
'in Y_range; move the starting point to next point in Y_range; go to Step 2.

    j% = Int(Y_range / binWidth)
    numberOfCycles(j%) = numberOfCycles(j%) + 0.5
    remainingPoints = remainingPoints - 1
    For i% = i_point% To remainingPoints%
       peakValleys(i%) = peakValleys(i% + 1)
    Next i%
    GoTo Step2

Step6:
'Count each range that has not been previously counted as one-half cycle.
    For i% = 1 To remainingPoints - 1
       j% = Int(Abs(peakValleys(i) - peakValleys(i + 1)) / binWidth)
       numberOfCycles(j) = numberOfCycles(j) + 0.5
    Next i%
End Sub
```

B.3   Data Synthesis

Data synthesis is the creation of artificial time series for some specific purpose. For example, hourly data may be needed for simulation models, but real hourly data may not be available. High frequency, turbulent wind speed with certain specific characteristics may be needed as input to an aeroelastic code. Such high-frequency data is seldom available. The accepted way to obtain data for such purposes is through the application of a data synthesis algorithm.

The type of algorithm that is used for data synthesis depends on the application and the type of other data (if any) that may be available. Of the various methods available, three of the most commonly used are: (1) Auto-regressive moving average (ARMA), (2) Markov chains, and (3) the Shinozuka method. A fourth method, known as the Sandia method, is a variant of the Shinozuka. It is used to produce multiple, partially correlated time series.

B.3.1 Autoregressive Moving Average (ARMA)

The ARMA method is able to synthesize a time series that has the following specified characteristics: (1) mean, $\bar{x}$, (2) standard deviation, σ, and (3) autocorrelation at a particular lag. The ARMA model uses a random number generator and will have the same probability distribution as the random numbers do. The data does not have specified frequency domain characteristics.

Various forms of the ARMA method are available (see Box and Jenkins, 2008, for details). The simplest, and most commonly used method (ARMA 1), is defined, first of all, by the following equation for a normalized time series:

$$y_n = R(1)y_{n-1} + N(n) \tag{B.22}$$

where

$N(n)$ = normally distributed noise term with zero mean and standard deviation, $\sigma_{noise} = \sqrt{(1 - R^2(1))}$.
$R(1)$ = the value of the autocorrelation at a lag of 1.

The final time series, x_n, with the prescribed mean, $\bar{x}$, and standard deviation σ_x is then given by:

$$x_n = \sigma_x y_n + \bar{x} \tag{B.23}$$

B.3.2 Markov Chains

Wind data, especially long term, hourly data, can be synthesized via Markov chains (see Kaminsky *et al.*, (1990) and Manwell *et al.*, (1999)). The approach used is a two-step process: (1) generation of a transition probability matrix (TPM) and (2) use of the TPM, together with a random number generator.

The method assumes that any time series can be represented by a sequence of "states." The number of states should be chosen so that there not so few that the generated time series appears too discontinuous, nor so many so that calculations are unnecessarily burdensome. The Markov TPM is a square matrix, whose dimension is equal to the number of states into which a time series is to be divided. The value in any given position in the matrix is the probability that the next point in the time series will fall (i.e., will make a transition) into the jth state, given that the present point is in the ith state. One feature of a TPM is that the sum of all the values in a given row should be equal to 1.0. This corresponds to the observation that any succeeding point must lie in one of the states. An additional consideration is that there must be a known relation between the state number and the value of that state. This value is normally the midpoint of the state. For example, suppose a time series with values between 0 and 50 were to be represented by 10 states. The first state would be represented by 2.5, the second by 7.5, etc.

The most intuitive way, and most common, to generate a Markov TPM is to start with a time series of data. Each successive pair of points is examined to see which states both points are in. By tallying these, the number of times, and hence probability, that there is a "transition" from state *i* to state *j* can be determined. The result of this analysis for transitions from each state to each other state yields all the values needed to fill in a TPM.

Generating a TPM without using an initial time series is also possible, although it is considerably less intuitive. In addition, it is possible to do so in such a way that the probability density function of the data generated with use of the TPM will be equal (given a sufficient number of points) to a target probability density function. More information on this method may be found in McNerney and Richardson (1992).

In either case, the resulting TPM can be used to generate a time series that will have a mean, standard deviation, and probability density function close to that of the original data or the target pdf. (They will not be exactly the same, as discussed below.) The time series will have an exponentially decreasing autocorrelation, but will not necessarily be equal to that of the original data (or the target) for all lags.

A time series is generated by first assuming a starting value. This can be any number corresponding to a real state. A random number generator is then used to select the next point, based on weightings that are proportional to the probabilities in the row determined by the present state. For example, suppose a 5 × 5 TPM is being used, and that the state transition probabilities in the row of interest are 0, 0.2, 0.4, 0.3, and 0.1. Then, there would be no chance of the next number being in state 1, a 20% chance that it would be in state 2, a 40% chance it would be in state 3, etc. Subsequent points are generated in an analogous way.

In practice, when a finite-length data set is synthesized, its mean and standard deviation may not be exactly equal to that of the original data or the target. Typically, scaling is then undertaken to produce a final time series with the expected summary statistics.

It is also the case that certain more deterministic characteristics in the final data set may be desired, such as seasonal or diurnal patterns in annual data. These patterns may be approximately accounted for by generating and then concatenating separate time series for each month of the year and then adding the diurnal effect by suitable scaling of the resultant time series.

B.3.3 Shinozuka Method

The so-called Shinozuka method (Shinozuka and Jan, 1972) is often used to synthesize turbulent wind speed data, with a specified mean, standard deviation, and spectral characteristics corresponding to the specified psd. In this method, the range of frequencies of interest (between an upper value f_u and lower value f_l in the psd is divided into N intervals, resulting in an interval between frequencies of $\Delta f = (f_u - f_l)/N$ and interval midpoint frequencies of f_i. Sine waves are generated at each midpoint frequency. The sine waves are then all weighted by the square root of the psd times Δf at that frequency to give the correct variance to the result. A random phase angle, ϕ_i, is used for each sine wave as well. The sum of all the sine waves results in a fluctuating time series with zero mean, $\tilde{x}_n$ as shown in Equation (B.24).

$$\tilde{x}_n = \sqrt{2\Delta f}\sum_{i=1}^{N}\sqrt{S(f_i)}\sin(2\pi f_i n + \phi_i) \tag{B.24}$$

The final result is the desired mean, $\bar{x}$, plus the fluctuating term as shown in Equation (B.25).

$$x_n = \tilde{x}_n + \bar{x} \tag{B.25}$$

It should be noted that when the number of desired points exceeds the number of frequencies, the time series will begin to repeat itself. For that reason, in their original formulation, Shinozuka and Jan proposed to add a certain amount of noise, or "jitter" given by $\alpha\Delta f$ to prevent the repetition. In that case, the expression for the zero-mean time series is:

$$\tilde{x}_n = \sqrt{2\Delta f}\sum_{i=1}^{N}\sqrt{S(f_i)}\sin(2\pi (f_i + \alpha\Delta f) n + \phi_i) \tag{B.26}$$

This can suffice in some situations, but the psd is also changed. Ideally, the number of frequencies should equal the number of desired data points and there should be no need for jitter. See Jeffries et al. (1991) for more details.

B.3.4 Sandia Method for Partially Correlated Time Series

It is sometimes of interest to synthesize multiple turbulent wind speed time series that are partially correlated with each other. This is particularly the case when a field of time series across the rotor plane is desired. Such partial correlations can take into account the observation that slow variations in wind speed are often spatially rather large, whereas rapid fluctuations tend to be rather small. Data of this type are needed as input to structural dynamics codes such as were discussed in Chapter 8.

One way to synthesize partially correlated time series data is with the Sandia method which was proposed by Veers (1988). It is essentially an extension of the Shinozuka method described above. In the Sandia method, the coherence function is used to produce the appropriate correlation as a function of spacing between the time series. The time series are generated by linear combinations of independent sinusoids. The method uses a spectral matrix $S(f_i)$ and a triangular transformation matrix $H(f_i)$, each of dimension $M \times M$ where M is the number of time series being produced. Matrix elements are updated for each frequency f_i being used. The elements of $S(f_i)$ are derived from the target psd $S(f_i)$, which is the same for all the time series, and an assumed coherence function, γ_{jk}. These are shown in Equation (B.27).

$$S_{jj} = S(f_i)$$
$$S_{jk} = \gamma_{jk} S(f_i) \tag{B.27}$$

The most convenient form of the coherence function is the exponential one, which was introduced in Chapter 6. The relation between the spectral matrix and the transformation matrix is shown in Equation (B.28). The matrix $H(f_i)$ can be thought of as the matrix equivalent of the square root of $S(f_i)$ in Equation (B.24).

$$S(f_i) = \mathbf{H}(f_i)\mathbf{H}^T(f_i) \tag{B.28}$$

The elements of the transformation matrix are found from a recursive relationship, as shown in Equation (B.29).

$$H_{11} = S_{11}^{1/2}$$
$$H_{21} = S_{21}/H_{11}$$
$$H_{22} = (S_{22} - H_{21}^2)^{1/2}$$
$$H_{31} = S_{31}/H_{11}$$
$$\vdots \tag{B.29}$$
$$H_{jk} = \left(S_{jk} - \sum_{l=1}^{k-1} H_{jl}H_{kl} \right)/H_{kk}$$
$$H_{kk} = \left(S_{kk} - \sum_{l=1}^{k-1} H_{kl}^2 \right)^{1/2}$$

The transformation matrix is then used to produce the contribution of each frequency to each of the desired time series. The partial correlation is obtained through the phase angle of the sinusoids. In particular, the matrix $H(f_i)$ is used to multiply an M dimensional vector X of sine waves of frequency f_i with independent phase angles ϕ_i where $i \le M$. The nth term of the M time series (corresponding to time $n\Delta t$) may be expressed as a vector $\tilde{x}_n$ as shown in Equation (B.30). The reader is referred to Veers (1988) for more details.

$$\tilde{\mathbf{x}}_n = \sqrt{2\Delta f} \sum_{j=1}^{N} \sum_{i=1}^{M} \mathbf{H}(f_i)\mathbf{X} \tag{B.30}$$

where

$$\mathbf{X} = \begin{bmatrix} \sin(2\pi f_i n\Delta t + \phi_1) \\ \sin(2\pi f_i n\Delta t + \phi_2) \\ \vdots \\ \sin(2\pi f_i n\Delta t + \phi_M) \end{bmatrix}$$

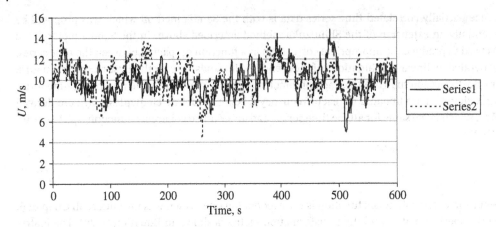

Figure B.7   Partially correlated, synthesized wind speed time series

For example, consider the synthesis of two 600 seconds (10 minutes) time series using a von Karman spectrum with the following characteristics: $\overline{U} = 10$ m/s, $\sigma_u = 0.15$, length scale = 100 m, separation between time series = 30 m, and coherence constant $A = 50$. The frequency range is 0.002 to 1 Hz, and 600 frequencies are used. The two-time series are shown in Figure B.7. As expected, the two-time series are generally similar but are not identical. Changing the separation would result in time series that appear correspondingly more or less correlated, depending on the separation.

References

Ariduru, S. (2004) *Fatigue Life Calculation by Rainflow Cycle Counting Method.* MSc Thesis, Middle East Technical University, http://etd.lib.metu.edu.tr/upload/12605614/index.pdf Accessed 12/8/23.

Bendat, J. S. and Piersol, A. G. (2010) *Random Data: Analysis and Measurement Procedures*, 4th edition. John Wiley & Sons, Chichester.

Box, G. E. P. and Jenkins, G. M. (2008) *Time Series Analysis: Forecasting and Control*, 4th edition. John Wiley & Sons, Chichester.

Jeffries, W. Q., Infield, D and Manwell, J. F. (1991) Limitations and recommendations regarding the Shinozuka method for simulating wind data. *Wind Engineering*, **15** (3), 147–154, Multi-Science, Brentwood, UK.

Kaminsky, F. C., Kirchhoff, R. H. and Syu, C. Y. (1990) A statistical technique for generating missing data from a wind speed time series. *Proceedings of Wind Power '90*, American Wind Energy Assoc., Washington, DC.

McNerney, J. and Richardson, R. (1992) The statistical smoothing of power delivered from utilities by multiple wind turbines. *IEEE Transactions on Energy Conversion*, **7** (4), 644–647.

Manwell, J. F., Rogers, A., Hayman, G., Avelar, C. T. and McGowan, J. G. (1999). *Hybrid2- A Hybrid System Simulation Model: Theory Manual.* Department of Mechanical Engineering, Univ. of Mass., Amherst, MA.

Veers, P. (1988) *Three-Dimensional Wind Simulation.* Sandia National Laboratory Report SAND88-0152 UC-261

Shinozuka, M. and Jan, C.-M. (1972) Digital simulation of random process and its application. *Journal of Sound and Vibration* **25** (1), 111–128, Elsevier, Amsterdam.

Appendix C

Notes on Probability Distributions

In the field of wind energy, there is much variability. Much of this variability can be quantified in terms of probability distributions. The following provides a summary of some of the most relevant distributions.

C.1   Basics

Probability concerns the behavior of random variables, which are variables whose numerical value is the outcome of some random process. For example, the random variable for wind speed could be X, but at any given time the actual wind speed could take on a range of values x. By convention random variables are capitalized; realizations have lower cases. Probability distributions are characterized by a probability density function, PDF, and a cumulative distribution function, CDF.

The PDF can be used to find the probability that something occurs between two values a and b:

$$\Pr(a \leq x \leq b) = \int_a^b f(x)\, dx \tag{C.1}$$

The integral over all values is equal to 1

$$\int_{-\infty}^{\infty} f(x)\, dx = 1 \tag{C.2}$$

The CDF is the integral of the PDF from $-\infty$ to some value x which is less than ∞. It gives the fraction of the time that a value is *less* than x. The CDF is always between 0 and 1 and will approach 1 as $x \to \infty$

$$F(x) = \int_{-\infty}^{x} f(u)\, du = 1 \tag{C.3}$$

The moment of a PDF is given by

$$\mu_n = \int_{-\infty}^{\infty} (x-a)^n f(x)\, dx \tag{C.4}$$

where a is a constant. The first moment μ_1, which is typically referred to as just μ, is the mean, or average. In this case, $a = 0$. For higher moments, a is often taken to be the average.

In many cases, the most convenient way to find the probability that a variable lies between two values a and b is to take the difference of two CDFs:

$$\Pr(a \leq x \leq b) = F(b) - F(a) \tag{C.5}$$

Wind Energy Explained: On Land and Offshore, Third Edition.
James F. Manwell, Emmanuel Branlard, Jon G. McGowan, and Bonnie Ram.
© 2024 John Wiley & Sons Ltd. Published 2024 by John Wiley & Sons Ltd.

C.1.1 Quantile

Quantiles are divisions in the range of a probability distribution such that each interval has equal probability. These are named according to the number of intervals; for example, the intervals in five divisions are called quintiles.

C.2 Normal Distribution

The PDF of the normal probability distribution (sometimes called the Gaussian) is:

$$f(x) = \frac{1}{\sigma\sqrt{2\pi}}e^{-\frac{1}{2}\left(\frac{x-\mu}{\sigma}\right)^2} \tag{C.6}$$

where μ = mean and σ = standard deviation.

The normal CDF is:

$$F(x) = \frac{1}{\sigma\sqrt{2\pi}}\int_{-\infty}^{x} e^{-\frac{1}{2}\left(\frac{u-\mu}{\sigma}\right)^2} du = \frac{1}{2}\left[1 + erf\left(\frac{(x-\mu)}{\sigma\sqrt{2}}\right)\right] \tag{C.7}$$

where u is a dummy variable, and $erf()$ is the error function. Note that the error function can be found easily in Excel using ERF().

C.3 Standard Normal Distribution

The standard normal distribution is a normal distribution with a mean of zero and a variance of 1.0. The PDF of the standard normal distribution is

$$f(x) = \frac{1}{\sqrt{2\pi}}e^{-\frac{1}{2}x^2} \tag{C.8}$$

The CDF of the standard normal distribution is:

$$F(x) = \frac{1}{\sqrt{2\pi}}\int_{-\infty}^{x} e^{-\frac{u^2}{2}} du \tag{C.9}$$

The CDF of the standard normal distribution is also given the symbol $\Phi()$ and can be related to the error function as shown by:

$$\Phi(x) = \frac{1}{\sqrt{2\pi}}\int_{-\infty}^{x} e^{-\frac{u^2}{2}} du = \frac{1}{2}\left[1 + erf\left(\frac{x}{\sqrt{2}}\right)\right] \tag{C.10}$$

The inverse of the standard normal cumulative distribution $\Phi^{-1}(p)$ is useful in many situations. It is also called the quantile function and is defined by:

$$\Phi^{-1}(p) = \sqrt{2}erf^{-1}(2p-1) \tag{C.11}$$

It is worth noting that the inverse error function $erf^{-1}(z)$ is related to the inverse of the gamma distribution, which is shown in Equation (C.12). The parameters α and β depend on the situation.

$$f_\Gamma(x; \alpha, \beta) = \frac{1}{\beta^\alpha \Gamma(\alpha)}x^{\alpha-1}e^{-x/\beta} \tag{C.12}$$

In Excel, the inverse error function is given by:

$$erf^{-1}(z) = SQRT(GAMMA.INV(z, 0.5, 1)) \text{ for } z \geq 0.5$$
$$= -SQRT(GAMMA.INV((1-z), 0.5, 1)) \text{ for } z < 0.5$$

Normal Distribution PDF

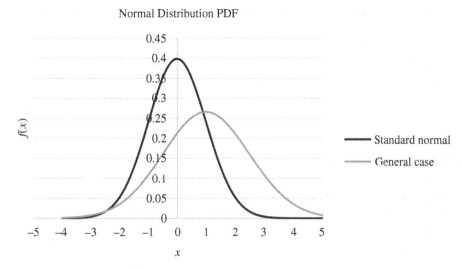

Figure C.1 Illustration of general case of normal distribution and standard normal distribution

CDF of Standard Normal Distribution

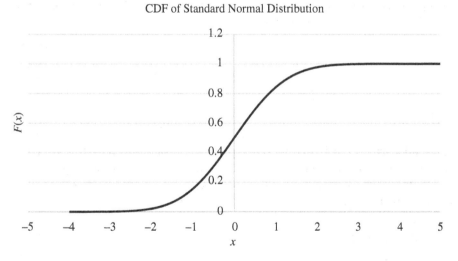

Figure C.2 Cumulative distribution function of standard normal distribution

$erf^{-1}(z) = $ SQRT(GAMMAINV(z,α,β)), where GAMMAINV is the inverse gamma distribution, $\alpha = 0.5$ and $\beta = 1$. (See http://www2.latech.edu/~sajones/REU/Learning%20Exercises/Tutorial%20on%20Statistical%20Testing.docx for more information on the gamma distribution and its relation to the inverse error function.) If $z < 0.5$.

Figure C.1 illustrates a typical general case of the normal distribution and a standard normal distribution. Figure C.2 illustrates the CDF of a standard normal distribution.

C.4 Gamma Function

The gamma function $\Gamma()$ is useful in a number of situations, including with the Weibull distribution. It is defined by:

$$\Gamma(a) = \int_0^\infty z^{a-1}e^{-z}dz \qquad\qquad (C.13)$$

Many software packages, such as Excel, include the gamma function. It is also possible to use this approximation (Jamil, 1994):

$$\Gamma(a) = \left(\sqrt{2\pi a}\right)\left(a^{a-1}\right)\left(e^{-a}\right)\left(1 + \frac{1}{12a} + \frac{1}{288a^2} - \frac{139}{51840a^3} - \frac{571}{2488320a^4} \cdots\right) \tag{C.14}$$

Note that when n is an integer, then

$$\Gamma(n) = (n-1)! \tag{C.15}$$

C.5 Weibull Distribution

The Weibull distribution is used in many situations in which the important variable is always equal to or greater than zero. The PDF of the Weibull distribution is:

$$f(x) = (k/c)(x/c)^{k-1}e^{-(x/c)^k} \tag{C.16}$$

The CDF is

$$F(x) = 1 - e^{-(x/c)^k} \tag{C.17}$$

The Weibull is characterized by two parameters, the scale factor c, and shape factor k; they are related to the mean and standard deviation by the gamma function as follows:

$$\mu = c\,\Gamma(1 + 1/k) \tag{C.18}$$
$$\sigma^2 = c^2\,\Gamma(1 + 2/k) - \mu^2 \tag{C.19}$$

or equivalently

$$\sigma^2 = \mu^2\left[\frac{\Gamma(1 + 2/k)}{\Gamma^2(1 + 1/k)} - 1\right] \tag{C.20}$$

With the mean and standard deviation known c and k may be found in a variety of ways, including trial and error in which k is guessed and then used to predict σ until convergence. Once k is found, c follows directly from (C.18). For typical wind speeds, there are closed-form approximations that work well.

The nth moment of the Weibull distribution is:

$$m_n = \int_0^\infty x^n k c^{-k} x^{k-1} e^{-(x/c)^k} dx = c^n \Gamma(1 + n/k) \tag{C.21}$$

The expression for the moments of the Weibull in terms of the gamma function can be justified as follows. First note that the nth moment is by definition:

$$m_n = \int_0^\infty x^n k c^{-k} x^{k-1} e^{-(x/c)^k} dx = \int_0^\infty x^n \left(\frac{k}{c}\right)\left(\frac{x}{c}\right)^{k-1} e^{-(x/c)^k} dx \tag{C.22}$$

From the definition of the gamma function, $\Gamma(a) = \int_0^\infty z^{a-1}e^{-z}dz$. In this case, we are interested in $\Gamma(1 + n/k) = \int_0^\infty z^{n/k}e^{-z}dz$

Now let $z = (x/c)^k$ so $dz = \frac{k}{c}\left(\frac{x}{c}\right)^{k-1}dx$
From above and substituting

$$m_n = \int_0^\infty x^n e^{-z} dz \tag{C.23}$$

Rearranging z we have $x^n = c^n z^{n/k}$ so

$$m_n = c^n \int_0^\infty z^{n/k} e^{-z} dz \tag{C.24}$$

Therefore, it follows directly that

$$\boxed{m_n = c^n \, \Gamma(1 + n/k)} \text{ q.e.d.} \tag{C.25}$$

There are some special forms of the Weibull, depending on k, including the exponential and the Rayleigh, as discussed in Sections (C.6) and (C.7).

The inverse CDF of the Weibull is given by:

$$F^{-1}(p) = c(-\ln(1-p))^{1/k} \tag{C.26}$$

C.6 Exponential Distribution

When $k = 1$, the Weibull distribution is known as the exponential distribution.

The PDF of the exponential distribution is:

$$f(x) = c^{-1} e^{-(x/c)}$$

The CDF of the exponential distribution is

$$F(x) = 1 - e^{-(x/c)} \tag{C.27}$$

Since $k = 1$ the nth moment for the exponential is

$$m_n = \int_0^\infty c^n \left(\frac{1}{c}\right) e^{-(x/c)} dx = c^n \Gamma(1 + n) = c^n n! \tag{C.28}$$

It is also the case that for the exponential the mean, Weibull scale factor, and standard deviation are all the same, i.e., $\mu = \sigma = c$

C.7 Rayleigh Distribution

When $k = 2$, the Weibull distribution is the same as Rayleigh distribution:

$$f(x) = \frac{2}{c^2} x e^{-(x/c)^2} \tag{C.29}$$

The CDF is

$$F(x) = 1 - e^{-(x/c)^2} \tag{C.30}$$

It may be shown from the expression for the first moment that $c = \dfrac{2\mu}{\sqrt{\pi}}$. Accordingly, alternative forms of the Rayleigh are as follows.

The PDF becomes:

$$f(x) = \frac{\pi}{2} \left(\frac{x}{\mu^2}\right) e^{\left[-\frac{\pi}{4}\left(\frac{x}{\mu}\right)^2\right]} \tag{C.31}$$

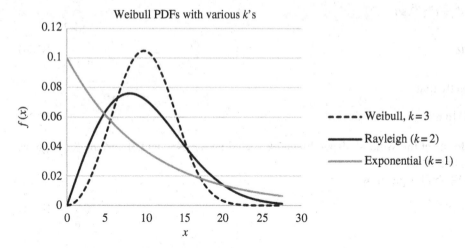

Figure C.3 Illustration of Weibull PDFs

The CDF of the Rayleigh is

$$F(x) = 1 - e^{\left[-\frac{\pi}{4}\left(\frac{x}{\mu}\right)^2\right]}$$
(C.32)

Since $k = 2$ from Equation (C.25) the nth moment for the Rayleigh is

$$m_n = c^n \, \Gamma(1 + n/2)$$
(C.33)

Thus, the mean $\mu = m_1$ is given by

$$\mu = c\,\Gamma(1.5) = \frac{2\mu}{\sqrt{\pi}}\Gamma(1.5) = \mu \text{ as expected}$$

Some Weibull PDFs with a mean of 10 and various k's are shown in Figure C.3.

C.8 Log Normal

The log-normal distribution is applicable to positive values of the variable x.
The PDF is:

$$f(x) = \frac{1}{x\sigma_{\ln x}\sqrt{2\pi}} e^{-\frac{(\ln x - \mu_{\ln x})^2}{2\sigma_{\ln x}^2}}$$
(C.34)

The CDF is:

$$F(x) = \frac{1}{2}\left[1 + erf\left(\frac{\ln x - \mu_{\ln x}}{\sigma_{\ln x}\sqrt{2}}\right)\right]$$
(C.35)

where

$$\mu_{\ln x} = \ln\left(\frac{\mu_x}{\sqrt{1 + \left(\frac{\sigma_x}{\mu_x}\right)^2}}\right)$$
(C.36)

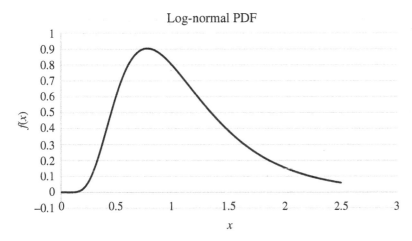

Figure C.4 Illustration of log-normal distribution

$$\sigma_{\ln x} = \sqrt{\ln\left(1 + \left(\frac{\sigma_x}{\mu_x}\right)^2\right)} \tag{C.37}$$

Note that $\mu_{\ln x}$ and $\sigma_{\ln x}$ are the mean and standard deviation of the variable x's natural logarithm, respectively, and μ_x and σ_x are the mean and standard deviation of the original variable x.

Conversely, it is worth noting the reverse of the above, namely:

$$\mu_x = \exp\left(\mu_{\ln x} + \frac{\sigma_{\ln x}^2}{2}\right) \tag{C.38}$$

$$\sigma_x = \sqrt{\left[\exp\left(\sigma_{\ln x}^2\right) - 1\right]\exp\left(2\mu_{\ln x} + \sigma_{\ln x}^2\right)} \tag{C.39}$$

Figure C.4 illustrates the log normal for $\mu_{\ln x} = 0$ and $\sigma_{\ln x} = 0.5$ (which corresponds to $\mu_x \cong 1.133$ and $\sigma_x \cong 0.6039$).

C.9 Gumbel Distribution

The Gumbel distribution is often used in assessing the likelihood of rare events.

Its PDF is:

$$f(x) = \frac{1}{\beta}\exp\left(\frac{-(x-\mu)}{\beta}\right)\exp\left(-\exp\left(\frac{-(x-\mu)}{\beta}\right)\right) \tag{C.40}$$

where

$\bar{x}$ = the mean of x, σ_x = standard deviation of x, $\beta = \left(\sigma_x\sqrt{6}\right)/\pi$, and $\mu = \bar{x} - 0.577\beta$

The CDF is:

$$F(x) = \exp\left(-\exp\left(\frac{-(x-\mu)}{\beta}\right)\right) \tag{C.41}$$

Figure C.5 illustrates the Gumbel PDF for a mean of 10 and a standard deviation of 5.

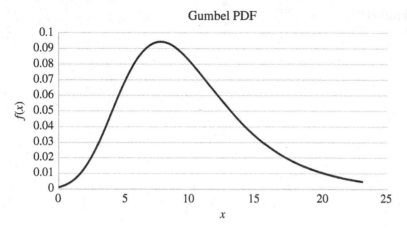

Figure C.5 Illustration of the Gumbel distribution

C.10   Joint Probability Distribution

A joint probability distribution gives the probability that each of a number of variables, $X_1, X_2,...X_n$, falls in a particular range. The joint PDF of two random variables X and Y is:

$$f_{X,Y}(x,y) = f_{Y|X}(y|x)f_X(x) = f_{X|Y}(x|y)f_Y(y) \tag{C.42}$$

where $f_{Y|X}(y|x)$ and $f_{X|Y}(x|y)$ are called *conditional* distributions and $f_X(x)$ and $f_Y(y)$ are called <u>marginal</u> distributions. Since these are probability distributions it is necessarily the case that:

$$\int_{-\infty}^{\infty} \int_{-\infty}^{\infty} f_{X,Y}(x,y)dxdy = 1 \tag{C.43}$$

The CDF is:

$$F_{X,Y}(x,y) = \int_{-\infty}^{X} \int_{-\infty}^{Y} f_{X,Y}(x,y)dxdy \tag{C.44}$$

When the variables X and Y are uncorrelated, Equation (C.42) becomes:

$$f_{X,Y}(x,y) = f_X(x)f_Y(y) \tag{C.45}$$

Equation (C.44) becomes:

$$F_{X,Y}(x,y) = \int_{-\infty}^{X} f_X(x)dx \int_{-\infty}^{Y} f_Y(y)dy \tag{C.46}$$

Joint distributions may be used, for example, to quantify the probability the wave height in the ocean falls within a given range of possible heights, given that the wind speed falls within a given range of speeds. In this situation, the wind speed would be characterized by the marginal distribution, the wave height by the conditional. See Chapter 6 for discussion of this topic.

Joint distributions of two variables are often depicted graphically by contours of constant probability density. Such contours are illustrated below for the case of joint normal distributions.

C.10.1   Bivariate Joint Normal Distribution

Multivariate joint distributions due not generally lend themselves to convenient, closed-form solutions. The joint normal bivariate (two variable) distribution is an exception to this and provides some useful examples.

For example, the joint normal uncorrelated PDF of two variables (with zero mean) is:

$$f_{X,Y}(x,y) = \frac{1}{2\pi\sigma_x\sigma_y} e^{-\frac{x^2}{2\sigma_x^2} - \frac{y^2}{2\sigma_y^2}} \tag{C.47}$$

The contours can be found from:

$$\frac{x^2}{\sigma_x^2} + \frac{y^2}{\sigma_y^2} = constant \tag{C.48}$$

In this case, the contours are ellipses centered around the origin with axes the same as the x axis and y axis. When the variables are correlated, the PDF is given by:

$$f_{X,Y}(x,y) = ce^{-q(x,y)} \tag{C.49}$$

where

$$c = \frac{1}{2\pi\sqrt{1-\rho^2}\sigma_x\sigma_y} \text{ and } q(x,y) = \frac{y^2}{2\sigma_y^2} + \frac{\left(x - \rho\frac{\sigma_x}{\sigma_y}y\right)^2}{2(1-\rho^2)\sigma_x^2} \tag{C.50}$$

The contours in this case are given by

$$\frac{x^2}{\sigma_x^2} - 2\rho\frac{xy}{\sigma_x\sigma_y} + \frac{y^2}{\sigma_y^2} = cons\tan t \tag{C.51}$$

Typical uncorrelated joint normal contours are illustrated in Figure C.6. In this example, $\sigma_x = 1$, $\sigma_y = 2$ and the constants $= 5$ and 10. The contours of correlated variables would have axes at an angle to the principal axes of the figure.

See http://athenasc.com/Bivariate-Normal.PDF

C.10.2 Correlated Distributions

For correlated variables, it is sometimes useful to divide the marginal variable into bins and make the assumption that within the bins the conditional variable and marginal variable are not correlated. This may be the case, for example, with mean wind and standard deviation or with mean wind and significant wave height. This approach is used in the IFORM method, which is discussed in Chapter 6.

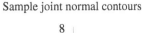

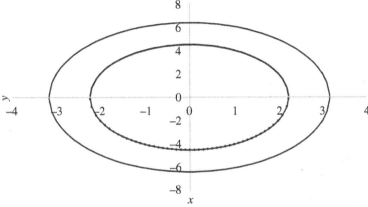

Figure C.6 Typical joint normal distribution contours

In some cases, analytic methods are used to account for partial correlations. One of these is the Plackett method, as described in C.10.3.

C.10.3 Plackett Model

A model developed by Plackett (1965) and adapted by Nerzic and Prevosto (2000) may be used to account for partial correlation of jointly distributed variables. In this case, the joint PDF is given by:

$$f_{X,Y}(x,y,\psi) = D(x,y,\psi)f_1(x)f_2(y) \tag{C.52}$$

where $f_1(x)$ and $f_2(y)$ are the marginal probability densities of x and y, and D is a dependency function defined as:

$$D(x,y,\psi) = \frac{\psi((\psi-1)[F_1(x) + F_2(y) - 2F_1(x)F_2(y)] + 1)}{\left([1 + (F_1(x) + F_2(y))(\psi-1)]^2 - 4\psi(\psi-1)F_1(x)F_2(y)\right)^{3/2}} \tag{C.53}$$

where $F_1(x)$ and $F_2(y)$ are the corresponding CDFs. When x and y are independent, $\psi = 1$; when they are correlated, ψ is a relatively large number. Figures C.7 and C.8 show examples for both x and y are Weibull distributed with

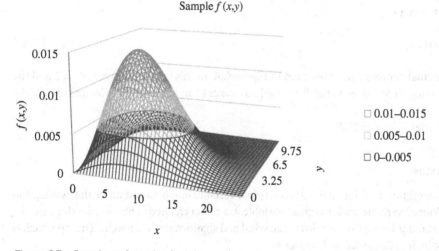

Figure C.7 Sample surface plot $f(x,y)$ for $\psi = 1$

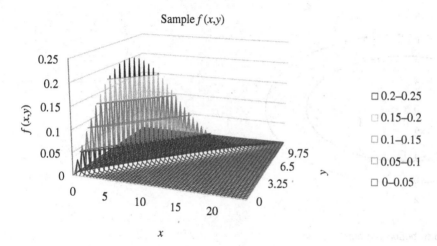

Figure C.8 Sample surface plot $f(x,y)$ for $\psi = 1000$

$k = 2$. In the case of x, $c = 10$ and for y, $c = 5$. In the first of these figures $\psi = 1$; in the second $\psi = 1000$. As can be seen, in the second situation, x and y are well correlated and the higher values of $f(x,y)$ are aligned along the line $y = 0.5x$.

References

Jamil, M. (1994) Wind Power Statistics and Evaluation of Wind Energy Density. *Wind Engineering*, **18**(5), 227–240.

Plackett, R. L. (1965) A Class of Bivariate Distributions. *Journal of the American Statistical Association*, **60**(310), 516–522, Taylor & Francis, Ltd. on behalf of the American Statistical Association, https://www.jstor.org/stable/2282685 Accessed 08/15/2018.

Nerzic, R. and Prevosto, M. (2000) Modelling of Wind and Wave Joint Occurrence Probability and Persistence Duration from Satellite Observation Data. *Proceedings of the Tenth (2000) International Offshore and Polar Engineering Conference*, Seattle, USA, May 28–June 2, 2000, https://www.researchgate.net/publication/267998233

Appendix D

Properties of Fundamental Support Structure Components

In the design of wind turbine support structures, there are a few types of components whose properties are useful to have on hand. These are hollow cuboids, cylinders, and frustums. Their most relevant properties are summarized in this appendix. In all cases, a uniform material density of ρ is assumed.

D.1 Cuboid

A cuboid is a box-like shape. Assume that a hollow cuboid has width W, length L, and height H, corresponding to x, y, and z axes. The thickness of the walls, top and bottom is t. Then, the volumes, mass, center of gravity, and mass moment of inertia about the y axis are:

$$V = WLH \qquad V_{in} = W_{in}L_{in}H_{in} \qquad V_{net} = V - V_{in} \tag{D.1}$$

$$m = m_{out} - m_{in} = \rho V_{net} = \rho(V - V_{in}) \tag{D.2}$$

$$h_{cg} = H/2 \tag{D.3}$$

$$
\begin{aligned}
J_{cg} &= \frac{1}{12}\rho\left[V\left(H^2 + W^2\right) - V_{in}\left(H_{in}^2 + W_{in}^2\right)\right] \\
&= \frac{1}{12}\left[m_{out}\left(H^2 + W^2\right) - m_{in}\left(H_{in}^2 + W_{in}^2\right)\right]
\end{aligned}
\tag{D.4}
$$

where $W_{in} = W - 2t$; $L_{in} = L - 2t$; $H_{in} = H - 2t$ and V_{net} is the total material volume.

D.2 Cylinder

A hollow cylinder with closed top and bottom with radius R_{out}, height H_{out}, and wall thickness t has the following volume, mass, center of gravity, and mass moment of inertia:

$$V = \pi R^2 H \qquad V_{in} = \pi R_{in}^2 H_{in} \tag{D.5}$$

$$V_{net} = V - V_{in} \tag{D.6}$$

$$m = m_{out} - m_{in} = \rho V_{net} = \rho \pi\left(R^2 H - R_{in}^2 H_{in}\right) \tag{D.7}$$

$$h_{cg} = H/2 \tag{D.8}$$

$$J_{cg} = \frac{1}{12}\rho\pi\left[R^2 H\left(3R^2 + H^2\right) - R_{in}^2 H_{in}\left(3R_{in}^2 + H_{in}^2\right)\right] \tag{D.9}$$

where $R_{in} = R - t$ and $H_{in} = H - 2t$. If the cylinder is open, then $H_{in} = H_{out}$.

Wind Energy Explained: On Land and Offshore, Third Edition.
James F. Manwell, Emmanuel Branlard, Jon G. McGowan, and Bonnie Ram.
© 2024 John Wiley & Sons Ltd. Published 2024 by John Wiley & Sons Ltd.

D.3 Frustum

A frustum is essentially the lower part of a cone. Consider a hollow frustum with open top and bottom which has an outer upper radius R_A, outer lower radius R_B, upper wall thickness t_A, lower wall thickness t_B, and height H. The volumes are defined by the outer and inner radii:

$$V = \frac{\pi H}{3}\left(R_A^2 + R_A R_B + R_B^2\right) \qquad V_{in} = \frac{\pi H}{3}\left(R_{A,in}^2 + R_{A,in}R_{B,in} + R_{B,in}^2\right) \tag{D.10}$$

$$V_{net} = V - V_{in} \tag{D.11}$$

The mass is then:

$$m = m_{out} - m_{in} = \rho V_{net} = \rho(V - V_{in}) \tag{D.12}$$

The centers of gravity $h_{cg,out}$ of the outer frustum and inner frustum $h_{cg,in}$ are given from Equation (D.13):

$$h_{cg,out} = \frac{H\left(3R_A^2 + 2R_A R_B + R_B^2\right)}{4\left(R_A^2 + R_A R_B + R_B^2\right)} \qquad h_{cg,in} = \frac{H\left(3R_{A,in}^2 + 2R_{A,in}R_{B,in} + R_{B,in}^2\right)}{4\left(R_{A,in}^2 + R_{A,in}R_{B,in} + R_{B,in}^2\right)} \tag{D.13}$$

The center of gravity of the frustum is found from:

$$h_{cg} = \frac{h_{cg,out}V - h_{cg,in}V_{in}}{V} \tag{D.14}$$

The mass moment of inertia of the frustum about the y axis through the center of gravity can be approximated numerically by dividing it into discs and applying Equation (D.15):

$$J_{cg} \cong \frac{\rho\pi H}{N_d}\sum_{i=1}^{N_d}\left[\left(R_i^4 - R_{in,i}^4\right)/4 + \left(h_i - h_{cg}\right)^2\right] \tag{D.15}$$

where i is the summation index and N_d is the number of discs.

An analytical equation for the mass moment of inertia of the frustum may be also found as follows. First note that a solid frustum could be considered the lower portion of a cone of height $h_1 = HR_B/(R_B - R_A)$. The upper part of the cone has a height $h_2 = h_1 - H$. It can then be shown that the mass moment of inertia $J_{cg,out}$ of the outer frustum is given by:

$$J_{cg,out} = \frac{\rho\pi}{3}\left\{ \begin{array}{l} h_1 R_B^2\left(\frac{3}{80}\left(h_1^2 + 4R_B^2\right) + \left(\frac{h_1}{4} - h_{cg,out}\right)^2\right) \\ -h_2 R_A^2\left(\frac{3}{80}\left(h_2^2 + 4R_A^2\right) + \left(\frac{h_2}{4} + H - h_{cg,out}\right)^2\right) \end{array}\right\} \tag{D.16}$$

The mass moment of inertia of the inner frustum $J_{cg,in}$ with the inner radii $R_{A,in}$ and $R_{B,in}$ is found analogously. The overall mass moment of inertia with respect to its center of gravity is then found by applying the parallel axis theorem. The result is:

$$J_{cg} = J_{cg,out} + m_{out}\left(h_{cg,out} - h_{cg}\right)^2 - \left(J_{cg,in} + m_{in}\left(h_{cg,in} - h_{cg}\right)^2\right) \tag{D.17}$$

Appendix E

Information Regarding Problems and Associated Files

Homework problems and associated data files to accompany this text are available on the internet site of the Wind Energy Center at the University of Massachusetts Amherst (http://www.umass.edu/windenergy/WindEnergy Explained). This site also contains the Wind Engineering MiniCodes which have been developed at the University of Massachusetts at Amherst. A number of these codes may be useful in solving problems. The Wind Engineering MiniCodes is a set of short computer codes for examining wind energy-related issues, especially in the context of an academic setting. The website is updated periodically.

Wind Energy Explained: On Land and Offshore, Third Edition.
James F. Manwell, Emmanuel Branlard, Jon G. McGowan, and Bonnie Ram.
© 2024 John Wiley & Sons Ltd. Published 2024 by John Wiley & Sons Ltd.

Index

Wind Energy Explained: On Land and Offshore, Third Edition.
James F. Manwell, Emmanuel Branlard, Jon G. McGowan, and Bonnie Ram.
© 2024 John Wiley & Sons Ltd. Published 2024 by John Wiley & Sons Ltd.